FOOD TOXICOLOGY

FOOD TOXICOLOGY

Edited by
Debasis Bagchi
Anand Swaroop

Foreword by Sidney J. Stohs

CRC Press
Taylor & Francis Group
Boca Raton London New York

CRC Press is an imprint of the
Taylor & Francis Group, an **informa** business

CRC Press
Taylor & Francis Group
6000 Broken Sound Parkway NW, Suite 300
Boca Raton, FL 33487-2742

Printed by CPI on sustainably sourced paper
Version Date: 20160829

International Standard Book Number-13: 978-1-4987-0874-6 (Hardback)

Visit the Taylor & Francis Web site at
http://www.taylorandfrancis.com

and the CRC Press Web site at
http://www.crcpress.com

Dedicated to my beloved father, the late Tarak Chandra Bagchi, MSc, AIC

Contents

Foreword

The World Health Organization estimates that over 350,000 people die of food poisoning around the world each year. Within the United States, the Centers for Disease Control and Prevention estimates that approximately one in every six Americans becomes ill each year as a result of food poisoning and foodborne diseases. In spite of numerous health and food-related regulations, approximately 130,000 people are hospitalized, while 3,000 people die annually in the USA as a result thereof. Many factors contribute to food-associated toxicities and poisonings.

Numerous steps may be involved in getting food from the farm or fishery to the dinner table, and as a consequence, contamination can occur at multiple sites along the food production chain. The industrialization and globalization of food production have introduced many additional opportunities for food to become contaminated with bacteria, viruses, molds, biological toxins, industrial toxins and wastes, pesticides, heavy metals, parasites, and plant and marine toxins.

The rapid, international production and shipping of food products around the world can result in widespread health problems when a food product is contaminated with a toxic or poisonous substance that can lead to delays in identifying the source and cause. In some cases, the consequences of food poisoning are rapid as may be the case for many bacterial and viral contaminations. Other cases of poisoning act more slowly and insidiously as is the case with drinking water contaminated with lead or a contaminated infant formula.

The consumption of foods is essential for life, and foods contain numerous essential nutrients as well as many other beneficial ingredients. However, it should be noted that the vast majority of the population believe that if something is "natural," it is safe, in spite of the fact that the most toxic substances known come from natural sources. "Natural" is a word for which there is no official definition in the United States, although countries such as England and Canada have official definitions. Just because they occur in nature, many food ingredients are called natural within the United States, but they would not meet the definition of "natural" in either Canada or England.

It should also be kept in mind that various groups of people are at higher risk and have a greater susceptibility to various toxins and food poisons. Most notable are children under the age of 5 years due to their rapid growth, pregnant women, adults over the age of 65 years due to their decreased defense mechanisms, and immunocompromised individuals.

This book, *Food Toxicology*, edited by Dr. Debasis Bagchi and Dr. Anand Swaroop, provides current information on many aspects of food-related poisoning. It addresses several controversial topics and provides current information regarding food-related toxicity topics that have been problematic for years.

Sidney J. Stohs, PhD, FACN, CNS, ATS, FAPhA, FASAHP

Preface

Food toxicology demonstrates how natural or synthetic poisons/toxicants in diverse food products cause harmful, detrimental, or adverse side effects in living organisms. The term *toxicity* highlights the degree to which a compound can cause toxic manifestation (poisonous effects) and bodily injury. However, the degree of toxicity or the propensity of the harmful effect depends on the dose, duration (time), and route of exposure. The route of exposure includes oral (ingestion), topical, intravenous (directly into a vein), intramuscular (into a muscle), subcutaneous (under the skin), intraperitoneal (within the membrane lining the organs of the abdomen), or nasal (for gases and aerosols).

It is important to note that food toxicology is a very important area because the entire world population depends on the food supply chain, and any contamination or toxic manifestation may cause serious widespread adverse health effects. Also, this book covers a diverse area of food items including the processing, preservation of health-promoting constituents in the food and their vital integrity, storage stability, warehouse storing, refrigeration, supply and transportation, distribution, and catering. This includes raw foods and their proper preservation, and unprocessed and processed foods and their proper packaging, in conjunction with a number of parameters including storage temperature, preservatives, humidity, and shelf-life.

This book covers all possible intricate areas of food toxicology. The first chapter discusses the importance of dose-response in food toxicology, while Chapter 2 highlights the aspects on the measurement of toxicants and toxicity. The third and fourth chapters discussed the aspects of food-associated carcinogenesis and food-derived chemical carcinogenesis. Dr. Ramesh demonstrates how lipids potentiate toxicant absorption in the fifth chapter. The sixth chapter discusses the intricate aspects of food allergy. Chapter 7 highlights the diverse pathogens associated with fruits and vegetables and discusses how the incidences of pathogen contamination can be prevented. Chapter 8 exemplifies the safety issues of food ingredients from plant cell, tissues, and organ cultures. Dr. Stoev summarizes food security and foodborne mycotoxicoses, risk assessment, prevention measures, and hazards in Chapter 9. Dr. Gerada discusses the detrimental effects of radionuclide exposure in foods in Chapter 10. Dr. April Neal-Kleuver and her diligent team have designed a vivid chapter on food-related toxicities in infants in Chapter 11. Chapter 12 considers naturally occurring toxicants as etiologic agents in foodborne diseases. Chapter 13 is dedicated to naturally occurring toxicants and antinutrients in plants and fungi. Drs. Hossain and Park provide an extensive review of mushroom toxins in Chapter 14. Chapter 15 on dioxin exposure via the food chain is addressed by Professor Sid Stohs, an eminently dedicated professor of toxicology. Chapters 16 through 18 extensively discussed the three most important heavy metal contaminants, namely, mercury, lead, and vanadium. Fluoride toxicity was extensively discussed in Chapter 19. The role of antioxidant additives in food preservation is considered in Chapter 20. Chapter 21 provides an overview of industrial food processing contaminants. Chapter 22 highlights how functional foods can target cancer stem cells and promote human health. Food packaging is a very critical aspect on food safety, and Chapter 23 is dedicated to this topic. Chapter 24 discusses the role of gut microbiome in human health and its modulation by environmental toxicants and therapeutic agents. The final chapter, Chapter 25, discusses various features including the pros and cons of genetically modified foods and how they will participate in the biotech revolution and human health in the future.

Finally, the foreword of this book was written by my respected and beloved guru (teacher) Sidney J. Stohs, PhD, FACN, CNS, ATS, FAPhA, FASHP, dean emeritus of the School of Pharmacy and Health Professions at Creighton University Medical Center, Omaha, Nebraska, who taught me the intricate aspects of toxicological sciences.

We sincerely acknowledge and thank all the contributing authors for their valuable contributions to this book, and we expect that the book will be useful to readers. The editors sincerely thank Namrata Bagchi, Tariq Ahmad and Randy Brehm for their continued technical support, suggestions and assistance.

Debasis Bagchi, PhD, MACN, CNS, MAIChE
Anand Swaroop, PhD, FACN

Editors

Debasis Bagchi, PhD, MACN, CNS, MAIChE, earned his PhD in medicinal chemistry in 1982. He is an adjunct professor in the Department of Pharmacological and Pharmaceutical Sciences at the University of Houston College of Pharmacy, Houston, Texas, and the chief scientific officer of Cepham Research Center, Piscataway, New Jersey. Dr. Bagchi is an adjunct faculty in Texas Southern University, Houston, Texas. He served as the senior vice president of Research & Development of InterHealth Nutraceuticals Inc., Benicia, California, from 1998 to February 2011, and then as the director of innovation and clinical affairs of Iovate Health Sciences, Oakville, Ontario, until June 2013. Dr. Bagchi received the Master of American College of Nutrition (ACN) Award in October 2010. He is the former chairman of International Society of Nutraceuticals and Functional Foods (ISNFF), former president of ACN, Clearwater, Florida, and former chair of the Nutraceuticals and Functional Foods Division of Institute of Food Technologists (IFT), Chicago, Illinois. He is serving as a distinguished advisor on the Japanese Institute for Health Food Standards (JIHFS), Tokyo, Japan. Dr. Bagchi is a member of the Study Section and Peer Review Committee of the National Institutes of Health, Bethesda, Maryland. He has 315 papers in peer-reviewed journals, 27 books, and 18 patents. He is also a member of the Society of Toxicology, a member of the New York Academy of Sciences, a fellow of the Nutrition Research Academy, and a member of the TCE stakeholder committee of the Wright Patterson Air Force Base, Ohio. Dr. Bagchi is the associate editor of the *Journal of Functional Foods*, *Journal of the American College of Nutrition*, and *Archives of Medical and Biomedical Research* and also serves as editorial board member of numerous peer-reviewed journals, including *Antioxidants & Redox Signaling*, *Cancer Letters*, *Toxicology Mechanisms and Methods*, and *The Original Internist.*

Anand Swaroop earned his MS in biochemistry and PhD in chemistry in 1997. He has gathered over two decades of diversified global management experience in phytopharmaceuticals, global regulations on nutraceuticals and functional foods, chemicals and pharmaceutical ingredients, outsourcing and supply chain management, business development, market research, and strategic planning. He has worked in various multinational pharmaceutical companies, including Alembic Pharmaceuticals Limited (Vadodara, India), Cadila Pharmaceuticals Limited (Ahmedabad, India), Nandesari Rasayanee Limited (Vadodara, India), and Dai-Ichi Laboratories Limited (Hyderabad, India), and Yag Mag Labs Pvt Limited (Hyderabad, India) in senior management positions. Currently, he is the founding president of a reputable group of companies including Cepham Inc. (Piscataway, New Jersey), Yag Mag Inc. (Piscataway, New Jersey), RALLIFE Inc. (Piscataway, New Jersey), and Cepham Life Sciences Inc. (Piscataway, New Jersey). Dr. Swaroop has introduced several cutting-edge nutraceutical ingredients for obesity, diabetes, metabolic syndrome, and versatile degenerative conditions in the United States and around the world. He has 12 peer-reviewed publications, two books with CRC Press/Taylor & Francis Group and Wiley-Interscience, and numerous patents. Dr. Swaroop is a member of the American Chemical Society, American Association of Pharmaceutical Scientists, American Society of Pharmacognosy, the New York

Academy of Sciences, the American College of Nutrition, the Council for Responsible Nutrition, the American Botanical Council, and the Natural Products Association. He has organized and chaired numerous sessions in the annual meetings of the Institutes of Food Technologists, the International Society of Nutraceuticals and Functional Foods, and other scientific meetings. He has delivered invited lectures and served as session chairs in various national and international scientific conferences and organized workshops and group discussion sessions.

Contributors

Mohammad Al-Mogbel
College of Applied Medical Sciences
and
Center for Molecular Diagnostics and Personalized Therapeutics
University of Ha'il
Ha'il, Saudi Arabia

Shrikant Anant
Department of Surgery
and
The University of Kansas Cancer Center
The University of Kansas Medical Center
Kansas City, Kansas

Anthony E. Archibong
Department of Physiology
Meharry Medical College
Nashville, Tennessee

Ibraheem Ashankyty
Center for Molecular Diagnostics and Personalized Therapeutics
University of Ha'il
Ha'il, Saudi Arabia

Samudra Prosad Banik
Department of Microbiology
Maulana Azad College
Kolkata, India

Leah D. Banks
Department of Biochemistry and Cancer Biology
Meharry Medical College
Nashville, Tennessee

Maheep Bhatnagar
Department of Zoology
Mohan Lal Sukhadia University
Udaipur, India

Rekha Bhatnagar
Department of Preventive and Social Medicine
Rabindra Nath Tagore Medical College
Udaipur, India

L.B. Borse
Department of Pharmacology
P.S.G.V.P.Mandal's College of Pharmacy
Shahada, India

S.L. Borse
Department of Pharmaceutical Chemistry
P.S.G.V.P.Mandal's College of Pharmacy
Shahada, India

Alessia Carocci
Department of Pharmacy—Drug Sciences
University of Bari
Bari, Italy

Alessia Catalano
Department of Pharmacy—Drug Sciences
University of Bari
Bari, Italy

Lauren Connelly
The Kraft Heinz Company
Glenview, Illinois

Prasad R. Dandawate
Department of Surgery
The University of Kansas Medical Center
Kansas City, Kansas

Vijayalaxmi S. Dandin
Department of Botany
Karnatak University
Dharwad, India

Katheryne Daughtry
Food, Bioprocessing and Nutrition Sciences
North Carolina State University
Raleigh, North Carolina

Jeffrey Fisher
Division of Biochemical Toxicology
National Center for Toxicological Research
Jefferson, Arkansas

Peter P. Fu
Division of Biochemical Toxicology
National Center for Toxicological Research
Jefferson, Arkansas

Koraljka Gall Troselj
Division of Molecular Medicine
Rudjer Boskovic Institute
Zagreb, Croatia

Giuseppe Genchi
Department of Pharmacy
Health and Nutritional Sciences
University of Calabria
Cosenza, Italy

Piya Das Ghatak
Department of Surgery
The Ohio State University Wexner Medical Center
Columbus, Ohio

Rituparna Ghosh
Department of Physiology
Bhairab Ganguly College
Kolkata, India

Jennifer Gill
Food, Bioprocessing and Nutrition Sciences
North Carolina State University
Raleigh, North Carolina

Tanja Matijevic Glavan
Division of Molecular Medicine
Rudjer Boskovic Institute
Zagreb, Croatia

Ellen Goertzen
Food, Bioprocessing and Nutrition Sciences
North Carolina State University
Raleigh, North Carolina

Alexander Gosslau
Department of Science
BMCC, The City University of New York
New York, New York

and

Department of Chemistry and Chemical Biology
Rutgers University
Piscataway, New Jersey

Francoise Gueraud
INRA Toxalim
Toulouse, France

Francisco Javier Guillén Gerada
LARUEX, Applied Physics Department
Faculty of Veterinary Sciences
University of Extremadura
Cáceres, Spain

N.A. Gujarathi
Department of Pharmaceutics
P.S.G.V.P.Mandal's College of Pharmacy
Shahada, India

Gabriel Keith Harris
Food, Bioprocessing and Nutrition Sciences
North Carolina State University
Raleigh, North Carolina

Kelly L. Harris
Department of Biochemistry and Cancer Biology
Meharry Medical College
Nashville, Tennessee

Kenneth J. Harris
Department of Biochemistry and Cancer Biology
Meharry Medical College
Nashville, Tennessee

Darryl B. Hood
Department of Neuroscience
Ohio State University
Columbus, Ohio

Ashfaque Hossain
College of Medicine
and
Center for Molecular Diagnostics and Personalized Therapeutics
University of Ha'il
Ha'il, Saudi Arabia

Md. Akil Hossain
College of Veterinary Medicine
Kyungpook National University
Daegu, South Korea

and

Veterinary Drugs and Biologics Division
Animal and Plant Quarantine Agency (QIA)
Gyeongsangbuk-do, South Korea

Ayushi Jain
Department of Biochemistry
King George's Medical University
Lucknow, India

and

Department of Zoology
Mohan Lal Sukhadia University
Udaipur, India

Elif Karlık
The Institute of Science and Engineering
Fatih University
Istanbul, Turkey

Raj K. Keservani
School of Pharmaceutical Sciences
Rajiv Gandhi Proudyogiki Vishwavidyalaya
Bhopal, India

Rajesh K. Kesharwani
Department of Biotechnology
National Institute of Technology
Warangal, India

Graziantonio Lauria
Department of Pharmacy
Health and Nutritional Sciences
University of Calabria
Cosenza, Italy

Stuart J. Lucas
Sabanci University Nanotechnology Research and Application Center
Sabanci University
Istanbul, Turkey

Abbas Ali Mahdi
Department of Biochemistry
King George's Medical University
Lucknow, India

Godfred Antony Menezes
College of Applied Medical Sciences
and
Center for Molecular Diagnostics and Personalized Therapeutics
University of Ha'il
Ha'il, Saudi Arabia

Hosakatte Niranjana Murthy
Research Center for the Development of Advanced Horticultural Technology
Chungbuk National University
Cheongju, Republic of Korea

and

Department of Botany
Karnatak University
Dharwad, India

April Neal-Kluever
Division of Food Contact Notifications
Center for Food Safety and Applied Nutrition
College Park, Maryland

P.S. Negi
Fruit and Vegetable Technology
CSIR—Central Food Technological Research Institute
Mysore, India

Bright I. Nwaru
Asthma UK Centre for Applied Research
Usher Institute of Population Health Sciences and Informatics
The University of Edinburgh
Edinburgh, United Kingdom

and

School of Health Sciences
University of Tampere
Tampere, Finland

Kee Yoeup Paek
Research Center for the Development of Advanced Horticultural Technology
Chungbuk National University
Cheongju, Republic of Korea

Seung-Chun Park
College of Veterinary Medicine
Kyungpook National University
Daegu, South Korea

So-Young Park
Research Center for the Development of Advanced Horticultural Technology
Chungbuk National University
Cheongju, Republic of Korea

Cameron Parsons
Food, Bioprocessing and Nutrition Sciences
North Carolina State University
Raleigh, North Carolina

Fabrice Pierre
INRA Toxalim
Toulouse, France

K.V. Ragavan
Food Engineering Department
CSIR—Central Food Technological Research Institute
Mysore, India

and

CSIR—Academy of Scientific and Innovative Research
Chennai, India

Aramandla Ramesh
Department of Biochemistry and Cancer Biology
Meharry Medical College
Nashville, Tennessee

Navin K. Rastogi
Food Engineering Department
CSIR—Central Food Technological Research Institute
Mysore, India

and

CSIR—Academy of Scientific and Innovative Research
Chennai, India

Sidhartha D. Ray
Division of Pharmaceutical Sciences
Manchester University
Ft. Wayne, Indiana

Y.S. Rohinishree
Fruit and Vegetable Technology
CSIR—Central Food Technological Research Institute
Mysore, India

Ana Sanches-Silva
Department of Food and Nutrition
National Institute of Health Dr. Ricardo Jorge, I.P.
Lisbon, Portugal

and

Studies Center of Animal Science (CECA)
University of Oporto
Oporto, Portugal

Chandan K. Sen
Department of Surgery
The Ohio State University Wexner Medical Center
Columbus, Ohio

Raquel Sendón
Department of Analytical Chemistry, Nutrition and Food Science
University of Santiago de Compostela
Santiago de Compostela, Spain

Anil K. Sharma
Department of Pharmaceutics
Delhi Institute of Pharmaceutical Sciences and Research
New Delhi, India

Aziz Sheikh
Asthma UK Centre for Applied Research
Usher Institute of Population Health Sciences and Informatics
The University of Edinburgh
Edinburgh, United Kingdom

Maria Stefania Sinicropi
Department of Pharmacy
Health and Nutritional Sciences
University of Calabria
Cosenza, Italy

Alok Kumar Srivastava
CSIR—Academy of Scientific and Innovative Research
Chennai, India

and

Food Safety and Analytical Quality Control Laboratory
CSIR—Central Food Technological Research Institute
Mysore, India

Stoycho D. Stoev
Department of General and Clinical Pathology
Faculty of Veterinary Medicine
Trakia University
Stara Zagora, Bulgaria

and

Faculty of Science
Department of Food Technology
University of Johannesburg
Johannesburg, South Africa

Sidney J. Stohs
School of Pharmacy and Health Professions
Creighton University Medical Center
Omaha, Nebraska

Dharmalingam Subramaniam
Department of Surgery
and
The University of Kansas Cancer Center
The University of Kansas Medical Center
Kansas City, Kansas

Aydın Türkeç
Department of Molecular Biology and Genetics
Uludağ University
Bursa, Turkey

M.C. Varadaraj
Microbiology and Fermentation Technology
CSIR—Central Food Technological Research Institute
Mysore, India

Yen-Ching Wu
Oak Ridge Institute for Science and Education
Research Participation Program
Center for Food Safety and Applied Nutrition
College Park, Maryland

Qingsu Xia
Division of Biochemical Toxicology
National Center for Toxicological Research
Jefferson, Arkansas

Neven Zarkovic
Division of Molecular Medicine
Rudjer Boskovic Institute
Zagreb, Croatia

1 Dose–Response
An Overview and Significance in Food Toxicology

Lauren Connelly, Katheryne Daughtry, Jennifer Gill, Ellen Goertzen, Cameron Parsons, and Gabriel Keith Harris

CONTENTS

1.1 INTRODUCTION

How safe is it to eat this food or this diet? This is the central question of food toxicology. To answer that simple question, we must consider the toxicity of food components, either as single components or as mixtures. We must also consider the methods used to determine toxicity. While all food components may be toxic at some level, food toxicology is primarily concerned with those that may have harmful effects at low doses. These effects may be acute (*apparent after only one exposure*) or chronic (*effects that are only observed after repeated exposures*). Acute effects would include the ingestion of bacterial toxins at a single meal, with an onset of symptoms occurring within hours or days. Chronic effects might include drinking water contaminated with arsenic, where symptoms only appear after months or years of consumption. In toxicology, something that may cause harm is referred to as a hazard. Risk is the probability of being harmed by a particular hazard. The determination of risk, better known as risk assessment, seeks to understand both the hazard and degree of exposure to it, in order to answer the "how safe is it?" question. Dose–response experiments involving cell culture and animal models are a key component of risk assessment: the estimation of how safe or unsafe a food might be. While the use of animals is viewed as essential in toxicology today, it raises controversy and questions about how well experimental results apply to humans. Our diets, the food we choose to eat on a daily basis, may be the ultimate dose–response experiment, since we eat and drink mixtures of thousands of compounds each day, each with a different toxic dose.

The Renaissance physician Paracelsus (1493–1541), often referred to as the Father of Toxicology, proposed the concept that "the dose makes the poison" (Borzelleca, 2000). This means that everything is toxic at a sufficiently high dose. For this reason, toxicologists do not ask whether or not something is toxic. Instead, they ask questions like "How toxic is this substance?" or "How risky is it?" Materials that most of us would consider poisons, such as arsenic, cyanide, or botulinum toxin, can cause harm or death at very low doses. For example, it has been estimated that 1 g, or less than one-quarter teaspoon, of botulinum toxin would be sufficient to kill one million people (Arnon, 2001). On the other end of the toxicity spectrum, even pure water can cause harm or death if an adult were to drink about 6 L (nearly 2 gal) in a period of several hours. The result of consuming too much water is called hyponatremia (low blood sodium), an osmotic imbalance that can cause brain swelling, seizures, and death (Farrell and Bower, 2003; Dolan et al., 2010). Since all food ingredients, including nutrients, may cause harm at some level, it is important to define these levels and remain well below them to avoid harming even sensitive individuals.

This chapter deals with dose–response, so it is important to define dose. When taking a "dose" of aspirin or other medications, it is common to think in terms of an amount, such as a tablet containing a certain number of milligrams of active compound. In toxicology, dose is typically defined as a concentration of a substance. Concentrations are ratios of mass to mass, mass to volume, or volume to volume. Substances may be quantified in terms of milligrams per kilogram (otherwise known as parts per million), grams per kilogram, grams per liter, etc. Why not just use an amount like 5 g or 20 mL? The reason is that variability in body weight, body composition, and metabolism affects how compounds distribute through the body. A small child may react very differently to the same 5 g of material than an adult. Part of this is due to age-based differences in metabolism, but a large component is due simply to differences in *dilution factors* between the child's smaller volume and the adult's larger volume. In this way, concentration is a more effective measure of dose than mass or volume alone. The term *dose–response* refers to the effect that a given concentration of a chemical substance has on the body in general or on specific target organs. These effects could be beneficial (as in the case of an appropriate dose of a medicine) or harmful (as in the case of a poison). Although dose–response relationships have been studied since the 1500s, there remains a vast amount of information yet to be learned. While toxicologists agree that the "dose makes the poison," debate continues about the relationship between the dose and the response. Is it linear, meaning that any dose causes some harm? Is there a threshold dose, below which no harm occurs for all toxins? Can tiny doses of toxic materials be beneficial? Why are some individuals especially sensitive to certain food components? What is the effect of consuming a mixture of many compounds (as with single foods or whole diets) instead of just one? This chapter seeks to address these questions by examining dose–response relationships in food.

1.2 DOSE–RESPONSE MODELS: DANGEROUS CURVES

Like other scientific disciplines, food toxicology attempts to understand the natural world using models, or simplifications of reality. This is a practical necessity since it is physically and financially impossible to study all hazards in all foods for all people. The shape of dose–response curves is essential to the understanding of a substance's toxicity (Calabrese et al., 1999 and Calabrese, 2003). While dose–response is typically curvilinear, it may be described in terms of either linear or nonlinear models, where dose appears on the *x*-axis and response appears on the *y*-axis. In the case of toxicology, the response may be harm or death. Linear models may result from examining only the linear portion of a dose–response curve, from linear extrapolation, or from data transformations designed to *linearize* a curve. Despite their limitations, linear models are preferred because of their ease of analysis and interpretation. Figure 1.1a shows a linear dose–response curve. This type of linear model assumes that there is no safe dose for a given substance and that any dose will cause some harm. These types of linear models are used to estimate risk related to cancer-causing agents (carcinogens), giving an added margin of safety. Threshold models (Figure 1.1b) assume that no harm occurs below a certain dose, but that there is a linear increase in harm after a *threshold dose* is surpassed. This threshold dose is referred to as "no observed adverse effect level" (NOAEL). The NOAEL could be thought of as the highest *safe* dose. NOAELs can be determined for either acute or chronic exposures of the same compound (the dose will be different for each type of exposure).

Most dose–response relationships in food toxicology are not linear. Plotting dose versus percent response for a sufficiently large population results in a curvilinear plot (Figure 1.1c). Dose is typically converted to a log scale, resulting in a sigmoid curve (Figure 1.1d). This conversion makes the center portion of the plot more linear, allowing for the determination of three important

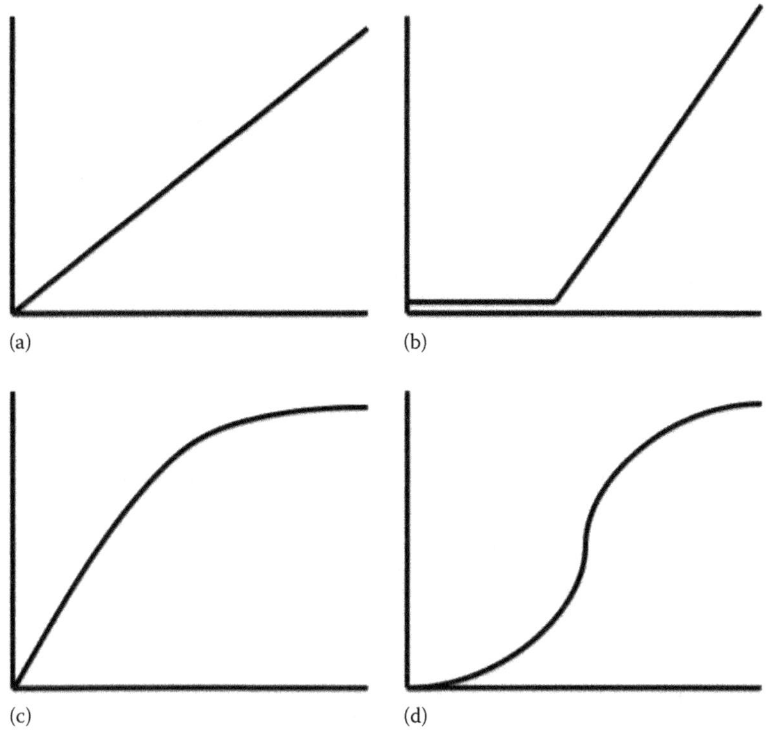

FIGURE 1.1 Dose–response curve shapes before and after transformation: (a) linear model, (b) threshold model, (c) curvilinear model, and (d) sigmoid curve. The *x*-axes represent dose (log dose in the case of the sigmoid curve). The *y*-axes represent response.

dose levels, the NOAEL, mentioned previously, the LOAEL, and the LD50. The LOAEL is the "lowest observed adverse effect level." The LOAEL is the lowest *unsafe* dose, or the minimum dose at which observable harm occurs. The LD50, or median lethal dose, describes the level at which 50% of a test population dies. LD50 testing is done in animal models and then extrapolated to humans, typically on a milligram per kilogram dose basis. Several species may be tested to estimate the human LD50. Animal to human extrapolations are typically made based on the most sensitive species. The LD50 is one of the most common ways of making direct "apples to apples" comparisons of the toxicity of different compounds. Probit (probability unit) transformations (Figure 1.2) are used to convert sigmoidal dose–response curves to straight lines, further easing LD50 calculations. This is accomplished by replacing percentage values on the *y*-axis with probit values ranging from two to eight. These probit values result from adding five to the three standard deviation units above and below the dose at which 50% of the population responds. Given the assumption of a Gaussian (bell curve) distribution of responses (Figure 1.3), the probit scale would cover dose–responses for 99.7% of a given population. On this scale, the LD50 is the dose corresponding to a probit value of five ($0 + 5 = 5$), since the mean, by definition, has a standard deviation of zero (Deshpande, 2002). Regarding the LD50, why would half a population die and the other half survive if given the same dose of a toxin? The reason for this is that, within a given population, there are individuals that are very susceptible to the toxin, individuals who are very resistant to the toxin, and everyone else in between. Susceptibility or resistance to particular toxins may result from genetic differences in drug-metabolizing enzymes, nutritional status, age, or other factors. When considering acute versus chronic exposures, there are different NOAELs, LOAELs, and LD50s for a given compound.

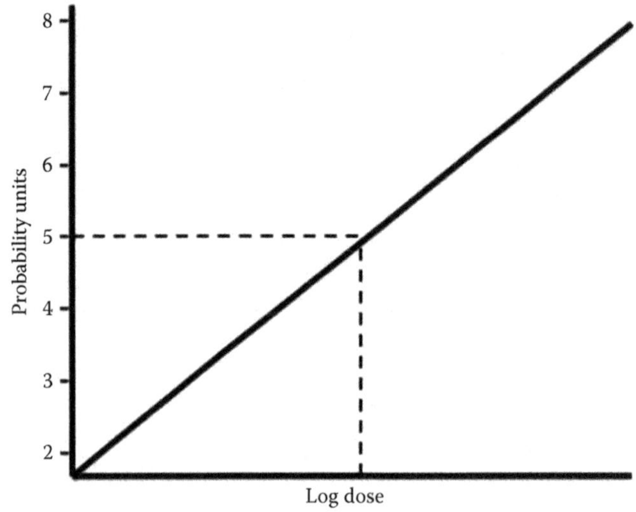

FIGURE 1.2 Probit transformation to convert sigmoid curves to linear responses. The *x*-axis represents log dose. The *y*-axis represents probit (probability unit) values.

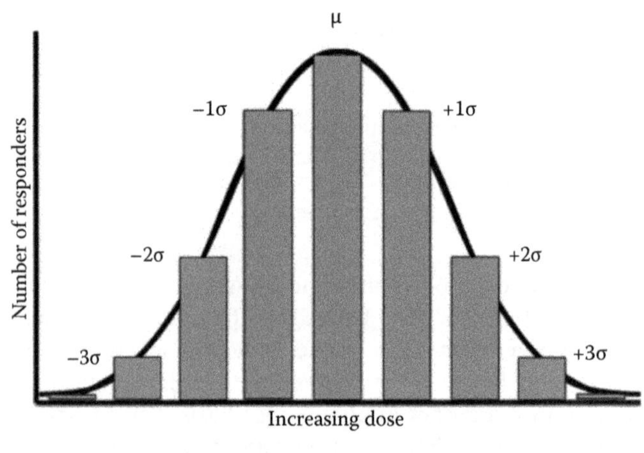

FIGURE 1.3 Gaussian distribution showing typical effects of increasing dose on the number of responders in a population, including mean dose and doses up to three standard deviations above or below the mean. Susceptible individuals may respond at doses up to three standard deviations below the mean. Resistant individuals may only respond at doses up to three standard deviations above the mean.

Since even nutrients such as vitamins and minerals may become toxic at very high levels, it is necessary to establish both upper and lower limits for their use. For essential nutrients, potential deficiencies and toxicities are estimated relative to dietary reference intakes. Dietary reference intakes take three different nutrient levels into account: estimated average requirements, recommended dietary allowances, and acceptable upper intake levels (USDA, 2010; DRIs, 2016). The recommended dietary allowances are calculated from estimated average requirements and represent intakes with two standard deviations above estimated average requirements, assuring that most of the population receives a sufficient level of essential nutrients but remains well below toxic levels. Acceptable upper intake levels correspond to the NOAELs described earlier (Taylor, 2008). Consuming more than the acceptable upper intake levels of a nutrient increases the risk of toxic effects.

For nonnutrient compounds, acceptable and tolerable daily intakes are used instead of dietary reference intakes. Acceptable daily intakes are defined as the amount of a food additive (based on a 60 kg average body weight) that can be safely consumed for a lifetime (Agget et al., 2015). Regarding food additives, an ongoing area of research is the investigation of possible links between artificial colors and attention-deficit disorder in children. While reactions have been reported when single doses of 50–100 mg have been administered in controlled trials, the question remains whether the doses normally consumed by children (reported to be approximately 60 mg per day in the United States) reach those levels (Stevens et al., 2013). Tolerable daily intakes are the amounts of nonnutrient, nonfood additive compounds that can be consumed over a lifetime without harm to the individual. Acceptable and tolerable daily intake values are generated from NOAELs based on the most sensitive species using a 100-fold margin of safety factor (Nasreddine and Parent-Massin, 2002).

Recently, the concept of linear and sigmoidal dose–response relationships has been called into question. An accumulating weight of evidence indicates that certain substances may produce *nonmonotonic* responses, meaning that higher doses do not uniformly produce greater effects. Examples of nonmonotonic curves would include those related to low-dose effects and to hormesis. Low-dose effects refer to the idea that compounds may produce very different effects at low doses than at high doses. Endocrine disruptors are compounds that interfere with the body's natural hormonal balance. Bisphenol A, a breakdown product of certain plastics, exerts estrogenic effects, even at very low doses. This could result in reproductive problems and may promote hormone-sensitive cancers. Endocrine disruptors are of great interest to toxicologists because they produce responses at very low concentrations and because responses are typically nonlinear (Vandenberg et al., 2012).

1.3 HORMESIS

One final set of nonlinear curves to consider are the J-shaped and U-shaped curves related to hormesis (Figure 1.4a and b). There is a large body of literature supporting the concept that low doses of otherwise toxic compounds can have beneficial effects. This concept is referred to as "hormesis." Dose–response curves may take on a J-shape or an inverted U-shape, depending on the endpoints measured. Generally speaking, J-shaped curves have some type of risk on the *y*-axis, such that risk decreases to a certain point as dose increases but then goes back up. The dose–response relationship between alcohol and stroke is an example of a J-shaped curve. Stroke risk drops as alcohol intake increases from zero to two or three *drinks* (one drink being defined as one ounce of alcohol) per day (Andréasson, 1998). From this perspective, two or three servings of alcohol per day appear to protect against stroke. When between four and five drinks per day are consumed, stroke risk is on par with that of nonconsumers. Beyond five drinks per day, stroke risk dramatically increases above that observed for both nonconsumers and moderate consumers of alcohol. This means that the risk of stroke is similar for those consuming either zero or five servings of alcohol per day. From a more general perspective, the concept of the J-shaped curve means that risk for a particular outcome may be similar at two different doses. It is important to point out that while two to three servings of alcohol may decrease stroke risk, those same two to three servings may also increase the risk of breast cancer (Scoccianti et al., 2014). This means that for the same dose of the same compound in the same person, risk may be decreased for one disease or condition and simultaneously increased for another. In contrast to J-shaped curves, inverted U-shaped curves typically have a type of function (such as cognitive function) as a dependent variable. An example of an *inverted U-shaped* curve might be caffeine intake and alertness. One or two cups of regular coffee might be expected to increase alertness. If three, four, or more cups of coffee were consumed at one sitting, nervousness, tremors, and other effects of excessive caffeine intake might be seen, decreasing overall alertness (mental function). Hormesis curves are now a well-accepted part of toxicology in general. The question is whether these curves can be applied to exposures, such as radiation or chemical carcinogens. Some researchers argue that they can (Calabrese, 2004, 2005).

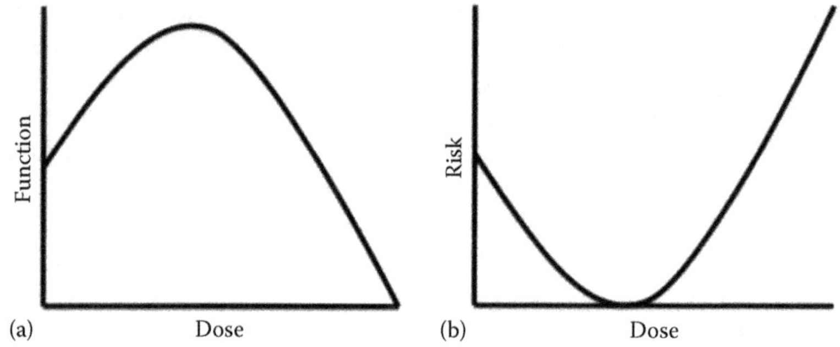

FIGURE 1.4 (a) J-shaped and (b) U-shaped hormesis curves. The x-axis represents dose in both cases, but the y-axis response depends on the type of curve. In the case of the J-shaped curve, the y-axis represents risk, while in the case of the U-shaped curve, the y-axis represents physical or mental function.

No discussion of dose–response would be complete without the mention of three critical factors: efficacy, potency, and time. Although frequently used interchangeably, efficacy and potency are quite different concepts in toxicology. Efficacy is the maximum effect that a substance can have, regardless of dose. Potency, in contrast, is the dose at which a certain effect occurs. In this way, an "efficacious" toxin would be one that could produce a specific injury, disease, or death at some dose, while a "potent" toxin is something that exerts its harmful effects at a low dose, relative to other compounds. Time is a critical factor in determining whether or not someone is harmed by a substance. Pharmacokinetic models examine the effects of time on tissue or blood levels of a compound. With a single dose, blood levels of an absorbed compound generally increase to a maximum level and fall over time (Figure 1.5a). Toxicologists are not only concerned with the peak levels of a compound but with the area under the curve, which represents the total amount of time an individual is exposed to varying levels of a toxin. The shape of the curve describing this rise and fall is determined by a number of factors, including digestibility, bioavailability, bioaccumulation, and excretion. This is, in turn, determined by the food matrix (solid vs. liquid). Digestibility is the percent of a compound that leaves the digestive tract and enters the bloodstream. Bioavailability, on the other hand, encompasses digestibility, xenobiotic metabolism, and the dose of compound that reaches target organs.

Xenobiotic metabolism is the enzymatic conversion of foreign compounds to ones that are more easily excreted from the body. This conversion typically occurs in two phases. Phase I enzymes, such as cytochrome P-450s, generally oxidize their metabolic targets to create more water-soluble metabolites. Phase II enzymes, on the other hand, tend to attach bulky groups to metabolites, both increasing

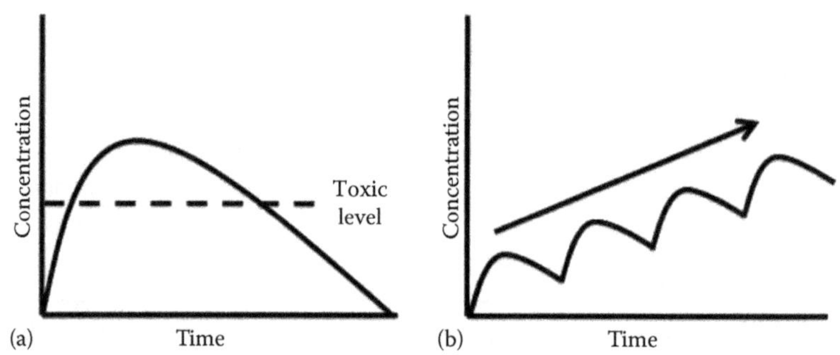

FIGURE 1.5 Pharmacokinetic curves. With a single dose of an absorbed compound, blood levels rise and fall over time (a) as compounds are metabolized, excreted, or sequestered. (b) With multiple doses of a compound, blood levels may rise over time if doses are spaced too closely or xenobiotic metabolism is slowed.

their water solubility and signaling that they should be excreted (Rose and Hodgson, 2004). While xenobiotic metabolism may occur in various parts of the body, the liver and kidneys are the primary sites. Some toxic substances may accumulate in bone or fat tissue rather than being excreted. Toxic effects may occur at a later time if toxins are liberated from storage (perhaps after significant weight loss). Depending on their chemical nature and the way in which they are metabolized, foodborne toxins may be excreted into urine, feces, or even sweat. Since foodborne toxins may be consumed repeatedly, it is important that pharmacokinetic curves be developed to include chronic exposure. Figure 1.5b shows that toxin levels may increase with multiple doses if toxins are not completely excreted prior to the next dose. If at any point in time toxin concentrations rise to sufficiently high levels, harm may occur. If levels rise higher, death may occur. Many foodborne toxins are naturally present in foods (e.g., aflatoxin in peanut butter or methanol in ripe fruit). Therefore, if it is impossible to avoid a toxin, it is essential to avoid reaching either harmful or fatal levels. All of the dose–response models described in this section have strengths and weaknesses. The key is to determine which model most accurately describes the effects of a given dose of a food toxin.

1.4 ANIMAL MODELS IN TOXICITY TESTING

So far, we have discussed a number of *dangerous curves*. How are these curves determined experimentally? The majority of knowledge regarding the relationship between dose and beneficial or harmful effects has been estimated from animal trials rather than from human studies. Most test animals are rodents (mice, rats, and guinea pigs). The use of animals, particularly nonhuman primates and dogs, has generated considerable controversy in recent years, given the experimental conditions to which animals may be subjected. Beyond concerns over animal welfare, questions have been raised about the relevance and need for animal studies. Variability in the response of different species to toxicants can be due to differences in compound digestibility, bioavailability, bioaccumulation, excretion, and target organ sensitivity, just as in humans (Dybing et al., 2002). For example, humans catalyze N-acetylation reactions very efficiently, whereas rats are slow acetylators and dogs are deficient in cytosolic N-acetyltransferase and therefore unable to catalyze the acetylation reaction. This means that choosing the *wrong* animal model will produce irrelevant results.

Despite the objections raised over their use, there are many physiological and anatomical similarities between laboratory animals and humans. For this reason, animals have been viewed as essential for the majority of toxicity-related studies, where causing adverse health effects in humans is not a viable research option. The use of laboratory animals to estimate toxicity in humans requires interspecies extrapolation so that the true effects of a toxin on humans may remain unknown (Oser, 1981). Based on concerns over animal testing, a general policy of *3 Rs* has been adopted. The 3 Rs stand for "refine, reduce, and replace" animal models (Flecknell, 2002). Refining animal studies would include the use of transgenic animals with genetic and physical traits closely mimicking those of humans. Reducing the use of animals may involve greater use of *in silico* (computer model) and *in vitro* models. One of the most promising avenues for limiting animal testing may be to use metabolomics to examine dose–response changes in toxicity-related biochemical pathways without reaching a *terminal* endpoint. This might be accomplished by using much lower doses than those that would cause harm or death and monitoring resulting disturbances in the normal metabolism. In this way, the same animal might be used for multiple tests, reducing the total number of animals used for toxicological studies. This approach seems most viable for acute toxicity testing. The question is how well this *microdosing* approach might work to predict chronic disease outcomes, such as cancer. Some researchers, rather than attempting to reduce the use of animals, have switched to the use of fish embryos, particularly zebrafish for toxicity testing. Will it be possible to completely replace animals for toxicity testing? The European Union completely banned animal testing for cosmetics in 2013 but continues to use them in toxicity research. For the present, it seems that the complete elimination of animals for toxicity testing is unlikely. In contrast, refining and reducing animal testing both appear to be achievable goals.

1.5 CODOSING

As stated earlier, the human diet is made up of mixtures of thousands of compounds, both natural and synthetic, each with a toxic dose. While these mixtures generally cause no harm, certain mixtures may increase health risk by a variety of mechanisms. This is referred to as "codosing effects." Grapefruit, alcohol, and even vitamins may all have codosing effects. Both whole grapefruit and grapefruit juice contain 6′,7′-dihydroxybergamottin, a naturally occurring furanocoumarin compound that inhibits an important xenobiotic-metabolizing enzyme known as cytochrome P450 3A4 or CYP3A4. The result is that the drug remains in its active form for a longer period of time and is excreted more slowly, simulating the effects of a higher dose (Mennen et al., 2005). If another dose of the same drug is given before the first is metabolized, toxic levels may be reached. Grapefruit is not the only food with codosing effects. The polyphenols found in limes, tea, wine, chocolate, soy, and other foods may affect xenobiotic-metabolizing enzymes and mineral bioavailability and, in the case of soy isoflavones, may have estrogenic effects (Bailey et al., 2003). While even large doses of vitamin E do not appear to be directly toxic, it can interfere with the bioavailability of other fat-soluble vitamins. High-dose vitamin E has been shown to result in decreased hepatic storage of vitamin A, impaired bone health, and coagulation, all of which are corrected by increased doses of vitamins A, D, and K (Combs, 2012). High dietary vitamin K may inhibit the activity of anticoagulant (blood-thinning) drugs. In addition to interactions between pharmaceuticals and specific food components, macronutrient ratios may also affect drug activity. High-protein, low-carbohydrate diets may decrease the effectiveness of the blood-thinning drug warfarin (Beatty et al., 2005). Given the risks it presents, codosing merits further study. Given the complexity of the diet and dietary components, metabolomics may be a good way to identify unknown codosing effects.

1.6 CONCLUSION

As new foods and new versions of familiar foods are introduced into the market every year, humans are exposed to a greater and greater variety of new substances. Worldwide travel and food distribution further increase this variety. The question is not whether foods contain potential toxins (they do), but whether toxin levels are high enough to cause harm in the short or the long term. That question applies not only to individuals but also to populations. When considering populations, it is important to keep in mind that some individuals will be especially sensitive to specific compounds. Identifying sensitive individuals and what dose of a particular substance may cause them harm is an area of continuing research in food toxicology. Some substances, like aflatoxin in peanuts or furanocoumarins in grapefruit, may be naturally occurring. Others may be added intentionally as preservatives, colors, or flavors. Microbial or chemical toxins may appear in the food supply due to contamination, poor storage, or processing conditions.

Debate continues about whether there are doses of some toxins, particularly carcinogens, below which no harm occurs. Since potentially carcinogenic compounds may be naturally present in food or produced during processing and storage, this will remain an important area of research in food toxicology. The concept of hormesis would suggest that small doses of even very toxic compounds may have beneficial effects. For a given dose of a given compound, there will be a range of responses in a population. Some very susceptible individuals may be harmed by the same dose that causes either no effect or even some benefit in other members of the population. Food is a complex mixture of chemical compounds and diet even more so. We know that certain food components may interact with one another, resulting in harm that would not occur if they were consumed separately. This is referred to as a "codosing" effect. As research continues, other interactions may be identified. As Paracelsus indicated over 500 years ago, the dose makes the poison.

While it is obviously unethical to conduct human dose–response studies to determine the acute and chronic toxicities of food components, debate continues regarding the ethics and efficacy of using animal models for toxicity testing. Is animal testing ethical? Is it really necessary? Does it yield useful information? Those that favor animal testing would say that the value to human life is worth the sacrifice of animal life. Those opposing animal testing believe that the suffering caused to animals is unethical, regardless of the potential benefits to humans. Animal testing is certainly a way to estimate the effects of specific compounds on complex organisms. On the other hand, running studies to determine LD50 and other values means that animals are sacrificed. Given the time and expense required to conduct a single animal study, animal models often use higher concentrations of toxic compounds than would be expected in the human diet. Extrapolation of the results from these *high-dose* studies to the lower concentrations in human diets has raised many questions about the resulting information. Concern about the use of animals for toxicity testing has led to the exploration of alternative methods. *In silico* and *in vitro* studies are now being used to an increasing degree in order to replace the use of animal models. Metabolomics (*the measurement of many metabolites at once*) may be an effective tool for reducing or replacing the use of animals, since it may help to identify the potential for toxicity even at very low doses, where no obvious harm occurs. With the advent of the Food Safety Modernization Act, microbial and chemical food safety is now front and center as a priority for government and industry. While it is impossible to completely eliminate risk in the food supply, dose–response studies are an important tool for the identification and avoidance of risk in food.

REFERENCES

Agget, P., Nordberg, G.F., and Nordberg, M. Essential metals: Assessing risk from deficiency and toxicity. In *Handbook on the Toxicology of Metals*, Nordberg, G.F. and Fowler, B.A., Eds., p. 283, Elsevier, San Diego, CA, 2015, Chapter 14.

Andréasson, S. Alcohol and J-shaped curves. *Alcoholism: Clinical and Experimental Research* 22(S7):359s–364s, 1998.

Arnon, S.S., Schechter, R., Inglesby, T.V. et al. Botulinum toxin as a biological weapon medical and public health management. *JAMA* 285(8):1059–1070, 2001.

Bailey, D.G., Dresser, G.K., and Bend, J.R. Bergamottin, lime juice, and red wine as inhibitors of cytochrome P450 3A4 activity: Comparison with grapefruit juice. *Clinical Pharmacology and Therapeutics* 73(6):529–537, 2003.

Beatty, S.J., Mehta, B.H., and Rodis, J.L. Decreased warfarin effect after initiation of high-protein, low-carbohydrate diets. *Annals of Pharmacotherapy* 39(4):744–747, 2005.

Borzelleca, J.F. Paracelsus: Herald of modern toxicology. *Toxicological Sciences* 53(1):2–4, 2000.

Calabrese, E.J. Hormesis: A revolution in toxicology, risk assessment and medicine *EMBO Report* 5(S1):S37–S40, 2004.

Calabrese, E.J. Paradigm lost, paradigm found: The re-emergence of hormesis as a fundamental dose response model in the toxicological sciences. *Environmental Pollution* 138(3):378–411, 2005.

Calabrese, E.J. and Baldwin, L.A. Reevaluation of the fundamental dose–response relationship: A new database suggests that the U-shaped, rather than the sigmoidal, curve predominates. *BioScience* 49(9):725–732, 1999.

Calabrese, E.J. and Baldwin, L.A. Hormesis: The dose-response revolution. *Annual Review of Pharmacology and Toxicology* 43(1):175–197, 2003.

Combs Jr., G.F. Vitamin E. *The Vitamins*. Academic Press, New York, 2012.

Deshpande, S.S. Measurements of toxicants and toxicity. In *Handbook of Food Toxicology*, Marcel Dekker, New York, 2002.

Dietary Reference Intakes (DRIs): Institute of Medicine. Tolerable upper intake levels, vitamins. USDA.gov. http://www.nal.usda.gov/fnic/DRI/DRI_Tables/recommended_intakes_individuals.pdf. Accessed on July 11, 2016.

Dietary Reference Intakes: Institute of Medicine. UL for vitamins and elements. USDA.gov. http://www.nal.usda.gov/fnic/DRI/DRI_Tables/UL_vitamins_elements.pdf. Accessed on July 11, 2016.

Dolan, L.C., Matulka, R.A., and Burdock, G.A. Naturally occurring food toxins. *Toxins* 2(9):2289–2332, 2010.

Dybing, E., Doe, J., Groten, J. et al. Hazard characterisation of chemicals in food and diet: Dose response, mechanisms and extrapolation issues. *Food and Chemical Toxicology* 40(2):237–282, 2002.

Farrell, D.J. and Bower, L. Fatal water intoxication. *Journal of Clinical Pathology* 56(10):803–804, 2003.

Flecknell, P. Replacement, reduction and refinement. *ALTEX* 19(2):73–78, 2002.

Mennen, L.I., Walker, R., Bennetau-Pelissero, C., and Scalbert, A. Risks and safety of polyphenol consumption. *The American Journal of Clinical Nutrition* 81(1):326S–329S, 2005.

Nasreddine, L. and Parent-Massin, D. Food contamination by metals and pesticides in the European Union. Should we worry? *Toxicology Letters* 127(1):29–41, 2002.

Oser, B.L. The rat as a model for human toxicological evaluation. *Journal of Toxicology and Environmental Health, Part A Current Issues* 8(4):521–542, 1981.

Rose, R.L. and Hodgson, E. Metabolism of toxicants. In *A Textbook of Modern Toxicology*, Hodgson, E., Ed., John Wiley & Sons, Hoboken, NJ, 2004.

Scoccianti, C., Lauby-Secretan, B., Bello, P.Y., Chajes, V., and Romieu, I. Female breast cancer and alcohol consumption: A review of the literature. *American Journal of Preventive Medicine* 46(3S1):S16–S25, 2014.

Stevens, L.J., Kuczek, T., Burgess, J.R., Stochelski, M.A., Arnold, L.E., and Galland, L. Mechanisms of behavioral, atopic, and other reactions to artificial food colors in children. *Nutritional Review* 71(5):268–281, 2013.

Taylor, CL. Framework for DRI development: Components known and components to be explored. 2008. http://www.nal.usda.gov/fnic/DRI/DRI_Process_Reports/Framework_DRI_Development.pdf accessed September 24, 2015.

Vandenberg, L.N., Colborn, T., Hayes, T.B. et al. Hormones and endocrine-disrupting chemicals: Low-dose effects and nonmonotonic dose responses. *Endocrine Reviews* 33(3):378–455, 2012.

2 Measurement of Toxicants and Toxicity

Alexander Gosslau

CONTENTS

2.1 OVERVIEW

The history of food toxicity might have started as early as Hippocrates made the statement "let food be thy medicine and medicine thy food," which presaged the modern science by over two millennia ago. Not until the development of modern biochemistry, molecular biology, cell culture techniques, computer science, and bioinformatics, it has been possible to identify and characterize potential toxicants in food (Paustenbach, 2000; Barlow et al., 2002; Eisenbrand et al., 2002; Houck and Kavlock, 2008; Sun et al., 2012; Ahuja and Sharma, 2014; Knudsen et al., 2015). Mechanistic insights gained by toxicity assessment in food using different models from *in vitro* biochemical, cell-based *in vitro*, animal *in vivo* to clinical settings were leading to a better food safety. The growing interest is reflected by stunning 6240 publications in PubMed as of January 2016 when combining "food, toxicity, review" with exploding numbers of over 100 reviews per year starting from 1995.

There are two different layers of measurement of toxicants and toxicity in food: (1) measurement of the effects of toxicants in different models ranging from *in vitro* biochemical systems, cell-based *in vitro* systems, and animal *in vivo* models to clinical settings analyzing systemic or organ-specific toxicity and (2) assessment of potential toxicants in food. These two layers are interrelated since the mechanistic knowledge gained by the assessment of the effects of toxicants is leading to the identification of potential toxicants in food. The majority of assessment systems for food toxicology were developed in the field of pharmacology (Sun et al., 2012; Ahuja and Sharma, 2014; McGonigle and Ruggeri, 2014; O'Brien, 2014). Pharmacology and nutritional science share common roots since many of the world's most commonly used drugs are derived from natural products (Newman and Cragg, 2007). The close relationship between pharmacology and nutritional science is illustrated by the term *nutraceutical* 1989 coined by Stephen L. DeFelice combining the words "nutrition" and "pharmaceutical" (Rajat et al., 2012).

The mechanisms of toxicants are multifactorial, interacting intrinsically and extrinsically with key molecules playing major roles in cell integrity, metabolisms, signaling pathways, gene expression, and translation. For a variety of toxicants, toxic effects appear to converge on the generation of electrophilic species (ES), leading to oxidative stress and chronic inflammation (Sies, 1993; Davies, 1995; Roberts et al., 2009; Reuter et al., 2010; Landskron et al., 2014). Oxidative reactions induced by toxicants lead to an accumulation of damages of macromolecules, thus harming cells, tissues, and organs. Therefore, toxicants play central roles in cell death, aging, and degenerative diseases such as Alzheimer's, Parkinson's, Huntington's, multiple sclerosis, myocardial infarction, arteriosclerosis, chronic inflammation, diabetes, rheumatoid arthritis, sterility, cataract, and many others (Floyd, 1990; Knight, 1998; Cottrell et al., 2000; Marnett, 2000; Kovacic and Jacintho, 2001; Roberts et al., 2009; Uttara et al., 2009; Reuter et al., 2010; Calle and Fernandez, 2012; Siti et al., 2015).

For *in vitro* assessment, a variety of biochemical systems have been developed to analyze damaging effects on integrity or activity of key biomolecules affecting cell integrity, metabolisms, signaling pathways, gene expression, and translation. The list is numerous and includes enzymes, receptors, membrane lipids, nucleic acids, and/or factors involved in gene expression (Eisenbrand et al., 2002; Jeffrey and Williams, 2005; Lee and Dordick, 2006; Houck and Kavlock, 2008; Niki, 2009; Mahadevan et al., 2011; Sun et al., 2012; Zanger and Schwab, 2013; Ahuja and Sharma, 2014; Anadon et al., 2014). On a cellular level, a variety of viability assays are routinely used to quantify toxic effects for extrapolation of a range of dosages used for maximal tolerated concentrations for not only *in vivo* animal models but also clinical settings (Eisenbrand et al., 2002; Houck and Kavlock, 2008; Sumantran, 2011; Astashkina et al., 2012; Sun et al., 2012; Yoon et al., 2012; Ahuja and Sharma, 2014). Inhibitory effects on cell viability as a measure of cytotoxicity due to necrosis, apoptosis, or autophagy can be discriminated by the use of specific assay systems (Sumantran, 2011; Astashkina et al., 2012; Ouyang et al., 2012). For more mechanistic insights, several other simple or complex cell-based *in vitro* systems were developed in combination with targeted *in vitro* analysis focusing on cell-specific key enzymes and receptor-dependent pathways.

Although the assessment of toxicity in animal *in vivo* rodent models still appears to be the golden standard, there is an increasing demand for alternative models. Acute toxicity tests currently still performed in rodent *in vivo* screening are believed to have little value in hazard identification for food chemicals and components with a few exceptions of chemicals showing anticholinesterase activity. Repeated lower dosing toxicity tests adapted from *in vitro* cell-based screening are considered to be more relevant for food toxicity screening as core tests (Barlow et al., 2002; Houck and Kavlock, 2008; McGonigle and Ruggeri, 2014). The limitations of traditional toxicity testing using rodents are characterized by high costs, low-throughput readouts, inconsistent responses, ethical issues, and concerns of extrapolability to humans (Barlow et al., 2002; Sun et al., 2012; Ahuja and Sharma, 2014; McGonigle and Ruggeri, 2014). Consequently, new strategies have been developed and the paradigm in toxicology switched from the traditional apical endpoint approach as determined in animal models to a mechanism-based approach by *in silico* methods (Valerio, 2009; Ahuja and Sharma, 2014; Anadon et al., 2014; Knudsen et al., 2015; Wilk-Zasadna et al., 2015).

In silico screening systems, a combination of *in vitro* cell-based models with computer-based algorithms are using a variety of different high-throughput and high-content screening technologies. Cell-specific biomarkers at gene, protein, or metabolite level can be measured by toxicogenomics, toxicoproteomics, or toxicometabolomics, respectively (Fabre et al., 2009; Merrick and Witzmann, 2009; George et al., 2010; Kiyosawa et al., 2010; Mahadevan et al., 2011; Ouedraogo et al., 2012; Ahuja and Sharma, 2014; Nishi et al., 2015). Data obtained throughout *in vitro* biochemical, cell-based, and animal models were leading to a mechanistic knowledge of systemic or organ-specific toxicity in humans and identification and use of specific surrogate biomarkers in clinical settings. The integration of food toxicology data obtained by in animal *in vivo* and human clinical settings enabled the establishment of alternative, highly predictable *in silico* models employing new focused cell-based bioassays as valuable tool for toxicity risk assessment in food. Advances of these *in silico* tools to assess toxicity in food have led to a significant reduction in the number of animals required for toxicological tests for a new active substance (Valerio, 2009; Mahadevan et al., 2011; Sun et al., 2012; Ahuja and Sharma, 2014; McGonigle and Ruggeri, 2014; Knudsen et al., 2015).

Although complex *in vitro* cell culture systems integrated in *in silico* systems provide unique mechanistic insights into *in vivo* toxicology more relevant to humans, they will never completely model the higher-level complexity of cross talk throughout different pathways present in an intact organism (Paustenbach, 2000; Barlow et al., 2002; Houck and Kavlock, 2008; Sun et al., 2012; Ahuja and Sharma, 2014; McGonigle and Ruggeri, 2014; Heinonen, 2015). Therefore, animal *in vivo* models for toxicology assessment are inevitable, particularly for new food molecules and thus remain a mainstay in food toxicology assessment. A large body of information about their responses were leading to various specific high-content animal models that may have the ability to simulate the genetic heterogeneity of the human population and evaluate reproductive and developmental toxicity, neurotoxicity, genotoxicity, carcinogenicity, immunotoxicity, food allergy, and endocrine disruption (Paustenbach, 2000; Barlow et al., 2002; Houck and Kavlock, 2008; Sun et al., 2012; Ahuja and Sharma, 2014; McGonigle and Ruggeri, 2014; O'Brien, 2014). In addition, the use of transgenic animals is considering common diseases and/or genetic polymorphisms within the general population that may be at particular risk following exposure to a food component (Barlow et al., 2002; Sun et al., 2012; Ahuja and Sharma, 2014; McGonigle and Ruggeri, 2014). Another development in toxicity assessment is the installation of alternative lower hierarchy surrogate animal models that offer an advantage in terms of ethical concerns and high throughput and genetic manipulation over traditional rodent models (Houck and Kavlock, 2008; Giacomotto and Segalat, 2010; Pandey and Nichols, 2011; Sun et al., 2012; Ahuja and Sharma, 2014). The value of using alternative submammalian vertebrate and invertebrate models became evident by the surprising discovery of the high degree of homology of genes between humans and zebrafish (*Danio rerio*), fruit flies (*Drosophila melanogaster*), or nematodes (*Caenorhabditis elegans*) (Barbazuk et al., 2000; Bier, 2005; Leung et al., 2008; Pandey and Nichols, 2011; Sun et al., 2012; Prussing et al., 2013; Chakravarthy et al., 2014; Raldua and Pina, 2014).

Overall, the achievements in food toxicology significantly improved the prediction rate of drug and food safety in dimensions unimagined only a decade ago (Olson et al., 2000; Houck and Kavlock, 2008; Andersen et al., 2010; Borchers et al., 2010; Sun et al., 2012; Ahuja and Sharma, 2014; McGonigle and Ruggeri, 2014). The deeper understanding of the molecular mode of action by which toxic effects are affecting key targets of biological pathways was enhancing predictivity and robustness of *in vitro* cell-based toxicity models, thus leading to the improvement of food safety. Despite those achievements, at present, there are still huge challenges to increase the rate of predictivity in various disciplines such as reproductive and developmental toxicity, neurotoxicity, genotoxicity, carcinogenicity, immunotoxicity, food allergy, and endocrine disruption (Kirkland et al., 2007; Houck and Kavlock, 2008; Kettenhofen and Bohlen, 2008; Mahadevan et al., 2011; Kraushaar et al., 2012; Sun et al., 2012; Ahuja and Sharma, 2014; McGonigle and Ruggeri, 2014).

2.2 MOLECULAR EFFECTS OF TOXICANTS

Although the mechanisms leading to toxic effects in humans are multifactorial, the majority of toxic effects appear to converge on the generation of free radicals. Different ES such as reactive oxygen species (ROS) or reactive nitrogen species (RNS) are oxidizing virtually all biomolecules. The accumulation of damages to membrane lipids, cellular proteins, carbohydrates, as well as nucleic acids is thus harming cells, tissues, and organs (Sies, 1993; Davies, 1995; Jones, 2006;

Franco et al., 2009; Roberts et al., 2009; Reuter et al., 2010; Landskron et al., 2014). Whereas a variety of toxicants generate ES directly, others are inducing a secondary response, leading indirectly to the generation of ES by immunocompetent leukocytes playing a key role in the inflammatory cascade (Reuter et al., 2010; Gosslau et al., 2011; Landskron et al., 2014). ES are also playing a role in the modulation of gene expression by interfering with transcription factors and/or DNA causing mutations and carcinogenesis. These and other observations strengthen the hypothesis that toxicants leading to oxidative stress and chronic inflammation play central roles in cell death, aging, and degenerative diseases such as Alzheimer's, Parkinson's, Huntington's, multiple sclerosis, myocardial infarction, arteriosclerosis, chronic inflammation, diabetes, rheumatoid arthritis, sterility, cataract, and many others (Floyd, 1990; Knight, 1998; Cottrell et al., 2000; Marnett, 2000; Kovacic and Jacintho, 2001; Roberts et al., 2009; Uttara et al., 2009; Reuter et al., 2010; Calle and Fernandez, 2012; Siti et al., 2015).

ROS comprise differently reduced oxygen species such as the superoxide anion radical ($^{\bullet}O_2^-$), hydrogen peroxide (H_2O_2), and the highly reactive hydroxyl radical ($^{\bullet}OH$), which is able to oxidize—and thus damage—virtually all macromolecular compounds of the cell (Davies, 1995; Alfassi, 1999; Finkel and Holbrook, 2000; Franco et al., 2009; Roberts et al., 2009). Importantly, several ROS members produced by endogenous cellular processes or exogenous sources are reacting with toxicants to produce even more reactive species. An important endogenous source for ROS generation is the mitochondrial inner membranes that give rise to superoxide anion radicals due to electron leakage from complexes I and III and transfer to O_2 (Sohal, 1993; Wallace, 1999; Cadenas and Davies, 2000; Finkel and Holbrook, 2000). ROS are also produced by oxidizing enzymes (e.g., xanthine oxidase, NADPH oxidase, lipoxygenase, and cyclooxygenase), detoxifying enzymes (cytochrome P450 system), or oxidation of fatty or amino acids within peroxisomes (Davies, 1995; Alfassi, 1999; Finkel and Holbrook, 2000; Franco et al., 2009; Roberts et al., 2009). Dismutation of these radicals leads to hydrogen peroxide as a fairly stable ROS member. Heavy metals (iron but also copper, chromium, or vanadium) as an important group of toxicants have led to the generation of the highly toxic hydroxyl radical using Fenton reaction ($H_2O_2 + Fe^{2+} \rightarrow {}^{\bullet}OH + OH^- + Fe^{3+}$) (Halliwell and Gutteridge, 1984; Davies, 1995; Franco et al., 2009; Roberts et al., 2009). Therefore, the amount of hydroxyl radicals formed in a cell depends not only on endogenous ROS generation but also on the amount of reduced metal ions for the Fenton reaction (Halliwell and Gutteridge, 1984; Stohs and Bagchi, 1995). The other highly reactive group of molecules consists of reactive nitrogen species (RNS): the signal molecule nitric oxide (NO) exists as NO^+, $NO^{\bullet}$, and NO^-, while the peroxynitrite ion ($ONOO^-$) is generated by the reaction of NO with the superoxide anion radical (Ischiropoulos et al., 1992; Jones, 2006; Franco et al., 2009). Phagocytes are capable of producing large quantities of ROS and RNS that are important for their antimicrobial and tumoricidal activities. Toxicants can interfere with the generation of ROS or RNS, potentially at different levels. Due to the mechanism of several toxicants to oxidize and thus damage biomolecules, the assessment of toxicity is focused on the measurements of oxidized products derived from lipids, proteins, nucleic acids, and carbohydrates (Sies, 1993; Davies, 1995; Eisenbrand et al., 2002; Jones, 2006; Franco et al., 2009; Roberts et al., 2009; Reuter et al., 2010; Landskron et al., 2014). Assessment of primary or secondary products induced by food toxicants is achieved throughout *in vitro* biochemical, cell-based, *in vivo* animal models or clinical settings as described in the following texts.

2.2.1 Lipid Peroxidation

Lipid peroxidation is the oxidative degradation of cell structures containing lipids that lead to cell damage and cell death. Membrane lipids, particularly the unsaturated fatty acids of the phospholipids, are major targets of ES induced by toxicants. A rearrangement of the double bonds results in lipid peroxidation, which is characterized by radical chain reactions with the production of a variety of breakdown products, including alcohols, ketones, alkanes, aldehydes, and ethers (Halliwell and Gutteridge, 1984; Buettner, 1993; Antunes et al., 1996; Dianzani and Barrera, 2008;

Franco et al., 2009; Niki, 2009). Lipid peroxidation in turn leads to membrane defects that influence membrane proteins and membrane permeability and causes—among other effects—an influx of calcium. Increased intracellular calcium enhances oxidative stress by activating NO synthase, xanthine oxidase, phospholipases, proteases, and endonucleases, thus creating a positive feedback in the ES generation. These events contribute to cell damage, necrosis, or apoptosis that lead to the onset of many degenerative diseases (Trump and Berezesky, 1995; Oyama et al., 1996; Franco et al., 2009; Ouyang et al., 2012).

2.2.2 Protein Oxidation

For proteins, thiol groups of cysteine residues are the most sensitive targets of ES. Redox-dependent modifications of intra- and intermolecular disulfide bonds lead to structural/functional changes and protein aggregation (Davies, 1995; Jeffrey and Williams, 2005; Jones, 2006; Kastle and Grune, 2011; Hohn et al., 2014). Tyrosine residues may also lead to dimerization when exposed to ES. Other amino acid residues sensitive to oxidative stress are histidine, proline, arginine, and lysine. Oxidation of peptide bonds can lead to protein fragmentation by way of peroxyl intermediates. Moreover, proteins may be cross-linked to lipids (e.g., lipofuscin), which together with oxidized proteins accumulate during exposure to toxicants (Stadtman, 1992; Davies, 1993; Kastle and Grune, 2011; Hohn et al., 2014). Altogether these ROS-induced damages cause malfunctioning of enzymes, transporters, signal transducers, or structural proteins. Several *in vitro* assessment systems are based on the measurement of oxidative modifications of these proteins.

2.2.3 DNA and RNA Oxidation

Nucleic acids are very delicate targets of ES induced by food toxicants. Damage of nucleic acids by ES results in single- and double-strand breaks, DNA–DNA, DNA–protein, DNA–lipid adducts, or numerous base modifications. Frequent modifications are 8-hydroxy-deoxyguanosine, 5-hydroxylmethyluracil, 8-hydroxydeoxyadenine, and thyminglycol (Davies, 1995, 2000; Marnett, 2000; Jeffrey and Williams, 2005; Hwang and Bowen, 2007; Moller et al., 2013; Li et al., 2014). Even in healthy cells that are not exposed to ES-generating toxicants there are an estimated 10^5–10^6 reactions of ROS with the DNA of a cell each day, which are not all eliminated by repair systems and thus lead to persistent mutations and an accumulation of DNA damage with age (Ames et al., 1993; Marnett, 2000). The increased probability to develop cancer with age is believed to be due to the accumulation of DNA mutations in mitotic cells partly through ES-generated DNA damage (Floyd, 1990; Ames et al., 1993; Croteau and Bohr, 1997; Knight, 1998; Marnett, 2000; Jeffrey and Williams, 2005; Hwang and Bowen, 2007). Mitochondrial DNA (mtDNA) is particularly susceptible to oxidative damage because of the absence of histones, an incomplete DNA repair system, and the generation of free radicals through electron leakage in the mitochondrial respiratory chain (Hwang and Bowen, 2007; Moller et al., 2013; Li et al., 2014).

2.2.4 Carbohydrate Oxidation

Toxicant-induced free radical formation also leads to the oxidation of carbohydrates. In hyperglycemia, typical of diabetes, food-derived ES are particularly harmful (Wright et al., 2006; Yan et al., 2009; Ramasamy et al., 2012). Glucose reacting with plasma proteins produces glycated proteins such as glycated hemoglobin (HbA1c) and advanced glycation end products (AGE) (Wright et al., 2006). The very reactive carbonyl species (RCS) methylglyoxal reacts with amino and sulfhydryl groups in proteins to form irreversible AGE, which bind to its receptor RAGE on endothelial cells and macrophages. Chronic inflammation then leads to cell damage and organ pathology (Wright et al., 2006; Yan et al., 2009; Ramasamy et al., 2012). Additionally, ROS and AGE activate transcription factor NFκB, thus increasing inflammation in a positive feedback regulation by inducing the transcription of inflammatory cytokines, chemokines, and adipokines, resulting in the production of even more ROS in a positive feedback loop (Ghosh and Hayden, 2008; Ramasamy et al., 2012). In the context of DNA damage, oxidation and fragmentation of deoxyribose fragments produced in DNA by free radical attack are believed to play a major role in mutations by blocking the action of DNA polymerase and DNA ligase (Marnett, 2000; Kirkland et al., 2007; Snyder, 2007; Mahadevan et al., 2011).

2.3 *IN VITRO* BIOCHEMICAL ASSESSMENT OF TOXICITY

The measurement of toxicity by numerous different *in vitro* systems is reflected by the molecular mechanisms of many toxicants converging on the generation of ES. Oxidation of lipids, proteins, nucleic acids, and carbohydrates is then leading to damage of cells, tissues, organs, and organisms (Sies, 1993; Davies, 1995; Jones, 2006; Franco et al., 2009; Roberts et al., 2009; Reuter et al., 2010; Landskron et al., 2014). Consequently, *in vitro* assessment of toxicology is focused on the integrity or activity of a variety of key biomolecules that play major roles in cell integrity, metabolisms, signaling pathways, gene expression, and translation. Testing whether a chemical modulates the activity of particular enzyme or binding affinities to a particular receptor or other biomolecule is the most direct way to gain mechanistic insights of action at the molecular level. There are different biochemical *in vitro* assays that comprise analysis of integrity or mutation of DNA and RNA, membrane lipids, binding and activity analysis of various receptors, enzymes involved in signaling transduction, drug, or neurotransmitter metabolism, among many others (Eisenbrand et al., 2002; Jeffrey and Williams, 2005; Lee and Dordick, 2006; Houck and Kavlock, 2008; Mahadevan et al., 2011; Sun et al., 2012; Zanger and Schwab, 2013; Ahuja and Sharma, 2014; Anadon et al., 2014).

Genotoxicity, the study of risk assessment of DNA-reactive toxicants in food, is of particular interest by virtue of the close correlation to carcinogenesis (Jeffrey and Williams, 2005; San, 2006; Custer and Sweder, 2008). To measure the potential for genotoxic activity of food compounds leading to mutations, traditionally the Ames bacterial reverse mutation test is used (McCann et al., 1975). The Ames test is based on several histidine-dependent *Salmonella* strains carrying different mutations in various genes of the histidine operon. The growth of *Salmonella* on minimal media agar plate is then used as a measure of toxicants in food sources, leading to mutations and potentially carcinogenesis (Mortelmans and Zeiger, 2000; Mahadevan et al., 2011; Ahuja and Sharma, 2014). Other methods measuring genotoxic potential in cell-based systems are discussed in the following texts.

Many enzyme and receptor-binding assays have been developed to examine specific mechanisms of action at the molecular level. Therefore, a variety of *in vitro* screening technologies are available that analyze the activities of different receptors (e.g., ion channels, G-protein-coupled receptors, tyrosine kinases, and nuclear receptors), signaling transduction enzymes (kinases, proteases, phosphatases, and phosphodiesterases), and enzyme-metabolizing drugs (e.g., cytochrome P450 monooxygenase) or neurotransmitters (e.g., acetylcholinesterase). As an *in vitro* prescreening tool for cardiotoxicity, the human ether-a-go-go-related gene (hERG) potassium ion channel is routinely used (Sanguinetti and Tristani-Firouzi, 2006; Taboureau and Jorgensen, 2011; Ferreiro et al., 2015). For a global assessment on neurotoxicity, acetylcholinesterase activity assays are frequently employed (Wille et al., 2010; Worek et al., 2012). An example to monitor the potential effects of toxicants interfering with anti-inflammatory drugs may be represented by high-throughput screening for microsomal prostaglandin E synthase activity (Andersson et al., 2012).

Xenobiotic metabolism is a commonly encountered problem during the development of new drug candidates; thus, it is also applicable to potential toxicants in food (Lee and Dordick, 2006; Houck and Kavlock, 2008; Patterson et al., 2010; Zanger and Schwab, 2013; O'Brien, 2014; Wilk-Zasadna et al., 2015). By far, the most important class of metabolic enzymes is cytochrome P450 monooxygenase (CYPs) drug-metabolizing phase I enzymes, but there are several other classes of biotransformation enzymes including uridine diphosphate glucuronosyltransferases (UGTs), glutathione S-transferases (GSTs), N-acetyltransferases (NATs), sulfotransferases (SATs), and methyltransferases (MTs) referred to phase II enzymes (Brandon et al., 2003; Hariparsad et al., 2006; Lee and Dordick, 2006; Foti et al., 2010; Zanger and Schwab, 2013). The majority of inhibition studies are using fluorescence or luminescence substrates toward recombinant CYPs and UGTs and high-throughput formats were developed, which involve protein-based microarrays (Eppinger et al., 2004; Trubetskoy et al., 2005; Hariparsad et al., 2006; Foti et al., 2010). Other approaches are using recombinant CYP450-metabolizing enzyme toxicology assay chips (MetaChips), which assess the toxicity of the generated metabolites by coupling with cell-based screening (Lee et al., 2005; Lee and Dordick, 2006). The abundance of *in vitro* biochemical testing platforms for CYP450 enzymes is understandable since many hepatoma-derived HepG2 cell lines used for hepatotoxicity screening lack functional expression of almost all the relevant human xenobiotic-metabolizing enzymes in the cytochrome P450 family (Donato et al., 2008; Patterson et al., 2010; Zanger and Schwab, 2013). High-throughput IC50 determinations and selectivity profiling in a microarray format using fluorescently tagged affinity label were also developed for kinetic analysis of various kinases (Horiuchi et al., 2006; Lee and Dordick, 2006; Ahuja and Sharma, 2014).

Lipid peroxidation is characterized by radical chain reactions with the production of a variety of breakdown products, which serve as oxidative markers in different biochemical *in vitro* assays (Halliwell and Gutteridge, 1984; Buettner, 1993; Antunes et al., 1996; Dianzani and Barrera, 2008; Franco et al., 2009; Niki, 2009). Generation of aldehydes, particularly malondialdehyde, is used by the thiobarbituric acid test to determine the degree of lipid peroxidation, designated as "thiobarbituric reactive substances" (TBARS) (Bernheim et al., 1948). Because of its sensitivity and simplicity, this method is considered as a first global measure of lipid peroxidation in a variety of chemical as well as biological materials (Liu et al., 1997; Niki, 2009). Another reliable and stable indicator of oxidative stress for the measurement of lipid peroxidation includes F2-isoprostanes. An increase in F2-isoprostanes during chronic inflammation has been observed (Montuschi et al., 2004; Milne et al., 2007). The *tert*-butyl hydroperoxide-initiated chemiluminescence is another method that has been successfully utilized to detect oxidative damage associated with experimental or pathological situations in subcellular fractions, tissue homogenates, or different organs (Gonzalez Flecha et al., 1991; Repetto, 2008).

2.4 CELL-BASED *IN VITRO* ASSESSMENT OF TOXICITY

The use of cellular models provides a much higher level of complexity than simple biochemical assays. Cell lines from a number of humans are easily accessible and a huge variety of different cell-based *in vitro* assays have been developed for the screening of the food toxicants (Eisenbrand et al., 2002; Houck and Kavlock, 2008; Sumantran, 2011; Astashkina et al., 2012; Sun et al., 2012; Yoon et al., 2012; Ahuja and Sharma, 2014; Heinonen, 2015). Usually, two basic approaches are applied: (1) a universal screening approach using one or a few cell lines to assess cell viability and (2) a target-organ-based approach, using a panel of different cells with more specialized functions using representative cell lines for different organs such as liver, heart, kidney lung, brain, or others.

2.4.1 Viability Assays

For general assessment of cytotoxicity, an indirect measure of cell viability is usually performed. There are several of those cellular bioassays that are routinely used for integrating different cytotoxicity endpoints such as membrane leakage or cellular activity (Sumantran, 2011; Astashkina et al., 2012; Ahuja and Sharma, 2014). Whereas the trypan blue, propidium iodide, crystal violet, or lactate dehydrogenase assays analyze membrane integrity based on exclusion, other viability assays such as the neutral red, alamar blue, or MTT assay are metabolic measures of cellular activity. Inhibitory concentration (IC) values obtained by viability assays are useful for initial dose selection in animals and human test setting (Eisenbrand et al., 2002; Sumantran, 2011; Astashkina et al., 2012; Sun et al., 2012; Yoon et al., 2012).

A prominent and reproducible method to analyze membrane integrity is the trypan blue exclusion method (Stoddart, 2011). Usually, the method is performed by cell counting using a hemocytometer, and the ratio of stained versus unstained cells is used as a reflection of membrane damage. Recently, an elegant quantitative measure of trypan blue staining has been introduced (Uliasz and Hewett, 2000). Similar to the trypan blue exclusion assay, the crystal violet assay is based on the growth rate reduction reflected by stained cells through the reaction of crystal violet with negatively charged cell components such as nucleic acids or peptidoglycans (Saotome et al., 1989; Itagaki et al., 1991). In addition to the measurement of dyes, the activity analysis of enzymes is used to determine the membrane integrity. Leakage of intracellular enzymes such as lactate dehydrogenase (LDH) or others into the extracellular medium is thus employed as an indicator of cell membrane damage (Decker and Lohmann-Matthes, 1988; Sumantran, 2011; Astashkina et al., 2012).

Other frequently used viability assays are based on cellular activity analysis such as the neutral red (NR), alamar blue, or MTT assay. The neutral red (NR) assay determines cell viability by endocytotic uptake of neutral red in lysosomes of uninjured cells (Borenfreund and Puerner, 1985). Therefore, a decrease in NR staining is a reflection of cytotoxic effects. In contrast to cellular endocytotic activity by NR, the MTT viability assay is based on mitochondrial activity by conversion of the tetrazolium salt MTT (3.(4.5-dimethylthiazol-2-yl)-2.5-diphenyltetrazolium bromide) by mitochondrial dehydrogenase activity to colored formazan (Berg et al., 1990). Similar to the MTT method, the alamar blue assay is based on redox reactions converting blue, oxidized resazurin to reduced resorufin that is red and highly fluorescent (Rampersad, 2012). It has been reported that the MTT method based on intracellular redox reactions may yield false-positive results for certain cell types when treated with different antioxidants (Berridge et al., 2005). Therefore, it appears to be advisable to include not only other viability assays besides the MTT but also the alamar blue method for cell-based toxicity screening (Gosslau et al., 2005).

Damage of mitochondria is a major contributor to organ toxicity, such as toxicity to liver, kidney, heart, muscle, and the central nervous system and mitochondrial dysfunction is increasingly implicated in a growing list of degenerative diseases (Wallace, 1999; Dykens and Will, 2008; Meyer et al., 2013). Therefore, the impact of toxicants on mitochondria as measured by the MTT method or by integrity analysis of mtDNA, electron transport chain or mitochondrial permeability transition pore, and ATP synthesis is of particular importance for toxicity assessment (Dykens and Will, 2008; Meyer et al., 2013; Ahuja and Sharma, 2014). For further differentiation between apoptosis and necrosis, there are several panels available using multiplexed endpoints as strong indicators of the type of cell death induced by the test extracts or compounds. Measurements may include caspase activity, phosphatidylserine exposure, DNA fragmentation, mitochondrial membrane potential, and Bax/Bcl-2 ratio, among many others (Sumantran, 2011; Astashkina et al., 2012; Ouyang et al., 2012).

2.4.2 Genotoxicity, Nutrigenomic, and Immunotoxicity Testing

Due to the nature of many toxicants to generate ES, it is not surprising that many of them are inducing oxidation and damage of DNA or RNA, thus leading to genotoxicity (Jeffrey and Williams, 2005; San, 2006; Custer and Sweder, 2008; Mahadevan et al., 2011; Chen et al., 2012). Since there is a strong correlation between genotoxicity and carcinogenicity, the aim of genotoxicity testing is to identify potentially carcinogenic food ingredients (Mahadevan et al., 2011; Chen et al., 2012; Ahuja and Sharma, 2014). Also, chronic inflammation is widely recognized not only as a major underlying cause of carcinogenesis but also various other degenerative conditions including cardiovascular diseases, Alzheimer's disease, arthritis, and diabetes. Therefore, cell-based models have been developed, which analyze the impact of food ingredients on inflammation. Using nutrigenomic screening, a method to analyze the influence of nutritional compounds on gene expression, potential anti-inflammatory effects can be analyzed by the modulation of inflammatory surrogate genes (e.g., interleukins, cyclooxygenase-2, tumor necrosis factor-α) by qPCR or microarray analysis (Gosslau et al., 2008; Gosslau et al., 2011; Gruber et al., 2012; Afman et al., 2014).

As a first layer for *in vitro* genotoxicity screening, the Ames reverse mutation test is routinely used in bacteria (McCann et al., 1975; Mahadevan et al., 2011). For cell-based genotoxic screening, various exploratory genetic toxicity assays are used to measure chromosomal damage in eukaryotic cells. Frequently used models are the mouse lymphoma tk gene mutation assay, the Chinese hamster ovary (CHO) chromosomal aberration assay, the micronucleus clastogenicity assay, and the Comet assay (Barlow et al., 2002; Mahadevan et al., 2011; Ahuja and Sharma, 2014). Interestingly, there has been a high false-positive rate (low specificity) in genotoxicity testing that might be addressed due to the recent discovery of noncovalent DNA interaction and interference of toxicants with critical DNA-metabolizing proteins such as topoisomerase and DNA polymerases (Kirkland et al., 2007; Snyder, 2007; Mahadevan et al., 2011; Moller et al., 2013; Li et al., 2014). The linkage among cell-based *in vitro*, *ex vivo*, and *in vivo* evaluation is evident by micronucleus test used for genotoxic screening of chromosomal damage in bone marrow polychromatic erythrocytes (Waters et al., 1994; Ahuja and Sharma, 2014). Also for immunotoxicity, there are several *ex vivo* test models available. Functional immunological tests include the activity of macrophages, natural killer cells, and immune responses of T and B cells (Barlow et al., 2002; Luebke, 2012; Hartung and Corsini, 2013). As for other organ-based *ex vivo* tissue or organ toxicology studies on heart, brain, lung, kidney, liver, or other organs, there is an advantage with the *in vitro* to *in vivo* extrapolation. Linking the data to *in vivo* cardiovascular, neurobehavioral, respiratory, renal, diabetes, P450 metabolism, or other studies will increase the predictivity frequency of *in vitro* cell-based models.

2.4.3 Complex Cellular Toxicity Assays

For viability, genotoxicity, immunotoxicity, or nutrigenomic testing, usually one cell line is employed. Recently, novel, more complex *in vitro* human cell–based systems are being increasingly used to model mammalian tissues in a more holistic approach (Andersen et al., 2010; Sun et al., 2012; Ahuja and Sharma, 2014; Anadon et al., 2014). Complex human *in vitro* cell-based models are advantageous through the elimination of interspecies extrapolation. A variety of parameters such as key signaling pathways, gene and protein expression levels, receptor activity, cytoskeletal and membrane integrity, energy status, morphology of cell organelles, cell movement, cell cycle status, and cell differentiation can be quantified on single cell level and compared throughout the panel of different cells employed to better represent *in vivo* physiology. The emergence of quantitative high-throughput screening and high-content methods has proved to be an efficient way to decompose those complex toxicological pathways of targeted organs (Lee and Dordick, 2006; Houck and Kavlock, 2008; Mahadevan et al., 2011; Sun et al., 2012; Ahuja and Sharma, 2014; O'Brien, 2014; Persson et al., 2014). Recently, there is an increase of high-content as well as high-throughput complex cellular *in vitro* models focusing on organ-specific toxicity such as hepatotoxicity (Chen et al., 2014; Gomez-Lechon et al., 2014; Persson et al., 2014), nephrotoxicity (Huang et al., 2014; Tiong et al., 2014), neurotoxicity (Smith, 2009; Crofton et al., 2012; Kadereit et al., 2012), cardiotoxicity (Kettenhofen and Bohlen, 2008; Peters et al., 2015), and respiratory toxicity (Nichols et al., 2013; Frohlich and Salar-Behzadi, 2014). By conducting quantitative *in vitro* to *in vivo* extrapolation, these models are leading to a better prediction of the *in vivo* response (Houck and Kavlock, 2008; Sun et al., 2012; Yoon et al., 2012; Ahuja and Sharma, 2014; Heinonen, 2015; Wilk-Zasadna et al., 2015). An even higher level of complexity is achieved with 3D organotypic models. Although still in its infancy, organotypic systems have a very high potential for predicting acute toxicity and can be used to answer fundamental biological questions and enable testing of novel therapeutic approaches, often using patient-derived cells (Dash et al., 2012; Shamir and Ewald, 2014; Knudsen et al., 2015).

2.4.4 Stem Cell Models

Recently, stem cell–based assays are also being discussed as source for various toxicological applications (Vojnits and Bremer, 2010; Liu et al., 2013; Ahuja and Sharma, 2014; Johnson and Hockemeyer, 2015). The Nobel Prize–winning discovery of how to reprogram ordinary somatic cells to behave like embryonic stem cells (Takahashi and Yamanaka, 2006), the so-called human-induced pluripotent stem cells (iPSCs), allows assays to consider an individual's genetic background and potential epigenetic influences that affect the variability of the toxicity response (Liu et al., 2013). By the use of comprehensive profiling via genomics, proteomics, transcriptomics, and metabolomics, stem cell *in vitro* models offer an unprecedented opportunity to test the effects of potential food toxicants in a very predictive and personalized manner (Vojnits and Bremer, 2010; Liu et al., 2013; Ahuja and Sharma, 2014; Johnson and Hockemeyer, 2015). Stem cell–based models are also of particular interest for toxicity measurements which either lack extrapolability in rodent models such as for genotoxicity, cardiotoxicity, respiratory toxicity or for different stages of disorders that largely remain unknown such as neurological disorders (depression, anxiety, Alzheimer's and Parkinson's disease), autoimmune diseases (multiple sclerosis, type 1 diabetes, asthma), systemic infection, cancer, and others (Kirkland et al., 2007; Kettenhofen and Bohlen, 2008; Mahadevan et al., 2011; Kraushaar et al., 2012; McGonigle and Ruggeri, 2014). Although still in early development, the long-term potential for application of stem cells in toxicity testing is very high. The increased use of human stem cell models may further reduce the usage of animals in safety and risk assessment studies and offers the potential to dramatically enhance our understanding and thus prediction of the molecular basis of toxicity (Vojnits and Bremer, 2010; Liu et al., 2013; Ahuja and Sharma, 2014; Johnson and Hockemeyer, 2015).

2.5 *IN VIVO* ANIMAL TOXICITY ASSESSMENT

While complex cell culture systems can provide unique insights into *in vivo* toxicology, they will never completely model the higher-order interactions present in an intact organism (Paustenbach, 2000; Barlow et al., 2002; Houck and Kavlock, 2008; Merlot, 2010; Sun et al., 2012; Ahuja and Sharma, 2014; McGonigle and Ruggeri, 2014; Heinonen, 2015). Therefore, the gold standard for toxicity assessment has been *in vivo* toxicology, where a particular molecule or complex food ingredients are dosed against animals. Various animal models have been developed to evaluate acute, subacute, and chronic effects of tested compounds on reproductive and developmental toxicity, neurotoxicity, genotoxicity, cardiotoxicity, hepatotoxicity, nephrotoxicity, carcinogenicity, immunotoxicity, food allergy, and endocrine disruption (Paustenbach, 2000; Barlow et al., 2002; Houck and Kavlock, 2008; Sun et al., 2012; Ahuja and Sharma, 2014; McGonigle and Ruggeri, 2014; O'Brien, 2014). The majority of animals used are rodents and the numbers of animals needed for testing are enormous with an estimation of 7000 animals and tens of millions of dollars for each test compound in the pharmaceutical industry (Erickson, 2011). Although the number of animals involved in food toxicity screening is decreasing, the number of compounds or food ingredients to be tested as well as the costs of the current *in vivo* assessment systems is exploding (Barlow et al., 2002; Garthoff, 2005; Andersen et al., 2010; Giacomotto and Segalat, 2010; Sun et al., 2012; Ahuja and Sharma, 2014; Heinonen, 2015; Knudsen et al., 2015). Traditional *in vivo* toxicity testing is also characterized by low-throughput readouts and ethical concerns of using such large quantities of animals raised by the animal protection lobby. Last, there are debates on whether the data could readily be extrapolated to human beings due to differences in not only the species sensitivity but also the level of the heterogeneity of the human population as reflected by a steady increase in reports on adverse events (Moore et al., 2008; McGonigle and Ruggeri, 2014). The high costs, low-throughput readouts, and ethical and extrapolability concerns led to an outcry for alternative strategies in toxicity testing methods (Garthoff, 2005; Sun et al., 2012; Knudsen et al., 2015).

Consequently, the paradigm of the high use of rodent for *in vivo* toxicity testing shifted in recent years with an approximate 50% reduction in the number of animals required for toxicological tests (Van Cauteren and Lumley, 1997; Houck and Kavlock, 2008; Sun et al., 2012; Ahuja and Sharma, 2014). This was paralleled by a focused use of specific animal models due to a better understanding of mechanisms. For example, the knowledge gained on the effects of toxicants on lipids stems from the similarity of lipid metabolism between mini pigs and humans, whereas the rat is an unsuitable model for the study of cholesterol levels (Barlow et al., 2002). Higher specificity was also achieved by the use of new genetically diverse *in vivo* high-content models simulating the genetic heterogeneity of the human population or transgenic animals modeling common diseases and/or genetic polymorphisms considering specific groups that may be at particular risk following exposure to a food component (Barlow et al., 2002; Andersen et al., 2010; Giacomotto and Segalat, 2010; Ahuja and Sharma, 2014; Knudsen et al., 2015).

In addition to enhancing the specificity of traditional animal models, the value of using alternative submammalian vertebrate and invertebrate models became evident by significant discoveries of homologous genes related to development, immune response, cancer, or related pathways in humans (Barbazuk et al., 2000; Bier, 2005; Leung et al., 2008; Pandey and Nichols, 2011; Sun et al., 2012; Prussing et al., 2013; Chakravarthy et al., 2014; Raldua and Pina, 2014). Mostly, three test platforms of model organisms are adapted to high-throughput screening that includes the fruit fly *D. melanogaster*, the nematode *C. elegans*, and the zebrafish *D. rerio*. The use of these or other alternative lower hierarchy surrogate animal models offers an advantage with relatively less ethical concerns and high throughput and genetic manipulation compared to traditional rodent models (Houck and Kavlock, 2008; Giacomotto and Segalat, 2010; Pandey and Nichols, 2011; Sun et al., 2012; Ahuja and Sharma, 2014).

The use of the fruit fly *D. melanogaster* has been recognized as a model organism in studies of genetics and developmental biology for over 100 years (Rubin and Lewis, 2000). The genome is fully sequenced, showing that nearly 75% of human disease-causing genes have a functional homolog in *D. melanogaster* (Bier, 2005; Pandey and Nichols, 2011; Prussing et al., 2013). Extensive genetic manipulation of the fruit fly such as knockouts or knockdown by RNAi is available for modifying fruit fly genetics (Bier, 2005; Drysdale and FlyBase, 2008; Pandey and Nichols, 2011). Fruit flies have the potential to be used for chemical-toxicity screens particularly for neurotoxicity due to the high degree of orthologs associated with genes known to be involved in neurodegenerative diseases (Nichols, 2006; Whitworth et al., 2006; Segalat, 2007; Pandey and Nichols, 2011; Prussing et al., 2013). The nematode *C. elegans* is easily cultured in the lab and widely used for biomedical research. More than 50% of human genes have functional orthologs in *C. elegans*, and all 959 somatic cells had been characterized with respect to lineage (Leung et al., 2008; Cho et al., 2014; Meyer and Williams, 2014). With regard to *D. melanogaster*, a variety of molecular tools provided the availability of a large number of transgenic strains suitable for a differential toxicity screening by high-throughput genomic studies (Leung et al., 2008; Ruan et al., 2009; Cho et al., 2014; Meyer and Williams, 2014). Since *C. elegans* is capable of rudimentary learning and many neurotransmitters are well conserved, it is well suited for neurotoxicity testing (Avila et al., 2012; McVey et al., 2012; Meyer and Williams, 2014). More recently, the zebrafish *D. rerio* has been used as a vertebrate model organism in research such as drug discovery and toxicology. The increased usage of zebrafish as *in vivo* model system reflects the striking similar toxicity profile between humans and zebrafish due to substantial physiological, anatomic, and genetic homologies (Barbazuk et al., 2000; Howitz et al., 2003; Hill et al., 2005; Raldua and Pina, 2014). The zebrafish model is also amenable to gene manipulation, is cheap, has a short generation time, and is particular suitable for high-throughput screening such as microarray and proteomic studies (Love et al., 2004; McGrath and Li, 2008; Chakravarthy et al., 2014). Since zebrafish larvae are transparent, they are ideal for studies on organ morphology by *in vivo* imaging techniques in addition to more detailed studies by immunohistochemistry or *in situ* hybridization (Howitz et al., 2003; McGrath and Li, 2008; Raldua and Pina, 2014).

2.6 TOXICITY ASSESSMENT IN HUMANS

Toxicity assessment in humans encompasses different fields such as clinical, forensic, environmental, and regulatory toxicologies. A systemic determination of toxicants in body tissues is usually obtained by biopsy or body fluids such as blood and urine. Clinical toxicology is mainly based on analytical toxicology analyzing genotoxicity, neurotoxicity, cardiotoxicity, hepatotoxicity, nephrotoxicity, carcinogenicity, immunotoxicity, food allergy, and/or endocrine disruption. Disturbances in these systems are setting the stage for a variety of disorders such as Alzheimer's, Parkinson's, Huntington's, multiple sclerosis, myocardial infarction, arteriosclerosis, chronic inflammation, diabetes, rheumatoid arthritis, sterility, cataract, and many others (Floyd, 1990; Knight, 1998; Cottrell et al., 2000; Marnett, 2000; Kovacic and Jacintho, 2001; Roberts et al., 2009; Uttara et al., 2009; Reuter et al., 2010; Calle and Fernandez, 2012; Siti et al., 2015).

A great deal of knowledge on toxicity in humans has been obtained by postmortem analysis of cells, tissues, and organs in the disciplines of molecular, cytotoxicologic, or anatomic toxicologic pathology (van Tongeren et al., 2011; Maeda et al., 2014). Forensic toxicology is much related to toxicologic pathology but focuses on the legal aspects (Dinis-Oliveira et al., 2010; Maeda et al., 2014). The discipline of environmental toxicology is related to studies of various chemical, biological, and physical agents, which are harmful to humans, whereas regulatory toxicology is concerned with risk assessment of food and potential toxicants (Andersen et al., 2010; Hennes, 2012). By virtue of advances in nanotechnology and its application in food industry, the newly created discipline of nanotoxicology investigates safety or potential hazards of nanoparticles (Borchers et al., 2010; Hubbs et al., 2013; Azhdarzadeh et al., 2015; Bitounis et al., 2016). Another dimension refers to genetically modified organisms (GMO) or food (GM) as the potential source of toxicity (Dona and Arvanitoyannis, 2009; Borchers et al., 2010; Ricroch, 2013). All the different disciplines of toxicity assessment in humans are not mutually exclusive but rather highly interconnected. The goal is to identify and understand the molecular mechanisms of toxicants causing adverse effects in order to ultimately prevent their intake, thus increasing food safety (van den Brandt et al., 2002; Borchers et al., 2010).

A major tool in clinical toxicology is the surrogate biomarker molecules employed as specific indicators of organ and tissue damages. As a consequence of the rapid development in biotechnology, an increase in specificity and sensitivity of detection levels led to a better predictivity of these biomarkers. Toxicological assessment of organ and tissue damages can be grouped in to two basic types of biomarkers indicating adverse biological effects on the organism: (1) biomarkers assessing function and integrity of cells and tissues and (2) biomarkers as a damage response of cells and tissues. Integration of data obtained by these two approaches as first line of toxicity assessment is usually validated by histology that may potentially lead to causative relationships relevant to degenerative diseases (Eisenbrand et al., 2002; Frank and Hargreaves, 2003; MacGregor, 2003; Bakhtiar, 2008; Louisse et al., 2012; Campion et al., 2013).

Typically, biomarkers assessing cellular integrity are tissue-specific cytoplasmic enzymes that leak from damaged or dying cells and can be monitored in blood or urine. Usually, a comprehensive clinical chemistry profile is performed employing a variety of tissue-specific biomarkers as first indicators for organ-specific damage such as liver, heart, kidney, brain, and muscles (Frank and Hargreaves, 2003; MacGregor, 2003; Gao et al., 2005; Bakhtiar, 2008; Wagner, 2008; Frederick and Bissell, 2012; Campion et al., 2013). The list of organ-specific biomarker is growing and may include aspartate aminotransferase (AST), alanine aminotransferase (ALT), alkaline phosphatase (ALP), glutamate dehydrogenase (GLDH), and glutathione *S*-transferase (GST-α) as indicators of liver damage. Creatine kinase (CK), myoglobin, and different isoforms of troponins usually serve in muscle damage. These enzymes as well as C-reactive protein (CRP) are considered to be global indicators for heart damage, but cardiac troponins T and I, cardiac natriuretic peptide (CNP), and ischemia-modified albumin (IMA) might be more specific. The occurrence of vascular endothelial growth factor (VEGF), caveolin-1, GRO/CINC-1, thrombospondin-1 (TSP-1), smooth muscle alpha actin, calponin, and transgelin is indicative of vascular injury. There are also several biomarkers indicating kidney damage such as kidney injury molecule (KIM), beta-2-microglobulin (B2M), neutrophil gelatinase–associated lipocalin (NGAL), glutathione *S*-transferase (GST-α), and cystatin C. Neuron-specific enolase (NSE) and neuregulin-1 (NRG-1) among many others might be considered as indicators for brain damage.

The other group of damage response biomarkers is based on the inducible cellular defense system. Due to the nature of most toxicants to be involved in ES-induced cell damage, the list of inducible biomarkers comprises antioxidant enzymes (e.g., superoxide dismutase, catalase, glutathione peroxidase, GST, heme oxygenase, and others), antioxidants (e.g., glutathione), metal-chelating proteins (e.g., transferrin, ferritin, ceruloplasmin, and metallothionein), repair enzymes for peroxidized lipids, proteins and DNA, stress proteins, transcription factors (e.g., p53, NFκB, GADD153), inflammatory (C-reactive protein, interleukin-6, tumor necrosis factor-alpha), immunoresponsive (lymphocyte/macrophage infiltration, leukocyte counts), and xenobiotic reactions (e.g., cytochrome p450 proteins such as CYP1A1 or CYP2E1), among many others (Stocker, 1990; Sies, 1993; Corso et al., 1994; Davies, 2000; MacGregor, 2003; Bakhtiar, 2008; Campion et al., 2013; Ahuja and Sharma, 2014).

As a result of recent developments in metabolomics, the clinical chemistry profile is enhanced by NMR technology to identify intermediary metabolites associated with a variety of pathologies and functional alterations, including renal and hepatic toxicities (MacGregor, 2003; Bakhtiar, 2008; Bouhifd et al., 2013). Furthermore, modern noninvasive techniques such as positron emission tomography (PET), fluorescence magnetic resonance imaging (fMRI), computerized tomography (CT), or single-photon emission computed tomography (SPECT) are increasingly considered for monitoring molecular biomarkers because of the potential for relatively direct clinical translation (Frank and Hargreaves, 2003; MacGregor, 2003; Rudin and Weissleder, 2003; Bakhtiar, 2008; Comley and Kallend, 2013; Matthews et al., 2013; Ahuja and Sharma, 2014).

2.7 PREDICTIVE *IN SILICO* TOXICITY MODELS

The assessment of differential effects of toxicants in different cells, tissues, and organs in the human body and even more throughout different species is a major challenge. In each system, the expression of either different or same key molecules plays critical roles in numerous biological pathways. Because of ethical but also extrapolability concerns, there has been a paradigm shift to reduce animal testing and move from the traditional endpoint testing to a mechanism-based approach by the use of predictive *in silico* toxicity assessment tools (Valerio, 2009; Ahuja and Sharma, 2014; Anadon et al., 2014; Knudsen et al., 2015; Wilk-Zasadna et al., 2015). *In silico* models are used to study pathways of subsequent cellular events, starting from a molecular initiating event through a sequential series of higher-order effects using complex *in vitro* cell-based models and computer algorithms (Valerio, 2009; Nicolotti et al., 2014; Knudsen et al., 2015; Wilk-Zasadna et al., 2015). As a basic principle, quantitative structure–activity relationships (QSARs) between a chemical structure and the biological effects give valuable insights into the molecular mechanisms of action of toxic substances. Correlation analysis with animal and clinical studies led to a high predictivity and robustness of complex cell-based *in vitro* models (Simon-Hettich et al., 2006; Sun et al., 2012; Knudsen et al., 2015; Wilk-Zasadna et al., 2015). The predictive value of QSARs can be greatly enhanced by quantitative *in vitro* to *in vivo* extrapolation when toxicokinetic data on xenobiotic biotransformation, chemical–chemical interactions, absorption, distribution, bioavailability, metabolism, and/or excretion of the substance under study are available (Barlow et al., 2002; Afshari et al., 2011; Louisse et al., 2012; Yoon et al., 2012; Knudsen et al., 2015; Meek and Lipscomb, 2015; Wilk-Zasadna et al., 2015). Through evaluation based on *in silico* approaches, a wealth of mechanistic information of adverse effects of food toxicants were gained and *in silico* models are increasingly recognized as predictive tools to analyze hepatotoxicity, cardiotoxicity, and nephrotoxicity, thus leading to a significant reduction in the number of animals used (Valerio, 2009; Combes, 2012; Louisse et al., 2012; Ouedraogo et al., 2012; Sutter et al., 2013; Wetmore et al., 2013; O'Brien, 2014; Wilk-Zasadna et al., 2015).

2.7.1 High-Throughput and High-Content Screening Methods

Thanks to the dramatic advances made in biotechnology during the last two decades, the majority of toxicity test systems are now analyzed by high-throughput or high-content screening technology (Lee and Dordick, 2006; Houck and Kavlock, 2008; Mahadevan et al., 2011; Sun et al., 2012; O'Brien, 2014; Persson et al., 2014). These newly emerging screening systems were originally developed by the pharmaceutical industry to identify bioactive compounds from huge corporate compound libraries (Gribbon and Sewing, 2005; Macarron, 2006; Sun et al., 2012). Whereas the high-throughput approach is based on quick screening of the biological activity of numerous compounds, high-content screening includes the measurement of many parameters in a single cell or tissue setting (Giuliano et al., 1997; Rausch, 2006; Houck and Kavlock, 2008; Sun et al., 2012; Ahuja and Sharma, 2014; O'Brien, 2014; Knudsen et al., 2015). Most of those systems are strongly driven by fluorescence-based bioassays or biosensor systems employing image analysis algorithm. These systems allow multiplexing fluorescent endpoints by the use of robotic-based screening in multiwell plates and automated liquid handling equipment settings (Houck and Kavlock, 2008; Banerjee et al., 2010; Sun et al., 2012; Ahuja and Sharma, 2014; Nishi et al., 2015). Due to the establishment of the different -omics technologies, the impact of toxicants on genes, proteins, or metabolites is analyzed by toxicogenomics, toxicoproteomics, or toxicometabolomics, respectively, in combination with high-throughput or high-content screening methods (Fabre et al., 2009; Merrick and Witzmann, 2009; George et al., 2010; Kiyosawa et al., 2010; Mahadevan et al., 2011; Ouedraogo et al., 2012; Ahuja and Sharma, 2014). Whereas toxicogenomics and toxicoproteomics are fairly established screening methods, toxicometabolomics has started to be integrated as profiling method in the panels for toxicity assessment only in the past few years (Bouhifd et al., 2013; Ahuja and Sharma, 2014). Now, high-throughput and high-content technology in combination with -omics technologies is routinely used throughout biochemical and cell-based *in vitro*, animal *in vivo* models to clinical and pathology analysis. These new techniques allow very quick screens of a huge number of compounds to yield important mechanistic information on numerous critical cell signaling pathways and cell health parameters with unprecedented quality and reproducibility. Former trade-offs of high-throughput and high-content technology due to high rates of false positives as well as false negatives have been recently addressed by the use of quantitative high-throughput screening (Huang et al., 2008; Xia et al., 2008; Sun et al., 2012).

2.8 CONCLUDING REMARKS

Previously, toxicology assessment relied heavily on the use of animal testing, particularly rodents, for the prediction of potential for toxicity in humans bearing the caveat of high costs, low-throughput readouts, ethical, and extrapolability concerns. Due to these concerns, and also an increase of test candidates, major advances in biotechnology by the use of high-throughput, high-content testing programs, -omics technologies, computational toxicology, as well as the establishment of QSAR prediction models, there has been shift in the paradigm of toxicity screening in the last two decades. Particularly, knowledge gained by QSAR provides insights into the molecular mechanisms of food molecules inducing toxicity. Integration and correlation analysis of toxicological data obtained by these new screening technologies throughout *in vitro* biochemical and cell-based models, *in vivo* experiments, as well as clinical assessment led to a better predictivity of complex *in vitro* cell-based bioassays. The advantage of complex human cell–based models is that there are no issues with interspecies extrapolation and complex toxicological endpoints can often be decomposed into few specific pathways in specific target organs. The evaluation of toxicants through complex *in vitro* human cell–based models evaluated by *in silico* methods is now being increasingly recognized as predictive tools to analyze toxicology. Nevertheless, there is an ongoing demand to increase the specificity and thus predictivity for a variety of test systems. The integration of data obtained by different QSAR-based *in silico* systems will pave the road toward the goal of higher specificity and accuracy aiming for a reduction in animal toxicological testing and human clinical trials. However, there will be a continuous need for traditional *in vivo* testing for novel toxicants to identify whole-animal and organ-level responses to integrate new data in the *in silico* software systems.

The establishment of *in silico* models as validation tools assessing food toxicity parallels a decrease in rodent animal models in favor of alternative animal models for use in food toxicology measurement. This trend was enabled by the discovery of a higher degree of evolutionary conservation of homologous genes between humans and lower vertebrate or invertebrates than assumed decades ago. Now, alternative lower surrogate animal models such as the zebrafish *D. rerio*, the nematode *C. elegans*, and the fruit fly *D. melanogaster* are routinely used. The use of these organisms offers an advantage in the ease of ethical concerns with genetic manipulation, organ toxicology, higher throughput, and costs over traditional rodent models, however, as with all animal models, with extrapolation issues related to species differences. Nevertheless, until prediction of the effects of food toxicants can be more reliably accomplished through computational *in silico* models, some degree of highly targeted testing using rodents or larger vertebrates may be needed, particularly for new food ingredients to give better mechanistic evaluations of *in vivo* responses.

Overall, the achievements in toxicology assessment in the last decades were significant. The deeper understanding of the molecular mode of action by which toxic effects are induced, particularly targets of key biological pathways, led to the identification of potential toxicants in food, thus enhancing food safety. Recently, organotypic systems and stem cell–based assays are discussed as very promising sources for various toxicological applications with a very high potential to predict acute toxicity, thus helping to further answer fundamental biological questions. The list of toxicants is heterogeneous and increasing and comprises different toxins or mycotoxins derived from pathogenic organisms or fungi, heavy metals (e.g., lead, mercury, cadmium, arsenic), pesticides (e.g., dichlorodiphenyltrichloroethane [DDT]), dioxins or dioxin-like compounds (e.g., polychlorinated dibenzofurans [PCDFs] and dibenzodioxins [PCDDs]), polychlorinated biphenyls (PCBs), polybrominated biphenyls (PBBs), polybrominated diphenyl ethers (PBDEs), polycyclic aromatic hydrocarbons (PAHs), polychlorinated naphthalenes, antibiotics, hormones or hormone-modifying agents, acrylamide, acrolein, phthalates, bisphenol A, semicarbazide, allergens, radionuclides, nitrosamines, and aromatic nitro compounds, among many others (Ames et al., 1993; Knight, 1998; Marnett, 2000; Paustenbach, 2000; Kovacic and Jacintho, 2001; Kroes et al., 2002; Borchers et al., 2010).

Although the achievements in the measurement of food toxicology leading to increased food safety are truly significant, there are numerous challenges in predicting and/or translating the effects of toxicants in food due to (1) cross talk of toxicants by interacting with different molecules throughout different toxicity pathways; (2) the fact that toxicity pathways will be perturbed differently due to cell-, tissue-, organ-, and even organism-type dependent variability in gene expression; (3) the aspect that some forms of toxicity are dependent on higher-order interactions of cells in tissues or organs; (4) the interaction of toxicants with other toxicants, drugs, or food matrices; (5) the aspects of digestion and absorption; (6) the aspects of sensitization or desensitization; (7) the aspects of bioavailability, biotransformation, and bioconversion; (8) the aspects of human variability such as heterogeneity, epigenetics, gender, size, and age; (9) the impact of toxicants on different nutritional health, disease status, or other environmental factors; (10) the aspects of xenometabolisms; and (11) the aspects of chemical characterization and normalization of test compounds.

REFERENCES

Afman, L., Milenkovic, D., Roche, H.M., 2014. Nutritional aspects of metabolic inflammation in relation to health-insights from transcriptomic biomarkers in PBMC of fatty acids and polyphenols. *Mol Nutr Food Res* 58, 1708–1720.

Afshari, C.A., Hamadeh, H.K., Bushel, P.R., 2011. The evolution of bioinformatics in toxicology: Advancing toxicogenomics. *Toxicol Sci* 120(Suppl. 1), S225–S237.

Ahuja, V., Sharma, S., 2014. Drug safety testing paradigm, current progress and future challenges: An overview. *J Appl Toxicol* 34, 576–594.

Alfassi, Z., 1999. *General Aspects of the Chemistry of Radicals*. John Wiley, Chichester, U.K.

Ames, B., Shigenaga, M., Hagen, T., 1993. Oxidants, antioxidants, and the degenerative diseases of aging. *Proc Natl Acad Sci USA* 90, 7915–7922.

Anadon, A., Martinez, M.A., Castellano, V., Martinez-Larranaga, M.R., 2014. The role of in vitro methods as alternatives to animals in toxicity testing. *Expert Opin Drug Metab Toxicol* 10, 67–79.

Andersen, M.E., Al-Zoughool, M., Croteau, M., Westphal, M., Krewski, D., 2010. The future of toxicity testing. *J Toxicol Environ Health B Crit Rev* 13, 163–196.

Andersson, S., Norman, M., Olsson, R., Smith, R., Liu, G., Nord, J., 2012. High-precision, room temperature screening assay for inhibitors of microsomal prostaglandin E synthase-1. *J Biomol Screen* 17, 1372–1378.

Antunes, F., Salvador, A., Marinho, H., Alves, R., Pinto, R., 1996. Lipid peroxidation in mitochondrial inner membranes. I. An integrative kinetic model. *Free Radic Biol Med* 21, 917–943.

Astashkina, A., Mann, B., Grainger, D.W., 2012. A critical evaluation of in vitro cell culture models for high-throughput drug screening and toxicity. *Pharmacol Ther* 134, 82–106.

Avila, D., Helmcke, K., Aschner, M., 2012. The *Caenorhabiditis elegans* model as a reliable tool in neurotoxicology. *Hum Exp Toxicol* 31, 236–243.

Azhdarzadeh, M., Saei, A.A., Sharifi, S., Hajipour, M.J., Alkilany, A.M., Sharifzadeh, M., Ramazani, F., Laurent, S., Mashaghi, A., Mahmoudi, M., 2015. Nanotoxicology: Advances and pitfalls in research methodology. *Nanomedicine (Lond)* 10, 2931–2952.

Bakhtiar, R., 2008. Biomarkers in drug discovery and development. *J Pharmacol Toxicol Methods* 57, 85–91.

Banerjee, P., Franz, B., Bhunia, A.K., 2010. Mammalian cell-based sensor system. *Adv Biochem Eng Biotechnol* 117, 21–55.

Barbazuk, W.B., Korf, I., Kadavi, C., Heyen, J., Tate, S., Wun, E., Bedell, J.A., McPherson, J.D., Johnson, S.L., 2000. The syntenic relationship of the zebrafish and human genomes. *Genome Res* 10, 1351–1358.

Barlow, S.M., Greig, J.B., Bridges, J.W., Carere, A., Carpy, A.J., Galli, C.L., Kleiner, J. et al., 2002. Hazard identification by methods of animal-based toxicology. *Food Chem Toxicol* 40, 145–191.

Berg, K., Hansen, M., Nielsen, S., 1990. A new sensitive bioassay for precise quantification of interferon activity as measured via the mitochondrial dehydrogenase function in cells (MTT-method). *APMIS* 98, 156–162.

Bernheim, F., Bernheim, M., Wilbur, K., 1948. The reaction between thiobarbituric acid and the oxidation products of certain lipids. *J Biol Chem* 174, 257–264.

Berridge, M.V., Herst, P.M., Tan, A.S., 2005. Tetrazolium dyes as tools in cell biology: New insights into their cellular reduction. *Biotechnol Annu Rev* 11, 127–152.

Bier, E., 2005. *Drosophila*, the golden bug, emerges as a tool for human genetics. *Nat Rev Genet* 6, 9–23.

Bitounis, D., Pourchez, J., Forest, V., Boudard, D., Cottier, M., Klein, J.P., 2016. Detection and analysis of nanoparticles in patients: A critical review of the status quo of clinical nanotoxicology. *Biomaterials* 76, 302–312.

Borchers, A., Teuber, S.S., Keen, C.L., Gershwin, M.E., 2010. Food safety. *Clin Rev Allergy Immunol* 39, 95–141.

Borenfreund, E., Puerner, J.A., 1985. Toxicity determined in vitro by morphological alterations and neutral red absorption. *Toxicol Lett* 24, 119–124.

Bouhifd, M., Hartung, T., Hogberg, H.T., Kleensang, A., Zhao, L., 2013. Review: Toxicometabolomics. *J Appl Toxicol* 33, 1365–1383.

Brandon, E.F., Raap, C.D., Meijerman, I., Beijnen, J.H., Schellens, J.H., 2003. An update on in vitro test methods in human hepatic drug biotransformation research: pros and cons. *Toxicol Appl Pharmacol* 189, 233–246.

Buettner, G., 1993. The pecking order of free radicals and antioxidants. *Arch Biochem Biophys* 300, 535–543.

Cadenas, E., Davies, K., 2000. Mitochondrial free radical generation, oxidative stress, and aging. *Free Radic Biol Med* 29, 222–230.

Calle, M.C., Fernandez, M.L., 2012. Inflammation and type 2 diabetes. *Diabetes Metab* 38, 183–191.

Campion, S., Aubrecht, J., Boekelheide, K., Brewster, D.W., Vaidya, V.S., Anderson, L., Burt, D. et al., 2013. The current status of biomarkers for predicting toxicity. *Expert Opin Drug Metab Toxicol* 9, 1391–1408.

Chakravarthy, S., Sadagopan, S., Nair, A., Sukumaran, S.K., 2014. Zebrafish as an in vivo high-throughput model for genotoxicity. *Zebrafish* 11, 154–166.

Chen, M., Bisgin, H., Tong, L., Hong, H., Fang, H., Borlak, J., Tong, W., 2014. Toward predictive models for drug-induced liver injury in humans: Are we there yet? *Biomark Med* 8, 201–213.

Chen, M., Zhang, M., Borlak, J., Tong, W., 2012. A decade of toxicogenomic research and its contribution to toxicological science. *Toxicol Sci* 130, 217–228.

Cho, A., Shin, J., Hwang, S., Kim, C., Shim, H., Kim, H., Kim, H., Lee, I., 2014. WormNet v3: A network-assisted hypothesis-generating server for *Caenorhabditis elegans*. *Nucleic Acids Res* 42, W76–W82.

Combes, R.D., 2012. In silico methods for toxicity prediction. *Adv Exp Med Biol* 745, 96–116.

Comley, R.A., Kallend, D., 2013. Imaging in the cardiovascular and metabolic disease area. *Drug Discov Today* 18, 185–192.

Corso, A.D., Cappiello, M., Mura, U., 1994. Thiol dependent oxidation of enzymes: The last chance against oxidative stress. *Int J Biochem Cell Biol* 26, 745–750.

Cottrell, D., Blakely, E., Borthwick, G., Johnson, M., Taylor, G., Brierley, E., Ince, P., Turnbull, D., 2000. Role of mitochondrial DNA mutations in disease and aging. *Ann NY Acad Sci* 908, 199–207.

Crofton, K.M., Mundy, W.R., Shafer, T.J., 2012. Developmental neurotoxicity testing: A path forward. *Congenit Anom (Kyoto)* 52, 140–146.

Croteau, D., Bohr, V., 1997. Repair of oxidative damage to nuclear and mitochondrial DNA in mammalian cells. *J Biol Chem* 272, 25409–25412.

Custer, L.L., Sweder, K.S., 2008. The role of genetic toxicology in drug discovery and optimization. *Curr Drug Metab* 9, 978–985.

Dash, A., Blackman, B.R., Wamhoff, B.R., 2012. Organotypic systems in drug metabolism and toxicity: Challenges and opportunities. *Expert Opin Drug Metab Toxicol* 8, 999–1014.

Davies, K., 1993. Protein modification by oxidants and the role of proteolytic enzymes. *Biochem Soc Trans* 21, 346–353.

Davies, K., 1995. *Oxidative Stress: The Paradox of Aerobic Life*. Portland Press, London, U.K.

Davies, K., 2000. Oxidative stress, antioxidant defenses, and damage removal, repair, and replacement systems. *IUBMB Life* 50, 279–289.

Decker, T., Lohmann-Matthes, M.L., 1988. A quick and simple method for the quantitation of lactate dehydrogenase release in measurements of cellular cytotoxicity and tumor necrosis factor (TNF) activity. *J Immunol Methods* 115, 61–69.

Dianzani, M., Barrera, G., 2008. *Pathology and Physiology of Lipid Peroxidation and Its Carbonyl Products*. Transworld Research Network, Kerala, India.

Dinis-Oliveira, R.J., Carvalho, F., Duarte, J.A., Remiao, F., Marques, A., Santos, A., Magalhaes, T., 2010. Collection of biological samples in forensic toxicology. *Toxicol Mech Methods* 20, 363–414.

Dona, A., Arvanitoyannis, I.S., 2009. Health risks of genetically modified foods. *Crit Rev Food Sci Nutr* 49, 164–175.

Donato, M.T., Lahoz, A., Castell, J.V., Gomez-Lechon, M.J., 2008. Cell lines: A tool for in vitro drug metabolism studies. *Curr Drug Metab* 9, 1–11.

Drysdale, R., FlyBase, C., 2008. FlyBase: A database for the *Drosophila* research community. *Methods Mol Biol* 420, 45–59.

Dykens, J.A., Will, Y., 2008. *Drug-Induced Mitochondrial Dysfunction*. Wiley, Hoboken, NJ.

Eisenbrand, G., Pool-Zobel, B., Baker, V., Balls, M., Blaauboer, B.J., Boobis, A., Carere, A. et al., 2002. Methods of in vitro toxicology. *Food Chem Toxicol* 40, 193–236.

Eppinger, J., Funeriu, D.P., Miyake, M., Denizot, L., Miyake, J., 2004. Enzyme microarrays: On-chip determination of inhibition constants based on affinity-label detection of enzymatic activity. *Angew Chem Int Ed Engl* 43, 3806–3810.

Erickson, B.E., 2011. Modernizing toxicity tests. *Chem Eng News* 89, 25–26.

Fabre, N., Anglade, I., Vericat, J.A., 2009. Application of toxicogenomic tools in the drug research and development process. *Toxicol Lett* 186, 13–17.

Ferreiro, S.F., Carrera, C., Vilarino, N., Louzao, M.C., Santamarina, G., Cantalapiedra, A.G., Botana, L.M., 2015. Acute cardiotoxicity evaluation of the marine biotoxins OA, DTX-1 and YTX. *Toxins (Basel)* 7, 1030–1047.

Finkel, T., Holbrook, N., 2000. Oxidants, oxidants stress and the biology of ageing. *Nature* 408, 239–247.

Floyd, R., 1990. Role of oxygen free radicals in carcinogenesis and brain ischemia. *FASEB J* 4, 2587–2597.

Foti, R.S., Wienkers, L.C., Wahlstrom, J.L., 2010. Application of cytochrome P450 drug interaction screening in drug discovery. *Comb Chem High Throughput Screen* 13, 145–158.

Franco, R., Sanchez-Olea, R., Reyes-Reyes, E.M., Panayiotidis, M.I., 2009. Environmental toxicity, oxidative stress and apoptosis: Menage a trois. *Mutat Res* 674, 3–22.

Frank, R., Hargreaves, R., 2003. Clinical biomarkers in drug discovery and development. *Nat Rev Drug Discov* 2, 566–580.

Frederick, D.L., Bissell, M.G., 2012. Overview of progress in clinical toxicology testing. *Clin Lab Med* 32, 353–359.

Frohlich, E., Salar-Behzadi, S., 2014. Toxicological assessment of inhaled nanoparticles: Role of in vivo, ex vivo, in vitro, and in silico studies. *Int J Mol Sci* 15, 4795–4822.

Gao, J., Garulacan, L.A., Storm, S.M., Opiteck, G.J., Dubaquie, Y., Hefta, S.A., Dambach, D.M., Dongre, A.R., 2005. Biomarker discovery in biological fluids. *Methods* 35, 291–302.

Garthoff, B., 2005. Alternatives to animal experimentation: The regulatory background. *Toxicol Appl Pharmacol* 207, 388–392.

George, J., Singh, R., Mahmood, Z., Shukla, Y., 2010. Toxicoproteomics: New paradigms in toxicology research. *Toxicol Mech Methods* 20, 415–423.

Ghosh, S., Hayden, M.S., 2008. New regulators of NF-kappaB in inflammation. *Nat Rev Immunol* 8, 837–848.

Giacomotto, J., Segalat, L., 2010. High-throughput screening and small animal models, where are we? *Br J Pharmacol* 160, 204–216.

Giuliano, K.A., DeBiasio, R.L., Dunlay, R.T., Gough, A., Volosky, J.M., Zock, J., Pavlakis, G.N., Taylor, D.L., 1997. High-content screening: A new approach to easing key bottlenecks in the drug discovery process. *J Biomol Screen* 2, 249–259.

Gomez-Lechon, M.J., Tolosa, L., Conde, I., Donato, M.T., 2014. Competency of different cell models to predict human hepatotoxic drugs. *Expert Opin Drug Metab Toxicol* 10, 1553–1568.

Gonzalez Flecha, B., Llesuy, S., Boveris, A., 1991. Hydroperoxide-initiated chemiluminescence: An assay for oxidative stress in biopsies of heart, liver, and muscle. *Free Radic Biol Med* 10, 93–100.

Gosslau, A., Chen, M., Ho, C., Chen, K., 2005. A methoxy derivative of resveratrol analogue selectively induced activation of the mitochondrial apoptotic pathway in transformed fibroblasts. *Br J Cancer* 92, 513–521.

Gosslau, A., Li, S., Ho, C.T., Chen, K.Y., Rawson, N.E., 2011. The importance of natural product characterization in studies of their anti-inflammatory activity. *Mol Nutr Food Res* 55, 74–82.

Gosslau, A., Sisk, S., Huang, M.-T., Ho, C.-T., Chen, K.Y., 2008. Case study of an anti-inflammatory ingredient discovered via nutrigenomic screening. *J Nutrigenet Nutrigenomics* 1, 76–77.

Gribbon, P., Sewing, A., 2005. High-throughput drug discovery: What can we expect from HTS? *Drug Discov Today* 10, 17–22.

Gruber, L., Lichti, P., Rath, E., Haller, D., 2012. Nutrigenomics and nutrigenetics in inflammatory bowel diseases. *J Clin Gastroenterol* 46, 735–747.

Halliwell, B., Gutteridge, J., 1984. Oxygen toxicity, oxygen radicals, transition metals and disease. *Biochem J* 219, 1–14.

Hariparsad, N., Sane, R.S., Strom, S.C., Desai, P.B., 2006. In vitro methods in human drug biotransformation research: Implications for cancer chemotherapy. *Toxicol In Vitro* 20, 135–153.

Hartung, T., Corsini, E., 2013. Immunotoxicology: Challenges in the 21st century and in vitro opportunities. *ALTEX* 30, 411–426.

Heinonen, T., 2015. Better science with human cell-based organ and tissue models. *Altern Lab Anim* 43, 29–38.

Hennes, E.C., 2012. An overview of values for the threshold of toxicological concern. *Toxicol Lett* 211, 296–303.

Hill, A.J., Teraoka, H., Heideman, W., Peterson, R.E., 2005. Zebrafish as a model vertebrate for investigating chemical toxicity. *Toxicol Sci* 86, 6–19.

Hohn, A., Jung, T., Grune, T., 2014. Pathophysiological importance of aggregated damaged proteins. *Free Radic Biol Med* 71, 70–89.

Horiuchi, K.Y., Wang, Y., Diamond, S.L., Ma, H., 2006. Microarrays for the functional analysis of the chemical-kinase interactome. *J Biomol Screen* 11, 48–56.

Houck, K.A., Kavlock, R.J., 2008. Understanding mechanisms of toxicity: Insights from drug discovery research. *Toxicol Appl Pharmacol* 227, 163–178.

Howitz, K.T., Bitterman, K.J., Cohen, H.Y., Lamming, D.W., Lavu, S., Wood, J.G., Zipkin, R.E. et al., 2003. Small molecule activators of sirtuins extend Saccharomyces cerevisiae lifespan. *Nature* 425, 191–196.

Huang, J.X., Blaskovich, M.A., Cooper, M.A., 2014. Cell- and biomarker-based assays for predicting nephrotoxicity. *Expert Opin Drug Metab Toxicol* 10, 1621–1635.

Huang, R., Southall, N., Cho, M.H., Xia, M., Inglese, J., Austin, C.P., 2008. Characterization of diversity in toxicity mechanism using in vitro cytotoxicity assays in quantitative high throughput screening. *Chem Res Toxicol* 21, 659–667.

Hubbs, A.F., Sargent, L.M., Porter, D.W., Sager, T.M., Chen, B.T., Frazer, D.G., Castranova, V. et al., 2013. Nanotechnology: Toxicologic pathology. *Toxicol Pathol* 41, 395–409.

Hwang, E.S., Bowen, P.E., 2007. DNA damage, a biomarker of carcinogenesis: Its measurement and modulation by diet and environment. *Crit Rev Food Sci Nutr* 47, 27–50.

Ischiropoulos, H., Zhu, L., Beckman, J., 1992. Peroxynitrite formation from macrophage-derived nitric oxide. *Arch Biochem Biophys* 298, 446–451.

Itagaki, H., Hagino, S., Kato, S., Kobayashi, T., Umeda, M., 1991. An in vitro alternative to the Draize eye-irritation test: Evaluation of the crystal violet staining method. *Toxicol In Vitro* 5, 139–143.
Jeffrey, A.M., Williams, G.M., 2005. Risk assessment of DNA-reactive carcinogens in food. *Toxicol Appl Pharmacol* 207, 628–635.
Johnson, J.Z., Hockemeyer, D., 2015. Human stem cell-based disease modeling: Prospects and challenges. *Curr Opin Cell Biol* 37, 84–90.
Jones, D.P., 2006. Redefining oxidative stress. *Antioxid Redox Signal* 8, 1865–1879.
Kadereit, S., Zimmer, B., van Thriel, C., Hengstler, J.G., Leist, M., 2012. Compound selection for in vitro modeling of developmental neurotoxicity. *Front Biosci (Landmark Ed)* 17, 2442–2460.
Kastle, M., Grune, T., 2011. Protein oxidative modification in the aging organism and the role of the ubiquitin proteasomal system. *Curr Pharm Des* 17, 4007–4022.
Kettenhofen, R., Bohlen, H., 2008. Preclinical assessment of cardiac toxicity. *Drug Discov Today* 13, 702–707.
Kirkland, D., Pfuhler, S., Tweats, D., Aardema, M., Corvi, R., Darroudi, F., Elhajouji, A. et al., 2007. How to reduce false positive results when undertaking in vitro genotoxicity testing and thus avoid unnecessary follow-up animal tests: Report of an ECVAM Workshop. *Mutat Res* 628, 31–55.
Kiyosawa, N., Manabe, S., Yamoto, T., Sanbuissho, A., 2010. Practical application of toxicogenomics for profiling toxicant-induced biological perturbations. *Int J Mol Sci* 11, 3397–3412.
Knight, J., 1998. Free radicals: Their history and current status in aging and disease. *Ann Clin Lab Sci* 28, 331–346.
Knudsen, T.B., Keller, D.A., Sander, M., Carney, E.W., Doerrer, N.G., Eaton, D.L., Fitzpatrick, S.C. et al., 2015. FutureTox II: In vitro data and in silico models for predictive toxicology. *Toxicol Sci* 143, 256–267.
Kovacic, P., Jacintho, J., 2001. Mechanisms of carcinogenesis: Focus on oxidative stress and electron transfer. *Curr Med Chem* 8, 773–796.
Kraushaar, U., Meyer, T., Hess, D., Gepstein, L., Mummery, C.L., Braam, S.R., Guenther, E., 2012. Cardiac safety pharmacology: From human ether-a-gogo related gene channel block towards induced pluripotent stem cell based disease models. *Expert Opin Drug Saf* 11, 285–298.
Kroes, R., Muller, D., Lambe, J., Lowik, M.R., van Klaveren, J., Kleiner, J., Massey, R. et al., 2002. Assessment of intake from the diet. *Food Chem Toxicol* 40, 327–385.
Landskron, G., De la Fuente, M., Thuwajit, P., Thuwajit, C., Hermoso, M.A., 2014. Chronic inflammation and cytokines in the tumor microenvironment. *J Immunol Res* 2014, 149185.
Lee, M.Y., Dordick, J.S., 2006. High-throughput human metabolism and toxicity analysis. *Curr Opin Biotechnol* 17, 619–627.
Lee, M.Y., Park, C.B., Dordick, J.S., Clark, D.S., 2005. Metabolizing enzyme toxicology assay chip (MetaChip) for high-throughput microscale toxicity analyses. *Proc Natl Acad Sci USA* 102, 983–987.
Leung, M.C., Williams, P.L., Benedetto, A., Au, C., Helmcke, K.J., Aschner, M., Meyer, J.N., 2008. *Caenorhabditis elegans*: An emerging model in biomedical and environmental toxicology. *Toxicol Sci* 106, 5–28.
Li, Z., Malla, S., Shin, B., Li, J.M., 2014. Battle against RNA oxidation: Molecular mechanisms for reducing oxidized RNA to protect cells. *Wiley Interdiscip Rev RNA* 5, 335–346.
Liu, J., Yeo, H., Doniger, S., Ames, B., 1997. Assay of aldehydes from lipid peroxidation: Gas chromatograph-mass spectrometry compared to thiobarbituric acid. *Anal Biochem* 245, 161–166.
Liu, W., Deng, Y., Liu, Y., Gong, W., Deng, W., 2013. Stem cell models for drug discovery and toxicology studies. *J Biochem Mol Toxicol* 27, 17–27.
Louisse, J., Verwei, M., Woutersen, R.A., Blaauboer, B.J., Rietjens, I.M., 2012. Toward in vitro biomarkers for developmental toxicity and their extrapolation to the in vivo situation. *Expert Opin Drug Metab Toxicol* 8, 11–27.
Love, D.R., Pichler, F.B., Dodd, A., Copp, B.R., Greenwood, D.R., 2004. Technology for high-throughput screens: The present and future using zebrafish. *Curr Opin Biotechnol* 15, 564–571.
Luebke, R., 2012. Immunotoxicant screening and prioritization in the twenty-first century. *Toxicol Pathol* 40, 294–299.
Macarron, R., 2006. Critical review of the role of HTS in drug discovery. *Drug Discov Today* 11, 277–279.
MacGregor, J.T., 2003. The future of regulatory toxicology: Impact of the biotechnology revolution. *Toxicol Sci* 75, 236–248.
Maeda, H., Ishikawa, T., Michiue, T., 2014. Forensic molecular pathology: Its impacts on routine work, education and training. *Leg Med (Tokyo)* 16, 61–69.
Mahadevan, B., Snyder, R.D., Waters, M.D., Benz, R.D., Kemper, R.A., Tice, R.R., Richard, A.M., 2011. Genetic toxicology in the 21st century: Reflections and future directions. *Environ Mol Mutagen* 52, 339–354.

Marnett, L., 2000. Oxyradicals and DNA damage. *Carcinogenesis* 21, 361–370.

Matthews, P.M., Coatney, R., Alsaid, H., Jucker, B., Ashworth, S., Parker, C., Changani, K., 2013. Technologies: Preclinical imaging for drug development. *Drug Discov Today Technol* 10, e343—e350.

McCann, J., Choi, E., Yamasaki, E., Ames, B.N., 1975. Detection of carcinogens as mutagens in the Salmonella/microsome test: Assay of 300 chemicals. *Proc Natl Acad Sci USA* 72, 5135–5139.

McGonigle, P., Ruggeri, B., 2014. Animal models of human disease: Challenges in enabling translation. *Biochem Pharmacol* 87, 162–171.

McGrath, P., Li, C.Q., 2008. Zebrafish: A predictive model for assessing drug-induced toxicity. *Drug Discov Today* 13, 394–401.

McVey, K.A., Mink, J.A., Snapp, I.B., Timberlake, W.S., Todt, C.E., Negga, R., Fitsanakis, V.A., 2012. *Caenorhabditis elegans*: An emerging model system for pesticide neurotoxicity. *J Environ Anal Toxicol* 2, 125–135.

Meek, M.E., Lipscomb, J.C., 2015. Gaining acceptance for the use of in vitro toxicity assays and QIVIVE in regulatory risk assessment. *Toxicology* 332, 112–123.

Merlot, C., 2010. Computational toxicology-A tool for early safety evaluation. *Drug Discov Today* 15, 16–22.

Merrick, B.A., Witzmann, F.A., 2009. The role of toxicoproteomics in assessing organ specific toxicity. *EXS* 99, 367–400.

Meyer, D., Williams, P.L., 2014. Toxicity testing of neurotoxic pesticides in *Caenorhabditis elegans*. *J Toxicol Environ Health B Crit Rev* 17, 284–306.

Meyer, J.N., Leung, M.C., Rooney, J.P., Sendoel, A., Hengartner, M.O., Kisby, G.E., Bess, A.S., 2013. Mitochondria as a target of environmental toxicants. *Toxicol Sci* 134, 1–17.

Milne, G.L., Sanchez, S.C., Musiek, E.S., Morrow, J.D., 2007. Quantification of F2-isoprostanes as a biomarker of oxidative stress. *Nat Protoc* 2, 221–226.

Moller, P., Danielsen, P.H., Jantzen, K., Roursgaard, M., Loft, S., 2013. Oxidatively damaged DNA in animals exposed to particles. *Crit Rev Toxicol* 43, 96–118.

Montuschi, P., Barnes, P.J., Roberts, L.J., 2nd, 2004. Isoprostanes: Markers and mediators of oxidative stress. *FASEB J* 18, 1791–1800.

Moore, K.M., Duddy, A., Braun, M.M., Platt, R., Brown, J.S., 2008. Potential population-based electronic data sources for rapid pandemic influenza vaccine adverse event detection: A survey of health plans. *Pharmacoepidemiol Drug Saf* 17, 1137–1141.

Mortelmans, K., Zeiger, E., 2000. The Ames Salmonella/microsome mutagenicity assay. *Mutat Res* 455, 29–60.

Newman, D.J., Cragg, G.M., 2007. Natural products as sources of new drugs over the last 25 years. *J Nat Prod* 70, 461–477.

Nichols, C.D., 2006. *Drosophila melanogaster* neurobiology, neuropharmacology, and how the fly can inform central nervous system drug discovery. *Pharmacol Ther* 112, 677–700.

Nichols, J.E., Niles, J.A., Vega, S.P., Cortiella, J., 2013. Novel in vitro respiratory models to study lung development, physiology, pathology and toxicology. *Stem Cell Res Ther* 4(Suppl. 1), S7.

Nicolotti, O., Benfenati, E., Carotti, A., Gadaleta, D., Gissi, A., Mangiatordi, G.F., Novellino, E., 2014. REACH and in silico methods: An attractive opportunity for medicinal chemists. *Drug Discov Today* 19, 1757–1768.

Niki, E., 2009. Lipid peroxidation: Physiological levels and dual biological effects. *Free Radic Biol Med* 47, 469–484.

Nishi, K., Isobe, S., Zhu, Y., Kiyama, R., 2015. Fluorescence-based bioassays for the detection and evaluation of food materials. *Sensors* (*Basel*) 15, 25831–25867.

O'Brien, P.J., 2014. High-content analysis in toxicology: Screening substances for human toxicity potential, elucidating subcellular mechanisms and in vivo use as translational safety biomarkers. *Basic Clin Pharmacol Toxicol* 115, 4–17.

Olson, H., Betton, G., Robinson, D., Thomas, K., Monro, A., Kolaja, G., Lilly, P. et al., 2000. Concordance of the toxicity of pharmaceuticals in humans and in animals. *Regul Toxicol Pharmacol* 32, 56–67.

Ouedraogo, M., Baudoux, T., Stevigny, C., Nortier, J., Colet, J.M., Efferth, T., Qu, F. et al., 2012. Review of current and "omics" methods for assessing the toxicity (genotoxicity, teratogenicity and nephrotoxicity) of herbal medicines and mushrooms. *J Ethnopharmacol* 140, 492–512.

Ouyang, L., Shi, Z., Zhao, S., Wang, F.T., Zhou, T.T., Liu, B., Bao, J.K., 2012. Programmed cell death pathways in cancer: A review of apoptosis, autophagy and programmed necrosis. *Cell Prolif* 45, 487–498.

Oyama, Y., Okazaki, E., Chikahisa, L., Nagano, T., Sadakata, C., 1996. Oxidative stress-induced increase in intracellular Ca^{2+} and Ca^{2+}-induced increase in oxidative stress: An experimental model using disassociated rat brain neurons. *Jpn J Pharmacol* 72, 381–385.

Pandey, U.B., Nichols, C.D., 2011. Human disease models in *Drosophila melanogaster* and the role of the fly in therapeutic drug discovery. *Pharmacol Rev* 63, 411–436.

Patterson, A.D., Gonzalez, F.J., Idle, J.R., 2010. Xenobiotic metabolism: A view through the metabolometer. *Chem Res Toxicol* 23, 851–860.

Paustenbach, D.J., 2000. The practice of exposure assessment: A state-of-the-art review. *J Toxicol Environ Health B Crit Rev* 3, 179–291.

Persson, M., Loye, A.F., Jacquet, M., Mow, N.S., Thougaard, A.V., Mow, T., Hornberg, J.J., 2014. High-content analysis/screening for predictive toxicology: Application to hepatotoxicity and genotoxicity. *Basic Clin Pharmacol Toxicol* 115, 18–23.

Peters, M.F., Lamore, S.D., Guo, L., Scott, C.W., Kolaja, K.L., 2015. Human stem cell-derived cardiomyocytes in cellular impedance assays: Bringing cardiotoxicity screening to the front line. *Cardiovasc Toxicol* 15, 127–139.

Prussing, K., Voigt, A., Schulz, J.B., 2013. *Drosophila melanogaster* as a model organism for Alzheimer's disease. *Mol Neurodegener* 8, 35.

Rajat, S., Manisha, S., Sunil, K., 2012. Nutraceuticals: A review. *Intern Res J Pharm* 3, 95–99.

Raldua, D., Pina, B., 2014. In vivo zebrafish assays for analyzing drug toxicity. *Expert Opin Drug Metab Toxicol* 10, 685–697.

Ramasamy, R., Yan, S.F., Schmidt, A.M., 2012. Advanced glycation endproducts: From precursors to RAGE: Round and round we go. *Amino Acids* 42, 1151–1161.

Rampersad, S.N., 2012. Multiple applications of Alamar Blue as an indicator of metabolic function and cellular health in cell viability bioassays. *Sensors (Basel)* 12, 12347–12360.

Rausch, O., 2006. High content cellular screening. *Curr Opin Chem Biol* 10, 316–320.

Repetto, M.G., 2008. *Clinical Use of Chemiluminescence Assays for the Determination of Systemic Oxidative Stress*. Transworld Research Network, Kerala, India.

Reuter, S., Gupta, S.C., Chaturvedi, M.M., Aggarwal, B.B., 2010. Oxidative stress, inflammation, and cancer: How are they linked? *Free Radic Biol Med* 49, 1603–1616.

Ricroch, A.E., 2013. Assessment of GE food safety using '-omics' techniques and long-term animal feeding studies. *N Biotechnol* 30, 349–354.

Roberts, R.A., Laskin, D.L., Smith, C.V., Robertson, F.M., Allen, E.M., Doorn, J.A., Slikker, W., 2009. Nitrative and oxidative stress in toxicology and disease. *Toxicol Sci* 112, 4–16.

Ruan, Q.L., Ju, J.J., Li, Y.H., Liu, R., Pu, Y.P., Yin, L.H., Wang, D.Y., 2009. Evaluation of pesticide toxicities with differing mechanisms using *Caenorhabditis elegans*. *J Toxicol Environ Health A* 72, 746–751.

Rubin, G.M., Lewis, E.B., 2000. A brief history of *Drosophila's* contributions to genome research. *Science* 287, 2216–2218.

Rudin, M., Weissleder, R., 2003. Molecular imaging in drug discovery and development. *Nat Rev Drug Discov* 2, 123–131.

San, R.H.C., 2006. *Preclinical Genotoxicity Testing-Past, Present, and Future*. Informa Healthcare, New York.

Sanguinetti, M.C., Tristani-Firouzi, M., 2006. hERG potassium channels and cardiac arrhythmia. *Nature* 440, 463–469.

Saotome, K., Morita, H., Umeda, M., 1989. Cytotoxicity test with simplified crystal violet staining method using microtitre plates and its application to injection drugs. *Toxicol In Vitro* 3, 317–321.

Segalat, L., 2007. Invertebrate animal models of diseases as screening tools in drug discovery. *ACS Chem Biol* 2, 231–236.

Shamir, E.R., Ewald, A.J., 2014. Three-dimensional organotypic culture: Experimental models of mammalian biology and disease. *Nat Rev Mol Cell Biol* 15, 647–664.

Sies, H., 1993. Strategies of antioxidant defense. *Eur J Biochem* 215, 213–219.

Simon-Hettich, B., Rothfuss, A., Steger-Hartmann, T., 2006. Use of computer-assisted prediction of toxic effects of chemical substances. *Toxicology* 224, 156–162.

Siti, H.N., Kamisah, Y., Kamsiah, J., 2015. The role of oxidative stress, antioxidants and vascular inflammation in cardiovascular disease (a review). *Vascul Pharmacol* 71, 40–56.

Smith, R.A., 2009. Twenty-first century challenges for in vitro neurotoxicity. *Altern Lab Anim* 37, 367–375.

Snyder, R.D., 2007. Assessment of atypical DNA intercalating agents in biological and in silico systems. *Mutat Res* 623, 72–82.

Sohal, R., 1993. Aging, cytochrome oxidase activity, and hydrogen peroxide release by mitochondria. *Free Radic Biol Med* 14, 583–588.

Stadtman, E., 1992. Protein oxidation and aging. *Science* 257, 1220–1224.

Stocker, R., 1990. Induction of haem oxygenase as a defense against oxidative stress. *Free Radic Res Commun* 9, 101–112.

Stoddart, M.J., 2011. Cell viability assays: Introduction. *Methods Mol Biol* 740, 1–6.
Stohs, S., Bagchi, D., 1995. Oxidative mechanisms in the toxicity of metal ions. *Free Radic Biol Med* 18, 321–326.
Sumantran, V.N., 2011. Cellular chemosensitivity assays: An overview. *Methods Mol Biol* 731, 219–236.
Sun, H., Xia, M., Austin, C.P., Huang, R., 2012. Paradigm shift in toxicity testing and modeling. *AAPS J* 14, 473–480.
Sutter, A., Amberg, A., Boyer, S., Brigo, A., Contrera, J.F., Custer, L.L., Dobo, K.L. et al., 2013. Use of in silico systems and expert knowledge for structure-based assessment of potentially mutagenic impurities. *Regul Toxicol Pharmacol* 67, 39–52.
Taboureau, O., Jorgensen, F.S., 2011. In silico predictions of hERG channel blockers in drug discovery: From ligand-based and target-based approaches to systems chemical biology. *Comb Chem High Throughput Screen* 14, 375–387.
Takahashi, K., Yamanaka, S., 2006. Induction of pluripotent stem cells from mouse embryonic and adult fibroblast cultures by defined factors. *Cell* 126, 663–676.
Tiong, H.Y., Huang, P., Xiong, S., Li, Y., Vathsala, A., Zink, D., 2014. Drug-induced nephrotoxicity: Clinical impact and preclinical in vitro models. *Mol Pharm* 11, 1933–1948.
Trubetskoy, O.V., Gibson, J.R., Marks, B.D., 2005. Highly miniaturized formats for in vitro drug metabolism assays using vivid fluorescent substrates and recombinant human cytochrome P450 enzymes. *J Biomol Screen* 10, 56–66.
Trump, B., Berezesky, I., 1995. Calcium-mediated cell injury and cell death. *FASEB J* 9, 219–228.
Uliasz, T.F., Hewett, S.J., 2000. A microtiter trypan blue absorbance assay for the quantitative determination of excitotoxic neuronal injury in cell culture. *J Neurosci Methods* 100, 157–163.
Uttara, B., Singh, A.V., Zamboni, P., Mahajan, R.T., 2009. Oxidative stress and neurodegenerative diseases: A review of upstream and downstream antioxidant therapeutic options. *Curr Neuropharmacol* 7, 65–74.
Valerio, L.G., Jr., 2009. In silico toxicology for the pharmaceutical sciences. *Toxicol Appl Pharmacol* 241, 356–370.
Van Cauteren, H., Lumley, C.E., 1997. Harmonization of international toxicity testing guidelines for pharmaceuticals. Contributions to refinement and reduction in animal use. *Eur Biomed Res Assoc Bull* 11, 4–9.
van den Brandt, P., Voorrips, L., Hertz-Picciotto, I., Shuker, D., Boeing, H., Speijers, G., Guittard, C. et al., 2002. The contribution of epidemiology. *Food Chem Toxicol* 40, 387–424.
van Tongeren, S., Fagerland, J.A., Conner, M.W., Diegel, K., Donnelly, K., Grubor, B., Lopez-Martinez, A. et al., 2011. The role of the toxicologic pathologist in the biopharmaceutical industry. *Int J Toxicol* 30, 568–582.
Vojnits, K., Bremer, S., 2010. Challenges of using pluripotent stem cells for safety assessments of substances. *Toxicology* 270, 10–17.
Wagner, J.A., 2008. Strategic approach to fit-for-purpose biomarkers in drug development. *Annu Rev Pharmacol Toxicol* 48, 631–651.
Wallace, D., 1999. Mitochondrial diseases in man and mouse. *Science* 283, 1482–1488.
Waters, M.D., Stack, H.F., Jackson, M.A., Bridges, B.A., Adler, I.D., 1994. The performance of short-term tests in identifying potential germ cell mutagens: A qualitative and quantitative analysis. *Mutat Res* 341, 109–131.
Wetmore, B.A., Wambaugh, J.F., Ferguson, S.S., Li, L., Clewell, H.J., 3rd, Judson, R.S., Freeman, K. et al., 2013. Relative impact of incorporating pharmacokinetics on predicting in vivo hazard and mode of action from high-throughput in vitro toxicity assays. *Toxicol Sci* 132, 327–346.
Whitworth, A.J., Wes, P.D., Pallanck, L.J., 2006. *Drosophila* models pioneer a new approach to drug discovery for Parkinson's disease. *Drug Discov Today* 11, 119–126.
Wilk-Zasadna, I., Bernasconi, C., Pelkonen, O., Coecke, S., 2015. Biotransformation in vitro: An essential consideration in the quantitative in vitro-to-in vivo extrapolation (QIVIVE) of toxicity data. *Toxicology* 332, 8–19.
Wille, T., Thiermann, H., Worek, F., 2010. Development of a high-throughput screening for nerve agent detoxifying materials using a fully-automated robot-assisted biological assay. *Toxicol In Vitro* 24, 1026–1031.
Worek, F., Eyer, P., Thiermann, H., 2012. Determination of acetylcholinesterase activity by the Ellman assay: A versatile tool for in vitro research on medical countermeasures against organophosphate poisoning. *Drug Test Anal* 4, 282–291.
Wright, E., Jr., Scism-Bacon, J.L., Glass, L.C., 2006. Oxidative stress in type 2 diabetes: The role of fasting and postprandial glycaemia. *Int J Clin Pract* 60, 308–314.
Xia, M., Huang, R., Witt, K.L., Southall, N., Fostel, J., Cho, M.H., Jadhav, A. et al., 2008. Compound cytotoxicity profiling using quantitative high-throughput screening. *Environ Health Perspect* 116, 284–291.

Yan, S.F., Yan, S.D., Ramasamy, R., Schmidt, A.M., 2009. Tempering the wrath of RAGE: An emerging therapeutic strategy against diabetic complications, neurodegeneration, and inflammation. *Ann Med* 41, 408–422.

Yoon, M., Campbell, J.L., Andersen, M.E., Clewell, H.J., 2012. Quantitative in vitro to in vivo extrapolation of cell-based toxicity assay results. *Crit Rev Toxicol* 42, 633–652.

Zanger, U.M., Schwab, M., 2013. Cytochrome P450 enzymes in drug metabolism: Regulation of gene expression, enzyme activities, and impact of genetic variation. *Pharmacol Ther* 138, 103–141.

3 A Review on Food-Associated Carcinogenesis

Koraljka Gall Troselj, Francoise Gueraud, Tanja Matijevic Glavan, Fabrice Pierre, and Neven Zarkovic

CONTENTS

3.1 CARCINOGENESIS

Nutrition is thought to play an important role in cancer development. Cancer represents the second leading cause of death in most high-income countries and a leading cause of death in less economically developed countries (Ferlay et al., 2015; Mosby et al., 2012).

Pulmonary cancer, breast cancer, and colorectal cancer (CRC) are the most commonly diagnosed types of cancer worldwide (Ferlay et al., 2015). On the molecular level, malignant tumors are characterized by numerous genetic and epigenetic aberrations. Genetic aberrations include inherited germ line mutations and induced somatic mutations such as point mutations (the substitution, addition, or deletion of a single nucleotide base), gene copy number variations, and gene/chromosomal rearrangements. Germ line mutations are passed from generation to generation through the germ cells. Somatic mutation represents genetic alteration acquired by a cell that can be passed to the progeny of the mutated cell in the course of mitosis.

Carcinogenesis is defined as the initiation of a tumor development. Carcinogens are substances that can lead to cancer initiation through disturbing various cellular processes and, in many cases, inducing DNA damage. A carcinogen can also be defined as a physical, chemical, or biological agent or a combination of agents that can induce cancer. Genotoxic carcinogens are agents that interact with DNA directly or after being metabolically activated. These interactions result in DNA mutations. Nongenotoxic carcinogens are compounds that can cause tumor initiation without damaging the DNA. The mechanisms involved are (1) receptor-mediated and nonreceptor-mediated endocrine modulation, (2) induction of cellular oxidative stress, (3) induction of tissue-specific toxicity and inflammation, (4) induction of cytotoxicity and immune suppression, and (5) induction of intracellular communication through gap junctions (Hernandez et al., 2009).

Epigenetic changes include histone modifications, DNA methylation changes, and alterations of non-coding RNA expression (Pogribny and Rusyn, 2013; Sharma et al., 2010). In a tumor cell, epigenetic changes may lead to global genomic hypomethylation and gene-specific hyper- or hypomethylation associated with different histone modification patterns. Demethylation of DNA can be a consequence of a deficiency of the universal methyl donor *S*-adenosyl-L-methionine (Pogribny et al., 2006) or may be caused by altered expression and/or activity of DNA methyltransferases (DNMTs) (Jones and Baylin, 2007). There are two ways in which global demethylation of DNA may contribute to tumorigenesis: (1) through elevation in mutation rates, activation of normally silenced tumor promoting genes, and loss of imprinting (Chen et al., 1998) and (2) through demethylation of repetitive DNA sequences and sequelae of events leading to chromosomal instability (Ehrlich, 2009). Gene-specific loss of DNA methylation in cancer cells is commonly found in the promoter region of oncogenes and in the Imprinted Control Regions of imprinted genes. On the other hand, DNA hypermethylation associated with the promoter of tumor suppressor genes commonly leads to their inactivation in cancer cells. Posttranslational modifications of histones are crucial for maintaining chromatin and genome stability. They include methylation, acetylation, phosphorylation, ubiquitination, sumoylation, biotinylation, and ADP-ribosylation (Cohen et al., 2011). The loss of histone H4 lysine 20 trimethylation (H4K20m3) and H4 lysine 16 acetylation (H4K16ac) is the common change in lymphomas and colon cancer tissues (Fraga et al., 2005). miRNAs can influence tumor initiation by inhibiting tumor suppressor genes or activating oncogenes (Kent and Mendell, 2006).

The International Agency for Research on Cancer (IARC) categorizes substances in the following order: group 1, carcinogenic to humans; group 2A, probably carcinogenic to humans; group 2B, possibly carcinogenic to humans; group 3, not classifiable as to carcinogenicity in humans; and group 4, probably not carcinogenic. According to the current listing, there are 116 agents in group 1, 73 in group 2A, 287 in group 2B, 503 in group 3, and 1 agent in group 4 (IARC). When assessing a carcinogenic agent, the following factors must be considered: (1) latency period (the duration between the exposure and the onset of cancer), (2) metabolism of the chemical agent (there are differences between the effect of a carcinogen that may be species specific and organ specific), (3) dose–response relationship (the probability of cancer initiation is proportional to dosage and duration of exposure to carcinogen), (4) the presence of benign tumor (the cancer formation is sometimes difficult to assess so benign tumors can also indicate the influence of hazardous substance), (5) cocarcinogenesis (cocarcinogens do not have carcinogenic activity but enhance the carcinogenic potential of a carcinogen), and (6) susceptibility of the host (the response to a substance can depend on the species, strain, age, sex, diet, health, etc.) (Warshawsky and Landolph, 2005).

In terms of food toxicology, the most important carcinogenic factors are chemical and biological. Based on the mechanism of action, chemical factors have been divided by Weisburger into three categories: (1) ultimate carcinogens (for direct action), which induce cancer without previous metabolic activation (nitrosamines, epoxides, ethyleneamines, and β-propiolactone); (2) procarcinogens, which need to be metabolically activated (aminozoic colorants, aromatic hydrocarbons, aflatoxins, aromatic amines, and urethane); and (3) cocarcinogens, which are chemical substances that cannot induce cancer per se but enhance the carcinogenic effect of other substances. Biological factors include carcinogens in plants (ptaquiloside, pyrrolizidine alkaloids, reserpine, sanguinarine, nicotine, caffeine, acronycine safrole, alkenylbenzene compounds, essential oils), mycotic toxins (aflatoxins and toxins produced by the *Penicillium* genus), and *Streptomyces* toxins (Baba and Landolph, 2007).

There are more than 25,000 different bioactive food components. More than 500 of these have been identified as possible modifiers of carcinogenesis. The selection of dietary constituents may positively or negatively modify cancer risk and tumor behavior (Milner, 2008). The list of major dietary carcinogenic factors and the types of cancer they induce and/or promote are shown in Table 3.1.

3.2 CARCINOGENIC RISK OF FOOD CONTAMINANTS

Data from GLOBOCAN 2012, produced by the IARC, estimate 14.1 million new cancer cases and 8.2 million deaths caused by cancer worldwide in 2012 (Torre et al., 2015). Exposure to a wide variety of harmful substances, not necessarily naturally present in the environment, counts for at least two-thirds of all cancer cases in Western countries. These substances reflect a lifestyle that includes smoking, alcohol consumption, and a Western style diet. The potential for carcinogenic actions by some food contaminants began receiving a considerable attention after the Second World War (Shubik, 1975). These investigations relayed mostly on epidemiological human studies and experiments using various animal models; hence, estimating human exposure to these contaminants and assessing the carcinogenic potency of such contaminants became a must. However, keeping in mind that food contaminants represent a multitude of compounds, there is another approach of analyses establishing a link between a specific type of food and contaminants with carcinogenic potential. The World Cancer Research Fund (WCRF), in its 2007 report on "Food, Nutrition, Physical Activity, and the Prevention of Cancer: A Global Perspective" (WCRF/AICR, 2007), compiled the then available epidemiological data relating to food and cancer.

That study represents an important milestone for this chapter. The report itself has been used to specify foods and contaminants associated with increased carcinogenic potential, categorized in the groups (Table 3.2). These associations are highlighted only when epidemiologically justified and based on relevant scientific literature revealing possible carcinogenic mechanisms. As a result, this information system evolves as new data are produced daily and are limited only because some critical contaminants may have been omitted from the list due to insufficient data. In addition to the food ingredients themselves, we want to stress the importance of pesticides, heavy metals, and persistent organic pollutants (POPs) found in many types of food, as well as the food contaminants belonging to group 1 or 2A, according to aforementioned IARC classification (IARC, 2015).

3.2.1 Drinks and Additives

In the *drinks* group (water, fruit juice, soft drinks, hot drinks), inorganic arsenic in drinking water has been determined by WCRF to be associated with lung, skin, and kidney and bladder cancers in a *convincing*, *probable*, and *limited–suggestive* manner, respectively. The level of contamination of drinking water with arsenic is region dependent, with the highest values recorded in China (Taiwan), Chile, Argentina, Bangladesh, and India (Ganga Plain) (IARC, 2004). More recently, European Food Safety Authority (EFSA) considered the possibility of a risk in dietary exposure to inorganic arsenic for some consumer groups, such as high consumers of rice and algae-based products (EFSA, 2009a). Inorganic arsenic and some other heavy metals (cadmium and chromium) are classified as IARC group 1 agents, *carcinogenic to humans*.

Some additives, color compounds, aroma, flavors, and compounds migrating from food package materials, are present in certain foodstuffs. The use of some of these has been criticized because they can cause cancer in animals when given in high doses. Their use at very low doses has been controlled by international and national regulatory bodies, which adds to their safety.

Processed meat and red meat have been recently classified by IARC as carcinogenic and probably carcinogenic to humans (Group1 and Group 2A), respectively (Bouvard et al., 2015).

TABLE 3.1
Major Dietary Carcinogenic Factors and the Types of Cancer They Induce

Types of Cancer Induced	Carcinogenic Factors from the Diet	Active Agent*	Group (IARC Classification)
Breast cancer	Poultry		
	Fat (postmenopausal breast cancer)		
	Alcohol (postmenopausal breast cancer)	Acetaldehyde	1
	Pesticides		
	Meat (heating >180°C)	PhIP[a] (HAA[b])	2B
Central nervous system cancers	Pesticides		
Colorectal cancer	Red meat (heating >180°C)	HAA, PAH[c]	2B
	Hemoglobin (heme)		
	Processed meat	Sodium nitrite (N-nitrosamines)	2A
	Low-fiber diet		
	Refined grains		
	Sugar		
	Fat		
	Alcohol		1
	Water fluoridation	Acetaldehyde	3
	Pesticides	Fluor	
Endometrial cancer	Poultry		
	Diet with high GL[d] but not high GI[e] (obese women)		
Gastric cancer	Pesticides		
Head and neck cancer	Alcohol (men)	Acetaldehyde	1
	Water fluoridation	Fluor	3
	Salted fish (Asia)	*N*-nitrosodimethylamine	2A
Hepatocellular cancer	Aflatoxines		1
Lymphoid tissue carcinomas	Pesticides		
	Food contamination/soft drinks	Benzene	1
Liver cancer	Alcohol (women)	Acetaldehyde	1
	Capsaicin		
	Safrole	Safrole	2B
	Water fluoridation	Fluor	3
	Aflatoxins	Aflatoxin B1	1
	Water contamination	1,2-Dichloropropane	1
	Food/water contamination	Polyvinyl chloride	3
Lung cancer	Fat		
	β-Carotene		
	Arsenic compounds	Arsene	3
	Pesticides		
Ovarian cancer	Pesticides		

(*Continued*)

TABLE 3.1 (*Continued*)
Major Dietary Carcinogenic Factors and the Types of Cancer They Induce

Types of Cancer Induced	Carcinogenic Factors from the Diet	Active Agent*	Group (IARC Classification)
Pancreatic cancer	Pesticides		
Testicular cancer	Pesticides		
Urinary organs cancer	Water fluoridation	Fluor	3
	Water contaminant	Trichloroethylene	1
	Pesticides	THM[f]	
	Arsenic compounds	Arsene	3
Multiple sites (unspecified)	Food contamination	Dioxins	1
		PCBs[g]	1
	Heated carbohydrate-rich food	Acrylamide	2A

Sources: Milner, J.A., *Cancer Lett.*, 269, 189, 2008; Mosby, T.T. et al., *Anticancer Res.*, 32, 4171, 2012.
[a] PhIP, 2-amino-1-methyl-6-phenylimidazo[4,5-*b*]pyridine.
[b] HAA, heterocyclic aromatic amine.
[c] PAH, polycyclic aromatic hydrocarbon.
[d] GL, glycemic load.
[e] GI, glycemic index.
[f] THM, trihalomethane.
[g] PCB, polychlorinated biphenyl.
* Empty lines indicate the absence of a specified agent.

3.2.2 Cereals, Roots, Tubers, and Plantain Foods

Cereals and peanuts represent foods that are most commonly contaminated by aflatoxins—fungal toxins—coming from *Aspergillus flavus* and *Aspergillus parasiticus*. Among *cereals, roots, tubers, and plantain* foods, the WCRF considers the presence of aflatoxins associated with an increased risk for liver cancer (hepatocellular carcinoma (HCC)), in a *convincing* manner, where "convincing" represents the judgment of the highest evidence. Recent data, based on the meta-analysis of 17 eligible studies conducted in high-exposure areas—China, Taiwan, and Sub-Saharan Africa—showed that aflatoxin multiplicatively interacts with hepatitis B virus (HBV) to induce HCC (Yu et al., 1997). The data allowed a predicted reduction of 23% in HCC incidence, if exposure to aflatoxin is kept at nondetectable levels (Liu et al., 2012). Aflatoxins are highly potent carcinogenic compounds (Kew, 2013). In 1961, the effect of 20% Brazilian groundnut meal in a purified rat diet was shown to be hepatocarcinogenic, as 9 of 11 animals fed with this diet for six months developed multiple liver tumors (Lancaster et al., 1961). According to IARC classification, aflatoxins (of which aflatoxin B1 [AFB1] exerts the strongest carcinogenic and mutagenic effect) belong to group 1 (IARC, 1993). Their bioactivation occurs in the liver, with the formation of a genotoxic epoxide derivative. Specifically, the AFB1 was shown to be metabolized in the liver by cytochrome P450 3A4 into a genotoxic metabolite that can bind to DNA, forming an adduct. During DNA replication, this adduct induces point mutation (transversion G-T) that can lead to cancer (Bailey et al., 1996). Aflatoxin B1 was also shown to induce oxidative DNA damage, including the formation of 8-oxodeoxyguanosine (8-oxodG), as shown through its time- and dose-dependent increase in rat hepatic DNA, following exposure to AFB1 (Shen et al., 1995).

It has been shown that glutathione-S-transferase μ 1 (GSTM1) and glutathione-S-transferase θ1 play a predominant role in AFB1 detoxification. Genes that code for these enzymes are

TABLE 3.2
Classification of Food and Food Contaminants Associated with Increased Carcinogenic Potential according to the Level of Evidence

Level of Evidence	Criteria	Upgrading Factors
Convincing	Robust enough to be highly unlikely modified by future studies • Evidence from more than one study type and at least 2 independent cohort studies • No substantial unexplained discrepancy between studies or in different populations • Good-quality studies • Presence of a *dose–response* in the association • Experimental evidence that human exposure can lead to cancer outcome	NA
Probable	Robust enough to justify goals and recommendations aimed at reducing cancer incidence • Evidence from at least 2 independent cohort studies or at least 5 case-control studies • No substantial unexplained discrepancy between studies • Good-quality studies • Biological plausibility	Upgrading factors • Presence of a dose–response • Large effect size (relative risk of 2.0 or more) • Evidence from randomized trials in humans • Reproducible and robust experimental evidence
Limited–suggestive	Not robust enough to be convincing or probable but evidence of a consistent direction of effect; does not generally justify recommendations • Evidence from at least 2 independent cohort studies or at least 5 case-control studies • Some unexplained heterogeneity that may be present • Biological plausibility	

highly polymorphic. These variations also include homozygous deletion polymorphism, present, for GSTM1, in a significant part of the population (20%–67%). There were many studies associating the *GSTM1*-null genotype with cancer risk, but the results seem to be inconclusive (Chen et al., 1996). The possible explanation for this lack of convincing correlation throughout the literature may be the result of a compensatory activity of GSTM2, as recently shown (Bhattacharjee et al., 2013). At the level of DNA, the previously mentioned transversion, G-T, has been found to be a hot spot for *p53* mutation, induced by AFB1. The occurrence of this mutation is common at codon 249, which seems to be a preferential site for AFB1 adduct formation (Besaratinia et al., 2009). Some other mycotoxins such as fumonisin B1 and ochratoxin A were reported to be carcinogenic in rodents exhibiting both cancer-initiating and cancer-promoting effects. Although classified as members of the IARC group 2B, they are not focused on by WCRF.

3.2.3 Seaweed, Fish, and Seafood

Based on occupational studies, the IARC has classified cadmium in group 1. However, the EFSA analysis of the data related to general population exposure to cadmium, pointed out a statistically significant association with an increased risk of lung, endometrial, bladder, and breast cancers. For the non-smoking population, the major source of cadmium is in the food. The highest cadmium concentrations were detected in seaweed, fish, and seafood (EFSA, 2009b). However, the food groups that most significantly contributed to dietary cadmium intake, primarily because of their high consumption levels, are cereals, cereal products, and vegetables.

3.2.4 Vegetables, Fruits, Pulses, Nuts, Seeds, Herbs, and Spices

EFSA estimated that certain population subgroups (vegetarians, children, smokers, and people living in highly contaminated areas) may exceed the tolerable weekly intake (2.5 μg/kg) of cadmium by two-fold. For that reason, the report concluded that the current population exposure to cadmium should be reduced (EFSA, 2009b). Despite the presence of carcinogenic contaminants such as cadmium or lead in various foods of plant origin, in addition to carcinogenic compounds found in herbs such as aristolochic acid (classified in category 1 by IARC), no increase in cancer risk is associated with *vegetables, fruits, pulses, nuts, seeds, herbs, and spices*, according to WCRF.

This food group would also be at the highest risk for the presence of pesticide residue, but there are insufficient epidemiological data associated with cancer risk. Likely reason for this may be very low concentrations of these contaminants, insufficient to exert carcinogenic effects. Most organochlorine pesticides have been replaced by less toxic organophosphorous and carbamates. However, some residues of those banned persistent compounds can still be found in food. Of note, the organophosphorus insecticides malathion and diazinon and the herbicide glyphosate have been very recently upgraded into IARC group 2A (IARC, 2015).

3.2.5 Food of Animal Origin: Meat, Poultry, Fish, and Eggs

The residue of veterinary drugs may also be found in foods of animal origin. Although the use of drugs that appear to be carcinogenic in animals has been banned, hormones, such as estrogens and testosterone used in cattle to promote growth, are still used in some countries, including the United States and Canada. These hormones are considered multisite carcinogens and belong to IARC group 1. POPs, a family of lipophilic compounds that are characterized by a resistance to environmental degradation, may accumulate in lipid-rich consumable animal tissues. Some of them, such as polychlorinated biphenyls and some dioxins, are listed in IARC group 1. While there is no evidence for an association between *milk, dairy products, fats and oils, sugar, and salt* and cancer, red and processed meats belonging to the *meat, poultry, fish, and eggs,* WCRF group show a *convincing* evidence for an increased risk of CRC. The evidence is *limited–suggestive* for the association of red meat (beef, goat, lamb, and pork) with esophageal, lung, pancreatic, and endometrial carcinomas. The same level of evidence is valid for the association of processed meat (defined as meat preserved by smoking, curing, salting, or with preservative [nitrites]) with esophageal, lung, pancreatic (WCRF/AICR, 2012), stomach, and prostate cancers. The consumption of Cantonese-style salted fish is associated with an increased risk of nasopharyngeal cancer in a *probable* manner. A fish preserved by this traditional method is classified as *carcinogen to human* (IARC group 1). It represents a mixture of various compounds, including a high concentration of carcinogenic *N*-nitrosocompound precursors.

Foods containing iron, in addition to smoked, grilled, or barbecued foods of animal origin, have been shown to associate with colorectal and stomach cancers, respectively, in a *limited–suggestive* manner. Processed meat contains both iron and nitrites as preservatives. Nitrite can form *N*-nitrosocompounds (nitrosamines and nitrosamides) upon reaction with amino acid degradation products. Some of these *neoformed* (newly formed) compounds are carcinogenic to humans (Mirvish, 1995). Nitrites arise from processed meat consumption or from plant nitrates, which are subsequently converted into nitrites within the body.

Heterocyclic amines (HCAs) and polycyclic aromatic hydrocarbons (PAHs) are two groups of chemicals that are formed, respectively, by cooking at very high temperatures or grilling meats and fishes over a direct flame. Considered as *thermal processing contaminants*, these compounds are carcinogenic to humans. The PAH benzo[a]pyrene belongs to IARC group 1 and HCA (2-amino-3-methylimidazo[4,5-*f*]-quinoline—IQ) belongs to IARC group 2A. Both compounds need to undergo enzyme-catalyzed activation to become mutagenic and carcinogenic (Sugimura, 2002). PAHs can (self-)induce this bioactivation.

Cooked fish and poultry may display high concentrations of HCAs, but they are not associated with an increased cancer risk due to the lack of heme iron. While iron is necessary for increased proliferation of both malignant and normal cells, it was shown to accumulate in colon carcinoma and corresponding metastases, even causing anemia (Prutki et al., 2006). On the other hand, heme iron was shown to be associated with an increased risk of proximal colon cancer, especially among women who drink (Lee et al., 2004) and in animal models by catalyzing the genesis of harmful food products, based on oxidation (Bastide et al., 2015). Heme iron can bind oxygen and catalyze the peroxidation of several cellular or dietary components, including dietary lipids.

3.2.6 Fat Metabolites: Newly Formed Food Processing Contaminants

One of the possible mechanisms linking heme iron and CRC could be the formation of secondary oxidation products of polyunsaturated fatty acids, mostly alkenals, such as 4-hydroxynonenal (HNE). Some of these compounds were reported to be cytotoxic, genotoxic, and mutagenic *in vitro* (Bastide et al., 2015). This is detailed in subchapter.

Similarly, acrylamide is a *thermally induced contaminant* that is formed in carbon-rich food heated at high temperature, similarly to the related, fat-originating acrolein. Acrolein is generally considered as an environmental pollutant and a toxic product of oxidized, thermally treated fat. Acrylamide is mainly found in foodstuffs not associated with an increased risk of cancer: fried potatoes (French fries and potato crisps), cereals (bread, biscuits), and coffee. It is generated as the result of the Maillard reaction between amino acids, mainly asparagine, and reducing sugars (Arvanitoyannis and Dionisopoulou, 2014). On the basis of occupational exposure, acrylamide has been classified in group 2A. More recently, dietary exposure to acrylamide in food was reported (Tareke et al., 2002) to be of concern for the general population with respect to its genotoxic and carcinogenic effects.

Since *neoformed* compounds may be absorbed, their existence in the organism may explain the probable association between red or processed meat consumption and non-digestive cancers (lung, endometrium, or prostate cancer).

In conclusion, consumers are exposed to a wide variety of contaminants with carcinogenic potential. They are present in animal and plant foods at very low doses. However, it is very difficult to precisely associate the exposure to these compounds with cancer incidence, except for a limited number of compounds, as already described. There are efforts to decrease exposure, while there is strong need to have more accurate assessment of the balance between the risk and potential benefits of their presence in food. Data on carcinogens need to be persistently verified, keeping in mind that *harmful exposure* represents, in majority of cases, variable contact with an assortment of compounds, which make toxicity and carcinogenicity studies difficult to perform and interpret.

Finally, carcinogenesis—as a multilevel process during prolonged period of time—certainly may be influenced by various potential food contaminants. For that reason, it is reasonable to be very careful in extrapolating the data obtained in experimental models to a daily routine for humans.

3.3 OXIDATIVE STRESS AND CARCINOGENESIS

Complete carcinogens can cause cancer as they possess both cancer-initiating and cancer-promoting activities. Among them are various carcinogenic chemicals and ionizing radiation that often cause oxidative stress, characterized by the excess of reactive oxygen species (ROS), which play an important role mostly in the promotional phase of carcinogenesis (Cerutti, 1985). The superoxide anion radical O_2^- and some organic peroxides not only are *tumor promoters* but may also be, similar to hyperbaric oxygen, *weak* complete carcinogens (Heston and Pratt, 1959). Due to the importance of DNA damage in carcinogenesis, any chemical capable of reacting with DNA, modifying its structure and function, may be considered as potentially carcinogenic. Polycyclic hydrocarbons are complete chemical carcinogens whose harmful activities are correlated with their ability to form free radicals (Szent-Gyorgyi et al., 1960). At least four different radicals might be formed from benzo(a)pyrene (BaP): the benzo(a)pyrene anion radical (BaP^-), the benzo(a)pyrene cation radical (BaP^+), the radical species formed by heating BaP (related to the radicals formed in cigarette smoke), and the oxygenated BaP radicals. Based on the activity of these radicals, the binding of BaP to DNA has been implicated as the first step of carcinogenesis caused by BaP.

The DNA from various cancer tissues (human renal cell carcinomas, breast carcinomas, squamous cell carcinoma of the lungs, and HCCs) contains increased amounts of one of the major oxidatively modified DNA base products—7,8-dihydro-8-oxo-2′-deoxyguanosine (8-OH-dG), which is a mutation-prone (G–C to A–T transversion) modification (Malins and Haimanot, 1991; Toyokuni et al., 1995). It is known that hydroxyl radical, singlet oxygen, or a photodynamic action is responsible for the formation of 8-OH-dG (Kasai et al., 1992). Higher amounts of 8-OH-dG in cancer cells rather than in the adjacent nontumorous tissue were reported for several tumors such as human renal cell carcinomas, breast carcinomas, squamous cell carcinoma of the lungs, and HCCs (Malins and Haimanot, 1991). Its increased presence in the DNA of malignant cells (when compared to nontumorous surrounding tissue) indicates that the DNA of malignant cells is continuously exposed to oxidative stress at a higher extent than that of the nonmalignant counterpart cells.

The chromosome breaking (clastogenic) effect of the "tumor promoter" phorbol myristate acetate is also mediated by the production of ROS (Emerit and Cerutti, 1981). Additionally, carcinogenic activities of tumor promoters are associated with a decrease in the cellular antioxidant defense mechanisms, which include some key enzymes in defending against oxidative stress, such as catalase (CAT) and superoxide dismutase (SOD) (Halliwell and Gutteridge, 1989).

Lowered activities of CuZnSOD are often, but not always, seen in (malignant) tumors. This phenomenon joined with increased ROS production points out that the decreased antioxidant protection not only might explain essential steps in carcinogenesis but also should be considered in the broader context of well-known features of cancer cells in general (Oberley and Oberley, 1984). Accordingly, increased production of ROS should be observed in cancer cells by comparing with normal cells, also being supported by iron, which is also required for cell growth and multiplication. Iron is a transition metal that triggers Fenton's reaction, which is needed for forming the most harmful hydroxyl radical from SOD-generated hydrogen peroxide. Similarly, iron-containing carcinogens often cause the production of the lipid peroxidation product HNE, indicating the involvement of lipid peroxidation in the signaling pathway of ROS-mediated carcinogenic activity of the iron ions (Toyokuni et al., 1994).

Subcellular fractions of cancer cells (cell cultures or animal tumors) show low levels of lipid peroxidation (Halliwell and Gutteridge, 1989). In concert with that are novel findings on the low content of one of the major lipid peroxidation products, aldehyde HNE, in cancer, by comparison with the normal tissue, especially stromal cells (Biasi et al., 2002; Negre-Salvayre et al., 2010). So far, exact reasons for the differential HNE metabolism or lipid peroxidation between the normal and malignant cells are not fully understood. However, these observations are very important and will be discussed in more details in the following paragraphs.

The overall metabolism of ROS in cancer cells is a research area that has yet to be intensively pursued. It is still an open question whether cancer cells produce larger amounts of ROS than nonneoplastic cells or whether the antioxidant systems are suppressed in cancer. In support of the first possibility are findings of large amounts of hydrogen peroxide (H_2O_2) produced *in vitro spontaneously* by various malignant cells (Szatrowski and Nathan, 1991). It is likely that 8-OH-dG is then being produced due to the hydroxyl radical generated from H_2O_2 in the presence of metal ions (such as iron) (Gutteridge et al., 1982). Hydroxyl radical generation can be confirmed in cancer cells by a spectrum of modified DNA bases induced by the hydroxyl radical attack.

On the other hand, the expression of the gene encoding for one of the major antioxidative enzymes, that is, anti-ROS defensive enzymes, namely, CAT, is reduced at least in some (hepatoma) cancer cells (Sato et al., 1992). Similarly, tumors are often low in SOD activity, while the metabolism of the ROS-scavenger glutathione and the enzymes involved in its metabolism are variable in different types of cancer and could be associated with their different sensitivities to cytostatic drugs and HNE (Borovic et al., 2007; Sun, 1990).

It seems probable that both mechanisms of involvement of oxidative stress in cancer development are relevant. That is, cancer cells may produce larger amounts of ROS than nonneoplastic cells, while the antioxidant systems may also be suppressed in cancer. Such unfavorable conditions should actually cause the death of cells, but, unfortunately, cancer survives such oxidative stress. It can also be assumed that unusual resistance to oxidative stress accompanying malignant transformation could be one of the reasons for the relative resistance of some cancer cells to chemotherapy based on ROS production (O'Donnell-Tormey et al., 1985).

The relevance of persistent oxidative stress in cancer could be in the continuous activation of transcription factors leading to intensive proliferation of the cells through altered expression of c-fos, c-jun, and c-myc genes (Schreck et al., 1991). Additionally, ROS can specifically damage proteinase inhibitors (alpha-1-proteinase inhibitors, alpha-1-macroglobulin, alpha-1-plasmin, etc.) by oxidizing methionine residues of the active sites, which would then allow the further progression of cancer (Johnson and Travis, 1979). Thus, it is obvious that ROS play a relevant role not only in cancer induction and promotion but also in its progression. Hence, cancer itself represents a condition of oxidative stress.

Cancer cells are sensitive to HNE, also known as the "second messenger of free radicals," which can not only have toxic effects on cancer cells but also regulate their proliferation and differentiation (Zarkovic, 2003), as will be presented in the next section.

3.4 LIPID PEROXIDATION AND MALIGNANCIES OF GASTROINTESTINAL TRACT

Environmental factors strongly contribute to gastrointestinal types of cancer, particularly CRC (WCRF/AICR, 2012). Various harmful substances are present in foods comprising a "Western diet." Recent meta-analyses showed that a high consumption of red or processed meat, characteristic of that diet, is associated with a modest but significant risk of some gastrointestinal malignancies (Chan et al., 2011). Namely, strong association is shown for CRC and is also suggested for esophageal, pancreatic, and stomach cancers. Among the compounds that can induce harmful (mutagenic, carcinogenic) effects are heme-induced lipid peroxidation end products (Bastide et al., 2015). A body of evidence suggests that oxidative stress and resulting lipid peroxidation are involved in various and numerous pathological states, including cancer (Barrera, 2012). This evidence is based on epidemiological studies supported by numerous experimental studies.

3.4.1 Epidemiological Evidence

3.4.1.1 Colorectal Cancer

CRC is the third most common type of cancer worldwide. With 1.36 million cases recorded in 2012, CRC accounts for around 10% of all diagnosed malignant neoplasms (Ferlay et al., 2015; WCRF/AICR, 2011). There are positive results from meta-analyses studying the association of consuming red and/or processed meat with CRC incidence, undertaken for the 2011 WCRF report (WCRF/AICR, 2011). These data confirm the results presented in the 2007 report, with a *convincing* level of evidence. This is the highest level for the WCRF panel. The report from 2011 was based on the meta-analysis of 24 cohort studies (Chan et al., 2011). The dose–response meta-analysis showed a significant increase of risk for developing CRC with the consumption of red meat (an increment of 100 g/day). The increase is 14% for CRC, 25% for colon, and 31% for rectum. These results were confirmed in the meta-analysis of Alexander et al. (2011), demonstrating an increase of CRC risk by 12% with red meat intake (comparison between the highest and lowest intakes among 25 cohort studies). Epidemiological (Lee et al., 2004) and experimental (Bastide et al., 2015) studies identified heme iron as the most significant factor in promoting CRC by red meat and showed that lipid peroxide end products play a central role in molecular mechanisms of action.

3.4.1.2 Pancreatic Cancer

Pancreatic cancer is the thirteenth most common type of cancer worldwide. This is one of the most aggressive and lethal types of cancer. About 338,000 cases were recorded in 2012, with a higher incidence in men than in women (WCRF/AICR, 2012). Pancreatic cancer is the burden of developed countries, suggesting, as CRC does, a positive association with Western dietary habits. Experimental studies have implicated oxidative stress in the pathogenesis of both chronic pancreatitis and pancreatic cancer (Kodydkova et al., 2013). Similarly, the concept of antioxidant supplementation as a preventive approach for pancreatic cancer has been proposed (Yu and Kim, 2014).

Regarding epidemiological data, results from the meta-analyses associating the consumption of red and/or processed meat with pancreas cancer risk, carried out for the 2012 WCRF report, were in favor of an increased risk at the level of evidence qualified as *suggested*. The dose–response meta-analysis for an increase of 100 g/day of red meat (8 cohort studies) and 50 g/day of processed meat (7 cohort studies) showed a significantly increased risk in men: 43% and 21%, respectively. However, the risk of cancer was not significantly increased in women. These results were confirmed in a recent meta-analysis (Larsson and Wolk, 2012) showing a significantly increased risk of pancreatic cancer with the consumption of red meat (for an increment of 120 g/day, 11 cohort studies) and of processed meat (for an increment of 50 g/day, 7 cohort studies): 29% and 19%, respectively.

3.4.1.3 Esophageal and Gastric Cancers

Esophageal and gastric cancers with 456,000 and 952,000 newly diagnosed cases in 2012 represent the eighth and fifth most commonly diagnosed types of cancer, respectively (Torre et al., 2015). Contrary to colorectal and pancreatic cancer, esophageal and stomach cancers occur mainly in less-developed countries. Nevertheless, in the United States and Europe, a positive association has been observed with red and processed meat for esophageal cancer and with processed meat for stomach cancer.

Three meta-analysis studies have shown the link between esophageal cancer and red or processed meat intake (50 g/day). For the consumption of red meat, Choi's study (Choi et al., 2013) (inclusion of 19 studies, cohort studies, and case-control studies) showed an increase of 42%. Huang's studies (6 case-control studies and 3 cohort studies) (Huang et al., 2013) and Salehi's studies (2 cohort studies and 15 case-control studies) (Salehi et al., 2013) found an increase of 31% and 40%, respectively. The dose–response meta-analyses for processed meats represented by these studies showed a significantly increased risk for the highest versus the lowest intake, thus being in favor of a *suggested* increased risk.

The results from the meta-analyses related to the consumption of processed meats and gastric cancer risk, carried out for the 2007 WCRF report, show an increased risk at the "suggested" level of evidence The meta-analysis related to dose–response for an increase of 20 g/day of processed meat (nine case-control studies) showed a significantly increased risk of 13% (WCRF/AICR, 2007). These results were confirmed in the recent meta-analysis of D'Elia, showing an increase of 24% (seven cohort studies) (D'Elia et al., 2012).

3.5 MECHANISMS OF CARCINOGENESIS IN GASTROINTESTINAL TRACT

Heme-induced lipid peroxidation has been proposed by Sesink et al. (1999) as the basis for an important hypothesis explaining the link between red and processed meat and increased risk for malignancies of the gastrointestinal tract, especially CRC. Recently, it has been proven that the reactive aldehydes produced by heme iron–induced lipid peroxidation play a prominent role in the promotion of colon cancer in animal models (Bastide et al., 2015). Significant amount of these aldehydes, or their metabolites, were found not only in the meat- and fat-containing food but also in the feces and urine of animals and humans having a heme iron–rich diet (Gasc et al., 2007; Pierre et al., 2006, 2007). Among these aldehydes, alkenals, that is, the aldehydes bearing a double bond between carbons 2 and 3, are of particular interest because of their reactivity toward cellular macromolecules: DNA and proteins (Esterbauer et al., 1991; Uchida, 2003).

The most bioreactive alkenals, that is, α,β-unsaturated aldehydes, are able to make covalent adducts with DNA bases (Chung et al., 1996). Such exocyclic adducts significantly contribute to cellular DNA damage (Winczura et al., 2012). The reactivity of alkenals to DNA depends on the alkyl chain length. Acrolein, the alkenal with the shortest chain length, was shown to be the most reactive, while its exocyclic adducts are mutagenic. Similarly, HNE preferentially induces p53 point mutation AGG249AGT (Hu et al., 2002; Xing et al., 2007). HNE was also shown to provoke DNA lesions in primary cultures of liver cells; sister chromatid exchanges at low dosage and micronuclei and chromosomal aberrations at higher dosage (Eckl, 2003; Eckl et al., 1993). *In vivo*, exocyclic adducts of another alkenal—4-oxo-nonenal—were increased in the intestine of the COX-2 overexpressing C57BL/6JAPC^{min} mice, an animal model developed for research related to inflammation and intestinal cancer (Williams et al., 2006).

Beside DNA, these alkenals can covalently bind to proteins and modify their function. Adduction is possible through Michael addition on the double bond or by Schiff base formation with the aldehyde function (Esterbauer et al., 1991). Protein adducts with lipid oxidation products closely relate to oxidative stress and have been shown to be increased under various pathologic conditions associated with inflammation and/or degeneration, not only carcinogenesis. Whether these adducts are the cause for or a consequence of the oxidative stress/inflammation process is still a matter of debate. HNE was reported to be involved in the regulation of several cellular pathways that control cellular growth and cellular defensive mechanism, as well as inflammatory reaction. Its effect is largely cell and tissue specific. It also depends on its concentration. In cancer, the amount of HNE protein adducts is highly dependent on the origin of cancer and its stage. Nevertheless, in colon carcinogenesis, levels of HNE- and acrolein-protein adducts were found to be inversely related to the malignancy. (Biasi et al., 2002; Negre-Salvayre et al., 2010; Zarkovic et al., 2006). This decrease has been attributed to a better metabolizing capacity of cancer cells toward lipid oxidation products, resulting in their removal (Canuto et al., 1993). Concerning colon cancer, *in vitro* studies showed that the single Apc mutation, a predisposing event for majority of colon cancers, rendered the cells strongly resistant to toxic effects induced by HNE (Pierre et al., 2007). This gives a clear survival advantage to the precancerous cells subjected to oxidative conditions. This phenomenon exists in the colon when oxidation-prone polyunsaturated fatty acids and prooxidant compounds such as heme iron are present in the diet, acting as a positive selection factor for precancerous cells and inducing cancer promotion. This selective advantage can be explained by the better metabolizing capacity of preneoplastic cells toward HNE (Baradat et al., 2011), as described earlier, for cancer cells. This could be due to the activation of the cellular defense Nrf2/antioxidant response element (ARE) pathway in preneoplastic cells (Shelton and Jaiswal, 2013). HNE itself has been shown to induce this cellular pathway (Huang et al., 2012), which contributes to inactivation/excretion of HNE. Of importance is that in colon carcinogenesis, HNE might have suppressive effects interacting with TGFβ-1 (Biasi et al., 2006). Namely, this cytokine suppresses proliferation of normal colon epithelium.

To achieve such suppressive effects, cells have to express type I TGFβ1 receptor, the expression of which is enhanced by HNE. At lower HNE levels, cells do not express enough type I receptor for the TGFβ1 and, as a consequence, their proliferation rate is increased. Accordingly, TGFβ1 and HNE can jointly induce apoptosis in colon carcinoma cells (Biasi et al., 2006), thus suppressing the colon carcinoma growth.

The gastrointestinal tract represents the first barrier against the products of lipid peroxidation originating from dietary products. HNE can increase the permeability of this barrier (Cindric et al., 2013), thus exerting not only multiple harmful but also some important growth-regulating effects, as mentioned earlier. Hence, these compounds may exert a Darwinian pressure selection on preneoplastic cells, making the missing link between the heme iron–containing red and processed meats and the increased incidence of colon cancer. *In vivo* studies need to be undertaken to confirm this effect. In parallel, more studies on potential antioxidant and/or Nrf2-targeting compounds, such as curcumin, might represent good prevention to counteract deleterious effects while preserving beneficial cancer-suppressing activities.

3.6 DIETARY ANTIOXIDANTS AND CANCER

3.6.1 Molecular Mechanisms of Action as the Must for Targeted Prevention

When considering molecular pathways that can be triggered by food contaminants and are relevant for inflammation and cancer, one must be aware of how complex various interactions are or may be. Signaling pathways related to inflammation are commonly centered around transcription regulation mediated by the nuclear factor kappaB (NF-κB). Various inflammatory mediators, including ROS, serve as NF-κB activators. They activate IκB kinase (IKK) complex that consists of two catalytic subunits, IKKα and IKKβ, together with the regulatory unit, IKKγ (NEMO), the key master regulator of the NF-κB-mediated inflammatory response (Karin and Delhase, 2000). The canonical pathway, dependent primarily on IKKβ and NEMO, is responsible for the phosphorylation of IκBα. It makes it a target for ubiquitination and proteasomal degradation (Li et al., 1999). As a consequence (in a simplified scenario), NF-κB dimmers translocate into the nucleus, bind to DNA, and transcriptionally activate specific target genes (Figure 3.1; Karin, 2006).

This *specificity* is particularly important in the context of carcinogenesis because NF-κB activates proproliferative and/or antiapoptotic genes (Ben-Neriah and Karin, 2011). Some of these genes, that is, COX-2 and cyclin D1, are commonly considered *a measure* of NF-κB activation. Of importance, COX-2 is the key molecule in the enzymatic conversion of arachidonic acid to prostaglandin H2, which is rapidly converted into a strong mediator of tumor promotion, PGE2 (Nakanishi and Rosenberg, 2013). While undetectable in most quiescent cells, COX-2 becomes detectable upon cellular stimulation by cytokines, mitogens, growth factors, and tumor promoters (Rao and Reddy, 2004).

Toxic effects of some food contaminants closely relate to abnormally elevated ROS and the pathogenic outcome known as oxidative stress. The key pathway needed for ROS elimination, KEAP1 (Kelch-like ECH-associated protein 1)-NRF2 ((erythroid-derived-2)-like-2), represents the best-understood model involved in protecting against environmental and xenobiotic stress and is considered to be the master pathway among redox-dependent antioxidant response mechanisms (Kensler et al., 2007). In oxidative stress, redox-sensitive thiol groups in KEAP1 at Cys-151, Cys-273, and Cys-288 become modified, leading to a failure in NRF2 docking and ubiquitination escape (Zhang and Hannink, 2003). As a consequence, NRF2, accumulated in the cytoplasm, translocates into the nucleus, binds with small Maf proteins, and activates cis-acting enhancer sequences: AREs/electrophile response element. AREs are present in the promoters of many phase II detoxification and antioxidant genes, including NAD(P)H dehydrogenase, quinones 1 and 2 (NQO1 and NQO2), and heme oxygenase (HO-1) (Rushmore et al., 1991). It is well known that Nrf2-null mice have defects in detoxifying carcinogens, including aflatoxin B1 (Kwak et al., 2001).

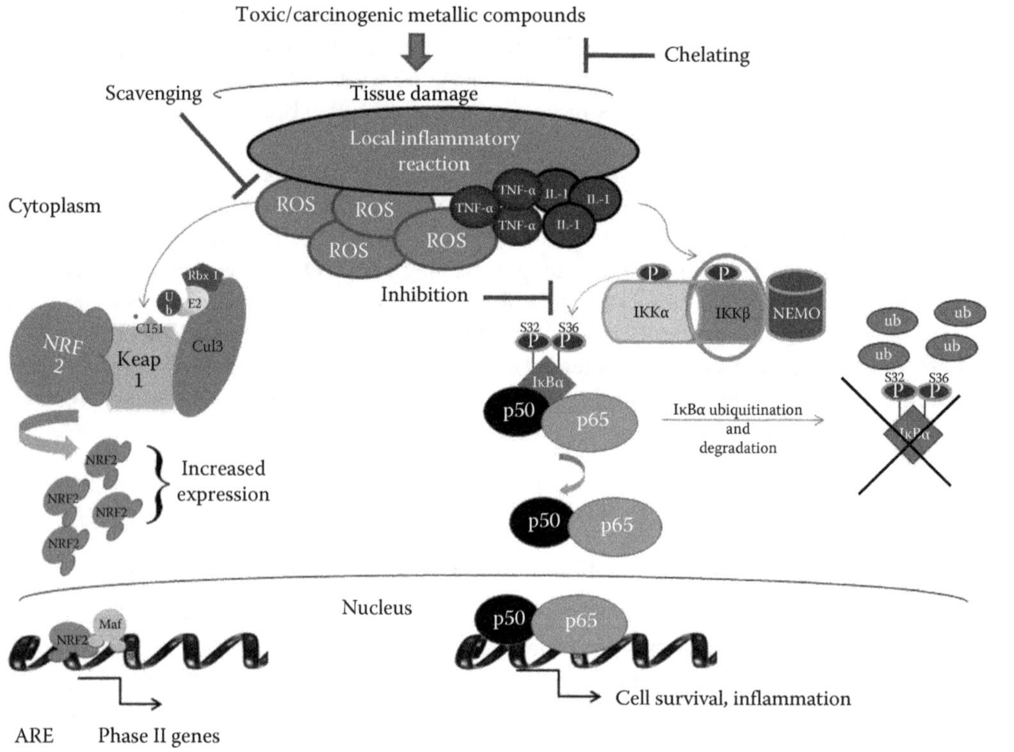

FIGURE 3.1 **(See color insert.)** Multiple protective effects of curcumin related to some toxic food contaminants. The desirable effects of curcumin, indicated by text written in red, include chelating of potentially carcinogenic heavy metals, scavenging of ROS, inhibition of inflammatory signaling, and increased expression of adaptive defense system against oxidative stress.

Ideally, in the context of exposure to food contaminants, a universal "protective approach" would selectively (only in those cells that are exposed to a harmful agent) suppress NF-κB-mediated inflammatory reaction and increase the activity of the KEAP1-NRF2 signaling pathway. Should we strive for a specific type of protective diet or for a smart protectant? Perhaps we should try both.

In the era of significant public health problems related to a prolonged lifetime accompanied with severe diseases (of which cancer represents part of the problem), various concepts of *healthy lifestyle* including *healthy diet* have been proposed. A Mediterranean diet, with its eight major attributes (relatively high monounsaturated to saturated fat ratio; moderate ethanol consumption; high consumption of legumes, cereals [particularly bread], fruits, and vegetables; low consumption of meat and meat products; and moderate consumption of milk and dairy products), appears to be favorable for a reduced cancer risk in general and protective against some types of cancer (i.e., gastric cancer) in particular (Buckland et al., 2010; La Vecchia, 2009). It is estimated that up to 25% of the incidence of CRC, 15% of the incidence of breast cancer, and 10% of the incidence of prostate, pancreas, and endometrial cancer could be prevented if the populations of the highly developed Western countries would shift to a traditional Mediterranean diet (Trichopoulou et al., 2000). In some aspects, an increasing amount of epidemiological data is not necessarily uniform and should be interpreted with caution. For example, recent results from a Swedish cohort study (45,222 women classified according to the Nordic food index: whole grain bread, oatmeal, apples/pears, cabbages, root vegetables, and fish/shellfish) do not support inverse association between consumption of these ingredients and CRC (Roswall et al., 2015). A large American study including more than 100,000 subjects did not show a correlation between cancer mortality and whole grain intake (28 g/day; Wu et al., 2015). These results should be put in the context of genetic variants, recognizing that each variant in gene-coding regions has biological impact on protein structure and function specific for individual subjects and/or certain population (Zhang et al., 2012). Variations in the genome significantly contribute to species variability, which is also related to the previously mentioned metabolic activation of food contaminants. It is also the reason for our current inability to predict person-specific response to xenobiotics.

This is only one of the many reasons for trying to complement these studies with exploring molecular mechanisms of action. In that context, some questions are of importance: How protective these and some other types of diets are or can be against food contaminants, in relation to NF-κB silencing and KEAP1-NRF2 stimulation? Does the same mechanism of action apply to every kind of cell, irrespective of its origin?

Sulforaphane is a potent isothiocyanate derivative that can be found in broccoli and other cruciferous vegetables. Although it does not prevent the entrance of NF-κB in the nucleus, it prevents its binding to COX-2 promoter (shown *in vitro* model: human bladder cancer) (Shan et al., 2009). In mammary epithelial cells stimulated with a prototypic tumor promoter 12-*O*-tetradecanoylphorbol-13-acetate, the final effect (COX-2 silencing) of sulforaphane was the same, but not necessarily the mechanism of action as, in these cells, nuclear localization of NF-κB was attenuated (Kim et al., 2014). Thus, the specific response does depend on the cell type.

The field of research related to naturally occurring compounds as health protectors is an emerging area in the field of human medicine. Among many compounds showing beneficial properties that have been extensively studied, curcumin is one that became the global focus of many research groups and clinical trials. Research on curcumin became especially prominent after 1995, the year when its suppressive properties on NF-κB-signaling pathway were first published (Singh and Aggarwal, 1995).

3.6.2 Curcumin: Indian Saffron

The natural source of curcumin is the rhizome of *Curcuma longa*, a native plant in Southwest India. The rhizome itself contains three curcuminoides (curcumin [diferuloylmethane], demethoxycurcumin, and bisdemethoxycurcumin). The best studied among them, curcumin, on average constitutes 2% of powdered turmeric. Its maximum daily intake in Nepal and India has been estimated to be between 1.5 and 2.0 g, corresponding to 50 and 75 mg of curcumin, respectively (Eigner and Scholz, 1999). It has been known in Traditional Chinese and Ayurvedic Indian medicine for centuries. It has limited systemic bioavailability due to its rapid resorption, conjugation in liver, and intestinal excretion. However, even when present in plasma at very low concentrations, curcumin shows potential for modifying many cellular functions.

The first research paper showing its anti-inflammatory properties, closely related to its anticancerogenic potential, was published in 1988 (Huang et al., 1988). These features have been further studied in various models because they are of particular interest in all fields of molecular medicine. It took more than 20 years to show that curcumin has a strong modifying effect on the epigenome. An exposure of DNA from a leukemia MV4-11 cell line to various concentrations of curcumin (1–30 μM) for 72 h resulted in a 15%–20% decrease in DNA methylation when 3.0 and 30.0 μM curcumin was applied (this effect was absent with 1 μM curcumin) (Liu et al., 2009). This DNA hypomethylating effect was shown to be important for curcumin's influence on NRF2 transcription. Already at 2.5 μM concentration, curcumin decreases promoter methylation of the Nrf2 in an experimental *in vitro* model of TRAP prostate cancer. As a consequence, transcription of the Nrf2 and its target gene, NQO1, significantly increases, primarily as the consequence of reduced methyltransferase activity and without significant change in transcription of the DNMT genes (Khor et al., 2011).

There is much evidence showing that, in addition to NF-κB inhibition, curcumin's actions contribute to an increase of cellular's defending mechanisms and NRF2. Good examples are the well-documented curcumin's hepatoprotective effects against environmental and occupational toxins.

Particularly well documented are its various protective effects against heavy metals (Garcia-Nino and Pedraza-Chaverri, 2014). Its enol form was shown to be a strong chelator of positively charged metal ions. The first research paper on this topic has shown its inhibitory effect on iron-catalyzed lipid peroxidation (Sreejayan and Rao, 1994). Additionally, as shown in rats fed by sodium arsenite (15 mg/kg BW), curcumin present in their diet (15 mg/kg BW) prevented both lipid peroxidation and the decrease in GSH content. While sodium arsenite caused a significant decrease in the activities of glutathione S-transferase (GST), SOD, and CAT in plasma and different organs (liver, lung, brain, kidney), there was a significant recovery relating to GST, SOD, and CAT in response to the combined application of curcumin and sodium arsenite (El-Demerdash et al., 2009). Similar experiments on rats have shown that curcumin can antagonize arsenic-induced hepatic oxidative stress through the upregulation of Nrf2 and induction of NQO1 and HO-1 proteins (Gao et al., 2013). Similar mechanism of action relating to curcumin's hepatoprotection was also shown in rats fed by aflatoxin B1 (25 μg/kg BW). In these animals, an intake of 200 mg/kg BW of curcumin during the duration of the experiment led to a decrease in alanine aminotransferase, aspartate aminotransferase, and gamma-glutamyl transferase and level of lipid peroxidation. At the same time, the presence of curcumin in the diet was related to increased levels of GSH, SOD, and CAT (El-Agamy, 2010).

Since the first report describing curcumin's antioxidative properties, in 1987, there have been many research articles focusing on this aspect of its activity (Shalini and Srinivas, 1987; Srinivasan, 2014) joined with research exploring its radical scavenging properties, originally discovered in 1994 (Sreejayan and Rao, 1994). Currently, curcumin is known for its potent scavenging activity against a variety of ROS, including the aforementioned O_2^-, HO, and peroxyl radical (ROO) (Fujisawa et al., 2004; Reddy and Lokesh, 1994).

For all these properties, curcumin has been recognized as a molecule with great potential for cancer chemoprevention, even more so because it was shown to be safe in animal models. Humans show individual tolerability and doses below 8 g/day were generally well tolerated (Cheng et al., 2001). Only a few exceptions, manifested as diarrhea, nausea, skin rush, and headaches occured in some individuals taking 8–12 g of curcumin daily (Cheng et al., 2001; Lao et al., 2006). Its poor bioavailability is a well-known obstacle. High doses, of up to 8 g/day, were shown to result in an apparently steady-state level of conjugated form in plasma (22–41 ng/mL; 59–111 nM), achieved by the third day of treatment in patients with pancreatic cancer (Cheng et al., 2001). These numbers are relatively low when compared with peak plasma levels of 0.41–1.75 μM, after administration of 4–8 g of curcumin daily for three months (Dhillon et al., 2008). The difference may be the consequence of variations in the methods used for the preparation of the plasma samples and the sensitivity of the measurement itself (Gall Troselj and Novak Kujundzic, 2014).

According to the database ClinicalTrial.Gov, a service of the U.S. National Institutes of Health, there are currently 46 worldwide (United States, 26; Canada, 1; Central America, 2; EU, 7; Middle East, 3; unknown, 9) clinical studies with known status and, in various stages, focusing on curcumin and cancer (https://clinicaltrials.gov/ct2/results?term=curcumin+cancer&no_unk=Y). Among them, few are designed in a way that would reveal the strength of curcumin as a chemopreventive agent. A majority strive to explore curcumin in the treatment of cancer, commonly in combination with various conventional cytostatic drugs. The rationale for this combined application is curcumin's inhibitory effect on NF-κB-signaling pathway that is activated as a part of cellular defense mechanism against cytostatic therapy. While this approach seems very promising, more clinical data are needed for achieving conclusions with a significant level of confidence.

CONCLUSIONS

While the organs of the gastrointestinal tract are the ones exposed to the highest doses of food associated carcinogenes, the harmful effects of these compounds affect the whole body system. Carcinogenesis in general, as well as carcinogenesis caused by food carcinogens, is associated with acute and/or chronic oxidative stress. However, pathophysiological association of oxidative stress and carcinogenesis also opens possibilities for preventive and even therapeutic use of beneficial, healthy food components denoted as nutraceuticals.

Therefore, we expect that the rapid progress of research in this field, from molecular level over experimental model studies up to clinical use and dietary practice, might result in a major breakthrough in our efforts to prevent and combat some of the most disastrous types of cancer of modern humankind.

ACKNOWLEDGMENTS

The authors thank Aaron Etra for his careful revision of the chapter and Tea Keckes for her technical support.

REFERENCES

Alexander, D.D., D.L. Weed, C.A. Cushing, and K.A. Lowe. 2011. Meta-analysis of prospective studies of red meat consumption and colorectal cancer. *Eur J Cancer Prev.* 20:293–307.

Arvanitoyannis, I.S. and N. Dionisopoulou. 2014. Acrylamide: Formation, occurrence in food products, detection methods, and legislation. *Crit Rev Food Sci Nutr.* 54:708–733.

Baba, A.I. and J.R. Landolph. 2007. *Comparative Oncology*. The Publishing House of the Romanian Academy, Bucharest, Romania.

Bailey, E.A., R.S. Iyer, M.P. Stone, T.M. Harris, and J.M. Essigmann. 1996. Mutational properties of the primary aflatoxin B1-DNA adduct. *Proc Natl Acad Sci USA*. 93:1535–1539.

Baradat, M., I. Jouanin, S. Dalleau et al. 2011. 4-Hydroxy-2(E)-nonenal metabolism differs in Apc(+/+) cells and in Apc(Min/+) cells: It may explain colon cancer promotion by heme iron. *Chem Res Toxicol.* 24:1984–1993.

Barrera, G. 2012. Oxidative stress and lipid peroxidation products in cancer progression and therapy. *ISRN Oncol.* 2012:137289.

Bastide, N.M., F. Chenni, M. Audebert et al. 2015. A central role for heme iron in colon carcinogenesis associated with red meat intake. *Cancer Res.* 75:870–879.

Ben-Neriah, Y. and M. Karin. 2011. Inflammation meets cancer, with NF-kappa B as the matchmaker. *Nat Immunol.* 12:715–723.

Besaratinia, A., S.I. Kim, P. Hainaut, and G.P. Pfeifer. 2009. In vitro recapitulating of TP53 mutagenesis in hepatocellular carcinoma associated with dietary aflatoxin B1 exposure. *Gastroenterology.* 137:1127–1137, 1137 e1121–1125.

Bhattacharjee, P., S. Paul, M. Banerjee et al. 2013. Functional compensation of glutathione S-transferase M1 (GSTM1) null by another GST superfamily member, GSTM2. *Sci Rep.* 3:2704.

Biasi, F., L. Tessitore, D. Zanetti et al. 2002. Associated changes of lipid peroxidation and transforming growth factor beta1 levels in human colon cancer during tumour progression. *Gut.* 50:361–367.

Biasi, F., B. Vizio, C. Mascia et al. 2006. c-Jun N-terminal kinase upregulation as a key event in the proapoptotic interaction between transforming growth factor-beta1 and 4-hydroxynonenal in colon mucosa. *Free Radic Biol Med.* 41:443–454.

Borovic, S., A. Cipak, A. Meinitzer et al. 2007. Differential sensitivity to 4-hydroxynonenal for normal and malignant mesenchymal cells. *Redox Rep.* 12:50–54.

Bouvard, V., D. Loomis, K.Z. Guyton et al. 2015. Carcinogenicity of consumption of red and processed meat. *Lancet Oncol.* 16:1599–1600.

Buckland, G., A. Agudo, L. Lujan et al. 2010. Adherence to a Mediterranean diet and risk of gastric adenocarcinoma within the European Prospective Investigation into Cancer and Nutrition (EPIC) cohort study. *Am J Clin Nutr.* 91:381–390.

Canuto, R.A., G. Muzio, M. Maggiora, M.E. Biocca, and M.U. Dianzani. 1993. Glutathione-S-transferase, alcohol dehydrogenase and aldehyde reductase activities during diethylnitrosamine-carcinogenesis in rat liver. *Cancer Lett.* 68:177–183.

Cerutti, P.A. 1985. Prooxidant states and tumor promotion. *Science.* 227:375–381.

Chan, D.S., R. Lau, D. Aune et al. 2011. Red and processed meat and colorectal cancer incidence: Meta-analysis of prospective studies. *PLoS One.* 6:e20456.

Chen, C.J., M.W. Yu, Y.F. Liaw et al. 1996. Chronic hepatitis B carriers with null genotypes of glutathione S-transferase M1 and T1 polymorphisms who are exposed to aflatoxin are at increased risk of hepatocellular carcinoma. *Am J Hum Gen.* 59:128–134.

Chen, R.Z., U. Pettersson, C. Beard, L. Jackson-Grusby, and R. Jaenisch. 1998. DNA hypomethylation leads to elevated mutation rates. *Nature*. 395:89–93.

Cheng, A.L., C.H. Hsu, J.K. Lin et al. 2001. Phase I clinical trial of curcumin, a chemopreventive agent, in patients with high-risk or pre-malignant lesions. *Anticancer Res*. 21:2895–2900.

Choi, Y., S. Song, Y. Song, and J.E. Lee. 2013. Consumption of red and processed meat and esophageal cancer risk: Meta-analysis. *World J Gastroenterol*. 19:1020–1029.

Chung, F.L., H.J. Chen, and R.G. Nath. 1996. Lipid peroxidation as a potential endogenous source for the formation of exocyclic DNA adducts. *Carcinogenesis*. 17:2105–2111.

Cindric, M., A. Cipak, E. Zapletal et al. 2013. Stobadine attenuates impairment of an intestinal barrier model caused by 4-hydroxynonenal. *Toxicol In Vitro*. 27:426–432.

Cohen, I., E. Poreba, K. Kamieniarz, and R. Schneider. 2011. Histone modifiers in cancer: Friends or foes? *Genes Cancer*. 2:631–647.

D'Elia, L., G. Rossi, R. Ippolito, F.P. Cappuccio, and P. Strazzullo. 2012. Habitual salt intake and risk of gastric cancer: A meta-analysis of prospective studies. *Clin Nutr*. 31:489–498.

Dhillon, N., B.B. Aggarwal, R.A. Newman et al. 2008. Phase II trial of curcumin in patients with advanced pancreatic cancer. *Clin Cancer Res*. 14:4491–4499.

Eckl, P.M. 2003. Genotoxicity of HNE. *Mol Aspects Med*. 24:161–165.

Eckl, P.M., A. Ortner, and H. Esterbauer. 1993. Genotoxic properties of 4-hydroxyalkenals and analogous aldehydes. *Mutat Res*. 290:183–192.

EFSA Panel on Contaminants in the Food Chain. 2009a. Scientific opinion on arsenic in food. *EFSA J*. 7:1351.

EFSA Panel on Contaminants in the Food Chain. 2009b. Cadmium in food—Scientific opinion of the Panel on Contaminants in the Food Chain. *EFSA J*. 980:1–139.

Ehrlich, M. 2009. DNA hypomethylation in cancer cells. *Epigenomics*. 1:239–259.

Eigner, D. and D. Scholz. 1999. Ferula asa-foetida and Curcuma longa in traditional medical treatment and diet in Nepal. *J Ethnopharmacol*. 67:1–6.

El-Agamy, D.S. 2010. Comparative effects of curcumin and resveratrol on aflatoxin B-1-induced liver injury in rats. *Arch Toxicol*. 84:389–396.

El-Demerdash, F.M., M.I. Yousef, and F.M.E. Radwan. 2009. Ameliorating effect of curcumin on sodium arsenite-induced oxidative damage and lipid peroxidation in different rat organs. *Food Chem Toxicol*. 47:249–254.

Emerit, I. and P.A. Cerutti. 1981. Tumour promoter phorbol-12-myristate-13-acetate induces chromosomal damage via indirect action. *Nature*. 293:144–146.

Esterbauer, H., R.J. Schaur, and H. Zollner. 1991. Chemistry and biochemistry of 4-hydroxynonenal, malonaldehyde and related aldehydes. *Free Radic Biol Med*. 11:81–128.

Ferlay, J., I. Soerjomataram, R. Dikshit et al. 2015. Cancer incidence and mortality worldwide: Sources, methods and major patterns in GLOBOCAN 2012. *Int J Cancer*. 136:E359–E386.

Fraga, M.F., E. Ballestar, A. Villar-Garea et al. 2005. Loss of acetylation at Lys16 and trimethylation at Lys20 of histone H4 is a common hallmark of human cancer. *Nat Genet*. 37:391–400.

Fujisawa, S., T. Atsumi, M. Ishihara, and Y. Kadoma. 2004. Cytotoxicity, ROS-generation activity and radical-scavenging activity of curcumin and related compounds. *Anticancer Res*. 24:563–569.

Gall Troselj, K. and R.N. Kujundzic. 2014. Curcumin in combined cancer therapy. *Cur. Pharm Design*. 20:6682–6696.

Gao, S., X.X. Duan, X. Wang et al. 2013. Curcumin attenuates arsenic-induced hepatic injuries and oxidative stress in experimental mice through activation of Nrf2 pathway, promotion of arsenic methylation and urinary excretion. *Food Chem Toxicol*. 59:739–747.

Garcia-Nino, W.R. and J. Pedraza-Chaverri. 2014. Protective effect of curcumin against heavy metals-induced liver damage. *Food Chem Toxicol*. 69:182–201.

Gasc, N., S. Tache, E. Rathahao, J. Bertrand-Michel, V. Roques, and F. Gueraud. 2007. 4-hydroxynonenal in foodstuffs: Heme concentration, fatty acid composition and freeze-drying are determining factors. *Redox Rep*. 12:40–44.

Gutteridge, J.M., D.A. Rowley, and B. Halliwell. 1982. Superoxide-dependent formation of hydroxyl radicals and lipid peroxidation in the presence of iron salts. Detection of 'catalytic' iron and anti-oxidant activity in extracellular fluids. *Biochem J*. 206:605–609.

Halliwell, B. and J.M.C. Gutteridge. 1989. *Free Radicals in Biology and Medicine*. Clarendon Press, Oxford, U.K.

Hernandez, L.G., H. van Steeg, M. Luijten, and J. van Benthem. 2009. Mechanisms of non-genotoxic carcinogens and importance of a weight of evidence approach. *Mutat Res*. 682:94–109.

Heston, W.E. and A.W. Pratt. 1959. Effect of concentration of oxygen on occurrence of pulmonary tumors in strain A mice. *J Natl Cancer Inst*. 22:707–717.

Hu, W., Z. Feng, J. Eveleigh et al. 2002. The major lipid peroxidation product, trans-4-hydroxy-2-nonenal, preferentially forms DNA adducts at codon 249 of human p53 gene, a unique mutational hotspot in hepatocellular carcinoma. *Carcinogenesis*. 23:1781–1789.

Huang, M.T., R.C. Smart, C.Q. Wong, and A.H. Conney. 1988. Inhibitory effect of curcumin, chlorogenic acid, caffeic acid, and ferulic acid on tumor promotion in mouse skin by 12-o-tetradecanoylphorbol-13-acetate. *Cancer Res*. 48:5941–5946.

Huang, W., Y.J. Han, J.P. Xu, W. Zhu, and Z.S. Li. 2013. Red and processed meat intake and risk of esophageal adenocarcinoma: A meta-analysis of observational studies. *Cancer Causes Control*. 24:193–201.

Huang, Y., W.G. Li, and A.N.T. Kong. 2012. Anti-oxidative stress regulator NF-E2-related factor 2 mediates the adaptive induction of antioxidant and detoxifying enzymes by lipid peroxidation metabolite 4-hydroxynonenal. *Cell Biosci*. 2:40.

IARC. 1993. Working Group on the Evaluation of Carcinogenic Risks to Humans. Aflatoxins. World Health Organization, International Agency for Research on Cancer, Lyon, France.

IARC. 2004. Working Group on the Evaluation of Carcinogenic Risks to Humans. Some drinking-water disinfectants and contaminants, including arsenic. World Health Organization, International Agency for Research on Cancer, Lyon, France.

IARC. 2015. Agents classified by the IARC monographs. IARC, Lyon, France. http://monographs.iarc.fr/ENG/Classification/latest_classif.php. Accessed on June 30, 2016.

Johnson, D. and J. Travis. 1979. The oxidative inactivation of human alpha-1-proteinase inhibitor. Further evidence for methionine at the reactive center. *J Biol Chem*. 254:4022–4026.

Jones, P.A. and S.B. Baylin. 2007. The epigenomics of cancer. *Cell*. 128:683–692.

Karin, M. 2006. Nuclear factor-kappaB in cancer development and progression. *Nature*. 441:431–436.

Karin, M. and M. Delhase. 2000. Hallmarks of cancer: The next generation. *Semin Immunol*. 12:85–98.

Kasai, H., Z. Yamaizumi, F. Yamamoto et al. 1992. Photosensitized formation of 8-hydroxyguanine (7,8-dihydro-8-oxoguanine) in DNA by riboflavin. *Nucleic Acids Symp Ser*. 187:181–182.

Kensler, T.W., N. Wakabayashi, and S. Biswal. 2007. Cell survival responses to environmental stresses via the Keap1-Nrf2-ARE pathway. *Annu Rev Pharmacol Toxicol*. 47:89–116.

Kent, O.A. and J.T. Mendell. 2006. A small piece in the cancer puzzle: MicroRNAs as tumor suppressors and oncogenes. *Oncogene*. 25:6188–6196.

Kew, M.C. 2013. Aflatoxins as a cause of hepatocellular carcinoma. *J Gastrointest Liver Dis*. 22:305–310.

Khor, T.O., Y. Huang, T.Y. Wu, L.M. Shu, J. Lee, and A.N.T. Kong. 2011. Pharmacodynamics of curcumin as DNA hypomethylation agent in restoring the expression of Nrf2 via promoter CpGs demethylation. *Biochem Pharmacol*. 82:1073–1078.

Kim, H.N., D.H. Kim, E.H. Kim et al. M.H. 2014. Sulforaphane inhibits phorbol ester-stimulated IKK-NF-kappa B signaling and COX-2 expression in human mammary epithelial cells by targeting NF-kappa B activating kinase and ERK. *Cancer Lett*. 351:41–49.

Kodydkova, J., L. Vavrova, B. Stankova, J. Macasek, T. Krechler, and A. Zak. 2013. Antioxidant status and oxidative stress markers in pancreatic cancer and chronic pancreatitis. *Pancreas*. 42:614–621.

Kwak, M.K., P.A. Egner, P.M. Dolan et al. 2001. Role of phase 2 enzyme induction in chemoprotection by dithiolethiones. *Mutat Res*. 480–481:305–315.

La Vecchia, C. 2009. Association between Mediterranean dietary patterns and cancer risk. *Nutr. Rev*. 67:S126–S129.

Lancaster, M., J.M. Philp, and F.P. Jenkins. 1961. Toxicity associated with certain samples of groundnuts. *Nature*. 192:1095.

Lao, C.D., M.T. Ruffin, D. Normolle et al. 2006. Dose escalation of a curcuminoid formulation. *BMC Complement Altern Med*. 6:10.

Larsson, S.C. and A. Wolk. 2012. Red and processed meat consumption and risk of pancreatic cancer: Meta-analysis of prospective studies. *Br J Cancer*. 106:603–607.

Lee, D.H., K.E. Anderson, L.J. Harnack, A.R. Folsom, and D.R. Jacobs, Jr. 2004. Heme iron, zinc, alcohol consumption, and colon cancer: Iowa Women's Health Study. *J Natl Cancer Inst*. 96:403–407.

Li, Z.W., W.M. Chu, Y.L. Hu et al. 1999. The IKK beta subunit of I kappa B kinase (IKK) is essential for nuclear factor kappa B activation and prevention of apoptosis. *J Exp Med*. 189:1839–1845.

Liu, Y., C.C. Chang, G.M. Marsh, and F. Wu. 2012. Population attributable risk of aflatoxin-related liver cancer: Systematic review and meta-analysis. *Eur J Cancer*. 48:2125–2136.

Liu, Z.F., Z.L. Xie, W. Jones et al. 2009. Curcumin is a potent DNA hypomethylation agent. *Bioorg Med Chem Lett*. 19:706–709.

Malins, D.C. and R. Haimanot. 1991. Major alterations in the nucleotide structure of DNA in cancer of the female breast. *Cancer Res.* 51:5430–5432.

Milner, J.A. 2008. Nutrition and cancer: Essential elements for a roadmap. *Cancer Lett.* 269:189–198.

Mirvish, S.S. 1995. Role of N-nitroso compounds (NOC) and N-nitrosation in etiology of gastric, esophageal, nasopharyngeal and bladder cancer and contribution to cancer of known exposures to NOC. *Cancer Lett.* 93:17–48.

Mosby, T.T., M. Cosgrove, S. Sarkardei, K.L. Platt, and B. Kaina. 2012. Nutrition in adult and childhood cancer: Role of carcinogens and anti-carcinogens. *Anticancer Res.* 32:4171–4192.

Nakanishi, M. and D.W. Rosenberg. 2013. Multifaceted roles of PGE2 in inflammation and cancer. *Semin Immunopathol.* 35:123–127.

Negre-Salvayre, A., N. Auge, V. Ayala et al. 2010. Pathological aspects of lipid peroxidation. *Free Radic Res.* 44:1125–1171.

O'Donnell-Tormey, J., C.J. DeBoer, and C.F. Nathan. 1985. Resistance of human tumor cells in vitro to oxidative cytolysis. *J Clin Invest.* 76:80–86.

Oberley, L.W. and T.D. Oberley. 1984. The role of superoxide dismutase and gene amplification in carcinogenesis. *J Theor Biol.* 106:403–422.

Pierre, F., G. Peiro, S. Tache, A.J. et al. 2006. New marker of colon cancer risk associated with heme intake: 1,4-dihydroxynonane mercapturic acid. *Cancer Epidemiol Biomarkers Prev.* 15:2274–2279.

Pierre, F., S. Tache, F. Gueraud, A.L. Rerole, M.L. Jourdan, and C. Petit. 2007. Apc mutation induces resistance of colonic cells to lipoperoxide-triggered apoptosis induced by faecal water from haem-fed rats. *Carcinogenesis.* 28:321–327.

Pogribny, I.P., S.A. Ross, C. Wise et al. 2006. Irreversible global DNA hypomethylation as a key step in hepatocarcinogenesis induced by dietary methyl deficiency. *Mutat Res.* 593:80–87.

Pogribny, I.P. and I. Rusyn. 2013. Environmental toxicants, epigenetics, and cancer. In *Epigenetic Alterations in Oncogenesis.* A.R. Karpf, ed., Vol. 754, pp. 215–232, Springer, Berlin, Germany.

Prutki, M., M. Poljak-Blazi, M. Jakopovic, D. Tomas, I. Stipancic, and N. Zarkovic. 2006. Altered iron metabolism, transferrin receptor 1 and ferritin in patients with colon cancer. *Cancer Lett.* 238:188–196.

Rao, C.V. and B.S. Reddy. 2004. NSAIDs and chemoprevention. *Curr Cancer Drug Targets.* 4:29–42.

Reddy, A.C.P. and B.R. Lokesh. 1994. Studies on the inhibitory effects of curcumin and eugenol on the formation of reactive oxygen species and the oxidation of ferrous iron. *Mol Cell Biochem.* 137:1–8.

Roswall, N., Y. Li, C. Kyro, S. et al. 2015. No association between adherence to a healthy nordic food index and colorectal cancer: Results from a Swedish Cohort Study. *Cancer Epidemiol Biomarkers Prev.* 24:755–757.

Rushmore, T.H., M.R. Morton, and C.B. Pickett. 1991. The antioxidant responsive element. Activation by oxidative stress and identification of the DNA consensus sequence required for functional activity. *J Biol Chem.* 266:11632–11639.

Salehi, M., M. Moradi-Lakeh, M.H. Salehi, M. Nojomi, and F. Kolahdooz. 2013. Meat, fish, and esophageal cancer risk: A systematic review and dose-response meta-analysis. *Nutr Rev.* 71:257–267.

Sato, K., K. Ito, H. Kohara, Y. Yamaguchi, K. Adachi, and H. Endo. 1992. Negative regulation of catalase gene expression in hepatoma cells. *Mol Cell Biol.* 12:2525–2533.

Schreck, R., P. Rieber, and P.A. Baeuerle. 1991. Reactive oxygen intermediates as apparently widely used messengers in the activation of the NF-kappa B transcription factor and HIV-1. *Embo J.* 10:2247–2258.

Sesink, A.L., D.S. Termont, J.H. Kleibeuker, and R. Van der Meer. 1999. Red meat and colon cancer: The cytotoxic and hyperproliferative effects of dietary heme. *Cancer Res.* 59:5704–5709.

Shalini, V.K. and L. Srinivas. 1987. Lipid peroxide induced dna damage - protection by turmeric (Curcuma-longa). *Mol Cell Biochem.* 77:3–10.

Shan, Y., K. Wu, W. Wang et al. 2009. Sulforaphane down-regulates COX-2 expression by activating p38 and inhibiting NF-kappaB-DNA-binding activity in human bladder T24 cells. *Int J Oncol.* 34:1129–1134.

Sharma, S., T.K. Kelly, and P.A. Jones. 2010. Epigenetics in cancer. *Carcinogenesis.* 31:27–36.

Shelton, P. and A.K. Jaiswal. 2013. The transcription factor NF-E2-related Factor 2 (Nrf2): A protooncogene? *Faseb J.* 27:414–423.

Shen, H.M., C.N. Ong, B.L. Lee, and C.Y. Shi. 1995. Aflatoxin B1-induced 8-hydroxydeoxyguanosine formation in rat hepatic DNA. *Carcinogenesis.* 16:419–422.

Shubik, P. 1975. Potential carcinogenicity of food additives and contaminants. *Cancer Res.* 35:3475–3480.

Singh, S. and B.B. Aggarwal. 1995. Activation of transcription factor Nf-kappa-b is suppressed by curcumin (diferulolylmethane). *J Biol Chem.* 270:24995–25000.

Sreejayan, N. and M.N.A. Rao. 1994. Curcuminoids as potent inhibitors of lipid-peroxidation. *J Pharm Pharmacol.* 46:1013–1016.

Srinivasan, K. 2014. Antioxidant potential of spices and their active constituents. *Crit. Rev Food Sci Nutr.* 54:352–372.

Sugimura, T. 2002. Food and cancer. *Toxicology.* 181–182:17–21.

Sun, Y. 1990. Free radicals, antioxidant enzymes, and carcinogenesis. *Free Radic Biol Med.* 8:583–599.

Szatrowski, T.P. and C.F. Nathan. 1991. Production of large amounts of hydrogen peroxide by human tumor cells. *Cancer Res.* 51:794–798.

Szent-Gyorgyi, A., I. Isenberg, and S.L. Baird. 1960. On the electron donating properties of carcinogens. *Proc Natl Acad Sci USA.* 46:1444–1449.

Tareke, E., P. Rydberg, P. Karlsson, S. Eriksson, and M. Tornqvist. 2002. Analysis of acrylamide, a carcinogen formed in heated foodstuffs. *J Agric Food Chem.* 50:4998–5006.

Torre, L.A., F. Bray, R.L. Siegel, J. Ferlay, J. Lortet-Tieulent, and A. Jemal. 2015. Global cancer statistics, 2012. *CA Cancer J Clin.* 65:87–108.

Toyokuni, S., K. Okamoto, J. Yodoi, and H. Hiai. 1995. Persistent oxidative stress in cancer. *FEBS Lett.* 358:1–3.

Toyokuni, S., K. Uchida, K. Okamoto, Y. Hattori-Nakakuki, H. Hiai, and E.R. Stadtman. 1994. Formation of 4-hydroxy-2-nonenal-modified proteins in the renal proximal tubules of rats treated with a renal carcinogen, ferric nitrilotriacetate. *Proc Natl Acad Sci USA.* 91:2616–2620.

Trichopoulou, A., P. Lagiou, H. Kuper, and D. Trichopoulos. 2000. Cancer and Mediterranean dietary traditions. *Cancer Epidemiol Biomarkers Prev.* 9:869–873.

Uchida, K. 2003. 4-Hydroxy-2-nonenal: A product and mediator of oxidative stress. *Prog Lipid Res.* 42:318–343.

Warshawsky, D. and J.R. Landolph. 2005. *Molecular Carcinogenesis and the Molecular Biology of Human Cancer.* CRC Press, Boca Raton, FL.

WCRF/AICR. 2007. *WCRF: Food, Nutrition, Physical Activity, and the Prevention of Cancer: A Global Perspective.* World Cancer Research Fund and American Institute for Cancer Research, Washington, DC.

WCRF/AICR. 2011. *WCRF: The Associations between Food, Nutrition and Physical Activity and the Risk of Colorectal Cancer.* World Cancer Research Fund and American Institute for Cancer Research, Washington, DC.

WCRF/AICR. 2012. *WCRF: Continuous Update Project Report. Food, Nutrition, Physical Activity, and the Prevention of Pancreatic Cancer.* World Cancer Research Fund and American Institute for Cancer Research, Washington, DC.

Williams, M.V., S.H. Lee, M. Pollack, and I.A. Blair. 2006. Endogenous lipid hydroperoxide-mediated DNA-adduct formation in min mice. *J Biol Chem.* 281:10127–10133.

Winczura, A., D. Zdzalik, and B. Tudek. 2012. Damage of DNA and proteins by major lipid peroxidation products in genome stability. *Free Radical Res.* 46:442–459.

Wu, H., Q. Qi, R.M. van Dam et al. 2015. Association between dietary whole grain intake and risk of mortality: Two large prospective studies in US men and women. *JAMA Intern Med.* 175:373–384.

Xing, D., L. Sun, R.I. Cukier, and Y. Bu. 2007. Theoretical prediction of the p53 gene mutagenic mechanism induced by trans-4-hydroxy-2-nonenal. *J Phys Chem B.* 111:5362–5371.

Yu, J.H. and H. Kim. 2014. Oxidative stress and cytokines in the pathogenesis of pancreatic cancer. *J Cancer Prev.* 19:97–102.

Yu, M.W., J.P. Lien, Y.H. Chiu, R.M. Santella, Y.F. Liaw, and C.J. Chen. 1997. Effect of aflatoxin metabolism and DNA adduct formation on hepatocellular carcinoma among chronic hepatitis B carriers in Taiwan. *J Hepatol.* 27:320–330.

Zarkovic, K., K. Uchida, D. Kolenc, L. Hlupic, and N. Zarkovic. 2006. Tissue distribution of lipid peroxidation product acrolein in human colon carcinogenesis. *Free Radic Res.* 40:543–552.

Zarkovic, N. 2003. 4-hydroxynonenal as a bioactive marker of pathophysiological processes. *Mol Aspects Med.* 24:281–291.

Zhang, D.D. and M. Hannink. 2003. Distinct cysteine residues in Keap1 are required for Keap1-dependent ubiquitination of Nrf2 and for stabilization of Nrf2 by chemopreventive agents and oxidative stress. *Mol Cell Biol.* 23:8137–8151.

Zhang, L., Y.F. Pei, R. Hai, Y. Lin, and H.W. Deng. 2012. Testing rare variants for association with diseases: A Bayesian marker selection approach. *Ann Hum Genet.* 76:74–85.

4 Food Chemical Carcinogens*

Sources and Mechanism of Exogenous DNA Adduct Formation

Peter P. Fu and Qingsu Xia

CONTENTS

* In: Food Chemical Carcinogens. Debasis Bagchi (ed.), Taylor & Francis Publisher.

4.1 INTRODUCTION

Cancer is the second leading cause of death in the United States, largely ascribed to exposure to chemical carcinogens in our daily life. Plants are used by humans daily in many different ways, including food and dietary supplements. Many phytochemicals produced as secondary metabolites are highly toxic to humans, including cytotoxic, genotoxic, photocytotoxic, photogenotoxic, and tumorigenic (Doll and Peto, 1981; Smith and Culvenor, 1981; Food, Nutrition and the Prevention of Cancer, 1997; Ames and Gold, 1998; Stegelmeier et al., 1999; Fu et al., 2002, 2013; Fu and Xia, 2014).

The major sources of human exposure to chemical carcinogens include natural foodborne chemicals, food pyrolysates produced by cooking, food additives, dietary supplements, food contaminants, environmental pollutants, drugs, cigarette smoke, and cosmetics. Among these, epidemiological studies have indicated that diet is the most important factor, accounting for about one-third of human cancer risk (Ames and Gold, 1998). The total amount of natural carcinogens ingested from food every day is much higher than that from environmental sources (Ames and Gold, 1998). The human cancer incidence can be reduced if exposure to foodborne carcinogens can be reduced.

There are three major sources of foodborne carcinogens: natural carcinogenic chemicals, food pyrolysates from cooking, and dietary supplements. Food contaminated with tumorigenic chemicals, such as heavy metals and pesticide residues, is relatively a minor source.

Many foodborne phytochemicals, plant extracts, and cooking pyrolysates have been found to be tumorigenic in experimental rodents, and mechanisms of cancer induction have been determined, among which the metabolic activation pathway involving the exogenous DNA adduct formation can provide insight of the mechanism at a molecular level.

We have been interested in determining the mechanism of cancer induction by carcinogenic chemicals present in food, drugs, and the environment. We have previously published review articles and book chapters on this research field, including pyrrolizidine alkaloids, Chinese herbal medicines, and dietary supplements (Fu et al., 2002, 2004; Fu and Xia, 2015); the toxicological effects of ginkgo biloba (Chan et al., 2007); the toxicity of kava kava (Fu et al., 2008); the phototoxicity of herbal plants, herbal products, and nitro-polycyclic aromatic hydrocarbons (Fu et al., 2013; Xia et al., 2013a); and quality assurance and safety protection of herbal dietary supplements (Fu et al., 2007, 2009; Fu and Xia, 2014). In this chapter, we briefly introduce the sources of foodborne chemical carcinogens and the induction of tumors by foodborne carcinogens mediated by exogenous DNA adduct formation. Using the mechanistic study of polycyclic aromatic hydrocarbons (PAHs) as an example, the challenge in determining the mechanism mediated by exogenous DNA adduct formation is also briefly described. Our laboratory has studied the mechanisms of several classes of genotoxic foodborne chemical carcinogens, including PAHs, nitro-PAHs, pyrrolizidine alkaloids, and herbal dietary supplements. Many examples described in this paper are from the authors' own work. Due to the scope of this chapter, the chemical carcinogenesis by tobacco smoking is not included.

4.2 HUMAN FOOD CARCINOGENS

There are two organizations that determine and classify human chemical carcinogens. The International Agency for Research on Cancer (IARC) is an agency of the World Health Organization (WHO). IARC classifies carcinogenic chemicals into the following groups:

Group 1	Carcinogenic to humans
Group 2A	Probably carcinogenic to humans
Group 2B	Possibly carcinogenic to humans
Group 3	Not classifiable due to its carcinogenicity to humans
Group 4	Probably not carcinogenic to humans

To date, IARC has published 113 volumes of IARC Monographs on the Evaluation of Carcinogenic Risks to Human. There are a series of food chemical carcinogens that have been listed in the IARC Monographs. These food chemical carcinogens are summarized in Table 4.1.

The U.S. National Toxicology Program (NTP), founded in 1978, has been conducting long-term tumorigenicity bioassays of environmental toxic chemicals, foodborne plant chemical constituents, and food pyrolysates, many of which have been determined to be rodent carcinogens. The Report on Carcinogens is a congressionally mandated, science-based, public health document. The Report on Carcinogens classifies human carcinogens into two categories: (1) known to be human carcinogen and (2) reasonably anticipated to be human carcinogen. Many food-borne chemical carcinogens have been listed in the Report on Carcinogens (NTP 13th Report of Carcinogens, 2014).

4.3 SOURCES OF FOOD CARCINOGENS

4.3.1 Chemical Carcinogens as Food Constituents

Edible plants consumed as foods account for about 60% of the human diet. It was estimated that Americans eat 5,000–10,000 different natural chemicals. Many plant secondary metabolites are toxic and tumorigenic (Hartmann, 2004; Fu et al., 2007). For example, 5-methoxypsoralen, 8-methoxypsoralen (8-MOP), safrole, estrole, cycasin, capsaicin, allyl isothiocyanate, and limonene found in our diet are carcinogenic (Figure 4.1). The IARC has determined that many plant secondary metabolites are human chemical carcinogens (Table 4.1).

N-Nitrosamines are a family of foodborne chemical carcinogens generated by acid-catalyzed reaction of nitrite with secondary amines in stomach (Mirvish, 1975; Wiestler et al., 1987). Humans are also exposed to *N*-nitrosamines in food. The general structure of *N*-nitrosamines is shown in Figure 4.1. The NTP classified 15 *N*-nitrosamines, for example, *N*-methyl-*N*′-nitro-nitrosoguanidine, *N*-nitrosodi-*N*-butylamine, *N*-nitrosodiethanolamine, *N*-nitrosodiethylamine, *N*-nitrosodimethylamine, *N*-nitrosodi-*N*-propylamine, *N*-nitroso-*N*-ethylurea, 4-*N*-nitrosomethylamino)-1-(3-pyridyl)-1-butanone, *N*-nitroso-*N*-methylurea, *N*-nitrosomethylvinylamine, *N*-nitrosomorpholine, *N*-nitrosonornicotine, *N*-nitrosopiperidine, *N*-nitrosopyrrolidine, and *N*-nitrososarcosine as “reasonably anticipated to be human carcinogen” in the Report on Carcinogens (NTP 13th Report on Carcinogens, 2014).

The IARC listed 19 *N*-nitrosamines in IARC Monographs, among which 2 are listed as group 1 human carcinogens, 5 as group 2A human carcinogens, and 11 as group 2B human carcinogens (Table 4.1). Many *N*-nitrosamines are produced in the process of food preparation. For example, salted fish prepared in a Chinese style containing several *N*-nitrosamines is listed by IARC as a group 1 human carcinogen. *N*-Nitrosodimethylamine (dimethylnitrosamine), listed in group 2A, is present in foodstuffs, especially those cooked, smoked, or cured.

Pyrrolizidine alkaloids are the most widespread phytochemicals and have been identified in over 6000 plants. Many pyrrolizidine alkaloids are hepatotoxic, genotoxic, and tumorigenic and are among the first naturally occurring carcinogens identified in plants (Fu et al., 2001, 2002, 2004, 2007; Chen et al., 2010; Mei et al., 2004, 2005, 2010). There are numerous reports of human poisoning caused by pyrrolizidine alkaloids (Mattocks, 1968; Prakash et al., 1999; Stegelmeier et al., 1999). Humans are exposed to toxic pyrrolizidine alkaloids through eating pyrrolizidine alkaloid–containing food (Prakash et al., 1999), contaminated staple food, herbal teas, herbal medicines, and herbal dietary supplements (Fu et al., 2002, 2007). Pyrrolizidine alkaloids have been found as contaminants in many human food sources, for example, wheat, milk, honey, bird eggs, and tea, and exposure through these sources is suspected to have caused worldwide human health problems (Fuet et al., 2004; Bodi et al., 2014; Fu and Xia, 2015).

TABLE 4.1
Human Food Chemical Carcinogens Classified in IARC Monographs

Agent	Group	Year
Food constituents		
Acrylamide	2A	1994
Aloe vera, whole leaf extract	2B	2015
Anthraquinone	2B	2013
Aristolochic acid	1	2012
Plants containing aristolochic acid	1	2012
Betel quid	1	2012
Cycasin	2B	1987
Furan	2B	1995
Ginkgo biloba extract	2B	2015
Goldenseal root powder	2B	2015
Kava extract	2B	2015
Safrole	2B	1987
5-Methoxypsoralen	2A	1987
8-Methoxypsoralen plus UV light	1	2012
Monocrotaline	2B	1987
Riddelliine	2B	2002
Nitrosamines		
N′-Nitrosonornicotine (NNN)	1	2012
NNK[a]	1	2012
N-Ethyl-*N*-nitrosourea	2A	1987
MNNG[b]	2A	1987
N-Methyl-*N*-nitrosourea	2A	1987
N-Nitrosodiethylamine	2A	1987
N-Nitrosodimethylamine	2A	1987
N-Methyl-*N*-nitrosourethane	2B	1987
N-Nitrosodi-*N*-butylamine	2B	1987
N-Nitrosodi-*N*-propylamine	2B	1987
N-Nitrosomethylethylamine	2B	1987
N-Nitrosomethylvinylamine	2B	1987
N-Nitrosomorpholine	2B	1987
N-Nitrosopiperidine	2B	1987
N-Nitrosopyrrolidine	2B	1987
N-Nitrososarcosine	2B	1987
N-Nitrosodiethanolamine	2B	2000
3-(*N*-Nitrosomethylamino)propionitrile	2B	2004
Salted fish, Chinese style	1	2012
Polycyclic aromatic hydrocarbons		
Cyclopenta[*cd*]pyrene	2A	2010
Dibenz[*a,h*]anthracene	2A	2010
Dibenzo[*a,l*]pyrene	2A	2010
Dibenz[*a,j*]acridine	2A	2013
Benz[*j*]aceanthrylene	2B	2010
Benz[*a*]anthracene (BA)	2B	2010
Benzo[*a*]pyrene (BaP)	1	2012
Benzo[*b*]fluoranthene	2B	2010

(*Continued*)

TABLE 4.1 (*Continued*)
Human Food Chemical Carcinogens Classified in IARC Monographs

Agent	Group	Year
Benzo[*j*]fluoranthene	2B	2010
Benzo[*k*]fluoranthene	2B	2010
Benzo[*c*]phenanthrene	2B	2010
Chrysene	2B	2010
Dibenzo[*a,h*]pyrene	2B	2010
Dibenzo[*a,i*]pyrene	2B	2010
5-Methylchrysene	2B	2010
Naphthalene	2B	2002
Dibenz[*a,h*]acridine	2B	2013
Dibenz[*c,h*]acridine	2B	2013
7*H*-Dibenzo[*c,g*]carbazole	2B	2013
Nitropolycyclic aromatic hydrocarbons		
1-Nitropyrene	2A	2013
6-Nitrochrysene	2A	2013
5-Nitroacenaphthene	2B	1987
4-Nitropyrene	2B	2013
3,7-Dinitrofluoranthene	2B	2013
Dinitrofluoranthene	2B	2013
1,3-Dinitropyrene	2B	2013
1,6-Dinitropyrene	2B	2013
1,8-Dinitropyrene	2B	2013
2-Nitrofluorene	2B	2013
Heterocyclic aromatic amines		
Glu-P-1[c]	2B	1987
Glu-P-2[d]	2B	1987
IQ[e]	2A	1993
MeIQ[f]	2B	1993
MeIQx[g]	2B	1993
PhIP[h]	2B	1993

[a] 4-(*N*-Nitrosomethylamino)-1-(3-pyridyl)-1-butanone.
[b] *N*-Methyl-*N*′-nitro-*N*-nitrosoguanidine.
[c] 2-Amino-6-methyldipyrido[1,2-*a*:3′,2′*d*]imidazole.
[d] 2-Aminodipyrido[1,2-*a*:3′,2′-d]imidazole.
[e] 2-Amino-3-methylimidazo[4,5-*f*]quinolone.
[f] 2-Amino-3,4-dimethylimidazo[4,5-*f*]quinolone.
[g] 2-Amino-3,8-dimethylimidazo[4,5-*f*]quinoxaline.
[h] 2-Amino-1-methyl-6-phenylimidazo[4,5-*b*]pyridine.

To date, there are more than 15 pyrrolizidine alkaloids that can induce tumors in rodents. Structures of these carcinogenic pyrrolizidine alkaloids are shown in Figure 4.2. Among these carcinogens, both lasiocarpine and riddelliine have been tested by the NTP as rodent carcinogens (NTP, 1978, 2003). The NTP classified riddelliine as "reasonably anticipated to be human carcinogen" in the 13th Report on Carcinogens (2014). The IARC listed monocrotaline and riddelliine as group 2B possible human carcinogens in 1987 and 2002, respectively (Table 4.1).

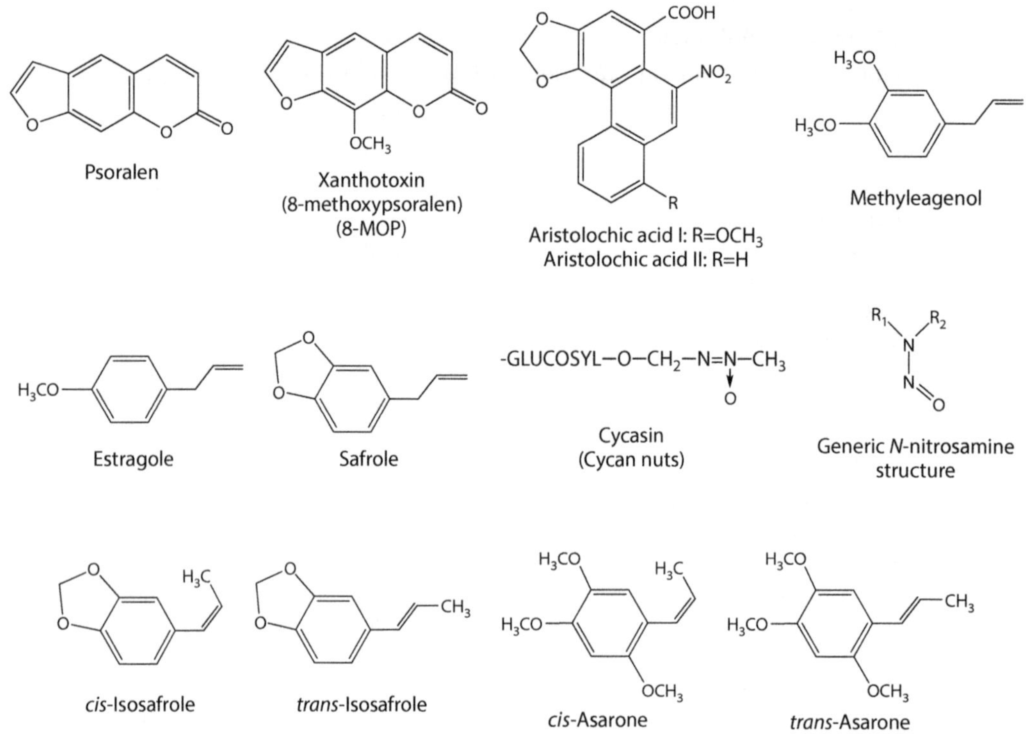

FIGURE 4.1 Structures of some representative natural food carcinogens.

4.3.2 Food Contaminants through Cooking Process

4.3.2.1 Polycyclic Aromatic Hydrocarbons

PAHs are first known as a class of genotoxic and tumorigenic environmental pollutants produced from incomplete combustion of natural organic materials, such as fossil fuel (Gelboin and T'so, 1978; IARC, 1983; Fang et al., 2004a,b; Fu et al., 2012). The study of carcinogenesis of PAHs has a long history and started in 1775 when Percivall Pott found a high occurrence of scrotal cancer in chimney sweeps and suspected that the cause was due to chemicals contained in the soots (Pott, 1775). This was one of the first epidemiological studies on chemical carcinogenesis. In 1918, Yamagiwa and Ichikawa painted coal tar on rabbits' ears and induced skin cancer, which represented the first study using experimental animals for the study of induction of cancer by chemicals. In 1933, Cook, Hewett, and Hieger isolated the carcinogen benzo[a]pyrene from coal tar, confirming that PAHs are the principal carcinogenic chemical constituents in soots and coal tar is responsible for cancer induction. The follow-up research that focused on the structure–activity relationship lasted several decades.

It was not until 1967 that PAHs were detected in food from grilling meat (Lijinsky and Ross 1967). It was found that high levels of PAHs were formed during food processing or cooking but were not present in raw food (WHO, 1998; Phillips, 1999; SCF, 2002). It was then established that the main source of human exposure to PAHs is food, accounting for more than 90% of total exposure, not due to exposure to environmental matters (WHO, 1998; Phillips, 1999; SCF, 2002).

The formation of PAHs from cooking occurs predominantly through pyrolysis of fat at a temperature of 200°C and much above, particularly with the food directly in contact with the flame (SCF, 2002). The highest level of PAHs was formed when cooking under high temperature, lack of water, and a long duration of cooking (WHO, 1998). Therefore, cooking methods with grilling,

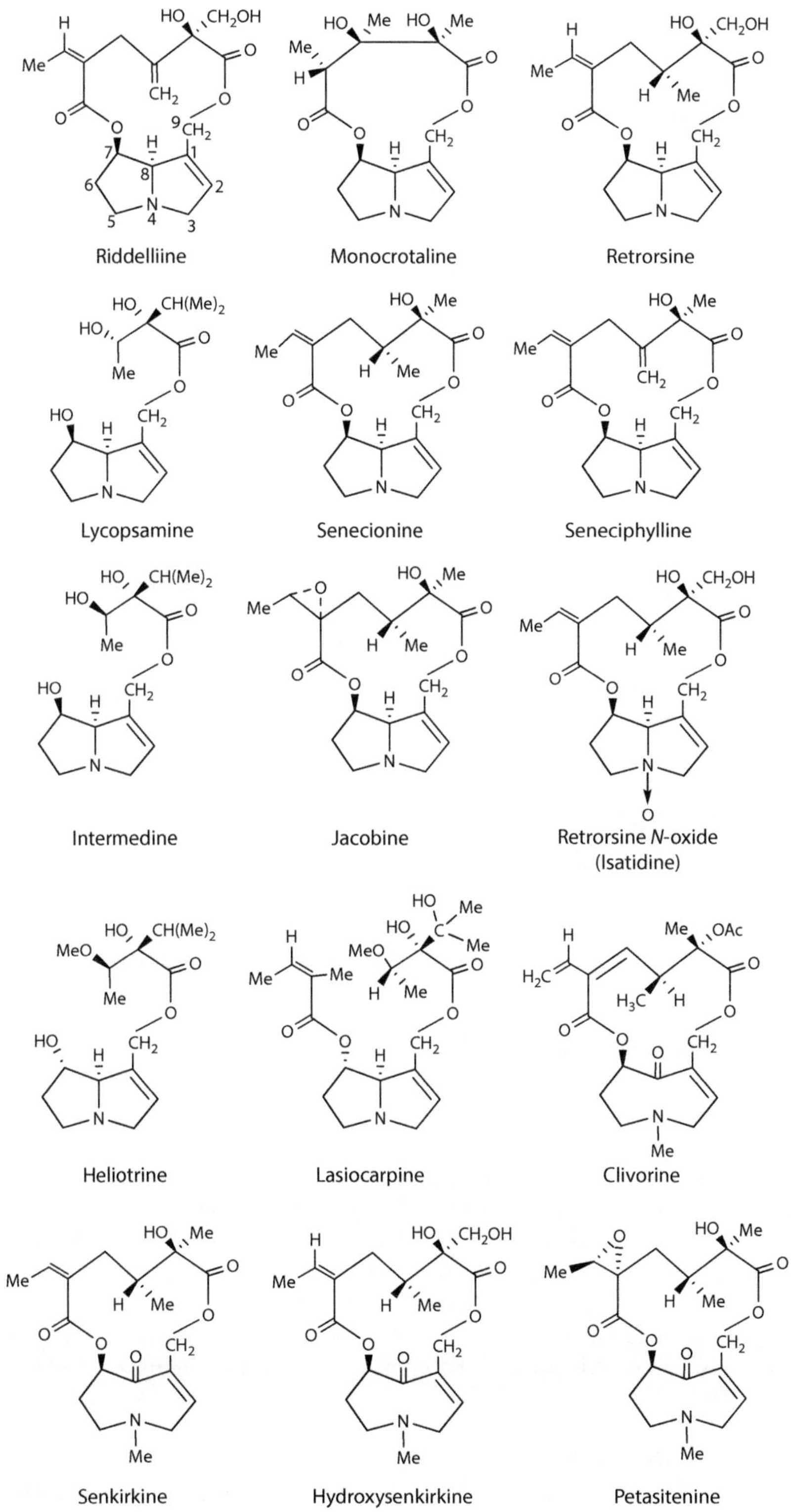

FIGURE 4.2 Structures of tumorigenic pyrrolizidine alkaloids and a pyrrolizidine alkaloid *N*-oxide, retrorsine *N*-oxide.

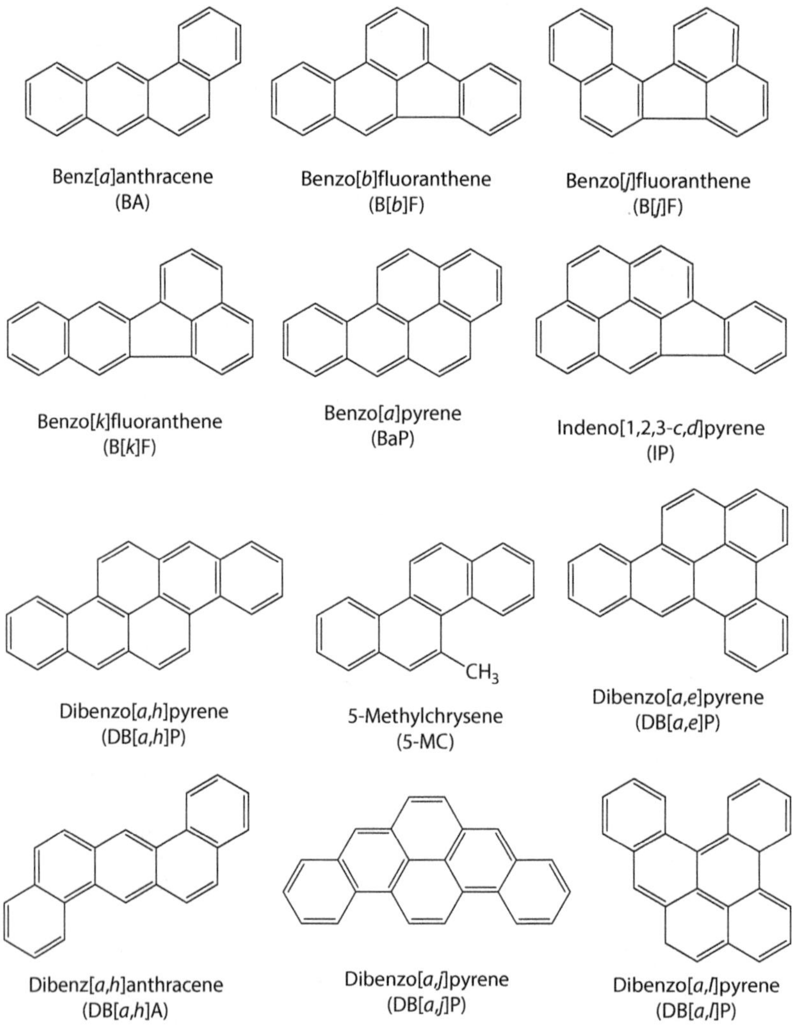

FIGURE 4.3 Structures of the 12 PAHs listed in the NTP 13th Report on Carcinogens as reasonably anticipated to be human carcinogens.

roasting, barbecuing, or smoking generate hundreds of PAHs in a high quantity, whereas cooking by steaming produces PAHs in a trace quantity (Chen and Lin, 1997).

IARC classified 19 PAHs as human carcinogens, among which benzo[a]pyrene (BaP) is a group 1 human carcinogen, cyclopenta[*cd*]pyrene, dibenz[*a,h*]anthracene, dibenzo[*a,l*]pyrene, and dibenz[*a,j*]acridine as group 2A probably carcinogenic to humans, and another 14 as group 2B possibly carcinogenic to humans (Table 4.1). Also, the U.S. NTP listed 12 PAHs in Report on Carcinogens as reasonably anticipated to be human carcinogens (Figure 4.3; NTP 13th Report on Carcinogens, 2014).

4.3.2.2 Nitro-PAHs (Nitroarenes)

During the energy crisis in the early 1970s, diesel fuel was studied to assess whether or not it can be a good substituted energy source. Compared with gasoline, the use of diesel fuel may be safer, because its combustion generates less quantity of PAHs, although nitrated gas was produced. The extensive study on the environmental formation and mutagenicity of nitro-PAHs began in the

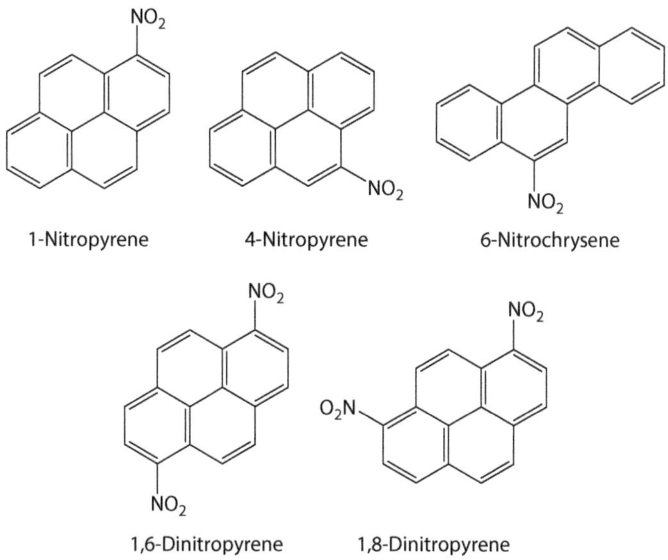

FIGURE 4.4 Structures of the five nitro-PAHs listed in the NTP 13th Report on Carcinogens as reasonably anticipated to be human carcinogens.

late 1970s when Pitts and coworkers detected nitro-PAHs in the atmosphere (Pitts et al., 1978). Since then, many potent mutagenic nitro-PAHs were detected in the environment (Fu, 1990; Fu and Herreno-Saenz, 1999).

The first report on the detection of nitro-PAHs in food was in 1986. Kinouchi et al. (1986) detected 1-nitropyrene in grilled chicken. Schlemitz and Pfannhauser (1996) identified a number of nitro-PAHs in vegetables, smoked and grilled foods, oil, tea, coffee, and spices. Many nitro-PAHs are potent mutagenic, and a number of them are also rodent carcinogens (Chae et al., 1997). The IARC listed 1-nitropyrene and 6-nitrochrysene as group 2A carcinogens and another 8 nitro-PAHs as group 2B possibly human carcinogens (Table 4.1). To date, the NTP classified 5 nitro-PAHs, 1,6-dinitropyrene, 1,8-dinitropyrene, 6-nitrochrysene, 1-nitropyrene, and 4-nitropyrene, as "reasonably anticipated to be human carcinogen" in Report on Carcinogens (2014; Figure 4.4).

4.3.2.3 Heterocyclic Aromatic Amines

The first report on the detection of heterocyclic amines in cooking food was in 1977 by Sugimura and coworker (Nagao et al., 1977). They found that pyrolysis of protein and amino acids at high temperature generated heterocyclic amines (or termed *food pyrolysates*) (Nagao et al., 1977). Subsequently, a series of foodborne heterocyclic amines were found to be potent mutagenic in Ames *Salmonella typhimurium* assays. To date, a number of heterocyclic amines generated from the pyrolysis of protein-containing foods have been determined to be carcinogenic in rodents. The structures and names of 10 carcinogenic heterocyclic amines are shown in Figure 4.5. Among these heterocyclic amines, 2-amino-3,4-dimethylimidazo[4,5-*f*] quinolone (MeIQ), 2-amino-3,8-dimethylimidazo[4,5-*f*]quinoxaline (MeIQx), 2-amino-3-methylimidazo[4,5-*f*] quinolone (IQ), and 2-amino-1-methyl-6-phenylimidazo[4,5-*b*] pyridine (PhIP) are listed as "reasonably anticipated to be human carcinogen" in Report on Carcinogens. In addition to these four heterocyclic amines, IARC also listed 2-amino-6-methyldipyrido[1,2-*a*:3′,2′d]imidazole (Glu-P-1) and 2-aminodipyrido[1,2-*a*:3′,2′-*d*]imidazole (Glu-P-2) as group 2B possibly human carcinogens (Table 4.1).

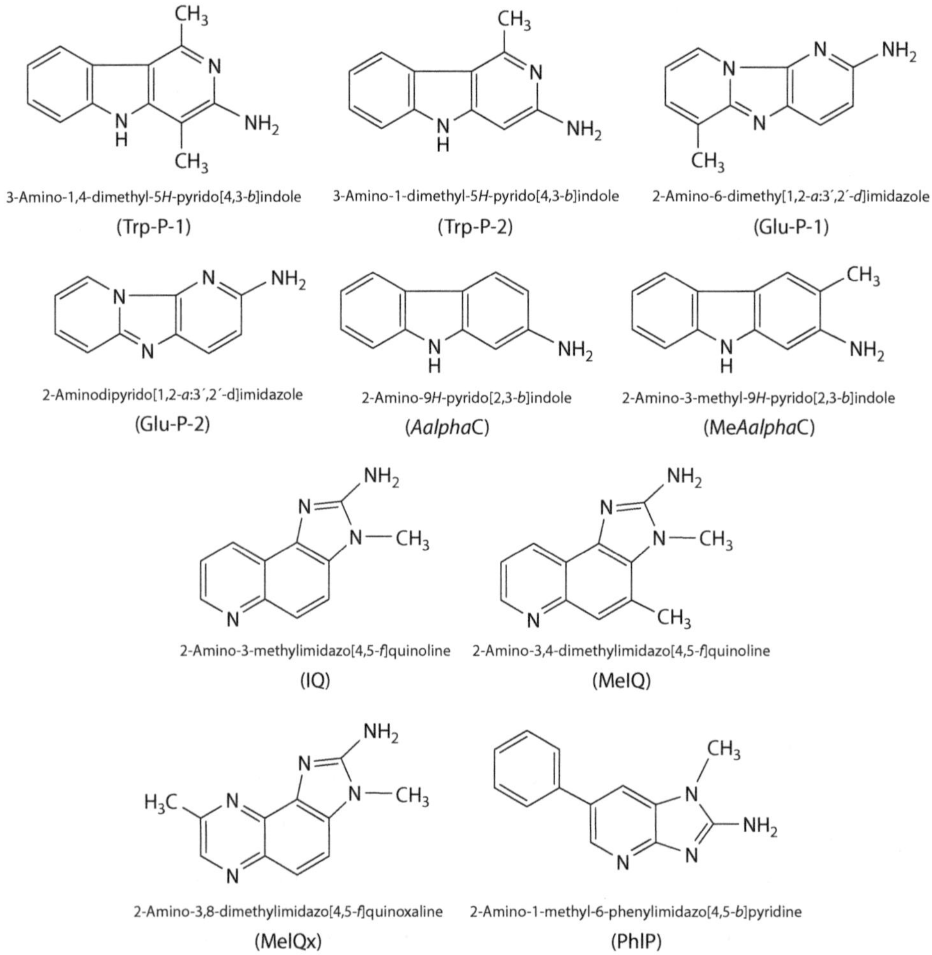

FIGURE 4.5 Structures of heterocyclic amines determined as rodent carcinogens.

4.3.2.4 Acrylamide, Glycidamide, and Furan

In 2002, Tareke et al. (2002) detected acrylamide in carbohydrate-rich foods, such as commercial potato products and breads prepared by high-temperature frying, baking, or roasting. Acrylamide was produced during frying and baking through the Maillard reaction of foods (Stadler et al., 2002).

A study by the U.S. Food and Drug Administration (FDA) identified furan, a liver carcinogen, in a variety of thermally treated foods (coffee, baked or fried cereal products, canned) and jarred foods (canned soup, baby food, and infant formula) (Becalski et al., 2005; FDA, 2005). The major pathway of furan generation in food is the thermal degradation of carbohydrates such as glucose, lactose, and fructose (Maga, 1979).

Acrylamide and furan were classified as probably carcinogenic to human (group 2A) and possibly carcinogenic to human (group 2B) by IARC in 1994 and 1995, respectively. The U.S. NTP listed acrylamide, glycidamide, and furan as "reasonably anticipated to be human carcinogen" in Report on Carcinogens.

4.3.3 Herbs Dietary Supplements

During the last several decades, the human use of botanicals/herbal products, mainly herbal dietary supplements and functional food, has been progressively increasing worldwide (Fu et al., 2009). During 2003–2006, about 20% of American adults used botanical supplements (Bailey et al., 2011).

As estimated by the American Herbal Products Association in 2004, there were about 3,000 plant species used in as many as 50,000 different products sold as herbal dietary supplements in the United States (Zurer and Hanson, 2004; Fu and Xia, 2014). It was reported that in 1999, the U.S. population consumed about US$31 billion for dietary supplements and functional foods. Their popular use in very large quantities becomes a third major source of foodborne carcinogen source. Unfortunately, numerous cases of toxic effects in humans by consumption of herbal products have been reported (Fu and Xia, 2014, 2015). In 1994, the U.S. Congress passed the Dietary Supplement Health and Education Act (DSHEA) that amended the U.S. Federal Food, Drug, and Cosmetic ACT (FFDCA) and created a new regulatory category, safety standard, and other rules for the U.S. FDA to regulate dietary supplements. Because of this act, dietary supplements belong to the food category; and consequently, many herbal plants and herbal products previously used merely as herbs now are listed as foods.

Since 1999, the U.S. NTP has been conducting long-term tumorigenicity bioassays on a large number of herbal plant extracts, herbal dietary supplements, and their functionally active chemical constituents nominated by the public and federal agencies. The nominated herbs and active ingredients are among the most commonly sold products in the United States and are suspected to possess a high potential toxic activity. Among the tested herbal chemical constituents, kava extract, ginkgo biloba extract, aloe vera extract, quercetin, and riddelliine have been determined to induce tumors in rodents.

4.3.4 Photocarcinogens

Many food pyrolysates, edible and herbal plants, herbal products, and their chemical components have been shown to be highly phototoxic and photogenotoxic (Fu et al., 2013). For instance, the two furocoumarins, psoralen and xanthotoxin (8-MOP) (Figure 4.1), used for photodynamic therapy can increase skin tumor incidence, such as squamous cell carcinoma, basalioma, and melanoma (IARC, 1986; Stern et al., 1997). Both psoralen and 8-MOP are listed as group 1 human carcinogens by IARC (Table 4.1).

It is noteworthy that PAHs and nitro-PAHs can absorb light energy to form photoexcited intermediates, which react with molecular oxygen to produce reactive oxygen species (ROS) and other free radicals, resulting in lipid peroxidation and DNA damage, including DNA strand breakage, oxidation to 8-oxo-2′-deoxyguanosine, and DNA adduct formation. Thus, PAHs and nitro-PAHs can be photocarcinogenic (Fu et al., 2012; Xia et al., 2013a).

4.4 MECHANISMS OF CANCER INITIATION BY CHEMICAL CARCINOGENS

Chemical carcinogens are conventionally classified as genotoxic and "nongenotoxic" depending on the mechanisms leading to cancer formation. Genotoxic chemical carcinogens are defined as chemicals that exert carcinogenicity through mechanisms involving the formation of covalently bound carcinogen–DNA (exogenous DNA) adducts that lead to mutation. Nongenotoxic chemical carcinogens are those that exert carcinogenicity through secondary mechanisms that may include (1) formation of DNA adducts from endogenous lipid peroxidation or oxidative stress; (2) epigenetic mechanisms, at least including hypomethylation, hypermethylation, and modulation of microRNAs; (3) modulation of sex hormones; and (4) induction of peroxisome proliferation.

The mechanisms of different classes of foodborne genotoxic chemical carcinogens have been extensively studied. These include PAHs (Jerina et al., 1978; Fu et al., 1978b), nitro-PAHs (Fu, 1990; Fu and Herreno-Saenz, 1999), food pyrolysates (heterocyclic aromatic amines) (Nagao et al., 1977; Howard et al., 1983; Beland and Kadlubar, 1985), *N*-nitrosamines (Wiseman et al., 1987), aromatic amines (Wiseman et al., 1987), and allylarenes including estragole and isosafrole (Wiseman et al., 1987).

There are two types of genotoxic carcinogens. The first type of genotoxic carcinogens, representing the majority of chemical carcinogens, requires metabolic activation, catalyzed by specific cytochrome P450. The resulting ultimate carcinogenic metabolites are electrophiles, capable of binding to the nucleophilic sites of cellular DNA, resulting in the production of exogenous DNA adducts and leading to mutations and tumor initiation. Mechanistic studies of this activation pathway at least include metabolism, DNA adduct formation, *ras*-oncogene activation, and tumorigenicity determination. The second type of genotoxic carcinogens is that the parent chemical carcinogens themselves, such as *N*-ethyl-*N*-nitrosourea, are electrophiles, can act like alkylating agents, and are capable of binding to DNA directly without requiring metabolic activation. To date, many foodborne carcinogens have been determined to initiate tumors through the activation pathway mediated by exogenous DNA adduct formation.

4.5 MECHANISM OF EXOGENOUS DNA ADDUCT FORMATION

Initiation of the cancer formation through a genetic mechanism requires that the ultimate carcinogenic metabolites must be electrophiles, capable of binding to cellular DNA to produce exogenous DNA adducts leading to mutations. An ultimate carcinogenic metabolite in general possesses a low electron density center that binds to the nucleophilic atom of the cellular DNA molecule *in vivo*. This binding may proceed through an S_N1 mechanism, an S_N2 mechanism, or a combination of both. To form the exogenous DNA adduct through an S_N2 mechanism, the leaving group released from the ultimate carcinogenic metabolite and the addition of the DNA molecule to the metabolite take place simultaneously. If this leaving group is located at a chiral carbon center, the configuration of the resulting covalent chemical bond that connects the metabolite and the DNA is inverted. In an S_N1 mechanism, the leaving group of the ultimate carcinogenic metabolite is removed first, resulting in the formation of a carbocation or a nitrenium ion as a strong electrophilic center. The configuration of the covalent bond linking the metabolite and DNA is racemized. Thus, the analysis of the ratio of the two enantiomers can determine that the mechanism is an S_N1, S_N2, or a combination of both. The most common electrophilic regions of ultimate carcinogenic metabolites can be classified into the following classes.

4.5.1 EPOXY RING

Examples include PAH diol epoxides and glycidamide, the ultimate carcinogenic metabolite of acrylamide. An ultimate carcinogenic metabolite carrying an epoxy-ring moiety may take place by an S_N2 mechanism through the electrophilic attack on the DNA molecule by the epoxy moiety, or through an S_N1 mechanism, initiated via the epoxy-ring opening, forming a carbocation that binds to DNA, or a combination of both. In an S_N2 mechanism, the nucleophile (DNA) usually attacks the less-hindered site of the epoxy-ring moiety.

4.5.2 ELECTROPHILIC NITROGEN

A nitrenium ion intermediate can be generated from the oxidative metabolism of carcinogenic aromatic amines and heterocyclic aromatic amines or from the reductive metabolism of nitro-PAHs and other chemical carcinogens that contain a nitro functional group, such as aristolochic acids.

4.5.3 ALKYL SULFATE ESTER (SULFURIC ACID ESTER)

7-Hydroxymethyl-12-methylbenz[*a*]anthracene is a carcinogenic metabolite of 7,12-dimethylbenz[*a*]anthracene (DMBA). 7-Sulfoxymethyl-12-methylbenz[*a*]anthracene is a metabolite carrying a sulfate ester moiety (Watabe et al., 1982). In 1982, Watabe et al. determined that tumor initiation induced by 7-hydroxymethyl-12-methylbenz[*a*]anthracene is mediated via the

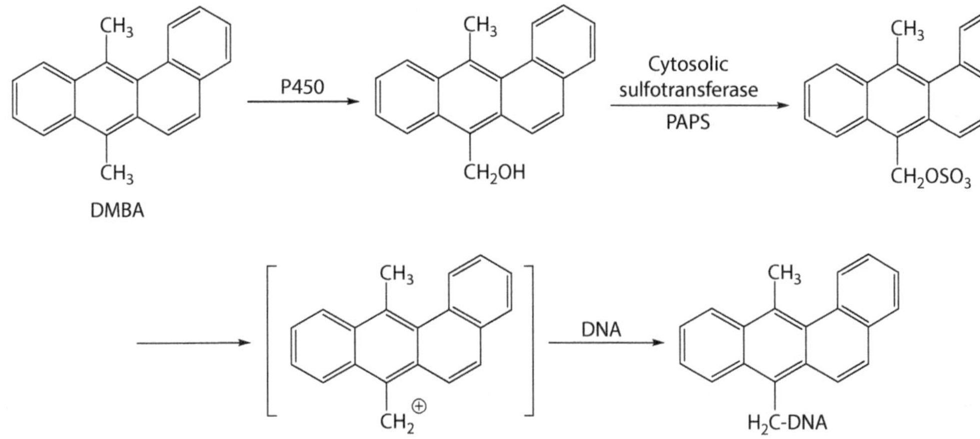

FIGURE 4.6 The proposed metabolic activation pathway of 7-hydroxymethyl-12-methylbenz[*a*]anthracene leading to the formation of 7-sulfoxymethyl-12-methylbenz[*a*]anthracene and tumor initiation.

transformation catalyzed by liver cytosolic sulfotransferase to the reactive 7-sulfoxymethyl-12-methylbenz[a]anthracene (Figure 4.6; Watabe et al., 1982; Chou et al., 1998).

1′-Hydroxysafrole and 1′-hydroxyestragole are metabolites of safrole and estragole, respectively. Enzymatic sulfation of these metabolites generated 1′-sulfoxy-safrole and 1′-sulfoxy-estragole. Upon dissociation of the sulfoxyl group, the resulting carbocation (at the 1′-position of safrole and estragole) binds to DNA to form exogenous DNA adducts leading to tumor initiation (Wiseman et al., 1987).

4.5.4 Ester Group at the Allylic or Benzylic Position

Taking mono- and diester dehydropyrrolizidine alkaloids as examples, the ester group at the allylic C7 or C9 position of the necine base is easily released from the necine ring to form a carbocation at the C7 or C9 position (Fu et al., 2014). In general, in some macrocyclic diester pyrrolizidine alkaloids (e.g., retrorsine, senecionine, monocrotaline, and riddelliine) (Figure 4.2), the allylic ester group at the C7 position separated from the necine ring to form a carbocation at the C7 position is favored than the ester group at the C9 position. This is because the resulting secondary carbocation (at the C7 position) is more stable than the primary carbocation formed at the C9 position (Fu et al., 2010; Jiang et al., 2015).

4.6 METABOLIC ACTIVATION OF FOODBORNE CHEMICAL CARCINOGENS MEDIATED BY EXOGENOUS DNA ADDUCT FORMATION

There are several classes of foodborne chemical carcinogens whose metabolic activation is mediated by exogenous DNA adduct formation. In this section, we present the metabolic activation of PAHs, pyrrolizidine alkaloids, nitro-PAHs, heterocyclic amines, acrylamide, and aristolochic acids that are mediated by exogenous DNA adduct formation. The other foodborne chemical carcinogens, such as *N*-nitrosamines, are not included for discussion.

4.6.1 PAHs

In the early 1960s, it was not known whether or not metabolism is required for tumor initiation. In the late 1960s, Millers' results brought the concept that chemical carcinogens in most cases are biologically inert and metabolism is required to generate reactive metabolites that bind to cellular macromolecules to initiate tumorigenesis (Miller, 1970). Since then, numerous mechanistic studies on PAHs by various laboratories have indicated that PAHs are genotoxic carcinogens (Gelboin and Ts'o, 1978; Conney, 1982). In the early 1970s, PAH K-region oxides, such as benzo[*a*]pyrene 4,5-oxide, were identified as metabolites formed *in vitro*. PAH K-region oxides were studied with the expectation that PAH K-region oxides can bind to cellular DNA and are responsible for tumor induction.

Borgen et al. (1973) discovered that the metabolism of BaP *trans*-7,8-dihydrodiol, a metabolite of benzo[*a*]pyrene (BaP), by hamster liver microsomes resulted in the level of covalent binding to DNA 10-fold greater than that of BaP. This represents the first evidence in the field of PAH carcinogenesis that a metabolite is more biologically active than the carcinogenic PAH substrate. Based on this finding, Sims et al. (1974) proposed that the ultimate metabolite of BaP should be BaP diol-epoxide and BaP *trans*-7,8-dihydrodiol is a proximate carcinogenic metabolite of BaP. A series of extensive studies followed, and the results confirmed that PAH diol epoxides, if formed, are ultimate carcinogenic metabolites responsible for PAH-induced tumorigenicity (Gelboin and Ts'o, 1978; Conney, 1982).

A series of PAH *trans*-dihydrodiols and PAH diol epoxides were studied for their biological and toxicological activities on a structure–activity relationship basis (Harvery and Fu, 1978; Jerina et al., 1978; Conney, 1982). Molecular orbital theoretical predictions were also conducted (Fu et al., 1978b, 1980; Jerina et al., 1978; Yeh et al., 1978). Based on theoretical predictions and experimental evidence, Jerina et al. proposed the *bay*-region theory, stating that if more than one diol-epoxide metabolite is formed from a PAH, the one with the epoxy ring located at the *bay* region of the molecule displays the most potent biological activity, forms the highest level of covalently bound DNA adducts, and is responsible for tumor initiation caused by this PAH (Jerina et al., 1978; Conney, 1982). The metabolic activation pathway involving the diol-epoxide formation represents one of the most important findings in the field of PAH carcinogenesis.

BaP is the most studied PAH on the metabolic activation leading to carcinogenesis. The metabolic activation of BP involves the initial cytochrome P450–catalyzed enzymatic oxidation to BaP 7,8-epoxide, which is converted to BaP *trans*-7,8-dihydrodiol catalyzed by epoxide hydrolase (Figure 4.7). Enzymatic epoxidation of BaP *trans*-7,8-dihydrodiol, catalyzed by cytochrome

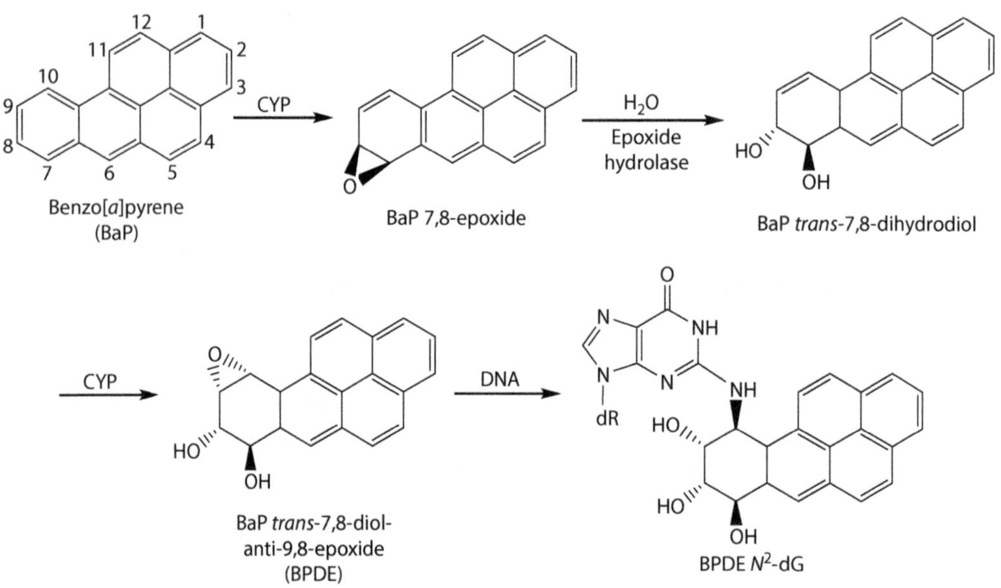

FIGURE 4.7 The metabolic activation pathway of benzo[a]pyrene leading to the formation of BPDE N^2-dG adduct and tumor initiation.

P450, produces the ultimate carcinogenic metabolite BaP *trans*-7,8-dihydroxy-*anti*-9,10-epoxy-7,8,9,10-tetrahydrobenzo[*a*]pyrene (BPDE). BPDE reacts with DNA to give a number of DNA adducts, of which the major DNA adduct is 10-(deoxyguanosin-N^2-yl)-7,8,9-trihydroxy-7,8,9,10-tetrahydrobenzo[*a*]pyrene (BPDE N^2-dG) (Figure 4.7; Gelboin and Ts'o, 1978; Beland et al., 2005). This DNA adduct has been detected and quantified in the lung tissue of mice administered with coal tar and human lung DNA samples. This BPDE N^2-dG adduct has been served as a biomarker of human exposure to PAHs (Beland et al., 2005). To date, it is commonly accepted that the formation of PAH diol-epoxide-derived DNA adducts is the principal and most important metabolic activation pathway of PAHs leading to cancer initiation.

There are several other metabolic activation pathways that have also been studied or proposed. 9-Hydroxybenzo[*a*]pyrene 4,5-oxide, a metabolite of BaP, was proposed to be capable of binding to DNA, although its binding to DNA has never been reported (King et al., 1976). As described previously, 7-sulfoxymethyl-12-methylbenz[*a*]anthracene, an activated metabolite of DMBA, can bind to DNA (Figure 4.6; Watabe et al., 1982; Chou et al., 1998). Cavalieri, Rogen, and coworkers proposed that one-electron oxidation of PAHs can produce exogenous depurinated DNA adducts and thus is a metabolic activation pathway of PAHs, including BaP and DMBA, leading to tumor initiation. This activation pathway was supported by the observation that the formation of DMBA–DNA adducts in mouse skin was mediated by radical cations (RamaKrishna et al., 1992). Penning et al. (1996) reported that the metabolic formation of PAH quinones generated ROS that can damage DNA, protein, and cell membrane.

4.6.2 Pyrrolizidine Alkaloids

Although it was established in 1954 that pyrrolizidine alkaloids can induce liver tumors in experimental animals (Schoental et al., 1954), the mechanism of tumor induction was not determined until in 2001 (Yang et al., 2001). Riddelliine, a tumorigenic pyrrolizidine alkaloid, was shown to induce liver tumors through a genotoxic mechanism mediated by 6,7-dihydro-7-hydroxy-1-hydroxymethyl-5*H*-pyrrolizine (DHP)-DNA adducts (Yang et al., 2001; Fu et al., 2010). The levels of DHP-DNA adducts correlated closely with the tumorigenic potencies of different doses of riddelliine fed to rats (Yang et al., 2001). A study with human liver microsomes indicates that the metabolic pattern and DNA adduct profiles obtained from human liver microsomal metabolism are similar to those formed in rat liver *in vitro* and *in vivo*. These results indicate that riddelliine could be highly genotoxic to humans (Xia et al., 2003).

High-performance liquid chromatography electrospray ionization tandem mass spectrometry (HPLC-ES-MS/MS) analysis showed that four DNA adducts were formed in rats dosed with riddelliine. With the synthesized DHP-DNA adducts available and the structures fully elucidated, the structures of these four DNA adducts were elucidated as (1) a pair of epimers of 7-hydroxy-9-(deoxyguanosin-N^2-yl)dehydrosupinidine adducts (termed as DHP-dG-3 and DHP-dG-4) as the predominant adducts and (2) a pair of epimers of 7-hydroxy-9-(deoxyadenosin-N^6-yl)dehydrosupinidine adducts (termed as DHP-dA-3 and DHP-dA-4 adducts) (Fu et al., 2010; Zhao et al., 2012).

A subsequent animal study determined that the rats dosed individually to seven liver carcinogens, riddelliine, retrorsine, monocrotaline, lasiocarpine, heliotrine, clivorine, and senkirkine, all generated DHP-dG-3, DHP-dG-4, DHP-dA-3, and DHP-dA-4 adducts in the liver (Xia et al., 2013b). These DNA adducts were not formed in the liver of rats dosed retronecine, lycopsamine, or platyphylline, which are not hepatocarcinogens or vehicle control (Xia et al., 2013b). In addition, there was a correlation between the liver tumor potency and the level of DNA adduct formation. This set of DNA adducts (DHP-dG-3, DHP-dG-4, DHP-dA-3, and DHP-dA-4) is a common biological biomarker of pyrrolizidine alkaloid–induced liver tumor formation and exposure of carcinogenic pyrrolizidine alkaloids (Figure 4.8; Xia et al., 2013b). This is the first time that a set of exogenous DNA adducts is commonly formed from a family of chemical carcinogens.

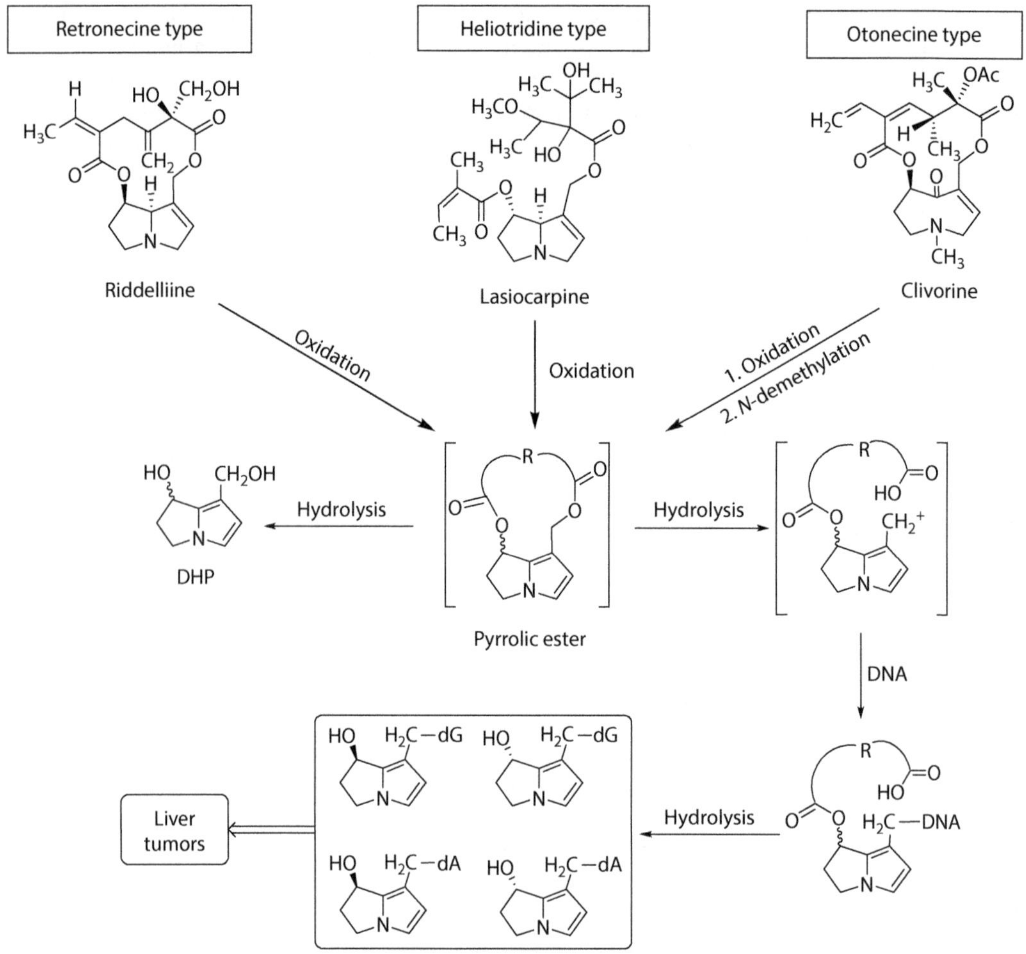

FIGURE 4.8 The proposed general mechanism leading to the DHP-derived DNA adduct formation from the metabolism of three types of carcinogenic pyrrolizidine alkaloids.

4.6.3 Nitro-PAHs

Similar to PAHs, nitro-PAHs require metabolic activation in order to exert tumorigenicity (Beland et al., 1985; Fu, 1990; Fu and Herreno-Saenz, 1999). Upon metabolic activation, the reactive metabolites covalently bind to cellular DNA to form exogenous DNA adducts to initiate mutation and tumor formation. Metabolism involves ring oxidation, reduction of the nitro group (nitroreduction), and a combination of both ring oxidation and nitroreduction. Nitroreduction proceeds through three consecutive enzymatic steps: (1) reduction of the nitro group to the corresponding nitroso-PAH metabolites, (2) further reduction to the *N*-hydroxyamino-PAHs, and (3) further reduction to the corresponding amino-PAHs (Figure 4.9; Beland et al., 1985; Fu, 1990; Fu and Herreno-Saenz, 1999). The *N*-hydroxyamino-PAHs are electrophiles, capable of binding to DNA *in vivo*. Certain *N*-hydroxyamino-PAHs are weak electrophiles, requiring further activation through enzymatic O-esterification (e.g., acetylation), to form the corresponding *N*-acetylamino-PAHs to bind to DNA (Figure 4.9; Fu, 1990; Fu and Herreno-Saenz, 1999). The metabolic activation pathway of 1-nitropyrene leading to the formation of *N*-(deoxyguanosin-8-yl)-1-aminopyrene is shown in Figure 4.9.

In general, nitroreduction of nitro-PAHs is the major activation pathway and the corresponding *N*-(deoxyguanosin-8-yl)amino-PAH adducts are responsible for most, if not all, nitro-PAH-induced mutations and tumor induction (Fu, 1990; Fu and Herreno-Saenz, 1999).

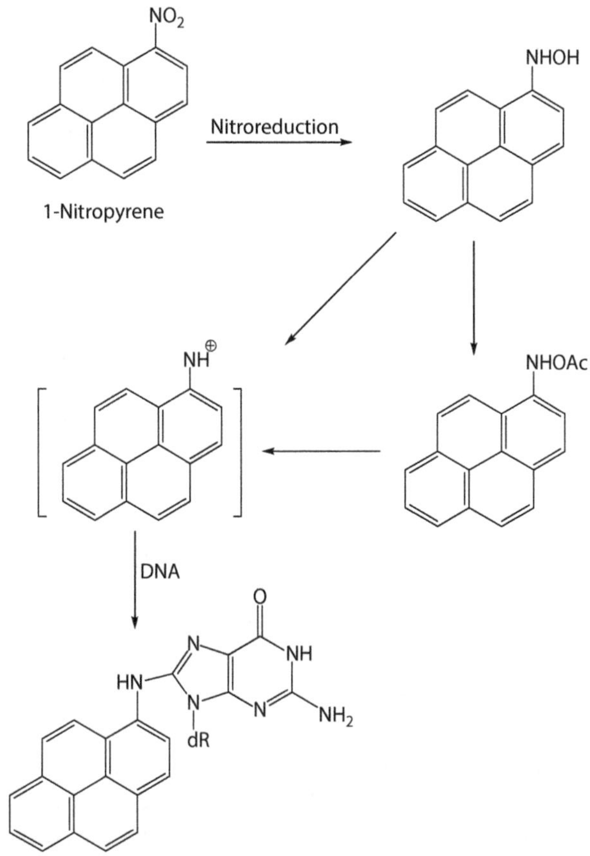

FIGURE 4.9 The metabolic activation pathway of 1-nitropyrene leading to the formation of *N*-(deoxyguanosin-8-yl)-1-aminopyrene.

4.6.4 Heterocyclic Aromatic Amines

The metabolic activation of food pyrolysates (heterocyclic aromatic amines) also produces *N*-hydroxyamino and/or *N*-acetoxyamino metabolites as ultimate carcinogenic metabolites that bind to DNA *in vivo* leading to mutations and tumor induction. The resulting DNA adducts are *N*-(deoxyguanosin-8-yl)amino-aza-PAH adducts (Howard et al., 1983; Beland and Kadlubar, 1985). The difference of the metabolic activation between nitro-PAHs and heterocyclic amines is the initial enzymatic step. The first step of the metabolic activation of nitro-PAH is nitroreduction, while the initial step in the metabolism of heterocyclic amines is enzymatic oxidation of the amino moiety. The DNA adducts responsible for induction of tumors by heterocyclic aromatic amines are also *N*-(deoxyguanosin-8-yl)-1-amino-type exogenous DNA adducts.

4.6.5 Acrylamide

Acrylamide, a foodborne carcinogen, is metabolically activated, catalyzed by cytochrome P450 2E1, to glycidamide, which bears an epoxy ring (Figure 4.10; Von Tungeln et al., 2012). Glycidamide reacts with DNA *in vivo* to generate two depurinated DNA adducts, *N*7-(2-carbamoyl-2-hydroxyethyl) guanine and *N*3-(2-carbamoyl-2-hydroxyethyl)adenine that are responsible for acrylamide-induced tumor formation (Figure 4.10).

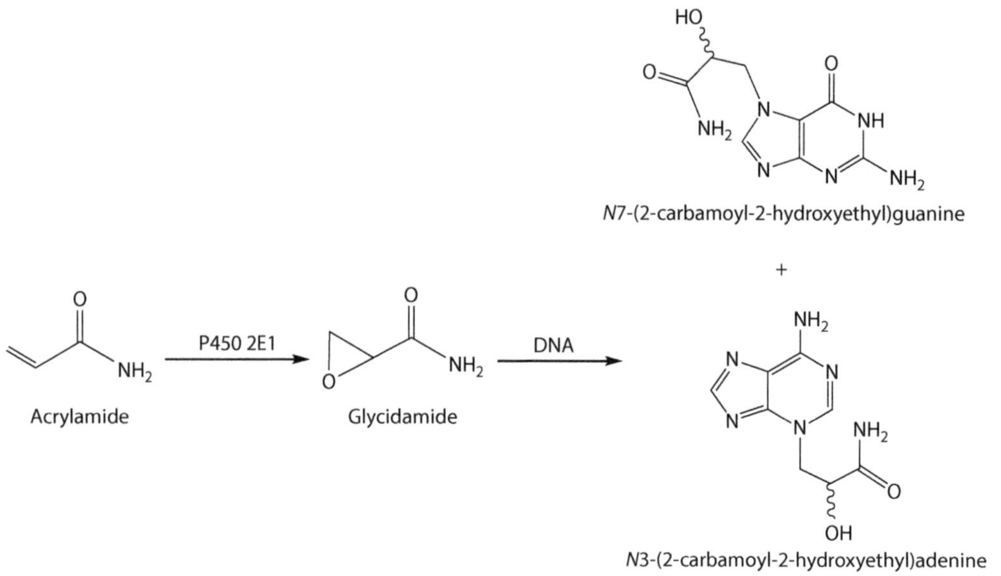

FIGURE 4.10 The metabolic activation of acrylamide to glycidamide and the formation of *N*7-(2-carbamoyl-2-hydroxyethyl)guanine and *N*3-(2-carbamoyl-2-hydroxyethyl)adenine leading to tumor induction.

4.6.6 Aristolochic Acids

Aristolochic acids contain two chemical components, aristolochic acid I (3,4-methylenedioxy-8-methoxy-10-nitrophenanthrene-1-carboxylic acid) and aristolochic acid II (3,4-methylenedioxy-10-nitrophenanthrene-1-carboxylic acid; Figure 4.1). Both aristolochic acids are constituents in the plant species of the genus *Aristolochia* (NTP, 2008). Because of causing urothelial cancer in humans, the IARC classified the aristolochic acid–containing plant species of the genus *Aristolochia* as human carcinogens (Group I) in 2002 (IARC, 2002, 2015). The NTP in 2011 classified aristolochic acids as "known to be human carcinogens."

Mechanistic studies revealed that both aristolochic acid I and aristolochic acid II undergo enzymatic nitroreduction to form cyclic amino nitrenium ions that react with DNA to form aristolochic acid–dG and aristolochic acid–dA adducts responsible for the induction of urothelial cancer (Figure 4.11; NTP, 2008).

4.7 CHALLENGES IN STUDYING MECHANISM MEDIATED BY EXOGENOUS DNA ADDUCT FORMATION

The study of chemical carcinogenesis centers on elucidating the mechanisms by which chemical carcinogens induce tumors. At the beginning, the goal is to determine the metabolic activation pathway and identify ultimate carcinogenic metabolites. In most cases, ultimate carcinogenic metabolites are phase I metabolites. Identification of ultimate carcinogenic metabolites from *in vivo* metabolism is complicated by phase II conjugation reactions. This can be illustrated by the metabolism of DMBA *in vitro* and *in vivo*. While Chou and Yang (1979) identified a total of 37 metabolites from rat liver microsomal metabolism of DMBA, only four metabolites were identified *in vivo*. Therefore, liver microsomal metabolism *in vitro* is commonly used for the study of metabolic activation of a chemical carcinogen.

The metabolic activation mediated by exogenous DNA adduct formation is biologically the pathway closely associated with tumor initiation. The study of this pathway is highly challenging. After metabolism, each metabolite has to be purified and the structure identified. Ultimate carcinogenic

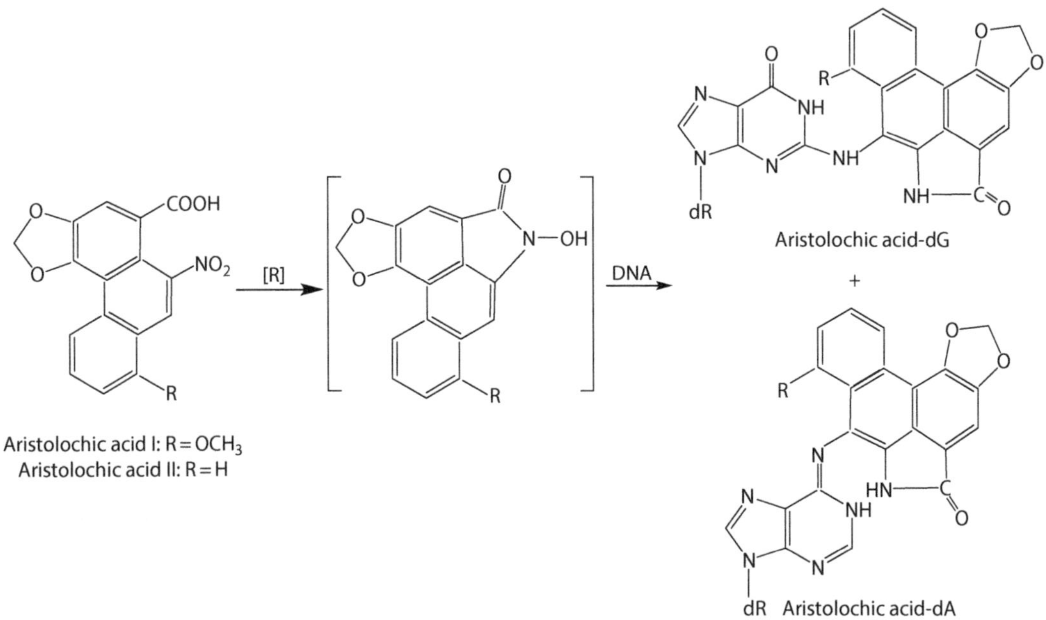

FIGURE 4.11 The metabolic activation of aristolochic acid I and aristolochic acid II leading to the formation of aristolochic acid–dG and aristolochic acid–dA adducts.

metabolites in most cases are produced in a small quantity from the metabolism and are highly unstable in aqueous medium, facilely hydrolyzed and decomposed, and therefore cannot be isolated from any *in vivo* or *in vitro* systems. In general, not too many metabolites are commercially available. In order to determine their genotoxic activities, such as mutagenicity, DNA-binding capability, and/or other genotoxicity, a sufficient amount of each metabolite has to be obtained either from the metabolism or prepared by multistep organic synthesis. Furthermore, tumorigenicity bioassays of both the substrate and metabolites have to be studied so that ultimate carcinogenic metabolites can be determined.

Also, exogenous DNA adducts formed *in vivo* or *in vitro* are always in a tiny quantity. The best strategy to identify their formation is through organic synthesis, by reacting synthetically prepared ultimate carcinogenic metabolite with DNA. The purified and structurally identified exogenous DNA adducts are then used as synthetic standards to identify and quantify, by LC/MS/MS analysis or other analytical approaches, the exogenous DNA adducts formed *in vitro* and *in vivo*. In addition, for quantitation, isotopically labeled DNA adducts have to be prepared as well. Furthermore, for the detection and quantitation of exogenous DNA adducts formed *in vivo* or *in vitro*, a new LC/MS/MS method, or other analytical methodology, has to be developed as well. The study of benzo[*a*]pyrene (BaP) represents an excellent example to illustrate how an activation pathway is mediated by exogenous DNA adduct formation.

Immediately after the finding by Borgen et al. (1973) and the proposal by Sims et al. (1974), both the proximate carcinogenic metabolite BaP *trans*-7,8-dihydrodiol and the ultimate carcinogenic metabolites BaP diol epoxides (BaP *trans*-7,8-diol-*syn*-epoxide and BaP *trans*-7,8-diol-*anti*-epoxide) were synthesized (McCaustland and Engel, 1975; Yagi and Jerina, 1975; Beland and Harvey, 1976; Fu and Harvey, 1977; Yagi et al., 1977; Harvey and Fu, 1978). As shown in Figure 4.12, it requires multiple steps to synthesize these reactive metabolites. The task is novel, time-consuming, labor-intensive, and challenging. As BaP *trans*-7,8-dihydrodibenzoate, BaP *trans*-7,8-dihydrodiol, and BaP diol-epoxides (Figure 4.12) are all highly unstable, for the synthesis of these compounds, new synthetic methods had to be developed (Fu and

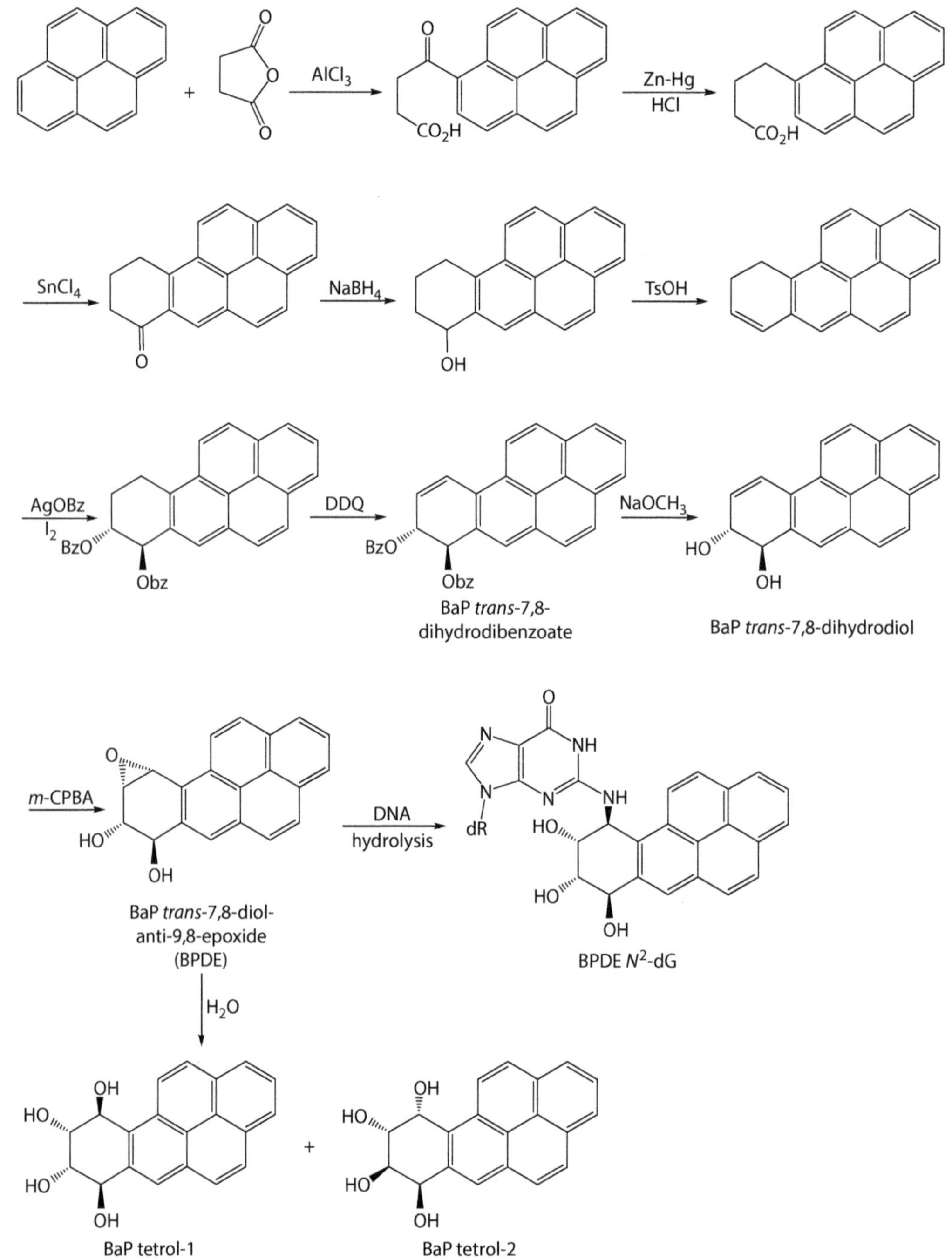

FIGURE 4.12 The synthesis of BaP *trans*-7,8-dihydrodiol, BaP *trans*-7,8-diol-*anti*-9,10-epoxide, and dG-BPDE adduct.

Harvey, 1977). For example, BaP *trans*-7,8-dihydrodibenzoate prepared by dehydrogenation of 7,8,9,10-tetrahydro-BaP *trans*-7,8-dibenzoate (Figure 4.12) via the standard method involving bromination with *N*-bromosuccinimide (NBS) followed by thermal or base-catalyzed dehydrobromination gave erratic results, low yield, impure product, and product readily decomposed on storage. The search for alternative methods led to the finding that dehydrogenation can be conveniently accomplished by DDQ, providing the pure BaP *trans*-7,8-dihydrodibenzoate in an excellent yield (Fu and Harvey, 1977). With BaP diol epoxides available in a large quantity, the exogenous DNA adducts, including BPDE-N^2dG (N^2dG-BPDE), were synthesized (Figure 4.12).

To study the structure–activity relationship of PAH diol epoxides, new synthetic methods were required for the synthesis of a series of structurally related PAH *trans*-dihydrodiols and PAH diol epoxides used for biological/toxicological assays (Fu and Harvey, 1977, 1978, 1979; Harvey et al., 1977; Fu et al., 1978a,b, 1981, 1993; Harvery and Fu, 1978; Wu et al., 1993).

New analytical methodologies have to be developed for the identification and quantitation of exogenous DNA adducts formed *in vivo*, including human tissues. The method development is always label-intensive and highly challenging. Several different analytical methodologies have been developed for the detection and quantitation of BPDE-N^2dG adduct in biological and human samples, including immunoassay, ^{32}P-postlabeling, and HPLC-ES-MS/MS (Beland et al., 2005).

Additional studies are also required to confirm that the activation pathway mediated by exogenous DNA adduct formation *in vivo* is indeed responsible for tumor induction. The studies include the correlation between the level of DNA adduct formation *in vivo* and the tumor potency of the tested carcinogen, the mutation type (point mutation) and profile, and the formation of specific type of *ras*-oncogene activation.

4.8 PERSPECTIVES

Diet is the major source of human cancer formation. Reducing the consumption of carcinogen-containing food can definitely decrease human cancer incidence. Accordingly, it is important to determine the sources of foodborne carcinogens, assess the level of human exposure, determine the tumor potency, and elucidate the mechanisms of tumor formation on a molecular basis.

The study of chemical carcinogenesis centers on determining the mechanism by which chemicals alter the normal cells into the cancerous cells *in vivo*. The cancer formation caused by carcinogenic chemicals involves three stages: tumor initiation, tumor promotion, and tumor progression. The study of tumor initiation is to elucidate the molecular mechanism to understand how the chemicals induce tumors. The next stage (tumor promotion) and third stage (tumor progression) are to study the pathological changes of the cells. The study of chemical carcinogenesis is an interdisciplinary-oriented research field with the initial study focused on the elucidation of metabolic activation pathways, among which the pathway mediated by exogenous DNA adduct formation is most important. Consequently, mechanism-based understanding on the cancer induction caused by foodborne chemical carcinogens is an important study for the prevention of human cancer incidence caused by the exposure to foodborne carcinogens.

To date, the IARC listed about 70 foodborne chemical carcinogens in IARC Monographs (1–113 volumes), accounting for about 15% of the total chemical carcinogens listed as human carcinogens (group 1), probably human carcinogens (group 2A), and possibly human carcinogens (group 2B) (Table 4.1). Among the listed foodborne chemical carcinogens, there are only about 16 belonging to food constituents or dietary supplements (Table 4.1). It is highly possible that many more foodborne chemicals, particularly herbal plant products including dietary supplements and functional food, will be found to be carcinogenic in humans.

There are more than 660 different pyrrolizidine alkaloids and *N*-oxide derivatives present in more than 6000 plants worldwide (Fu et al., 2004). Pyrrolizidine alkaloid–containing plants are probably the most common poisonous plants affecting livestock, wildlife, and humans. To date, IARC listed only riddelliine and monocrotaline as group 2B possible human carcinogens. Based on the published tumorigenicity data obtained from experiments using rodent animals, it is anticipated that additional pyrrolizidine alkaloids, as well as pyrrolizidine alkaloid–containing edible plants, would belong to possible or probable human carcinogens.

Nanomaterials used commercially for novel applications for humans are increasing exponentially. It is important to determine whether or not any of the nanomaterials used in food products are carcinogenic; and if carcinogenic, the mechanism of action has to be elucidated.

ACKNOWLEDGMENTS

The authors thank Dr. Frederick A. Beland for critical review of this manuscript. This article is not an official guidance or policy statement of the U.S. FDA. No official support or endorsement by the U.S. FDA is intended or should be inferred.

REFERENCES

Ames, B.N. and Gold, L.S. 1998. The prevention of cancer. *Drug Metab. Rev.* 30:201–223.

Bailey, R.L., Gahche, J.J., Lentino, C.V. et al. 2011. Dietary supplement use in the United States, 2003–2006. *J. Nutr.* 141:261–266.

Becalski, A., Forsyth, D., Casey, V. et al. 2005. Development and validation of a headspace method for determination of furan in food. *Food Addit. Contam.* 22:535–540.

Beland, F.A., Churchwell, M.I., Von Tungeln, L.S. et al. 2005. High performance liquid chromatography electrospray ionization tandem mass spectrometry for the detection and quantitation of benzo[*a*]pyrene-DNA adducts. *Chem. Res. Toxicol.* 18:1306–1315.

Beland, F.A. and Harvey, R.G. 1976. The isomeric 9,10-epoxide of trans-7,8-dihydroxy-7,8-dihydrobenzo[a]pyrene. *J. Chem. Soc. Chem. Commun.* 84–85.

Beland, F.A., Heflich, R.H., Howard, P.C. et al. 1985. The *in vitro* metabolic activation of nitro polycyclic aromatic hydrocarbons. In: *Polycyclic Hydrocarbons and Carcinogenesis*, pp. 371–396. R.G. Harvey (ed.), ACS Symposium Series 283, American Chemical Society, Washington, DC.

Beland, F.A. and Kadlubar, F.F. 1985. The formation and persistence of arylamine DNA adducts *in vivo*. *Environ. Health Perspect.* 62:18–30.

Bodi, D., Ronczka, S., Gottschalk, C. et al. 2014. Determination of pyrrolizidine alkaloids in tea, herbal drugs and honey. *Food Addit. Contam.: Part A* 31(11):1884–1895.

Borgen, A., Darvey, H., Castagnoli, N. et al. 1973. Metabolic conversion of benzo[a]pyrene by Syrian hamster liver microsomes and binding of metabolites to deoxyribonucleic acid. *J. Med. Chem.* 16:502–506.

Chae, Y.-H., Upadhyaya, P., Ji, B.-Y. et al. 1997. Comparative metabolism and DNA binding of 1-, 2-, and 4-nitropyrene in rats. *Mutation Res.* 376:21–28.

Chan, P.-C., Xia, Q., and Fu, P.P. 2007. Ginkgo biloba leave extract: Biological, medicinal, and toxicological effects. *J. Environ. Sci Health, Part C* 25:211–244.

Chen, B.H. and Lin, Y.S. 1997. Formation of PAHs during processing of duck meat. *J. Agric. Food Chem.* 45:1394–1403.

Chen, T., Mei, N., and Fu, P.P. 2010. Genotoxicity of pyrrolizidine alkaloids. *J. Appl. Toxicol.* 30:183–196.

Chou, H.-C., Ozawa, S., Fu, P.P. et al. 1998. Metabolic activation of methyl-hydroxylated derivatives of 7,12-dimethylbenz[a]anthracene by human liver steroid sulfotransferase. *Carcinogenesis* 19:1071–1076.

Chou, M.W. and Yang, S.K. 1979. Combined reversed-phase and normal-phase high performance liquid chromatography in the purification and identification of 7,12-dimethylbenz[*a*]anthracene metabolites. *J. Chromat.* 185:635–654.

Conney, A.H. 1982. Induction of microsomal enzymes by foreign chemicals and carcinogenesis by polycyclic aromatic hydrocarbons: G.H.A. Clowes Memorial Lecture. *Cancer Res.* 42:4875–4917.

Doll, R. and Peto, R. 1981. The cause of cancer: Quantitative estimates of avoidable risks of cancer in the United States today. *J. Nat. Cancer Inst.* 66:1191–1198.

Fang, G.-C., Chang, C.-N., Wu, Y.-S. et al. 2004a. Characterization, identification of ambient air and road dust polycyclic aromatic hydrocarbons in central Taiwan, Taichung. *Sci. Total Environ.* 327:135–146.

Fang, G.-C., Wu, Y.-S., Fu, P.P. et al. 2004b. Polycyclic aromatic hydrocarbons in the ambient air of suburban and industrial regions of central Taiwan. *Chemosphere* 54:443–452.

FDA. 2005. Exploratory data on furan in food: Individual food products, Last updated: 6/15/05. http://www.fda.gov/Food/FoodSafety?FoodContaminantsAdlteration/ChemicalContaminants/Furan/UCMO78439.

Food, Nutrition and the Prevention of Cancer: A Global Perspective. World Cancer Research Fund in Association with American Institute for Cancer Research, Washington, DC, 1997.

Fu, P.P. 1990. Metabolism of nitro-polycyclic aromatic hydrocarbons. *Drug Metab. Rev.* 22:209–268.

Fu, P.P., Beland, F.A., and Yang, S.K. 1980. Cyclopenta-polycyclic aromatic hydrocarbons: Potential carcinogens and mutagens. *Carcinogenesis* 1(8):725–727.

Fu, P.P., Chiang, H.-M., Xia, Q. et al. 2009. Quality assurance and safety of herbal dietary supplements. *J. Environ. Sci. Health, Part C* 27:91–119.

Fu, P.P., Chou, M.W., Churchwell, M. et al. 2010. High-performance liquid chromatography electrospray ionization tandem mass spectrometry for the detection and quantitation of pyrrolizidine alkaloid-derived DNA adducts in vitro and in vivo. *Chem. Res. Toxicol.* 23:637–652.

Fu, P.P., Chou, M.W., Xia, Q. et al. 2001. Genotoxic pyrrolizidine alkaloids and pyrrolizidine alkaloid N-oxides—Mechanisms leading to DNA adduct formation and tumorigenicity. *Environ. Carcinogen. Ecotoxicol. Rev.* 19:353–386.

Fu, P.P. and Harvey, R.G. 1977. Synthesis of the diols and diolepoxides of carcinogenic hydrocarbons. *Tetrahedron Lett.* 2059–2062.

Fu, P.P. and Harvey, R.G. 1978. Synthesis of the chrysene bay region anti-diolepoxide from chrysene. *J.C.S. Chem. Commun.* 585–586.

Fu, P.P. and Harvey, R.G. 1979. Synthesis of the dihydro diols and diol epoxides of chrysene from chrysene. *J. Org. Chem.* 44:3778–3784.

Fu, P.P., Harvey, R.G., and Beland, F.A. 1978b. Molecular orbital theoretical prediction of the isomeric products formed from reactions of arene oxides and related metabolites of polycyclic aromatic hydrocarbons. *Tetrahedron* 34:857–866.

Fu, P.P. and Herreno-Saenz, D. 1999. Nitro-polycyclic aromatic hydrocarbons: A class of genotoxic environmental pollutants. *J. Environ. Sci. Health, Part C* 17:1–43.

Fu, P.P., Lai, C.C., and Yang, S.K. 1981. Synthesis of cis and trans 7,8-dihydrodiols of 7 methylbenzo[a]pyrene. *J. Org. Chem.* 46:220 222.

Fu, P.P., Lee, H.M., and Harvey, R.G. 1978a. Novel synthesis of the dihydroarene precursors of carcinogenic arene dihydrodiols and diolepoxides. *Tetrahedron Lett.* 551–554.

Fu, P.P., Wu, Y.S., Von Tungeln, L.S. et al., 1993. Synthesis of 3 nitrobenzo[a]pyrene bay region trans 7,8 diol anti 9,10 epoxide and the corresponding N2 deoxyguanosine adduct. *Chem. Res. Toxicol.* 6:603–608.

Fu, P.P. and Xia, Q. 2014. Assessment of safety and quality assurance of herbal dietary supplements. In: *Nutraceutical and Functional Food Regulations in the United States and Around the World*, 2nd edn., pp. 151–168. D. Bagchi (eds.), Elsevier/Academic Press, New York, Chapter 10.

Fu, P.P. and Xia, Q. 2015. Pyrrolizidine alkaloids: Toxic phytochemicals found in food. In: *Food Poisoning: Outbreaks, Bacterial Sources and Adverse Health Effects*, pp. 1–33. P.C. Ray (ed.), Nova Science Publishers, Hauppauge, NY.

Fu, P.P., Xia, Q., Chou, M.W. et al. 2007. Detection, hepatotoxicity, and tumorigenicity of pyrrolizidine alkaloids in Chinese herbal plants and herbal dietary supplements. *J. Food Drug Anal.* 15:400–415.

Fu, P.P., Xia, Q., Guo, L. et al. 2008. Toxicity of kava kava. *J. Environ. Sci. Health C: Environ. Carcinog. Ecotoxicol. Rev.* 26:89–112.

Fu, P.P., Xia, Q., Hwang, H.-M. et al. 2014. Mechanism of nanotoxicity—Generation of reactive oxygen species. *J. Food Drug Anal.* 22:64–75.

Fu, P.P., Xia, Q., Lin, G., and Chou, M.W. 2004. Pyrrolizidine alkaloids—Genotoxicity, metabolism enzymes, metabolic activation, and mechanisms. *Drug Metab. Rev.* 36:1–55.

Fu, P.P., Xia, Q., Sun, X. et al. 2012. Phototoxicity and environmental transformation of polycyclic aromatic hydrocarbons—Light-induced reactive oxygen species, lipid peroxidation, and DNA damage. *J. Environ. Sci. Health, Part C* 30:1–41.

Fu, P.P., Xia, Q., Zhao, Y. et al. 2013. Phototoxicity of herbal plants and herbal products. *J. Environ. Sci. Health, Part C* 31:213–255.

Fu, P.P., Yang, Y.-C., Xia, Q. et al. 2002. Pyrrolizidine alkaloids—Tumorigenic components in Chinese herbal medicines and dietary supplements. *J. Food Drug Anal.* 10:198–211.

Gelboin, H.V. and Ts'o, P.O.P. 1978. *Polycyclic Hydrocarbons and Cancer: Chemistry, Molecular Biology and Environment*, Vol. 1, pp. 173–188. Academic Press, New York.

Hartmann, T. 2004. Plant-derived secondary metabolites as defensive chemicals in herbivorous insects: A case study in chemical ecology. *Planta.* 219:1–4.

Harvey, R.G. and Fu, P.P. 1978. Synthesis and reactions of diol epoxides and related metabolites of carcinogenic hydrocarbons. In: *Polycyclic Hydrocarbons and Cancer: Chemistry, Molecular Biology and Environment*, Vol. 1, pp. 133–166. H.V. Gelboin and P.O.P. Ts'o (eds.), Academic Press, New York.

Harvey, R.G., Fu, P.P., Cortez, C. et al. 1977. Synthesis of a biologically active D-ring diolepoxide of the potent carcinogen 7,12-dimethylbenz[α]anthracene. *Tetrahedron Lett.* 3533–3536.

Howard, P.C., Heflich, R.H., Evans, F.E. et al. 1983. Formation of DNA adducts formed *in vitro* and in *Salmonella typhimurium* upon metabolic reduction of the environmental mutagen 1-nitropyrene. *Cancer Res.* 43:2052–2058.

IARC. 1983. *Polynuclear Aromatic Compounds. Part I: Chemical, Environmental and Experimental Data.* International Agency for Research on Cancer, Lyon, France.

IARC. 1986. Some naturally occurring and synthetic food components, furocoumarins and ultraviolet radiation. *Int. Agency Res. Cancer, Monogr. Eval. Carcinogenic Risk Chem. Hum.* 40:1–415.

IARC. 2002. Some traditional herbal medicines, some mycotoxins, naphthalene and styrene. In: *IARC Monographs on the Evaluation of Carcinogenic Risk of Chemicals to Humans.* IARC, Lyons, France.

IARC. 2015. Some drugs and herbal products. In: *IARC Monographs on the Evaluation of Carcinogenic Risks to Humans.* IARC. Lyons, France.

Jerina, D.M., Yagi, H., Lehr, R.E. et al. 1978. The Bay-region theory of carcinogenesis by polycyclic aromatic hydrocarbons. In: *Polycyclic Hydrocarbons and Cancer: Chemistry, Molecular Biology and Environment,* Vol. 1, pp. 173–188. H.V. Gelboin and P.O.P. Ts'o (eds.), Academic Press, New York.

Jiang, X., Wang, S., Zhao, Y. et al. 2015. Absolute configuration, stability, and mechanism of interconversion of epimeric phenylthiohydantoin products derived from dehydropyrrolizidine alkaloids *J. Food Drug Anal.* 23:318–326.

King, H.W.S., Thompson, M.H., and Brookes, P. 1976. The role of 9-hydroxybenzo[a]pyrene in the microsome mediated binding of benzo[a]pyrene to DNA. *Int. J. Cancer* 18:339–344.

Kinouchi, T., Tsutsui, H., and Ohnishi, Y. 1986. Detection of 1-nitropyrene in yakitori (grilled chicken). *Mutation Res.* 171:105–113.

Lijinsky, W. and Ross, A.E. 1967. Production of carcinogenic polynuclear hydrocarbons in the cooking of food. *Fd Cosmet. Toxicol.* 5:343–347.

Maga, J.A. 1979. Furan in foods. *Crit. Rev. Food Sci. Nutr.* 11:35–400.

Mattocks, A.R. 1968. Toxicity of pyrrolizidine alkaloids. *Nature* 217:723–728.

McCaustland, D.F. and Engel, J.F. 1975. Metabolites of polycyclic aromatic hydrocarbons. II. Synthesis of 7,8-dihydroxybenzo[a]pyrene-7,8-diol and 7,8-dihydroxybenzo[a]pyrene-7,8-epoxide. *Tetrahydron Lett.* 2549–2552.

Mei, N., Chou, M.W., Fu, P.P. et al. 2004. Differential mutagenicity of riddelliine in liver endothelial and parenchymal cells of transgenic big blue rats. *Cancer Lett.* 215:151–158.

Mei, N., Guo, L., Fu, P.P. et al. 2005. Mutagenicity of comfrey (*Symphytum officinale*) in rat liver. *Br. J. Cancer* 92:873–875.

Mei, N., Guo, L., Fu, P.P. et al. 2010. Metabolism, genotoxicity, and carcinogenicity of comfrey. *J. Toxicol. Environ. Health, B Crit Rev.* 13:509–526.

Miller, J.A. 1970. Carcinogenesis by chemicals: An overview—G. H. A. Clowes Memorial Lecture. *Cancer Res.* 30:559–576.

Mirvish, S.S. 1975. Formation of N-nitroso compounds: Chemistry, kinetics, and in vivo occurrence. *Toxicol Appl. Pharmacol.* 31:325–351.

Nagao, M., Honda, M., Seino, Y. et al. 1977. Mutagenicities of protein pyrolysates, *Cancer Lett.* 2:335–339.

NTP. 1978. Bioassay of lasiocarpine for possible carcinogenicity. *Natl. Cancer Inst. Carcinog. Tech. Rep. Ser.* 39:1–66.

NTP. 2003. Toxicology and carcinogenesis studies of riddelliine (CAS No. 23246-96-0) in F344/N rats and B6C3F1 mice (gavage studies). *Natl. Toxicol. Program Tech. Rep. Ser.* 508:1–280.

NTP. 2008. *Report on Carcinogens Background Document for Aristolochic Acids.* National Toxicology Program, Research Triangle Park, NC. http://ntp.niehs.nih.gov/files/Aristolochic_Acids_(FINAL-02Sep08)_Redo2(3).pdf.

NTP. 2014. 1*3th Report on Carcinogens.* Research Triangle Park, NC: U.S. Department of Health and Human Services, Public Health Service. http://ntp.niehs.nih.gov/pubhealth/roc/roc13/.

Penning, T.M., Ohnishi, S.T., Ohnishi, T., and Harvey, R.G. 1996. Generation of reactive oxygen species during the enzymatic oxidation of polycyclic aromatic hydrocarbon trans-dihydrodiols catalyzed by dihydrodiol dehydrogenase. *Chem. Res. Toxicol.* 9:84–92.

Phillips, D.H. 1999. PAHs in the diet. *Mutat. Res.* 443:139–147.

Pitts, J.N. Jr., Van Cauwenberghe, K.A., Grosjean, D. et al. 1978 Atmospheric reactions of polycyclic aromatic hydrocarbons: Facile formation of mutagenic nitro derivatives. *Science* 202:515–519.

Pott, P. 1775. CancerScroti. In: *Chirurgical Observations Relative to the Cataract, the Polypus of the Nose, the Cancer of the Scrotum, the Different Kinds of Ruptures, and the Modification of the Toes and Feet,* pp. 63–68. Hawes, Clarke, Collins, London, U.K.

Prakash, A.S., Pereira, T.N., Reilly, P.E., and Seawright, A.A. 1999. Pyrrolizidine alkaloids in human diet. *Mutat. Res.* 443:53–67.

RamaKrishna, N.V.S., Cavalieri, E.L., Rogan, E.G. et al. 1992. Synthesis and structure determination of the adducts of the potent carcinogen 7,12-dimethylbenz[a]anthracene and deoxyribonucleosides formed by electrochemical oxidation: Models for metabolic activation by one-electron oxidation *J. Am. Chem. Soc.* 114:1863–1874.

Schlemitz, S. and Pfannhauser, M. 1996. Monitoring of nitropolycyclic aromatic hydrocarbons in food using gas chromatography. *Z. Lebensm. Unters. Forsch.* 203:61–64.

Schoental, R., Head, M.A., and Peacock, P.R. 1954. Senecio alkaloids: Primary liver tumours in rats as a result of treatment with (1) a mixture of alkaloids from *S. jacobaea* lin.; (2) retrorsine; (3) isatidine. *Br. J. Cancer* 8:458–465.

Scientific Committee on Foods of EC (SCF). 2002. Opinion of the Scientific Committee on Food in the risk to human health of PAHs in food. SCF, Brussels, Belgium.

Sims, P., Grover, P.L., Swaisland, A. et al. 1974. Metabolic activation of benzo[a]pyrene proceeds by a diol-epoxide. *Nature* (*London*) 252:326–328.

Smith, L.W. and Culvenor, C.C. 1981. Plant sources of hepatotoxic pyrrolizidine alkaloids. *J. Nat. Prod.* 44:129–152.

Stadler, R.H., Blank, I., Varga, N. et al. 2002. Acrylamide from Maillard reaction products. *Nature* 419:449–450.

Stegelmeier, B.L., Edgar, J.A., Colegate, S.M. et al. 1999. Pyrrolizidine alkaloid plants, metabolism and toxicity. *J. Nat. Toxins* 8:95–116.

Stern, R.S., Nichols, K.T., and V¨akev¨a, L.H. 1997. Malignant melanoma in patients treated for psoriasis with methoxsalen (psoralen) and ultraviolet A radiation (PUVA). *New Engl. J. Med.* 336:1041–1045.

Tareke, E., Rydrerg, P., Karlesson, P. et al. 2002. Analysis of acrylamide, a carcinogen formed in heated foodstuff. *J. Agric. Food Chem.* 50:4998–5006.

Von Tungeln, L.S., Doerge, D.R., Gambo da Costa, G. et al. 2012. Tumorigenicity of acrylamide and its metabolite glycidamide in the neonatal mouse bioassay. *Int. J. Cancer* 131:2008–2015.

Watabe, T., Ishizuka, T., Isobe, M., and Ozawa, N. 1982. A 7-hydroxymethyl sulfate ester as an active metabolite of 7,12-dimethylbenz[alpha]anthracene. *Science* 215:403–405.

Wiestler, O.D., von Deimling, A., von Hofe, E. et al. 1987. Interorgan shift of nitrosamine metabolism by dietary ethanol. Mechanism and models in toxicology. *Arch. Toxicol. Suppl.* 11:53–65.

Wiseman, R.W., Miller, E.C., Miller, J.A., and Liem, A. 1987. Structure-activity studies of the hepatocarcinogenicities of alkenylbenzene derivatives related to estragole and safrole on administration to preweanling male C57BL/6J × C3H/HeJ F1 mice. *Cancer Res.* 47:2275–2283.

World Health Organization (WHO). 1998. Environmental Health Criteria 202, Selected non-heterocyclic PAHs. WHO, Geneva, Switzerland. Available from: http://www.inchem.org/documents/ehc/ehc/ehc202.htm.

Wu, Y.S., Lai, J.S., and Fu, P.P. 1993. Synthesis of the biologically reactive bay region diolepoxide of the mutagenic environmental contaminant 1 nitrobenzo[a]pyrene. *J. Org. Chem.* 58:7283–7285.

Xia, Q., Chou, M.W., Kadlubar, F.F., Chan, P.C., and Fu, P.P. 2003. Human liver microsomal metabolism and DNA adduct formation of the tumorigenic pyrrolizidine alkaloid, riddelliine. *Chem. Res. Toxicol.* 16:66–73.

Xia, Q., Yin, J.J., Zhao, Y. et al. 2013a. UVA photoirradiation of nitro-polycyclic aromatic hydrocarbons—Formation of reactive oxygen species and lipid peroxides. *Int. J. Environ. Res. Public Health* 10:1062–1084.

Xia, Q., Zhao, Y., Von Tungeln, L.S. et al. 2013b. Pyrrolizidine alkaloids derived DHP-DNA adducts are a common biological biomarker of pyrrolizidine alkaloid-initiated tumorigenicity. *Chem. Res. Toxicol.* 26:1384–1396.

Yagi, H. and Jerina, D.M. 1975. A general synthetic method for non-K-region arene oxides. *J. Am. Chem. Soc.* 97:3185–3187.

Yagi, H., Thakker, D.R., Hernandez, O. et al. 1977. Synthesis and reactions of the highly mutagenic 7,8-diol-9,10-epoxides of the carcinogen benzo[a]pyrene. *J. Am. Chem. Soc.* 99:1604–1612.

Yang, Y.C., Yan, J., Doerge, D.R., Chan, P.C., Fu, P.P., and Chou, M.W. 2001. Metabolic activation of the tumorigenic pyrrolizidine alkaloid, riddelliine, leading to DNA adduct formation in vivo. *Chem. Res. Toxicol.* 14:101–109.

Yamagiw, K. and Ichikawa, K. 1918. Experimental study of the pathogenesis of carcinoma. *J. Cancer Res.* 3:1–21.

Yeh, C.Y., Fu, P.P., and Beland, H.R.G. 1978. Application of CNDO/2 theoretical calculations to interpretation of the chemical reactivity and biological activity of the syn and anti diolepoxides of benzo[α]pyrene. *Bioorg. Chem.* 7:497–506.

Zhao, Y., Xia, Q., Gamboa da Costa, G., Yu, H., Cai, L., and Fu, P.P. 2012. Full structure assignments of pyrrolizidine alkaloid DNA adducts and mechanism of tumor initiation. *Chem. Res. Toxicol.* 25:1985–1996.

Zurer, P. and Hanson, D. 2004. Chemistry puts herbal supplements to the test. *Chem. Eng. News* 82:16.

5 Absorption of Food Toxicants
An Overview on Polycyclic Aromatic Hydrocarbons

Kelly L. Harris, Leah D. Banks, Kenneth J. Harris, Anthony E. Archibong, Darryl B. Hood, and Aramandla Ramesh

CONTENTS

5.1 INTRODUCTION

There has been a burgeoning public awareness about the consumption of healthy, nutritious food and staying fit. Furthermore, there is a mounting concern about food safety from the standpoint of microbial contamination and contamination by a variety of chemicals. These chemicals have been shown to occur as natural toxicants in produce (glycoalkaloids, glucosinolates, saponins, illudane sesquiterpenes, etc.) or those used during pre- and postharvest of agricultural crops, fruits, and vegetables (agrochemicals). Other chemical contaminants include feed supplements for livestock and industrial chemicals (chlorinated hydrocarbons, polycyclic aromatic hydrocarbons [PAHs], heterocyclic amines, phthalates, heavy metals, etc., as reviewed in Creaser and Purchase [1991] and D'Mello [2003]). Therefore, from a food safety and regulatory perspective, monitoring of foodstuffs for chemical residues is warranted. From a human health standpoint, the processes by which these chemicals are absorbed through the gastrointestinal (GI) tract and cause toxicity and/or cancer are also important. Because there are several hundreds of food toxicants belonging to various chemical classes with differing physicochemical properties and environmental loadings, it is beyond the scope of this chapter to deal with each chemical class and address their absorption processes and the factors that influence these processes. Therefore, in this treatise, we will focus on the absorption of PAH family of toxicants. These compounds that are environmental and dietary toxicants are also implicated as causative agents for lung, breast, and colon cancers and cause toxicity to the nervous, reproductive, developmental, and cardiovascular systems (for comprehensive reviews refer to WHO, 2010; Ramesh et al., 2011).

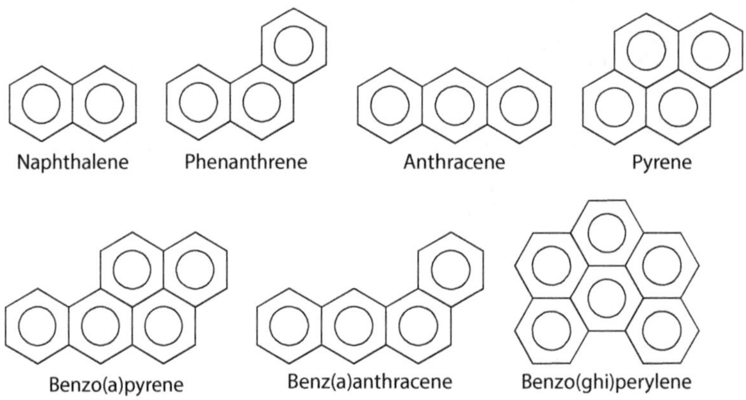

FIGURE 5.1 Some representative polycyclic aromatic hydrocarbon compounds.

PAHs are a large family of environmental toxicants characterized by two or more fused aromatic rings. These compounds are formed as a result of pyrolytic (incomplete combustion of organic materials during industrial and other human activities such as coal-fired power generation, refuse burning, wild fires, etc.) and petrogenic (drilling for petroleum, crude oil deliveries, etc.) processes. Other sources for PAH contamination of the environment and exposure to humans include industrial and automobile emissions, hazardous waste sites, arson, and warfare (IPCS, 1998; Ramesh et al., 2004; Hood et al., 2009). Being semivolatile, sparingly water-soluble, lipophilic, and high-molecular-weight compounds, these pollutants are pervasive in the environment and depending on their loading may persist in soil and sediments as mixtures (Ramesh et al., 2011). A few representative PAHs that are encountered in the environment are shown in Figure 5.1.

5.2 DIETARY PAH EXPOSURE AND INTAKE ESTIMATES

As indicated earlier, humans are exposed to PAHs through several routes that include air, water, food, skin contact with soil, and sediment and in occupational settings. For the general population, food ingestion is the major route of exposure compared to inhalation of polluted air through cigarette smoke, vehicle exhaust, and factory emissions (Butler et al., 1993). Studies conducted on human exposure to benzo(a)pyrene (B(a)P), a prototypical member of the PAH family, revealed that the range and magnitude of dietary exposures (2–500 ng/day) were larger than those for inhalation (10–50 ng/day; Lioy et al., 1988). Diet contributes to more than 70% of nonoccupational exposure to PAHs in nonsmokers (Beckman Sundh et al., 1998; Phillips, 1999). Intake of PAHs through water was reported to vary from 0.0002 to 0.12 μg/day (assuming a daily consumption of 2 L of water/person/day [WHO, 2003]), which is significantly below their estimated intake from food.

PAH contamination of food arises from a number of sources, including environment and food processing techniques and methods used for analysis. From a food safety perspective, the contamination of food products of animal origin (milk, butter, cheese, curd, meat, etc.) by PAHs has received considerable attention as these products are consumed on a daily basis. For unprocessed food such as vegetables, tubers, fruits, and grains, atmosphere and soil are prime sources of PAH contamination (Wickstrom et al., 1986). While accumulation of PAHs in foods of terrestrial animal origin such as livestock is due to consumption of contaminated pasture and vegetation (Hartemann, 2004; Ramesh et al., 2012), PAHs in edible fish and shellfish are a result of contamination of fresh and coastal waters. Food processed through smoking (Reinik et al., 2007), charcoal cooking (Kazerouni et al., 2001), combustion gas heating (Klein et al., 1993), and refining (Guillén and Sopelana, 2003) also contributes substantially to the intake of PAHs. Additionally, the cooking time, amount of fat, and oil influence the formation of PAHs (Vainiotalo and Matveinen 1993; Perez et al., 2002). The concentrations of PAHs found in products of plant and animal origin have extensively been reviewed in Ramesh et al. (2004) and Domingo and Nadal (2015). Global dietary intake of PAHs is shown in Table 5.1. The intake ranged from 0.02 to 25.9 μg/person/day.

The increasing contamination of the environment by these PAHs (WHO, 2010; Ramesh et al., 2011) and their environmental persistence (half-life time of 100 days in soil; Jurjanz et al., 2008) have raised awareness on possible risks to humans. The toxicity that ensues from the intake of PAHs through diet and subsequent digestion, particularly if the doses are high, depends on the following two factors: the bioaccessibility and bioavailability of the PAH test compounds. The amount of PAHs released into the lumen of the GI tract during digestion is considered the "bioaccessible fraction" (Oomen et al., 2002; Armstrong et al., 2007). Subsequent to absorption of a PAH compound, the fraction that is absorbed into the systemic circulation and subsequently causes toxicity in target organs/tissues is considered as the "bioavailable fraction" (NAS/NRC, 2002). The factors that influence bioaccessibility and subsequently the bioavailability of PAHs are the physicochemical properties of these compounds, their residue levels in dietary items (or doses if laboratory animal models are involved), nature of the dietary items (plant based, animal based, fat, fiber, etc.), and physiological status of the organism (reviewed in Ng et al., 2010; Harris et al., 2013).

TABLE 5.1
Estimated Dietary Intakes of Polycyclic Aromatic Hydrocarbon in Various Countries[a]

Country	Estimated Intake (μg/person/day)	Reference
Australia[b]	0.03	Benford et al. (2006)
Austria	0.39	Benford et al. (2006)
Brazil	0.87	Rojo Camargo et al. (2011)
China	371	Li et al. (2014)
Czech Republic	0.19	Benford et al. (2006)
Denmark	0.02	Benford et al. (2006)
Finland	2.34	Benford et al. (2006)
France	0.104	Veyrand et al. (2013)
Germany	0.19	Benford et al. (2006)
Greece	2.630	Voutsa and Samara (1998)
India	11.0	Pandey et al. (2004)
Italy[b]	1.7	Cirillo et al. (2010)
Japan[b]	0.09	Suzuki and Yoshinaga (2007)
Netherlands	5.2–17.1	DeVos et al. (1990)
New Zealand[b]	0.16	Benford et al. (2006)
Nigeria[b]	25.9	Iwegbue and Bassey (2013)
Norway	0.02	Benford et al. (2006)
South Korea	1.10	Moon et al. (2010)
Spain	12.04	Marti-Cid et al. (2008)
Sweden	0.27	Abramsson-Zetterberg et al. (2014)
United Kingdom	0.06	Benford et al. (2006)
United States[b]	0.05	Sinha et al. (2005)

[a] The variety of food items considered for intake estimation, analytical methods and the polycyclic aromatic hydrocarbons used for computing dietary intake, and the years in which samples were collected and analyzed vary.

[b] Data pertain to benzo(a)pyrene intake only.

5.3 ABSORPTION OF PAHs THROUGH DIETARY INGREDIENTS

Following ingestion, PAHs present in food items are acted upon by salivary analyses and lipases in the oral cavity. Subsequent to mastication, the ingested food passes into the stomach via the esophagus. In the stomach, the food is exposed to gastric juices containing hydrochloric acid secreted from the stomach cells. From the stomach, the chyme (partially digested food, hydrochloric acid, and digestive enzymes) is carried to the duodenum. During this transit, the bile secreted from the gall bladder, pancreatin, pepsin, amylases, lipases, and protease (for cleaving carbohydrates, fats, and proteins in food) are added (Johnson, 2001). The chyme then passes from the duodenum to the ileum where it comes into contact with enterocytes that harbor microvilli. The microvilli provide a large surface area for absorption (Tso, 1994).

5.3.1 Uptake and Transport of PAHs

5.3.1.1 Fat

Transfer of PAHs into circulation occurs concurrently with lipids (Stavric and Klassen, 1994). In as much as lipids contribute 42% of dietary energy (Gurr, 1983), the role of dietary fat is important in the uptake of PAHs because this component possesses a hydrophobic domain that is not disrupted easily during lipolysis (Vasiluk et al., 2008).

5.3.1.1.1 Lymphatic Transport

Fatty acids of larger size (more than 10 carbon atoms) are transported into blood via the lymphatic system (Guillot and Vaugelade, 1993). Zaleski et al. (1991) have shown that orally administered B(a) P is sequestered in lipid droplets. Further studies have reported that lipid-soluble toxicants such as PAHs undergo absorption via the lymphatic route (Busbee et al., 1990; Hussain et al., 1996).

Fat in the diet is broken down during digestion into monoglycerides, fatty acids, cholesterol, and phospholipids, which are in micellar form in the intestinal lumen. In enterocytes, monoglycerides and free fatty acids are converted back into triacylglycerols (TAGs). The lipids that were processed in the intestine accumulate in the vesicles of the endoplasmic reticulum and are generally referred to as chylomicrons. Chylomicrons comprise 80%–90% TAGs, 10% phospholipids, 3% cholesterol, and 2% protein (Hobson and Hobson, 1986). TAGs are involved in the absorption of dietary cholesterol (Iqbal and Hussain, 2009) and serve as a natural solvent and delivery vehicle for PAHs during their passage from the stomach to the duodenum for their subsequent absorption in the ileum (Alingst and Shen, 2007). TAGs are emulsified by interacting with bile salts and phospholipids, which ensure efficient absorption (Thomson et al., 1989). The amphiphilic nature of bile salts promotes the diffusion of lipids through the unstirred layer (Weber and Lanno, 2001), which surrounds the intestinal membrane and acts as a major barrier to absorption of lipids and/or lipophilic toxicants. It has been reported that PAHs are transferred via chylomicron formation within the enterocytes for subsequent transfer into lymph (Busbee et al., 1990). Rahman et al. (1986) used a rat model to demonstrate that bile salt–facilitated micellar solubilization drives the transport of PAHs such as phenanthrene, anthracene, 7,12-dimethylbenz(a)anthracene (DMBA), and B(a)P across the unstirred water layer to enterocytes. Enterocytes are involved in the formation of lymph containing chylomicrons that are secreted into interstitial space. From there, chylomicrons are taken up by the surrounding mesenteric lymphatic system. Gershkovich and Hoffman (2005) found a linear correlation between the uptake of B(a)P by chylomicrons under *ex vivo* conditions and intestinal lymphatic bioavailability.

The uptake of BaP, a prototypical PAH toxicant, has been studied extensively. This toxicant like other chemicals is transported from chylomicrons to lipoproteins after entering systemic circulation. Studies revealed that 70% of B(a)P in blood is partitioned into plasma, out of which 95% is partitioned into plasma lipoproteins, specifically apolipoprotein B (primary apolipoprotein of chylomicrons and low-density lipoproteins), which is the main carrier for the transport of B(a)P to target organs (Polyakov et al., 1996). Both *in vitro* and *in vivo* studies reveal that the rate of B(a)P transfer between chylomicrons and lipoproteins depends on the rate of chylomicrons clearance and metabolism (Vauhkonen et al., 1980). The PAHs transferred through chylomicrons into lymph allow these chemicals to bypass the liver and first pass elimination (Hussain et al., 1996) underscoring the importance of chylomicrons in lymphatic transport of PAHs. The processes involved in absorption and eventual fate of PAHs in the body resulting in target organ toxicities/cancer are schematically shown in Figure 5.2.

With increased lipid burden, prolonged transport, absorption, and increased bioavailability of PAHs are expected. A direct relationship between PAH absorption and fat absorption was demonstrated by the studies of Laurent et al. (2001). When pigs were administered [^{14}C] phenanthrene or [^{14}C] B(a)P in milk, the time-course plasma concentrations of these chemicals showed 5–6-hour peak concentrations postadministration, corresponding with the period of maximum fat absorption.

On the other hand, Laher et al. (1984) reported similar bioaccessibility, bioavailability, and lymphatic transport of B(a)P, when ingested in two doses of dietary lipid. The ability of enterocytes to

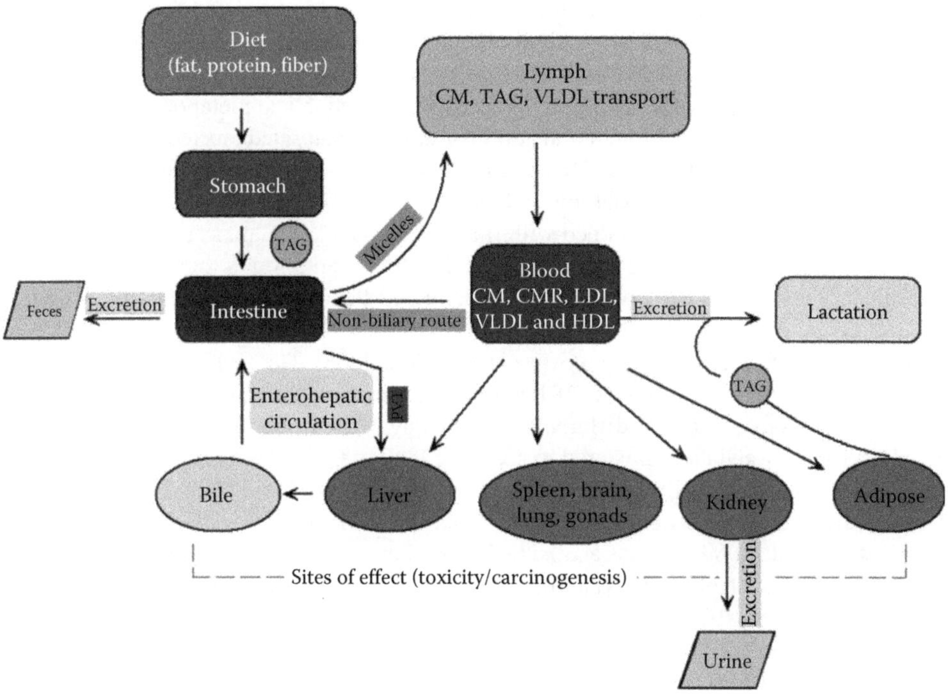

FIGURE 5.2 Schematic representation of absorption and biological fate of dietary polycyclic aromatic hydrocarbon (PAH) toxicants. Oral intake of PAHs occurs through dietary items of both plant origin and animal origin, which comprise carbohydrate, fat, and protein as the principal dietary ingredients. The high-molecular-weight PAHs are packaged and carried through chylomicrons (CMs) and transported to the heart, lungs, kidneys, gonads, etc., through arterial blood flow (lymphatic transport). The low-molecular-weight PAHs are transported from the intestine via mesenteric blood vessels and portal vein to the liver, then to bile, and transported back to the intestine (portal transport) through enterohepatic circulation. The lipolytic enzymes hydrolyze the triglyceride cores in CMs resulting in the formation of chylomicron remnants (CMRs). The hepatocytes, adipocytes, splenocytes, and foam cells (macrophages) are involved in uptake of CMRs. The target organs harboring these cells are affected by prolonged exposure to dietary PAH parent compounds and/or metabolites culminating in disruption of cellular homeostasis causing toxicity or cancer. In addition to CMs, triacylglycerols (TAGs) and lipoproteins act as carriers of PAHs in the absorption process. While the more polar PAH products or even unchanged PAH parent compounds (at high doses) are excreted through urine and feces, the lipophilic PAHs in adipose stores are released through TAG into milk during lactation. CMs, chylomicrons; LDL, low-density lipoprotein; VLDL, very-low-density lipoprotein; HDL, high-density lipoprotein. (Modified from Harris, K.L. et al., *Expert. Opin. Drug Metab. Toxicol.*, 9, 1465, Copyright 2013, Informa Healthcare.)

adapt to varying doses of dietary fat and exposure frequencies (single or multiple exposures) obviates the need for large flux of chylomicrons for B(a)P transport (Laher et al., 1984). Specifically, B(a)P partitioning into lymph was independent of lymphatic fat. Along the same line, accordingly, Bowes and Renwick (1986) observed no change in the inducible levels of B(a)P hydroxylase and DNA binding in the intestine of guinea pigs that were fed normal or high-fat diets. Alternatively, Clinton and Visek (1989a) reported enhanced bioavailability of DMBA in rats fed high-fat diets.

Studies were conducted to determine if intake of B(a)P through dietary fat would have any influence on B(a)P metabolite formation. Gower and Willis (1986) reported a positive correlation between the amount of B(a)P metabolites produced by the intestine and quantity of dietary fat. Mice that were fed a diet rich in dietary fats in the normal human intake range generated B(a)P metabolites with increased dietary fats leading to an increased production of these metabolites (O'Neill et al. 1990a,b). A decrease in concentrations of glucuronide and sulfate conjugates of B(a)P in extrahepatic tissues of rats maintained on food-restricted diet compared to tissues of those fed high-fat diet was noted (Kwei et al., 1991). Rats that were maintained on high-fat or restricted-food diets showed an inverse relationship between the levels of TAGs and B(a)P metabolism in hepatocytes (Zaleski et al., 1991). Studies conducted in our laboratory are also in support of the view that the type of dietary fat administered will have a profound effect on the toxicant absorption and metabolism. Rodent models receiving B(a)P or fluoranthene through oral administration via saturated fats (coconut oil) registered greater metabolite concentrations and DNA adducts than rats that received these chemicals through unsaturated fat (peanut oil) in a subchronic (60 days) exposure regimen (Walker et al., 2007; Diggs et al., 2013). Similar results were obtained with B(a)P using olive oil (Banks et al., 2016).

Aside from using laboratory animal models, studies that used animals such as goats and sheep also provided evidence that PAH absorption is facilitated through fat. Specifically, a single oral administration of 5 μg/kg ^{14}C-Phe, ^{14}C-Pyr, or^{14}C-BaP to goats resulted in approximately 2% of the parent dose through lactational transfer (Grova et al., 2002, 2005). The molecular structure of B(a)P (five benzene rings) compared to those of Phe (three benzene rings) and Pyr (four benzene rings) may have been responsible for the differential processing of these chemicals by goats. Studies on 20 μg PAHs/kg body weight administered to goats for 28 days showed both PAH parent compounds and their hydroxylated metabolites in milk except for B(a)P (Grova et al., 2006; Lapole et al., 2007). Similar bioavailability results were obtained when castor oil or sunflower oil was used as delivery vehicles (Costera et al., 2009; Ounnas et al., 2009a). Studies conducted on sheep also showed appreciable lactational transfer of hydroxylated metabolites of radiolabeled 3-methylcholanthrene and B(a)P (West and Horton, 1976).

5.3.1.1.2 Portal Transport

Fatty acids of smaller size (about 2–6 carbon atoms) and medium size (about 6–10 carbon atoms) are transported into blood via the portal vein. Studies conducted by Laher et al. (1984) on lymph and bile duct–cannulated rats demonstrated that the transfer of about 80% of absorbed PAHs to the general circulation was via hepatic portal transport rather than via lymph. Another study by Kim et al. (2012) confirmed these findings. Their studies concluded that approximately twice the amount of PAHs entering into the general circulation cross through hepatic portal transfer rather than through chylomicron formation and transport through the lymphatic system. Therefore, for a great majority of ingested PAHs, their entry into the body via the portal vein to the liver and subsequently to the systemic circulation cannot be ruled out.

5.3.1.2 Protein

During digestion, proteins are broken down to di- and tripeptides that are amenable for absorption. The specific role of proteins in the uptake of PAHs is difficult to discern because diets rich in proteins are also rich in fats, and when humans consume different dietary items, it is difficult to exclude carbohydrates and lipids from proteins. Therefore, the factors that affect the uptake of PAHs through fats are also expected to play a similar role in the case of proteins.

Next to lipid, animal protein was found to bind strongly to B(a)P in a rat model and facilitate the free radical activation of B(a)P (O'Neill et al., 1990).

5.3.1.3 Fiber

Dietary fibers and associated compounds have been known to sequester toxicants by preventing their absorption through the GI tract, thereby limiting their access to the target organs of their action. Kawamura et al. (1988) reported that the absorption of B(a)P in a rat model was prevented by cellulose (30%) compared to soybean oil (39%) and triolein (51%). Clinton and Visek (1989b) demonstrated that 10% of wheat bran inhibited the B(a)P-induced B(a)P hydroxylase activity in the small intestine of Sprague-Dawley (SD) rats. Stavric and Klassen (1994) reported that foods rich in fibers reduce the uptake of B(a)P. Fergusion et al. (1995) reported that gum arabic, a soluble dietary fiber, reduced the adsorption of B(a)P due to alpha-cellulose, an insoluble dietary fiber (IDF). Both the IDF alpha-cell (bulk cellulose) and soluble fiber (pectin) were found to reduce the urinary excretion of 1-hydroxypyrene (a metabolite of pyrene) in SD rats presumably due to low bioaccessibility of this PAH compound (Viau et al., 2004). Similarly, Boki et al. (2007) found that the IDF held PAHs during their passage through the GI tract. Using artificial gastric and intestinal juices and adsorption isotherm approach, these authors found that the IDF from powdered barley leaves held PAHs such as B(a)P, benz(a)anthracene, chrysene, benzo(b)fluoranthene, and benzo(k)fluoranthene tightly. Kim et al. (2012) have shown that green tea extract inhibits the intestinal absorption of B(a)P through the lymphatic route and enhances the excretion of absorbed B(a)P via the biliary route in rats. The previous findings will have implications for dietary fiber protection against toxicant-induced colorectal cancer.

5.4 UTILITY OF VARIOUS MODELS TO STUDY PAH ABSORPTION

Animal models have widely been used to study the absorption and bioavailability of PAHs. Since soil contains higher concentrations of PAHs compared to other matrices, researchers have focused on soil-bound PAH bioavailability as a surrogate of diet-incorporated PAH bioavailability using rodents (van Schooten et al., 1997), mini pigs (Roos et al., 2004), ruminants (Ounnas et al., 2009b), and swine (Peters et al., 2015) as models. In addition to *in vivo* models, various *in vitro* digestion models mimicking human physiology have been developed to estimate the fraction of soil-bound PAHs that were bioaccessible under GI conditions (Khan et al., 2008). However, adopting those strategies for human dietary ingredients poses some issues. Soil is known to sequester PAHs due to clay, organic carbon, the presence of other toxicants, etc., whereas bile salts, liquid/solid ratio of food, and composition of food (fats, fibers, proteins, sugars, etc.) are known to influence the bioaccessibility in humans (Ounnas et al., 2009b). Additionally, PAH sequestration in dietary lipids and tissues only occurs when high amounts of lipids are ingested over a longer period of time. Furthermore, studies that have used some of these *in vitro* and *in vivo* models have not taken into consideration the role of the microbiota in PAH absorption in the GI tract. Interested readers may refer to a recent review by Williams et al. (2015) for information on the pros and cons of various human gut models available that could be used to study toxicant absorption.

5.5 CONSEQUENCES OF ALTERED ABSORPTION

The PAHs packaged into chylomicrons reach target tissues such as the lung, heart, brain, kidney, etc., through arterial circulation. The triglyceride core in chylomicrons is hydrolyzed in the peripheral circulation by lipoprotein lipase, resulting in the formation of chylomicron remnants (CMRs; Nebert and Dalton, 2006). CMRs are taken up by hepatocytes, adipocytes, spleen, and foam cells (Stary, 1987; Cooper, 1997), which are the organ sites that are targeted by PAHs.

Increased consumption of PAH-contaminated dietary items that are rich in fats such as barbecue meat contributes to colorectal cancers (Sinha et al., 2005). As the fat content and fat type facilitate chylomicron formation, an increased intake of these chemicals increases the likelihood of CMRs reaching other organs, which can contribute to lung and breast cancers as well (Moorthy et al., 2015). Another possibility for these chemicals partitioning into adipose tissue cannot be ruled out. Given their lipophilic nature (Hutcheon et al., 1983), this puts humans at risk of obesity-associated insulin resistance (Choi et al., 2015), atherosclerosis (Marinković et al., 2013), reduced fertility (Xia et al., 2009), poor birth outcomes (Jedrychowski et al., 2012), and impaired neurobehavioral function (Perera et al., 2015).

Altering the absorption of ingested PAHs may contribute to the reduction of bioavailable fraction of these toxicants, thereby reducing toxicity and cancer. Phytochemicals have been advocated as potent preventive agents against many diseases including cancer. The abundance of these chemicals in fruits, vegetables, and oils and their biologically active properties (Eprilati and Ginjom, 2012) render them beneficial for human health. Compounds such as chlorophyllins have been reported to reduce the uptake of dibenzo(a,l)pyrene from Caco-2 intestinal cells (Mata et al., 2004). Similarly, red raspberry extract was found to inhibit absorption of B(a)P and dibenzo(a,l)pyrene across the Calu-3 intestinal cells (Mahadevan et al., 2005).

5.6 CONCLUSIONS AND FUTURE RESEARCH DIRECTIONS

In summary, gaining an understanding of the absorption processes of PAHs and other food toxicants is necessary for the protection of humans from exposure to PAH-contaminated foods. At this time, it is speculated that the absorption of PAHs could be relatively low in fat-free or low-fat diets consumed by vegetarians. However, if a vegetarian meal contains one-third of its calories in the form of vegetable or seed oil, the diet will still have enough fat to solubilize the lipophilic toxicants. However, if a vegetarian diet contains a bulk of fiber to bind strongly to contaminants, it will aid in the facilitation of their excretion. Most of the PAH absorption and bioavailability studies were conducted on single PAH compounds, and we know that humans seldom ingest a particular PAH compound through diet. As the widely distributed PAHs differ in their molecular weight, physicochemical properties, and persistence, extrapolations made with a single compound do not represent the true chemical fate and exposure risk to humans. Therefore, absorption studies with PAH mixtures are warranted. Great caution should be exercised in choosing not only the test compounds but also the animal model for conducting PAH absorption and bioavailability studies. This is important as the metabolic and transport pathways used by various species to process ingested PAH greatly differ. Another overlooked aspect in PAH absorption studies is the presence of microflora in the gut. As gut microflora is known to regulate fat storage and breakdown of dietary items and associated toxicants, studies need to be conducted on the effect of both pro- and prebiotics on the processing of PAH toxicants in foods and the mechanisms involved. These studies will provide useful information on prevention strategies to be adopted against diet and environment-induced diseases.

ACKNOWLEDGMENTS

The authors gratefully acknowledge the research funding from the National Institutes of Health through grant numbers 1RO1CA142845-O1A1 (AR), U54RR026140 (AEA), U54NS041071 (DBH), and S11ES014156 (DBH, AR, and AEA), which supported PAH research in their laboratories. They also appreciate the student support from NIH grants 1F31ESO2407901 (KLH), 5T32HL007735-20 (LDB), and 5R25GM059994-13 (KJH) received through Meharry.

REFERENCES

Abramsson-Zetterberg, L., Darnerud, P.O., and S. Wretling. 2014. Low intake of polycyclic aromatic hydrocarbons in Sweden: Results based on market basket data and a barbecue study. *Food Chem. Toxicol.* 74: 107–111.

Alingst, B. and D.D. Shen. 2007. Gastrointestinal absorption. In *Gastrointestinal Toxicology*, ed. K. Rozman and O. Hänninen, pp. 29–42. London, U.K.: Elsevier Science Publishers.

Armstrong, S.A., Van de Wiele, T., Jermida, J.J. et al. 2007. Aryl hydrocarbon bioaccessibility to small mammals from arctic plants using in vitro techniques. *Environ. Toxicol. Chem.* 26: 491–496.

Banks, L.D., Amoah, P., Niaz, M.S., Washington, MK., Adunyah, S.E., and A. Ramesh. 2016. Olive oil prevents benzo(a)pyrene [B(a)P]-induced colon carcinogenesis through altered B(a)P metabolism and decreased oxidative damage in Apc^{Min} mouse model. *J. Nutr. Biochem.* 28: 37–50.

Beckman Sundh, O., Thuvander, A., and C. Andersson. 1998. Preview of PAHs in food: Potential health effects and contents in food, Report 8, Swedish National Food Administration, Uppsala, Sweden.

Benford, D., Agudo, B., Carrington, C. et al. 2006. Polycyclic aromatic hydrocarbons. WHO Food Additive Series 55, Safety Evaluation of Certain Contaminants in Food. http://whqlibdoc.who.int/publications/2006/9241660554_PAH_eng.pdf. Downloaded on January 22, 2016.

Boki, K., Kadota, K., Takahashi, M., and M. Kitakouji. 2007. Uptake of polycyclic aromatic hydrocarbons by insoluble dietary fiber. *J. Health Sci.* 53: 99–106.

Bowes, S.G. and A.G. Renwick. 1986. The hepatic metabolism and biliary excretion of benzo(a)pyrene in guinea-pigs fed normal, high-fat or high-cholesterol diets. *Xenobiotica* 16: 531–542.

Busbee, D.L., Norman, J.O., and R.L. Ziprin. 1990. Comparative uptake, vascular transport, and cellular internalization of aflatoxin-B1 and benzo(a)pyrene. *Arch. Toxicol.* 64: 285–290.

Butler, J.P., Post, G.B., Lioy, P.J., Waldman, J.M., and A. Greenberg. 1993. Assessment of carcinogenic risk from personal exposure to benzo(a)pyrene in the total human environmental exposure study (THEES). *J. Air Waste Manag. Assoc.* 43: 970–977.

Choi, Y-H., Kim, J.H., and Y.C. Hong. 2015. Sex-dependent and body weight-dependent associations between environmental PAHs exposure and insulin resistance: Korean urban elderly panel. *J. Epidemiol. Commun. Health* 69: 625–631.

Cirillo, T., Montuori, P., Mainardi, P., Russo, I., Fasano, E., Triassi, M., and R. Amodio-Cocchieri, 2010. Assessment of the dietary habits and polycyclic aromatic hydrocarbon exposure in primary school children. *Food Addit. Contam. Part A Chem. Anal. Control Expos. Risk Assess.* 27: 1025–1039.

Clinton, S.K. and W.J. Visek. 1989a. Wheat bran and the induction of intestinal benzo(a)pyrene hydroxylase by dietary benzo(a)pyrene. *J. Nutr.* 119: 395–402.

Clinton, S.K. and W.J. Visek. 1989b. Hepatic and intestinal drug metabolizing enzymes and the tissue distribution and excretion of ^{14}C-7,12-Dimethylbenz(a)anthracene in rats fed diets varying in fat concentration. *Pharmacology* 39: 89–97.

Cooper, A.D. 1997. Hepatic uptake of chylomicron remnants. *J. Lipid Res.* 38: 2173–2192.

Costera, A., Feidt, C., Dziurla, M.A., Monteau, F., Le Bizec, B., and G. Rychen. 2009. Bioavailability of polycyclic aromatic hydrocarbons (PAHs) from soil and hay matrices in lactating goats. *J. Agric. Food Chem.* 57: 5352–5357.

Creaser, C. and R. Purchase. 1991. *Food Contaminants, Sources and Surveillance.* Cambridge, U.K.: The Royal Society of Chemistry, 205pp.

D'Mello, J.P.F. 2003. *Food Safety: Contaminants and Toxins.* Cambridge, MA: CABI Publishing, 452pp.

de Vos, R.H., van Dokkum, W., Schouten, A., and P. de Jong-Berkhout. 1990. Polycyclic aromatic hydrocarbons in Dutch total diet samples (1984–1986). *Food Chem. Toxicol.* 28: 263–268.

Diggs, D.L., Huderson, A.C., Harris, K.L., Myers, J.N., Banks, L.D., Rekhadevi, P.V., Niaz, M.S., and A. Ramesh. 2011. Polycyclic aromatic hydrocarbons and digestive tract cancers: A perspective. *J. Environ. Sci. Health C Environ. Carcinog. Ecotoxicol. Rev.* 29: 324–357.

Diggs, D.L., Myers, J.N., Banks, L.D., Niaz, M.S., Hood, D.B., Roberts, L.J. 2nd, and A. Ramesh. 2013. Influence of dietary fat type on benzo(a)pyrene [B(a)P] biotransformation in a B(a)P-induced mouse model of colon cancer. *J. Nutr. Biochem.* 24: 2051–2063.

Domingo, J.L. and M. Nada. 2015. Human dietary exposure to polycyclic aromatic hydrocarbons: A review of the scientific literature. *Food Chem. Toxicol.* 86: 144–153.

Eprilati, I. and I.R. Ginjom. 2012. Bioavailability of phytochemicals. In *Phytochemicals: A Global Perspective of Their Role in Nutrition and Health,* ed. D.V. Rao. Rijeka, Croatia: InTech Publishers.

Ferguson, L.R., Roberton, A.M., Watson, M.E., Triggs, C.M., and P.J. Harris. 1995. The effects of a soluble-fibre polysaccharide on the adsorption of carcinogens to insoluble dietary fibres. *Chem. Biol. Interact.* 95: 245–255.

Gershkovich, P. and A. Hoffman. 2005. Uptake of lipophilic drugs by plasma derived isolated chylomicrons: Linear correlation with intestinal lymphatic bioavailability. *Eur. J. Pharm. Sci.* 26: 394–404.

Gower, J.D. and E.D. Willis. 1986. The dependence on the rate of BP metabolism in the rat small intestinal mucosa on the composition of the dietary fat. *Nutr. Cancer.* 8: 151–161.

Grova, N., Feidt, C., Laurent, C., and G. Rychen. 2002. [^{14}C] milk, urine and faeces excretion kinetics in lactating goats after an oral administration of [^{14}C] polycyclic aromatic hydrocarbons. *Int. Dairy J.* 12: 1025–1031.

Grova, N., Monteau, F., Le Bizec, B., Feidt, C., Andre, F., and G. Rychen. 2005. Determination of phenanthrene and hydroxyphenanthrenes in various biological matrices at trace levels using gas chromatography-mass spectrometry. *J. Anal. Toxicol.* 29: 175–181.

Grova, N., Rychen, G., Monteau, F., Le Bizec, B., and C. Feidt. 2006. Effect of oral exposure to polycyclic aromatic hydrocarbons on goat's milk contamination. *Agron. Sustain. Dev.* 26: 195–199.

Guillén, M.D. and P. Sopelana. 2003. Polycyclic aromatic hydrocarbons in diverse foods. In *Food Safety: Contaminants and Toxins*, ed. J. P. F. D'Mello, pp. 175–197. CAB International, Wallingford, Oxon, UK.

Guillot, E., Vaugelade, P., Lemarchal, P. et al. 1993. Intestinal absorption and liver uptake of medium-chain fatty acids in non-anaesthetized pigs. *Br. J. Nutr.* 69: 431–442.

Gurr, M.I. 1983. The nutritional significance of lipids. In *Developments in Dairy Chemistry-2*, ed. P.F. Fox, pp. 365–417. London, U.K.: Applied Science Publishers Ltd.

Harris, K.L., Banks, L.D., Mantey, J.A., Huderson, A.C., and A. Ramesh. 2013. Bioaccessibility of polycyclic aromatic hydrocarbons: Relevance to toxicity and carcinogenesis. *Expert. Opin. Drug Metab. Toxicol.* 9: 1465–1480.

Hartemann, P. 2004. Risk assessment of organic xenobiotics in the environment. In *Pharmaceuticals in the Environment*, ed. Kummercerk, pp. 251–267. Heidelberg, Germany: Springer-Verlag.

Hobson, D.W. and V.L. Hobson. 1986. Normal and abnormal intestinal absorption by humans. In *Toxicology of the Gastrointestinal Tract*, ed. S.C. Gad, pp. 27–30. Boca Raton, FL: CRC Press.

Hood, D.B., Ramesh, A., and M. Aschner. 2009. Polycyclic aromatic hydrocarbons: Exposure from emission products and terrorist attacks on US targets: Implications for developmental central nervous system toxicity. In *Handbook of Toxicology of Chemical Warfare Agents*, ed. R.C. Gupta, pp. 229–243. London, U.K.: Academic Press.

Hussain, M.M., Kancha, R.K., Zhou, Z., Luchoomun, J., Zu, H., and A. Bakillah. 1996. Chylomicron assembly and catabolism: Role of apolipoproteins and receptors. *Biochem. Biophys. Acta* 1300: 151–170.

Hutcheon, D.E., Kantrowitz, J., Van Gelder, R.N., and E. Flynn. 1983. Factors affecting plasma benzo[a] pyrene levels in environmental studies. *Environ. Res.* 32: 104–110.

IPCS. 1998. Environmental Health Criteria 202: Selected non-heterocyclic polycyclic aromatic hydrocarbons. International Programme on Chemical Safety, World Health Organization, Lyon, France.

Iqbal, J. and M.M. Hussain. 2009. Intestinal lipid absorption. *Am. J. Physiol. Endocrinol. Metab.* 296:E1183–E1194.

Iwegbue, C.M.A. and F.I. Bassey. 2013. Concentrations and health hazards of polycyclic aromatic hydrocarbons in selected commercial brands of milk. *Food Measure* 7: 177–184.

Jedrychowski, W., Perera, FP., Tang, D., Stigter, L., Mroz, E., Flak, E., Spengler, J., Budzyn-Mrozek, D., Kaim, I., and R. Jacek. 2012. Impact of barbecued meat consumed in pregnancy on birth outcomes accounting for personal prenatal exposure to airborne polycyclic aromatic hydrocarbons: Birth cohort study in Poland. *Nutrition* 28: 372–377.

Johnson, L.R. 2001. Gastric secretion. In *Gastrointestinal Physiology*, 6th edn., ed. L.R. Johnson. St. Louis, MI: Mosby Publishing.

Jurjanz, S., Rychen, G., and C. Feidt. 2008. Dairy livestock exposure to persistent organic pollutants and their transfer to milk: A review. In *Impact of Pollution on Animal Products*, eds. B. Faye and Y. Sinyavskiy, pp. 63–83. Dordrecht, the Netherlands: Springer Science and Business Media.

Kawamura, Y., Kamata, E., Ogama, Y. et al. 1988. The effect of various foods on the intestinal absorption of benzo(a)pyrene in rats. *J. Food Hyg. Soc. Jpn.* 29: 21–25.

Kazerouni, N., Sinha, R., Hsu, C.H., Greenberg, A., and N. Rothman. 2001. Analysis of 200 food items for benzo[a]pyrene and estimation of its intake in an epidemiologic study. *Food Chem. Toxicol.* 39: 423–436.

Khan, S., Cao, Q., Lin, A.J., and Y.G. Zhu. 2008. Concentrations and bioaccessibility of polycyclic aromatic hydrocarbons in wastewater-irrigated soil using in vitro gastrointestinal test. *Environ. Sci. Pollut. Res. Int.* 15: 344–353.

Kim, J., Koo, S.I., and S.K. Noh. 2012. Green tea extract markedly lowers the lymphatic absorption and increases the biliary secretion of ^{14}C-benzo[a]pyrene in rats. *J. Nutr. Biochem.* 23: 1007–1011.

Klein, H., Speer, K., and E.H.F. Schmidt. 1993. Polycyclic aromatic hydrocarbons in raw coffee and roasted coffee. *Bundesgesundheitsblatt* 36: 98–100.

Kwei, G.Y., Zaleski J., Thurman R.G., and F.C. Kauffman. 1991. Enzyme activities associated with carcinogen metabolism in liver and nonhepatic tissues of rats maintained on high fat and food-restricted diets. *J. Nutr.* 121: 131–137.

Laher, J.M., G.A. Chernenko, and J.A. Barrowman. 1983. Studies of the absorption and enterohepatic circulation of 7,12-dimethylbenz[a]anthracene in the rat. *Can. J. Physiol. Pharmacol.* 61: 1368–1373.

Laher, J.M., Rigler M.W., Vetter R.D., Barrowman J.A., and J.S. Patton. 1984. Similar bioavailability and lymphatic transport of Benzo(a)pyrene when administered to rats in different amounts of dietary fat. *J. Lipid Res.* 25: 1337–1342.

Lapole, D., Rychen, G., Grova, N., Monteau, F., Le Bizec, B., and C. Feidt. 2007. Milk and urine excretion of polycyclic aromatic hydrocarbons and their hydroxylated metabolites after a single oral administration in ruminants. *J. Dairy Sci.* 90: 2624–2629.

Laurent, C., Feidt, C., Lichtfouse, E., Grova, N., Laurent, F., and G. Rychen. 2001. Milk-blood transfer of 14C-tagged polycyclic aromatic hydrocarbons (PAHs) in pigs. *J. Agric. Food. Chem.* 49: 2493–2496.

Li, J., Fan, R., Lu, S., Zhang, D., Zhou, Y., and Y. Ly. 2014. Exposure to polycyclic aromatic hydrocarbons could cause their oxidative DNA damage: A case study for college students in Guangzhou, China. *Environ. Sci. Pollut. Res. Int.* 22: 1770–1777.

Lioy, P.L., Waldman, J.M., Greenberg, A., Harkov, A., and C. Pietarinen. 1988. The total Human Environmental Exposure Study (THEES) to benzo(a)pyrene: Comparison of the inhalation and food pathways. *Arch. Environ. Health* 43: 304–312.

Mahadevan, B., Mata, J.E., Albershardt, D.J., Stevens, J.F., Pereira, C.B., Rodriguez-Proteau, R., and W.M. Baird. 2005. The effect of red raspberry extract on PAH transport across Calu-3 cell monolayer, an in vitro cell model. *Int. J. Cancer Prev.* 2: 129–141.

Marinković, N., Pasalić, D., and S. Potocki. 2013. Polymorphisms of genes involved in polycyclic aromatic hydrocarbons' biotransformation and atherosclerosis. *Biochem. Med. (Zagreb)* 23: 255–265.

Martí-Cid, R., Llobet, J.M., Castell, V., and J.L. Domingo. 2008. Evolution of the dietary exposure to polycyclic aromatic hydrocarbons in Catalonia, Spain. *Food Chem. Toxicol.* 46: 3163–3171.

Martorell, I., Perelló, G., Martí-Cid, R., Castell, V., Llobet, J.M., and J.L. Domingo. 2010. Polycyclic aromatic hydrocarbons (PAH) in foods and estimated PAH intake by the population of Catalonia, Spain: Temporal trend. *Environ. Int.* 36: 424–432.

Mata, J.E., Yu, Z., Gray, J.E., Williams, D.E., and R. Rodriguez-Proteau. 2004. Effects of chlorophyllin on transport of dibenzo(a,l)pyrene, 2-amino-1-methyl-6-phenylimidazo-[4,5-b]pyridine, and aflatoxin B1 across Caco-2 cell monolayers. *Toxicology* 196: 117–125,

Moon, H.B., Kim, H.S., Choi, M. et al. 2010. Intake and potential health risk of polycyclic aromatic hydrocarbons associated with seafood consumption in Korea from 2005 to 2007. *Arch. Environ. Contam. Toxicol.* 58: 214–221.

Moorthy, B., Chu, C., and D.J. Carlin. 2015. Polycyclic aromatic hydrocarbons: From metabolism to lung cancer. *Toxicol. Sci.* 145: 5–15.

National Academy of Sciences/National Research Council (NAS/NRC). 2002. *Bioavailability of Contaminants in Soils and Sediments: Processes, Tools, and Applications.* Washington, DC: National Academy Press.

Nebert, D.W. and T.P. Dalton. 2006. The role of cytochrome P450 enzymes in endogenous signalling pathways and environmental carcinogenesis. *Nat. Rev. Cancer.* 6: 947–960.

Ng, J.C., Juharz, A.L., Smith, E., and R. Naidu. 2010. Contaminant bioavailability and bioaccessibility. Part 1: A scientific and technical review. Technical Report Series, No. 14. Cooperative Research Centre for Contamination Assessment and Remediation of the Environment, Adelaide, South Australia, Australia.

O'Neill, I.K., Bingham, S., Povey, A.C., Brouet, I., and J.C. Bereziat. 1990a. Modulating effects in human diets of dietary fibre and beef, and of time and dose on the reactive microcapsule trapping of benzo(a)pyrene metabolites in the rat gastrointestinal tract. *Carcinogenesis* 11: 599–607.

O'Neill, I.K., Povey, A.C., Bingham, S., and E. Cordis. 1990b. Systematic modulation by human diet levels of dietary fibre and beef on metabolism and disposition of benzo(a)pyrene in the gastrointestinal tract of F344 rats. *Carcinogenesis* 11: 609–616.

Oomen, A.G., Hack, A., Minekus, M. et al. 2002. Comparison of five in vitro digestion models to study the bioaccessibility of soil contaminants. *Environ. Sci. Technol.* 36: 3326–3334.

Ounnas, F., Jurjanz, S., Dziurla, M.A., Guiavarc'h, Y., Feidt, C., and G. Rychen. 2009a. Relative bioavailability of soil-bound polycyclic aromatic hydrocarbons in goats. *Chemosphere* 77: 115–122.

Ounnas, F., Rychen, G., Feidt, C., and S. Jurjanz. 2009b. Bioavailability of soil-bound persistent organic pollutants in dairy ruminants: A review. In *Ruminant Physiology: Digestion, Metabolism and Effects of Nutrition on Reproduction and Welfare*, 1st edn., eds. Y. Chilliard, F. Glasser, Y. Faulconnier, F. Bocquier, and I. Veissier, pp. 302–303. Wageningen Academic Publishers, Wageningen, the Netherlands.

Pandey, M.K., Mishra, K.K., Khanna, S.K. et al. 2004. Detection of polycyclic aromatic hydrocarbons in commonly consumed edible oils and their likely intake in the Indian population. *J. Am. Oil Chem. Soc.* 81: 1131–1136.

Perera, F., Phillips, D.H., Wang, Y., Roen, E., Herbstman, J., Rauh, V., Wang, S., and D. Tang. 2015. Prenatal exposure to polycyclic aromatic hydrocarbons/aromatics, BDNF and child development. *Environ. Res.* 142: 602–608.

Perez, C., Lopez de Carain, A., and J. Bello. 2002. Modulation of mutagenic activity in meat samples after deep-frying in vegetable oils. *Mutagenesis* 17: 63–66.

Peters, R.E., Wickstrom, M., and S.D. Siciliano. 2015. The bioavailability of polycyclic aromatic hydrocarbons from different dose media after single and sub-chronic exposure in juvenile swine. *Sci. Total Environ.* 506–507: 308–314.

Phillips, D.H. 1999. Polycyclic aromatic hydrocarbons in the diet. *Mutat. Res.* 443: 139–147.

Polyakov, L.M., Chasovskikh, M.I., and L.E. Panin. 1996. Binding and transport of benzo(a)pyrene by blood plasma lipoproteins: The possible role of apolipoprotein B in this process. *Bioconjug. Chem.* 7: 346–400.

Rahman, A., Barrowman, J.A., and A. Rahimtula. 1986. The influence of bile on the bioavailability of polynuclear aromatic hydrocarbons from the rat intestine. *Can. J. Physiol. Pharmacol.* 64: 1214–1218.

Ramesh, A., Archibong, A., Hood, D.B., Guo, Z., and B.G. Loganathan. 2011. Global environmental distribution and human health effects of polycyclic aromatic hydrocarbons. In *Global Contamination Trends of Persistent Organic Chemicals*, eds. B.G. Loganathan and P.K.S. Lam, pp. 95–124. Boca Raton, FL: Taylor & Francis Publishers.

Ramesh, A., Archibong, A.E., Huderson, A.C., Diggs, D.L., Myers, J.N., Hood, D.B., Rekhadevi, P.V., and M.S. Niaz. 2012. Polycyclic aromatic hydrocarbons. In *Veterinary Toxicology*, ed. R.C. Gupta, pp. 797–809. London, U.K.: Academic Press.

Ramesh, A., Walker, S.A., Hood, D.B., Guillén, M.D., Schneider, K., and EH. Weyand. 2004. Bioavailability and risk assessment of orally ingested polycyclic aromatic hydrocarbons. *Int. J. Toxicol.* 23: 301–333.

Reinik, M., Tamme, T., Roasto, M., Juhkam, K., Tenno, T., and A. Kiis. 2007. Polycyclic aromatic hydrocarbons (PAHs) in meat products and estimated PAH intake by children and the general population in Estonia. *Food Addit. Contam.* 24: 429–437.

Rojo-Camargo, M.C., Antoniolli, P.R., Vicente, E., and S.A.V. Tfouni. 2011. PAHs in Brazilian commercial soybean oils and dietary exposure. *Food Addit. Contam. Part B* 4: 152–159.

Roos, P.H., Tschirbs, S., Hack, A., Welge, P., and M. Wilhelm. 2004. Different mechanisms of handling ingested polycyclic aromatic hydrocarbons in mammalian species: Organ-specific response patterns of CYP1A1-induction after oral intake of PAH-contaminated soils. *Xenobiotica* 34: 781–795.

Sinha, R., Kulldorff, M., Gunter, M.J., Strickland, P., and N. Rothman. 2005. Dietary benzo(a)pyrene intake and risk of colorectal adenoma. *Cancer Epidemiol. Biomarkers Prev.* 14: 2030–2034.

Stary, H.C. 1987. Macrophages, macrophage foam cells, and eccentric intimal thickening in the coronary arteries of young children. *Atherosclerosis* 64: 91–108.

Stavric, B. and R. Klassen. 1994. Dietary effects on the uptake of benzo(a)pyrene. *Food Chem. Toxicol.* 32: 727–734.

Suzuki, K. and J. Yoshinaga. 2007. Inhalation and dietary exposure to polycyclic aromatic hydrocarbons and urinary 1-hydroxypyrene in non-smoking university students. *Inter. Archiv. Occup. Environ. Health* 81: 115–121.

Thomson, A.B., Keelan, M., Garg, M.L. et al. 1989. Intestinal aspects of lipid absorption: In review. *Can. J. Physiol. Pharmacol.* 67: 179–191.

Tso, P. 1994. Intestinal lipid absorption. In *Physiology of the Gastrointestinal Tract*, 3rd edn., ed. L.R. Johnson, Vol. 2. New York: Raven Press.

Vainiotalo, S. and K. Matveinen. 1993. Cooking fumes as a hygienic problem in the food and catering industries. *Am. Ind. Hyg. Assoc. J.* 54: 376–382.

van Schooten, F.J., Moonen, E.J., van der Wal, L., Levels, P., and J.C. Kleinjans. 1997. Determination of polycyclic aromatic hydrocarbons (PAH) and their metabolites in blood, feces, and urine of rats orally exposed to PAH contaminated soils. *Arch. Environ. Contam. Toxicol.* 33: 317–322.

Vasiluk, L., Pinto, L.J., and W.S. Tsang. 2008. The uptake and metabolism of benzo[a]pyrene from a sample food substrate in an in vitro model of digestion. *Food Chem. Toxicol.* 46: 610–618.

Vauhkonen, M., T. Kuusi, and P.K. Kinnunen. 1980. Serum and tissue distribution of benzo(a)pyrene from intravenously injected chylomicrons in rat in vivo. *Cancer Lett.* 11: 113–119.

Veyrand, B., Sirot, V., Durand, S., Pollono, C., Marchand, P., Dervilly-Pinel, G., Tard, A., Leblanc, J.C., and B. Le Bizec. 2013. Human dietary exposure to polycyclic aromatic hydrocarbons: Results of the second French Total Diet Study. *Environ. Int.* 54: 11–17.

Viau, C., Zaqui, C., and S. Charhonneau. 2004. Dietary fibers reduce the urinary excretion of 1-hydroxypyrene following intravenous administration of pyrene. *Toxicol. Sci.* 78: 15–19.

Voutsa, D. and C. Samara. 1998. Dietary intake of trace elements and polycyclic aromatic hydrocarbons via vegetables grown in an industrial Greek area. *Sci. Total Environ.* 218: 203–216.

Walker, S.A., Addai, A.B., Mathis, M., and A. Ramesh. 2007. Effect of dietary fat on metabolism and DNA adduct formation after acute oral exposure of F-344 rats to fluoranthene. *J. Nutr. Biochem.* 18: 236–249.

Weber, L.P. and R.P. Lanno. 2001. Effect of bile salts, lipid, and humic acids on absorption of benzo[a] pyrene by isolated channel catfish (Ictalurus punctatus) intestine segments. *Environ. Toxicol. Chem.* 20: 1117–1124.

West, C.E. and B.J. Horton. 1976. Transfer of polycyclic hydrocarbons from diet to milk in rats, rabbits and sheep. *Life Sci.* 19: 1543–1552.

WHO. 2003. Selected nitro- and nitro-oxy-polycyclic aromatic hydrocarbons. *Environmental Health Criteria*. Geneva, Switzerland: World Health Organization.

WHO. 2010. IARC Monographs on the Evaluation of Carcinogenic Risks to Humans. Vol. 92. Lyon, France: International Agency for Research on Cancer; Some non-heterocyclic polycyclic aromatic hydrocarbons and some related exposures.

Wickstrom, K., Pyysalo, H., Plaami-Heikkila, S. et al. 1986. Polycyclic aromatic compounds (PAC) in leaf lettuce. *Z. Lebensm. Unters. Forsch.* 183: 182–185.

Williams, C.F., Walton, G.E., Jiang, L., Plummer, S., Garaiova, I., and G.R. Gibson. 2015. Comparative analysis of intestinal tract models. *Annu. Rev. Food Sci. Technol.* 6: 329–350.

Xia, Y., Zhu, P., Han, Y., Lu, C., Wang, S., Gu, A., Fu, G., Zhao, R., Song, L., and X. Wang. 2009. Urinary metabolites of polycyclic aromatic hydrocarbons in relation to idiopathic male infertility. *Hum. Reprod.* 24: 1067–1074.

Zaleski, J., Kwei, G.Y., Thurman, R.G., and F.C. Kauffman. 1991. Suppression of benzo(a)pyrene metabolism by accumulation of triacylglycerols in rat hepatocytes: Effect of high fat and food-restricted diet. *Carcinogenesis* 12: 2073–2079.

6 Overview of Food Allergy in Children and Adults

Bright I. Nwaru and Aziz Sheikh

CONTENTS

6.1 INTRODUCTION

Food allergy has been defined "as an adverse health effect arising from a specific immune response that occurs reproducibly on exposure to given food" (Boyce et al. 2010; Burks et al. 2012; Muraro et al. 2014a; Sampson et al. 2014). An adverse reaction to foods without any underlying immunologic mechanisms is called "nonallergic food hypersensitivity," but those with underlying immunologic mechanisms are grouped as *food allergy*; *food hypersensitivity* is used to refer to adverse reactions to foods, regardless of whether or not immunologic mechanisms are involved (Johansson et al. 2001, 2004). Food allergy thus manifests from a sequel of immunological activities that result in intolerance of particular offending foods (Boyce et al. 2010; Burks et al. 2012; Longo et al. 2013; Muraro et al. 2014a; Sampson et al. 2014). In the majority of cases, food allergy is primarily mediated by immunoglobulin E (IgE), but some food allergies may result from aberrant cell-mediated immunity against food proteins in the absence of specific IgE antibodies (Longo et al. 2013). In the case of IgE-mediated food allergy, there is an imbalance in the T helper cell 1 (Th1)/Th2 ratio, with a skewing toward Th2, allowing for the excessive IgE production and relative deficiency of Th1 (Boyce et al. 2010; Burks et al. 2012; Longo et al. 2013; Muraro et al. 2014a; Sabra et al. 2003; Sampson et al. 2014). IgE-mediated reactions to foods typically have rapid onset of symptoms after consumption of an offending food, and reactions generally involve the skin, gastrointestinal tract, cardiovascular system, and respiratory tract (Boyce et al. 2010; Burks et al. 2012; Muraro et al. 2014a; Sampson et al. 2014). Non-IgE-mediated food allergy, on the other hand, is exclusively Th1-cell-mediated activity and commonly presents with a spectrum of gastrointestinal disorders, including protein-induced enterocolitis, proctocolitis, and enteropathy syndromes, primarily affecting infants and young children (Boyce et al. 2010; Burks et al. 2012; Jyonouchi 2012; Longo et al. 2013; Muraro et al. 2014a; Sabra et al. 2003; Sampson et al. 2014). The pathogenesis of some conditions, such as eosinophilic esophagitis, eosinophilic gastroenteritis, and atopic eczema, very commonly results from processes involving both IgE- and non-IgE-mediated mechanisms and hence has been described as mixed IgE- and non-IgE-mediated food allergy (Boyce et al. 2010; Burks et al. 2012; Jyonouchi 2012; Longo et al. 2013; Muraro et al. 2014a; Sabra et al. 2003; Sampson et al. 2014).

In contrast to respiratory allergic diseases (i.e., asthma and allergic rhinitis) that were observed to have risen in prevalence during the latter half of the twentieth century, the incidence and prevalence of food allergy are believed to have risen during the past 10–15 years (Boyce et al. 2010; Nwaru et al. 2014a; Prescott and Allen 2011). This has led some investigators to describe food allergy as the second wave of the allergy epidemic and appears to indicate that different risk factors may be at play (Nwaru et al. 2014a). Global estimates of the incidence, prevalence, and overall burden of food allergy have been elusive, and this has been attributed to the variations in definition and diagnosis of food allergy across settings (Boyce et al. 2010; Longon et al. 2013; Nwaru et al. 2014a; Rona et al. 2007; Sampson 2005; Sampson et al. 2014; Sicherer and Sampson 2014). Depending on the definition used across studies, it is generally estimated that around 10% of children and adults may be affected by food allergy from the general population; these estimates however vary across settings and may be much more less when assessment is performed using double-blind, placebo-controlled food challenge (DBPCFC), which is currently the gold standard tool for the assessment of food allergy (Allen and Koplin 2012; Chafen et al. 2010; Longo et al. 2013; Osborne et al. 2011; Rona et al. 2007; Sampson 2005; Sicherer and Sampson 2014; Zuberbier et al. 2004). Despite the believe that the incidence of food allergy is increasing, reliable data to clearly understand the time trends are limited (Allen and Koplin 2012; Burks et al. 2012; Chafen et al. 2010; Rona et al. 2007; Sampson 2005; Sicherer and Sampson 2014; Zuberbier et al. 2004), but it is clear that healthcare utilization due to food allergy has increased in recent years (Longo et al. 2013). Although about 170 foods have now been identified to potentially trigger IgE-mediated food allergic reactions (Boyce et al. 2010; Burks et al. 2012), about 90% of allergic reactions to foods in children are caused by only eight foods (see Box 6.1): tree nuts appear to be the most common foods among adults, although these also vary across settings (Allen and Koplin 2012; Boyce et al. 2010; Burks et al. 2012).

Clearly, food allergy represents an evolving public health problem, with increasing morbidity, poor quality of life, substantial healthcare and societal burden, and potential fatalities in cases of food-induced anaphylaxis (Allen and Koplin 2012; Boyce et al. 2010; Burks et al. 2012; Chafen et al. 2010; Longo et al. 2013; Muraro et al. 2014a; Osborne et al. 2011; Prescott and Allen 2011; Rona et al. 2007; Sampson 2005; Sampson et al. 2014; Sicherer and Sampson 2014; Zuberbier et al. 2004). Currently, there is no cure for food allergy; while the majority of food allergies may be outgrown, some—particularly, peanut, tree nuts, and seafood allergies—may remain lifelong disorders. Therefore, current efforts toward evidence-based preventive and management strategies are well placed. In this chapter, drawing from key evidence-based guidelines from the U.S. National Institutes of Allergy and Infectious Diseases (NIAID) (Boyce et al. 2010); the Joint Task Force on Practice Parameters representing the American Academy of Allergy, Asthma & Immunology (AAAAI),

BOX 6.1 COMMON FOOD ALLERGENS CAUSING FOOD ALLERGY IN CHILDREN AND ADULTS

Food Allergens Common in Children	Food Allergens Common in Adults
• Cow's milk	• Peanut
• Egg	• Tree nuts
• Wheat	• Fish
• Soy	• Shellfish
• Peanut	

the American College of Allergy, Asthma & Immunology (ACAAI), and the Joint Council of Allergy, Asthma & Immunology (JCAAI) (Sampson et al. 2014); World Allergy Organization (WAO) (Johansson et al. 2001, 2004); European Academy of Allergy and Clinical Immunology (EAACI) (Muraro et al. 2014a); the International Consensus (ICON) on food allergy (Burks et al. 2012); and rigorously conducted systematic reviews of the evidence and relevant population-based studies, we present an overview of the epidemiology, diagnosis, prevention, and management of food allergy in children and adults and discuss future directions in research and clinical practice. We focus primarily on IgE-mediated food allergies because their epidemiology, natural history, diagnosis, and management are more clearly understood than those of non-IgE-mediated and mixed IgE- and non-IgE-mediated food allergies.

6.2 EPIDEMIOLOGY OF FOOD ALLERGY IN CHILDREN AND ADULTS

For epidemiologic purposes, the ultimate goal of food allergy diagnosis/assessment is to derive estimates to describe and understand the burden of food allergy in the population, including estimates of the incidence, prevalence, and quality of life associated with food allergy. For these purposes, various approaches have been employed to measure food allergy, including clinical history, serum allergen sensitization, and oral food challenge (OFC); patient-reported symptoms based on questionnaires are a common approach used and these are of interest in epidemiologic studies (Chafen et al. 2010; Nwaru et al. 2014a–c; Rona et al. 2007; Zuidmeer et al. 2008). This provides the basis to gain a comprehensive picture of the various scenarios of the burden of food allergy, both from the patient perspective and by using more objective clinical judgment and confirmatory tests; this then will allow comparing and contrasting of the estimates provided by the different assessment methods (Leung et al. 2014). Previous systematic reviews on the epidemiology of food allergy have demonstrated that the prevalence estimates as provided by patient symptom report, serum IgE sensitization, and OFC vary between these assessment approaches, usually with self-report having the highest prevalence, followed by IgE sensitization and OFC having the least (Chafen et al. 2010; Nwaru et al. 2014a,b; Rona et al. 2007; Zuidmeer et al. 2008;). Understandably, estimates provided by self-reported questionnaires are highest because they will usually include all adverse reactions to foods, including nonallergic food hypersensitivity, non-IgE-mediated food allergy, and IgE-mediated food allergy, or a mixture of these, and are influenced by local culture, language, knowledge, and beliefs (Kung et al. 2014; Lee et al. 2013; Sicherer and Sampson 2010). In contrast, both IgE sensitization and OFC include only IgE-mediated food allergy and hence have lower prevalence of food allergy. Going forward, researchers need to be explicit on the assessment approaches employed and, where possible, present estimates of the epidemiological burden of food allergy using various approaches (Nwaru et al. 2014c).

6.2.1 INCIDENCE

There is a paucity of data on the actual incidence of food allergy across populations (Nwaru et al. 2014a). Most of the available incidence estimates describe the cumulative incidence of food allergy, primarily among children (Nwaru et al. 2014a). A recent systematic review of the epidemiology of food allergy in Europe reported a cumulative incidence of 6% based on clinical history or positivity to OFC and 5% based on clinical history or positivity to DBPCFC by the age of 3; for OFC-proven food allergy alone, the cumulative incidence was 1.5% by 1 year, 3.3% by 4 years, and 3.6% by 6 years (Nwaru et al. 2014a). These estimates are however insufficient to gain a comprehensive picture of the incidence of the new onset food allergy, because while they may be representative of the European region, they apply only to the pediatric population. Although most new onset food allergies, particularly cow's milk, egg, wheat, soy, and peanut, occur during infancy and early childhood, there are however indications that new onset food allergies, particularly tree nuts, fish, and shellfish, are manifesting in adulthood, and the frequency of these may be increasing (Chaudhry and

Oppenheimer 2012; Jansen et al. 1994; Kagan 2003; Kivity 2012; Nwaru et al. 2014c). However, given a paucity of longitudinal data to map out the developmental trajectories of food allergy from infancy to adulthood, we currently do not know what proportion of the food allergy manifesting in adulthood reflects (1) persistence of childhood allergy, (2) actual new onset disease in adulthood, or (3) a combination of both (Nwaru et al. 2014c).

6.2.2 Prevalence

Global estimates of the prevalence of food allergy are lacking. The frequency of food allergy varies by age, local diet, and beliefs, and, in addition to these, differences in definitions and assessment methods make it challenging to derive overall global estimates of the prevalence of food allergies (Boyce et al. 2010; Kung et al. 2014; Lee et al. 2013; Nwaru et al. 2014a; Rona et al. 2007; Sampson 2005; Sampson et al. 2014; Sicherer and Sampson 2014). A recent survey by the WAO to examine the global patterns of the burden of food allergy in children in 89 countries concluded that whereas the prevalence of OFC-proven food allergy can be estimated as high as 10% in developed countries and around 7% in developing countries, lack of comparable quality data on food allergy across regions limited the derivation of an overall estimate of the prevalence of food allergy (Prescott et al. 2013). In previous systematic reviews of population-based studies with no limitation to any particular world region (Chafen et al. 2010; Rona et al. 2007; Zuidmeer et al. 2008), pooled prevalence estimates of self-report of any food allergy were around 12% for children (0–16 years) and around 13% for adults (>16 years) (Sicherer and Sampson 2010). For specific food allergies, there were wide variations across studies included in the reviews: among children, prevalence of self-reported food allergy ranged from 1.2% to 17% for milk allergy, 0.2% to 7% for egg allergy, 0% to 2% for peanut and fish allergy, and 0% to 10% for shellfish allergy (Rona et al. 2007). The corresponding estimates for measurements based on serum IgE sensitization, skin prick test (SPT), or OFC were between 0% and 5%. In general, these prevalence estimates were higher in children than in adults, except for fish, shellfish, and tree nut allergies; but only few of the studies included were undertaken in adults (Chafen et al. 2010; Rona et al. 2007; Zuidmeer et al. 2008). While these estimates may represent the most comprehensive available global prevalence of food allergy, all studies included in the reviews however came from Western countries, that is, North America, Europe, and Australia.

Although the number of studies undertaken in Africa to estimate the prevalence and overall burden of food allergy has grown over the years, findings from a recent review indicate that available studies from the region are however limited to provide reliable estimates, considering that a majority of the studies were out in selected populations (e.g., asthma population or those with other atopic diseases) and hence do not provide a clear picture of the population-based prevalence of food allergy (Kung et al. 2014). Interestingly, however, the studies highlight that, apart from the common food-causing allergy in Western countries, other novel food allergens are now being observed and are on the increase in the African region, including pineapple in Ghana, okra in Nigeria, and mopane worm in Botswana (kung et al. 2014). Reliable prevalence data are also lacking in Central and South America and the Middle East (Prescott et al. 2013).

In Asia, substantial numbers of studies have been carried out, providing somewhat reliable estimates of the prevalence of food allergy, mostly in children (Lee et al. 2013). The prevalence of any self-reported food allergy is estimated to range between 3% and 11%, whereas OFC-proven food allergy ranges from 1% to around 7% (Lee et al. 2013). Shellfish is the most common food allergen in Asia with an estimated prevalence of about 5% in children, whereas peanut is very uncommon with less than 1% prevalence based on positive OFC. The prevalence of egg allergy in children is up to 4%, cow's milk allergy is less than 4%, and fish allergy stands at about 2%. The prevalence of food allergy in Australia is one of the highest in the world, with an estimated prevalence of 10% of infants being positive for OFC-proven positivity to any food allergen, 9% positive for egg allergy, 3% for peanut allergy, and 1% for sesame allergy (Osborne et al. 2011).

A recent systematic review has provided the most contemporaneous and reliable estimates of food allergy by age groups and geographical regions in Europe, although a majority of the studies were undertaken in children (see summary estimates from the study in Table 6.1; Nwaru et al. 2014a,b). The pooled prevalence of any self-reported food allergy was 5.9% in all age groups (children 6.9%, adults 5.1%); for specific food allergens, the prevalence in all age groups ranged from 0.6% for fish to 2.3% for cow's milk (from <1% for shell fish to 3.8% for cow's milk in children; from 0.8% for soy to 6.7% for peanut allergy in adults). For food allergy assessed by convincing clinical history, the prevalence was 2.6%, but this was only available in children. The prevalence of any OFC-proven food allergy in all age groups was 0.9% (children 1%, adults 0.9%); for specific food allergens, the prevalence was <1% in all age groups as well as in children and adults. While few studies have been undertaken in Eastern and Southern Europe, compared to studies undertaken in Northern and Western Europe, the prevalence of food allergy varied across the different European regions, with Northern Europe generally having the highest prevalence estimates (Nwaru et al. 2014a,b).

The prevalence of food allergy widely varies geographically and racially across the United States (Gupta et al. 2011, 2012; Lau et al. 2012; Liu et al. 2010; Warren et al. 2013); however, based on estimates from nationally representative surveys, the prevalence of clinical food allergy (based on a combination of clinical criteria and IgE sensitization) in all age groups is estimated to be 2.5% (1.3% for peanut, 0.4% milk, and 0.2% egg; Liu et al. 2010). In other nationally representative samples, the prevalence of self-reported food allergy among children has been estimated at 8% for any food allergy, 2% peanut, 1.7% milk, 1.4% shellfish, 1% tree nuts, 0.8% egg, 0.5% fish, and 0.4% wheat and soy (Gupta et al. 2011, 2012; Lau et al. 2012; Warren et al. 2013). Among adults, the prevalence of any self-reported food allergy is estimated at 0.7%, shellfish 0.4%, peanut 0.2%, tree nuts 0.1%, and fish 0.05% (Agarwal and Wang 2014). Overall, there is a paucity of data on the prevalence estimates of food allergy in adults in the United States.

6.2.3 Time Trends

The incidence, prevalence, and overall burden of food allergy are seen to have dramatically increased during the last 10–15 years (Allen and Koplin 2012; Nwaru et al. 2014c; Prescott and Allen 2011). While anecdotal evidence points to an increase, there has been a paucity of reliable longitudinal data to clearly map out the time trends of the burden of food allergy over the years (Nwaru et al. 2014c). With regard to the incidence of food allergy, the recent systematic review from Europe, based on a study from the United Kingdom, showed that the incidence of peanut allergy remained broadly stable between 2001 and 2005 (Kotz et al. 2011). For the same period however, the prevalence of physician-diagnosed peanut allergy doubled (Kotz et al. 2011), which could probably be as a result of less spontaneous resolution of peanut allergy and/or as a result of better recognition or record of diagnosis. Similar observation was made with regard to admission rates for any food allergy between 1990 and 2004 (Nwaru et al. 2014a), probably due to changes in disease severity or changes in clinical practice such as decrease in the threshold for admission as a result of greater public and professional understanding of the risk associated with food allergy. A recent study from Sweden also found an indication of increasing trend in pediatric emergency visits as a result of food allergy (primarily due to peanut and tree nuts) between 2001 and 2010 (Johnson et al. 2014).

Two studies from Asia evaluated time trends in the prevalence of OFC-proven food allergy between 1999 and 2009 (China) and self-reported food allergy between 1995 and 2000 (South Korea) in children: while OFC-proven food allergy doubled from 3.5% in 1999 to 7.7% in 2009, self-reported food allergy remained fairly stable between 1995 and 2000 (10.9% to 8.9% in 6–12-year-olds and 11.3% to 12.6% in 12–15-year-olds) (Hu et al. 2010; Kung et al. 2014). Mullins used changes in demand for a specialist for food allergy services as a proxy to evaluate changes in the prevalence of food allergy in Australia between 1995 and 2006 among 0–5-year-old children and reported a 12-fold increase during this period (Mullins 2007). Finally, in the United States, in a nationally

TABLE 6.1

Point Prevalence of Food Allergy in Children and Adults in Europe Based on Patient-Reported, Clinical History, and OFC/DBPCFC-Confirmed Estimates

	Self-Report % (95% CI)		Convincing Clinical History % (95% CI)		OFC/DBPCFC[a] Positivity % (95% CI)	
Type of Food Allergy	**Children (≤17 Years)[b]**	**Adults (18+ Years)**	**Children (≤17 Years)[b]**	**Adults (18+ Years)**	**Children (≤17 Years)[b]**	**Adults (18+ Years)**
Any food	6.86 (6.58–7.15)	5.05 (4.82–5.28)	2.60 (2.08–3.12)	NA[c]	0.99 (0.80–1.18)	0.89 (0.75–1.04)
Cow's milk	3.75 (3.32–4.19)	2.14 (1.47–2.81)	1.58 (1.21–1.90)	NA[c]	2.00 (1.48–2.52)	0.21 (0.06–0.47)
Egg	2.46 (2.07–2.84)	1.09 (0.64–1.54)	1.03 (0.75–1.32)	NA[c]	0.75 (0.46–1.04)	0.09 (0.04–0.14)
Wheat	2.67 (2.08–3.25)	3.15 (2.16–4.14)	0.30 (0.02–0.59)	NA[c]	0.16 (0.03–0.36)	0.00 (0.00–0.40)
Soy	1.30 (0.80–2.00)	0.60 (0.30–1.40)	NA[c]	NA[c]	0.50 (0.20–1.40)	0.00 (0.00–0.40)
Peanut	1.97 (1.75–2.18)	6.73 (5.30–8.15)	1.55 (1.18–1.93)	NA[c]	0.57 (0.23–0.91)	0.06 (0.00–0.13)
Tree nuts	1.81 (1.60–2.01)	2.50 (1.97–3.03)	0.05 (0.00–0.18)	NA[b]	0.45 (0.08–0.81)	0.45 (0.08–0.81)
Fish	1.10 (0.80–1.50)	0.95 (0.46–1.43)	0.05 (0.00–0.23)	NA[c]	0.12 (0.00–0.32)	0.15 (0.10–0.40)
Shellfish	0.68 (0.53–0.83)	1.34 (0.99–1.69)	NA[c]	NA[c]	0.12 (0.06–0.29)	NA[c]

Sources: Nwaru, B.I. et al., *Allergy*, 69, 62, 2014a; Nwaru, B.I. et al., *Allergy*, 69, 992, 2014b.

[a] OFC, open food challenge; DBPCFC, double-blind, placebo-controlled food challenge.

[b] For the specific food allergens, estimates for children were presented in age groups of 0–1, 2–5, and 6–17 years; estimates presented in this table are the highest estimates reported for any of these age groups.

[c] NA, no available data.

representative sample of children, as reported by parents or guardians, there was an 18% increase in the prevalence of self-reported food allergy between 1997 and 2007, tripling in ambulatory care visits, and a 72% increase in hospital discharges for food allergy (Branum and Lukacs 2009).

6.2.4 Natural History

The natural course of food allergy is now well characterized for many foods, and this has shown that it varies with the type of food, age, pathophysiology of the food allergen, and type of resultant disorder; for example, it is known that food allergy–related eosinophilic esophagitis shows relatively poor chance of resolution (Boyce et al. 2010; Burks et al. 2012; Longo et al. 2013; Muraro et al. 2014a; Sampson et al. 2014; Sicherer and Sampson 2014). Overall, resolution of food allergy occurs more quickly in children than in adults for most foods (Boyce et al. 2010; Burks et al. 2012; Longo et al. 2013; Muraro et al. 2014a; Sampson et al. 2014; Sicherer and Sampson 2014). The spontaneous resolution for cow's milk, egg, wheat, and soy tends to follow similar patterns: in general, at least 60% of children tolerate these foods by the age of 10 (Boyce et al. 2010; Burks et al. 2012; Longo et al. 2013; Muraro et al. 2014a; Sampson et al. 2014; Sicherer and Sampson 2014). Peanut, tree nuts, fish, and shellfish are typically considered persistent allergies, with a high risk of anaphylaxis and an average of about 20% tolerance by teenage years (Boyce et al. 2010; Burks et al. 2012; Longo et al. 2013; Muraro et al. 2014a; Sampson et al. 2014; Sicherer and Sampson 2014). A possible marker of tolerance to a particular food is usually seen by a reduction in serum-specific IgE (sIgE) levels or the size of the SPT wheal over time (Boyce et al. 2010; Burks et al. 2012; Longo et al. 2013; Muraro et al. 2014a; Sampson et al. 2014; Sicherer and Sampson 2014). However, some food allergies that occur during adulthood may persist regardless of a reduction in sIgE levels over time (Boyce et al. 2010; Burks et al. 2012; Longo et al. 2013; Muraro et al. 2014a; Sampson et al. 2014; Sicherer and Sampson 2014).

6.2.5 Risk Factors

Risk factors identified for the development of food allergy are both genetic and environmental, although findings regarding several environmental factors have generally been contradictory (Burks et al. 2012; Lack 2012; Longo et al. 2013). There is evidence that a child with parents or siblings with peanut allergy has up to sevenfold increased risk of developing peanut allergy, whereas for monozygotic twins the risk of developing peanut allergy increases to 64% (Hourihane et al. 1996; Lack 2012; Sicherer et al. 2000). A family history of other atopic diseases or the presence of other atopic diseases, in particular atopic eczema, asthma, and allergic rhinitis, is a risk factor for the developing food allergy (Dhami and Sheikh 2015; Koplin et al. 2012; Lack 2012; Nwaru et al. 2014a). The propensity to developing food allergy is associated with age, the risk being higher during childhood than in later life, particularly for cow's milk, egg, wheat, and soy allergies, but for peanut, tree nuts, fish, and shellfish, the risk appears to be somewhat associated with increasing age (Nwaru et al. 2014a). Sex has been defined as an important risk factor for the development of food allergy: while boys are at a greater risk than girls during childhood, a reversal of risk is seen during adulthood so that women are at a greater risk than men (Koplin et al. 2012; Lack 2012; Nwaru et al. 2014a). Ethnicity has also been consistently noted as an important risk factor for food allergy, particularly in the United States, with non-Hispanic black populations at a greater risk than white populations (Lack 2012; Liu et al. 2010; Warren et al. 2013). Geographical residence and affluence are also key risk factors for food allergy: a higher risk has been noted in Westernized or more affluent settings than in non-Westernized or less affluent settings; a higher risk has also been reported in urban compared to rural dwellings (Gupta et al. 2011, 2012; Lack 2012; Liu et al. 2010; Warren et al. 2013).

Changes in dietary intake have been proposed to be associated with the increasing pattern of food allergy, and, in this regard, key dietary factors believed to play a role have included vitamin D, dietary fat, antioxidants, and the Mediterranean dietary pattern (Anandan et al. 2009; Julia et al. 2015;

Lack 2012; Nurmatov et al. 2011). Definite clinical trials are now underway to confirm the putative effect of each of these dietary factors, and it is anticipated that the results from the ongoing studies will clarify our understanding of their role in the development of food allergy (Clark et al. 2012; Litonjua et al. 2014; Meldrum et al. 2011; Sewell et al. 2013). The role of breastfeeding and infant formulas in the development of food allergy has remained controversial (Agostoni et al. 2008; Fiocchi et al. 2006; Greer et al. 2008; Prescott and Tang 2005); the ethical dilemma in randomizing breastfeeding has precluded the evaluation of its effectiveness through formal randomized clinical trials (Greer et al. 2008). For the timing of introduction of complementary foods, much of the emerging evidence from observational studies indicates that early introduction of foods is associated with a decreased risk of developing food allergy; few recent studies also indicate that increased dietary diversity during infancy may decrease the risk of food allergy in children (Agostoni et al. 2008; Fiocchi et al. 2006; Greer et al. 2008; Nwaru et al. 2010, 2013; Poole et al. 2006; Prescott and Tang 2005; Roduit et al. 2014; Zutavern et al. 2008). Specific to peanut allergy, evidence from a recent randomized controlled clinical trial (RCT) shows that early introduction of peanut compared to avoidance of peanut led to upto 80% relative reduction in the incidence of OFC-proven peanut allergy in infants with high risk of peanut allergy (Du Toit et al. 2015); this finding has now led to an interim guidance of early introduction of peanut for infants at a high risk of peanut allergy by the leading expert groups (Fleischer et al. 2015). Evidence for other risk factors, including cesarean section delivery, maternal smoking, socioeconomic status, antibiotics during early life, and obesity, is unclear.

6.3 PREVENTION OF FOOD ALLERGY

Strategies for the prevention of food allergy emphasize primarily on modifying early environmental exposure to potential allergens (Boyce et al. 2010; Burks et al. 2012; De Silva et al. 2014a; Muraro et al. 2014a,b; Sampson et al. 2014). While most of the existing strategies have focused on dietary exposure prenatally, during breastfeeding and infancy, very little is known about food allergy prevention during childhood and adulthood, although bacillus Calmette–Guérin (BCG) vaccinations (Linehan et al. 2014), fish oil (Anandan et al. 2009) and vitamin (Mamsjo et al. 2009) supplements have been evaluated in these later age groups, but no real benefits have been found for them (De Silva et al. 2014a; Muraro et al. 2014b). Prenatally and during lactation, key strategies have concerned dietary restrictions and use of supplements, whereas during infancy, in addition to these, strategies have included duration of breastfeeding, use of hydrolyzed infant formulas, and timing of introduction of solid foods (Arshad et al. 1992; Boyce et al. 2010; Burks et al. 2012; De Silva et al. 2014a; Muraro et al. 2014a,b; Sampson et al. 2014). There is no strong evidence supporting the effectiveness of dietary restrictions for primary prevention of food allergy, and the current recommendation is that normal diet without any restrictions should be eaten during pregnancy and lactation by mothers of high-risk children (Arshad et al. 1992; Boyce et al. 2010; Burks et al. 2012; De Silva et al. 2014a; McLean and Sheikh 2010; Muraro et al. 2014b; Sampson et al. 2014). Similarly, there is currently no strong evidence to support the use of supplements, particularly, pre- and probiotics, for food allergy prevention (De Silva et al. 2014a; Muraro et al. 2014b).

Although the benefits of breastfeeding have been widely articulated for many generations, evidence is weak regarding its role in primary prevention of food allergy; however, current recommendation remains that high-risk children should be exclusively breastfed for at least 4–6 months of life (Agostoni et al. 2008; De Silva et al. 2014a; Fiocchi et al. 2006; Greer et al. 2008; Muraro et al. 2014b; Prescott and Tang 2005). For babies who are unable to be breastfed or the breast milk is insufficient, hypoallergenic formulas, particularly extensively hydrolyzed whey or casein formulas or partially hydrolyzed infant formulas, are recommended to be beneficial for the first 4 months of life (De Silva et al. 2014a; Muraro et al. 2014b). Evidence supporting the use of soy-based formulas is lacking, and hence, their use is not recommended for primary prevention of food allergy (De Silva et al. 2014a; Muraro et al. 2014b). During infancy, it has been thought that delaying the introduction of complementary foods for the first 4–6 months is beneficial (Agostoni et al. 2008; Fiocchi et al. 2006; Greer et al. 2008; Prescott and Tang 2005), but evidence from several observational studies (Nwaru et al. 2010, 2013; Poole et al. 2006; Roduit et al. 2014; Zutavern et al. 2008), which has now been confirmed in an RCT (Du Toit et al. 2015) regarding peanut allergy, shows that early complementary feeding for high-risk children may confer some benefits of reduced risk of food allergy. While these preventive strategies have been demonstrated in high-risk populations, their role in normal-risk populations has not shown any replicated benefits (De Silva et al. 2014a; Muraro et al. 2014b).

6.4 DIAGNOSIS OF FOOD ALLERGY

For both children and adults, the clinical diagnosis of food allergy is based on a careful and detailed medical history and a focused physical examination (Boyce et al. 2010; Burks et al. 2012; Longo et al. 2013; Muraro et al. 2014a; Sampson et al. 2010, 2014; Sicherer and Sampson 2010, 2014). This approach cannot however completely distinguish between IgE-mediated and non-IgE-mediated food allergy and thus provides only a preliminary basis for identifying the type of food-induced allergic reaction and the potential food triggering such reaction (Boyce et al. 2010; Burks et al. 2012; Longo et al. 2013; Muraro et al. 2014a; Sampson et al. 2014; Sicherer and Sampson 2010, 2014). A detailed medical history should therefore determine the possible causal food, the form it was consumed, the quantity consumed, time course of reaction, nature of reaction, and cofactors (Boyce et al. 2010; Burks et al. 2012; Longo et al. 2013; Muraro et al. 2014a; Sampson et al. 2014; Sicherer and Sampson 2010, 2014). The aspect of physical examination that accompanies a medical history aims to identify signs in the individual presenting with symptoms that are consistent with an allergic reaction (Boyce et al. 2010; Burks et al. 2012; Longo et al. 2013; Muraro et al. 2014a; Sampson et al. 2014; Sicherer and Sampson 2010, 2014). As medical history and physical examination lack sufficient sensitivity and specificity to make a definite diagnosis of IgE-mediated food allergy, further laboratory evaluations are usually required in order to identify the presence of allergen-specific IgE (Boyce et al. 2010; Burks et al. 2012; Longo et al. 2013; Muraro et al. 2014a; Sampson et al. 2014; Sicherer and Sampson 2010, 2014). For this purpose, the SPT and immunoassays of serum sIgE levels are used and these have been helpful in identifying the most common allergens (Boyce et al. 2010; Burks et al. 2012; Longo et al. 2013; Muraro et al. 2014a; Sampson et al. 2014; Sicherer and Sampson 2010, 2014). The size of the SPT wheal and increased concentration of serum sIgE provide some indications of the likelihood of clinical reaction and can demonstrate the presence of specific foods that may be provoking IgE-mediated food-induced allergic reactions (Boyce et al. 2010; Burks et al. 2012; Longo et al. 2013; Muraro et al. 2014a; Sampson et al. 2014; Sicherer and Sampson 2010, 2014). However, neither the size of the SPT wheal nor the concentration of serum sIgE alone can be considered sufficiently diagnostic of food allergy as neither correlates with nor is predictive of the severity of the allergic reaction to food; thus, they must be combined with medical history when ascertaining IgE-mediated food allergy (Boyce et al. 2010; Burks et al. 2012; Longo et al. 2013; Muraro et al. 2014a; Sampson et al. 2014; Sicherer and Sampson 2010, 2014). While both wheal size and serum concentration of sIgE may vary by age, SPT wheal also varies by diurnal variation, site of the body, skin reactivity, and the device and SPT reagents used (Boyce et al. 2010; Burks et al. 2012; Longo et al. 2013; Muraro et al. 2014a; Sampson et al. 2014; Sicherer and Sampson 2010, 2014).

In cases where there is discordance between the medical history and the results of allergen-specific SPT and sIgE tests, an OFC is indicated to help clarify the diagnosis (Boyce et al. 2010; Burks et al. 2012; Longo et al. 2013; Muraro et al. 2014a; Sampson et al. 2014; Sicherer and Sampson 2010, 2014). OFC can be undertaken as either an open or a blinded procedure; in many situations, open OFC with an objective unequivocal reaction is sufficient to diagnose food allergy; if however the results are equivocal, then blinded OFC should be performed (Boyce et al. 2010; Burks et al. 2012; Longo et al. 2013; Muraro et al. 2014a; Sampson et al. 2014; Sicherer and Sampson 2010, 2014).

Blinded OFC may be single blinded or double blinded. If either the open OFC or single-blind OFC is negative, food allergy can be ruled out, but if either elicits symptoms that correlate with medical history and is supported by laboratory tests, then the diagnosis of food allergy is confirmed (Boyce et al. 2010; Burks et al. 2012; Longo et al. 2013; Muraro et al. 2014a; Sampson et al. 2014; Sicherer and Sampson 2010, 2014).

The DBPCFC is the most specific test and the gold standard for diagnosing food allergy (Boyce et al. 2010; Burks et al. 2012; Longo et al. 2013; Muraro et al. 2014a; Sampson et al. 2014; Sicherer and Sampson 2010, 2014). However, because of the costs in administering it, inconvenience, time, labor-intensiveness, safety concerns (particularly in the very young), and an associated risk of severe reactions and anaphylaxis, DBPCFC is less often used and hence the open or single-blind OFCs are more commonly performed in clinical practice (Boyce et al. 2010; Burks et al. 2012; Lack 2012; Longo et al. 2013; Muraro et al. 2014a; Sampson et al. 2014; Sicherer and Sampson 2010, 2014). DBPCFC needs to be performed in a health facility with on-site medical supervision and the resources to monitor and manage any resulting acute reactions including anaphylaxis (Boyce et al. 2010; Burks et al. 2012; Longo et al. 2013; Muraro et al. 2014a; Sampson et al. 2014; Sicherer and Sampson 2010, 2014). Guidelines are available on the procedures of performing DBPCFC, including accurate selection of patients, safety criteria, type and quantity of food allergens to administer, timings and doses, periods of observation, and other criteria (Boyce et al. 2010; Burks et al. 2012; Longo et al. 2013; Muraro et al. 2014a; Perry et al. 2004; Sampson et al. 2012, 2014; Sicherer and Sampson 2010, 2014). There is now growing interest in the use of the basophil activation test (BAT) technique to diagnose the presence of food allergy, particularly among those with food allergy sensitization despite the absence of any oral exposure to food allergens (Santos et al. 2014, 2015a,b). BAT is now increasingly being used and aimed to overcome the challenges of safety concerns and costs of employing the DBPCFC (Santos et al. 2014, 2015a,b).

An evolving and promising approach to the diagnosis of food allergy relates to the various developments in molecular technologies that hopefully will improve the diagnosis of IgE-mediated food allergy (Ballmer-Weber and Hoffmann-Sommergruber 2011; Bindsley-Jensen et al. 2004; Canonica et al. 2013; Caubet et al. 2011; Eller and Bindsley-Jensen 2013; Fiocchi et al. 2011; Glaumann et al. 2012; Muraro et al. 2014a; Nicolau and Custovic 2011). In this context, IgE responses to complex allergen extracts are not the focus, instead specific IgE responses are evaluated against individual allergenic molecules (using component-resolved diagnosis) or the epitopes of those allergens (using epitope mapping or profiling; Ballmer-Weber and Hoffmann-Sommergruber 2011; Bindsley-Jensen et al. 2004; Canonica et al. 2013; Caubet et al. 2011; Eller and Bindsley-Jensen 2013; Fiocchi et al. 2011; Glaumann et al. 2012; Muraro et al. 2014a; Nicolau and Custovic 2011). The molecular diagnostic tools may have the potential to enhance the specificity and sensitivity of serum IgE responses of selected foods, facilitate the determination of food allergy phenotypes, enhance the opportunity to compile a patient-tailored risk profile to specific food allergies, be able to distinguish between primary and secondary sensitizers, and be undertaken either in single test formats or in a microarray by simultaneously evaluating a range of allergens (Ballmer-Weber and Hoffmann-Sommergruber 2011; Bindsley-Jensen et al. 2004; Canonica et al. 2013; Caubet et al. 2011; Eller and Bindsley-Jensen 2013; Fiocchi et al. 2011; Glaumann et al. 2012; Muraro et al. 2014a; Nicolau and Custovic 2011). Currently, the major allergens in different types of food allergies have now been determined and the results of their sensitivity and specificity are judged to be excellent (Ballmer-Weber and Hoffmann-Sommergruber 2011; Bindsley-Jensen et al. 2004; Canonica et al. 2013; Caubet et al. 2011; Eller and Bindsley-Jensen 2013; Fiocchi et al. 2011; Glaumann et al. 2012; Muraro et al. 2014a; Nicolau and Custovic 2011); these include, for example, Ara h 2 for peanut allergy (Eller and Bindsley-Jensen 2013; Glaumann et al. 2012; Muraro et al. 2014a; Nicolau and Custovic 2011), omega-5-gliadin for wheat allergy (Morita et al. 2009; Muraro et al. 2014a), rGly m 4 for soy milk allergy (Fiocchi et al. 2011; Muraro et al. 2014a), and allergies of a number of nuts, tree nuts, fruits, vegetables, and fish (Ballmer-Weber and Hoffmann-Sommergruber 2011; Muraro et al. 2014a). However, well-designed RCTs are being awaited to confirm these diagnostic successes

(Muraro et al. 2014a). The costs of undertaking these approaches and the need for experience in interpreting their findings may limit the wider application. Furthermore, as these molecular tests focus only on allergen sensitization, which in themselves do not imply clinical responsiveness, they need to be evaluated alongside clinical history and physical examination of each patient in order to confirm food allergy positivity (Ballmer-Weber and Hoffmann-Sommergruber 2011; Canonica et al. 2013; Caubet et al. 2011; Eller and Bindsley-Jensen 2013; Fiocchi et al. 2011; Glaumann et al. 2012; Morita et al. 2009; Nicolau and Custovic 2011).

6.5 MANAGEMENT OF FOOD ALLERGY

The management of food allergy involves the following two aspects: (1) management of acute reactions and (2) long-term management (Boyce et al. 2010; Burks et al. 2012; De Silva et al. 2014b; Muraro et al. 2014a; Sampson et al. 2014; Sheikh et al. 2011). Of particular concern is that reactions to food allergens may lead to anaphylaxis, a potentially life-threatening systemic allergic disorder. While risk may vary across subgroups of patients, careful and immediate assessment and management of risk of severe reactions are crucial (Boyce et al. 2010; Burks et al. 2012; De Silva et al. 2014b; Muraro et al. 2014a; Sampson et al. 2014; Sheikh et al. 2011). The first-line treatment for acute management of food allergy is the use of epinephrine (Boyce et al. 2010; Burks et al. 2012; De Silva et al. 2014b; Muraro et al. 2014a; Sampson et al. 2014). However, other pharmacological treatments, known as second- or third-line treatments, are commonly used and are available for the management of acute reactions from food allergy. Among these treatments, H1-antihistamines are the most commonly used class of drugs and are believed to confer some benefits, but evidence supporting their use is currently weak (Boyce et al. 2010; Burks et al. 2012; De Silva et al. 2014b; Muraro et al. 2014a; Sampson et al. 2014; Sheikh et al. 2011). Prophylactic use of antihistamines may mask symptoms of anaphylaxis, thereby causing delay in treatment with adrenaline (epinephrine) (Boyce et al. 2010; Burks et al. 2012; De Silva et al. 2014b; Muraro et al. 2014a; Sampson et al. 2014; Sheikh et al. 2011). Although other medications are sometimes used in the management of acute food allergy reactions, there is no robust evidence to demonstrate their benefits (Boyce et al. 2010; Burks et al. 2012; De Silva et al. 2014b; Muraro et al. 2014a; Sampson et al. 2014; Sheikh et al. 2011).

Long-term management strategies for food allergy focus on preventing any recurrence or worsening of symptoms, and, along this line, allergen avoidance, preparation for prompt treatment of reactions, and increasing patient education have been emphasized (Boyce et al. 2010; Burks et al. 2012; De Silva et al. 2014b; Muraro et al. 2014a; Sampson et al. 2014; Sheikh et al. 2011). Dietary avoidance of potential offending foods is key in managing food allergy and aims at eliminating the specific culprit food allergens and requires careful monitoring and regular dietetic reviews to avoid negative consequences that may impair on the quality of life of patients (Boyce et al. 2010; Burks et al. 2012; De Silva et al. 2014b; Muraro et al. 2014a; Sampson et al. 2014; Sheikh et al. 2011). Education is said to be "the key pillar" of effective long-term management of food allergy (Muraro et al. 2014a). Patients, their caregivers, relatives, and friends should therefore be educated about avoidance strategies, including risky situations; cross-contact with other allergens; food labeling; food choice strategies when traveling, in a restaurant, and at school; and the relevant legislation covering food labeling and manufacturing (Boyce et al. 2010; Burks et al. 2012; Muraro et al. 2014a; Sampson et al. 2014; Sheikh et al. 2011). Education should also include symptom recognition, indication for specific treatment, and self-administration of recommended doses of treatment (e.g., epinephrine) (Boyce et al. 2010; Burks et al. 2012; Muraro et al. 2014a; Sampson et al. 2014). Short-acting beta agonists are recommended to be included in the management plan of patients with coexistence of asthma (Muraro et al. 2014a). The evidence is weak regarding the use of pharmacological treatments for the long-term management of food allergy: evidence for prophylactic use of mast cell stabilizers is mixed; as well insufficient evidence on the role of H1-antihistamines (De Silva et al. 2014b; Muraro et al. 2014a). The role of allergen-specific immunotherapy has been emphasized and appears promising for the long-term management of some types of food allergies

(Boyce et al. 2010; Burks et al. 2012; De Silva et al. 2014b; Muraro et al. 2014a; Sampson et al. 2014; Scadding 2013); however, while its effectiveness has been confirmed for disorders such as allergic rhinitis and asthma, its role in food allergy has not been well defined but remains an area of vigorous research and debate (Calderon et al. 2007; Nurmatov et al. 2012, 2014; Romantsik et al. 2014; Jutel et al. 2015). For instance, issues regarding the routes of administration (subcutaneous vs. sublingual immunotherapy) are still being debated (De Silva et al. 2014b). Other areas of current controversy for allergen-specific immunotherapy for food allergy include alternative up-dosing and maintenance schedules; the role of adjuvant therapy in improving safety and its preparation; and the role of peptide, recombinant, and allergoid immunotherapy.

6.6 CONCLUSIONS AND FUTURE DIRECTIONS

IgE-mediated food allergy currently affects less than 10% of children and adults and up to 90% of the disorders are caused primarily by eight foods, including cow's milk, egg, wheat, soy, peanut, tree nuts, fish, and shellfish. The epidemiology, natural history, diagnosis, prevention, and management of food allergy are well defined in children, but there is lack of data to better understand these in the adult population. Furthermore, while the prevalence and overall burden of IgE-mediated food allergy are believed to have increased during the last 10–15 years, reliable estimates to clearly map out the disease burden have primarily come from Western countries and some parts of Asia, and, in most cases, existing data across settings are not well longitudinally constructed to allow a clear time-trend analysis of the burden of food allergy. There is currently a paucity of data from Africa, Central and South America, Eastern Europe, and Middle East. There is therefore the need to increase research and initiate major data informed policies and investments that will allow the derivation of population-based and longitudinal data that will allow better understanding of the burden of food allergy across settings and age groups. A starting point in achieving this will be to take advantage of the various extensive routinely collected healthcare data sets and registries now available in many countries, for example, the United Kingdom and Nordic countries. The emerging informatics capabilities across board will definitely play a key role here in advancing the probing of these data sets and uncovering the various nuances that will allow us to more reliably estimate the burden of food allergy. The ongoing EuroPrevall study has provided a good foundation for advancing these perspectives, showing that broad-based collaborative work is indispensable for success, involving epidemiologists, genetics, statisticians, health economists, healthcare services researchers, and data scientists (de Blok et al. 2007; Kummeling et al. 2009).

In addition to the lack of data from several world regions, one of the key challenges to derivation of global estimates of food allergy is the lack of standardized diagnosis and definition of food allergy. Currently, as recommended by expert groups, the diagnosis of food allergy is performed by (1) careful and detailed medical history and a focused physical examination, (2) allergen-specific SPT and immunoassays of serum IgE, and (3) use of OFC, the DBPCFC being the gold standard of food allergy diagnosis. Advancing this field of inquiry will require a standardized and globally harmonized definition and diagnosis criteria. In this respect, definitions and diagnosis criteria jointly formulated by expert bodies in the area should be more proactively adopted across settings. The recent developments in molecular approaches to the diagnosis of food allergy through the application of component-resolved diagnosis and epitope mapping appear to hold some promises for improvement in the diagnosis of food allergy, but their clinical responsiveness still needs to be evaluated before their wider application in clinical care. Nevertheless, these developments in molecular diagnostic techniques, in combination with ongoing developments in omics techniques (Di Girolamo et al. 2015; Nakamura and Teshima 2013), will remain crucial for future transformation of the field and will have important implications in relation to the effectiveness and safety of immunotherapy going forward as we will in effect be moving much closer to a precision management perspective (Morjaria et al. 2011; Wölbing and Biedemann 2013).

The recent evidence confirming that early introduction of peanut to the diet of infants at high risk of developing peanut allergy reduces the risk of peanut is promising for the field of allergy prevention and has already had immediate impact in influencing some preliminary changes in policies. While replication of these novel findings, having now been demonstrated in only one clinical trial of highly selected population, will strengthen the evidence base with regard to the prevention of peanut allergy, similar work also needs to be undertaken with regard to effective preventive strategies for other common allergens, in particular, egg, wheat, soy, tree nuts, fish, and shellfish. In addition, it is important that the long-term preventive effects of this strategy be carefully evaluated. Currently, most preventive strategies for food allergy have been evaluated among the infant population and very little is known during childhood and adulthood. The increasing onset of new food allergies during adulthood presents the impetus for catalyzing research that will also uncover evidence-based preventive strategies for young adults and adults.

More work is still needed to ascertain the putative effectiveness of various pharmacological treatments being used for the management of acute and long-term food allergies, for example, H1-antihistamines. Although it is promising and has been shown to be effective in the management of asthma and allergic rhinitis, there remain unresolved controversies on the role of allergen-specific immunotherapy in the long-term management of food allergy. Although it is pretty clear clinically that these will help to improve cutaneous manifestations of food allergy, they are however far less effective in dealing with respiratory and cardiovascular manifestations. More work is therefore needed in this front, particularly to ascertain the most appropriate routes of administration (subcutaneous vs. sublingual immunotherapy), alternative up-dosing and maintenance schedules, the role of adjuvant therapy in improving safety, and its preparation.

While education remains the key pillar for the long-term management of food allergy, it is known that this strategy has limited sustained impact on adolescent population, who, given their psychosocial transitions and environments, remain disproportionately at a greater risk of severe food allergy outcomes, including anaphylaxis and fatality (Akeson et al. 2007; Gallagher et al. 2011; Marrs and Lack 2013; Monks et al. 2010; Nwaru and Sheikh 2015; Sampson et al. 2006; Worth et al. 2013). Transition into adolescence is known to carry greater likelihood for risk-taking decisions that may affect proper care among adolescents with food allergy; consequently, current educational management strategies appear not comprehensive enough for this age group (Nwaru and Sheikh 2015). There is therefore an urgent need to further develop long-term food allergy management approaches that go beyond the current emphasis on education and that are more closely tailored toward adolescents, taking into cognizance their everyday developmental transition challenges (Nwaru and Sheikh 2015). A recent proposed tripartite management framework, developed mainly within the context of management of food-triggered anaphylaxis, is a further step in this regard, emphasizing consideration of the following three key perspectives: (1) challenges of adolescents' developmental transition, (2) shortcomings of current management strategies among adolescent population, and (3) drawing on adolescents' social network (Nwaru and Sheikh 2015).

Finally, we now have promising preventive and curative approaches in sight for food allergy; continuously streamlining these efforts alongside the opportunity to be able to stratify patients using CRD will enable us to continue with major breakthroughs in reducing the burden of food allergy.

REFERENCES

Agarwal S, Wang J. 2014. Prevalence and characteristics of food allergy in urban minority adults. *Ann Allergy Asthma Immunol* 112: 471–478.

Agostoni C, Decsi T, Fewtrell M. 2008. Complementary feeding: A commentary by the ESPGHAN Committee on Nutrition. *J Pediatr Gastroenterol Nutr* 46: 99–110.

Akeson N, Worth A, Sheikh A. 2007. The psychosocial impact of anaphylaxis on young people and their parents. *Clin Exp Allergy* 37: 1213–1220.

Allen JK, Koplin JJ. 2012. The epidemiology of IgE-mediated food allergy and anaphylaxis. *Immunol Allergy Clin N Am* 32: 35–50.

Anandan C, Nurmatov U, Sheikh A. 2009. Omega 3 and 6 oils for primary prevention of allergic disease: Systematic review and meta-analysis. *Allergy* 64: 840–848.

Arshad SH, Matthews S, Gant C, Hide DW. 1992. Effect of allergen avoidance on development of allergic disorders in infancy. *Lancet* 339: 1493–1497.

Ballmer-Weber BK, Hoffmann-Sommergruber K. 2011. Molecular diagnosis of fruit and vegetable allergy. *Curr Opin Allergy Clin Immunol* 11: 229–235.

Bindslev-Jensen C, Ballmer-Weber BK, Bengtsson U et al. 2004. Standardization of food challenges in patients with immediate reactions to foods—Position paper from the European Academy of Allergology and Clinical Immunology. *Allergy* 59: 690–697.

Branum AM, Lukacs SL. 2009. Food allergy among children in the United States. *Pediatrics* 124: 1549–1555.

Burks AW, Tang M, Sicherer S et al. 2012. ICON: Food allergy. *J Allergy Clin Immunol* 129: 906–920.

Calderon MA, Alves B, Jacobson M, Hurwitz B, Sheikh A, Durham S. 2007. Allergen injection immunotherapy for seasonal allergic rhinitis. *Cochrane Database Syst Rev* 1:CD001936.

Canonica GW, Ansotegui IJ, Pawankar R et al. 2013. A WAO-ARIA-GA2LEN consensus document on molecular-based allergy diagnostics. *World Allergy Organ J* 6: 17.

Caubet JC, Kondo Y, Urisu A, Nowak-Wegrzyn A. 2011. Molecular diagnosis of egg allergy. *Curr Opin Allergy Clin Immunol* 11: 210–215.

Chafen JJ, Newberry SJ, Riedl MA et al. 2010. Diagnosing and managing common food allergies: A systematic review. *JAMA* 303: 1848–1856.

Chaudhry RQ, Oppenheimer JJ. 2012. Update on food allergy in adults. *Curr Allergy Asthma Rep* 12: 311–320.

Clark J, Craig L, McNeill G. Smith N, Norrie J, Devereux G. 2012. A novel dietary intervention to optimize vitamin E intake of pregnant women to 15 mg/day. *J Acad Nutr Diet* 112: 297–301.

de Blok BM, Vlieg-Boerstra BJ, Oude Elberink JN et al. 2007. A framework for measuring the social impact of food allergy across Europe: A EuroPreval state of the art paper. *Allergy* 62: 733–737.

De Silva D, Geromi M, Halken S et al. 2014a. Primary prevention of food allergy in children and adults: Systematic review. *Allergy* 69: 581–589.

De Silva D, Geromi M, Panesar SS et al. 2014b. Acute and long-term management of food allergy: Systematic review. *Allergy* 69: 159–167.

Dhami S, Sheikh A. 2015. Estimating the prevalence of aero-allergy and/or food allergy in infants, children and young people with moderate-to-severe atopic eczema/dermatitis in primary care multi-centre, cross-sectional study. *J R Soc Med* 108: 229–236.

Di Girolamo F, Muraca M, Mazzina O, Lante I, Dahdah L. 2015. Proteomic applications in food allergy: Food allergenomics. *Curr Opin Allergy Clin Immunol* 15: 259–266.

Du Toit G, Roberts G, Sayre PH et al. 2015. Randomised trial of peanut consumption in infants at risk for peanut allergy. *N Engl J Med* 372: 803–813.

Eller E, Bindslev-Jensen C. 2013. Clinical value of component-resolved diagnostics in peanut-allergic patients. *Allergy* 68: 190–194.

Fiocchi A, Assa'ad A, Bahna S. 2006. Food allergy and the introduction of solid foods to infants: A consensus document. Adverse Reactions to Foods Committee; American College of Allergy, Asthma and Immunology. *Ann Allergy Asthma Immunol* 97: 10–20.

Fiocchi A, Bouygue GR, Albarini M, Restani P. 2011. Molecular diagnosis of cow's milk allergy. *Curr Opin Allergy Clin Immunol* 11: 216–221.

Fleischer DM, Sicherer S, Greenhawt M et al. 2015. Consensus communication on early peanut introduction and the prevention of peanut allergy in high-risk infants. *J Allergy Clin Immunol* 136: 258–261.

Gallagher M, Worth S, Cunningham-Burley S, Sheikh A. 2011. Epinephrine auto-injector use in adolescents at risk of anaphylaxis: A qualitative study in Scotland, UK. *Clin Exp Allergy* 41: 869–877.

Glaumann S, Nopp A, Johansson SG, Rudengren M, Borres MP, Nilsson C. 2012. Basophil allergen threshold sensitivity, CD-sens, IgE-sensitization and DBPCFC in peanut-sensitized children. *Allergy* 67: 242–247.

Greer FR, Sicherer SH, Burks AW; American Academy of Pediatrics, Committee on Nutrition and Section on Allergy and Immunology. 2008. Effects of early nutritional interventions on the development of atopic disease in infants and children: The role of maternal dietary restriction, breastfeeding, and timing of introduction of complementary foods, and hydrolyzed formulas. *Pediatrics* 121: 183–191.

Gupta RS, Springston EE, Smith B, Warrier MR, Pongracic J, Holl JL. 2012. Geographic variability of childhood food allergy in the United States. *Clin Pediatr* 51: 856–861.

Gupta RS, Springston EE, Warrier MR et al. 2011. The prevalence, severity, and distribution of childhood food allergy in the United States. *Pediatrics* 128: e9–e17.

Hourihane JO, Dean TP, Warner JO. 1996. Peanut allergy in relation to heredity, maternal diet, and other atopic diseases: Results of a questionnaire survey, skin prick testing, and food challenges. *BMJ* 313: 518–521.

Hu Y, Chen J, Li H. 2010. Comparison of food allergy prevalence among Chinese infants in Chongqing, 2009 versus 1999. *Pediatr Int* 52: 820–824.

Jansen JJ, Kardinaal AF, Huijbers G, Vlieg-Boerstra BJ, Martens BP, Ockhuizen T. 1994. Prevalence of food allergy and intolerance in the adult Dutch population. *J Allergy Clin Immunol* 93: 446–456.

Johansson SGO, Bieber T, Dahl R et al. 2004. Revised nomenclature for allergy for global use: Report of the Nomenclature Review Committee of the World Allergy Organization, October 2003. *J Allergy Clin Immunol* 113: 832–836.

Johansson SGO, Hourihane JOB, Bousquet J et al. 2001. A revised nomenclature for allergy: An EAACI position statement from the EAACI nomenclature task force. *Allergy* 56: 813–824.

Johnson J, Malinovschi A, Alving K, Lidholm J, Borres MP, Nordvall L. 2014. Ten-year review reveals changing trends and severity of allergic reactions to nuts and other foods. *Acta Paediatr* 103: 862–867.

Julia V, Macia L, Dombrowicz D. 2015. The impact of diet on asthma and allergic diseases. *Nature Rev Immunol* 15: 308–322.

Jutel M, Agache I, Bonini S et al. 2015. International consensus on allergy immunotherapy. *J Allergy Clin Immunol* 136: 556–568.

Jyonouchi H. 2012. Non-IgE mediated food allergy—Update of recent progress in mucosal immunity. *Inflam Allergy Drug Targets* 11: 382–396.

Kagan RS. 2003. Food allergy: An overview. *Environ Health Perspect* 111: 223–225.

Kivity S. 2012. Adult-onset food allergy. *Isr Med Assoc J* 14: 70–72.

Koplin JJ, Dharmage SC, Ponsonby AL et al. 2012. Environmental and demographic risk factors for egg allergy in a population-based study of infants. *Allergy* 67: 1415–1422.

Kotz D, Simpson CR, Sheikh A. 2011. Incidence, prevalence, and trends of general practitioner-recorded diagnosis of peanut allergy in England, 2001 to 2005. *J Allergy Clin Immunol* 127: 623–630.

Kummeling I, Mills EN, Clausen M et al. 2009. The EuroPrevall surveys on the prevalence of food allergies in children and adults: Background and study methodology. *Allergy* 1493–1497.

Kung SJ, Steenhoff AP, Gray C. 2014. Food allergy in Africa: Myth or reality? *Clinic Rev Allerg Immunol* 46: 241–249.

Lack G. 2012. Update on risk factors for food allergy. *J Allergy Clin Immunol* 129: 1187–1197.

Lau CH, Springston EE, Smith B, Pongracic J, Holl JL, Gupta RS. 2012. Parent report of childhood shellfish allergy in the United States. *Allergy Asthma Proc* 33: 474–480.

Lee AJ, Thalayasingam M, Lee BW. 2013. Food allergy in Asia: How does it compare? *Asia Pac Allergy* 3: 3–14.

Leung PS, Shu SA, Chang C. 2014. The changing geoepidemiology of food allergies. *Clin Rev Allerg Immunol* 2014; 46: 169–179.

Linehan MF, Nurmatov U, Frank TL, Niven RM, Baxter DN, Sheikh A. 2014. Does BCG vaccination protect against childhood asthma? Final results from the Manchester Community Asthma Study retrospective cohort study and updated systematic review and meta-analysis. *J Allergy Clin Immunol* 133: 688–695.

Litonjua AA, Lange NE, Carey VJ et al. 2014. The Vitamin D Antenatal Asthma Reduction (VDAART): Rationale, design, and methods of a randomized, controlled trial of vitamin D supplementation in pregnancy for the primary prevention of asthma and allergies in children. *Contemp Clin Trials* 38: 37–50.

Liu AH, Jaramillo R, Sicherer SH et al. 2010. National prevalence and risk factors for food allergy and relationship to asthma: Results from the National Health and Nutrition Examination Survey 2005–2006. *J Allergy Clin Immunol* 126: 798–806.

Longo G, Berti I, Burks AW, Kraus B, Barbi E. 2013. Ige-mediated food allergy in children. *Lancet* 382: 1656–1664.

Marmsjo K, Roselund H, Kull I et al. 2009. Use of multivitamin supplements in relation to allergic disease in 8-y-old children. *Am J Clin Nutr* 90: 1693–1698

Marrs T, Lack G. 2013. Why do few food-allergic adolescents treat anaphylaxis with adrenaline?—Reviewing a pressing issue. *Pediatr Allergy Immunol* 24: 222–229.

McLean S, Sheikh A. 2010. Does avoidance of peanuts in early life reduce the risk of peanut allergy? *BMJ* 340: c424.

Meldrum SJ, D'Vaz N, Dunstan J, Mori TA, Prescott SL. 2011. The Infant Fish Oil Supplementation Study (IFOS): Design and research protocol of a double-blind, randomised controlled n-3 LCPUFA intervention trial in term infants. *Contemp Clin Trials* 32: 771–778.

Monks H, Gowland MH, MacKenzie H et al. 2010. How do teenagers manage their food allergies? *Clin Exp Allergy* 2010; 40: 1533–1540.

Morita E, Matsuo H, Chinuki Y, Takahashi H, Dahlström J, Tanaka A. 2009. Food-dependent exercise-induced anaphylaxis—Importance of omega-5 gliadin and HMW-glutenin as causative antigens for wheat-dependent exercise-induced anaphylaxis. *Allergol Int* 58: 493–498.

Morjaria JB, Proiti M, Polosa R. 2011. Stratified medicine in selecting biologics for the treatment of severe asthma. *Curr Opin Allergy Clin Immunol* 11: 58–63.

Mullins RJ. 2007. Paediatric food allergy trends in a community-based specialist allergy practice, 1995–2006. *MJA* 186: 618–621.

Muraro A, Halken S, Arshad SH et al. 2014b. EAACI Food Allergy and Anaphylaxis Guidelines. Primary prevention of food allergy. *Allergy* 69: 590–601.

Muraro A, Werfel T, Hoffman-Sommergruber K et al. 2014a. EAACI food allergy and anaphylaxis guidelines: Diagnosis and management of food allergy. *Allergy* 69: 1008–1025.

Nakamura R, Teshima R. 2013. Proteomics-based allergen analysis in plants. *J Proteomics* 93: 40–49.

NIAD-Sponsored Expert Panel, Boyce JA, Assa'ad A et al. 2010. Guidelines for the diagnosis and management of food allergy in the United States: Report of the NIAD-Sponsored Expert Panel. *J Allergy Clin Immunol* 126: S1–S58.

Nicolaou N, Custovic A. 2011. Molecular diagnosis of peanut and legume allergy. *Curr Opin Allergy Clin Immunol* 11: 222–228.

Nurmatov U, Devereux G, Sheikh A. 2011. Nutrients and foods for the primary prevention of asthma and allergy: Systematic review and meta-analysis. *J Allergy Clin Immunol* 127: 724–733.

Nurmatov U, Devereux G, Worthh A, Healy L, Sheikh A. 2014. Effectiveness and safety of orally administered food allergies: A systematic review and meta-analysis. *Br J Nutr* 111: 12–22.

Nurmatov U, Venderbosch I, Devereux G, Simons FE, Sheikh A. 2012. Allergen-specific oral immunotherapy for peanut allergy. *Cochrane Database Syst Rev* 9: CD009014.

Nwaru BI, Erkkola M, Ahonen S et al. 2010. Age at introduction of solid foods during the first year and allergic sensitization at age 5. *Pediatrics* 125: 50–59.

Nwaru BI, Hickstein L, Panessar SS et al. 2014a. The epidemiology of food allergy in Europe: A systematic review and meta-analysis. *Allergy* 69: 62–75.

Nwaru BI, Hickstein L, Panessar SS et al. 2014b. Prevalence of common food allergies in Europe: A systematic review and meta-analysis. *Allergy* 69: 992–1007.

Nwaru BI, Muraro A, Sheikh A. 2014c. Charting a research agenda for understanding the epidemiology of food allergy in adults in Europe. *Allergy* 69: 975–977.

Nwaru BI, Sheikh A. 2015. Anaphylaxis in adolescents: A potential tripartite management framework. *Curr Opin Allergy Clin Immunol* 15: 344–349.

Nwaru BI, Takkinen HM, Niemela O et al. 2013. Introduction of complementary foods in infancy and atopic sensitization at the age of 5 years: Timing and food diversity in a Finnish birth cohort. *Allergy* 68: 507–516.

Osborne NJ, Koplin JJ, Martin PE et al. 2011. Prevalence of challenge-proven IgE-mediated food allergy using population-based sampling and predetermined challenge criteria in infants. *J Allergy Clin Immunol* 127: 668–676.

Perry TT, Matsui EC, Kay Conover-Walker M, Wood RA. 2004. The relationship of allergen-specific IgE levels and oral food challenge outcome. *Allergy Clin Immunol* 114: 144–149.

Poole JA, Barriga K, Leung DY, Hoffman M, Eisenbarth GS, Rewers M. 2006. Timing of initial exposure to cereal grains and the risk of wheat allergy. *Pediatrics* 117: 2175–2182.

Prescott S, Allen KJ. 2011. Food allergy: Riding the second wave of the allergy epidemic. *Pediatr Allergy Immunol* 22: 155–160.

Prescott SL, Pawankar R, Allen KJ et al. 2013. A global survey of changing patterns of food allergy burden in children. *World Allergy Organ J* 6: 21.

Prescott SL, Tang MLK. 2005. The Australian Society of Clinical Immunology and Allergy position statement: Summary of allergy prevention in children. *Med J Aust* 181: 464–467.

Roduit C, Frei R, Depner M et al. 2014. Increased food diversity in the first year of life is inversely associated with allergic diseases. *J Allergy Clin Immunol* 133: 1056–1064.

Romantsik O, Bruschettini M, Tosca MA, Zappettini S, Della Casa Alberighi O, Calevo MG. 2014. Oral and sublingual immunotherapy for egg allergy. *Cochrane Database Syst Rev* 11: CD010638.

Rona RJ, Keil T, Summers C et al. 2007. The prevalence of food allergy: A meta-analysis. *J Allergy Clini Immunol* 120: 638–646.

Sabra A, Bellani JA, Rais JM, Castro HJ, de Inocencio JM, Sabra S. 2003. IgE and non-IgE food allergy. *Ann Allergy Asthma Immunol* 90: 71–76.

Sampson HA. 2005. Food allergy—Accurately identifying clinical reactivity. *Allergy* 60: 19–24.

Sampson HA, Aceves S, Bock A et al. 2014. Food allergy: A practice parameter update—2014. *J Allergy Clin Immunol* 134: 1016–1025.e43

Sampson HA, Gerth van Wijk R, Bindslev-Jensen C et al. 2012. Standardizing double-blind, placebo-controlled oral food challenges: American Academy of Allergy, Asthma & Immunology-European Academy of Allergy and Clinical Immunology PRACTALL consensus report. *J Allergy Clin Immunol* 130: 1260–1274.

Sampson MA, Muñoz-Furlong A, Sicherer SH. 2006. Risk-taking and coping strategies of adolescents and young adults with food allergy. *J Allergy Clin Immunol* 117: 1440–1445.

Santos AF, Douri A, Bécares N et al. 2014. Basophil activation test discriminates between allergy tolerance in peanut-sensitized children. *J Allergy Clini Immunol* 134: 645–652.

Santos AF, James LK, Bahnson HT et al. 2015a. IgG4 inhibits peanut-induced basophil and mast cell activation in peanut-tolerant children sensitized to peanut major allergens. *J Allergy Clini Immunol* 135: 1249–1256.

Santos AF, Du Toit G, Douri A et al. 2015b. Distinct parameters of the basophil activation test reflect the severity and threshold of allergic reactions to peanut. *J Allergy Clini Immunol* 135: 179–186.

Scadding G. 2013. Laying the foundation for specific immunotherapy. In *Landmark Papers in Allergy: Seminal Papers in Allergy with Expert Commentaries*, eds. Sheikh A, Platts-Mills T, Worth A. Oxford, U.K.: Oxford University Press.

Sewell DA, Hammersley VS, Devereux G et al. 2013. Investigating the effectiveness of the Mediterranean diet in pregnant women for the primary prevention of asthma and allergy in high-risk infants: Protocol for a pilot randomized controlled trial. *Trials* 14: 173.

Sheikh A, Worth A, Gallagher M. 2011. Allergic problems. In *Practical General Practice: Guidelines for Effective Clinical Management*, eds. Khot A, Polmear A, pp. 585–592. Edinburgh, Scotland: Elsevier.

Sicherer SH, Furlong TJ, Maes HH, Desnick RJ, Sampson HA, Gelb BD. 2000. Genetics of peanut allergy: A twin study. *J Allergy Clin Immunol* 106: 53–56.

Sicherer SH, Sampson HA. 2010. Food allergy. *J Allergy Clin Immunol* 125: S116–S125.

Sicherer SH, Sampson HA. 2014. Food allergy: Epidemiology, pathogenesis, diagnosis, and treatment. *J Allergy Clin Immunol* 133: 291–307.

Warren CM, Jhaveri S, Warrier MR, Smith B, Gupta RS. 2013. The epidemiology of milk allergy in US children. *Ann Allergy Asthma Immunol* 110: 370–374.

Worth A, Regent L, Levy M, Ledford C, East M, Sheik A. 2013. Living with severe allergy: An Anaphylaxis Campaign national survey of young people. *Clin Transl Allergy* 3: 2.

Wölbing F, Biedermann T. 2013. Anaphylaxis: Opportunities of stratified medicine for diagnosis and risk assessment. *Allergy* 68: 499–508.

Zuberbier T, Edenharter G, Worm M et al. 2004. Prevalence of adverse reactions to food in Germany—A population study. *Allergy* 59: 338–345.

Zuidmeer L, Goldhahn K, Rona RJ et al. 2008. The prevalence of plant food allergies: A systematic review. *Allergy Clin Immunol* 121: 1210–1218, e4.

Zutavern A. Brockow I, Schaaf B et al. 2008. Timing of solid food introduction in relation to eczema, asthma, allergic rhinitis, and food and inhalant sensitization at the age of 6 years: Results from the prospective birth cohort study LISA. *Pediatrics* 121: e44–e52.

7 Pathogens Associated with Fruits and Vegetables

Contamination, Incidences, and Prevention

Y.S. Rohinishree, M.C. Varadaraj, and P.S. Negi

CONTENTS

7.1 INTRODUCTION

Codex Alimentarius Commission defines food as "any substance, whether processed, semi-processed or raw, which is intended for human consumption." On the other hand, food may be defined as a substance of essential nutrients, which provides nutritional support for the body. Most of the foods are from plant origin, but animal, fungi, mushrooms, and blue-green algae have also been used as foods.

A recent, WHO report examined the effects of 31 agents that spread foodborne diseases (FBDs), including bacteria, viruses, parasites, toxins, and chemicals, and found that about 10% of the world's population become ill after eating contaminated food resulting in about 0.42 m death, which include 0.125 m children below 5 years of age (WHO, 2015). Ensuring food safety is very important in the context of changing food habits and the globalization of our food supply. Because of globalization, there is a need to strengthen food safety systems among all the countries to ensure safe food for one and all. The burden of FBD mainly encompasses microbial, parasitic, and chemical (including biotoxic) contamination of food. However, FBD is also closely associated with an absence of essential elements in the diet, food allergies, food issues such as avian influenza, bovine spongiform encephalopathy, and infections transmitted through direct contact with food.

FBDs are a serious global problem and have devastating health and economic consequences in both developed and developing countries. The regional and global burden of FBD was recently estimated (WHO, 2015), classifying hazards as important cause of FBD in all regions (e.g., nontyphoidal *Salmonella enterica*) and some of local nature (e.g., aflatoxin). A higher economic burden of FBD on developing countries as compared to developed countries has been reported (Sudershan Rao et al., 2012). Many FBDs can also be transmitted through water and air and also through infected animals, pets, and human beings. Health experts estimate that the annual cost of all FBDs in the United States is \$5–\$6 billion in direct medical expenses and lost productivity (Kirch, 2008), and economic burdens on affected people are even higher in India on per-capita basis (Sudershan Rao et al., 2010). The large burden from food points toward the importance of food safety (WHO, 2015).

FBDs resulting from the pathogenic microorganisms can be classified into five types, namely, infection, intoxication, metabolic food disorders, allergies, and idiosyncratic illness. Foodborne infection is due to consumption of contaminated live microorganisms, which establishes in the intestinal tract. Foodborne intoxication results from the ingestion of food contaminated with preformed toxins by microorganisms. Infection and intoxication can affect all individuals, whereas metabolic food disorders, allergies, and idiosyncratic illness depend on individual health conditions and affect only immunocompromised individuals (Taylor et al., 1999). The five factors collectively termed as *foodborne illness risk factors* have been identified as the most significant contributing factors directly related to food safety, which include food from unsafe sources, inadequate cooking, improper holding temperature, contaminated equipment, and poor personal hygiene (Food and Drug Administration, 2009).

The increase in the production and advancement in postharvest handling has increased the consumption of fresh fruits and vegetables in the daily diet. Raw fresh fruits and vegetables are rich sources of vitamins, minerals, nutrients, and fibers for human consumption and can be consumed as precut and ready-to-eat (RTE) salads. However, in recent years, the number of outbreaks of foodborne illnesses linked to the consumption of fruits and vegetables has increased significantly (WHO, 2015). The cause for the outbreak of illness due to consumption of raw fruits and vegetables is mainly viral or bacterial pathogens contaminating the produce in the field through soil, feces, and water used for irrigation; through the application of manure, biosolids, pesticides, and fertilizers; and through dust, insects, and animals. During harvesting and postharvest operation, equipment, transport containers, insects, and dust can contaminate raw fruits and vegetables, and food handlers in food service establishments can also contaminate them (Tournas, 2005). It has been estimated that half of the postharvest losses of the world's production of fruits and vegetables due to spoilage occur during storage and transport. Therefore, efforts to solve the food safety problems must be linked to fresh produce handling.

7.2 FOODBORNE PATHOGENS ASSOCIATED WITH FRESH AND PROCESSED FRUITS AND VEGETABLES

Fresh fruits and vegetables are often considered as safe food as compared to foods from animal origin, as they usually harbor nonpathogenic epiphytic microorganisms. Unfortunately, recent outbreaks associated with fruits and vegetables have been associated with pathogenic microorganisms (e.g., *Norovirus* outbreak due to frozen raspberry fruit in Scandinavian countries in 2006; *Escherichia coli* O157:H7 associated with spinach in the Unites States in 2007; *Salmonella* outbreak with tomatoes in 2008; and outbreak due to consumption of fresh-cut produce, FDA, 2015). During growth and minimal processing, pathogenic microorganisms may enter into the fresh produce, and they may become potential source of foodborne illness. Contamination can be transiently present on the surface or it may penetrate deep inside through damaged sites on the surface of the fresh produce. Numerous pathogens have been recovered from fresh fruits and vegetables; however, only a few of them are included in foodborne illness (Table 7.1). Not only bacteria present in fresh-cut fruits and vegetables but also pathogens like viruses and protozoans may be responsible for foodborne illness caused by fruits and vegetables.

The pathogenic organisms of public health importance that may be transmitted through contaminated foods are bacteria, viruses, protozoa, trematodes, cestodes, and nematodes. Among these, the most important organisms that cause FBDs include bacteria like *Aeromonas* spp., *Bacillus cereus*, *Brucella* spp., *Campylobacter jejuni*, *Clostridium botulinum*, *Clostridium perfringens*, *E. coli*, *Helicobacter pylori*, *Listeria monocytogenes*, *Salmonella typhi* and *Salmonella paratyphi*, *Shigella* spp., *Staphylococcus aureus*, *Vibrio cholera*, and *Vibrio parahaemolyticus* (Rao et al., 1989; Aktar et al., 2014).

7.2.1 *Aeromonas* Species

Cytotoxic strains of *Aeromonas* have been isolated from a wide range of vegetables such as seed sprouts, lettuce, green salads, mixed raw vegetables, parsley, and carrots (Table 7.1); however, outbreaks associated with *Aeromonas* have not been reported. The infection of *A. hydrophila* and *A. sobria* to human has not been completely confirmed; however, they are considered as potential infectious agents (Wadstrom and Ljungh, 1991). The pathogen can grow rapidly on raw vegetables and seed sprouts at refrigerated temperature (Table 7.2). Controlled or modified atmosphere storage used frequently to extend the shelf life of fresh fruits and vegetables does not significantly affect the growth of *A. hydrophila*.

7.2.2 *Campylobacter* Species

Campylobacter has been isolated from a variety of fresh produce items from farmer's markets in Canada and from mushrooms from retail markets in the United States. *Campylobacter* was also found on fresh produce due to cross-contamination in the field with the poultry manure. *Campylobacter enteritis* has also been associated with lettuce or salads (Table 7.1) and *C. jejuni* can

TABLE 7.1
Reported Outbreaks of Foodborne Illnesses and Pathogens Associated with Fruits and Vegetables

Vegetables/Fruits	Microorganisms
Lettuce	*A. hydrophila* or *A. caviae*, *Escherichia coli*, *L. monocytogenes*, *Shigella sonnei*, *Salmonella* spp., *hepatitis A virus*, *C. jejuni*, *C. enteritis*
Spinach	*L. monocytogenes*, *E. coli*, *coliform*, *Salmonella*
Tomato	*Hepatitis A virus*, *L. monocytogenes*, *mesophilic microorganisms*, *E. coli* O157:H7, *Salmonella* Baildon, *S.* Javiana, *S.* Montevideo
Radish	*E. coli* O157:H7, *Salmonella*
Broccoli, cauliflower, green beans, kidney beans, bean sprouts	*L. monocytogenes*, *Yersinia enterocolitica*
Cabbage/cabbage salad	*E. coli*, *L. monocytogenes*, *Clostridium botulinum* (type A), *V. cholerae*
Onion	*Salmonella* spp., *L. monocytogenes*, *Giardia lamblia*, *S. aureus*
Green onions	*Cryptosporidium parvum*, *Shigella flexneri 6A*
Carrots	*Yersinia enterocolitica*, Enterotoxigenic *E. coli*
Coriander	*Salmonella*, *L. monocytogenes*
Chopped garlic in oil	*C. botulinum types A and B*
Mixed vegetables	*Vibrio cholerae*
Vegetable salads	*S. aureus*, *hepatitis A virus*
Watermelon	*Salmonella* Javiana, *S.* Oranienburg, *S. miami*, *coliforms*, *Shigella sonnei*
Cantaloupe	*E. coli* O157:H7, *Salmonella chester*, *S. oranienburg*, *S. poona*, *S. saphra*
Raspberries	*Cyclospora cayetanensis*, *Calicivirus*, *hepatitis A virus*
Strawberries	*Hepatitis A virus*
Apple	*Cryptosporidium parvum*, *E. coli* O157:H7, *S. typhi*, *S. typhimurium*
Orange	Enterotoxigenic *E. coli*; *S.* Gaminera, *S.* Hartford, and *S. rubislaw*; *S. typhimurium*
Orange juice (reconstituted)	*Hepatitis A virus*; *Salmonella typhi*
Fruit salad	*E. coli* O157:H7, *Norwalk virus*
Alfalfa	*E. coli* O157:H7, *E. coli* O157:NM; *Salmonella* Bovismorbificans, *S. enteriditis*, *S. havana/cubana/tennessee*

Source: Compiled from Harris, L.J. et al., *Comprehens. Rev. Food Sci. Food Safety*, 2, 78, 2003; Barth, M. et al., Microbiological spoilage of fruits and vegetables, in Sperber, W.H. and M.P. Doyle (Eds.), *Compendium of the Microbiological Spoilage of Foods and Beverages*, Food Microbiology and Food Safety, Springer Science, 2009, pp. 135–183; FDA, Analysis and evaluation of preventive control measures for the control and reduction/elimination of microbial hazards on fresh and fresh-cut produce: Chapter IV, in *Outbreaks Associated with Fresh and Fresh-Cut Produce. Incidence, Growth, and Survival of Pathogens in Fresh and Fresh-Cut Produce*, http://www.fda.gov/Food/FoodScienceResearch/SafePracticesforFoodProcesses/ucm091265.htm2015.

survive on watermelon and papaya slices; however, they require sufficient time to be a risk to the consumer (Table 7.2). *Campylobacter jejuni* and *Campylobacter coli* are reported to be the chief causes of bacterial enteritis (Castillo and Escartin, 1994).

7.2.3 *ESCHERICHIA COLI*

Enterotoxigenic *E. coli* is a common cause of traveler's diarrhea, and raw vegetables are frequently associated with the travelers' diarrhea. *E. coli* O157:H7 is recognized as an important enterotoxigenic foodborne pathogen. *E. coli* contamination has been reported from garden salad made from iceberg and romaine lettuce, endive, and shredded carrots and tossed salad containing onions,

TABLE 7.2
Survival and Growth of Pathogenic Organism on Fresh Fruits and Vegetable Produce

Pathogen	Produce Type	pH	Temp. (°C)	Initial Counts (log_{10} CFU or Spores/g or mL or Slice or Cube)	Incubation Time	Final Counts (log_{10} CFU or Spores/g or mL or Slice or Cube)	References
Aeromonas hydrophila	Broccoli	NR	4.0–15	4.0–5.1	4–14 days	From ~6.0 to ~8.2	Berrang et al. (1989a)
A. hydrophila	Cauliflower	NR	4.0–15	3.5–4	4–14 days	From ~5.9 to ~8.5	Berrang et al. (1989a)
A. hydrophila	Vegetable salads (mixed vegetable)	6	4–15.0	3	24 h to 9 days	From ~3.5 to ~7.9	Garcia-Gimeno et al. (1996)
A. hydrophila (2 strains)	Sprouts, mung beans	ND	ND	5	5–7 days	From ~6.8 to ~9.2	Aytac and Gorris (1994)
Escherichia coli O157:H7 (5 strains)	Carrot (shredded)	7.1	5.0–21	2.5–5.3	3–14 days	From <1 to 6.42	Abdul-Raouf et al. (1993)
E. coli O157:H7 (5 strains)	Cucumber (sliced)	7.3	5.0–21.0	2.3–5.1	3–10 days	From <1 to 5.4	Abdul-Raouf et al. (1993)
E. coli O157:H7	Lettuce (shredded)	~7.4	5.0–21.0	5.3	3–14 days	From 3.1–8.8	Abdul-Raouf et al. (1993)
E. coli O157:H7	Iceberg lettuce (shredded)	NR	4.0–15	3.4 to ~3.9	4–18 days	From 2.9–7.6	Li et al. (2001)
E. coli O157:H7	Apple, cider	3.6–4.0	8.0–26	2.0–6.0	3–15 days	From <1.0 to 5.5	Janisiewicz et al. (1999); Zhao et al. (1993)
E. coli O157:H7	Seeds, alfalfa	ND	ND	3.0–6.9	5 days–54 weeks	From <0.3 to 8.0	Taormina and Beuchat (1999)
Salmonella (5 serotypes)	Cantaloupe, cubes	6.67	5.0–23.0	2	24 h	From ~1.6 to ~7.2	Golden et al. (1993)
Salmonella (5 serotypes)	Honeydew, cubes	5.95	5.0–23.0	2	24 h	From ~1.8 to ~8.0	Golden et al. (1993)
Salmonella (5 serotypes)	Watermelon, cubes	5.9	5.0–23.0	2	24 h	From ~1.9 to 8.6	Golden et al. (1993)
Salmonella chester	Apple (Golden delicious, disks)	4.1	8.0–20.0	~5.5	66 h	From ~5.3 to ~10.4	Liao and Sapers (2000)
S. typhi	Papaya, cubes	3.59–5.69	25–27	2.9–3.0	6 h	From 3.8 to 4.3	Fernandez Escartin et al. (1989)
S. typhi	Jicama (cubes)	3.3–5.97	25–27	3.1–3.2	6 h	3.4–4.7	Fernandez Escartin et al. (1989)
S. typhi	Sprouts, alfalfa	ND	ND	2.5–4.8	6–15 days	2.4–3.7	Castro-Rosas and Escartin (2000)

(*Continued*)

TABLE 7.2 (*Continued*)
Survival and Growth of Pathogenic Organism on Fresh Fruits and Vegetable Produce

Pathogen	Produce Type	pH	Temp. (°C)	Initial Counts (log_{10} CFU or Spores/g or mL or Slice or Cube)	Incubation Time	Final Counts (log_{10} CFU or Spores/g or mL or Slice or Cube)	References
Salmonella (5 serotypes)	Sprouts alfalfa	ND	ND	7.7	11 days	6.9	Weissinger et al. (2001)
S. stanley	Seeds, alfalfa	ND	ND	2.5	8–9 weeks	0.9–1.81	Jaquette et al. (1996)
Salmonella (2 serotypes)	Orange, (Hamlin, peeled)	6.0–6.5	4.0–24.0	4.4	1–14 days	From ~2.5 to ~7.6	Pao et al. (1998)
S. baildon (4 strains)	Tomato (diced)	4.39–4.52	4.0–30.0	0.4–3.4	72 h–12 days	From <1.0 to 8.1	Weissinger et al. (2000)
S. baildon	Iceberg lettuce (shredded)	6.06–7.00	4	0.3–3.3	12 days	From <1.0 to 1.8	Weissinger et al.,2000
S. montevideo (3 strains)	Tomato (cut slices)	4.31–4.52	25	3.4–7.4	12 h	From ~7.5 to ~9.5	Wei et al. (1995)
S. montevideo (3 strains)	Tomato (mature green, wounded)	4.33–4.52	25	3.0–7.0	24 h	From ~4.8 to 7.0	Wei et al. (1995)
S. montevideo (3 strains)	Tomato (ripe, chopped)	4.1	5.0–30.	4.4	22 h–96 h	~4.2–7.0	Zhuang et al. (1995)
S. montevideo	Tomato (whole, mature green)	NR	10–30.0	~1.4	1–18 days	From ~2.2 to ~5.0	Zhuang et al. (1995)
S. montevideo	Iceberg lettuce (pieces)		4	~4.1	4	~4.4	Lin et al. (2000)
Salmonella (3 serotypes)	Tomato (cut, small pieces)	4.0–4.40	22–30.0	1.1	24 h	6.3–8.4	Asplund and Nurmi (1991)
Shigella (3 species)	Papaya, cubes	5.69	25–27	2.0–2.4	6 h	3.8–4.2	Fernandez Escartin et al. (1989)
Shigella flexneri	Watermelon, cubes	NR	22–26	2.8	6 h	4.5	Fernandez Escartin et al. (1989)
S. flexneri (1 strain)	Coleslaw	NR	4	~5.1	16 days	~4.3	Rafii and Lunsford (1997)
S. flexneri (1 strain)	Cabbage (chopped)	NR	4	6.5	26 days	3.1	Rafii and Lunsford (1997)
S. flexneri (1 strain)	Onion (chopped)	NR	4	6.7	12 days	5.3	Rafii and Lunsford (1997)
S. flexneri (1 strain)	Carrot salad	NR	5	~6.7	11 days	~2.5	Rafii and Lunsford (1997)
S. flexneri (1 strain)	Green pepper (chopped)	NR	4	6.7	12 days	4.3	Rafii and Lunsford (1997)
S. sonnei	Parsley (chopped leaves)	NR	4.0–21	3.2–6.48	1–14 days	From 0.4 to ~9.2	Wu et al. (2000)
S. sonnei	Cabbage (shredded)	6.8	24±2	2.6–3.8	1–7 days	From <1.0 to 7.0	Satchell et al. (1990)
S. sonnei	Watermelon, whole	NR	22.0–30.0	2	2–4 days	~9.0	Fredlund et al. (1987)

(*Continued*)

TABLE 7.2 (*Continued*)
Survival and Growth of Pathogenic Organism on Fresh Fruits and Vegetable Produce

Pathogen	Produce Type	pH	Temp .(°C)	Initial Counts (log_{10} CFU or Spores/g or mL or Slice or Cube)	Incubation Time	Final Counts (log_{10} CFU or Spores/g or mL or Slice or Cube)	References
S. sonnei	Lettuce (shredded)	NR	5.0–22.0	3.1	12 h–7 days	1.0–6.3	Davis et al. (1988)
Shigella (3 species)	Jicama (cubes)	5.97	25–27	2.5–2.6	6 h	3.3–4.0	Fernandez Escartin et al. (1989)
S. dysenteriae (1 strain)	Jicama (cubes)	5.97	25–27	1.5	6 h	2.2	Fernandez Escartin et al. (1989)
Listeria monocytogenes (2 strains)	Orange, (Hamlin, peeled)	6.0–6.5	8.0–24.0	~2.8–3.9	1–14 days	From ~3.3 to ~5.5	Pao et al. (1998)
L. monocytogenes (4 strains)	Tomato (chopped)	4.1	October 21	4.5–5.0	8–10 days	From ~1.0 to 4.2	Beuchat and Brackett (1991)
L. monocytogenes (4 strains)	Tomato (whole, cherry)	NR	October 21	3.4–3.7	2–20 days	3.5–5.5	Beuchat and Brackett (1991)
L. monocytogenes (2 strains)	Asparagus	NR	2.0–20	4.0–5.0	3–24 days	4.5–9.0	Castillejo Rodriguez et al. (2000)
L. monocytogenes (2 strains)	Broccoli	NR	4.0–15	4.0–5.6	10–21 days	From ~4.0 to ~8.8	Berrang et al. (1989b)
L. monocytogenes (5 strains)	Butternut squash	6.3	4–10.0	3	2–4 days	From ~6.5 to ~8.5	Farber et al. (1998)
L. monocytogenes (1 strain)	Cabbage (shredded)	NR	5	3.0–4.1	2–64 days	From <1.3 to 8.0	Beuchat et al. (1986); Kallander et al. (1991)
L. monocytogenes	Carrot	NR	4.0–15.0	<1.0–3.0	7–24 days	From <1.0 to 5.8	Beuchat and Brackett (1990); Farber et al. (1998)
L. monocytogenes (2 strains)	Cauliflower	NR	4.0–15.0	2.8–4.0	8–21 days	From ~3.0 to ~7.1	Berrang et al. (1989b)
L. monocytogenes (5 strains)	Coleslaw	6.6	4.0–10.0	2.5	9 days	From ~4.0 to ~5.0	Farber et al. (1998)
L. monocytogenes (5 strains)	Endive (chicory)	NR	3.0–10.0	3.3–4.8	7–10 days	From ~2.5 to ~7.0	Aytac and Gorris (1994); Carlin and Nguyen-the (1994)
L. monocytogenes (5 strains)	Rutabaga	6.3	4.0–10.0	3	9 days	From ~4.0 to ~6.0	Farber et al. (1998)
Clostridium botulinum (10 strains)	Broccoli	6.9	4.0–21	2	3–30 days	NR	Larson et al. (1997

(*Continued*)

TABLE 7.2 (*Continued*)
Survival and Growth of Pathogenic Organism on Fresh Fruits and Vegetable Produce

Pathogen	Produce Type	pH	Temp. (°C)	Initial Counts ($\log_{10}$ CFU or Spores/g or mL or Slice or Cube)	Incubation Time	Final Counts ($\log_{10}$ CFU or Spores/g or mL or Slice or Cube)	References
C. botulinum nonproteolytic	Butternut squash	NR	5.0–15.0	2.0–3.0	7–21 days	NR	Austin et al. (1998)
C. botulinum proteolytic	Butternut squash	NR	15.0–25.0	2	3–14 days	NR	Austin et al. (1998)
C. botulinum (10 strains, 5 type A and 5 type B)	Cabbage (shredded)	6.2	4.4–25	0.8–3.0	4–37 days	2.2	Solomon et al. (1990); Petran et al. (1995)
C. botulinum (10 strains)	Carrot (sliced)	6.3	4.0–21.0	2.0–3.0	4–78 days	NR	Larson et al. (1997); Austin et al. (1998)
C. botulinum nonproteolytic (8 strains)	Coleslaw	NR	January 05	3	21 days	NR	Austin et al. (1998)
C. botulinum proteolytic (10 strains)	Coleslaw	NR	15–25	2	21 days	NR	Austin et al. (1998)
C. botulinum	Green beans	6.3	4.0–21.0	2	7–35 days	NR	Larson et al. (1997)
C. botulinum proteolytic (10 strains)	Mushroom (enoki)	NR	6.0–27.0	1.0–3.0	2–21 days	No toxin	Malizio and Johnson (1991)
C. botulinum proteolytic	Onion	NR	25	3	6 days	—	Austin et al. (1998)
C. botulinum (10 strains)	Lettuce (intact)	6.2	4.0–21.0	2	6–50 days	ND	Larson et al. (1997)
C. botulinum nonproteolytic (8 strains)	Mixed salad	NR	5.0–25.0	1.0–3.0	4–21 days	From 3.7 to >4.5	Austin et al. (1998)
C. botulinum (12 strains)	Romaine lettuce (shredded)	NR	4.4–21	2	28 days	2.0–2.1	Petran et al. (1995)
C. botulinum nonproteolytic (8 strains)	Romaine lettuce	NR	5.0–10	3	21 days	NR	Austin et al. (1998)
C. botulinum proteolytic (10 strains)	Romaine lettuce	NR	15–25	1.0–2.0	21 days	NR	Austin et al. (1998)
Vibrio cholerae	Sprouts, alfalfa	ND	ND	2.5–3.1	6–15 days	0.5–3.7	Castro-Rosas and Escartin (2000)

NR- not reported
ND- not detected

TABLE 7.3
Reported Foodborne Incidences Associated with Fruits and Vegetables

Year	Pathogens	Cases	Regions	Food
2011	*E. coli* O26	14	North America	Raw clover sprouts
2011	*E. coli* O104:H4	4000+	Europe	Fenugreek seeds
2008	*Salmonella*	1442	North America	Fresh peppers, tomatoes
2007	*Salmonella*	51	Europe, North America	Fresh basil
2007	*Shigella*	175	Australia, Europe	Alfalfa sprouts
2007	*Salmonella*	45	Europe	Alfalfa sprouts
2007	*Staphylococcus aureus*	100	Madhya Pradesh, India	Potato ball
2006	*Salmonella*	206	North America	Fresh spinach
2006	*Salmonella*	20+	Europe	Arugula
1996	*Salmonella* paratyphi A var durazzo	33	Maharashtra, India	Vegetarian food
1990	*Vibrio fluvialis*	14	Pune, India	Vegetarian food

Sources: Compiled from Lynch, M.F. et al., *Epidemiol. Infect.*, 137, 307, 2009; Wallace, C.A. et al., *Food Safety for the 21st Century: Managing HACCP and Food Safety throughout the Global Supply Chain*, Wiley- Blackwell: A John Wiley & Sons, Ltd, Publications, 2011; Marler, B., http://www.marlerblog.com/, 2012; Fule et al., *Indian J. Med. Res.*, 103, 74, 1996; Thekdi et al., *J. Diarrh. Dis. Res.*, 8, 163, 1990; Nema, V.R. et al., *Int. J. Food Microbiol.*, 117, 29, 2007.

carrots, zucchini, peppers, broccoli, mushrooms, and tomatoes (Beuchat, 1996). *E. coli* was also isolated from mixed vegetables, cilantro, coriander, and celery in Mexico City; however, outbreaks have also been linked to lettuce, unpasteurized apple cider, cantaloupe, and sprouts (Table 7.2). *Escherichia coli* O157:H7 grows rapidly in several types of raw fruits and vegetables, particularly when stored at 12°C or above (Table 7.2). Packaging under modified atmosphere is reported not to affect the survival or growth of *E. coli* O157:H7 (Janisiewicz et al., 1999; Table 7.2).

7.2.4 *Helicobacter pylori*

Helicobacter pylori, one of the most common bacteria causing infections and affecting approximately 50% of the world's population, was first isolated from human gastric mucosa in 1982 (Warren and Marshall, 1983). Earlier, this bacterium was classified as *Campylobacter pylori*; later in 1989, it was included under a new genus, *Helicobacter*, and renamed as *Helicobacter pylori* (Goodwin et al., 1989). The bacteria has been classified as a Class I carcinogen by the World Health Organization (Aruin, 1997).

Fruits and vegetables can easily be contaminated by *H. pylori* from irrigation water, soil, animal manure, and human stool (Hopkins et al., 1993; Sasaki et al., 1999; Mazari-Hiriart et al., 2001; Safaei et al., 2011) or by a vector such as the fly (Grubel et al., 1998); they are prone to cross-contamination in processing stages (Yahaghi et al., 2014). The possible routes of transmission of *H. pylori* are from person to person, milk, water, vectors, latrogens and zoonotic vectors, and waterborne transmission; transmission from nonhuman primates and domestic animals such as cats, pigs, and housefly has been reported (Brown, 2000). Studies show that contamination from irrigation water and animal manure are the main sources of *Helicobacter pylori*. In Chile, sewage-contaminated irrigation water used for vegetables cultivation and the consumption of harvested raw vegetables were the key factors for the transmission of enteric pathogens (Hopkins et al., 1993). The inefficient sanitation practices for removing or killing pathogens on raw fruits and vegetables may increase the chances of their contamination by *H. pylori*.

Although increased consumption of fruits and/or vegetables was found to significantly reduce the risk of *H. pylori* infection (Fontham et al., 1995; Goodman et al., 1997), Goodman et al. (1996) and Hopkins et al. (1993) reported that consumption of raw or uncooked vegetables was related to the increased risk of *H. pylori* infection. It is reported that the acquisition rate of *H. pylori* is more rapid in developing counties than in developed countries (Mazari-Hiriart et al., 1992; Pounder and Ng, 1995) probably due to poor hygienic practices (Brown, 2000). It is reported that the food prepared under poor sanitary conditions was also responsible for transmission of *H. pylori* in Lima, Peru (Begue et al., 1998), and consumption of uncooked lettuce or food from street vendors was a risk factor for *H. pylori* infection in the southern Colombian Andes (Goodman et al., 1996).

7.2.5 *Listeria monocytogenes*

Listeria monocytogenes is widely distributed on raw fruits and vegetables (Table 7.1), and salad vegetables play a role in distributing the pathogen from natural habitats to the human food supply (Beuchat, 1996). It can grow on fresh produce, as well as on fresh-cut fruits, asparagus, broccoli, butternut squash, coleslaw and cauliflower, rutabaga stored at temperature 4°C, lettuce stored at 5°C, and chicory endive stored at 6.5°C (Table 7.3). Controlled atmosphere storage does not have any influence on growth rates of *L. monocytogenes.*

Listeria monocytogenes causes relatively mild gastroenteritis in healthy adults, but severe gastroenteritis has been reported in susceptible individuals including pregnant women, neonates, and immune compromised individuals. The infective dose has not been clearly established, although it is thought to be relatively low among susceptible individuals. Carrot juice is reported to inhibit its growth due to the presence of phytoalexins, which act as antimicrobial agent (Beuchat and Brackett, 1990; Nguyen-the and Lund, 1991, 1992; Beuchat et al., 1994; Beuchat and Doyle, 1995). Documented outbreaks associated with this organism linked to fresh produce have been limited, and contaminated vegetable salads served with these foods were raw celery, coleslaw, tomatoes, and lettuce.

7.2.6 *Salmonella*

Salmonella is a fecal organism–associated foodborne pathogen with significant morbidity and mortality rate in all age groups. Salmonellosis has been linked to several fruits and vegetables including tomatoes, seed sprouts, cantaloupe, mamey, apple juice, and orange juice (Beuchat, 2002). Their growth on fresh fruits and vegetables was influenced by temperature (T >10°C–12°C), whereas *S. sonnei* and *S. flexneri* were able to grow even at low temperatures of 6°C–7°C (ICMSF, 1996). However, *Salmonella* is less commonly associated with fresh fruits because of their low pH, and outbreaks involving fruits and vegetables have been documented mostly in cantaloupe and sprouts. Several large outbreaks of salmonellosis have been attributed to fresh produce (raw tomatoes) involving *Salmonella* Javiana in 1992, *Salmonella* Montevideo in 1993, and *Salmonella* Baildon in 2000.

The genus *Salmonella* has over 2700 serotypes (Doyle, 1990; Beuchat, 1996; Tauxe et al., 1997; Francis et al., 1999), and the presence of several *Salmonella* serotypes capable of causing human infection has been reported from fresh produce. In Spain and the Netherlands, about 7.5%–65% of vegetables sampled in field and retail shops were contaminated with *Salmonella* spp. (Nguyen-the and Carlin, 1994; Pirovani et al., 2000). The growth of *Salmonella* was influenced by temperature, and significant reduction was observed below 15°C with no growth below 7°C (ICMSF, 1996). However, *S. hadar* was reported to recover from shredded cabbage that was stored at 4°C for 10 days and grew normally at 12°C–20°C (Piagentini et al., 1997). They were also able to grow on damaged, chopped, and sliced tomatoes (pH 4.1–4.5) stored at 20°C–30°C (Table 7.2).

7.2.7 *Shigella* Species

The genus *Shigella* consists of four species, namely, *S. dysenteriae*, *S. boydii*, *S. sonnei*, and *S. flexneri*, which are pathogenic to humans even at low dose of less than 100 cells. The infection (shigellosis) can be transmitted through consumption of contaminated water and foods in particular fruits and vegetable salads, and shigellosis was associated with tomatoes, seed sprouts, cantaloupe, mamey, apple juice, orange juice (Beuchat, 2002), lettuce, scallions, and parsley (Cook et al., 1995; Martin et al., 2003). *Shigella sonnei* can survive at 5°C for 3 days without decrease in number and recover rapidly showing 1000-fold increase in growth when transferred at 22°C. *Shigella* can grow in shredded cabbage and chopped parsley when stored at 24°C (Table 7.2). Populations of *S. sonnei*, *S. flexneri*, and *S. dysenteriae* show a substantial increase in growth within 4–6 h at 22°C–27°C on the surface of freshly cut cubes of papaya (pH 5.69) and jicama (pH 5.97) (Table 7.2).

The available reports of foodborne illness outbreaks from *Shigella* are from the developed countries such as the United States and the United Kingdom. A lettuce processing facility was the common source of product responsible for outbreaks caused by *S. sonnei* that occurred in Texas. In another outbreak of *S. sonnei*, gastroenteritis was associated with eating shredded lettuce (Table 7.1), where a worker in the plant was the source of contamination and the method of processing allowed contamination of the lettuce. Two outbreaks of *S. flexneri* infection have been linked to the consumption of fresh green onions in midwestern United States (Beuchat, 1996). The onions were traced to a farm in Mexico, where they probably got contaminated at harvest or during packing. Although the association of fresh fruits and vegetables with *Salmonella* was reported from underdeveloped countries like Nigeria, so far no foodborne outbreaks have been reported (Lennox et al., 2015).

7.2.8 *Staphylococcus aureus*

S. aureus is one of the foodborne pathogens, commonly present on the surface of the skin of warm-blooded animals. This organism has been detected on fresh produce and RTE vegetable salads (Table 7.1), probably contaminated through food handlers. Enterotoxigenic *S. aureus* does not compete well with other microorganisms normally present on fresh produce, so incipient spoilage caused by nonpathogenic microbiota would likely overgrow, making produce less vulnerable to be the cause of staphylococcal foodborne illness.

An outbreak of staphylococcal foodborne illness was associated with canned mushrooms, and the possible contamination occurred prior to processing as mushrooms were stored under ambient conditions in plastic bags with salt. Due to anaerobic conditions of plastic bags, normal spoilage microbiota may have been inhibited, but *S. aureus* was able to survive and produce thermostable enterotoxins, suggesting that staphylococcal outbreaks associated with raw produce potentially occur only at favorable conditions. *S. aureus* has been shown to grow on peeled Hamlin oranges (Table 7.2) stored at 24°C and was able to survive up to 14 days at 4°C–8°C.

7.2.9 *Yersinia enterocolitica*

Yersinia enterocolitica has been isolated from several raw vegetables, and infection has been associated with the consumption of carrots and raw vegetables. The incidence of foodborne illness from *Yersinia* was higher in root and leafy vegetables than for tomatoes or cucumbers. The pathogen can survive and grow at refrigeration temperatures commonly used during transport and storage of fresh produce. In France, out of 58 samples, 27% of grated carrots were contaminated with *Yersinia* (Catteau et al., 1985) and 7% of the samples contained *Y. enterocolitica* serotypes. Fifty percent of prepared raw vegetables destined for school meals, stored at refrigerated conditions for up to 5 days, contained *Yersinia* species. *Y. enterocolitica* was the only species isolated from grated carrots, whereas *Y. intermedia* and *Y. kristensenii* were mainly isolated from lettuce.

7.2.10 Spore-Forming Pathogenic Bacteria

Spore-forming pathogenic bacteria such as *Bacillus cereus*, *Clostridium botulinum*, or *Clostridium perfringens* are commonly present in soil and can contaminate fresh fruits and vegetables. These organisms were able to grow in vegetables packed under modified atmospheric conditions and germinate their spores, which can become a serious threat to public health. An outbreak related to *Bacillus cereus* has been associated with consumption of mixed seed sprouts and the organism was subsequently shown to be present at relatively high levels in a variety of seeds sold for sprouting (Portnoy et al., 1976).

Clostridium botulinum requires relatively high water activity, pH greater than 4.6, relatively warm temperatures, and anaerobic conditions to grow and produce toxin. Botulism has been linked to consumption of cooked vegetables than to fresh produce and the outbreaks are associated to the butternut squash, sliced onions, cabbage and garlic in oil, coleslaw (Solomon et al., 1990), and mushrooms (Malizio and Johnson, 1991). Coleslaw was prepared from packaged, shredded cabbage mixed with coleslaw dressing of pH 3.5, and packed in a modified atmosphere. A survey revealed that 12 of 88 cabbages obtained from supermarkets contained *C. botulinum* spores and botulinum toxin was formed in shredded cabbage packed under atmosphere containing reduced oxygen and stored at 22°C–25°C for 4–6 days, whereas botulinum toxin was produced in mushrooms at an anaerobic environment, which was wrapped with polyvinyl chloride wrapped packages and stored at 20°C for 3–4 days (Sugiyama and Yang, 1975). Proteolytic strains of *Clostridium botulinum* grew and produced toxin in vacuum-packaged Enoki mushrooms at 15°C–27°C. *Clostridium perfringens* was associated with an outbreak by the consumption of salad (Table 7.2).

7.2.11 Viruses

Hepatitis A virus, calicivirus, and Norwalk-like viruses (NLV) have been associated with the foodborne illness by the consumption of produce and the contamination via infected food handlers during final preparation. Outbreaks have been associated with frozen raspberries or frozen strawberries, lettuce, melons, salads, watercress, diced tomatoes, and fresh-cut fruit (Table 7.1). Several enteroviruses (poliomyelitis, enteroviruses, hepatitis A, rotavirus, and Coxsackie viruses) can survive depending on pH, moisture content, and temperature in a variety of raw vegetables and vegetable salads. These observations indicate that salad vegetables can serve as vehicles for the transmission of viral pathogens also to humans.

7.2.12 Protozoan Parasites

Parasites are organisms that derive nutrient from other host organisms, and they are transmitted by water, soil, and from person to person. Parasites are single to multicellular organisms such as protozoa to helminthes. Protozoa such as *Cryptosporidium* spp., *Giardia intestinalis*, *Cyclospora cayetanensis*, and *Toxoplasma gondii*, roundworms such as *Trichinella* spp. and *Anisakis* spp., and tapeworms such as *Diphyllobothrium* spp. and *Taenia* spp. are the most common foodborne parasites.

Sensitive and reliable methods for the detection of parasites contamination in raw produce are lacking and therefore, incidences of outbreak associated with parasites are not available except in Costa Rica. In Costa Rica, a total of 640 samples from eight different vegetables were surveyed for the presence of *Cryptosporidium* oocysts, fecal coliforms, and generic *E. coli* (Monge and Chinchilla, 1996). *E. coli* was found at populations from 10^1 (tomato) to $10^5/10^6$ MPN/g (cilantro leaves/cilantro roots), and *Cryptosporidium* oocysts were found in at least one of 80 samples of each vegetable except cabbage with maximum number in cilantro leaves and cilantro roots. The highest contamination rates were observed in the rainy season, and probably produce was contaminated through irrigation water.

7.3 CONTAMINATION

Microbial pathogens may contaminate fresh produce of fruits and vegetables by a wide variety of mechanisms. Foodborne illness occurs due to contamination of pathogens at any point in the food-handling chain and gets aggravated by improper handling and storage during production, harvest, processing, and transporting, as well as in retail and food service establishments.

Various sources are involved in contamination during preharvest, harvest, and postharvest stages of fruits and vegetables. In preharvest stage, sources of contamination include application of fertilizer (organic wastes), waters for irrigation contaminated with fecal material, direct contamination by livestock, pesticide and fungicide dusts, wild animals and birds, field workers, and harvesting equipments (Beuchat, 2006; Matthews, 2013, 2014). Pathogenic bacteria such as *E. coli* O157:H7, *Salmonella*, and *L. monocytogenes* are mainly associated with produce outbreaks known to contaminate the environment through animal hosts (Ivanek et al., 2006; Franz and van Bruggen, 2008), and they are able to survive long term in agricultural waters, soil amendments, and soil (Franz et al., 2005; Scott et al., 2006; Franz and van Bruggen, 2008).

7.3.1 Preharvest

7.3.1.1 Soil

Soil is a rich reservoir for a variety of microorganisms including pathogenic and nonpathogenic microorganisms. Nonpathogenic microorganisms degrade dead plants and animals in the environment and help in mineralization of soil. Soil is also a reservoir of foodborne pathogens, such as *Bacillus cereus*, *Clostridium botulinum*, and *Clostridium perfringens* (Lund, 1986), and *Listeria monocytogenes* has been isolated from noncultivated soil (Welshimer and Donker-Voet, 1971). These microorganisms may degrade the tissue of plants and damage the fruits and vegetables, which may expose them to further microbial attack during transportation and storage.

7.3.1.2 Water

Water is one of the primary requirements for plant growth and mainly used for irrigation of plants. Quality of water mainly depends on the source, and it will vary from surface water or potable water to underground water. Water may also be a source of contamination of enteric pathogenic microorganisms such as protozoa, bacteria, and viruses (Park et al., 2012), and the occurrence of *L. monocytogenes, Salmonella,* and viruses has been reported (Castillo Martín et al., 1994; Nguyen-the and Carlin, 1994, 2000; Park et al., 2012).

The transfer of pathogenic microorganisms from irrigation water to fruits and vegetables will depend on the type of irrigation technique used and on the nature of the produce (NACMCF, 1999a). Spray irrigation is expected to increase the risk of contamination in comparison to drip irrigation or flooding (Matthews, 2014). Leafy vegetables provide large surfaces for contact with water and for the attachment of microorganisms. *E. coli* O157:H7 was able to survive on lettuce leaves for 20 days that were spray irrigated, and the level of pathogen increased with the repeated irrigation with contaminated water (Solomon et al., 2002, 2003). Overhead sprinkle irrigation was also associated with increased enteric pathogen contamination of butter head lettuce (Van der Linden et al., 2013). Repeated spray irrigation with a low dose (approximately 3.5 log CFU mL^{-1}) of *E. coli* O157:H7 resulted in internalization of the pathogen in spinach and parsley leaves, but not in lettuce leaves (Erickson et al., 2014) and these pathogens were able to persist for 2 days even after postharvest, suggesting that irrigation immediately prior to harvest presents an increased microbiological risk.

Untreated sewage water is often used for irrigation of fruits and vegetables in most of the developing countries, and they are the rich source of enteric human pathogens. In Pachuca City, Hidalgo, Mexico, untreated sewage water contaminated RTE vegetable salads with fecal coliforms (99%), *E. coli* (85%), and diarrheagenic *E. coli* (7%) (Castro-Rosas et al., 2012). The treated sewage water used for irrigation should be of good quality, and the level of fecal coliforms contamination in wastewater for irrigation of plants should be less than 1000 cfu mL^{-1} or the most probable number (MPN) should not exceed 100 mL^{-1} (World Health Organization/Food and Agriculture Organization, 2006).

River water used for irrigation can also harbor pathogenic microorganisms. In South Africa, river water used for irrigation of vegetable farms was contaminated with fecal coliforms and *E. coli* up to 1.6×10^4 cfu mL^{-1} and 3.5×10^3 cfu mL^{-1}, respectively (Gemmell and Schmidt, 2012). In Australia, pond and tidal creek water samples are reported to harbor *E. coli* O157, *Campylobacter jejuni*, and *Salmonella* (Ahmed et al., 2009). *Salmonella* spp. was also isolated from 57 out of 72 surface water samples collected in Georgia, United States (Haley et al., 2009). In Canada, river and canal water samples also contain *E. coli* O157:H7 and *Salmonella* (Gannon et al., 2004). *Salmonella* Newport was associated with an outbreak of salmonellosis due to consumption of tomatoes irrigated using contaminated pond water (Greene et al., 2008).

7.3.1.3 Organic Fertilizers

Organic fertilizers, such as animal manures and slurries (Beuchat, 1996; Natvig et al., 2002), abattoir wastes (Avery et al., 2005), compost of human and animal origin, and sewage sludge (Al-Ghazali and Al-Azawi, 1990), are commonly used for fruits and vegetables cultivation particularly in organic cultivation. The fecal matters of these fertilizers are the potential risk for the contamination of human pathogenic microorganisms such as viruses, bacteria, and parasites. Enterobacteriaceae members such as *Salmonella*, *Shigella*, *Yersinia*, *E. coli*, and *Campylobacter* were found in the intestinal tract of a wide range of domestic, wild, and companion animals. About one-fifth of organic fertilizers and sewage sludge samples were contaminated with *L. monocytogenes* in Belgium and Finland (Husu, 1990; Van Renterghem et al., 1991; Strauch, 1991), and *L. monocytogenes* was also detected in sewage effluent treatment plants in Italy and the Netherlands (Dijkstra, 1989; Bernagozzi et al., 1994). Foodborne outbreaks linked with contamination of *L. monocytogenes* on cabbage in Canada, and *Salmonella* and *E. coli* O157:H7 on apples used to make apple juice in the United States (Tauxe et al., 1997; Nguyen-the and Carlin, 2000) were suspected to be due to the application of contaminated manure.

7.3.1.4 Environment and Climatic Conditions

The survivability, prevalence, and transfer of enteric pathogens in soil, soil amendments, irrigation water, and type of crops may individually or collectively are influenced by environmental conditions or geographic location (Liu et al., 2013). As the ambient temperature increased, reduction in the survivability of enteric pathogens like *E. coli* O157 and *Salmonella* in manure and soil was observed (Mukherjee et al., 2006; Semenov et al., 2007; Danyluk et al., 2008). However, better survival of *E. coli* O157:H7 and *S. typhimurium* in manure and manure-amended soil in temperate regions as compared to tropical climatic conditions is reported (Ongeng et al., 2011). Greater intensity of rain events may lead to fruits and vegetable contamination with pathogens by splashing manure and soil particles onto edible portions of the crop (Cevallos-Cevallos et al., 2012), and with the increase in moisture content pathogens enhance their survivability (Warriner et al., 2009). Seasonality has also been demonstrated to affect microbial populations and survival of pathogens associated with fruits and vegetables (Ailes et al., 2008; Haley et al., 2009; Oliveira et al., 2012; Abakpa et al., 2013). Collectively, these studies indicate that a wide range of climatic and geographic variables must be considered for developing strategies to enhance the microbial safety of crops in the field.

7.3.2 Harvest/Postharvest

During harvesting, fruits and vegetables can get cross-contaminated with pathogenic microorganisms through contact with harvesting equipment, knives, human handling or gloves, and transport containers such as bins, boxes, and buckets, wild and domestic animals, air, transport vehicles, ice, or water (Beuchat and Doyle, 1995; Matthews, 2013).

An outbreak of *Shigella flexneri* and hepatitis A was reported due to contamination from infected people working on the fields or in the packaging facility (Beuchat and Doyle, 1995). Clean, well-designed, and well-maintained equipment are less likely to cause damage to fresh produce and to introduce spoilage and pathogenic microorganisms (Brackett, 1992), whereas unhygienic practices in storage with the presence of rodents, birds, and insects increase the risk of contamination with foodborne pathogens (FDA, 1998).

During harvesting of lettuce, trimming and coring in the field have been widely adopted to reduce the costs of shipping and waste disposal. However, this practice of core removal with human handling damages the tissue and increases microbial infiltration through the cut edges and leads to microbial growth and contamination (Food and Agriculture Organization/World Health Organization, 2008). On the other hand, coring knives can get contaminated by worker's hand after contact with pathogen-contaminated soil and can transfer *E. coli* O157:H7 to lettuce heads (Taormina et al., 2009). Widely used harvesting machines may also contaminate the vegetables and fruits with pathogenic microorganisms (Fallon et al., 2011; Buchholz et al., 2012). Accordingly, harvesting tools and machines should be kept clean through periodic washing and sanitization to prevent the risk of cross-contamination. Harvesting at the appropriate time and keeping the harvested product under controlled environmental conditions can help retard growth of postharvest spoilage and pathogenic microorganisms (Brackett, 1992).

7.4 OVERVIEW OF FOODBORNE ILLNESS INCIDENCES ASSOCIATED WITH FRUITS AND VEGETABLES

Food poisoning associated with fresh-cut fruits and vegetables is emerging due to increased consumption and increased consumer awareness of their nutritional values and convenience. Globalization of fresh produce supply chain has resulted in the round-the-year availability of the fresh produce; however, it requires more time between harvest and consumption. This situation demands proper storage conditions and transport to longer distances; failure may lead to incidences of FBD. Processing of fresh produce involves peeling, cutting, and coring, which removes natural plant protection barriers and provides suitable medium for microbial growth and also increases the possibility of cross-contamination.

The Center for Science in the Public Interest database shows that fresh produce outbreaks accounted for about 13% of foodborne outbreaks during 1990–2005 (deWaal and Bhuiya, 2007). The 2006 outbreak linked to *E. coli* O157:H7–contaminated spinach that resulted in 205 confirmed cases and three deaths served as a catalyst for research efforts to ensure the safety of leafy greens (CDC, 2006). The response to these incidents has led to the adoption of new recommendations intended to improve produce safety. Sprout-related disease outbreaks have also been reported in Japan, the United Kingdom, Finland, Denmark, Sweden, and Canada and have involved alfalfa, cress, radish, and mung bean sprouts (Puohiniemi et al., 1997; Taormina et al., 1999; Taylor et al., 2002).

During the 1996 summer in Japan, more than 6000 cases of *E. coli* O157:H7 infection were reported (Guiterrez, 1997). The largest outbreak resulted in four deaths and more than 4000 schoolchildren were affected due to consumption of raw radish sprouts in and around Sakai City, although the mechanism of sprout contamination was not determined.

In the United States, fresh produce is considered as the second most important food product (22%) associated with foodborne outbreaks, followed by poultry (18%), beef (16%), and eggs (13%). Between 1995 and 1998, nine outbreaks involving more than 1234 cases of foodborne illness caused by *Salmonella* or *E. coli* O157:H7 were reported at Missouri, Michigan, California, Washington, Arizona, and Nevada due to consumption of fresh vegetable sprouts (NACMCF, 1999a), and in most cases, alfalfa or clover seeds were involved. The most frequently implicated produce types were salad, lettuce, juice, melon, and sprouts, and the outbreaks have increased from 2 incidences/year

TABLE 7.4
Human Pathogens Recovered from Vegetables

Pathogens	Products
Aeromonas	Alfalfa sprouts, asparagus broccoli, cauliflower, celery, lettuce, pepper, spinach
Bacillus cereus	Alfalfa sprouts, watercress, mustard sprouts, soybean sprouts
Clostridium jejuni	Green onions, lettuce, potato, pepper, spinach
Clostridium botulinum	Cabbage, pepper, garlic, potato, carrots
E. coli O26	Raw clover sprouts
E. coli O104:H4	Fenugreek seeds
E. coli O157:H7	Alfalfa sprouts, cabbage, celery, watercress, lettuce, cabbage
Listeria monocytogenes	Beans sprouts, cabbage, chicory, eggplant, lettuce, potatoes, radish, lettuce
Salmonella	Alfalfa sprouts, cabbage, celery, watercress, lettuce, cabbage fresh peppers, tomatoes, fresh basil, arugula, artichokes, beet leaves, cauliflower, eggplant, endive, fennel, green onions, mung bean sprouts, mustard cress, pepper, salad greens, spinach
Shigella	Alfalfa sprouts, celery, lettuce, green onions
Staphylococcus	Alfalfa sprouts, carrot, lettuce, onion sprouts, radish
Vibrio cholera	Cabbage, lettuce
Enteric viruses (NLV, *hepatitis* A)	Lettuce, green onions, watercress
Protozoa	Lettuce, green onions, onion

Sources: Compiled from Beuchat, L.R., *J. Food Protect.*, 59, 204, 1996; FDA (Food and Drug Administration), Survey of domestic fresh produce: interim results, July 31, 2001, available from: http://www.fda.gov/Food/FoodSafety/Product-SpecificInformation/FruitsVegetablesJuices/GuidanceComplianceRegulatoryInformation/ucm118844.htm, accessed March 3, 2010.

in the 1970s to 7 incidences/year in the 1980s and 16 incidences/year in the 1990s. In a survey of about 1028 outbreak incidences regarding fruits and vegetables, *Salmonella*, *Shigella*, and *E. coli* O157:H7 were found to be the most dominant causative agents (Shivapalasingam et al., 2004). The most probable human pathogens associated with vegetables have been listed in Table 7.4.

In Europe, fresh-cut fruits and vegetables are rarely involved in foodborne outbreaks, when compared to other sources (EFSA, 2007). Foodborne illness alert notifications for fresh fruits and vegetables remained stable during 2000–2005 at about 9%, but in 2006, about 72 alert notifications were reported related to the presence of mycotoxins and pesticide residues following pathogenic microorganisms especially *Salmonella* (EFSA, 2007).

In India, the magnitude of the problem of FBD is not clear, and the majority of outbreaks are unreported. In case of reported outbreaks, most are uninvestigated or unrecognized and only noticed after major health or economic damage has occurred. Control of the outbreaks and investigation to prevent the future outbreaks and strengthening of food safety programs becomes difficult in such situation. The reported bacterial FBD outbreaks in India during 1980–2009 indicated the occurrence of 24 cases involving 1130 persons, and bacteria causing these outbreaks were *S. aureus*, *Vibrio*, *Salmonella*, *E. coli*, and *Yersinia enterocolitica* (Sudershan Rao et al., 2012). Rohinishree and Negi (2011) also found high level of contaminants in street food sold in Mysuru, India, and they identified *S. xylosus* as the major contaminant among *Staphylococcus*. The presence of toxigenic genes was observed among these *Staphylococci* (Rohinishree and Negi, 2012). *S. aureus* toxin was responsible for food poisoning among 31 persons in Haryana, India, after consumption of sweet meat made of coconut (Mandokhot and Garg, 1986), 42 persons in an army unit in India due to consumption of meat (Nayar et al., 1993), and more than 100 children in Madhya Pradesh, India, due to consumption of potato balls deep fried in vegetable oil (Nema et al., 2007).

Viruses are the second most important foodborne pathogen next to *Salmonella* in Europe and were linked to about 10.2% of outbreaks reported in 2006. The main viruses contaminating fresh produce were identified as adenovirus, norovirus, enterovirus, hepatitis A virus (HAV), and rotavirus (EFSA, 2007).

In the United States, approximately 67% of foodborne illness was reported from viruses only, and norovirus was found to be the most predominant virus responsible for the illness. Outbreaks from viruses associated with fruits and vegetables were reported from coleslaw, tossed salads (White, 1986), green salads (Griffin et al., 1982), green onions (Wheeler et al., 2005), fresh-cut fruits (Harwaldt et al., 1994), small soft fruits such as raspberries (Gaulin et al., 1999; Calder et al., 2003; Colterelle et al., 2005; Korsager et al., 2005), and potato salad (Patterson, 1997).

In North America, outbreak due to protozoans was reported during the late 1990s from fresh raspberries imported from Guatemala. The raspberries were contaminated with protozoans *Cyclospora cayetanensis* (Herwaldt, 2000). Although protozoans such as *Cryptosporidium parvum* and *Giardia lamblia* are more likely to cause waterborne illness, they are also associated with green onions (Quinn et al., 1997) and basil (Lopez et al., 2001) outbreaks.

7.5 RESPONSES OF PATHOGENS TO VARIOUS FOOD ENVIRONMENTS

Foodborne pathogens are exposed to different physical and chemical environments during production, processing, preservation, storage, transportation, and consumption. As a result, organisms may become stressed and show several changes of altered characteristics. The term *stress* describes the response against any agent or treatment that causes *injury* to the microorganisms. This gives a more subtle meaning that is "not necessarily causing physical damage but altering organism behavior" (Hurst, 1984). Though food is a metabolically supportive and rich source of nutrients for the growth of microorganisms, intrinsic factors in foods such as water activity (a_w), pH, oxidation–reduction potential, and competitive exclusion by protective cultures also cause some stress (Archer, 1996). Other processing stresses in food may include exposure to acids, bases, bioactive antimicrobial peptides, oxidants, osmotic pressure, heating, freezing, thawing, and the presence of innate and supplemented antimicrobial compounds, which often acts as bacteriostatic or bactericidal agents (Miller et al., 2000).

In the last 50 years, three different terminologies have been applied to describe bacterial altered character when exposed to some physical and chemical stress conditions, namely, sublethal injury, viable but nonculturable (VBNC) state, and stress adaptation (Andrew and Russell, 1984; Hurst and Nasim, 1984; Ray, 1989; Colwell and Grimes, 1995; Yousef and Juneja, 2003). Depending on the type of stress and its effect on bacterial cells, they are categorized as minor to moderate, severe, extreme, and eventually lethal (Storz and Hengge-Aronis, 2000). Bacterial cells exposed to minor stress completely adapt to the changed conditions and multiply with normal growth rate. Low levels of stress may cause a transient adaptation (adaptive response) with a temporary physiological change that often results in increased stress tolerance (Yousef and Juneja, 2003). During lethal conditions, stress is experienced only by a fraction of cell population accompanied by stress gene responses leading to adaptive mutations that might improve survival and results in the death of only few cells (Archer, 1996; Storz and Hengge-Aronis, 2000). Moderate stress may result in a continuum of injury, ranging from mild to severe (Stephens et al., 1997; Mackey, 2000).

Studies have been conducted to determine pathogen responses and adaptation to stress. When one or two generations of *E. coli* cells were exposed to pH 5.0, subsequently they were able to survive better at pH 3 and 4 (but not pH <2), and when *E. coli* cells were exposed to temperature 50°C, they adapted resistance to survive at 60°C but not at >72°C (Buchanan and Edelson, 1999). Similarly, acid-adapted *L. monocytogenes* was able to survive at pH 3.5 and also developed resistance against nisin. When *L. monocytogenes* cells were exposed to 0.1% H_2O_2, they developed cross-resistance against subsequent exposure to 0.5% H_2O_2, 5% ethanol, 7% NaCl, pH 5, and 45°C as compared to nonexposed control cells.

Increase in D value of pathogens represents heat resistance of an organism, and the studies were conducted both *in vivo* and *in vitro.* When *S. aureus*, *L. monocytogenes*, *Salmonella* serovars, and *E. coli* O157:H7 cells were suspended initially at low pH fruit juices (orange, apple, or grape juices, pH 3.5–3.9) and subsequently heated at 56°C, a significant increase in the D value of all the organisms was observed (Mazzotta, 2001). This indicates that the organisms develop cross-resistance to subsequent stresses and may need more effective treatments to control the foodborne pathogens during processing and preservation.

Earlier studies have indicated that stress responses of the pathogens are mediated by changing the lipid profile of the cytoplasmic or cell membrane, to maintain the fluid state. Thus, during low- or high-temperature stress, the cell membrane was accumulated with low molecular weight and unsaturated fatty acids, whereas at normal growth conditions, accumulation with high molecular weight and saturated fatty acids was observed (Booth, 1999). Lately, stress adaptation of bacteria is reported to be mediated through the synthesis of shock or stress response proteins, some of which are specific for specific stress, whereas others are nonspecific and expressed in response to one or more stresses. Synthesis of stress response proteins protects the cell structures, DNA, and many enzymes. Synthesis of stress response proteins in large quantities is mediated through the cascade of stress-related gene system or global gene regulator. Expression of stress-related genes is initiated by sigma factors synthesized by specific genes.

In *Staphylococcus* spp., predicted growth curve analysis revealed that both wild type (WT) and *sigB*$^-$ (mutant) staphylococcal cultures were able to grow normally at 37°C. While heat (42.5°C) and alkaline stresses (pH 10) have a moderate effect on WT and *sigB*$^-$ staphylococcal strains, in contrast a lethal effect of saline stress (8% NaCl) on *sigB*$^-$ staphylococcal strains was observed, though *Staphylococcus* spp. have the ability to grow up to 20% of NaCl. WT and *sigB* mutant strains of *Staphylococci* produced SEA and SEB toxins even at stress conditions in synthetic medium (BHI) and in food model systems used to study the effect of acid and heat stresses (Rohinishree, 2015). Preservation techniques, if not used to optimum level, can enhance the expression of the toxigenic genes and toxin production by *Staphylococci* (Rohinishree, 2015; Rohinishree and Negi, 2016).

7.6 FOOD SAFETY AND EFFECTIVE MANAGEMENT STRATEGIES

Food safety is a growing concern for consumers and professionals in the field of food and food processing sectors. There is a growing demand for food safety information at the international, national, and local levels. Occurrence of foodborne illness worldwide aroused the concern and anxiety of the public about food safety. Most of these foodborne illnesses are due to mishandling and improper preparation/storage of food by food handlers. In order to ensure food safety and prevent food poisoning, all food businesses including food service organizations should comply with the existing food regulations. Besides regulations, WHO encourages governments, industry, and consumer organizations to disseminate five key messages: cleanliness, separation of raw and cooked foods, proper preparation, use of safe water and raw materials, and proper food handling as key to FBD prevention, which are very important for food hygiene and ensuring safety of foods.

7.6.1 Prevention of Contamination

To some extent, contamination of fruits and vegetables from pathogenic microorganisms can be controlled during preharvest, harvest, and postharvest stages. As an initiative to ensure the safety of fruits and vegetables and provide assurance about health of the people, U.S. Food and Drug Administration (USFDA) and U.S. Department of Agriculture (USDA) developed some guidance on good agricultural practices (GAPs) and good manufacturing practices (GMPs) about growing, harvesting, washing, sorting, packing, and transporting most fruits and vegetables sold to consumers in unprocessed, minimally processed, and processed forms.

7.6.1.1 Preharvest

Selection of organic manure is a primary factor, and comprehensive guidelines on treatment and time of its application to minimize contamination of crops are available. In the United Kingdom, annually over 90 million tonnes of animal waste are used as organic manure, and they have developed guidelines as Safe Sludge Matrix (ADAS) (Anon, 2001) and the Codes of Good Agricultural Practice. The Safe Sludge Matrix states that even when enhanced-treated sludge is applied to land, a 10-month harvest interval is necessary, whereas a 30-month harvest interval is required for salad crops if conventionally treated sludge is used. These intervals should be sufficient to ensure the microbial quality of produce at harvest. In the United States, Environment Protection Agency Part 503 Biosolids Rule (Anon, 1993) and Canadian Ministry of the Environment guidelines (Anon, 1996) have applied similar recommendations for organic manure. In general, increasing the delay between the application of organic fertilizers and harvest could reduce the occurrence of foodborne pathogens on fruits and vegetables.

Soil is a rich source of microorganisms of all kinds including pathogenic or nonpathogenic microorganisms, spore formers or nonspore formers. Survival of microorganisms in the soil is influenced by several factors, for example, soil type, soil humidity and its variation, temperature, and microflora (Bryan, 1977; Tierney et al., 1977; Dowe et al., 1997). Soil is microbiologically processed during composting or aeration and also heated to 50°C–60°C temperature at the time of treatment. If this process is managed carefully, it will kill nonspore-forming foodborne pathogens (Strauch, 1991). Usually, vegetative foodborne pathogenic bacteria and viruses decline within a few days after their introduction into the soil (Van Renterghem et al., 1991; Dowe et al., 1997) or on the plant surfaces (Ward and Irving, 1987); however, in few cases, they may survive in low numbers for several weeks or months (Al-Ghazali and Al-Azawi, 1990; Dowe et al., 1997).

Water is one of the main sources for microbial contamination. Water from sewage plants is being used for irrigation purpose and sewage water contains all types of human pathogens. Water used for irrigation should be of suitable microbiological quality, and the presence or absence of pathogens/indicator microorganisms must be monitored to determine the quality of agricultural waters applied to crops (Gerba, 2009). Recycling of water for agricultural purposes has been implemented in several countries, and hygienic treatment of sewage water before use is a must to reduce the risk of contaminating the crop (Castillo Martín et al., 1994). The safety of treated sewage water depends on the efficiency and reliability of treatments to inactivate pathogens. The establishment of appropriate guidelines for irrigation water is difficult (Matthews, 2013, 2014).

Pathogens present on hands of workers can be transferred to fruits or vegetables or on food contact surfaces during harvest (Todd et al., 2008, 2009), and the lack of suitable sanitary hand-washing facilities in the production area can potentially create a hygienic problem. These unhygienic practices transfer enteric viruses such as hepatitis A, and bidirectional transfer of *S. typhimurium* between bare and gloved hands was reduced on hands by using the combination of hand-washing and application of alcohol-based hand gel (Jimenez et al., 2007), which can minimize the risk of cross-contamination by hands or gloves. Appropriate hand-washing and bathroom facilities should be provided at work sites, without creating a potential hazard for the contamination of nearby crops (NACMCF, 1999a). Before transport to the processing facility, harvested fruits and vegetables may get contaminated through soil, as observed for baby greens, which are commonly harvested into bins, and the bins are often directly placed onto soil and stacked on top of each other, which leads to the transfer of contaminants during stacking. A change in harvesting practices is suggested to avoid contamination, and the bins should not be placed on the ground nor should they be stacked after filling (Matthews, 2014).

7.6.1.2 Postharvest

Cutting, sorting, washing and sanitizing, cooling, packaging, storing, and transporting are processing operations that provide many opportunities for cross-contamination. After harvesting, fresh fruits and vegetables undergo immediate surface sanitation and rapid cooling to slow the product metabolism that reduces the growth of spoilage microorganisms. Reduction in the rate of metabolism also reduces the product respiration, which, in turn, reduces the rate of deterioration, or perishability, of the crop (Kader, 1992).

In many instances, cooling and sanitation of product are accomplished simultaneously through one or more washings with chilled sanitizing water. Chlorine is mostly used as a sanitizing agent in the form of sodium hypochlorite, calcium hypochlorite, or chlorine gas, by maintaining a constant chlorine dose of 50–75 ppm (Rickhow and Singer, 1985).

Rapid cooling of fresh fruits and vegetables after harvest is one of the important factors for maintaining the quality. Cooling technique not only decreases the rate of respiration but also extends the ripening process and indirectly reduces the growth of enteric pathogens (Buchholz et al., 2010; Matthews, 2014). Methods such as room cooling (e.g., tomatoes), forced air cooling (e.g., tomatoes and peppers), hydrocooling (e.g., leafy greens, carrots, asparagus), and vacuum cooling (e.g., leafy greens, pepper) will be used depending on the type of the commodity. For leafy green vegetables, hydrocooling and vacuum cooling methods are generally used (Buchholz et al., 2010), and these cooling processes may lead to cross-contamination or internalization of pathogens into the commodities. Foodborne pathogen *E. coli* O157:H7 was reported to filtrate in lettuce sample by vacuum cooling (Li et al., 2008). Some foodborne pathogens are able to survive and grow at refrigeration temperature and may become the cause for foodborne illnesses following cooling operations (Koseki and Isobe, 2005; Hsu et al., 2006; Khalil and Frank, 2010).

Packaging is one of the most important steps for the long-term preservation of fresh fruits and vegetables until it reaches the consumer. Bags, crates, hampers, baskets, cartons, bulk bins, and palletized containers are convenient containers for handling, transporting, and marketing fresh produce. More than 1500 different types of packages are used for fruits and vegetables in the United States and the number increases as the industry introduces new packaging materials and concepts (Robertson, 2003). The key features of packaging should be its proper design to contain and protect the produce from mechanical damage and poor environmental conditions during handling and distribution, and it must contain sufficient information to identify the produce.

A large variety of packaging materials are used for the fresh commodities, and they include wood, bamboo, rigid and foam plastic, solid cardboard, and corrugated fiber board. Selection of packaging material and structure mainly depends on the perforation of material, the distance to its destination, the value of the product, and the requirement of the market.

After harvesting fruits and vegetables, they continue to undergo biological advancement such as respiration and ripening. This causes gradual changes in the quality of produce. During respiration, the product consumes oxygen and liberates carbon dioxide, water, and heat and breaks down the carbohydrates, which in turn affects the freshness, taste, and health of the produce. To maintain the quality of the produce during storage, a suitable packaging material and a modified atmosphere should be created (Zagory and Kader, 1988). Modified atmosphere packaging is the method used to extend the shelf life of perishable and semiperishable food products by altering the relative proportions of atmospheric gases that surround the produce, and modified atmosphere is often used in the packaging mechanism for fruits and vegetables.

Storage temperature is the most important environmental factor that influences the deterioration of harvested fruits and vegetables commodities. An optimal shelf life of most perishable horticultural commodities is at and around 0°C for temperate and 10°C for tropical growth regions. The rate of deterioration of perishables however increases twofold to threefold with every 10°C increase in temperature. Temperature has a significant effect on internal and external factors, which influence the quality of fresh produce and the growth of pathogens. Temperatures either above or below the optimal range can cause rapid deterioration of fresh produce, and mesophilic range temperature favors the growth of most of the pathogens. Therefore, temperature management is the most effective tool for extending the shelf life of fresh horticultural commodities and keep them safe for consumption.

Control atmosphere storage is different in its composition from that of normal atmosphere (78% nitrogen, 21% oxygen, 0.03% carbon dioxide, and traces of the noble gases), where the components of gases are precisely adjusted to specific concentrations and maintained throughout the storage and distribution of the perishable produce. Optimum oxygen and carbon dioxide concentrations lower respiration and ethylene production rates, reduce ethylene action, delay ripening and senescence, and retard the growth of decay-causing pathogens (Kader and Rolle, 2004).

Transportation and distribution are the most important and critical parameter for ensuring supply of quality fruits and vegetables. Fast transportation with minimum damage is very important for successful marketing, and all the advantages of proper growth, harvest, and handling can be nullified by improper transport. Transportation requirements may vary with respect to different fruit and vegetable commodities for their safe preservation. As far as possible, environmental conditions (temperature; relative humidity; concentrations of oxygen, carbon dioxide, and ethylene) should be optimized in transport vehicles with a special system of ventilation and maintenance of proper temperature and relative humidity to maintain the quality of the fresh produce (Kader, 1992).

7.6.2 Food Safety Management Systems

A food safety management system (FSMS) is a network of interrelated elements that ensure the safety of food for human consumption. Programs, plans, policies, procedures, objectives, methods, practices, processes, goals, controls, roles, responsibilities, relationships, documents, records, and resources are the key elements of FSMS.

The Codex Alimentarius Commission was created by the joint efforts of FAO and WHO in 1963 to develop food standards, guidelines, and related texts. Later in 1969, The Codex Alimentarius Commission developed Recommended International Code of Practice—General Principles of Food Hygiene (GHP). In 2005, the ISO (International Organization for Standardization) brought out FSMSs (ISO 22000:2005) to harmonize food safety at global level. FSMSs consists of five basic key elements and provides an excellent tools for professionals in the food supply chain (van Schothorst, 2004), which includes GMP/GHP, hazard analysis critical control point (HACCP), management element/system, statutory and regulatory requirements, and communication.

GMPs for handling, packing, storage, and transportation of fresh produce include operations such as produce cleaning and pretreatment, packing, storage and transportation, equipment cleaning, and sanitation. Food laws and regulations have been developed by adapting food safety system, investigating FBD outbreaks, and implementing international food laws and regulations.

HACCP system is a process control system that identifies where hazards might occur in the food production process and puts into place stringent actions to prevent the hazards from occurring. By strictly monitoring and controlling each step of the process, there is less chance for hazards to occur. In food processing, HACCP is an integral part to identify and control microbiological, chemical, and physical contaminants. At consumer level by following proper storage, handling, cooking, and cleaning procedures, one can implement HACCP-like practices. The HACCP approach has resulted in improved consumer protection and has promoted domestic and international food trade. A classical HACCP system is based on written work plan, and this plan must be specific to facilities, customers, equipment, process, and operations. Several food service organizations do not consider HACCP system feasible due to the multiplicity of food products, lack of standardized methods, lack of systematic production planning, and lack of expertise to develop the HACCP system. Based on the microbiological criteria on food, the National Advisory Committee created seven basic principles for HACCP system, which include (1) administration of a hazard analysis, (2) determination of critical control points (CCP), (3) establishment of critical limits for each CCP, (4) establishment of monitoring procedures for CCP, (5) identification of corrective actions, (6) verification of a system work, and (7) establishment of procedures for records maintenance and documentation.

7.7 FOODBORNE DISEASE SURVEILLANCE SYSTEMS AND FOOD RECALL

Estimating FBDs with accuracy at global level is a very difficult task as in most of the countries FBDs are not properly recorded. The variety and extent of FBDs are too vast to get accurate data on their incidence or prevalence. Many times, the incidences of FBDs do not come to the notice of health authorities. WHO estimates that in developing countries, only 1% of the FBD episodes are notified, whereas in developed countries, this figure is 10% (Bhat and Nageswara Rao, 1987).

Lack of integrated food safety framework poses danger to the food supply chain. Microbiological contamination of food and resulting foodborne illness are routine incidence in underdeveloped countries, though majority of the foodborne illness cases go unreported and unrecognized. Foodborne illness concerns related to food safety in different food chains may be mitigated through strict monitoring and surveillance throughout the world. Correct evaluation of the burden of foodborne infections and outbreak identification is not possible without proper monitoring and surveillance. Strengthening the surveillance systems to prevent outbreak requires intervention of international agencies, awareness, and strict legislation in the countries of world.

Surveillance system is defined as the systematic collection, analysis, and interpretation of data required for the planning, implementation, and evaluation of public health practice. Many surveillance systems have been used for the determination of occurrence of FBDs, and they rely on data available from local and state health agencies. Surveillance systems play a major role in detecting and preventing the outbreak of foodborne illnesses. The main objectives of surveillance systems are to (1) estimate the burden of disease in order to determine the magnitude of the problem, (2) monitor trends whether the situation is improving or worsening, (3) detect outbreaks to determine insistent action, (4) assess control programs to obtain information on performance, and (5) generate data to be used for risk analysis and ensure safety of food supplies (Teutsch and Churchill, 2000).

Foodborne surveillance systems are categorized into no formal system, syndromic surveillance, and laboratory-based surveillance or integrated food chain surveillance. *No formal surveillance system* is typical for countries where there is political instability, recent history of war, or struggle for extreme poverty. In this case, the public health system will not be a priority or generally outside agencies undertake certain aspects of surveillance. Even though large or unusual outbreaks may be detected and investigated by outside agencies such as NGOs, no such data are available.

Syndromic surveillance systems are generally used to target and investigate the potential disease cases (e.g., influenza). This system monitors data through school absenteeism, emergency calls, hospitals, over-the-counter drug sale records, Internet searches, and other data sources to detect unusual patterns. When an activity is noticed in any of the disease-monitoring systems, epidemiologists and public health professionals are alerted that there may be an issue. Based on the available data information related to outbreaks with the seasonal variation at local level or at national level and the number of reported cases, outbreak will be defined as at risk and high risk among the populations.

Laboratory-based surveillance and integrated food chain surveillance systems are the preferred FBD surveillance system, which rapidly detects the outbreak strains. Laboratory-based surveillance system works systematically by collecting, analyzing, and interpreting the disseminated data available based on laboratory-confirmed infections for public health action and classifying the disease based on standard case definitions. Clinical laboratories are key sources of isolates and random isolates are routinely sent to reference laboratories for confirmation or subtyping. To distinguish strains, subtyping is an important method to detect the outbreak and sources of infection. Subtyping methods include (1) serotyping, (2) phage typing, (3) antibiotic resistance profiling, and (4) molecular typing, for example, pulsed-field gel electrophoresis (PFGE). Taking a case of the identification of *Salmonella*, this common foodborne pathogen has more than 2500 serotypes and each one of them has its own biology and epidemiology (serotype Typhi causes typhoid fever, serotype Enteritidis is commonly transmitted by eggs, and serotype Typhimurium is transmitted by a wide variety of animal food sources), and all these serotypes can be identified by using these subtyping methods.

7.7.1 Foodborne Surveillance in the United States

Most of the surveillance systems are provided by CDC in the United States and they are as follows:

Foodborne Disease Active Surveillance Network (FoodNet): *FoodNet is an active sentinel surveillance system that* provides information of incidence and trends in foodborne and diarrheal diseases. Ten state health departments in the United States are participating and working in collaboration with CDC, USDA, and FDA. The FoodNet surveillance covers 15% of the U.S. population and conducts population-based surveillance for foodborne pathogens that include seven bacteria and two parasitic infections. FoodNet surveillance also conducts studies and analyzes the epidemiologic design, which in turn helps public health officials to better understand the epidemiology of FBDs in the United States.

Annual reports of FoodNet are the summaries of information collected through active population-based surveillance for infections caused by any of nine pathogens transmitted commonly through food. A preliminary version of this report is published annually in CDC's MMWR. The FoodNet final report becomes available after current census information becomes available. FoodNet reports and studies are also available online for viewing of the general public.

National Antimicrobial Resistance Monitoring System for Enteric Bacteria (NARMS): Most of the foodborne pathogens are resistant to antibiotics and NARMS conducts examination to detect, respond, and prevent antimicrobial resistance among foodborne bacteria in humans, retail meat, and animals with the collaboration of the CDC, Center for Veterinary Medicine in the FDA, and USDA. NARMS annual reports are published and can be accessed online.

National Electronic Norovirus Outbreak Network (CaliciNet): Noroviruses, a group of caliciviruses, are the most important viruses responsible for the majority of the foodborne illness in the United States. With the joint venture of CDC, state and local public health laboratories developed and implemented an electronic norovirus outbreak surveillance network (CaliciNet), which was launched in March 2009. The aim of the CaliciNet is to provide a better understanding of noroviruses to control its spread. This network compares norovirus sequences to identify emerging norovirus strains and to rapidly link norovirus outbreaks with a common food source. Periodic outbreak investigation reports are published weekly in CDC and *MMWR* and posted online. Since 2009, norovirus outbreaks, regardless of mode of transmission (i.e., foodborne or person-to-person), are reported to the National Outbreak Reporting System (NORS), which is linked to CaliciNet.

National Molecular Subtyping Network for Foodborne Disease Surveillance (PulseNet): PulseNet is a national network of local, state, territorial, agricultural, and federal laboratories coordinated by CDC and the Association of Public Health Laboratories (APHL). Application of standardized pulsed field gel electrophoresis (PFGE) methods revolutionized the detection of foodborne bacterial pathogens and outbreak by comparing the uploaded PFGE patterns with the available patterns of pathogens isolated from humans, animals, and foods. It helps in identifying a possible outbreak, and it can be prevented. It helps to investigate the possible linkages between the pathogens (e.g., outbreaks), especially when the outbreak occurs in multiple sites across the country. Publications and presentations of PulseNet studies are available online.

National Notifiable Diseases Surveillance System (NNDSS): NNDSS is a reporting surveillance system, in which healthcare providers and laboratory professionals report the individual cases of disease when selected pathogens are identified in patient specimens or specific clinical syndromes are recognized. Local public health agencies report the outbreak of diseases to the state or territorial public health agency, which in turn voluntarily submits the information to the NNDSS. The reported FBDs include botulism, hemolytic uremic syndrome, listeriosis, salmonellosis, shiga toxin–producing *E. coli* infections, and vibriosis. Annual reports of nationally notifiable infectious diseases are published in the *MMWR* Summary of Notifiable Diseases, United States. Early release of state/territory counts are published in CDC's weekly publication.

National Outbreak Reporting System (NORS): Earlier, NORS system was called as electronic Foodborne Outbreak Reporting System (eFORS; http://www.cdc.gov/outbreaknet/nors/). NORS collects reports of foodborne outbreak from enteric bacterial, viral, parasitic, and chemical agents from state, local, and territorial public health agencies, which in turn reports to CDC. The NORS surveillance team analyzes available data to better understand the impact of foodborne outbreaks on human health, pathogens, foods, settings, and other contributing factors involved in the illness. Earlier in 2009, the system was using modules for reporting enteric disease outbreaks that are transmitted through water, person-to-person contact, or direct contact with animals. Outbreak surveillance data are published in the *MMWR*. State-specific information about outbreaks, a Foodborne Outbreak Online Database, is available online.

Contributing factor surveillance (Environmental Health Specialists Network [EHS-Net]): State and local public health agency investigators collect the information about factors responsible for foodborne outbreaks by food control officials and reports to CDC. To investigate that the environmental factors are one of the contributing factors for FBDs, CDC established EHS-Net in 2000. The FDA, the Food Safety Inspection Service in the USDA, CDC, and environmental health specialists and epidemiologists are the participants for EHS-Net. An electronic reporting EHS-Net was developed in 2011 by CDC to report the contributing factors and other root causes of foodborne outbreaks. Publications and presentations of EHS-Net are available online.

Public Health Laboratory Information System (PHLIS): PHLIS is an electronic reporting system that reports for *Salmonella* and *Shigella* isolates of human sources. State Public Health Laboratory Directors and State and Territorial Epidemiologists use this system. Its annual reports are published regularly and can be found online.

7.7.2 Foodborne Surveillance in Europe

European and EU-led initiatives for surveillance led to the establishment of the European Food Safety Authority (EFSA) and the European Centre for Disease Control and Prevention (ECDC) and the introduction of legislation to revise Zoonoses Directive and development of multiannual National Control Plans (NCP) in member states. The ECDC was established in 2005 to identify, access, and communicate against current and emerging infectious diseases on human health. The overall goal of the ECDC is to reduce and prevent the incidence of communicable diseases in Europe. A significant development in ECDC was to facilitate a strategic and coordinating approach to surveillance, which harmonizes case definitions of infectious diseases for the development of the European Surveillance System (TESSy) and the Threat Tracking Tool (TTT or 3T). The information provided by these tools can be used by decision-makers, professionals, and healthcare workers in member states and European organizations. Earlier food and waterborne diseases were investigated by an International surveillance network ENTERNET, which was founded by EU, and later a routine activity was undertaken by ECDC. ECDC collects the data of infectious diseases such as salmonellosis, VTEC, and *Campylobacter* infections on a quarterly basis and gastrointestinal pathogens on annual basis and analyzes them in collaboration with EFSA.

The EFSA established in 2002 works with the collaboration of national authorities and provides scientific advice regarding direct or indirect impact on food and feed safety. It is a part of comprehensive legislation to improve EU food safety and ensures a high-level confidence and protection on EU food supply. The role of EFSA is to assess and communicate on all the risks associated with the food chain. A part of EFSA collects, identifies, and analyzes the scientific data in the case of food crisis and provides scientific support to the European Commission.

Every year member states and other reporting countries submit data on zoonoses, zoonotic agents, antimicrobial resistance, microbiological contaminants, and foodborne outbreaks to EFSA and ECDC. EFSA Zoonoses Unit, in collaboration with ECDC, analyzes the data and publishes annual Community Summary Report on zoonotic agents in animal and food. The aim of the EFSA is to harmonize data collection and provide information to support risk management by intriguing effective decisions to protect humans against zoonotic infections.

7.7.3 Foodborne Surveillance in Australia and New Zealand

Food Safety Authority of Australia and New Zealand (FSANZ) and other government agencies in Australia and New Zealand monitor the food supply to ensure that it is safe and that foods comply with standards for microbiological contaminants, pesticide residue limits, and chemical contamination. FSANZ coordinates a Bi-National Food Surveillance and Enforcement Strategy that involves food/health agencies in Australia and New Zealand to discuss and share information about monitoring and surveillance of the food chain in Australia and New Zealand. FSANZ acts as the central point for the collection of food surveillance data from public health units in Australia and New Zealand. These data include the results of general compliance testing and specially targeted surveys conducted in the various jurisdictions.

FSANZ monitors the safety of the food supply through the Implementation Subcommittee for Food Regulation. FSANZ also collects food surveillance data, including the results of general compliance testing and targeted surveys conducted in jurisdictions from public health units across Australia and New Zealand. FSANZ publishes surveys through Food Surveillance News e-mail alerts. Occasionally, state and territory governments undertake targeted surveys of food to ensure it is safe and complies with food standards.

7.7.4 Foodborne Surveillance in African Region

In 1998, the Integrated Disease Surveillance and Response was established with the collaboration of regional office and partners in African regions. This system addresses cholera, bloody diarrhea, and diarrhea with dehydration among children under 5 years old. The WHO Global Foodborne Infections Network improved the system by providing training for isolation, identification, and typing of *Salmonella* spp., *Campylobacter* spp., *Vibrio cholerae*, *Vibrio* species non-cholera, and *Shigella* of human and food sample isolates.

7.7.5 Foodborne Surveillance in China

China is the world's largest producer of agricultural products by value. The agricultural productions as well as the manufacture of food packaging, containers, chemical additives, drug production, and business regulations are directly under the control of Chinese government. Quality and safety regulations for agricultural production are covered in the Agriculture Law of the People's Republic of China. The first comprehensive Food Safety Law (FSL) was passed by China's National People's Congress Standing Committee, which came into effect on June 1, 2009 (China, Peoples Republic of FAIRS Subject Report, 2009). The regulatory body was rebranded and restructured as the China Food and Drug Administration (CFDA) in March 2013, which replaced a large group of overlapping regulators. The segmented regulatory model supplemented by supervision of individual foods was converted to a centralized regulatory model by CFDA, which has integrated various regulatory resources and has unified the supervision and management of food production, distribution, and consumption.

The FSL describes Surveillance and Assessment of Food Safety Risks, which says that a national surveillance system for food safety risks will be established to monitor FBDs, food contamination, and other food-related hazards. The executive department of health under the State Council will in a timely manner adjust the surveillance plan on food safety risks upon verification of the information with other relevant authorities. A national assessment mechanism for food safety risks will be established to assess the risks on biological, chemical, and physical hazards in foods and food additives. The FSL also states that the executive department of health under the State Council, upon discovering any possible safety problem through food safety risk surveillance or reports of other parties, will immediately organize inspection and food safety risk assessment. Food safety risk assessment results will be the scientific basis for developing and modifying food safety standards as well as regulating food safety (Wu and Zhu, 2014).

7.7.6 Foodborne Surveillance in India

In India, effective food control system is undermined by plethora of fragmented legislation, multiple jurisdictions, and weakness in surveillance, monitoring, and enforcement, which neither assures safety, hygiene, or quality. The Government of India in 2006 had enacted the Food Safety and Standards Act to regulate food manufacture, storage, distribution, sale, and import under a single independent statutory authority the Food Safety and Standards Authority of India (FSSAI). The Indian Food Processing Policy, calls for certifying authorities to be vigilant in enforcing standards at least in case of processed foods, training of quality enforcement staff in HACCP, GMP and GHP requirement, well-equipped and certified laboratory requirements, harmonization of national standards with Codex standards, and relevant science-based standards relevant to Indian conditions (Anon, 2005).

FBD outbreak information (2002–2010) is available in the comprehensive food safety information website (FoodHACCP.com) and also in the website www.indiaenvironmentportal.org.in managed by the Centre for Science and Environment. But none of the food poisoning cases were investigated and reported in the scientific journals.

CDC also collaborates with the Government of India, Indian institutions, and international organizations to strengthen the monitoring system in India. CDC has developed foodborne/acute diarrheal disease surveillance in India with the goal to reduce and control the number of acute diarrheal disease illnesses and economic costs. The Government of India had also declared year 2008–2009 as "Food Safety and Quality Year" to ensure safe, healthy, and sustainable food supply and bring about awareness among consumers.

7.7.7 Food Recall

A food recall is defined as an action taken to remove foods that may pose a safety risk from the sale, distribution, or consumption to consumers. The food can be removed from the market at any stage of the food chain, including that possessed by consumers. Food recall is a fundamental tool in the management of risks in response to food safety events and emergencies in developed countries. A food recall may be initiated as a result of a report or complaint from a variety of sources—manufacturers, wholesalers, retailers, government agencies, and consumers. Recalls are conducted by food businesses to ensure that potentially hazardous or unsafe foods are not consumed. Microbiological contamination is the primary cause of food recall, but it may be due to chemical contamination. Food recall is mandatory for consumer protection in industrialized countries and commonly practiced by multinational food industries.

The globalization poses new challenges in the conduct of key activities associated with food recalls, such as the trace-forward and trace-back activities required for a food suspected or confirmed to be unsafe. The FAO/WHO International Food Safety Authorities Network (INFOSAN) coordinates communication regarding contaminated food that has been internationally distributed to allow for food recalls in importing countries. Such alerts are communicated to national INFOSAN Emergency Contact Points in each country for their attention and action as required.

7.8 IMPROVING THE SAFETY AND QUALITY OF FRESH FRUIT AND VEGETABLES

One of the important factors to ensure safety and quality of fresh fruit and vegetables is by following GMPs for handling, packing, storing, and transporting fresh produce. The harvested produce must be cleaned properly and treated to prevent its contamination. It must be packed in suitable packaging material and stored under proper storage conditions to avoid contamination. Transportation vehicle must be cleaned properly before loading fruits and vegetables. Hygienic conditions should be maintained by the way of equipment cleaning and sanitation during processing.

Enforcement of food laws and regulation is another important factor to ensure safety and quality of fresh fruit and vegetables. Food laws enacted to ensure food safety must be compiled with to ensure the safety of consumers. There must be a proper system to investigate FBD outbreaks by way of reporting the incidence, identifying the causal organism, and tracing its origin to prevent the future outbreaks. All the international food laws and regulations must be complied when export of product is done.

Food safety and quality assurance issues must be addressed to ensure safety and quality of fresh fruit and vegetables. The product must be safe to consume and free from chemical and microbiological contaminants, and its quality in terms of color, texture, flavor, and nutritive value must be assured. Quality attributes of the product should match to those defined by legal limits. The product must be categorized to proper grades and standards before marketing.

7.9 CONCLUSION

Ensuring food safety starts with the production of food at the farm level. The fruits and vegetables must be grown in accordance with proper recommendations, and misuse of agrochemicals, including pesticides, fungicides, and growth hormones, should be avoided. The microbial and chemical risks could be introduced at the farm level, and therefore, GAPs should be applied to reduce them. Ensuring food safety requires due attention not only during growing but also during harvesting, transporting, processing, storing, and finally food preparing and storing by consumers; therefore, a proper care must be taken by all involved.

The burden due to FBD is not fully understood, probably due to the absence of surveillance system in most parts of the world. Though the socioeconomic impact of FBDs is very high, there are limited effective measures to reduce their incidences. Nonexistence of infrastructure and nonadherence with the existing food laws are the significant factors to make our food supply unsafe, which is further intensified by low level of public awareness on health and economic consequences of foodborne illnesses. In developing countries, illiteracy, coupled with poverty and corruption, and the absence of proper legislation make food safety even more complicated.

To prevent foodborne outbreaks in the future, strengthening of food safety policies and collection of epidemiological data should be aimed. In developing world, there is a need to increase awareness on food spoilage, FBD, and effective food control measures. The continuous monitoring of FBDs coupled with surveillance program can help in reducing incidences of foodborne outbreaks worldwide.

REFERENCES

Abakpa, G.O., V.J. Umoh, J.B. Ameh, and S.E. Yakubu. 2013. Microbial quality of irrigation water and irrigated vegetables in Kano State, Nigeria. *International Food Research Journal* 20:2933–2938.

Abdul-Raouf, U.M., L.R. Beuchat, and M.S. Ammar. 1993. Survival and growth of *Escherichia coli* O157:H7 on salad vegetables. *Applied and Environmental Microbiology* 59(7):1999–2006.

Ahmed, W., S. Sawant, F. Huygens, A. Goonetilleke, and T. Gardner. 2009. Prevalence and occurrence of zoonotic bacterial pathogens in surface waters determined by quantitative PCR. *Water Research* 43:4918–4928.

Ailes, E.C., J.S. Leon, L.A. Jaykus et al. 2008. Microbial concentrations on fresh produce are affected by postharvest processing, importation, and season. *Journal of Food Protection* 71:2389–2397.

Akhtar, S., M.R. Sarker, and A. Hossain. 2014. Microbiological food safety: A dilemma of developing societies. *Critical Reviews in Microbiology* 40(4):348–359.

Al-Ghazali, M.R. and S.K. Al-Azawi. 1990. *Listeria monocytogenes* contamination of crops grown on soil treated with sewage sludge cake. *Journal of Applied Bacteriology* 69:642–647.

Andrew, M.H.E. and A.D. Russell Eds. 1984. *The Revival of Injured Microbes.* Academic Press, Orlando, FL.

Anon. 1993. United States Environmental Protection Agency Rule 503 in biosolids generation, use, and disposal in the United States. Washington, DC: US Environmental Protection Agency. Available from http://www.epa.gov/epaoswer/non-hw/compost/biosolid.pdf. Accessed April 22, 2007.

Anon. 1996. *Ministry of the Environment Guidelines for the Utilization of Biosolids and Other Wastes on Agricultural Land.* Toronto, Ontario, Canada: Ministry of the Environment. Available from http://www.ene.gov.on.ca/envision/gp/3425e.pdf. Accessed April 22, 2007.

Anon. 2001. The safe sludge matrix, 3rd edn. Wolverhampton, U.K.: ADAS. Downloadable from http://www.adas.co.uk/media_files/Publications/SSM.pdf. Accessed January 18, 2007.

Anon. 2005. Food processing policy. *Indian Food Industry* 24:16–19.

Archer, D. L. 1996. Preservation microbiology and safety: Evidence that stress enhances virulence and triggers adaptive mutations. *Trends in Food Science and Technology* 7:91–95.

Aruin, L.I. 1997. *Helicobacter pylori* infection is carcinogenic for humans. *Arkhiv Patologii* 59(3):74–78.

Asplund, K. and E. Nurmi. 1991. The growth of salmonellae in tomatoes. *International Journal of Food Microbiology* 13(2):177–182.

Austin, J.W., K.L. Dodds, B. Blanchfield, and J.M. Farber. 1998. Growth and toxin production by *Clostridium botulinum* on inoculated fresh-cut packaged vegetables. *Journal of Food Protection* 61(3):324–328.

Avery, L.M., K. Killham, and D.L. Jones. 2005. Survival of *E. coli* O157: H7 in organic wastes destined for land application. *Journal of Applied Microbiology* 98:814–822.

Aytac, S.A. and L.G.M. Gorris. 1994. Survival of *Aeromonas hydrophila* and *Listeria monocytogenes* on fresh vegetables stored under moderate vacuum. *World Journal of Microbiology and Biotechnology* 10:670–672.

Barth, M., T.R. Hankinson, H. Zhuang, and F. Breidt. 2009. Microbiological spoilage of fruits and vegetables. In W.H. Sperber and M.P. Doyle, Eds., *Compendium of the Microbiological Spoilage of Foods and Beverages.* Food Microbiology and Food Safety, Springer Science + Business Media, LLC, pp. 135–183.

Begue, R.E., J.L. Gonzales, H. Correa-Gracian et al. 1998. Dietary risk factors associated with the transmission of *Helicobacter pylori* in Lima, Peru. *American Journal of Tropical Medicine and Hygiene* 59:637–640.

Bernagozzi, M., F. Bianucci, R. Sachetti, and P. Bisbini. 1994. Study of the prevalence of *Listeria* spp. in surface water. *Zentralblatt Fur Hygiene Und Umweltmedizin* 196:237–244.

Berrang, M.E., R.E. Brackett, and L.R. Beuchat. 1989a. Growth of *Listeria monocytogenes* on fresh vegetables stored under controlled atmosphere. *Journal of Food Protection* 52(10):702–705.

Berrang, M.E., R.E. Brackett, and L.R. Beuchat. 1989b. Growth of *Aeromonas hydrophila* on fresh vegetables stored under a controlled atmosphere. *Applied and Environmental Microbiology* 55(9):2167–2171.

Beuchat, L.R. 1996. Pathogenic microorganisms associated with fresh produce. *Journal Food Protection* 59(2):204–216.

Beuchat, L.R. 2002. Ecological factors influencing survival and growth of human pathogens on raw fruits and vegetables. *Microbes and Infection* 14:413–423.

Beuchat, L.R. 2006. Vectors and conditions for pre-harvest contamination of fruits and vegetables with pathogens capable of causing enteric diseases. *British Food Journal* 108:38–53.

Beuchat, L.R. and R.E. Brackett. 1990. Survival and growth of *Listeria monocytogenes* on lettuce as influenced by shredding, chlorine treatment, modified atmosphere packaging and temperature. *Journal of Food Science* 55(3):755–758:870.

Beuchat, L.R. and R.E. Brackett. 1991. Behavior of *Listeria monocytogenes* inoculated into raw tomatoes and processed tomato products. *Applied and Environmental Microbiology* 57(5):1367–1371.

Beuchat, L.R., R.E. Brackett, D.Y. Hao, and D.E. Conner. 1986. Growth and thermal inactivation of Listeria monocytogenes in cabbage and cabbage juice. *Canadian Journal of Microbiology* 32:791–795.

Beuchat, L.R., R.E. Brackett, and M.P. Doyle. 1994. Lethality of carrot juice to *Listeria monocytogenes* as affected by pH, sodium chloride and temperature. *Journal of Food Protection* 57(6):470–474.

Beuchat, L.R. and Doyle, M.P. 1995. Survival and growth of *Listeria monocytogenes* in foods treated or supplemented with carrot juice. *Food Microbiology* 12:73–80.

Bhat, R.V. and N.R. Rao. 1987. Foodborne diseases in India. *Indian Journal of Pediatrics* 54 (4):553–562.

Booth, L.R. 1999. Adaptation to extreme environments. In Lengeler, J.W., G. Drews, and H.G. Schlegel, Eds., *Biology of the Prokaryotes.* New York: Blackwell Science, p. 652.

Brackett, R.E. 1992. Shelf stability and safety of fresh produce as influenced by sanitation and disinfection. *Journal of Food Protection* 10(55):808–814.

Brown, L.M. 2000. *Helicobacter pylori:* Epidemiology and routes of transmission. *Epidemiologic Reviews* 22(2):283–297.

Bryan, F.L. 1977. Diseases transmitted by foods contaminated by wastewater. *Journal of Food Protection* 40:45–56.

Bryan, F.L. and M.P. Doyle. 1995. Health risk and consequence of *Salmonella* and *Campylobacter jejuni* in raw poultry. *Journal of Food Protection* 58:326–344.

Buchanan, R.L. and Edelson, S.G. 1999. pH dependent stationary-phase acid resistance response of enterohaemorrhagic *Escherichia coli* in the presence of various acidulants. *Journal of Food Protection* 62:211.

Buchholz, A.L., G.D. Davidson, and E.T. Ryser. 2010. Microbiology of fresh and processed vegetables. In Shinha, N.K., Y.H. Hui, E.O. Evranuz, M. Siddiq, and J. Ahmed, Eds., *Handbook of Vegetables and Vegetable Processing.* Hoboken, NJ: Wiley-Blackwell, pp. 159–181.

Buchholz, A.L., G.R. Davidson, B.P. Marks, E.C.D. Todd, and E.T. Ryser. 2012. Quantitative transfer of Escherichia coli O157:H7 to equipment during small-scale production of fresh-cut leafy greens. *Journal of Food Protection* 75:1184–1197.

Calder, L., G. Simmons, C. Thornley, P. Taylor, K. Pritchard, G. Greening, and J. Bishop. 2003. An outbreak of hepatitis A associated with consumption of raw blueberries. *Epidemiology and Infection* 131:745–751.

Carlin, F. and C. Nguyen-The. 1994. Fate of *Listeria monocytogenes* on four types of minimally processed green salads. *Letters in Applied Microbiology* 18:222–226.

Castillejo Rodriguez, A.M., E. Barco-Alcala, R.M. Garcia-Gimeno, and G. Zurera-Cosano. 2000. Growth modelling of *Listeria monocytogenes* in packaged fresh green asparagus. *Food Microbiology* 17:421–427.

Castillo, M.A. and E.F. Escartin. 1994. Survival of *Campylobacter jejuni* on sliced watermelon and papaya. *Journal of Food Protection* 57(2):166–168.

Castillo, M.A., J.C. Jordán, A.M. Fernández, B.G.V. Ruiz, J.H. Ruiz, J.L. Sorinas, R.N. Vargas-Machuca, and J.P. Muñoz. 1994. Health assessment criteria for projects direct reuse of treated urban wastewater, Government of Andalusia, Seville.

Castro-Rosas, J., J.F. Cerna-Cortes, E. Mendez-Reyes, D. Lopez-Hernandez, C.A. Gomez-Aldapa, and T. Estrada-Garcia. 2012. Presence of faecal coliforms, *Escherichia coli* and diarrheagenic *E. coli* pathotypes in ready-to-eat salads, from an area where crops are irrigated with untreated sewage water. *International Journal of Food Microbiology* 156:176–180.

Castro-Rosas, J. and E.F. Escartin. 2000. Survival and growth of *Vibrio cholerae* O1, *Salmonella typhi*, and *Escherichia coli* O157:H7 in alfalfa sprouts. *Journal of Food Science* 65(1):162–165.

Catteau, M., C. Krembel, and G. Wauters. 1985. Yersinia in raw vegetables. *Science of Aliments* 5:103–106.

CDC. 2006. Ongoing multistate outbreak of *Escherichia coli* serotype O157:H7 infections associated with consumption of fresh spinach—United States, September 2006. *Morbidity Mortality Weekly Report* 55:1045–1046.

Cevallos-Cevallos, J.M., M.D. Danyluk, G.Y. Gu, G.E. Vallad, and A.H.C. van Bruggen. 2012. Dispersal of *Salmonella* Typhimurium by rain splash onto tomato plants. *Journal of Food Protection* 75:472–479.

China, Peoples Republic of FAIRS Subject Report. 2009. Food Safety Law of the People's Republic of China. Available from http://apps.fas.usda.gov/gainfiles/200903/146327461.pdf. Accessed July 4, 2016.

Colwell, R.R. and D.J. Grimes, Eds. 1995. *Nonculturable Microorganisms in the Environment.* Washington, DC: ASM Press.

Cook, K.A., T. Boyce, C. Langkop, K. Kuo, M. Swartz, and D. Ewert. 1995. Scallions and shigellosis: A multistate outbreak traced to imported green onions. *Proceedings of the 44th Annual Conference of the Epidemic Intelligence Service.* Centres for Disease Control and Prevention, Atlanta, GA, p. 36.

Cotterelle, B., C. Drougard, C. Rolland, M. Becamel, M. Boudon, S. Pinede, O. Traoré, and K. Balay. 2005. Outbreak of Norovirus infection associated with the consumption of frozen raspberries, France, March 2005. *Eurosurveillance Weekly* 10: E050428.1. Available from http://www.eurosurveillance.org/ew/2005/050428.asp. Accessed July 4, 2016.

Danyluk, M.D., M. Nozawa-Inoue, K.R. Hristova, K.M. Scow, B. Lampinen, and L.J. Harris. 2008. Survival and growth of *Salmonella* Enteritidis PT 30 in almond orchard soils. *Journal of Applied Microbiology* 104:1391–1399.

Davis, H., J.P. Taylor, J.N. Perdue, G.N. Stelma Jr., J.M. Humphreys Jr., R. Rowntree III, and K.D. Greene. 1988. A shigellosis outbreak traced to commercially distributed shredded lettuce. *American Journal of Epidemiology* 128(6):1312–1321.

DeWaal, C.S. and F. Bhuiya. 2007. Outbreaks by the numbers: Fruits and vegetables. 1990–2005. Available from http://www.cspinet.org/foodsafety/IAFPPoster.pdf. Accessed March 3, 2010.

Dijkstra, R. 1989. Ecology of *Listeria. Microbiology. Alimentarius Nutrition* 7:353–359.

Dowe, M.J., E.D. Jackson, J.G. Mori, and C.R. Bell. 1997. *Listeria monocytogenes* survival in soil and incidence in agricultural soils. *Journal of Food Protection* 60:1201–1207.

Doyle, M.P. 1990. Fruit and vegetable safety: Microbiological considerations. *Horticultural Science* 25:1478–1482.

Erickson, M.C, C.C. Webb, L.E. Davey, A.S. Payton, I.D. Flitcroft, and M.P. Doyle. 2014. Internalization and fate of *Escherichia coli* O157:H7 in leafy green phyllosphere tissue using various spray conditions. *Journal of Food Protection* 77:713–721.

European Food Safety Authority (EFSA). 2007. The community summary report on trends and sources of zoonoses, zoonotic agents, antimicrobial resistance and foodborne outbreaks in the European Union in 2006. *The EFSA Journal* 130:23–352. Available from http://www.efsa.europa. eu/EFSA/DocumentSet/Zoon_report_2006_en,0.pdf. Accessed July 4, 2016.

European Food Safety Authority (EFSA). 2009. The community summary report on foodborne outbreaks in the European Union in 2007. Available from http://www.efsa.europa.eu/en/efsajournal/doc/271r.pdf. Accessed July 4, 2016.

Fallon, S.D., D.V. Rios, and J.M. Fonseca. 2011. Scrutinizing the sanitizing process involving aqueous chlorine from field to fresh-cut iceberg lettuce. *Acta Horticulture* 906:37–42.

Fang, F.C., S.J. Libby, N.A. Buchmeier, P.C. Loewen, and J. Switala. 1992. The alternative sigma factor *katF* (*rpoS*) regulates *Salmonella* virulence. *Proceedings of the National Academy of Science of the United States of America* 89:11978–11982.

FAO/ WHO. 2008. Microbiological hazards in fresh fruits and vegetables. Food and Agriculture Organization of the United Nations, Rome, Italy/World Health Organization, Geneva, Switzerland.

Farber, J.M., S.L. Wang, Y.Cai, and S. Zhang. 1998. Changes in populations of *Listeria monocytogenes* inoculated on packaged fresh-cut vegetables. *Journal of Food Protection* 61(2):192–195.

FDA. 1998. Guide to minimise the microbial food safety hazards for fresh fruits and vegetables: Guidance for industry. Center for Food Safety and Applied Nutrition. Washington DC: FDA, p. 20204.

FDA (Food and Drug Administration). July 31, 2001. Survey of domestic fresh produce: Interim results. Available from http://www.fda.gov/Food/FoodSafety/Product-SpecificInformation/FruitsVegetablesJuices/GuidanceComplianceRegulatoryInformation/ucm118844.htm. Accessed March 3, 2010.

FDA. 2009. Food Code: Annex 4—Management of food safety practices—Achieving active managerial control of foodborne illness risk factors. U.S. Food and Drug Administration, Silver Spring, MD, USA.

FDA. 2015. Analysis and evaluation of preventive control measures for the control and reduction/elimination of microbial hazards on fresh and fresh-cut produce: Chapter IV. *Outbreaks Associated with Fresh and Fresh-Cut Produce: Incidence, Growth, and Survival of Pathogens in Fresh and Fresh-Cut Produce.* Available from http://www.fda.gov/Food/FoodScienceResearch/SafePracticesforFoodProcesses/ucm091265.htm. Accessed July 4, 2016.

Fernandez Escartin, E.F., A. Castillo Ayala, and J. Saldana Lozano. 1989. Survival and growth of *Salmonella* and *Shigella* on sliced fresh fruit. *Journal of Food Protection* 52(7):471–472.

Fontham, E. T., B. Ruiz, A. Perez et al. 1995. Determinants of *Helicobacter pylori* infection and chronic gastritis. *American Journal of Gastroenterology* 90:1094–1101.

Francis, G. A., C. Thomas, and D. O'Beirne 1999. The microbiological safety of minimally processed vegetables. *International Journal of Food Science and Technology* 34:1–22.

Franz, E. and A.H. van Bruggen. 2008. Ecology of *E. coli* O157:H7 and *Salmonella enterica* in the primary vegetable production chain. *Critical Review of Microbiology* 34:143–161.

Franz, E., A.D. van Diepeningen, O.J. de Vos, and A.H. van Bruggen. 2005. Effects of cattle feeding regimen and soil management type on the fate of *Escherichia coli* O157:H7 and *Salmonella enterica* serovar *Typhimurium* in manure, manure-amended soil, and lettuce. *Applied Environmental Microbiology* 71:6165–6174.

Fredlund, H., E. Back, L. Sjoberg, and E. Tornquist. 1987. Watermelon as a vehicle of transmission of Shigellosis. *Scand ford Journal of Infectious Diseases* 19:219–221.

Fule, R.P., K.V. Ingole, S.V. Jalgaonkar, and B.U. Moon. 1996. Outbreak of food poisoning due to Salmonella paratyphi A var durazzo in Yavatmal (Maharashtra) in May 1995. *Indian Journal of Medical Research* 103:74–76.

Gannon, V.P.J., T.A. Graham, S. Read, K. Ziebell, A. Muckle, and J. Mori. 2004. Bacterial pathogens in rural water supplies in southern Alberta, Canada. *Journal of Toxicological and Environment Health* 67:1643–1653.

Garcia-Gimeno, R.M., M.D. Sanchez-Pozo, M.A. Amaro-Lopez, and G. Zurera-Cosano. 1996. Behaviour of *Aeromonas hydrophila* in vegetable salads stored under modified atmosphere at 4 and 15°C. *Food Microbiology* 13:369–374.

Gaulin, C.D., D. Ramsay, P. Cardinal, and M.A.D. Halevyn. 1999. Epidemic of gastroenteritis of viral origins associated with eating imported raspberries. *Canadian Journal of Public Health* 90:37–40.

Gemmell, M.E. and S. Schmidt. 2012. Microbiological assessment of river water used for the irrigation of fresh produce in a sub-urban community in Sobantu, South Africa. *Food Research International* 47:300–305.

Gerba, C.P. 2009. The role of water and water testing in produce safety. In Fan, X., B.A. Niemira, C.J. Doona, F.E. Feeherry, and R.B. Gravani, Eds., *Microbial Safety of Fresh Produce.* Hoboken, NJ: Wiley-Blackwell, pp. 129–142.

Gilbert, P. 1984. The revival of microorganisms sublethally injured by chemical inhibitors. In M.H.E. Andrew and A.D. Russell, Eds., *The Revival of Injured Microbes*. London, U.K.: Academic Press, pp. 175–197.

Golden, D.A., E.J. Rhodehamel, and D.A. Kautter. 1993. Growth of *Salmonella* spp. in cantaloupe, watermelon, and honeydew melons. *Journal of Food Protection* 56(3):194–196.

Goodman, K.J., P. Correa, H.J.T. Aux, H. Ramirez, J.P. DeLany, O.G. Pepinosa, M.L. Quinones, and T.C. Parra. 1996. *Helicobacter pylori* infection in the Colombian Andes: A population-based study of transmission pathways. *American Journal of Epidemiology* 144:290–299.

Goodman, K.J., P. Correa, A.H. Tengana et al. 1997. Nutritional factors and *Helicobacter pylori* infection in Colombian children. *Journal of Paediatrics Gastroenterology and Nutrition* 25:507–515.

Goodwin, C.S., J.A. Armstrong, T. Chilvers, M. Peters, M.D. Collins, L. Sly, W. McConnell, and W.E.S. Harper. 1989. Transfer of *Campylobacter pylori* and *Campylobacter mustelae* to *Helicobacter* gen. nov. as *Helicobacter pylori* comb. nov. and *Helicobacter mustelae* comb, nov., respectively. *International Journal of Systematic Bacteriology* 39:397–405.

Gould, R.M. and W.J. Lennarz. 1970. Metabolism of phosphatidyl glycerol and lysyl phosphatidylglycerol in *Staphylococcus aureus*. *Journal of Bacteriology* 104:1135–1144.

Greene, S.K., E.R. Daly, E.A. Talbot, L.J. Demma, S. Holzbauer, and N.J. Patel. 2008. Recurrent multistate outbreak of *Salmonella* Newport associated with tomatoes from contaminated fields. *Epidemiological Infection* 136:157–165.

Griffin, M.R., J.J. Surowiec, D.I. Mccloskey et al. 1982. Foodborne Norwalk virus. *American Journal of Epidemiology* 115:178–184.

Griffin, P.M. and R.V. Tauxe. Updated 2001 June 20. Surveillance for outbreaks of *Escherichia coli* 0157:H7 infection summary of 1998 data. Atlanta, GA: Center for Disease Control, National Center for Infectious Diseases, Division of Bacterial and Mycotic Diseases.

Grubel, P., L. Huang, N. Masubuchi et al. 1998. Detection of *Helicobacter pylori* DNA in houseflies (*Musca domestica*) on three continents. *Lancet* 352:788–789.

Gruber, T.M. and C.A. Gross. 2003. Multiple σ subunits and the partitioning of bacterial transcription space. *Annual Review of Microbiology* 57:441–466.

Gutierrez, E. 1997. Japan prepares as O157 strikes again. *Lancet* 349:1156.

Haley, B.J., D.J. Cole, and E.K. Lipp. 2009. Distribution, diversity, and seasonality of waterborne salmonellae in a rural watershed. *Applied Environmental Microbiology* 75:1248–1255.

Harris, L.J., J.N. Farber, L.R. Beuchat, M.E. Parish, T.V. Suslow, E.H. Garrett, and F.F. Busta. 2003. Outbreaks associated with fresh produce: Incidence, growth, and survival of pathogens in fresh and fresh-cut produce. *Comprehensive Reviews in Food Science and Food Safety* 2:78–141.

Herwaldt, B.L. 2000. *Cyclospora cayetanensis*: A review, focusing on the outbreaks of cyclosporiasis in the 1990s. *Clinical Infectious Disease* 31(4):1040–1057.

Herwaldt, B.L., J.F. Lew, C.L. Moe, D.C. Lewis, C.D. Humphrey, S.S. Monroe, E.W. Pon, and R.I. Glass. 1994. Characterization of a variant strain of norwalk virus from a foodborne outbreak of gastroenteritis on a cruise ship in Hawaii. *Journal of Clinical Microbiology* 32:861–866.

Hopkins, R.J., P.A. Vial, C. Ferreccio et al. 1993. Seroprevalence of *Helicobacter pylori* in Chile: Vegetables may serve as one route of transmission. *Journal of Infectious Diseases* 168(1):222–226.

Hsu, W.Y., A. Simonne, and P. Jitareerat. 2006. Fates of seeded *Escherichia coli* O157:H7 and *Salmonella* on selected fresh culinary herbs during refrigerated storage. *Journal of Food Protection* 69:1997–2001.

http://www.foodstandards.gov.au/science/surveillance/Pages/default.aspx. Accessed July 4, 2016.

http:/www.ecdc.europa.eu. Accessed July 4, 2016.

Hurst, A. 1977. Bacterial injury: A review. *Canadian Journal of Microbiology* 23:935–944.

Hurst, A. 1984. Revival of vegetative bacteria after sublethal heating. In M.H.E. Andrew and A.D. Russell, Eds., *The Revival of Injured Microbes*. London, U.K.: Academic Press, pp. 77–103.

Hurst, A., G.S. Hendry, A. Hughes, and B. Pale. 1976. Enumeration of sub-lethally heated Staphylococci in some dried foods. *Canadian Journal of Microbiology* 22:677–683.

Hurst, A. and Nasim, A. 1984. *Repairable Lesions in Microorganisms*. Orlando, FL: Academic Press.

Husu, J.R. 1990. Epidemiological studies on the occurrence of *Listeria monocytogenes* in the faeces of dairy cattle. *Journal of Veterinary Medicine B* 37:276–282.

ICMSF. 1996. *Shigella.* In T.A. Roberts, A.C. Baird-Parker, and R.B. Tompkin, Eds. *Microorganisms in Food 5: Microbiological Specifications of Food Pathogens.* London, U.K.: Blackie Academic and Professional, pp. 280–298.

Ivanek, R., Y.T. Grohn, and M. Wiedmann. 2006. *Listeria monocytogenes* in multiple habitats and host populations: Review of available data for mathematical modeling. *Foodborne Pathogen Diseases* 3:319–336.

Janisiewicz, W.J., W.S. Conway, M.W. Brown, G.M. Sapers, P. Fratamico, and R.L. Buchanan. 1999. Fate of *Escherichia coli* O157:H7 on fresh-cut apple tissue and its potential for transmission by fruit flies. *Applied Environmental Microbiology* 65(1):1–5.

Jaquette, C.B., L.R. Beuchat, and B.E. Mahon. 1996. Efficacy of chlorine and heat treatment in killing *Salmonella* stanley inoculated onto alfalfa seeds and growth and survival of the pathogen during sprouting and storage. *Applied and Environmental Microbiology* 62(6):2212–2215.

Jimenez, M., J.H. Siller, J.B. Valdez, A. Carrillo, and C. Chaidez. 2007. Bidirectional *Salmonella enterica* serovar *Typhimurium* transfer between bare/glove hands and green bell pepper and its interruption. *International Journal of Environment and Health Research* 17:381–388.

Kader, A.A. 1992. Postharvest biology and technology: An overview. In A. Kader, Ed., *Postharvest Technology of Horticultural Crops.* Berkeley, CA: University of California Division of Agriculture and Natural Resources, Pub. 3311, pp. 15–20.

Kader, A.A. and R.S. Rolle. 2004. The role of post-harvest management in assuring the quality and safety of horticultural produce. Rome. FAO Agricultural Surveillance Bulletin 152:51.

Kallander, K.D., A.D. Hitchins, G.A. Lancette, J.A. Schmieg, G.R. Garcia, H.M. Solomon, and J.N. Sofos. 1991. Fate of *Listeria monocytogenes* in shredded cabbage stored at 5 and 25°C under a modified atmosphere. *Journal of Food Protection* 54(4):302–304.

Khalil, R.K. and J.F. Frank. 2010. Behavior of *Escherichia coli* O157:H7 on damaged leaves of spinach, lettuce, cilantro, and parsley stored at abusive temperatures. *Journal of Food Protection* 73:212–220.

Kirch, W. 2008. Microbiological hazards, food safety. Encyclopedia of Public Health, p. 457.

Korsager, B., B. Hede, H. Boggild, B. Bottiger, and K. Molbak, K. 2005. Two outbreaks of Norovirus infections associated with the consumption of imported frozen raspberries, Denmark. *Eurosurveillance Weekly* 10: E050623.1. Available from http://www.eurosurveillance.org/ew/2005/050623.asp. Accessed July 4, 2016.

Koseki, S. and S. Isobe. 2005. Prediction of pathogen growth on iceberg lettuce under real temperature history during distribution from farm to table. *International Journal of Food Microbiology* 104:239–248.

Kumar, V. 2012. Incidence of *Salmonella* sp. and *Listeria monocytogenes* in some salad vegetables which are eaten raw: A study of Dhanbad city, India. *International Journal of Engineering Science and Research* 2:2277–2685.

Larson, A.E, E.A. Johnson, C.R. Barmore, and M.D. Hughes. 1997. Evaluation of the botulism hazard from vegetables in modified atmosphere packaging. *Journal of Food Protection* 60(10):1208–1214.

Lennox, J.A., M. Eja, U. Edeghor, C. Okoro, and E.C. Okpako. 2015. Incidence of *Salmonella* and *Shigella* species on some selected fruits and vegetables obtained from open area markets in Calabar Metropolis. *International Journal of Current Microbiology and Applied Sciences* 4 (5):262–268

Li, H., M. Tajkarimi, and B.I. Osburn. 2008. Impact of vacuum cooling on *Escherichia coli* O157:H7 infiltration into lettuce tissue. *Applied and Environmental Microbiology* 74:3138–3142.

Li, Y., R.E. Brackett, J. Chen, and L.R. Beuchat. 2001. Survival and growth of *Escherichia coli* 0157:H7 inoculated onto cut lettuce before or after heating in chlorinated water, followed by storage at 5°C or 15°C. *Journal of Food Protection* 64(3):305–309.

Liao, C.H. and G.M. Sapers. 2000. Attachment and growth of *Salmonella chester* on apple fruit and in vivo response of attached bacteria to sanitizer treatments. *Journal of Food Protection* 63(7):876–883.

Lin, C.M., J. Kim, W.X. Du, and C.I. Wei. 2000. Bactericidal activity of isothiocyanate against pathogens on fresh produce. *Journal of Food Protection* 63(1):25–30.

Liu, C., N. Hofstra, and E. Franz. 2013. Impacts of climate change on the microbial safety of pre-harvest leafy green vegetables as indicated by *Escherichia coli* O157 and *Salmonella* spp. *International Journal of Food Microbiology* 163:119–128.

Lopez, A.S., D.R. Dodson, M.J. Arrowood et al. 2001. Outbreak of cyclosporiasis associated with basil in Missouri in 1999. *Clinical Infectious Diseases* 32:1010–1017.

Lund, B.M. 1986. Anaerobes in relation to foods of plant origin. In E.M. Barnes and G.C. Mead, Eds., *Anaerobic Bacteria in Habitats Other than Man.* Oxford, U.K.: Blackwell Scientific Publications, pp. 351–372.

Lynch, M.F., R.V. Tauxe, and C.W. Hedberg. 2009. The growing burden of foodborne outbreaks due to contaminated fresh produce: Risks and opportunities. *Epidemiological Infection* 137:307–315.

Mackey, B.M. 2000. Injured bacteria. In A.M. Lund, T.C. Baird-Parker, and G.W. Gould, Eds., *The Microbiological Safety and Quality of Food*. Gaithersburg, MD: Aspen Publishers, Inc., pp. 315–341.

Malizio, C.J. and E.A. Johnson. 1991. Evaluation of the botulism hazard from vacuum-packaged enoki mushrooms [*Flammulina velutipes*]. *Journal of Food Protection* 54(1):20–21.

Mandokhot, V.V. and S.R. Garg. 1986. Market quality of Khoa Burfi and Pera: A critical review. *Journal of Food Science and Technology* 22:299.

Marler, B. 2012. Six state outbreak of Shiga Toxin-producing *E. coli* O26 infections linked to raw clover sprouts at Jimmy John's restaurants. htpp://www.marlerblog.com/. Accessed July 4, 2016.

Martin, D.L., P.S. Mead., L. Suarez, and L. Slutsker. 2003. Food related illness and death in the United States. *Emerging Infectious Diseases* 5:607–625.

Matthews, K.R. 2013. Sources of enteric pathogen contamination of fruits and vegetables: Future directions of research. *Stewart Postharvest Review* 9:1–5.

Matthews, K.R. 2014. Leafy vegetables. In K.R. Matthews, G.M. Sapers, and C.P. Gerba, Eds., *The Produce Contamination Problem: Causes and Solutions*. Waltham, MA: Elsevier, pp. 187–205.

Mazari-Hiriart, M., Y. Lopez-Vidal, G. Castillo-Rojas, S. Ponce de Leon, and A. Cravioto. 2001. *Helicobacter pylori* and other enteric bacteria in freshwater environments in Mexico City. *Archives of Medical Research* 32(5):458–467.

Mazari-Hiriart, M., H.M. Ponce-Mitchell, Y.Y. Li, P.J. Hu, Q. Liu, M. Chen, G.G. Du, Z.J. Wang, A. Lee, and S.L. Hazell. 1992. Epidemiology of *Helicobacter pylori* in southern China: Identification of early childhood as the critical period for acquisition. *Journal of Infectious Diseases* 166:149–153.

Mazzotta, A.S. 2001. Thermal inactivation of stationary phase and acid-adapted *Escherichia coli* O157:H7, *Salmonella* and *Listeria monocytogenes* in fruit juices. *Journal of Food Protection* 64:315.

Merson, M.H., G.K. Morris, D.A. Sack, J.E. Wells, J.C. Feeley, R.B. Sack, W.B. Creech, A.Z. Kapikian, and E.J. Gangarosa. 1976. Travelers' diarrhea in Mexico. *New England Journal of Medicine* 294:1299–1305.

Miller, K.J., J.R. Shebuski, and O. Vilhelmsson. 2000. Effects of growth at low water activity on the thermal tolerance of *Staphylococcus aureus*. *Journal of Food Protection* 63:1277–1281.

Minnikin, D.E. and H. Abdolrahimzadeh. 1974. Effect of pH on the proportions of polar lipids, in chemostat cultures of *Bacillus subtilis*. *Journal of Bacteriology* 120:999–1003.

Monge, R. and M. Chinchilla. 1996. Presence of Cryptosporidium oocysts in fresh vegetables. *Journal of Food Protection* 59(2):202–203.

Mukherjee, A., S. Cho, J. Scheftel, S. Jawahir, K. Smith, and F. Diez-Gonzalez. 2006. Soil survival of *Escherichia coli* O157:H7 acquired by a child from garden soil recently fertilized with cattle manure. *Journal of Applied Microbiology* 101:429–436.

NACMCF (National Advisory Committee on Microbiological Criteria for Foods). 1999a. Microbiological safety evaluations and recommendations on sprouted seeds. *International Journal of Food Microbiology* 52:123–153.

NACMCF (National Advisory Committee on Microbiological Criteria for Foods). 1999b. Microbiological safety evaluations and recommendations on fresh produce. *Food Control* 10:117–143.

Natvig, E.E., S.C. Ingham, B.H. Ingham, L.R. Cooperband, and T.R. Roper. 2002. *Salmonella enterica* serovar *Typhimurium* and *Escherichia coli* contamination of root and leaf vegetables grown in soils with incorporated bovine manure. *Applied and Environmental Microbiology* 68:2737–2744.

Nayar, K.R., S.K. Krishnan, and S.P. Gorthi.1993. Epidemiological investigation of an outbreak of food poisoning. *Medical Journal of Armed Forced in India*. 49:257–260.

Nema, V., R. Agrawal, D.V. Kamboj, A.K. Goel, and L. Singh.2007. Isolation and characterization of heat resistant enterotoxigenic *Staphylococcus aureus* from a food poisoning outbreak in Indian subcontinent. *International Journal of Food Microbiology* 117(1):29–35.

Nguyen-the, C. and B.M. Lund. 1991. The lethal effect of carrot on *Listeria* species. *Journal of Applied Bacteriology* 70:479–488.

Nguyen-the, C. and B.M. Lund. 1992. An investigation of the antibacterial effect of carrot on *Listeria monocytogenes*. *Journal of Applied Bacteriology* 73:23–30.

Nguyen-the, C. and F. Carlin. 1994. The microbiology of minimally processed fresh fruits and vegetables. *Critical Review of Food Science and Nutrition* 34:371–401.

Nguyen-the, C. and F. Carlin. 2000. Fresh and processed vegetables. In B.M. Lund, T.C. Baid-Parker, and G.W. Gould, Eds. *The Microbiological Safety and Quality of Foods*. Gaithersburg, MD: Aspen Publication, pp. 620–684.

Oliveira, M., I. Vinas, J. Usall, M. Anguera, and M. Abadias. 2012. Presence and survival of *Escherichia coli* O157:H7 on lettuce leaves and in soil treated with contaminated compost and irrigation water. *International Journal of Food Microbiology* 156:133–140.

Ongeng, D., C. Muyanja, A.H. Geeraerd, D. Springael, and J. Ryckeboer. 2011. Survival of *Escherichia coli* O157:H7 and *Salmonella enterica* serovar *Typhimurium* in manure and manure-amended soil under tropical climatic conditions in Sub-Saharan Africa. *Journal of Applied Microbiology* 110:1007–1022.

Pao, S., G.E. Brown, and K.R. Schneider. 1998. Challenge studies with selected pathogenic bacteria on freshly peeled hamlin orange. *Journal of Food Science* 63(2):359–362.

Park, S., B. Szonyi, R. Gautam, K. Nightingale, J. Anciso, and R. Ivanek. 2012. Risk factors for microbial contamination in fruits and vegetables at the preharvest level: A systematic review. *Journal of Food Protection* 75:2055–2081.

Patterson, W., P.T. Haswell, and J. Green. 1997. Outbreak of small round structured virus gastroenteritis arose after kitchen assistant vomited. *Canada Communicable Disease Report* 27:R101–R103.

Petran, R.L., W.H. Sperber, and A.R. Davis. 1995. *Clostridium botulinum* toxin formation in romaine lettuce and shredded cabbage: Effect of storage and packaging conditions. *Journal of Food Protection* 58:624–627.

Piagentini, A.M., M.E. Pirovani, D.R. Guemes, J.H. Di Pentima, and M.A. Tessi. 1997. Survival and growth of *Salmonella hadar* on minimally processed cabbage as influenced by storage abuse conditions. *Journal of Food Science* 62 (3):616–618.

Pirovani, M.E., D.R. Guemes, J.H. Di Pentima, and M.A. Tessi. 2000. Survival of *Salmonella* hadar after washing disinfection of minimally processed spinach. *Letters in Applied Microbiology* 31:143–148.

Portnoy, B.L., J.M. Goepfert, and S.M. Harmon. 1976. An outbreak of *Bacillus cereus* food poisoning resulting from contaminated vegetable sprouts. *American Journal of Epidemiology* 103(6):589–594.

Pounder, R.E. and D. Ng. 1995. The prevalence of *Helicobacter pylori* infection in different countries. *Alimentary Pharmacology and Therapeutics* 9(Suppl. 2):33–39.

Puohiniemi, R., T. Heiskanen, and A. Siitonen. 1997. Molecular epidemiology of two international sprout borne *Salmonella* outbreaks. *Journal of Clinical Microbiology* 35:2487–2491.

Quinn, K., G. Baldwin, P. Stepak, K. Thorburn, C. Bartleson, M. Goldoft, J. Kobayashi, and P. Stehr-Green. 1997. Civilian outbreak of adenovirus acute respiratory disease–South Dakota (Reprinted from *MMWR*, vol 47, pg 565–567, 1998). *Journal of the American Medical Association* 280:595–596.

Rafii, F. and P. Lunsford. 1997. Survival and detection of *Shigella flexneri* in vegetables and commercially prepared salads. *Journal of AOAC International* 80(6):1191–1197.

Rao, N.R., P. Sudhakar, R.V. Bhat, and C.P. Gupta. 1989. A study of recorded cases of foodborne diseases at Hyderabad during 1984 and 1985. *Journal of Tropical Medicine and Hygiene* 92:320–324.

Ray, B. Ed. 1989. *Injured Index and Pathogenic Bacteria*. Boca Raton, FL: CRC Press.

Rickhow, D.A. and P.C. Singer. 1985. Mechanisms of organic halide formation during fulvic acid chlorination and implications with respect to preozonation. In R.V. Jolley, R.J. Bull, W.P. Davis, S. Katz, M.H. Roberts, and V.A. Jacobs, Eds., *Water Chlorination—Chemistry, Environmental Impact and Health Effects*, Vol. 5. Chelsea, MI: Lewis Publishers, pp. 1229–1257

Robertson, G.L. 2003. *Food Packaging Principles and Practices*. New York: Marcel and Deckker, Inc.

Rohinishree, Y.S. 2015. Molecular characterization of toxigenic *Staphylococcus* species for stress responses under simulating conditions of food processing in model food systems PhD thesis submitted to the University of Mysore, Mysore, India.

Rohinishree, Y.S. and P.S. Negi. 2011. Detection, identification and characterization of staphylococci in street vend foods. *Food and Nutrition Sciences* 2:304–313.

Rohinishree, Y.S. and P.S. Negi. 2012. Multiplex reverse transcription polymerase chain reaction to study the expression of virulence and stress response genes in *Staphylococcus aureus*. *Journal of Food Science* 77: M95–M101.

Rohinishree, Y.S. and P.S. Negi. 2016. Effect of licorice extract on cell viability, biofilm formation and exotoxin production by *Staphylococcus aureus*. *Journal of Food Science and Technology* 53:1092–1100.

Safaei, H.G., E. Rahimi, A. Zandi, and A. Rashidipour. 2011. *Helicobacter pylori* as a zoonotic infection: The detection of *H. pylori* antigens in the milk and faeces of cows. *Journal of Research in Medical Sciences* 16(2):184–187.

Sasaki, K., Y. Tajiri, M. Sata et al. 1999. *Helicobacter pylori* in the natural environment. *Scandinavian Journal of Infectious Diseases* 31(3):275–279.

Satchell, F.B., P. Stephenson, W.H. Andrews, L. Estela, and G. Allen G. 1990. The survival of *Shigella sonnei* in shredded cabbage. *Journal of Food Protection* 53(7):558–562.

Scott, L., P. McGee, J.J. Sheridan, B. Earley, and N. Leonard. 2006. A comparison of the survival in feces and water of *Escherichia coli* O157:H7 grown under laboratory conditions or obtained from cattle feces. *Journal of Food Protection* 69:6–11.

Semenov, A.V., A.H.C. van Bruggen, L. van Overbeek, A.J. Termorshuizen, and A.M. Semenov. 2007. Influence of temperature fluctuations on *Escherichia coli* O157:H7 and *Salmonella enterica* serovar *Typhimurium* in cow manure. *FEMS Microbiology and Ecology* 60:419–428.

Sivaplasingham, S., C.R. Friedman, L. Cohen, and R.V. Tauxe. 2004. Fresh produce: A growing cause of outbreaks of foodborne illness in the United States, 1973 through 1997. *Journal of Food Protection* 67:2342–2353.

Solomon, E.B., H.J. Pang, and K.R. Matthews. 2003. Persistence of *Escherichia coli* O157:H7 on lettuce plants following spray irrigation with contaminated water. *Journal of Food Protection* 66:2198–2202.

Solomon, E.B., C.J. Potenski, and K.R. Matthews. 2002. Effect of irrigation method on transmission to and persistence of *Escherichia coli* O157:H7 on lettuce. *Journal of Food Protection* 65:673–676.

Solomon, H.M., D.A. Kautter, T. Lilly, and E.J. Rhodehamel. 1990. Outgrowth of *Clostridium botulinum* in shredded cabbage at room temperature under modified atmosphere. *Journal of Food Protection* 53(10):831–833.

Stephens, P., J.A. Joynson, K.W. Davies et al. 1997. The use of automated growth analyzer to measure recovery times of single heat-injured. *Salmonella* cells. *Journal of Applied Microbiology* 83:445–455.

Storz, G. and R. Hengge-Aronis. 2000. Oxidative stress. In G. Storz and R. Hengge-Aronis, Eds., *Bacterial Stress Responses*. Washington, DC: ASM Press, pp. 47–59.

Strauch, D. 1991. Survival of microorganisms and parasites in excreta, manure and sewage sludge. *Review Scientific Techniques International Office of Epizootics* 10:816–846.

SudershanRao, V., R.N. Kumar, L. Kashinath, V. Bhaskar, and K. Polasa. 2010. Economic impact of a foodborne disease outbreak in Hyderabad—A case study. *Indian Journal of Nutrition and Dietetics* 47:246–251.

SudershanRao, V., R.N. Kumar, and K. Polasa. 2012. Foodborne diseases in India. *British Food Journal* 114(5):661–680.

Sudhakar, P., N.R. Rao, R.V. Bhat, and C.P. Gupta. 1988. The Economic impact of food borne disease out break due to *Staphylococcus aureus Journal of Food Protection* 51:898–900.

Sugiyama, H. and K.H. Yang. 1975. Growth potential of *Clostridium botulinum* in fresh mushrooms packaged in semipermeable plastic film. *Applied Microbiology* 30:964–969.

Szabo, E.A., K.J. Scurrah, and J.M. Burrows. 2000. Survey for Psychrotrophic bacterial pathogens in minimally processed lettuce. *Letters in Applied Microbiology* 30(6):456–460.

Taormina, P.J. and L.R. Beuchat. 1999. Comparison of chemical treatments to eliminate enterohemorrhagic *Escherichia coli* O157:H7 on alfalfa seeds. *Journal of Food Protection* 62(4):318–324.

Taormina, P.J., L.R. Beuchat, M.C. Erickson, L. Ma, G. Zhang, and M.P. Doyle. 2009. Transfer of *Escherichia coli* O157:H7 to iceberg lettuce via simulated field coring. *Journal of Food Protection* 72:465–472.

Taormina, P.J., L.R. Beuchat, and R. Slusker. 1999. Infections associated with eating seed sprouts: An international concern. *Emerging Infectious Diseases* 5:629–634.

Tauxe, R., H. Kruse, C. Hedberg, M. Potter, J. Madden, and K. Wachsmuth. 1997. Microbiological hazards and emerging issues associated with produce. A preliminary report to the National Advisory Committee on Microbiologic Criteria for Foods. *Journal of Food Protection* 60:1400–1408.

Taylor, E., J. Bates, D. Kenyon, M. Maccaferri, and J. Thomas. 2002. Modern molecular methods for characterisation and diagnosis of seed-borne fungal pathogens. *Journal of Plant Pathology* 83:75–81.

Taylor, S.L., S.L. Hefle, and A. Munoz-Furlong. 1999. Food allergies and avoidance diets. *Nutrition Today* 34:15–22.

Teutsch, S.M. and R.E. Churchill. 2000. *Principles and Practice of Public Health Surveillance*, 2nd edn. Oxford, U.K.: Oxford University Press.

Thekdi, R.J., A.G. Lakhani, V.B. Rale, and M.V. Panse. 1990. An outbreak of food poisoning suspected to be caused by *Vibrio fluvialis*. *Journal of Diarrhoeal Disease Research* 8:163–165.

Tierney, J.T., R. Sullivan, and E.P. Larkin. 1977. Persistence of Poliovirus 1 in soil and on vegetables grown in soil previously flooded with inoculated sewage sludge or effluent. *Applied Environmental Microbiology* 33:109–113.

Todd, E.C., J.D. Greig, C.A. Bartleson, and B.S. Michaels. 2008. Outbreaks where food workers have been implicated in the spread of foodborne disease. Part 5. Sources of contamination and pathogen excretion from infected persons. *Journal of Food Protection* 71:2582–2595.

Todd, E.C., J.D. Greig, C.A. Bartleson, and B.S. Michaels. 2009. Outbreaks where food workers have been implicated in the spread of foodborne disease. Part 6. Transmission and survival of pathogens in the food processing and preparation environment. *Journal of Food Protection* 72:202–219.

Tournas, V.H. 2005. Moulds and yeasts in fresh and minimally processed vegetables and sprouts. *International Journal of Food Microbiology* 99:71–77.

Van der Linden, I., B. Cottyn, M. Uyttendaele, G. Vlaemynck, M. Heyndrickx, and M. Maes. 2013. Survival of enteric pathogens during butterhead lettuce growth: Crop stage, leaf age, and irrigation. *Foodborne Pathogens and Diseases* 10:485–491.

Van Renterghem, B., F. Huysman, R. Rygole, and W. Verstraete. 1991. Detection and prevalence of *Listeria monocytogenes* in the agricultural ecosystem. *Journal of Applied Bacteriology* 71:211–217.

van Schothorst, M. 2004. *A Simple Guide to Understanding and Applying the Hazard Analysis Critical Control Point Concept*, 3rd edn. Europe, Brussels: Concise Monograph Series, ILSI.

Wadstrom, T. and A. Ljungh. 1991. Aeromonas and Plesiomonas as food and waterborne pathogens. *International Journal of Food Microbiology* 12(4):303–311.

Wallace, C.A., W.H. Sperber, and S.E. Mortimore. 2011. *Food Safety for the 21st Century: Managing HACCP and Food Safety throughout the Global Supply Chain.* Wiley-Blackwell/John Wiley & Sons, Ltd., Publications, London, U.K.

Ward, B.K. and L.G. Irving. 1987. Virus survival on vegetables spray irrigated with wastewater. *Water Research* 21:57–63.

Warren, J.R. and B. Marshall. 1983. Unidentified curved bacilli on gastric epithelium in active chronic gastritis. *Lancet* 1:1273–1275.

Warriner, K., A. Huber, A. Namvar, W. Fan, and K. Dunfield. 2009. Recent advances in the microbial safety of fresh fruits and vegetables. *Advances in Food and Nutrition Research* 57:155–208.

Wei, C.I., T.S. Huang, J.M. Kim, W.F. Lin, M.L. Tamplin, and J.A. Bartz. 1995. Growth and survival of *Salmonella montevideo* on tomatoes and disinfection with chlorinated water. *Journal of Food Protection* 58(8):829–836.

Weissinger, W.R., W. Chantarapanont, and L.R. Beuchat. 2000. Survival and growth of *Salmonella* Baildon in shredded lettuce and diced tomatoes, and effectiveness of chlorinated water as a sanitizer. *International Journal of Food Microbiology* 62:123–131.

Weissinger, W.R., K.H. McWatters, and L.R. Beuchat. 2001. Evaluation of volatile chemical treatments for lethality to *Salmonella* on alfalfa seeds and sprouts. *Journal of Food Protection* 64(4):442–450.

Welshimer, H.J. and J. Donker-Voet. 1971. *Listeria monocytogenes* in nature. *Applied Microbiology* 21:516–519.

Wesche, A.M., J.B. Gurtler, B.P. Marks, and E.T. Ryser. 2009. Stress, sublethal injury, resuscitation and virulence of bacterial foodborne pathogens. *Journal of Food Protection* 72(5):1121–1138.

Wheeler, C., T.M. Vogt, G.L. Armstrong et al. 2005. An outbreak of hepatitis A associated with green onions. *New England Journal of Medicine* 353:890–897.

White, K.E. 1986. A foodborne outbreak of Norwalk virus gastroenteritis: Evidence for post-recovery transmission. *American Journal of Epidemiology* 124:120–126.

WHO. 2015. *WHO Estimates of Global Burden of Foodborne Diseases: Foodborne Disease Burden Epidemiology Reference Group 2007–2015.* Geneva, Switzerland: World Health Organization Press.

WHO. 2006. Guidelines for the safe use of wastewater, excreta and grey water, Vol. 2. Wastewater Use in Agriculture. Geneva, Switzerland: World Health Organization Press. p222. Available from http://www.who.int/water_sanitation_health/wastewater/gsuweg2/en/. Accessed July 4, 2016.

Wu, F.M., M.P. Doyle, L.R. Beuchat, J.G. Wells, E.D. Mintz, and B. Swaminathan. 2000. Fate of *Shigella sonnei* on parsley and methods of disinfection. *Journal of Food Protection* 63(5):568–572.

Wu, L. and D. Zhu. 2014. Food safety in China: a comprehensive review. Boca Raton, FL: CRC Press/Taylor & Francis Group.

Yahaghi, E., F. Khamesipour, F. Mashayekhi, F.S. Dehkordi, M.H. Sakhaei, M. Masoudimanesh, and M. K. Khameneie. 2014. *Helicobacter pylori* in vegetables and salads: Genotyping and antimicrobial resistance properties. *BioMed Research International.* Available from http://dx.doi.org/10.1155/2014/757941. p.11.

Yousef, A.E. and V.K. Juneja, Eds. 2003. *Microbial Stress Adaptation and Food Safety.* Boca Raton, FL: CRC Press.

Zagory, D. and A.A. Kader. 1988. Modified atmosphere packaging of fresh produce. *Food Technology* 42(9):70–77.

Zhao, T., M.P. Doyle, and R.E. Besser. 1993. Fate of enterohemorrhagic *Escherichia coli* O157:H7 in apple cider with and without preservatives. *Applied Environmental Microbiology* 59(8):2526–2530.

Zhuang, R.Y., L.R. Beuchat, and F.J. Angulo. 1995. Fate of *Salmonella montevideo* on and in raw tomatoes as affected by temperature and treatment with chlorine. *Applied Environmental Microbiology* 61(6):2127–2131.

8 Safety Issues of Food Ingredients from Plant Cell, Tissue, and Organ Cultures

An Explication

Hosakatte Niranjana Murthy, Kee Yoeup Paek, So-Young Park, and Vijayalaxmi S. Dandin

CONTENTS

8.1 INTRODUCTION

Plants constitute the major part of our traditional foods and also render a wide range of phytochemicals that are included in our food stuffs at low levels (Stafford 1991). In recent years, knowledge about the effects of these commercially valuable phytochemicals on health and disease risk is accumulating as they have a vital impact on all systems of the body and can modulate different functions of these systems (Meltzer 2010). An extensive array of phytochemicals are used as food ingredients such as flavors, colorants, essential oils, sweeteners, antioxidants, and nutraceuticals, and they are derived from conventional primary or secondary food sources (Kroes and Walker 2004) or from plant cell, tissue, and organ culture (PCTOC; Murthy et al. 2014a–i). PCTOCs are the important alternative sources for the production of phytochemicals that are used as food ingredients, and recently, novel strategies in PCTOC have been established to improve the biomass and phytochemical production. A remarkable advancement in bioreactor technologies has eased the large-scale cultivation of plant cells and organs for the production of phytochemicals (Murthy et al. 2014c–h; Paek et al. 2009). However, the use of plant cell, tissue, and organ cultures for the production of food ingredients impels new issues related to safety and efficacy of the food ingredients as these culture methods employ different natural and synthetic growth regulators as well as various physical and chemical elicitors for the enhancement of biomass and metabolite production. During the extraction of food ingredients from raw materials, scrupulous purification methods are also not followed, which may lead to the formation of complex mixtures and even the toxic by-products. Hence, the issues concerned with the extraction of food ingredients from raw materials have provoked the regulatory agencies across the world to consider the safety standards for PCTOC products (Murthy et al. 2015a).

The basic technical functions of the food ingredients in food as outlined jointly by the Food and Drug Administration (FDA) and the International Food Information Council suggest to improve or maintain safety, freshness, nutritional value, taste, texture, and appearance (Rulis and Levitt 2009). A resolution taken by various regulatory agencies for safety assessment of PCTOC-derived products is based on the substantial equivalence of these ingredients with their whole plant counterparts (Fu 1999). The food ingredients derived from PCTOC include an infinitely large set of potential substances; unless their use is limited in some sensible way through a premarket approval, it is difficult to produce safe foods for protecting public health effectively. At present, standard specifications on safety, efficacy, standardization and documentation of botanicals and herbal preparations (Abdel-Rahaman et al. 2011; Kroes and Walker 2004), natural food additives (Anonymous 2012a,b, 2007), and genetically modified crops (Anonymous 2008; Konig et al. 2004) have been proposed. On the other hand, till date, such specifications for safety assessment have not been suggested to PCTOC products as food ingredients. Here, the safety issues related to food ingredients derived from PCTOC, the safety considerations during the initiation, proliferation of cultures, harvesting the raw material, and biosafety evaluation of PCTOC raw materials and the products are addressed and elaborated.

8.2 PLANT CELL, TISSUE, AND ORGAN CULTURE SYSTEMS AS SOURCES OF FOOD INGREDIENTS

Few decades ago, plant cell and tissue cultures were virtually considered as the potential sources for high-value secondary products, and many of its targets were pharmaceuticals such as digoxin, quinine, morphine, and taxol (Stafford 1991), subsequently, with the advent in understanding of the phytonutrient value, and also in response to the increasing commercial demand for the food ingredients, the research focus was directed to generate such food ingredients from PCTOC.

8.2.1 Novel Food Ingredients from Small-Scale Cultures

Plant cell and organ culture systems are rich reservoirs of phytochemicals, which are used as food ingredients. PCTOC systems have been shaped to produce such compounds in small-scale cultures such as suspension, shoot, adventitious root, and hairy root cultures. This technology has been applied to investigate the production of flavor ingredients; however, it has not yielded commercially viable production methods (Hallagan et al. 1999). PCTOC technology used for the production of flavor compounds is remarkably complex. While continuing progress has been achieved, product yields remain low (Scragg 1997). Although various cell lines have been established in various systems including strawberry (*Fragaria* sp.), *Mentha*, *Citrus*, and *Vanilla*, numerous biochemical, genetic, and engineering obstacles must be overcome if commercialization of flavor ingredients using this technology is to be realized (Westcott 1989). The coloring properties and desirable biological activities of betalains have made it interesting for the *in vitro* production of these food colorants. Suspension cultures of *Rubia cordifolia* have proven to be a rich source of anthraquinones, whereas cultures of *Daucus carota* have long been known to accumulate anthocyanin pigments under appropriate growth conditions, and this trait, together with the good growth characteristics of suspension cultures of *D. carota*, continues to make this a popular system for studies of regulation of the synthesis of anthocyanins (Ellis 1988). A vast range of essential oils have been obtained from different plant culture systems. A wide variation in the essential oil profiles of callus and suspension cultures has been reported. Culture age appears to be an important factor in determining whether a particular product is present, and the cell line used has a significant effect on the oil profile. Furthermore, the extraction method is of critical importance in determining essential oil profiles. However, the consistent finding is that, in the absence of precursors, undifferentiated cell cultures are usually capable of accumulating very low uneconomic levels of essential oils (Stafford 1991). The glycoside stevioside is a natural sweetening agent—300 times sweeter than glucose. Plant cell cultures of *Stevia rebaudiana*, the natural source plant of stevioside, and of *Digitalis purpurea* have been shown to transform steviol into stevioside; novel glycosides have been detected as products of biotransformations (Furuya 1978). PCTOC technology has also been applied for the production of pungent additives such as capsaicin (Johnson et al. 1990; Sudha and Ravishankar 2002, 2003) and piperin (Balbuena et al. 2009). Only low levels of compounds have been detected in undifferentiated cultures; however, various dietary supplements like saponins and ginkgolides have been successfully produced through PCTOC (Carrier et al. 1991; Choi et al. 2000).

8.2.2 Food Ingredients from Large-Scale Cultures

Various bioreactor systems are applied for the production of different phytochemicals in large scale. The most successful plant cell and organ culture processes to date have been based upon cell suspensions (Stafford 1991; Thanh et al. 2014), hairy roots (Guillon et al. 2006), adventitious roots (Murthy et al. 2008, 2016), and somatic embryos (Shohael et al. 2014). A number of plant cell suspensions derived from different species have been grown to very large volumes (Stafford 1991). A wide variety of bioreactor designs tested and used for plant cell cultures are stirred tank reactors, bubble column reactors, airlift reactors, and ebb and flow reactors (Murthy et al. 2014a). High yields of secondary products have been produced by cell suspensions of many cell lines, some of which have been obtained by the selection of highly pigmented callus cultures, as in case of shikonin (Stafford 1991). Adventitious and hairy root cultures have been efficiently established in various bioreactors like stirred tank, bubble column, siphon mist, trickle-bed nutrient mist, fluidized bed, and modified airlift bioreactors called balloon-type bubble bioreactors. There are a very few successful examples for bioreactor production of food ingredients: *Echinacea purpurea* adventitious root biomass of 40.5 kg was produced with the inoculation of 7 g L^{-1} of inoculum in 1000 L airlift bioreactors (Wu et al. 2007). The accumulation of 5 mg g^{-1} of dry weight (DW) of chlorogenic acid, 22 mg g^{-1} of DW of cichoric acid, and 4 mg g^{-1} of DW of caftaric acid was achieved with adventitious roots grown in 1000 L bioreactors. Similarly, an inoculation of 500 g of fresh weight of ginseng adventitious roots into 500 L balloon-type bubble bioreactors yielded 74.8 kg of root biomass at 8 weeks of culture. The saponin content obtained in the small-scale (20 L) and pilot-scale (500 L) bioreactors was 1% based on DW (Choi et al. 2000). Such experimental results have led to the establishment of pilot-scale bioreactors (up to 10,000 L) for the production of ginseng adventitious root biomass for commercial exploitation by CBN Biotech, Cheongju, Republic of Korea. Recently, adventitious root culture system of *Hypericum perforatum* for high-density production of root biomass and hypericin using balloon-type bubble bioreactors has been established (Cui et al. 2014). A trickle-bed nutrient mist bioreactor was used for the hairy root culture of *Beta vulgaris* (Dilorio et al. 1992) and *D. carota* (Kondo et al. 1989) to obtain betacyanins and anthocyanins, respectively. If difficulties encountered during undifferentiated cell cultures, submerged shoot, as well as hairy and adventitious root cultures during scaling-up are properly tackled through the engineering of bioreactors, these systems serve as efficient productive units for the production of quality food ingredients in large scale in future.

8.3 SAFETY CONSIDERATIONS FOR THE PLANT CELL, TISSUE, AND ORGAN CULTURE PROCESSES

Setting up of PCTOC processes for the production of safe food ingredients involves safety considerations such as the development of an efficient cell line, scaling-up process, production, and purification of the phytocompounds (Fu 1999). The phytocompounds that are used as food ingredients have been produced from various tissue culture systems such as cell suspension, shoot, adventitious root, embryo, transformed shoot, and hairy root cultures (Paek et al. 2009; Park et al. 2005b; Shohael et al. 2014; Than et al. 2014). Organ cultures are considered to be more stable than the cell cultures in terms of genetic stability and the production of phytochemicals (Choi et al. 2000; Murthy et al. 2014a; Paek et al. 2009).

8.3.1 Safety Considerations for the Initiation of Cultures

It is empirical to contemplate the following safety measures to initiate high-yielding, stable cultures to obtain uniform raw materials and quality food ingredients from PCTOC.

1. *Selection of the plant material*: Initiation of culture processes is carried out considering the fundamental information about the plant material such as Latin name (genus, species, and authority), common name(s), chemotype, and geographic origin as the selection of the most promising plant material depends on the age of the plant at the time of obtaining the plant part, geography, and soil conditions (Figure 8.1).
2. *PCTOC method of production of the raw material*: Selection of suitable explants for induction of callus, cell, and organ lines is necessary as selection of suitable explant is dependent on the kind of culture to be initiated and the part of the plant used, as the concentration of the active ingredients varies with part of the plant (Murthy et al. 2015a).
3. *Selection of cell or organ lines for the production of biomass*: Strain improvement begins with the choice of the parent plant with high contents of desired product to obtain high-producing cell lines. Further, the heterogeneity within the cell population can be screened to select lines capable of higher biosynthetic activity to accumulate the elevated level of metabolites. A strain of *Euphorbia milii* accumulated about sevenfold the level of anthocyanins produced by the parent culture after 24 selections (Yamamoto et al. 1982). Selection can be easily achieved if the product of interest is a pigment (Fujita et al. 1984).
4. *Optimization of culture conditions/parameters and regulation of in vitro environment*: Optimized culture conditions/parameters for *in vitro* cultivation such as medium, salt strength, growth regulator and its concentration, medium pH, temperature, illumination, light intensity, and quality have control over the biomass growth and production of the phytochemicals (Figure 8.1).

8.3.2 Safety Considerations for the Proliferation of Cultures

1. Numerous problems like relative unstability in productivity, high shear sensitivity, slow growth rate, and low oxygen requirements may arise during scaling-up for the proliferation of cultures in bioreactors (Murthy et al. 2014a). In case of root cultures, mechanical agitation causes wounding of hairy roots and leads to callus formation. With a product of sufficiently high value, it is feasible to use batch fermentation, harvest the roots, and extract the product. Hence, safety considerations for the enhancement of biomass and phytochemical production depend on the type of culture vessel (bioreactor) used, agitation, aeration, mode of operation, and bioreactor conditions. During growth and production cycles in small-scale bioreactors, all the parameters have to be standardized before shifting to large-scale cultures (Murthy et al. 2015a; Figure 8.1).

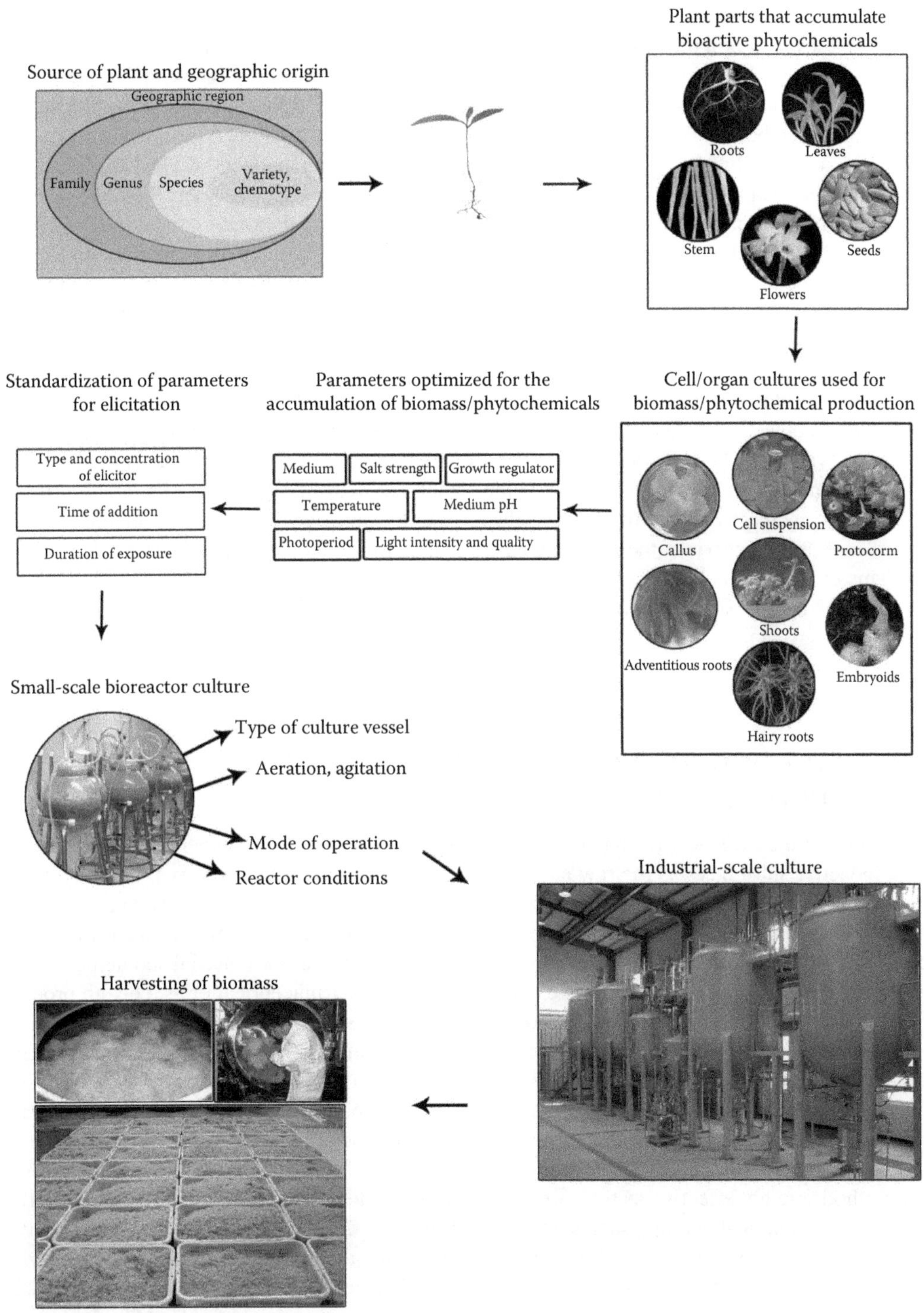

FIGURE 8.1 **(See color insert.)** A schematic depiction of the plant cell, tissue, and organ culture (PCTOC) processes for the production of PCTOC raw material.

2. Further, various techniques are used to increase the yield of products during the production stage of a cell or organ culture like addition of inducers or precursors, elicitation, and *in situ* product removal. A few considerations ought to be outlined concerning the following issues to obtain the best out of elicitation: elicitation methodology used, type and concentration of elicitor used, time of addition of elicitor and duration of exposure (Figure 8.1), the precise nature of a functional food in the health and legal frameworks, and plant secondary metabolism. Various types of chemical and physical elicitors have been tested and used successfully for the overaccumulation of bioactive compounds in PCTOC (Zhao et al. 2005). For example, addition of methyl jasmonate as an elicitor during the *Panax ginseng* adventitious root culture has been improved sevenfold accumulation of ginsenosides (Kim et al. 2004; Paek et al. 2009). In view of the biosafety of products, it is recommendable to avoid toxic elicitors such as heavy metals, detergents, xenobiochemicals, fungicides, herbicides, and other harmful chemicals. In addition, elicitors of biological origin (methyl jasmonate, salicylic acid) and physical elicitors (UV irradiation, differential light sources) can be used for the improvement in product accumulation (Murthy et al. 2015a; Figure 8.1).
3. *In situ* product removal techniques have been followed in the second phase of PCTOC with the objective of removing the products from the cultured cells and organs. The biomass types may be cells, adventitious roots, hairy roots, embryos, or shoots (Figure 8.1). Application of *in situ* product removal makes the recovery of products easier, and it often leads to increased productivity. Two-phase systems have also been used to facilitate the release and recovery of the secondary compounds in the medium (Buitelaar et al. 1991). This technology helps to continuously remove the compounds from the medium and prevent the feedback repression of the synthesis (Holmes et al. 1997). Further, detailed positive experimental evidences are necessary for such options as such experiments may lead to the accumulation of toxic substances in the cultures.

8.4 SAFETY CONSIDERATIONS FOR THE PROCESSING OF THE PCTOC RAW MATERIAL

Preservation of the PCTOC raw material immediately after its harvest is imperative to prevent the loss of bioactive substances as the raw material produced from PCTOC possesses the initial moisture content of 70%–85%, which is responsible for high enzyme activity and relieved heat flow, and provides the congenial environment for the growth of microorganisms, which leads to the loss of bioactive substances. Further, to avoid microbial contamination, the raw material should be dried to a low moisture content (<10%) for storage and utilization as pharmaceutical or as a health product. Drying is a process in which water is removed from the moist raw material by using the input of heat energy. Drying is a complex mechanism that involves combined heat and mass transfer within the biological raw material. Drying is an effective method for the preservation of the stable and safe PCTOC raw material as it reduces the effect of water and extends the shelf life of the raw material. Many conventional thermal methods such as airflow drying, vacuum drying, and freeze-drying have been under practice till date for the dehydration of the PCTOC raw material (Figure 8.2). However, these methods are not effective as they are characterized by low drying rates and long dehydration periods. In conventional heating, such as hot air and infrared drying, thermal energy is transmitted from the surface of the raw material to the inner core along a temperature gradient. These drying processes are responsible for a low drying rate that takes longer duration for drying, which may lead to undesirable thermal degradation of the ingredients. For example, oven drying (at 60°C) caused a marked loss of volatiles in sage, whereas freeze-drying preserved the same and also increased the absolute thymol content by 33% and deviated the volatile profile of the plant (Venskutonis 1997). Freeze-drying involves the removal of moisture from a frozen product without thawing that product. Temperature must be below the freezing point for that product, and the vapor pressure must

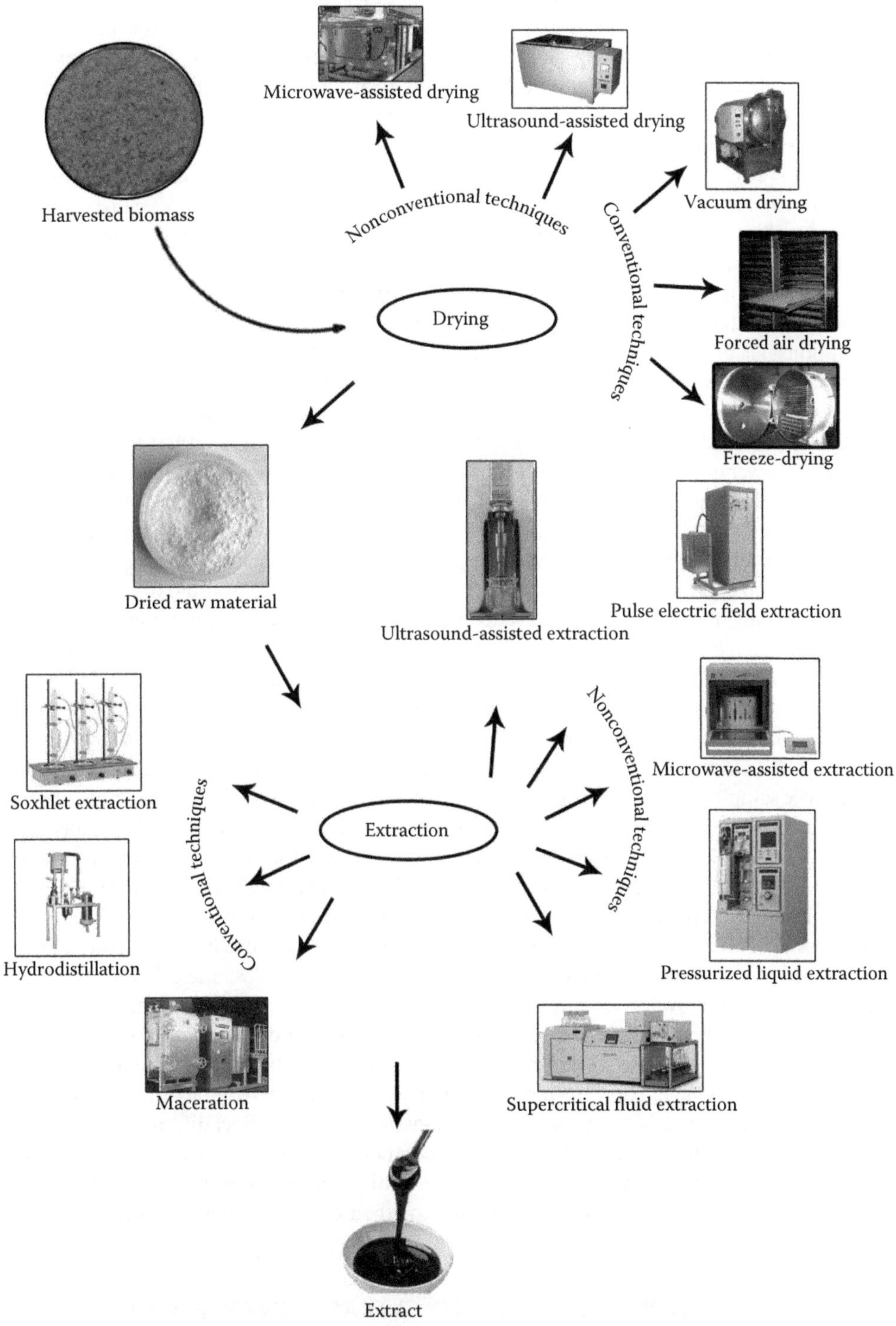

FIGURE 8.2 A schematic representation of the techniques used for drying of plant cell, tissue, and organ culture (PCTOC) raw material and extraction of phytochemicals from the PCTOC raw material.

be maintained at a very low level to permit moisture removal by sublimation. Because of the low temperature, low pressure, and low drying rate, freeze-drying is quite expensive compared to many other drying processes (Abascal et al. 2004). Conventional freeze-drying method is known to affect volatiles, for instance, freeze-drying alone substantially reduces the amount of monoterpenes such as 1, 8-cineole, linalool, and geraniol (Diaz-Maroto et al. 2002). Similarly, freeze-dried willow (*Salix spp.*) had lower concentrations of tannins compared with air-dried and vacuum-dried leaves, both of which had high concentration of metabolites (Orians 1995). In contrast, freeze-drying of purple cone flower (*E. purpurea*) led to a greater retention of cichoric acid compared to air drying. In vacuum drying, the product is usually placed inside a chamber where the pressure is reduced to create vacuum. Since the total pressure in the chamber is very low, the partial pressure of the water vapor in the chamber will also be very low. This low partial pressure causes a large partial pressure difference between the water in the product and the surrounding environment. Thus, water moves more readily from the product to the surrounding environment in the chamber. Drying under vacuum conditions permits the drying at a lower temperature, and this characteristic of vacuum drying is very important for the products that may suffer significant flavor changes at higher temperatures.

Recently, ultrasound-assisted drying and microwave-assisted drying methods have been followed for drying the botanical samples and PCTOC raw materials (Vilkhu et al. 2008; Figure 8.2). When low-frequency power ultrasound is applied to the biological material, it causes a rapid series of alternative compression and expansions of the material, whereas the liquid content in the material leads to the formation of bubbles that generate localized pressure and temperature increase, which leads to dehydration of the material (Simal et al. 1998). Ultrasonic pretreatment has been used for drying and dehydration of food and medicinal raw materials. Similarly, microwave-assisted drying is being used for drying of botanical samples. Microwave drying uses the transfer of electromagnetic waves (i.e., microwaves) through the material to generate heat by the oscillation of molecules. The heat generated is in a uniform rate within the entire volume of the material. Sharma and Prasad (2006) studied the microwave vacuum drying and hot air drying effects on drying of garlic. They noticed 80% reduction in the drying time without affecting the quality of the dried garlic with microwaves of 0.4 W g^{-1} to hot air at 60°C–70°C. On the other hand, they have reported a huge loss in garlic pungency with hot air drying in garlic samples. *Echinacea* roots (*E. purpurea* Moench) were dried to a final moisture content of 7.3% in an experimental rotating table microwave vacuum dryer (Anonymous 2000), roots were cut into 3 mm slices and 1.39 kWh of microwave energy per kg of fresh roots was applied for a drying time of 25 min, absolute pressure was 4650 Pa, and product temperature was maintained below 50°C. Roots were also subjected to freeze-drying (shelf temperature, 20°C; condenser temperature, 55°C; absolute pressure, 200 Pa) and hot air drying (air temperature 70°C, relative humidity 10%, air velocity 0.9 m s^{-1}, drying time 3.5 h) as a reference. The total alkamides were 2.85 mg g^{-1} DW for air drying, 3.07 mg g^{-1} DW for microwave vacuum drying, and 3.28 for freeze-drying. The microwave vacuum drying process retained higher concentrations of alkamides than air drying, but the quality of freeze-dried roots was better due to the lower product temperature encountered. In summary, conventional thermal methods of drying (airflow drying, vacuum drying, and freeze-drying) and modern methods of drying (ultrasound-assisted drying and microwave-assisted drying) affect different raw materials in a variety of ways for retaining the desired bioactive compounds in the raw material. They emphasize the need for more systematic studies to assist in determining the preferred method of drying for the preservation of various plant/PCTOC raw materials and their constituent compounds.

8.5 SAFETY CONSIDERATIONS FOR THE EXTRACTION OF FOOD INGREDIENTS FROM THE PCTOC RAW MATERIAL

The methods that have been developed for the isolation and production of bioactive compounds from botanical samples are also applicable for the extraction of useful phytochemicals from the PCTOC raw material (Murthy et al. 2015a). Several methods have been developed and improved over the years for the extraction of bioactive compounds from botanical samples (Azmir et al. 2013; Wang and Weller 2006). Traditional methods such as (1) soxhlet extraction, (2) maceration, and (3) hydrodistillation are commonly used for the extraction of bioactive constituents from plant materials (Figure 8.2). Generally, the raw material is exposed to different solvents in soxhlet and maceration techniques, which take up compound of interest such as flavors and coloring agents, whereas certain samples usually need centrifugation and filtration to remove the solid phase and extract could be used as additive, food supplement, or to be used as functional food after appropriate refinement. The important factors that affect the extraction process are matrix properties of the sample, solvent used, solvent temperature, pressure, and duration of treatment (Hernandez et al. 2009). Extraction efficiency of any conventional methods mainly depends on the choice and polarity of the solvents. Other factors such as molecular affinity between solute and solvent, mass transfer, use of cosolvent, human toxicity, and cost of extraction should be thoroughly studied for developing extraction procedures (Azmir et al. 2013). Hydrodistillation is another traditional method for the extraction of bioactive compounds in which water or steam is the main factor to free bioactive compounds from the plant material, and it involves hydrodiffusion, hydrolysis, and decomposition by heat. However, this method may not be suitable for the extraction of certain volatile compounds that may be lost due to high extraction temperature.

The major drawbacks of conventional methods of extraction are longer extraction period, low extraction selectivity, and thermal decomposition of thermolabile compounds. Various novel methods including (1) ultrasound-assisted extraction, (2) microwave-assisted extraction, (3) supercritical fluid extraction, (4) pulsed-electric field extraction, (5) enzyme-assisted extraction, and (6) accelerated solvent extraction have been developed for the extraction of botanicals, and these methods have effectively caused the reduction in the extraction time, decrease in solvent consumption, increase in the extraction yield, and enhancement of the quality of extracts (Azmir et al. 2013; Wang and Weller 2006; Figure 8.2). *Ultrasound-assisted extraction* involves the release of ultrasonic waves (20–100 MHz) that pass through the medium by creating compression and expansion of the matter. The process produces growth and collapse of bubbles (cavitation), and thus, the implosion of cavitation bubbles can hit the surface of the solid matrix and disintegrate the cells causing the release of desired compounds (Esclapez et al. 2011). Temperature, pressure, frequency, and time of sonication are factors that influence the release of bioactive compounds. Furthermore, moisture content of the sample, milling degree, particle size, and solvent are also important factors for obtaining efficient and effective extraction. Ultrasound-assisted extraction seems to be an effective extraction technique for the extraction of bioactive compound botanicals and the PCTOC raw material. For example, ultrasound-assisted extraction of ginseng saponins was approximately 3 times faster than the traditional solvent extraction. Furthermore, the ultrasound-assisted extraction technique was achieved at lower temperatures that are more favorable for thermally unstable compounds (Wu et al. 2001). Similarly, twofold higher carvone and limonene were extracted from caraway seeds when compared to that of conventional solvent extraction (Chemat et al. 2004). *Microwave-assisted extraction* involves the usage of electromagnetic field (300 MHz to 300 GHz), which is converted into heat during ionic conduction and dipole rotation mechanisms, during which there will be a separation of solutes from active sites of the sample matrix under increased pressure and temperature, which leads to the release of bioactive compounds in solvent. Microwave-assisted extraction was successfully used for rapid extraction of ginsenosides from ginseng roots within 15 min, whereas conventional solvent extraction needs 10 h (Shu et al. 2003). Dhobi et al. (2009) showed increased extraction efficiency of flavonolignan and silibinin from *Silybum marianum* through microwave-assisted extraction compared to that of soxhlet and maceration methods. The use of fluid (water or any solvents such as carbon dioxide, propane, butane, and ethylene) under high temperature and the pressure below supercritical conditions in extraction processes is generally referred to as *supercritical fluid extraction*. This technique has emerged as a useful tool to replace conventional extraction methods because extraction process is faster and produced higher yields; in addition, the use of solvents can be greatly reduced (Herrero et al. 2006). Higher amount of L-dopa was extracted from *Mucuna* using supercritical carbon dioxide when compared to conventional solvent extraction (Garcia et al. 2012). Supercritical extraction is reported to be a useful method for the extraction of plant flavors and fragrances (Capuzzo et al. 2013). *Pulsed-electric field technology* is the application of short pulses of high electric fields (in the order of 10–80 kV cm^{-1}) to the target material for short duration (micro to milliseconds) in different intervals. During this process, the cell membrane is weakened or destroyed and is useful in pressing, drying, and extraction (Angersbach et al. 2000). The effectiveness of pulsed-electric field extraction depends on the process parameters such as field strength, specific energy input, pulse number, treatment temperature, and properties of the material to be treated (Heinz et al. 2003). Twofold increment in polyphenol content was reported with the treatment of pulsed-electric field (0.5 kV cm^{-1}, 50 pulses) in grapes compared with conventional extraction methods (Balasa et al. 2006). *Enzyme-assisted extraction* is a method in which botanicals are treated with specific enzymes like cellulase, amylase, and pectinase during extraction. The enzymatic treatment materials are responsible for the breakage of cell wall, and hydrolysis of structural polysaccharides and lipid bodies leads to enhanced recovery of bioactive compounds. Landbo and Meyer (2001) showed improved release of phenolics from *Ribes nigrum* using various enzymes. *Accelerated solvent extraction* or *pressurized liquid extraction* is yet another method that involves the usage of solvents at high temperature (50°C–200°C) and pressure (1450–2175 psi) to ensure the rapid extraction rate of compounds (Mustafa and Turner 2011). A systematic study was carried out to evaluate variation in solvents and extraction procedures used for the extraction of isoflavones from soybeans. For the extraction, six different mixtures were evaluated: acetonitrile–water, ethanol–water, methanol–water, dimethysulfoxide–acetonitrile–water, dimethysulfoxide–ethanol–water and genapol–water. The assessment was done using six different extraction techniques: shaking, vertexing, sonication, stirring, soxhlet, and accelerated solvent extraction. The optimum isoflavone recoveries from soybean samples were obtained with dimethysulfoxide–ethanol–water (5:70:25, v/v/v) solvent mixing using accelerated solvent extraction (Luthria et al. 2007). In summary, the choice of extraction procedure depends on the nature of the raw material and the compounds to be isolated. The solvents used in extraction should be generally recognized as safe (GRAS), and the processes should be economical and environmental friendly.

8.6 BASIC GUIDELINES FOR SAFETY ASSESSMENT OF FOOD INGREDIENTS DERIVED FROM PCTOC

The PCTOC raw materials/products that are supposed to be used in food industry need thorough toxicological and safety evaluation (Figure 8.3). The quality and consistency of the product require a careful control in terms of product identity, potency, and purity because of concerns about impurities or toxins arising from the manufacturing process (microbial contamination, endotoxin, and process chemicals) and fidelity of desired bioactive compound(s). Biosafety measures of food ingredients obtained through PCTOC processes depend on the nature of the end product. For instance, if the PCTOC product is a flavoring agent used in the food industry, the Flavor and Extract Manufacturers Association (FEMA) safety model is applicable in the United States (Hallagan et al. 1999). If the PCTOC product is a food and dietary supplement, the FDA regulations such as "Dietary Supplement Health Education Act" (DESHA) are applied in the United States (Anonymous 2012a,b; Beru 1999; Rulis and Levitt 2009), and generally, the product should be declared as GRAS. The regulations for the approval of PCTOC products in European Union, Canada, and Japan also vary considerably (McIntyre 1999; Ushiyama 1999). Since the biosafety guidelines of the PCTOC raw materials/products vary in various countries, here, we are presenting succinct guidelines/criteria to be followed for the production of the PCTOC raw material for biosafety evaluation of raw materials/product for commercialization.

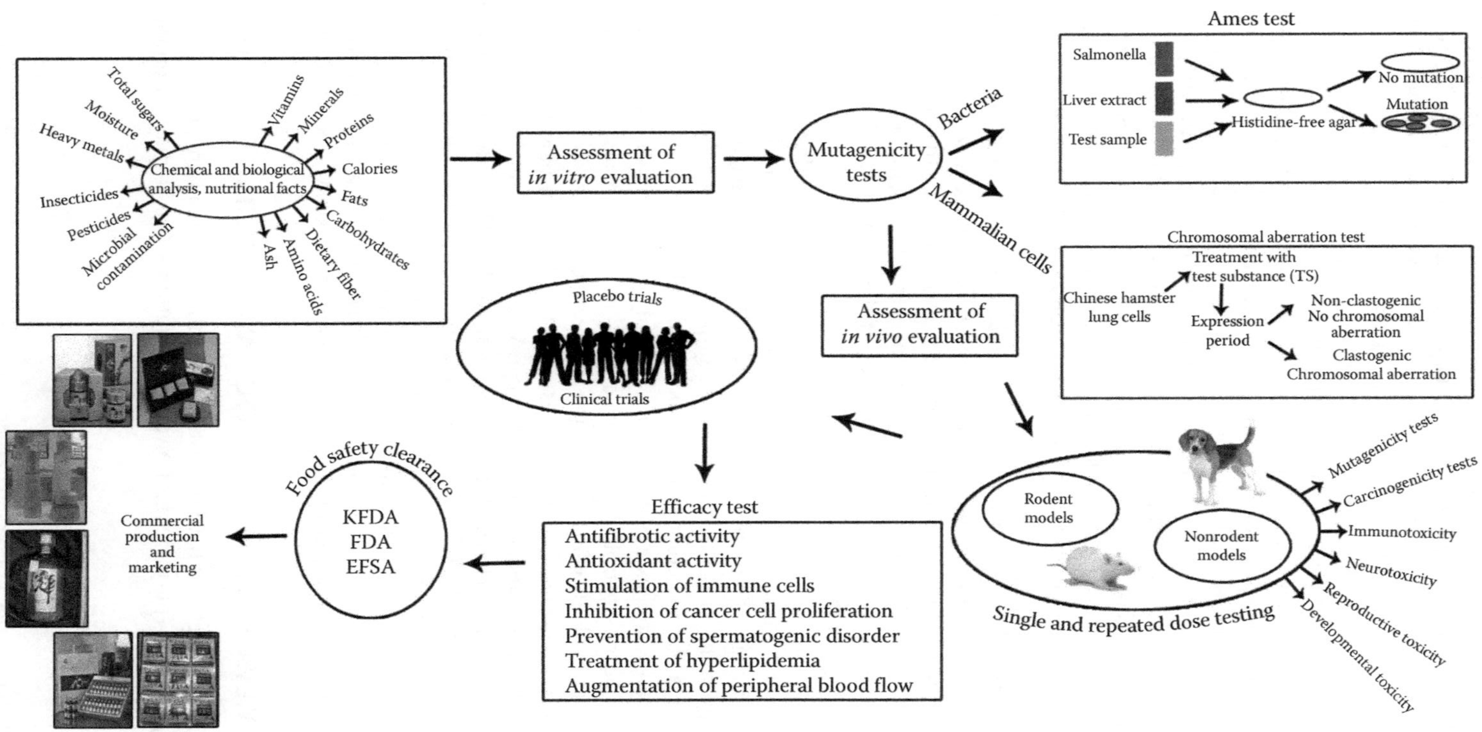

FIGURE 8.3 **(See color insert.)** A schematic depiction of the biosafety evaluation of the plant cell, tissue, and organ culture raw material/product.

Safety measures/evaluations of PCTOC raw material/product involve the following concrete steps:

1. *Biological, chemical, and nutritional analysis*: Assessment of microbial contaminants, heavy metals and pesticide/herbicide/insecticide contaminants, residual solvents and chemical/nutritional analysis including carbohydrates, proteins, fats, dietary fibers, minerals, ash, amino acids, vitamins, and calories of energy per gram material.
2. *Toxicological evaluation*: *In vitro* evaluation, point mutation with bacteria, and chromosomal mutation with mammalian cell lines (Figure 8.3).
3. *Toxicological evaluation with rodents and nonrodents*: Single- and repeated-dose assay and evaluation of mutagenicity, carcinogenicity, developmental toxicity, reproductive toxicity, immunotoxicity, neurotoxicity. During *in vivo* toxicological evaluation, selection of animal species, selection of doses, route of administration, duration of treatment/experiment, and number of animals used should be justified. Single-dose and multiple-dose tests involving developmental/reproductive toxicity tests should be carried out rigorously (Figure 8.3).
4. *Pharmacokinetic studies*: In the development of a compound as a single entity, pharmacokinetic studies can demonstrate the systemic exposure and can facilitate a comparison of dose response between animals and humans.
5. *Chemical identity, fidelity, and purity of target bioactive compound(s)*: To demonstrate the substantial equivalence of product derived from PCTOC process with a substance from conventional sources (e.g., an extract of natural plants), methods that are adopted should enable to have a "profile" or "fingerprint" of the substance to be produced by PCTOC method. For example, chromatographic profiles of the product may be used as a part of characterization of the product.
6. *Clinical studies*: For the use of PCTOC products as food ingredients, the primary evidence of safety and efficacy must be derived from human clinical trials due to limitations in the animal models. The most valuable clinical trials are designed to evaluate safety and efficacy by comparing the outcomes in a group of subjects receiving treatment to a group receiving a placebo-controlled treatment and followed for the same period of time. The highest-quality studies are those that randomly assign participants in the study of treatment and placebo-controlled groups where both the participants and investigators are unaware of the individuals who are receiving the treatment or placebo/control (Figure 8.3).
7. *Approval by the competent authorities*: Based on the thorough biosafety evaluation, the PCTOC raw material or products should be approved for commercialization by FDA in the United States; Korean Food and Drug Administration (KFDA) in Republic of Korea.
8. The safety assessment of PCTOC raw material or ingredients produced by genetically modified organs, such as hairy root and shooty teratomas, needs careful toxicological evaluation based on the recommendations proposed by Konig et al. (2004).
9. *Periodic batch testing to meet the specifications*: Periodic checking for identity and purity of target bioactive compound(s) should be carried out from batch to batch production.

8.7 BIOSAFETY EVALUATION OF PCTOC RAW MATERIAL: CASE STUDIES

Herein, two specific examples that have been studied for their biosafety are presented. In *P. ginseng* (ginseng) and *E. purpurea* (purple cone flower), adventitious root biomass is produced through PCTOC technology.

8.7.1 GINSENG

P. ginseng C.A. Meyer roots contain triterpene glycosides known as ginsenosides (saponins). Ginsenosides are divided into three groups based on their structure, that is, Rb group (protopanaxadiols including Rb1, Rb2, Rc, and Rd), Rg group (protopanaxatiols including Rg1, Re, Rf, and Rg2), and Ro group (oleanolic acid) (Murthy et al. 2014c; Park et al. 2005a). Other than saponins, ginseng roots also contain biophenols, polyacetylenes, sesquiterpenes, polysaccharides, peptidoglycans, fatty acids, and vitamins. Extracts and powder of the roots are used as health tonic and health food (Murthy et al. 2014c; Park et al. 2005a). Pharmacological effects of ginseng have been demonstrated in cancer, diabetes mellitus, and in disorders related to cardiovascular system, immune system, and central nervous system, including antistress and antioxidant activities (Murthy et al. 2014c; Park et al. 2005a).

Wild ginseng is a very rare commodity, and hence, Research Institute for the Development of Horticultural Technology, Chungbuk National University, Republic of Korea, has developed a standard cell and organ culture technology for large-scale biomass production of ginseng biomass using bioreactors (cells and adventitious roots) (Murthy et al. 2014c; Paek et al. 2009; Thanh et al. 2014). The biomass produced through PCTOC was subjected for in-depth biosafety analysis (Murthy and Paek 2014; Sivakumar et al. 2006), and the biomass of tissue-cultured mountain ginseng adventitious roots (TCMGARs) and their products have been approved for commercialization by KFDA (Murthy et al. 2014c, 2015a).

Many *in vitro* and *in vivo* studies and clinical trials (randomized, double-blinded, placebo-controlled clinical studies), pharmaceutical effects, efficacies, and biosafety of ginseng (Coleman et al. 2003; Kitts and Hu 2000) have been reviewed by Lee and Son (2011) and Choi et al. (2013), and ginseng is considered generally as safe with no serious adverse effects. Chemical evaluation and biosafety tests were also conducted on TCMGARs by Murthy and Paek (2014) and Sivakumar et al. (2006), and results revealed that the biomass of TCMGARs is free from microbial contamination, heavy metals, pesticides, and insecticides. *In vitro* reverse mutation (*Salmonella typhimurium and Escherichia coli*), chromosomal aberration (Chinese hamster lung cells), and micronucleus (mammalian polychromatic erythrocyte cells) tests did not show any abnormalities (Murthy and Paek 2014; Sivakumar et al. 2006). A 13-week repeated-dose tests on rodent and beagle dog confirmed the nontoxicity and biosafety at an average consumption level (Murthy and Paek 2014; Sivakumar et al. 2006). Chemical profile showed that TCMGARs are rich in varied ginsenosides especially Rb1, Rb2, Rc, Rd, Rg3, and Rh2 when compared to field-cultivated Korean ginseng and displayed high antioxidant (Ali et al. 2006), antidiabetic (Murthy et al. 2014j), hepatoprotective (Murthy et al. 2014i), and antifibrotic (Lim et al. 2007) activities. Varied efficacy tests of TCMGARs displayed the prevention of spermatogenic disorders (Park et al. 2006), inhibition of platelet aggregation, and thrombus formation (Jeon et al. 2008; Lee et al. 2011). Based on biosafety evaluation tests, KFDA and U.S. Food and Drug Administration (USFDA) have given a consent for commercial production of ginseng adventitious roots and their products to Research Centre for the Development of Advanced Horticultural Technology, Chungbuk National University, Republic of Korea, and CBN Biotech, 319–9 Songdae-ri, Ohchang-myeon, Cheongwon-gun, Chungcheongbuk-do, Republic of Korea. Various TCMGAR-based ginseng products like ginseng powder/syrup, wine, soap, and cosmetics are available in the market (Murthy et al. 2014a–c).

8.7.2 Purple Cone Flower

Purple cone flower (*E. purpurea*) is one of the top-selling phytomedicines in the United States and Europe (Brevoort 1998). It is used as an immunostimulant and acts by stimulating T-cell production, phagocytosis, lymphocytic activity, and cellular respiration by inhibiting hyaluronidase activity (Bauer et al. 1989). Caffeic acid derivatives (CADs), polysaccharides, and alkamides are the main phytochemicals that are responsible for biological activities of the purple cone flower (Bauer 1998). Among CADs, cichoric acid has been shown to act as an immunostimulant by stimulating phagocytosis (Wagner and Proksch 1985), and it has a potent antiviral activity against HIV (Barrett 2003).

Toxicological tests for chromosome aberrations and sister chromatid exchange in bacteria and animal cells have found no evidence of mutagenicity (Mengs et al. 1991). Animal experiments have failed to demonstrate a significant risk, and laboratory analysis of blood, urine, and organ specimens from animals treated with *Echinacea* products provides further evidence of safety (Mengs et al. 1991). Various biosafety tests proved *Echinacea* as a relatively safe herbal medicine (Melchart and Linde 1999; Percival 2000). However, allergic reactions to *Echinacea* have been reported (Barrett 2003), and no deaths have been reported.

Successful adventitious root cultures were established in *E. purpurea* by Murthy et al. (2014e,f) and Wu et al. (2007) using bioreactor systems for the production of potential bioactive compounds. Toxicological and biosafety evaluations of tissue-cultured *E. purpurea* adventitious roots (TCEPARs) were performed by Murthy et al. (2015b). Reverse mutation (*S. typhimurium and E. coli*) and chromosomal aberration (Chinese hamster lung cells) tests showed no significant mutagenicity. Furthermore, repeated 4-week oral dose tests performed in Sprague-Dawley rats did not show any notable change in the general behavior of the rats, in the gross appearance of internal organs, or in their mortality rate. No differences were observed between the control group and the treatment group in parameters such as absolute body weight, hematology, blood chemistry, and absolute and relative organ weights. The results of these toxicological studies indicate that TCEPARs are nontoxic and safe at an average dietary level. Further, the process of evaluation for large-scale production of TCEPARs and utilization is pending with KFDA.

8.8 CONCLUDING REMARKS

PCTOC systems have been modeled as apposite systems for the production of phytochemicals that are used as food ingredients. Although contamination-free culture systems have fashioned to explore these systems for the production of food compounds, the stability of the culture systems to produce only target compound free from any toxin has made the research opportunities still abound. The PCTOC can be modulated through metabolic engineering to direct metabolic flux for obtaining valuable ingredients without other potential substances that have an impact on the human systems, which may also aid in the production of food ingredients through PCTOC in a sensible way to produce safe foods for protecting public health effectively.

ACKNOWLEDGMENTS

Dr. H. N. Murthy is thankful to the Ministry of Education, Science and Technology, Republic of Korea, for the award of Brain Pool Fellowship (131S-4-3-0523); University Grants Commission, New Delhi [No. 41-423/2012(SR) and No. F.4-29/2015/DSA-1]; and Department of Atomic Energy [BRNS Project No. 2013/35/BRNS/20], Mumbai, for the financial assistance.

REFERENCES

Abascal, K., Ganora, L., and E. Yarnell. 2004. The effect of freeze-drying and its implications for botanical medicine: A review. *Phytother Res* 19: 655–660.

Abdel-Rehaman, A., Anyangwe, N., Carlacci, L. et al. 2011. The safety and regulation of natural products used as food and food ingredients. *Toxicol Sci* 123: 333–348.

Ali, M.B., Yu, K.W., Hahn, E.J., and K.Y. Paek. 2006. Methyl jasmonate and salicylic acid elicitation induces ginsenoside accumulation, enzymatic and non enzymatic antioxidant in suspension culture of *Panax ginseng* roots in bioreactors. *Plant cell Rep* 25: 613–620.

Angersbach, A., Heinz, V., and D. Knorr. 2000. Effects of pulsed electric fields on cell membranes in real food systems. *Innov Food Sci Emerg* 1: 135–149.

Anonymous. 2000. Process for drying medicinal plants. US Patent Number 6,128,831, October 10, 2000.

Anonymous. 2007. Guidance for industry and other stakeholders, Toxicological principles for the safety assessment of food ingredients. Redbook 2000 (updated July 2007) U.S. Department of Health and Human Services, Food and Drug Administration, Silver Spring, MD.

Anonymous. 2008. Safety and nutritional assessment of GM plants and derived food and feed: The role of animal feeding trials. *Food Chem Toxicol* 46: S2–S70.

Anonymous. 2012a. Safety evaluation of certain food additives. WHO Food Additives Series: 67, World Health Organization, Geneva, Switzerland.

Anonymous. 2012b. Guidance for submission for food additives evaluations. EFSA panel on food additives and nutrient sources added to food (ANS). *EFSA J* 10: 2760, 1–60.

Azmir, J., Zaidul, I.S.M., Rahman, M.M. et al. 2013. Techniques for extraction of bioactive compound from plant materials: A review. *J Food Eng* 117: 425–436.

Balasa, A., Toefl, S., and D. Knorr. 2006. Pulsed electric field treatment of grapes. Food Factory of the Future 3, Gothenburg, Sweden.

Balbuena, T.S., Santa-Catarina, C., Silvera V., Kato M.J., and E.I.S. Floh. 2009. *In vitro* morphogenesis and cell suspension culture establishment in *Piper solmsianum* DC. (Piperaceae). *Acta Bot Bras* 23: 229–236.

Barrett, B. 2003. Medicinal properties of Echinacea: A critical review. *Phytomed* 10: 66–86.

Bauer, R. 1998. Echinacea: Biological effects and active principles. In *Phytomedicines of Europe*, eds. L.D. Lawson and R. Bauer, pp. 140–157. American Chemical Society, Washington, DC.

Bauer, R., Remiger, P., Jurcic, K., and H. Wagner. 1989. Beeinflussung der phagozytose-aktivitat durch *Echinacea*-extrakete. *Z. Phytother* 10: 43–48.

Beru, N. 1999. Food ingredients from plant cell and tissue culture. In *Plant Cell and Tissue Culture for the Production of Food Ingredients*, eds. J.T. Fu, G. Sing, and W.R. Curtis, pp. 265–274. Academic/Plenum Publishers, New York.

Brevoort, P. 1998. The booming U.S. botanic market. *HerbalGram* 44: 33–45.

Buitelaar, R.M., Langenhoff, A.A.M., Heidstra, R., and J. Trampers. 1991. Growth and thiophene production by hairy root cultures of *Tagetes patula* in various two-liquid-phase bioreactors. *Enzyme Microb Techn* 13: 487–494.

Capuzzo, A., Maffei, M.E., and A. Occhipinti. 2013. Supercritical fluid extraction of plant flavors and fragrances. *Molecules* 18: 7194–7238.

Carrier, D., Chauret, N., Mancini, M. et al. 1991. Detection of ginkgolide A in *Ginkgo biloba* cell cultures. *Plant Cell Rep* 10: 256–259.

Chemat, S., Lagha, A., AitAmar, H., Bartels, V., and F. Chemat. 2004. Comparison of conventional and ultrasound-assisted extraction of carvone and limonene from carway seeds. *Flavour Frag J* 19: 188–195.

Choi, J., Kim, T.H., Choi, T.Y., and M.S. Lee. 2013. Ginseng for health care: A systematic review of randomized controlled trials in Korean literature. *Plos One* 8: 1–14.

Choi, S.M., Son, S.H., Yun, S.R., Kwon, O.W., Seon, J.H., and K.Y. Paek. 2000. Pilot-scale culture of adventitious roots of ginseng in a bioreactor system. *Plant Cell Tissue Org Cult* 62: 187–192.

Coleman, C.I., Hebert, J.H., and P. Reddy. 2003. The effects of *Panax ginseng* on quality of life. *J Clin Pharma Ther* 28: 5–15.

Cui, X.H., Murthy, H.N., and K.Y. Paek. 2014. Pilot-Scale culture of *Hypericum perforatum* L. adventitious roots in airlift bioreactors for the production of bioactive compounds. *Appl Biochem Biotech* 174: 784–792.

Dhobi, M., Mandal, V., and S. Hemalatha. 2009. Optimization of microwave assisted extraction of bioactive flavonolignan-silybinin. *J Chem Metrl* 3: 13–23.

Diaz-Maroto, M.C., Perez-Coello, M.S., and M.D. Cabezudo. 2002. Effect of different drying methods on the volatile components of parsley (*Petroselinum crispum* L.). *Eur Food Res Technol* 215: 227–230.

Dilorio, A.A., Cheetham, R.D., and P.J. Weathers. 1992. Growth of transformed roots in a nutrient mist bioreactor: Reactor performance and evaluation. *Appl Microbiol Biot* 37: 457–462.

Ellis, B.E. 1988. Natural products from plant tissue culture. *Nat Prod Rep* 5: 581–612.

Esclapez, M., Gracia Perez, J., Mulet, A., and J. Carcel. 2011. Ultrasound-assisted extraction of natural products. *Food Eng Rev* 3: 108–120.

Fu, T.J. 1999. Plant cell and tissue culture for food ingredient production, safety considerations. In *Plant Cell and Tissue Culture for the Production of Food Ingredients*, eds. J.T. Fu, G. Singh, and W.R. Curtis, pp. 237–250. Kluwer Academic/Plenum Publishers, New York.

Fujita, Y., Takahashi, S., and Y. Yamada. 1984. Selection of cell lines with high productivity of shikonin derivatives through protoplasts of *Lithospermum erythrorhizon. P Eur Cog Biotch* 1: 161–166.

Furuya, T. 1978. Biotransformation by plant cell cultures. In *Frontiers of Plant Tissue Culture,* ed. T.A. Thorpe, pp. 191–200. University of Calgary, Calgary, Alberta, Canada.

Garcia, V.A.d.S., Cabral, V.F., Zanoelo, E.F., da Silva, C., and L.C. Filho. 2012. Extraction of *Mucuna* seed oil using supercritical carbon dioxide to increase the concentration of L-Dopa in the defatted meal. *J Supercrit Fluid* 69: 75–81.

Guillon, S., Tremouillaux-Guiller, J., Pati, P.K., Rideau, M., and P. Gantet. 2006. Harnessing the potential of hairy roots: Dawn of a new era. *Trends Biotechnol* 24: 403–409.

Hallagan, J.B., Putnam, J.M., Higley, N.A., and R.L. Hall. 1999. The safety assessment of flavor ingredients derived from plant cell and tissue culture. In *Plant Cell and Tissue Culture for the Production of Food Ingredients,* eds. J.T. Fu, G. Sing, and W.R. Curtis, pp. 251–257. Academic/Plenum Publishers, New York.

Heinz, V., Toepfl, S., and C. Knorr. 2003. Impact of temperature on lethality and energy efficiency of apple juice pasteurization by pulsed electric fields treatment. *Innov Food Sci Emer Technol* 4: 167–175.

Hernandez, O., Ruiz-Matute, A.I., Olano, A., Moreno, F.J., and M.L. Sanz. 2009. Comparison of fractionation techniques to obtain prebiotic galactooligosaccharides. *Int Dairy J* 19: 531–536.

Herrero, M., Cifuentes, A., and E. Ibanez. 2006. Sub and supercritical fluid extraction of functional ingredients from different natural sources: Plants, food-by-products, algae and microalgae: A review. *Food Chem* 98: 136–148.

Holmes, P., Li, S.L., Green, K.D., Ford-Lloyd, B.V., and N.H. Thomas. 1997. Drip-tube technology for continuous culture of hairy roots with integrated alkaloid extraction. In *Hairy Roots, Culture and Applications*, ed. P.M. Doran, pp. 201–208. Harwood Academic Publisher, Amsterdam, the Netherlands.

Jeon, W.K., Yoo, B.K., Kim, Y.E. et al. 2008. Anti platelet activity of tissue-cultured mountain ginseng adventitious roots in human whole blood. *Food Sci Biotechnol* 17: 1197–1202.

Johnson, T.S., Ravishankar, G.A., and L.V. Venkataraman. 1990. *In vitro* capsaicin production by immobilized cells and placental tissues of *Capsicum annuum* L. grown in liquid medium. *Plant Sci* 70: 223–229.

Kim, Y.S., Hahn, E.J., Murthy, H.N., and K.Y. Paek. 2004. Adventitious root growth and ginsenoside accumulation in *Panax ginseng* as affected by methyl jasmonate. *Biotechnol Lett* 26: 1619–1622.

Kitts, D.D. and C. Hu. 2000. Efficacy and safety of ginseng. *Public Health Nutr* 3(4A): 473–485.

Kondo, O., Honda, H., Taya, M., and T. Kobayashi. 1989. Comparison of growth properties of carrot hairy root in various bioreactors. *Appl Microbiol Biotechnol* 32: 291–294.

Konig, A., Cockburn, A., Crevel, R.W.R. et al. 2004. Assessment of the safety of foods derived from genetically modified (GM) crops. *Food Chem Toxicol* 42: 1047–1088.

Kroes, R. and R. Walker. 2004. Safety issues of botanicals and botanical preparations in functional foods. *Toxicology* 198: 213–220.

Landbo, A.K. and A.S. Meyer. 2001. Enzyme-assisted extraction of antioxidative phenols from black current juice press residues (*Ribes nigrum*). *J Agric Food Chem* 49: 3169–3177.

Lee, I.S., Kim, S.K., Jeon, M.H., and W.K. Jeon. 2011. Ethyl acetate extract of tissue cultured mountain ginseng adventitious roots inhibits *in vitro* platelet aggregation in whole human blood and augments peripheral blood flow in mice. *J Ginseng Res* 35: 442–448.

Lee, N.H. and C.G. Son. 2011. Systematic review of randomized controlled trials evaluating the efficacy and safety of ginseng. *J Acupunct Meridian Stud* 4: 85–97.

Lim, H.K., Kim, Y.W., Lee, D.H., Kim, S.M., and M.J. Cho. 2007. The antifibrotic and antioxidant activities of hot water extract of adventitious root culture of *Panax ginseng* (ARCP). *J Appl Biol Chem* 50: 78–84.

Luthria, D.L., Biswas, R., and S. Natarajan. 2007. Comparison of extraction solvents and techniques used for the assay of isoflavones from soybean. *Food Chem* 105: 325–333.

McIntyre, K.E. 1999. Regulatory issues, Canadian perspective. In *Plant Cell and Tissue Culture for the Production of Food Ingredients*, eds. J.T. Fu, G. Sing, and W.R. Curtis, pp. 275–279. Academic/Plenum Publishers, New York.

Melchart, D. and K. Linde. 1999. Clinical investigations of Echinacea pharmaceuticals. In *Immunomodulatory Agent from Plants*, ed. H. Wagner, pp. 105–118. Birkhauser Verlag, Berlin, Germany.

Meltzer, H.M. 2010. Bioactive compounds through food, nutraceuticals or pills? In *Bioactive Compounds in Plants—Benefits and Risks for Man and Animals*, ed. A. Bernhoft, pp. 205–211 The Norwegian academy of Science and Letters, Oslo, Norway.

Mengs, U., Clare, C.B., and J.A. Poiley. 1991. Toxicity of *Echinacea purpurea*. Acute, subacute and genotoxicity studies. *Arzneimittelforschung* 41: 1076–1081.

Murthy, H.N., Dandin, V.S., Lee, E.J., and K.Y. Paek. 2014j. Efficacy of ginseng adventitious root extract on hyperglycemia in streptozotocin-induced diabetic rats. *J Ethanopharmacol* 153: 917–921.

Murthy, H.N., Dandin, V.S., and K.Y. Paek. 2016. Tools for biotechnological production of useful phytochemicals from adventitious root cultures. *Phytochem Rev* 15: 129–145.

Murthy, H.N., Dandin, V.S., and K.Y. Paek. 2014i. Hepatoprotective activity of ginsenosides from *Panax ginseng* adventitious roots against carbon tetrachloride treated hepatic injury in rats. *J Enthanophamcol* 158: 442–446.

Murthy, H.N., Dandin, V.S., Zhong, J.J., and K.Y. Paek. 2014b. Strategies for enhanced production of plant secondary metabolites from cell and organ cultures. In *Production of Biomass and Bioactive Compounds Using Bioreactor Technology*, eds. K.Y. Paek, H.N. Murthy, and J.J. Zhong, pp. 471–508. Springer, Dordrecht, the Netherlands.

Murthy, H.N., Georgiev, M.I., Kim, Y.S. et al. 2014c. Ginsenosides: Prospective for sustainable biotechnological production. *Appl Microbiol Biotechnol* 98: 6243–6254.

Murthy, H.N., Georgiev, M.I., Park, S.Y., Dandin, V.S., and K.Y. Paek. 2015a. The safety assessment of food ingredients derived from plant cell, tissue and organ cultures: A review. *Food Chem* 176: 426–432.

Murthy, H.N., Hahn, E.J., and K.Y. Paek. 2008. Adventitious roots and secondary metabolism. *Chinese J Biotechnol* 24: 711–716.

Murthy, H.N., Kim, Y.S., Georgiev, M.I., and K.Y. Paek. 2014h. Biotechnological production of eleutherosides: Current state and perspectives. *Appl Microbiol Biotechnol* 98: 7319–7329.

Murthy, H.N., Kim, Y.S., Jeong, C.S., Kim, S.J., Zhong, J.J., and K.Y. Paek. 2014d. Production of ginsenosides from adventitious root cultures of *Panax ginseng*. In *Production of Biomass and Bioactive Compounds Using Bioreactor Technology*, eds. K.Y. Paek, H.N. Murthy, and J.J. Zhong, pp. 625–654. Springer, Dordrecht, the Netherlands.

Murthy, H.N., Kim, Y.S., Park, S.Y., and K.Y. Paek. 2014e. Biotechnological production of caffeic acid derivatives from cell and organ cultures of *Echinacea* species. *Appl Microbiol Biotechnol* 98: 7707–7717.

Murthy, H.N., Kim, Y.S., Park, S.Y., and K.Y. Paek. 2014g. Hypericins: Biotechnological production from cell and organ cultures. *Appl Microbiol Biotechnol* 98: 9187–9198.

Murthy, H.N., Lee, E.J., and K.Y. Paek. 2014a. Production of secondary metabolites from cell and organ cultures: Strategies and approaches for biomass improvement and metabolite accumulation. *Plant Cell Tiss Organ* 118: 1–16.

Murthy, H.N. and K.Y. Paek. 2014. Food ingredients from plant cell and organ cultures: Biosafety and efficacy evaluations. In *Production of Biomass and Bioactive Compounds Using Bioreactor Technology*, eds. K.Y. Paek, H.N. Murthy, and J.J. Zhong, pp. 251–284. Springer, Dordrecht, the Netherlands.

Murthy, H.N., Park, S.Y., Lee, E.J., and K.Y. Paek. 2015b. Biosafety and toxicological evaluation of tissue-cultured *Echinacea purpurea* adventitious roots. *Kor J Hort Sci Technol* 33: 124–132.

Murthy, H.N., Wu, C.H., Cui, Y.Y., and K.Y. Paek. 2014f. Production of caffeic acid derivatives from adventitious root cultures of *Echinacea purpurea* (L.) Moench. In *Production of Biomass and Bioactive Compounds Using Bioreactor Technology*, eds. K.Y. Paek, H.N. Murthy, and J.J. Zhong, pp. 167–184. Springer, Dordrecht, the Netherlands.

Mustafa, A. and C. Turner. 2011. Pressurized liquid extraction as a green approach in food and herbal plant extraction: A review. *Analyt Chem Acta* 703: 8–18.

Orians, C.M. 1995. Preserving leaves for tannin and phenolic glycoside analyses: A comparison of methods using three willow taxa. *J Chem Ecol* 21: 1235–1243.

Paek, K.Y., Murthy, H.N., Hahn, E.J., and J.J. Zhong. 2009. Large scale culture of ginseng adventitious root cultures for production of ginsenosides. *Adv Biochem Eng/Biotechnol* 113: 151–176.

Park, D.J., Rhee, D.K., and Y.H. Lee. 2005a. Biological activities and chemistry of saponin form *Panax ginseng* C. A. Meyer. *Phytochem Rev* 4: 800–807.

Park, J.S., Hwang, S.Y., Lee, W.S., Yu, K.W, Paek, K.Y., and B.Y. Hwang. 2006. The therapeutic effect of tissue cultured root of wild *Panax ginseng* C. A. Meyer on spermatogenetic disorder. *Arch Pharm Res* 29: 800–807.

Park, S.Y., Ahn, J.W., Lee, W.Y., Murthy, H.N., and K.Y. Paek. 2005b. Mass propagation of *Eleutherococcus koreanum* plantlets via somatic embryogenesis from root cultures and accumulation of eleutherosides in regenerants. *Plant Sci* 168: 1221–1225.

Percival, S.S. 2000. Use of Echinacea in medicine. *Biochem Pharmacol* 60: 153–158.

Rulis, A.M. and J.A. Levitt. 2009. FDA's food ingredient approval process. Safety assurance based on scientific assessment. *Regul Toxicol Pharm* 53: 20–31.

Scragg, A.H. 1997. The production of aromas by plant cell cultures. *Adv Biochem Eng/Biotechnol* 55: 239–263.

Sharma, G.P. and S. Prasad. 2006. Optimization of process parameters for microwave drying of garlic cloves, *J Food Eng* 75: 441–446.

Shohael, A.M., Murthy, H.N., and K.Y. Paek. 2014. Pilot-scale culture of somatic embryos of *Eleutherococcus senticosus* in airlift bioreactors for the production of eleutherosides. *Biotechnol Lett* 36: 1727–1733.

Shu, Y.Y., Ko, M.Y., and Y.S. Chang. 2003. Microwave-assisted extraction of ginsenosides form ginseng root. *Microchem J* 74: 131–139.

Simal, S., Benedito, J., Saanchez, E.S., and C. Rosello. 1998. Use of ultrasound to increase mass transport rate during osmotic dehydration. *J Food Eng* 36: 323–336.

Sivakumar, G., Yu, K.W., Lee, J.S. et al. 2006. Tissue cultured mountain ginseng adventitious root TM: Safety and toxicology evaluation. *Eng Life Sci* 6: 372–383.

Stafford, A. 1991. The manufacture of food ingredients using plant cell and tissue cultures. *Trends Food Sci Technol* 2: 116–122.

Sudha, G. and G.A. Ravishankar. 2002. Influence of calcium channel modulators in capsaicin production by cell suspension cultures of *Capsicum frutescens*. Mill. *Curr Sci* 83: 480–484.

Sudha, G. and G.A. Ravishankar. 2003. Putrescine facilitated enhancement of capsaicin production in cell suspension cultures of *Capsicum frutescens*. *J Plant Physiol* 160: 339–346.

Thanh, N.T., Murthy, H.N., and K.Y. Paek. 2014. Optimization of ginseng cell culture in airlift bioreactors and developing the large-scale production system. *Ind Crop Prod* 60: 343–348.

Ushiyama, K. 1999. Regulations for plant cell culture derived products in Japan. In *Plant Cell and Tissue Culture for the Production of Food Ingredients*, eds. J.T. Fu, G. Singh, and W.R. Curtis, pp. 281–285. Kluwer Academic/Plenum Publishers, New York.

Venskutonis, P.R. 1997. The effect of drying on the volatile constituents of thyme (*Thymus vulgaris* L.) and sage (*Salvia officinalis* L.). *Food Chem* 53: 210–227.

Vilkhu, K., Mawson, R., Simons, L., and D. Bates. 2008. Applications and opportunities for ultrasound assisted extraction in the food industry—A review. *Innov Food Sci Emer Technol* 9: 161–169.

Wagner, H. and A. Proksch. 1985. Immunostimulatory drugs of fungi and higher plants. In *Economic and Medicinal Plant Research*, eds. H. Wagner, H. Hikino, and N.R. Farnsworth, pp. 113–153. Academic Press, Orlando, FL.

Wang, L. and C.L. Weller. 2006. Recent advances in extraction of nutraceuticals from plants. *Trend Food Sci Technol* 17: 300–312.

Westcott, R.J. 1989. Progress and prospects in plant biotechnology. In *Biotechnology Challenges for the Flavor and Food Industry*, eds. R.C. Lindsey and B.J. Willis, pp. 45–53. Elsevier Applied Science, New York.

Wu, C.H., Murthy, H.N., Hahn, E.J., and K.Y. Paek. 2007. Large-scale cultivation of adventitious roots of *Echinacea purpurea* in airlift bioreactors for the production of cichoric acid, chlorogenic acid and caftaric acid. *Biotechnol Lett* 29: 1179–1182.

Wu, J., Lin, L., and F. Chau. 2001. Ultrasound-assisted extraction of ginseng saponins from ginseng roots and cultured ginseng cells. *Ultrson Sonochem* 8: 347–352.

Yamamoto, T., Mizuguchi, R., and Y. Yamada. 1982. Selection of a high stable pigment producing strain in cultured *Euphorbia milli* cells. *Theor Appl Genet* 16: 113–116.

Zhao, J., Davis, L.C., and R. Verpoorte. 2005. Elicitor signal transduction leading to production of plant secondary metabolites. *Biotechnol Adv* 23: 283–333.

9 Food Security and Foodborne Mycotoxicoses, Risk Assessment, Preventive Measures, and Underestimated Hazard of Masked Mycotoxins or Joint Mycotoxin Interaction

Stoycho D. Stoev

CONTENTS

9.1 INTRODUCTION

Protecting humans or animals from various foodborne diseases has become an increasingly important task for many specialists in food safety issues in the field of animal and human health all over the world, especially with the important mission to ensure safe and healthy feeds or foods for animals and humans free from any dangerous contamination levels of various mycotoxins. While developed countries have well-developed infrastructures for monitoring food quality standards, people in developing countries are not protected by food quality monitoring and enforcement of safe standards within their countries. However, even in the case of introduction of such food safety standards within a particular developing country, it would often encourage the exportation of the best-quality crops, because the foods being exported are expected to comply with the food quality standards accepted in the countries importing such foods and thereby resulting in higher risk of mycotoxins exposure of the citizens in the developing countries, having in mind that only the best-quality crops leave the country, whereas the poor-quality foods usually remain for local domestic consumption (Stoev, 2013a). Therefore, stricter regulations in importing countries can be even considered as an additional health risk burden on the exporting country population since only the best-quality foods leave the country, whereas the commodities with higher levels of mycotoxin contamination usually remain for domestic consumption. In addition, the difficulty in meeting U.S. and EU mycotoxin safety standards possibly means the loss of an export market. Or, alternatively, it means exportation of the best-quality grains in order to keep the export market. By keeping the poorer-quality grain in the domestic market, local consumers bear the health burdens (Stoev, 2007).

Safety and quality control activities along the food supply chain are usually important to enhance the safety of various foods and feeds and to prevent foodborne diseases, but the same activities at each level require integration into a well-coordinated and integrated system. The existence of target professionals and knowledge in various scientific fields such as biochemistry, agriculture, veterinary medicine, environmental health science, food science, and technology, in addition to the upgrading of the respective target knowledge and skills via professional courses, is of crucial significance for implementing these safety and quality control activities in practice. However, only integrated approach to food safety, which includes systematic identification and assessment of hazards in food production, various means to control them, and introduction of hazard analysis and critical control point (HACCP) system, which can improve significantly the food safety management, could resolve all kind of existing problems in this field. The effective surveillance control and the respective enforcement of food safety laws and regulations is also compulsory to enhance the security of food and to reduce the number of foodborne diseases and the consequent social burdens of health care. However, if the regulations and food safety laws are too much restrictive, unjustified rejections of various commodities or raw food ingredients can lead to disastrous consequences for the producer or the distributor and to create some unjustified trade barriers (Rosner, 1998). Therefore, a harmonization of various national standards or an elaboration of common standards and regulations for mycotoxins contents in various food commodities or raw materials should be undertaken for better protection of the consumers all over the world, ensuring a global food safety and fair international trade (Stoev, 2007).

9.2 FOODBORNE MYCOTOXICOSES AND HAZARD OF JOINT MYCOTOXIN EXPOSURE

The contamination of foods/feeds with various mycotoxins is a serious threat to animals or humans all over the world, especially in developing countries, and can lead to many health problems and foodborne ailments or diseases. Mycotoxins are usually toxic secondary metabolites produced by certain fungi in agricultural products, which are susceptible to mold infestation. The cereals are usually invaded by fungi both in the field and after harvest, and this circumstance increases significantly the potential for multitoxin contamination. The production of mycotoxins is often unavoidable and depends on target environmental conditions during the plant growth and the subsequent food storage. Temperature stress is an important cause of fungi growth on crops in the field, and high moisture content (water activity) and temperature are associated with the growth of fungi in stored grain (Figures 9.1 and 9.2). These mycotoxins can occur in wheat, barley, rye, oats, triticale, maize, sorghum, etc., although some of them such as the fumonisins mainly occur in maize. The cocontamination of foods/feeds with known mycotoxins, in addition to some unknown mycotoxins, is occurring at a high rate (Stoev et al., 2010a,b). Although there are limited evidences on the definite effects of target mycotoxin or mycotoxin combinations in the course of various human/animal diseases, there are considerable amount of proofs supporting the association between mycotoxins and certain animal syndromes or livestock diseases including ergotism, various kinds of fusariotoxicoses such as alimentary toxic-aleukia in humans, porcine pulmonary edema, vulvovaginitis or rectal prolapse (Figure 9.3), equine leukoencephalomalacia or human esophageal carcinoma, various aflatoxicoses, mycotoxic porcine/chicken nephropathy (Figures 9.4 through 9.6), and stachybotryotoxicosis (Figure 9.7; Peraica et al., 1999; Stoev, 2007).

FIGURE 9.1 **(See color insert.)** An inappropriate store in rural area of Limpopo Province of South Africa, which would not be able to preserve feed from rain.

FIGURE 9.2 **(See color insert.)** An inappropriate manner for drying or preserving of maize flour in the rural area of Limpopo Province of South Africa.

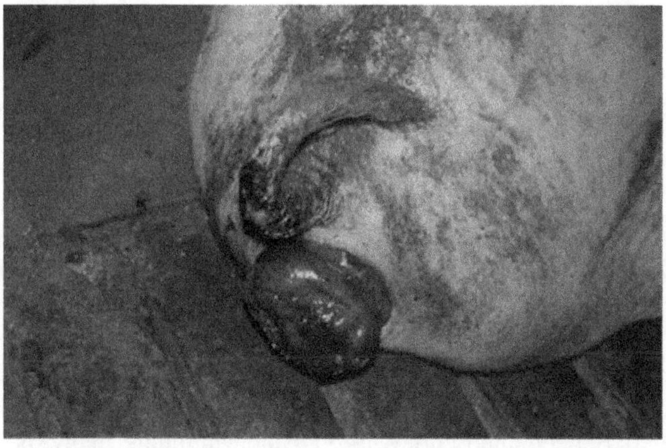

FIGURE 9.3 **(See color insert.)** A rectal prolapse in spontaneous case of fusariotoxicosis in pig from Bulgaria due to the estrogenic effect of mycotoxin zearalenone.

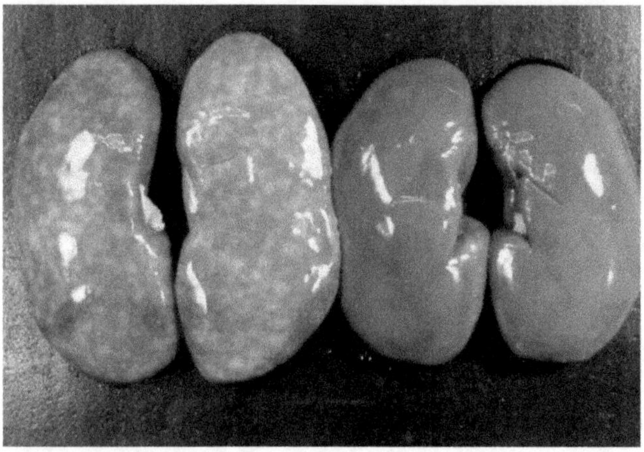

FIGURE 9.4 **(See color insert.)** Spontaneous case of mycotoxic porcine nephropathy identified at the slaughterhouse in Bulgaria. Enlarged and mottled appearance of kidneys from a 6- to 8-month-old pig (left) and normal kidneys in a pig of the same age (right).

FIGURE 9.5 **(See color insert.)** Nervous symptoms such as flexion of legs and sitting posture in chicks fed on moldy diet containing 790 ppb ochratoxin A and 2000–5000 ppb penicillic acid for 10 months.

Ergotism, caused by the fungus *Claviceps purpurea*, is provoked by many different mycotoxins such as ergometrine, ergine, ergotamine, ergosine, ergosecaline, ergocristine, ergocryptine, and some biogene amines such as histamine and acetylcholine, which are presented in fungal sclerotia—compact mass of hyphae cemented—which resembles a large, dark wheat grain. Mycotoxins significant for the development of ergotism are, however, ergometrine, ergotamine, and ergocristine. The disease is also known as St Anthony's fire, which is one of the oldest foodborne diseases in man causing hallucination symptoms and death of many people in France and other European countries during the Middle Ages (Betina, 1989). The cases of ergotism have been reported in a French village of Pont Saint Esprit in 1951 (Gabbai and Pourquier, 1951; Bennett and Bentley, 1999) and in Ethiopia in 1978 (Sibanda et al., 1997). The disease is still of public health importance, especially in the developing world (IPCS, 1990; Schneider et al., 1996). It is characterized mainly by dry gangrene of the extremities, loss of one or more limbs, in addition to some gastrointestinal symptoms (IPCS, 1990; Bennett and Bentley, 1999). Among animals, many species are sensitive to this foodborne mycotoxicosis, but the horses and ruminants are among the most sensitive species. The mycotoxins, responsible for ergotism, also increase the contraction of the uterus and provoke

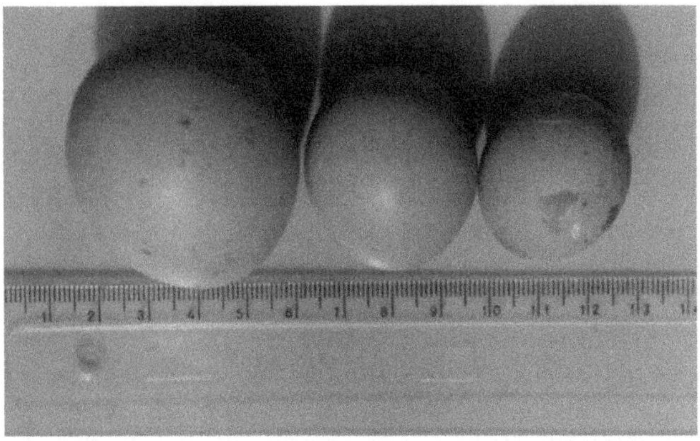

FIGURE 9.6 **(See color insert.)** Small size of eggs from laying hens exposed to 5 ppm OTA in a diet with the weight of the eggs of 15.8 g (center) and 25.8 g (right). Normal size of an egg from a laying hen without OTA treatment (left).

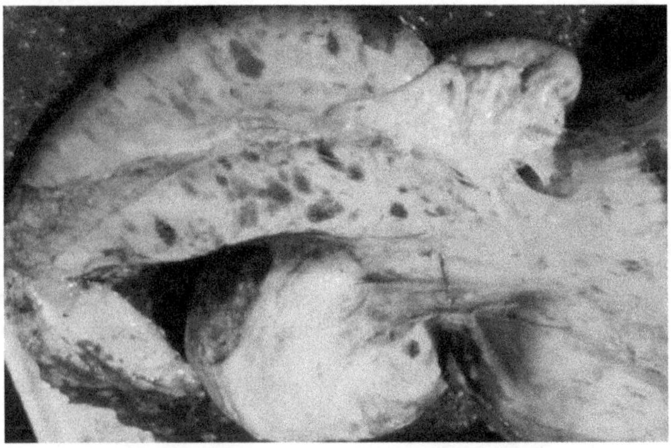

FIGURE 9.7 **(See color insert.)** Spontaneous case of stachybotryotoxicosis in cattle. Strong hemorrhagic diathesis on the serosa of the abomasus and rumen and ulcerative necrotic damages, which are translucent under the serosa.

many abortions and a prolapse of uterus, in addition to the vasoconstriction of blood vessels and the subsequent ischemic necroses in peripheral body parts such as the tail, ears, and crown of hooves (IPCS, 1990; Schneider et al., 1996).

Aflatoxins were discovered due to their devastating effect on turkey poults (turkey X disease) and some chicks in 1960 in England, when more than 100,000 turkey poults died (Allcroft and Carnaghan, 1962) as the target toxicity of these mycotoxins was found in the liver, which is usually enlarged and yellow. The so-called turkey X disease was eventually tied to high levels of aflatoxin in Brazilian peanut meal imported as a feed ingredient. Aflatoxin contamination is most common in African, Asian, and South American countries with warm and humid climates and also occurs in temperate areas of North America and Europe. The young animals and ducks or turkeys are usually the most sensitive species, and hemorrhagic diathesis, liver necrosis, enlarged gall bladder with edematous wall and yellowish staining of the tissue around it, proliferative processes in the hepatic interstitium, strong inflammatory necrotic changes in the digestive tract, and mild dystrophic changes in other organs were also often found in addition to liver damages, whereas in chronic course, the main damages are usually ascites, hydrothorax, icterus, cirrhosis of the liver or lung edema, but in the cattles, thickening of the skin around the mouth and neck and some papillomas on the mucosa of abomasus are also observed (Stoev, 2013b).

On the other hand, ochratoxin A was found as a major causal agent in mycotoxic nephropathy (MN) in pigs (Krogh et al., 1973; Stoev et al., 1998a) and probably Balkan endemic nephropathy (BEN) in humans, widely encountered in Balkan countries (Krogh, 1972; Stoev, 1998, 2008; Stoev and Denev, 2013). This nephropathy, however, was recently proved to have much complicated pathology (Figure 9.5) and etiology in some countries such as Bulgaria and South Africa (Stoev, 2008). In the same countries, spontaneous nephropathy in pigs (Stoev et al., 1998a–c) or chicks (Stoev et al., 2002a) described previously was recently found to be provoked by the combined action of several mycotoxins such as ochratoxin A, penicillic acid, and fumonisin B_1, in addition to not yet identified metabolite, having a strong synergistic interaction (Stoev et al., 2010a,b). A part of the same mycotoxins produced by storage fungi such as ochratoxin A and penicillic acid are secondary fungal metabolites encountered in feeds/food made mainly from cereals or fibrous plants, which is kept in storehouse conditions and increased humidity, whereas the other target mycotoxin fumonisin B_1 produced by field fungi mainly contaminates maize before harvesting (Stoev and Denev, 2013). Usually, the farms with such nephropathy problems had a history of inappropriate feed storage, but sometimes the problem came from certain feed plants whose grains, collected during moist and rainy days, had not been properly dried. Subsequently, all farms that used these plants as their main feed source had nephropathy problems and growth depression among their pigs, but after changing the source of feed supply, the problems with poor growth of pigs disappeared (Stoev et al., 1998a,c). The timely diagnosis of the same disease during the meat inspection at slaughterhouses is very important (Figure 9.4) because of the harmful effects, which can be observed in human kidneys after consuming the meat of animals with nephropathy problems and the respective mycotoxin content (Stoev and Denev, 2013). Therefore, the exposure of humans to the same hazardous and relatively heat stable target mycotoxins from chicken/pigs meat should be prevented (Stoev, 2008; Stoev and Denev, 2013). Such a growth depression, nervous signs (Figure 9.5), kidneys and liver damages, and a significant decrease in the weight of the eggs (Figure 9.6) can also be observed in ochratoxin A–exposed chicks or laying hens (Stoev, 2010b, Stoev et al., 2002d).

Maize molded by various *Fusarium* fungi was suspected to be a cause of horse or pig diseases, including the death in many horses more than 150 years ago, and this was subsequently associated with the presence of *Fusarium moniliforme* (*F. verticillioides*). However, the mycotoxins provoking the same diseases, the fumonisins, were not well known until the 1980s (Bezuidenhout et al., 1988; Gelderblom et al., 1988). We became aware about the significant hazard of these mycotoxins, especially of fumonisin B_1, after a number of outbreaks of equine leukoencephalomalacia (Marasas et al., 1976; Conkova et al., 2003) and porcine pulmonary edema killing many horses or pigs, fed on diets containing fumonisin-contaminated maize in the United States during 1989 and 1990 (Marasas, 2001). These damages in the horse brain and the pig lung appeared to be due to the alteration of vascular function and endothelial cell permeability provoked by disruption of sphingolipid metabolism (Ramasamy et al., 1995). On the other hand, fumonisin B_1 can decrease myocardial contractility in pigs, which is thought to be mediated by increased sphingosine inhibiting the L-type calcium channels in the myocardium (Haschek et al., 2009) and that is why this mycotoxin can subsequently complicate the pulmonary edema via the induced left-sided heart failure. Such a disease in myocardial contractility in humans was observed in 1980 by Dr. George Campbell, who practiced in rural hospitals in South Africa and observed a similar disease in different patients, which he called idiopathic congestive cardiomyopathy (ICC)—weakness of the heart due to influx of much blood in it. Only elderly people, whose main diets were made from home-produced maize and who drank a lot of homemade beer, have been sick, which clearly suggests that the main reason for this disease is probably fumonisin B_1, in addition to some other *Fusarium* mycotoxins or provoking factors (Stoev, 2007, 2013b).

Zearalenone is another *Fusarium* mycotoxin, which contaminated the moldy maize that was found to possess estrogenic activity in animals and to provoke swelling of the vulva and mammary glands, infertility, vulvovaginitis, vaginal, and/or rectal prolapse in swine (Shibamato and Bjeldanes, 1993; Shier, 1998; Friends et al., 1999; Stoev, 2013a,b; Figure 9.3). We should have in mind that hyperestrogenism (vulvovaginitis, prolapse of the rectum and vagina, etc.), due to the high content of zearalenone in feed, usually occurred after a longer period of exposure to molded feed, whereas the first sign of fusariotoxicosis caused by *Fusarium graminearum* and *Fusarium culmorum* is usually associated with cytotoxic (gastrointestinal changes and degenerative lesions in internal organs), neurotropic (paresis of hind limbs), emetic, and immunosuppressive (secondary bacterial infections) effect of deoxynivalenol (DON) (Stoev, 2007).

Another fusariotoxicosis is the one caused by *Fusarium sporotrichiella* and *F. poae*, but less attention is also given to *F. tricinctum* and *F. cladosporium*. The same fungi produce high levels of T-2 toxin and DAS (diacetoxyscirpenol), which have highly cytotoxic, immunosuppressive, and genotoxic activity, accumulate in the body, and are toxic in all animals and people (Stoev, 2007, 2013b). Intoxication usually occurs after feeding on feed made from grain, hay, or straw, wintering in the open. Poisoning in humans is also known as alimentary toxic aleukia (Lutsky and Mor, 1981). Toxins have a local irritant effect and cause serous-hemorrhagic inflammation, necrosis and ulceration in the digestive tract; dystrophy in the liver, kidney, heart, brain, and peripheral ganglia of the vegetative nervous system leading to some spasms and tremors of some muscles; and sometimes paresis of hind limbs. Damages are also seen in the blood vessel wall and a hemorrhagic diathesis is provoked (Stoev, 2013b).

Stachybotryotoxicosis caused by the fungus *Stachybotrys alternans* (*Stachybotrys atra*), which easily develops on moist substrates (hay, straw, oats) rich in cellulose and forms a black coating, is another problematic mycotoxicosis among animals provoked by several mycotoxins such as satratoxins, verrucarins, and roridins. The same mycotoxins can initially provoke only a local irritation in the mouth and nasal areas or swelling of the lips and fissures of the mouth, but they usually accumulate in the body, damage the vessels (leading to hemorrhages and edematous changes) and hematopoietic tissue (leading to leukopenia and thrombocytopenia), and also provoke degenerative changes in all internal organs as well as secondary (trophic) deep symmetrical areactive necroses and ulcers in the mouth and digestive tract (Schneider et al., 1979; Lefebvre et al., 1994; Stoev, 2007, 2013b; Figure 9.7). Fungus remains outside the substrate (soil) for a long time, which explains the stationarity of this mycotoxicosis.

It is also important to emphasize that the main concern for human health in regard to aflatoxins, ochratoxin A (Stoev, 2010a), and fumonisins in developed countries appeared to be their carcinogenic (Figures 9.8 and 9.9) or genotoxic and teratogenic effects (Figure 9.10) rather than their acute

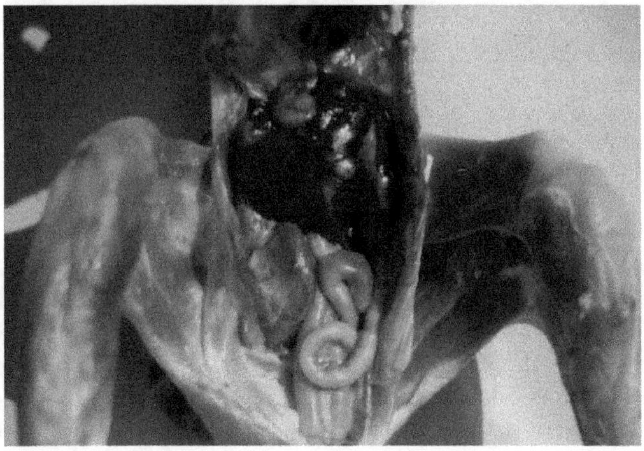

FIGURE 9.8 **(See color insert.)** Adenocarcinoma in the liver of a chick exposed to 5 ppm ochratoxin A for 10 months. Large gray-white neoplastic foci are seen in the liver and protruded significantly above its surface.

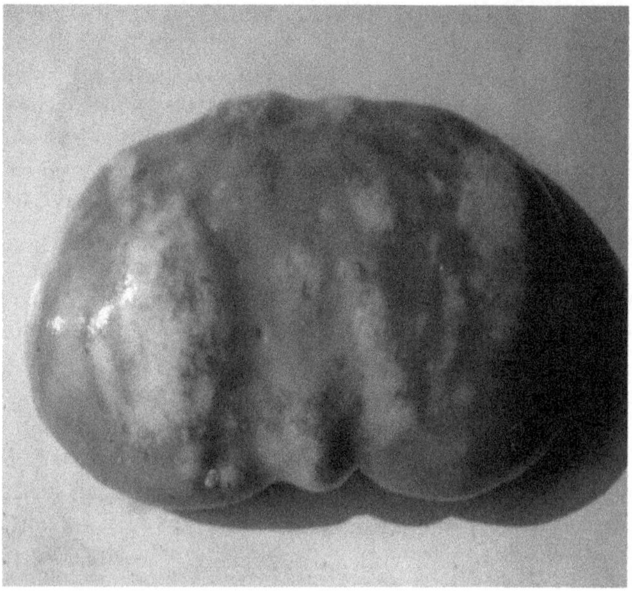

FIGURE 9.9 **(See color insert.)** Kidney with neoplastic tissue (fibroadenoma)—a spontaneous case of MPN in Bulgaria.

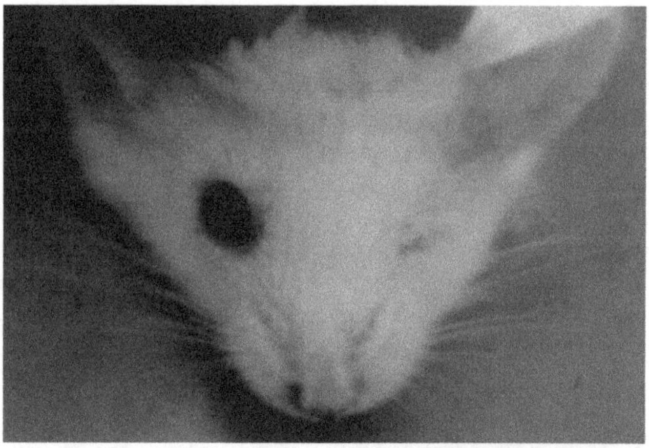

FIGURE 9.10 **(See color insert.)** Malformation (monophthalmia) in mouse, whose mother was exposed to 10 ppm ochratoxin A via the feed up to day eighth of the pregnancy.

effects (Stoev, 2013a,b). It is important to mention that fumonisin B_1 was first discovered in connection with esophageal cancer in South Africa (Marasas et al., 1988) and was also shown to be a liver carcinogen in rats (Gelderblom et al., 1992) and nephrotoxin (Bucci et al., 1998; Howard et al., 2001), and recently, the same mycotoxin is suspected to be involved in human and animal nephropathies in Bulgaria and South Africa (Stoev et al., 2010a,b) together with ochratoxin A and penicillic acid, as recently proved (Stoev and Denev, 2013).

This time there is a lack of sound evidences for the definitive relationship between mycotoxins and current human diseases in developing countries, which, however, does not necessarily imply that dietary exposure does not represent a real risk for the men. In these countries, many individuals are also chronically exposed to high levels of mycotoxins in their diet. Various epidemiological studies revealed a correlation between the level of fumonisin B_1 in corn, the amount of corn consumed in the diet, and the rate of esophageal cancer (Marasas et al., 1988; Sydenham et al., 1990; Dutton, 1996), and therefore, this mycotoxin has been clearly associated with human esophageal cancer in some countries such as China and South Africa. It was now proved that the highest level of esophageal cancer in South Africa (Marasas et al., 1988; Sydenham et al., 1990; Rheeder et al., 1992) and China (Chu and Li, 1994) occurred among those populations consuming the largest amount of corn with the highest level of fumonisin B_1 content. It is now alarming that pregnant women who are exposed to high levels of fumonisins in their diet have a high risk to bear a child with a birth defect of the brain or spinal cord. The last studies clearly suggest that maternal ingestion of high levels of fumonisin B_1 among women during early pregnancy may increase the risk of neural tube defects (NTDs) such as brain and spinal cord defects (van Waes et al., 2005; Missmer et al., 2006).

Lots of mycotoxins are known to have many different adverse effects such as cytotoxic, genotoxic, immunotoxic, carcinogenic, or teratogenic effects on animals and/or humans. Trichothecenes are reported to be immunotoxins (Sharma, 1993). Aflatoxins are not only hepatotoxins in animals or humans (Hgindu et al., 1982) but also carcinogens, genotoxins, and immunotoxins suppressing both cellular and humoral response and causing growth retardation in animals (Dirheimer, 1998; Kubena et al., 1998; Coulombe et al., 2005). Ochratoxins (mainly ochratoxin A) and fumonisins are proven nephrotoxins (Figure 9.3; Krogh, 1972; Krogh et al., 1973; Stoev et al., 1998a, 1999, 2001, 2002c, 2004, 2012), immunotoxins (Stoev et al., 2000a,b, 2002b), genotoxins (Dirheimer, 1998; Pfohl-Leszkowicz and Manderville, 2012), and carcinogens (Stoev, 2010a; Figures 9.4, 9.7, and 9.8).

It is important to emphasize that some experiments focused attention on the immunosuppression as the first expressed toxic effect of ochratoxin A, which may become evident clinically before nephropathy and the other biochemical changes. For the first time, a high susceptibility to natural infectious disease and/or secondary bacterial infections has been demonstrated in pigs fed on ochratoxin A–contaminated diet in levels often encountered in the field (Stoev et al., 2000b; Figure 9.11). The reported suppression of humoral and cellular immunity provoked by this mycotoxin, defined in principle (NNT, 1991), has been demonstrated in practice via natural development of secondary bacterial infections in pigs at only 1 ppm ochratoxin A in diet (Stoev et al., 2000b). It seems that humoral immunity has been affected to the extent of allowing the development of clinical disease in pigs at only 1 ppm ochratoxin A in the diet (Stoev et al., 2000b; Figure 9.12).

It was found that the repetitive exposures to the trichothecenes T-2 toxin can increase susceptibility to *Mycobacterium bovis*, *Salmonella typhimurium*, *Listeria monocytogenes*, and *Staphylococcus*

FIGURE 9.11 **(See color insert.)** Diphtheroid accretions on colon mucosa of a pig given 3 ppm ochratoxin A for 17 days and sick by salmonellosis.

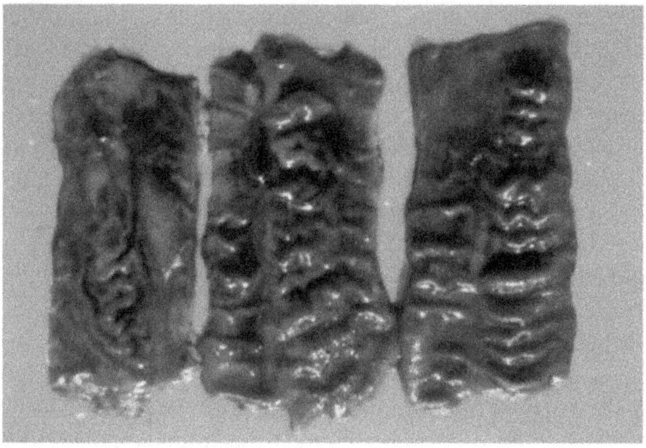

FIGURE 9.12 **(See color insert.)** Hemorrhagic dysentery in a pig 47 days after commencing a diet containing 1 ppm ochratoxin A.

aureus infections in animals or chickens (Boonchuvit et al., 1975; Ziprin et al., 1987; Tai and Pestka, 1988; Cooray and Jonsson, 1990; Oswald and Comera, 1998).

Some other authors observed a much heavier progression of *Pasteurella multocida* (Halloy et al., 2005) or porcine reproductive and respiratory syndrome virus infections in swine (Ramos et al., 2010; Pósa et al., 2011, 2013) compromised by feed intake of fumonisin B_1. This proinflammatory effect of fumonisin B_1 (Taranu et al., 2005) is probably due to the reduction of phagocytic activity of pulmonary macrophages (Halloy et al., 2005) and the increased permeability of pulmonary capillaries (Stoev et al., 2012). The immunosuppressive effect of fumonisin B_1 is supposed to be due to the accumulation of free sphingoid bases, which inhibits the proliferation of lymphocytes (Taranu et al., 2005). It has been reported that fumonisin B_1 given at diet levels of 10 ppm also possesses a significant immunosuppressive effect on humoral immune response in pigs, which was manifested with lower antibody titers at days 21 and 35 after immunization against the Morbus Aujeszky disease (Stoev et al., 2012).

Ochratoxin A, which possesses a strong suppressive effect on both cellular and humoral immune response in animals/chicks, has also been found to increase the susceptibility of chickens to coccidiosis (Huff and Ruff, 1982; Stoev et al., 2002b; Koynarski et al., 2007), salmonellosis (Elissalde et al., 1994; Fukata et al., 1996; Gupta et al., 2008), and colibacillosis (Kumar et al., 2003). The increased mortality among chicks/pigs fed on moldy diet containing ochratoxin A is probably due to the increased susceptibility to secondary bacterial enteric disease (Stoev et al., 2000b; Figures 9.11 and 9.12) or to a complicated and heavy progression of some often encountered parasitic diseases (Stoev et al., 2002b; Koynarski et al., 2007) or microbial infections (Dwivedi and Burns, 1985; Oswald and Comera, 1998; Stoev et al., 2000a,b).

9.3 RISK ASSESSMENT AND POSSIBLE HYGIENE CONTROL

Experience has shown that it is necessary to adopt measures aimed at guaranteeing that unsafe food is not placed on the market and at ensuring that systems exist to identify and respond to food safety problems to ensure the proper functioning of the internal market and to protect human health. Where food law is aimed at the reduction, elimination, or avoidance of a risk to health, the three interconnected components of risk analysis—risk assessment, risk management, and risk communication—provide a systematic methodology for the determination of effective, proportionate, and targeted measures or other actions to protect health (EC Regulation No. 178/2002). The "risk assessment" itself is a scientifically based process consisting of four steps, hazard identification, hazard characterization, exposure assessment, and risk characterization, whereas the "risk communication" addresses the exchange of information and opinions throughout the risk analysis process as regards hazards and risks, risk-related factors, and risk perceptions, among risk assessors, risk managers, consumers, feed and food businesses, and other interested parties. It is recognized that scientific risk assessment alone cannot, in some cases, provide all the information on which a risk management decision should be based and that other factors relevant to the matter under consideration should be taken into account including societal, economic, traditional, ethical and environmental factors, and the feasibility of controls (EC Regulation No. 178/2002).

To eliminate or minimize mycotoxin contents in the foods/feeds, some target agricultural practices and storage protocols that can effectively prevent their formation should be introduced. The fate of the particular mycotoxin at each stage of the process should also be traced in order to create a reliable control and to reduce the quantity of various mycotoxins that can reach the consumer. Unfortunately, a complete removal of mycotoxins is not always achievable. It is compulsory to protect the consumer from the effects of a particular mycotoxin in the food supply when a risk assessment of the contaminant indicates that its level of exposure is going to be unacceptable. Therefore, when the risk assessment of a particular mycotoxin is considered to be significant, legislation should be introduced together with some measures aimed to minimize the exposure of consumer to that mycotoxin. The setting maximum permissible limit presents a powerful tool for such an effective control in the food chain (Rosner, 1998). There are two main compounds taken into account in each risk assessment: the toxicologic effect of a particular mycotoxin and the estimate of mycotoxin exposure of the consumer. It is of a particular importance that all limits to be supported by reliable and extensive research to avoid unnecessary restrictions and subsequent economic loss (Stoev, 2013a).

The first limits for mycotoxins were set in the late 1960s for the aflatoxins. By the end of 2003, approximately 100 countries had developed specific limits for mycotoxins in foodstuffs and feedstuffs, and the number continues to grow. Such maximum permissible limits have been set for some mycotoxins by national or international organizations in EU or in some particular countries all over the world such as World Health Organization (WHO), Joint FAO/WHO Expert Committee on Food Additives (JECFA), and EU Scientific Committee for Food (SCF). The acquisition of more and more information about the content of various mycotoxins during various kinds of food processing may allow more lenient limits for raw ingredients to be set in the case of a significant decrease through the food processing chain. On the other hand, the profound knowledge of processing techniques and mycotoxin destroying or elimination can justify those food products that are more susceptible to mycotoxins contamination and therefore need to be exposed to more frequent monitoring than the other products (FAO, 1988, 1997; Tables 9.1 and 9.2). Such knowledge is of particular importance for mycotoxins such as deoxynivalenol, zearalenone, and fumonisins in maize that can be found in very high levels in raw cereals, but the same are significantly decreased during food processing. Therefore, any failure to take this information into account and subsequently to realize a sound balance between economic considerations and consumer risk could result in unnecessary penalizing of the cereal producers without improving consumer safety. In some particular cases, a kind of food processing such as the removal of bran from the final product might decrease significantly human exposure to some mycotoxins but simultaneously might decrease the nutritional and health benefits provided by the same final product. A controversial question then arises, whether the men should consume wholemeal bread with its known health benefits or white bread to ensure less intake of a particular mycotoxin (Scudamore and Banks, 2004).

The European Community (EC) has introduced maximum permissible limits for aflatoxins, ochratoxin A, patulin, and some other mycotoxins in specific products and is also actively involved in considering which other mycotoxins need regulation (WHO/FAO, 1995; FAO, 1997; Boutrif and Canet, 1998; Scudamore and Banks, 2004; Tables 9.1 and 9.2). The European Union has also accepted an official limit for ochratoxin A in cereals designed for direct consumption of about 5 μg OTA/kg, whereas for end consumer products and foods, the limit is 3 μg/kg (Rosner, 1998;

TABLE 9.1
Maximum Permitted Content of Mycotoxins in Feed Stuffs

Mycotoxins	Animal Feeds	Max. Content mg/kg (ppm)
Aflatoxin B1	– Complete feeding stuffs for cattle, sheep, and goats with exception of	0.02
	✓ Complete feeding stuffs for dairy animals	0.005
	✓ Complete feeding stuffs for calves and lambs	0.01
	– Complete feeding stuffs for pigs and poultry (except young animals)	0.02
Ochratoxin A	– Cereals and cereal products	0.25
	– Complete and complimentary feedstuffs with exception of	5
	✓ Complete and complimentary feedstuffs for pigs	0.05
	✓ Complete and complimentary feedstuffs for poultry	0.1
Deoxynivalenol	– Cereals without maize by-products	8
	– Maize by-products	12
	– Complete and complimentary feedstuffs with exception of	5
	✓ Complete and complimentary feedstuffs for pigs	0.9
	✓ Complete and complimentary feedstuffs for calves, lambs, and kids	2
Zearalenone	– Cereals without maize by-products	2
	– Maize by-products	3
	– Complete and complimentary feedstuffs for calves, dairy cattle, sheep (including lamb), and goats (including kids)	0.5
	– Complete and complimentary feedstuffs for piglets, gilts (young sows)	0.1
	– Complete and complimentary feedstuffs for sows and fattening pigs	0.25
Fumonisins $B_1 + B_2$	Maize by-products	60
	– Complete and complimentary feedstuffs for pigs, Equidae, rabbits	5
	– Complete and complimentary feedstuffs for poultry, calves, lambs, kids	20
	– Complete and complimentary feedstuffs for adult ruminants and mink	50
	– Complete and complimentary feedstuffs for fish	10
T-2 + HT-2	– Unprocessed barley (including malting barley) and maize	0.2
	– Unprocessed wheat, rye, and other cereals	0.1
	– Unprocessed oats (with husk)	1
	– Oat milling products (husks) for feed and compound feed	2
	– Other cereal products for feed and compound feed	0.5
	– Compound feed, with the exception of feed for cats	0.25
Rye ergot	– All feeding stuffs containing unground cereals	1000

Sources: EC Directives 2002/32/EC of the European Parliament and of the Council on undesirable substances in animal feed, May 7, 2002, pp. 1–27; EC Recommendation 2006/576/EC on the presence of deoxynivalenol, zearalenone, ochratoxin A, T-2 and HT-2 and fumonisins in products intended for animal feeding, August 17, 2006, pp. 1–3; EC Recommendations 2013/165/EU on the presence of T-2 and HT-2 toxin in cereals and cereal products, March 27, 2013, pp. 1–4.

Scudamore and Banks, 2004; Table 9.2). The control and regulation of nephrotoxic mycotoxins in animal feeds in European countries have been summarized by van Egmond (1989) and by Boutrif and Canet (1998) (Table 9.1). The official limits for ochratoxin A in cereals in some countries can be seen in various FAO reports and commonly ranged between 2 μg/kg (Switzerland) and 20 μg/kg (Czech Republic) but sometimes reached up to 50 μg/kg (Uruguay) (Boutrif and Canet, 1998). Various limits have also been accepted for the maximum permitted content of aflatoxin B_1 in animal feeding stuffs in EU (Table 9.1) and in some other particular countries (MAFF, 1982). It is a pity that many countries in the world have their own limits or legal regulations for the effective control of

TABLE 9.2
Maximum Permitted Content of Mycotoxins in Human Food

Mycotoxins	Foods	Max. Content mg/kg (ppm)
Aflatoxin B1	– Groundnuts and oilseeds subjected to physical treatment, before human consumption with exception of that for refined vegetable oil production	0.008
	– Almonds, pistachios, and apricot kernels subjected to physical treatment, before human consumption or use as ingredient in foodstuffs	0.012
	– Almonds, pistachios, and apricot kernels, intended for direct human consumption or use as ingredient in foodstuffs	0.008
	– Hazelnuts and Brazil nuts subjected to physical treatment, before human consumption or use as ingredient in foodstuffs	0.008
	– Hazelnuts and Brazil nuts for direct human consumption or use as ingredient in foodstuffs	0.005
	– Groundnuts and oilseeds or processed products thereof for direct human consumption or use as ingredient in foodstuffs, with the exception of crude vegetable oils and refined vegetable oils	0.002
	– Dried fruit subjected to physical treatment, before human consumption or use as ingredient in foodstuffs	0.005
	– Dried fruit and processed products thereof for direct human consumption or use as ingredient in foodstuffs	0,002
	– All cereals and processed cereal products excluding	0.002
	✓ Maize/rice subjected to physical treatment before consumption or use as ingredient in foodstuffs, and chillies, chilli powder, cayenne, white or black peppers, paprika, ginger, turmeric, and nutmeg	0.005
	✓ Processed cereal-based foods and baby foods for infants or children	0.0001
Aflatoxin M1	– Raw milk, heat-treated milk, and milk-based products	0.00005
	– Dietary foods for special medical purposes and infant milk	0.000025
Ochratoxin A	– Unprocessed cereals	0.005
	– Processed cereal products and cereals for direct human consumption	0.003
	– Dried vine fruit (currants, raisins, and sultanas) and soluble coffee	0.01
	– Roasted coffee beans and ground roasted coffee, excluding soluble coffee	0.005
	– Wine (including sparkling wine, excluding liqueur wine), wine products or cocktails and grape juice or nectar for direct human consumption	0.002
	– Processed cereal-based foods and baby foods for infants/children	0.0005
	– Chillies, chilli powder, cayenne, white or black peppers, paprika, ginger, turmeric, and nutmeg	0.015
	– Liquorice root, ingredient for herbal infusion	0.02
	– Liquorice extract, for use in food in beverages and confectionary	0.08
Patulin	– Fruit juices, fruit nectars, spirit drinks, cider, and other fermented drinks derived from apples or apple-containing juice	0.05
	– Solid apple products such as compote or apple puree for direct consumption	0.025
	– Baby foods other than processed cereal-based foods, apple juice, and solid apple products such as compote or apple puree for infants and children	0.01
Deoxynivalenol	– Unprocessed cereals other than durum wheat, oats, and maize	1.25
	– Unprocessed durum wheat, oats, and unprocessed maize, with the exception of unprocessed maize intended for wet milling	1.75
	– Cereals for direct human consumption such as flour, bran, pasta, and germ	0.75
	– Bread and bakery wares, pastries, biscuits, cereal snacks, or breakfast	0.5
	– Processed cereal-based foods and baby foods for infants and children	0.2

(*Continued*)

TABLE 9.2 (*Continued*)
Maximum Permitted Content of Mycotoxins in Human Food

Mycotoxins	Foods	Max. Content mg/kg (ppm)
Zearalenone	– Unprocessed cereals other than maize	0.1
	– Unprocessed maize with exception of maize intended for wet milling	0.35
	– Cereals for direct human consumption such as cereal flour, bran, and germ	0.075
	– Refined maize oil	0.4
	– Bread and bakery wares, pastries, biscuits, cereal snacks, or breakfast cereals, excluding maize snacks and maize-based breakfast cereals	0.05
	– Maize for direct human consumption such as snacks and breakfast cereals	0.1
	– Processed cereal-based foods and baby foods for infants and children	0.02
Fumonisins B_1+B_2	– Unprocessed maize, with exception of maize intended for wet milling	4
	– Maize and maize-based food for direct human consumption, excluding	1
	✓ Maize-based breakfast cereals and maize-based snacks	0.8
	✓ Processed maize-based foods or baby foods for infants and children	0.2
T-2 + HT-2	– Cereal grains for direct human consumption	
	✓ Oats	0.2
	✓ Maize	0.1
	✓ Other cereals	0.05
	– Cereal products for human consumption	
	✓ Oat bran and flaked oats	0.2
	✓ Cereal bran except oat bran, oat milling products other than oat bran and flaked oats, and maize milling products	0.1
	✓ Other cereal milling products	0.05
	✓ Breakfast cereals including formed cereal flakes	0.075
	✓ Bread and bakery wares, pastries, biscuits, cereal snacks, pasta	0.025
	✓ Cereal-based foods for infants and young children	0.015

Sources: EC Regulation 1881/2006 setting maximum levels for certain contaminants in foodstuffs, December 19, 2006, pp. 1–26; EC Recommendations 2013/165/EU on the presence of T-2 and HT-2 toxin in cereals and cereal products, March 27, 2013, pp. 1–4.

some mycotoxins in foods or feeds (Rosner, 1998), but there are still not internationally recognized limits. Such statutory limits for some mycotoxins in various commodities in each particular country are usually chosen depending on several important factors such as the existence of sufficient data for toxicological effect of mycotoxins, the existence of data on the occurrence of mycotoxins in various commodities, the availability of methods of sampling and analysis, the intercountry trade implications, or the existence of sufficient food supply (Boutrif and Canet, 1998), and the same may have important implications in international trade especially if the export of a commodity may represent a significant percentage of trade earnings of particular country (Stoev, 2013a). For example, in the developing countries, where food supplies are limited, drastic legal measures may lead to lack of food and to excessive prices. It is now clear that the food safety regulations such as specific tolerance levels usually have significant economic consequences and that different perceptions about appropriate trade-offs between health and economic losses are the source of potential conflict between countries.

These days, in spite of all difficulties mentioned earlier, some efforts should be undertaken for the development of internationally harmonized regulatory control measures for mycotoxins in order to protect public health and to ensure fair trade at the international level. It usually takes a lot of time to agree and introduce internationally recognized limits, so it may be much easier to introduce guideline limits as temporary measure when the risk for public health is really significant. In this point of view, JECFA has recently evaluated the most dangerous mycotoxins and provides a mechanism for assessing the toxicity of various additives, veterinary drugs, and other food contaminants such as mycotoxins (JECFA, 2001). The toxicological evaluation carried out by JECFA is based on the determination of no-observed-effect level (NOEL) in toxicological studies and the application of a safety factor and normally results in the estimation of a provisional tolerable weekly intake (PTWI). This approach of establishing maximum tolerated levels of mycotoxins in foods, however, does not apply for mycotoxins where carcinogenicity is the basis for concern as is the case with aflatoxins. In all these cases, a different approach such as "as low as reasonably achievable" (ALARA principle) or "as low as possible but technologically feasible and analytically detectable in the food ready for consumption" is necessary to be introduced (Rosner, 1998). Therefore, a program for the evaluation of carcinogenic effects of some mycotoxins on humans is undertaken by International Agency for Research on Cancer (IARC), and according to this program, mycotoxins can be distributed in the following groups: group 1 (carcinogenic to humans as many aflatoxins), group 2A (probably carcinogenic to humans), group 2B (possibly carcinogenic to humans as fumonisins or ochratoxin A), and group 3 (not classifiable due to its carcinogenicity to humans) (Castegnaro and McGregor, 1998). Various revisions of the criteria for this classification and reclassification of some mycotoxins are made periodically by IARC.

The hazards of different mycotoxins are often evaluated and reevaluated at various meetings of JECFA evaluated, and the respective reports issued after such meetings are widely distributed (JECFA, 2001; Scudamore and Banks, 2004). The quantitative evaluation of the degree of human exposure and risk likely to occur in the population is also an important subject in the work of JECFA (Boutrif and Canet, 1998). The maximum levels of mycotoxins and other chemical contaminants in foods and various commodities are usually considered by the Codex Committee on Food Additives and Contaminants in consultation with the Codex Commodity Committee, but the final decision is usually taken by the Codex Alimentarius Commission (CAC). Usually, the final conclusions are based on the scientific evaluation made by JECFA and some other relevant expert organizations (Boutrif and Canet, 1998; Table 9.3). The CAC, supported by FAO and WHO, aims to

TABLE 9.3
Tolerable Daily Intake Values of Mycotoxins

Mycotoxins	TDI per kg b.w. (μg/kg b.w.)
Aflatoxin B1	Less than 0.001–0.01 (the lower dose concerns carcinogenic effect)
Aflatoxin M1	0.01 (or 10 times lower carcinogenic effect than aflatoxin B1)
Ochratoxin	0.0002–0.014 (the lower dose concerns carcinogenic effect)
Patulin	0.4
Fumonisins	2
Deoxynivalenol (DON)	1
Zearalenone	0.2
T-2 or HT-2 toxins	0.06
T-2 + HT-2 toxins	0.1
Nivalenol	0.7

Sources: EC Regulation 1881/2006 setting maximum levels for certain contaminants in foodstuffs, December 19, 2006, pp. 1–26; EC Recommendations 2013/165/EU on the presence of T-2 and HT-2 toxin in cereals and cereal products, March 27, 2013, pp. 1–4.

facilitate world trade and protect the health of consumers through the development of international standards for foods and feeds.

The increase of scientific reports on ochratoxin A contamination in beverages and many kinds of foods or final food products provoked JECFA to assess all available information on the subject and to propose 112 ng/kg b.w. as a PTWI for ochratoxin A (WHO, 1991), which corresponds to about 16 ng/kg body weight per day. Subsequently, its PTWI was decreased to 100 ng/kg b.w., which corresponds to about 14 ng/kg body weight per day (JECFA, 1997; Table 9.3). In this particular case, JECFA bases its calculation of the PTWI mainly on the nephrotoxic effect of this mycotoxin and does not address the question of the toxin's carcinogenic effect. On the other hand, Kuiper-Goodman and Scott take into account that the carcinogenic effect is the most important effect and therefore base their analysis on this effect and find that tolerable daily intakes (TDIs) in such cases range from 0.2 to 4.2 ng/kg b.w. depending on the method used. With a big anxiety, we have to warn the people that the calculated average daily intakes of humans in the areas of Bulgaria with BEN from 26.8 ng/kg b.w. for 1988, 36.4 ng/kg b.w. for 1989, and 34.2 ng/kg b.w. for 1990, respectively (Stoev and Petkova-Bocharova, 1994; Stoev, 1998), exceed strongly the TDI of 14 ng/kg b.w., as proposed by JECFA (1997). Moreover, it can be seen clearly that the average daily intake in humans from the same endemic areas in Bulgaria exceeds strongly both TDIs (Stoev and Petkova-Bocharova, 1994; Stoev, 1998) calculated on the basis of cancerogenic effect (0.2–4.2 ng/kg b.w.) or even on the basis of nephrotoxic effect (14 ng/kg b.w.) of ochratoxin A (Kuiper-Goodman and Scott, 1989).

When considering the ochratoxin A contamination of various animal products, we can conclude that the current measures undertaken in some European countries are also very inappropriate or irrelevant (Stoev, 2008, 2013a). For example, in Denmark, there are regulations, according to which all "enlarged and/or mottled kidneys" are investigated for residues of ochratoxin A at slaughter time and all carcasses, whose kidneys contained ochratoxin A levels above 10 μg/kg, are subsequently condemned (Boutrif and Canet, 1998). Such regulations are not very clever or safe or even they are not satisfactory, because macroscopic changes in pig kidneys can be found only after 1–3 months ochratoxin A exposure via the feeds (Krogh et al., 1973; Stoev et al., 2001). Therefore, in spite of such toxicological investigations of "enlarged and/or mottled kidneys," ochratoxin A–contaminated pork may enter easily in the human food chain and thus represents a real public health hazard (Stoev et al., 1998c; Stoev, 2008, 2013a,b) for humans in these countries.

Having in mind the well-based assumption (Krogh, 1972; Stoev, 1998) as well as some recent proofs (Stoev, 2008, Stoev and Denev, 2013) that mycotoxins and especially ochratoxin A are involved in the etiology of BEN in humans, the possible exposure of humans to this very hazardous toxin from pork or chicken meat (by the way "feed–pork/chicken–food") should be effectively prevented. In this regard, a much better and clever procedure for preventing any dangerous exposure of humans to this hazardous mycotoxin from meat would be a toxicological analysis of a few blood samples of pigs/chicken from risky farms having nephropathy problems or suspected of mycotoxic nephropathy only several weeks (in pigs) or several days (in chicken) before slaughter and a change in the feed source for a week (pigs) or for 2–3 days (chicken), if it is necessary. Another possible measure would be the prolongation of the period of feed deprivation of pigs/chicken before slaughter (Stoev et al., 1998c, 2002a; Stoev, 2008). The short half-life of ochratoxin A in pigs (72–120 h) and especially in chickens (4 h; NNT, 1991) leads to a quick decrease of its concentration in blood and various tissues after changing the feed source or after prolonging the period of feed deprivation before slaughtering. In such a way, the loss of condemnation of pig/chicken production would be prevented, and a much better procedure (than toxicological investigations of "enlarged mottled kidneys" as accepted in Denmark) would be realized for preventing the human exposure to ochratoxin A from meat. The possible preventive measures in already slaughtered chicks should include condemnation and removing of the kidneys and liver, where ochratoxin A is accumulated, whereas in already slaughtered pigs, only the kidneys should be condemned and/or removed (Stoev et al., 2002a; Stoev, 2008).

9.3.1 Underestimated Hazard of Joint Mycotoxin Exposure and Mycotoxin Interaction

With a big concern, we have to pay attention to the circumstance that various mixtures of mycotoxins usually have at least an additive, if not synergistic toxic effect (Stoev, 2008; Stoev and Denev, 2013), and the same are often produced by the same fungus. For example, a potent synergistic effect was established between ochratoxin A and penicillic acid that are produced by the same ochratoxinogenic fungi, when the same mycotoxins were given simultaneously to pigs or chickens (Micco et al., 1991; Stoev et al., 1999, 2000a, 2001, 2004). It seems that the presence of multiple toxins in various foods and food products presents a new important concern and challenge since toxicological information on the effects of simultaneous exposure is still very scarce and limited (Stoev, 2008; Stoev et al., 2010a,b). It is expected that in a diverse human diet, exposure will be obviously to multiple mycotoxins at a low concentration on an intermittent rate over long periods of time (Stoev, 2008; Stoev and Denev, 2013). Unfortunately, at present, the ultimate effect of such a continuous exposure is still unknown or very limited, although there is some evidence of strong synergistic or additive effect between some mycotoxins such as ochratoxin A, penicillic acid, fumonisin B_1, and citrinin (Stoev et al., 2010a,b) that are found simultaneously in some feeds in farms with nephropathy problems (Stoev et al., 2010a,b) or between various couple of mycotoxins such as ochratoxin A and penicillic acid (Micco et al., 1991; Stoev et al., 1999, 2000a, 2001, 2004) or between ochratoxin A and fumonisin B_1 (Creppy et al., 2004; Klaric et al., 2007; Stoev et al., 2012) or between ochratoxin A and citrinin (Pfohl-Leszkowicz et al., 2008). Now, we are almost sure that simultaneous exposure to those mycotoxins even in low contamination levels might be an important factor for the development of chronic renal diseases in animals and humans, especially after long-term exposure as the same mycotoxins were recently found in high (fumonisin B_1 and penicillic acid) or low contamination levels (ochratoxin A and citrinin) in most of the feeds originated from farms with mycotoxic porcine or avian nephropathy in Bulgaria and South Africa (Stoev et al., 2010a,b). Therefore, these days, it is of particular importance to introduce new regulations and limits in regard to combined contamination of food by several mycotoxins having in mind their possible interaction and increased hazard for humans as well. Obviously, simultaneous analysis of food for co-occurrence of target mycotoxins should be initiated worldwide, and the real hazard of such co-occurrence for human health should be carefully assessed. Moreover, new critical limits at critical control points should be introduced by HACCP system in EU and in all over the world having in mind the interaction of target mycotoxins co-contaminated various foods and food products.

9.3.2 Hazard of Masked Mycotoxins for Human Health

Another big concern in risk assessment and possible hazard for humans are masked mycotoxins–contaminated various foods or food products (Stoev and Denev, 2013). The masked mycotoxins or mycotoxin derivatives cannot be usually detected or are underestimated by the conventional analytical methods, which are usually designed only for parent (native) mycotoxins.

Masked mycotoxins could represent a large amount of total mycotoxins and after ingestion could give back the native mycotoxins after microbial and/or animal metabolisms (Berthiller et al., 2013). Having in mind this circumstance, it can be supposed that the found levels of target mycotoxins in foods/feedstuffs are often actually higher, because masked mycotoxins have not been measured (Stoev and Denev, 2013).

The group of masked mycotoxins comprises both extractable conjugated and bound (nonextractable) varieties. Unfortunately, extractable conjugated mycotoxins can be detected by appropriate analytical methods only when their structure is known and analytical standards are available. The bound mycotoxins, however, are not directly accessible and have to be liberated or separated from the matrix by chemical or enzymatic treatment prior to chemical analysis (Berthiller et al., 2013).

Transformation of masked mycotoxins into parent molecules can be also undertaken in order to use the conventional analytical methods, and this usually involves hydrolysis (Sewram et al., 2003; Dall'Asta et al., 2009), the type of which should be carefully selected. Unfortunately, there is not a single hydrolysis method, which can be applicable to all masked mycotoxins. Masked mycotoxins might be less toxic than their parent compounds, but they might also be more toxic than their parent compounds, for example, when they are more bioavailable (Berthiller et al., 2013). With a big worry, we have to emphasize that currently there are no sufficient toxicokinetic and toxicodynamic studies available to evaluate the possible hazard or risk assessment of masked mycotoxins in comparison with their parent forms, and therefore, it is impossible to perform a proper risk assessment for these mycotoxins in food due to the lack of data on exposure and toxic properties. The recognition of the toxicological relevance of masked mycotoxins, in addition to the evaluation of the hazard of co-occurrence of target mycotoxins in food commodities, is a new big challenge that should be addressed by the respective regulatory bodies, food manufacturers, and monitoring authorities to protect the health of consumers and to evaluate human health hazard (Stoev and Denev, 2013).

In regard to different kinds of animals and chickens, we should have into account the circumstance that the guidance mycotoxin values for cereals and feeds have been determined for the most tolerant animal species and are therefore to be considered as a upper guidance value, whereas for the feeds intended for more sensitive animals, the lower guidance values should be applied by the manufacturers (EC Recommendation 2006/576/EC).

9.4 MANAGEMENT OF THE RISK OF MYCOTOXIN CONTAMINATION OF FOODS/FEEDS

Usually, cereals are invaded by fungi both in the field and after harvest and have the potential for multitoxin contamination. The formation of each particular mycotoxin depends on a number of factors but particularly on the climate, the type of cereal, drying at harvest, and storage conditions. Thus, cereals may be contaminated by any of the main mycotoxin groups that include aflatoxins, ochratoxin A, deoxynivalenol, and other related trichothecenes, zearalenone, fumonisins, or moniliformin. These mycotoxins can occur in wheat, barley, rye, oats, triticale, maize, sorghum, etc., although the fumonisins mainly occur in maize. Some of the mycotoxins are formed by the fungi encountered in the growing crop prior harvesting, and therefore, the development of the same fungi, mainly belonging to *Fusarium*, *Alternaria*, and *Aspergillus* genera, is often difficult to control. It is much easier to prevent the formation of those mycotoxins that arise during storage, mainly formed by species belonging to *Penicillium* or *Aspergillus* genera, which could be performed by drying crops at harvest to safe moisture contents as soon as possible. However, sometimes this is not easily done, or if it is done, the subsequent poor storage practice may result in mold growth and mycotoxin formation, which could have been prevented (Stoev, 2007).

The management of the risks associated with mycotoxin contamination in foods/feeds could be realized only by the use of an integrated system based on the HACCP approach and should involve strategies for prevention, control, good manufacturing practices, and quality control at all stages of production from the field up to the final consumer. Such HACCP-like approaches are now being applied to different mycotoxins, and the FAO has published a comprehensive manual on the application of the HACCP system for mycotoxin prevention and control (FAO, 2002). The HACCP system can be used to identify the steps or points at which mycotoxin contamination of feeds/foods might be prevented or mycotoxins might be removed and to identify the stages at which monitoring systems can be set up. For example, such approach is used in automated sorting and segregation of peanuts in order to remove those nuts highly contaminated with aflatoxins or in cleaning procedures for cereals prior to milling that removes dust, fungal spores, broken grains, and other debris that may contain high levels of mycotoxins (Dowell et al., 1990; FAO, 2002). The incidence of molds and mycotoxins occurrence can be decreased by the application of various preventative measures both before and after harvest including appropriate control measures, timely harvesting, cleanup, drying and storage practices, creating of plant cultures resistant to fungi infestation, management of insect infestation, crop rotation, and others (Doko et al., 1995; Bata et al., 2001).

9.4.1 Prevention of Mycotoxin Contamination of Foods/Feeds

The occurrence of molds and mycotoxins can be decreased by the application of a variety of preventative measures both before and after harvest including appropriate control measures, timely harvesting, cleanup, drying and storage practices, crop rotation, management of insect infestation, and creating of plant cultures resistant to fungi infestation. For example, Bt-corn (transgenic corn) products have less risk of fumonisin contamination depending on the types of insects attacking the corn. Genetic engineering has also been useful in the development of host resistance through the addition or enhancement of antifungal genes in order to create mold-resistant wheat cultivars or maize hybrids, etc. (Doko et al., 1995; Bata et al., 2001). Also, it is well known that mold contamination is more pronounced if wheat is sown after maize and vice versa. Applying multifield crop rotation in which, for example, rape, sugar beet, sunflower, or soya beans are present reduces the degree of mold infestation. It should be taken into account that fertilization with nitrogen increase plants, sensitivity to mold (Reid et al., 2001). On the other hand, fungicides, applied before blossoming, decrease *Fusarium* contamination and the respective mycotoxins production, whereas delayed harvesting favors *Fusarium* contamination (Peraica et al., 2002).

Another way to prevent the production of some mycotoxins such as ochratoxin A or aflatoxins is the prevention of the growth of the storage fungi. Some studies indicate that fungi in cereal grains stored at moisture contents lower than 15% generally do not produce ochratoxin A, which suggests that the moisture content in the cereal grains has to be decreased below 15% via various drying procedures before storage (Frohlich et al., 1991). Storage at higher moisture levels requires that grain be maintained under anaerobic conditions, which prevent the growth of fungi. Otherwise, a combination of mold inhibitors and sterilization techniques, possibly coupled with the inoculation of nontoxigenic competitive microflora, has to be applied to prevent the growth of ochratoxin A–producing fungi (Stoev, 2008).

On the other hand, Chelack and collaborators (1991a,b) have used radiation to control the growth of ochratoxin A–producing fungi. They demonstrated that gamma or electron beam irradiation is a highly effective means of destroying spores from some ochratoxin A–producing fungi such as *Aspergillus alutaceus* (Chelack et al., 1991a,b).

Leitao and collaborators (1990) demonstrated that phosphine (PH_3) was effective at inhibiting both fungal growth and sterigmatocystin production by *Aspergillus versicolor*, and this could be considered as another specific way to control the growth of mycotoxin production by fungi.

For foods with pH values from 5 to 6, such as the cereals or sorghum, the antimicrobial agents (food additives) that are able to prevent the growth and ochratoxin A production by *Aspergillus* and *Penicillium* species are methyl paraben or potassium sorbate (Tong and Draughon, 1985). Small concentrations of these compounds can inhibit completely the growth by both genera of fungi and their ochratoxin A production. At pH 4.5, as occurs in silage, fungal growth and ochratoxin A production were completely inhibited by 0.02% potassium sorbate, 0.7% methyl paraben, and 0.2% sodium propionate (Tong and Draughon, 1985). The same could also be used to suppress the growth of various other fungi, producing different mycotoxins.

Prevention through preharvest management is the best method for controlling mycotoxin contamination. Ideally, risks associated with mycotoxin hazards should be minimized in each phase of the food processing. Control parameters in processing commodities susceptible to mycotoxin contamination should include time of harvesting, temperature, moisture during storage and transportation, as well as humidity of air in the storehouses (Scott, 1998). Adequate storage conditions with optimal temperature and humidity of grain and storehouses may decrease significantly the growth of toxigenic molds (Peraica et al., 2002).

However, once mycotoxin contamination occurs, the hazards associated with various mycotoxins must be managed through postharvest procedures, if the product is to be used as human food or animal feed. Such measures, which can decrease mycotoxin content in food or feed, include various physical or chemical decontamination methods such as cleaning, segregation, electronic sorting, and steeping as well as various kinds of processing such as milling, which is usually used to manage mycotoxins in raw ingredients. Further baking, cooking, extrusion, malting, and brewing can additionally decrease mycotoxin content in the final product.

9.4.2 Physical Methods of Mycotoxin Decontamination of Foods/Feeds

The physical methods of mycotoxin decontamination (Scott, 1998) include various kinds of procedures such as cleaning, mechanical sorting and separation, washing, segregation, thermal inactivation, irradiation, and solvent extraction.

Cleaning procedures are used for all cereals prior to food processing, and this process includes removal of dust, broken grains, and other unwanted material and may involve the physical removal of some of the outer layers of grains by abrasion, for example, "scouring." The data of the effectiveness of cleaning for the aflatoxins vary, but an average reduction of about 40% in concentration was usually reported (Brekke et al., 1975a). Generally, separation of corn screenings can significantly reduce fumonisin and aflatoxin concentration (Broggli et al., 2002). Cleaning of grain can remove ergot alkaloids from rye and wheat (Scott, 1998). Some reduction by cleaning and scouring procedures is also reported for ochratoxin-contaminated grain (Scudamore and Banks, 2004). The highest concentrations of deoxynivalenol can be usually found in the smaller broken and shrivelled grains, and therefore, mean toxin concentrations could be reduced by discarding the small grains (Chelkowski and Perkowski, 1992). Substantial reduction of zearalenone occurs in barley by polishing, but cleaning of maize grains appeared to have little effect on zearalenone concentration as reported by Scudamore and collaborators (Scudamore et al., 1998, 1999). Similarly, the exclusion of rotten or poor-quality apples can significantly reduce the concentrations of patulin in apple juice (Scudamore and Banks, 2004).

The automated sorting and segregation of peanuts can remove damaged or inadequately developed nuts highly contaminated with aflatoxins, which are usually used for oil production (Dowell et al., 1990). Electronic sorting is another method for the decontamination of peanuts, which is based on the color of roasted, blanched peanuts (Pelletier and Reizner, 1992), whereas fluorescence sorting is used mainly for screening and decontamination of corn, cottonseed, and dried figs (Muller, 1983; Steiner et al., 1988).

Washing procedures by water or sodium carbonate solution is sometimes used for decreasing the concentration of some *Fusarium* mycotoxins as fumonisins, zearalenone, and deoxynivalenol in grains or corn cultures, but it might be a useful procedure only prior to wet milling, because of the eventual high price of the subsequent drying procedures (Scott, 1998).

Thermal inactivation is not a suitable method for decontamination, because most of the mycotoxins are heat stable. Partial decontamination can be only achieved by coffee roasting or by microwave treatment, which may destroy mainly aflatoxins in peanuts or trichothecenes in corn (Basappa and Shantha, 1996; Scott, 1998).

Different types of radiation such as γ-irradiation, x-rays, and ultraviolet light were explored for the decontamination of some mycotoxins such as T-2, deoxynivalenol, or aflatoxins as well as for the control of the growth of some fungi (Müller, 1983; Samarajeewa, 1991) but have some disadvantages, because radiation is effective only when applied to a thin layer of grain (Peraica et al., 2002). Fortunately, solar radiation is an inexpensive way of partial detoxification of coconuts, peanuts, corn, and sesame contaminated with aflatoxins, which is widespread in some tropical areas of world (Basappa and Shantha, 1996).

Solvent extraction by ethanol, isopropanol, or methoxymethane can be an effective way for the decontamination of some mycotoxins such as aflatoxins, but the high prices and possible solvent residues present barriers for commercial exploitation of such methods (Basappa and Shantha, 1996; Peraica et al., 2002).

9.4.3 Chemical Methods of Mycotoxin Decontamination of Foods/Feeds

There are also various chemicals such as acids (e.g., formic and propionic acid), bases (e.g., ammonia, sodium hydroxide), oxidizing agents (e.g., hydrogen peroxide, ozone), reducing agent (e.g., sodium bisulfite), and some other chemicals such as chlorine or formaldehyde, which have been studied for their effectiveness in mycotoxin decontamination (Peraica et al., 2002), but these usually leave some toxic metabolites or reduce the nutritional value of foods/feeds.

According to Laciakova and collaborators (1998), formic acid of 0.25% concentration degraded ochratoxin A after 3 h exposure, propionic and sorbic acids in 1% concentration after 24 h exposure, and benzoic acid in 0.5% concentration after 24 h exposure (Laciakova et al., 1998).

Hydrogen peroxide has been used for detoxifying peanuts, whereas calcium hydroxide or monomethylamine was used for detoxifying oil seeds and corn, contaminated, respectively, with aflatoxins or fumonisins (Scott, 1998). Sodium bisulfite is another chemical used for the detoxification of corn or dried figs contaminated with aflatoxins or deoxynivalenol (Scott, 1998).

Ammoniation and ozonation are other chemical methods that have received the most research attention as a practical solution to decontaminate aflatoxins- or fumonisins-contaminated feeds or peanuts, but such methods usually are not allowed within the EC for human foods (Peraica et al., 2002). Chelkowski and collaborators (1981) reported that the treatment of ochratoxin A–contaminated grain with ammonia reduced concentrations of this mycotoxin to undetectable levels. They concluded that ammoniation of grain not only detoxifies several mycotoxins, including fumonisins, but also inhibits mold growth (Chelkowski et al., 1981). Feeding studies have shown no toxic effects related to the ammoniation process, but there are some changes in the nutritional quality of the feed, such as a decrease in lysine and sulfur-containing amino acids (Scott, 1998). In addition, adequate aeration after ammoniation is necessary for the acceptance of the feed by animals.

Madsen and collaborators (1983) reported that the treatment of ochratoxin A–contaminated barley with 5% NH_3 for 96 h at 70°C or warming of grain to 105°C in the presence of 0.5% NaOH as well as autoclaving at 132°C for 0.5 h can destroy the main part of ochratoxin A in the barley, but such treatments are not practical and are not able to effectively reduce the mycotoxin contamination in grain.

Except ammoniation, many of the techniques proposed to remove mycotoxins are currently perceived as impractical, ineffective, and/or potentially unsafe for large-scale use (CAST, 1989).

9.4.4 Feed Additives and Antidotes Used Against Toxic Effects of Mycotoxins

There is also possibility of addition of various chemicals or feed additives with possibilities to fix and neutralize mycotoxins or to realize antidote effects against mycotoxin action (Stoev et al., 1999, 2000a, 2002d, 2004; Stoev, 2008, 2010b).

Some feed additives are designed to bind mycotoxins in gastrointestinal system and to reduce their bioavailability. Clay (hydrated sodium calcium aluminosilicate [HSCAS], for example, clay and others) and zeolitic minerals present a broad family of functionally diverse silicoaluminosilicates, which have shown promising effects on the binding of mycotoxins in gastrointestinal tract of animals, significantly decreasing their bioavailability and associated toxicities. HSCAS is very useful for preventing aflatoxicosis in farm animals and for reducing aflatoxins concentrations in milk but is not very effective for other mycotoxins such as ochratoxin A, T-2 toxin, fumonisins, or deoxynivalenol (Ramos et al., 1996; Kubena et al., 1998; Stoev, 2008).

Several different approaches have been used to reduce ochratoxin A absorption, including the use of HSCAS, bentonite, charcoal, and cholestyramine. The addition of HSCAS (1%) and bentonite

(1% and 10%) to a diet containing ochratoxin A had no effect on ochratoxin A concentration in swine blood, serum, tissues, and bile (Rotter et al., 1989). Bentonite, however, can effectively absorb aflatoxins (Ramos et al., 1996).

The addition of 1% activated charcoal to the diet, in contrast, caused a slight decrease in the concentration of ochratoxin A in swine blood, whereas 10% charcoal decreased the concentration of ochratoxin A in the blood, liver, kidney, spleen, and heart by 50%–80% (Marquardt and Frolich, 1992). However, the supplementation of diet with activated charcoal was considered as an impractical method of reducing ochratoxin A toxicity in chicks or pigs that were continuously exposed to this mycotoxin, because of its high cost and the possibility to reduce minerals and vitamins in domestic animals (Rotter et al., 1989).

Cholestyramine, in contrast to the nonspecific absorbent discussed previously, seems to be an effective absorbent of ochratoxin A, zearalenone, and fumonisins in the gastrointestinal tract of nonruminant animals. Cholestyramine is a commercial anion exchange resin that has been shown to reduce blood ochratoxin A concentrations by 50%, when it was included in 0.5% in a rat diet containing 1 ppm ochratoxin A (Marquardt and Frolich, 1992).

Another way to reduce toxicity of various mycotoxins is to elucidate a good understanding of the mechanisms of their toxicity. This approach includes an addition of specific antidotes or vitamins to the diet to prevent the specific toxic action of each mycotoxin (Stoev, 2008). For example, the ascorbic acid supplementation (300 mg/kg) of laying hen diet that contained 3 ppm ochratoxin A has been reported to have a good protective effect against ochratoxin action and can partially ameliorate its toxic effects (including the negative effect of OTA on the eggs production and on the weight of the eggs) (Haazele, 1992; Haazele et al., 1993).

The use of some other feed additives was also found to protect against the toxic effects of ochratoxin A, reducing significantly various farm losses from a decrease of weight gain in stock chicks (Stoev et al., 2000a, 2002d, 2004) and from a decrease of egg production in laying hens (Stoev, 2010b). For example, 5% total water extract of artichoke (*Cynara scolymus* L.), prepared as a steam infusion from dried leaves of artichoke and given to chicks in concentration of 5 mL/kg b.w. via the feed or water, has been found to have a good protective effect against the toxic effect of ochratoxin A and to improve its fast elimination from animals (Stoev et al., 1999, 2000a, 2002d, 2004). Other similar antidotes against the toxic effects of ochratoxin A such as Roxazyme-G (polyenzyme complement produced by the fungal genus *Trichoderma* given in concentration 0.2 g/kg feed) or Rosallsat (plant extract of bulbus *Allii Sativi* and seminum *Rosae canina* given *per os* in 0.6 mL/kg body mass daily as a supplement to the feed) were also found to protect against various adverse toxic effects of ochratoxin A in chicks (Stoev et al., 1999, 2002d). Such antidotes, used as supplements to the feeds, could be used as a practical approach for safely utilizing ochratoxin A–contaminated feed. In such a way, the rejection or condemnation of such feed will be avoided as well as there will be no need to eliminate this mycotoxin from the feed if its contamination levels are similar to these encountered in the practice (normally up to 1–2 ppm) (Stoev et al., 2002d).

In addition, the feeding of contaminated grain to animal species that are less susceptible to a particular mycotoxin is also a good measure to utilize mycotoxin-contaminated feeds. For example, feeding ochratoxin A–contaminated feeds to ruminants can safely utilize such feeds, because ruminants are less sensitive to this mycotoxin and are able to hydrolyze it in the rumen to its nontoxic form ochratoxin α (OTα) (Sreemannarayana et al., 1988).

9.4.5 Mycotoxin Decontamination via Food/Feed Processing

Processing can be defined as any chemical, biological, or physical treatment that is applied to a raw material to produce the final consumer product (Scudamore and Banks, 2004). That includes any procedure from dry and wet milling of grains, baking, extrusion, steaming, and brewing to feeding cereal-based complete feeds to animals to produce meat or milk. The stability of mycotoxins during processing may be affected by chemical or biological reactions and factors such as temperature, pH, moisture content, pressure, buffering conditions, and the presence of other constituents and enzymes. However, some important mycotoxins such as aflatoxins, ochratoxin A, deoxynivalenol, zearalenone, and fumonisins can usually survive processing of raw cereals to some extent and thus occur in consumer foods, but these may also undergo significant processing so that the concentration of particular mycotoxin in a food item that reaches the consumer may be considerably lower than in the raw harvested crop. Therefore, appropriate management of the commodity during processing becomes of prime importance.

Wet and dry milling are operations traditionally used for grain processing but can be also used for mycotoxin decontamination of food in some cases (in zearalenone, fumonisins, aflatoxins, trichothecenes, and ochratoxin A contamination). This method separates the grain into different fractions, and it is important to identify the fractions that remain toxic. In dry milling, mycotoxins are usually concentrated in the bran so that white flour from wheat (Osborne et al., 1996) or maize grits (e.g., Broggli et al., 2002) usually has significantly reduced mycotoxin levels. In wet milling of maize, significant amounts of most mycotoxins (especially fumonisins, zearalenone, and aflatoxins) are removed in the steep water, and subsequent processing results in very low residues in maize starch (Bennett and Andersen, 1978) although high concentrations may remain in the gluten or germ fractions that are mainly used for livestock (Scudamore et al., 1998). In dry milling, aflatoxins concentrate in the bran and offal fractions of wheat and germs as well as in the outer layers of maize while being reduced in flour or corn grits (Brekke et al., 1975b). In wet milling, a large percentage of aflatoxins are similarly removed in the steep water (Bennett and Anderson, 1978). Similarly, ochratoxin A tends to be concentrated in the outer bran layers of cereals, and this raises the possibility of redistribution during milling with the result that both reduction and increase in concentration can occur, depending on the milled fraction examined. The analysis of each milled fraction showed that high ochratoxin A concentrations were present in the bran and offal fractions, while those in the white flour fractions were much reduced unlike the wholemeal flour (Osborne et al., 1996). Therefore, ochratoxin A can be reduced during the production of bread by elimination of the bran and offal components although the high temperature of baking had minimal effect on the final concentrations (Scudamore and Banks, 2004). After milling, higher concentrations of deoxynivalenol and zearalenone were found in bran and shorts, while lower concentrations were found in straight grade (white) flour (Scudamore and Banks, 2004). Dry milling results in a redistribution of fumonisins in the different fractions with reduced amounts in milling grits and flour and increased concentrations in bran and germ (Broggli et al., 2002). Industry information indicates that dry milling results in fumonisin-containing fractions in descending order of the highest to lowest fumonisin levels as follows: bran, flour, meal, grits, and flaking grits (Broggli et al., 2002).

In some specific kinds of processings, such as the treatment of corn with lime water in the manufacture of tortillas, levels of aflatoxins can be considerably decreased (Scott, 1998). Refining of oil successfully eliminates any extracted aflatoxins. Similarly, addition of sodium chloride (salt) during the cooking of unshelled peanuts under pressure can decrease significantly aflatoxins content, whereas vitamin C in apple juice can slowly decrease the content of patulin (Scott, 1998; Peraica et al., 2002).

The stability of aflatoxins to heat in processes such as baking (Stoloff and Trucksess, 1981) and extrusion (Martinez and Monsalve, 1989) depends on temperature and pH so that higher temperatures or alkaline processes such as the use of leavening agents or tortilla production can reduce the content of aflatoxins (Price and Jorgensen, 1985, Abbas et al., 1988, De Arrola et al., 1988). The baking process itself is ineffective in destroying significant amounts of ochratoxin A. Deoxynivalenol is heat stable at 120°C and moderately stable at 180°C (deep frying) and decomposes within 30–40 minutes at 210°C (grilling) (Kamimura, 1989). Zearalenone also survives baking (Scudamore and Banks, 2004). The fumonisins are unstable during roasting but comparatively stable during baking and canning and therefore can be considered as moderately stable compounds (Castelo et al., 1998). Ergot alkaloids are much less stable and are destroyed on baking bread up to 100% (Scott, 1998).

Breakdown by enzymes depends on the process, but aflatoxins and most of other mycotoxins such as deoxynivalenol, zearalenone, and fumonisins can survive and occur in beer produced from maize or wheat (Scott and Lawrence, 1994; Scott, 1996; Scudamore and Banks, 2004). Malting of barley and brewing can partially destroy ochratoxin A, but this does not happen for deoxynivalenol or zearalenone (Scott, 1996). Regardless of the circumstance that ochratoxin A is a relatively stable mycotoxin, under certain situations such as high temperatures and acid or alkaline conditions and in the presence of enzymes, some breakdown can occur. The fermentation by yeasts has been found to be effective in destroying patulin and rubratoxin B (Scott, 1998). Rotter and collaborators (1990) demonstrated that inoculation of barley with a *Lactobacillus* species followed by ensiling (decreasing of pH) can reduce the concentration of ochratoxin A by approximately 50%, but ensiling of corn does not usually destroy aflatoxins, zearalenone, or deoxynivalenol (Rotter et al., 1990).

While processing tends to reduce the concentration of mycotoxins, there are some circumstances that can increase their levels. Such circumstances are poor storage of intermediate materials or end products, which can lead to mold growth and mycotoxin formation. For example, wheat flour, if not stored properly, could be contaminated by various storage mycotoxins such as ochratoxin A. On the other hand, the milling is a kind of processing, in which contamination levels of various mycotoxins often change in both directions, because mycotoxins are unequally distributed among the milled fractions, reflecting how the same originally formed in each individual grain. In this context, the concentrations of mycotoxins are usually found to be higher in some fractions, commonly bran or germ but reduced in flour (Scudamore and Banks, 2004).

9.5 CONCLUSIONS

It is obvious that human exposure to mycotoxins cannot be completely prevented, because they are natural contaminants. Target mycotoxins such as aflatoxin B_1, zearalenone, and ochratoxin A are much more dangerous and would represent safety hazard twice because of their transmission in milk of lactating cows of either parent toxin or toxic metabolites such as aflatoxin M_1 or α zearalenol (Galtier, 1998) as well as because of the transmission in eggs or meat such as ochratoxin A (Dailey et al., 1980; Galtier, 1998). Whenever any effort is made for improving the quality of foods or feeds, it should be clearly defined that there is a definite need for a better-quality product and that the respective community is prepared to bear any associated increase in the cost of the improved commodity. It becomes more and more clear that national programs for the monitoring, prevention, and control of mycotoxin contamination based on the assessment and analysis of the situation in each separate country are not enough and helpful now. Therefore, an implementation of carefully designed surveillance studies and modern internationally recognized biomonitoring methods measuring the exposure to mycotoxins of individuals should be introduced worldwide for profound evaluation of the factors that are compromising the quality of the products of the commodity system and leading to the production of molds and mycotoxins. A well-designed networking system is also necessary to be sustained in the future for appropriate dissemination of available information in addition to staff training at regional and international basis (Stoev, 2013a). The regulations enacted for mycotoxins in food and feed, and those under development, should be the result of sound cooperation between interested parties, drawn from science, consumers, industry, and policy makers to achieve a realistic protection of consumer. This is an incredibly difficult task as many factors have to be considered when deciding on regulatory standards. In addition to scientific factors, such as risk assessment and analytical accuracy or measuring the toxicological effects of a variety of different mycotoxin combinations as they occur in nature, economical and political factors such as the commercial interests of each country and the constant necessity of a sufficient food supply also play a role in the decision-making process. In spite of all these difficulties, these days, a harmonization of mycotoxin regulations and control procedures worldwide is required in order to facilitate international food trade and to ensure global food safety based on sound scientific knowledge about human health hazard and risk analysis.

ABBREVIATIONS

ALARA	As low as reasonably achievable
BEN	Balkan endemic nephropathy
CAC	Codex Alimentarius Commission
CCFAC	Codex Committee on Food Additives and Contaminants
HACCP	Hazard analysis and critical control point
JECFA	Joint FAO/WHO Expert Committee on Food Additives
IARC	International Agency for Research on Cancer
ICC	Idiopathic Congestive Cardiomyopathy
MN	Mycotoxic nephropathy
NOEL	No-observed-effect-level
TDI	Tolerable daily intake
PTWI	Provisional tolerable weekly intake
SCF	Scientific Committee for Food
WHO	World Health Organization

ACKNOWLEDGMENTS

This research has been supported in part through EC under Marie Curie Outgoing International Fellowship under the 6th framework and Marie Curie International Research Staff Exchange Scheme under 7th framework, Department of Science and Technology in South Africa, UK Royal Society Joint Project with Central and Eastern Europe, and Foundation of Ministry of science and education of Bulgaria via 5 Research projects.

REFERENCES

Abbas, H.K., Mirocha, C.J. Rosiles, R., Carvajal, M. (1988). Effect of Tortilla-preparation process on aflatoxin B_1 and B_2 in corn. *Mycotoxin Research* **4**: 33–36.

Allcroft, R., Carnaghan, R.B.A. (1962). Groundnut toxicity. *Aspergillus flavus* toxin (aflatoxin) in animal products: Preliminary communication. *The Veterinary Record* **74**: 863–864.

Basappa, S.C., Shantha, T. (1996). Methods for detoxification of aflatoxins in foods and feeds—A critical appraisal. *Journal of Food Science Technology* **33**: 95–107.

Bata, A., Rafai, P., Kovacs, S. (2001). Investigation and a new evaluation method of the resistance of maize hybrids grown in Hungary to *Fusarium* moulds. *Phytopathology* **149**: 107–111.

Bennett, G.A., Anderson, R.A. (1978). Distribution of aflatoxin and/or zearalenone in wet-milled corn products: A review. *Journal of Agricultural and Food Chemistry* **26**:1055–1060.

Bennett, J.W., Bentley, R. (1999). Pride and prejudice: The story of ergot. *Perspectives in Biology and Medicine* **42**: 333–355.

Berthiller, F., Crews, C., Dall'Asta, C., De Saeger, S., Haesaert, G., Karlovsky, P., Oswald, I., Seefelder, W., Speijers, G., Stroka, J. (2013). Masked mycotoxins: A review. *Molecular Nutrition & Food Research* **57**: 165–186.

Betina, V. (1989). Biological effects of mycotoxins. In: V. Betina (Ed.), *Mycotoxins: Chemical, Biological and Environmental Aspects*. Elsevier Science Publishers, Amsterdam, the Netherlands, pp. 42–58.

Bezuidenhout, S.C., Gelderblom, W.C.A., Gorst-Allman, C.P., Horak, R.M., Marasas, W.F.O., Spiteller, G., Vleggaar, R. (1988). Structure elucidation of the fumonisins, mycotoxins from *Fusarium moniliforme*. *Journal of the Chemical Society, Chemical Communications* **11**: 743–745.

Boonchuvit, B., Hamilton, P.B., Burmeister, H.R. (1975). Interaction of T-2 toxin with *Salmonella* infection in chickens. *Poultry Science* **54**: 1693–1696.

Boutrif, E., Canet, C. (1998). Mycotoxin prevention and control: FAO programes. *Revue de Medecine Veterinaire* **149**: 681–694.

Brekke, O.L., Peplinski, A.J., Griffin, E.L. Jr. (1975a). Cleaning trials for corn containing aflatoxin. *Cereal Chemistry* **52**: 198–204.

Brekke, O.L., Peplinski, A.J., Nelson, G.E.N., Griffin, E.L. Jr. (1975b). Pilot-scale dry milling of corn containing aflatoxin. *Cereal Chemistry* **52**: 205–211.

Broggli, L.E., Resnik, S.L., Pacin, A.M., Gonzalez, H.H.L., Cano, G., Taglieri, D. (2002). Distribution of fumonisins in dry-milled corn fractions in Argentina. *Food Additives and Contaminants* **19**: 465–469.

Bucci, T., Howard, P., Tolleson, W., Laborde, J, Hansen, D. (1998). Renal effects of fumonisin mycotoxins in animals. *Toxicologic Pathology* **26**: 190–194.

CAST (Council for Agricultural Science and Technology). (1989). Mycotoxins: Economic and health risk. Task Force Report 116. Ames, IA, pp. 1–91.

Castegnaro, M., McGregor, D. (1998). Carcinogenic risk assessment of mycotoxins. *Revue de Medecine Veterinaire* **149**: 671–678.

Castelo, M.M., Sumner, S.S., Bullerman, L.B. (1998). Stability of fumonisins in thermally processed corn products. *Journal of Food Protectection* **61**: 1030–1033.

Chelack, W.S., Borsa, J., Marquardt, R.R., Frohlich, A.A. (1991a). Role of competitive microbial flora in the radiation induced enhancement of ochratoxin production by *Aspergillus alutaceus* var alutaceus NRRL-3174. *Applied and Environmental Microbiology* **57**: 2492–2496.

Chelack, W.S., Borsa, J., Szekely, J.G., Marquardt, R.R., Frohlich, A.A. (1991b). Variants of Aspergillus alutaceus var alutaceus formerly *Aspergillus ochraceus*, with altered ochratoxin A production. *Applied and Environmental Microbiology* **57**: 2487–2491.

Chelkowski, J., Golinski, P., Godlewska, B., Radomyska, W., Szebiotko, K., Wiewiorowska, M. (1981). Mycotoxins in cereal grains. Part IV. Inactivation of ochratoxin A and other mycotoxins during ammoniation. *Nahrung* **25**: 631–637.

Chelkowski, J., Perkowski, J. (1992). Mycotoxins in cereal grain (part 15). Distribution of deoxynivalenol in naturally contaminated wheat kernels. *Mycotoxin Research* **8**: 27–30.

Chu, F.S., Li, G.Y. (1994). Simultaneous occurrence of fumonisin B_1 and other mycotoxins in moldy corn collected from the People's Republic of China in regions of high incidences of oesophageal cancer. *Applied and Environmental Microbiology* **60**: 847–852.

Conkova, E., Laciakova, A., Kovac, G., Seidel, H. (2003). *Fusarial* toxins and their role in animal diseases. *Veterinary Journal* **165**: 214–220.

Cooray, R., Jonsson, P. (1990). Modulation of resistance to mastitis pathogens by pre-treatment of mice with T-2 toxin. *Food and Chemical Toxicology* **28**: 687–692.

Coulombe, R.A., Guarisco, J.A., Klein, P.J., Hall, J.O. (2005). Chemoprevention of aflatoxicosis in poultry by dietary butylated hyodroxytoluene. *Animal Feed Science and Technology* **121**: 217–225.

Creppy, E.E., Chiarappa, P., Baudrimont, I., Borracci, P., Moukha, S., Carratù, M.R. (2004). Synergistic effects of fumonisin B_1 and ochratoxin A: Are "in vitro" cytotoxicity data predictive of "in vivo" acute toxicity. *Toxicology* **201**: 115–123.

Dailey, R.E., Reese, R.E., Brouwer, E.A. (1980). Metabolism of 14^{C}-zearalenone in laying hens. *Journal of Agricultural and Food Chemistry* **28**: 286–291.

Dall'Asta, C., Mangia, M., Berthiller, F., Molinelli, A., Sulyok, M., Schuhmacher, R., Krska, R., Galaverna, G., Dossena, A., Marchelli, R. (2009). Difficulties in fumonisin determination: The issue of hidden fumonisins. *Analytical and Bioanalytical Chemistry* **395**: 1335–1345.

De Arrola, M., Del, C., de Porres, E., de Cabrera, S., de Zepeda, M., Rolz, C. (1988). Aflatoxin fate during alkaline cooking of corn for tortilla preparation. *Journal of Agricultural and Food Chemistry* **36**: 530–533.

Dirheimer, G. (1998). Recent advances in the genotoxicity of mycotoxins. *Revue de Medecine Veterinaire* **149**: 605–616.

Doko, M., Rapior, S., Visconti, A., Schjoth, J.E. (1995). Incidence of levels of fumonisin contamination in maize by genotypes grown in Europe and Africa. *Journal of Agricultural and Food Chemistry* **43**: 429–434.

Dowell, F.E., Dorner, J.W., Cole, R.J., Davidson, J.I. (1990). Aflatoxin reduction by screening farmers stock peanuts. *Peanut Science* **17**: 6–8.

Dutton, M.F. (1996). Fumonisins, mycotoxins of increasing importance: Their nature and their effects. *Pharmacology & Therapeutics* **70**: 137–161.

Dwivedi, P., Burns, R.B. (1985). Immunosuppressive effects of ochratoxin A in young turkeys. *Avian Pathology* **14**: 213–225.

EC Directives 2002/32/EC of the European Parliament and of the Council on undesirable substances in animal feed, May 7, 2002, pp. 1–27.

EC Recommendation 2006/576/EC on the presence of deoxynivalenol, zearalenone, ochratoxin A, T-2 and HT-2 and fumonisins in products intended for animal feeding, August 17, 2006, pp. 1–3.

EC Recommendations 2013/165/EU on the presence of T-2 and HT-2 toxin in cereals and cereal products, March 27, 2013, pp. 1–4.

EC Regulation No. 178/2002 of the European Parliament and of the Council laying down the general principles and requirements of food law, establishing the European Food Safety Authority and laying down procedures in matters of food safety, January 28, 2002, pp. 1–40.

EC Regulation 1881/2006 setting maximum levels for certain contaminants in foodstuffs, December 19, 2006, pp. 1–26.

Elissalde, M.H., Ziprin, R.L., Huff, W.E., Kubena, L.F., Harvey, R.B. (1994). Effect of ochratoxin A on *Salmonella*-challenged broiler chicks. *Poultry Science* **73**: 1241–1248.

FAO. (1988). Nairobi+10 Report of the second Joint FAO/WHO/UNEP International Conference on Mycotoxins. Bangkok, Thailand, September 28–October 2, 1987, FAO, Rome, Italy.

FAO. (1997). Food and Nutrition Paper N 64, Worldwide Regulations for Mycotoxins, FAO, Rome, Italy.

FAO. (2002). Manual on the application of the HACCP system in mycotoxin prevention and control. Joint FAO/WHO Food Standards Programme FAO, FAO, Rome, Italy.

Friends, D.W., Trenholm, H.L., Thompson, B.K., Hartin, K.E., Fiser, P.S., Asem, E.K., Tsang, B.K. (1999). The reproductive efficiency of gilts fed very low levels of Zea. *Canadian Journal of Animal Science* **70**: 635–645.

Frohlich, A.A., Marquardt, R.R., Ominski, K.R. (1991). Ochratoxin A as a contaminant in the human food chain: A Canadian perspective. In: M. Castegnaro, R. Plestina, G. Dirheimer, I.N. Chernozemsky and H. Bartsch (Eds.), *Mycotoxins, Endemic Nephropathy and Urinary Tract Tumours*, Publication 115. IARC, Lyon, France, pp. 139–144.

Fukata, T., Sasai, K., Baba, E., Arakawa, A. (1996). Effect of ochratoxin A on *Salmonella typhimurium*-challenged layer chickens. *Avian Diseases* **40**(4): 924–926.

Gabbai, A., Pourquier, L. (September 1951). Ergot poisoning at Pont St. Esprit. *British Medical Journal* **15**: 650–651.

Galtier, P. (1998). Biological fate of mycotoxins in animals. *Revue de Medecine Veterinaire* **149**: 549–554.

Gelderblom, W., Marasas, W., Farber, E. (1992). The cancer initiating potential of the fumonisin B mycotoxins. *Carcinogenesis* **13**: 433–437.

Gelderblom, W.C.A., Jaskiewicz, K., Marasas, W.F.O., Theil, P.G., Horak, R.M., Vleggaar, R., Kriek, N.P.J. (1988). Fumonisins-novel mycotoxins with cancer-promoting activity produced by *Fusarium moniliforme*. *Applied and Environmental Microbiology* **54**: 1806–1811.

Gupta, S., Jindal, N., Khokhar, R.S., Asrani, R.K., Ledoux, D.R., Rottinghaus, G.E. (2008). Individual and combined effects of ochratoxin A and *Salmonella enterica* serovar *Gallinarum* infection on pathological changes in broiler chickens. *Avian Pathology* **37**: 265–272.

Haazele, F.M. (1992). Response to dietary ascorbic acid supplementation in laying hens: Effects of exposure to high temperature and ochratoxin ingestion, PhD thesis, University of Manitoba, Winnipeg, Manitoba, Canada.

Haazele, F.M., Guenter, W., Marquardt, R.R., Frohlich, A.A. (1993). Beneficial effects of dietary ascorbic acid supplement on hens subjected to ochratoxin A toxicosis under normal and high ambient temperatures. *Canadian Journal of Animal Science* **73**: 149–157.

Halloy, D.J., Gustin, P.G., Bouhet, S., Oswald, I.P. (2005). Oral exposure to culture material extract containing fumonisins predisposes swine to the development of pneumonitis caused by *Pasteurella multocida*. *Toxicology* **213**: 34–44.

Haschek, W.M., Rousseaux, C.G., Wallig, M.A. (2009). *Fundamentals of Toxicologic Pathology*. Academic Press, Elsevier, San Diego, CA, pp. 1–691.

Hgindu, A., Johnson, B.A., Kenya, P.R. (1982). An outbreak of acute hepatitis by aflatoxin poisoning in Kenya. *The Lancet* **319**: 1346–1348.

Howard, P., Warbritton, A., Voss, K., Lorenzen, R., Thurman, J., Kovach, R., Bucci, T. (2001). Compensatory regeneration as a mechanism for renal tubule carcinogenesis of fumonisin B_1 in F344/N/Nctr BR rat. *Environmental Health Perspectives* **109**: 309–314.

Huff, W.E., Ruff, M.D. (1982). *Eimeria acervulina* and *Eimeria tenella* infections in ochratoxin A-compromised broiler chickens. *Poultry Science* **61**: 685–692.

IPCS. (1990). Selected mycotoxins: Ochratoxins, trichothecenes, ergot. International Program of Chemical Safety (IPCS). Environmental Health Criteria No. 105. WHO, Geneva, Switzerland.

JECFA. (1997). Toxicological evaluation of certain food additives. WHO Food Additives Series, Forty-Ninth meeting of the Joint FAO/WHO Expert Committee on Food Additives, Geneva, Switzerland.

JECFA. (2001). Evaluation of certain mycotoxins that may contaminate food. 56th meeting of the Joint FAO/WHO Expert Committee on Food Additives, Geneva, Switzerland.

Kamimura, H. (1989). Removal of mycotoxins during food processing. In: S. Natori, K. Hashimoto, Y. Ueno (Eds.), *Mycotoxins and Phytotoxins'88*. Elsevier Science Publishers, B. V., Amsterdam, the Netherlands, pp. 169–176.

Klaric, M.S., Rumora, L., Ljubanovic, D., Pepeljnjak, S. (2007). Cytotoxicity and apoptosis induced by fumonisin B_1, beauvericin and ochratoxin A in porcine kidney PK15 cells: Effects of individual and combined treatment. *Archives of Toxicology* **82**: 247–255.

Koynarski, V., Stoev, S., Grozeva, N., Mirtcheva, T., Daskalov, H., Mitev, J., Mantle, P. (2007). Experimental coccidiosis provoked by *Eimeria acervulina* in chicks simultaneously fed on ochratoxin A contaminated diet. *Research in Veterinary Science* **82**: 225–231.

Krogh, P. (1972). Mycotoxic porcine nephropathy: A possible model for Balkan endemic nephropathy. In: *Proceedings of the Second International Symposium on Endemic Nephropathy*, Sofia, Bulgaria, pp. 266–270.

Krogh, P., Hald, B., Pederson, J. (1973). Occurrence of ochratoxin A and citrinin in cereals associated with mycotoxic porcine nephropathy. *Acta Pathologica Mcrobiologica Scandinavica Section B* **81**: 689–695.

Kubena, L.F., Harvey, R.B., Bailey, R.H., Buckley, S.A., Rottinghaus, G.E. (1998). Effects of a hydrated sodium calcium aluminosilicate (T-Bind^{TM}) on mycotoxicosis in young broiler chickens. *Poultry Science* **77**: 1502–1509.

Kuiper-Goodman, T., Scott, P.M. (1989). Risk assessment of the mycotoxin ochratoxin A. *Biomedical and Environmental Sciences* **2**: 179–248.

Kumar, A., Jindal, N., Shukla, C.L., Pal, Y., Ledoux, D.R., Rottinghaus, G.E. (2003). Effect of ochratoxin A on *Escherichia coli*-challenged broiler chicks. *Avian Diseases* **47**: 415–424.

Laciakova, A., Cabadaj, R., Conkova, E., Pastorova, B. (1998). Degradation of ochratoxin A by feed additives. *Revue de Medecine Veterinaire* **149**: 567.

Lefebvre, H.P., Le Bars, J., Legrand, C., Le Bars, P., Dossin, O., Toutain, P.L., Braun, J.P. (1994). Three cases of equine stachybotryotoxicosis. *Revue de Medecine Veterinaire* **145**: 267–269.

Leitao, J., Bailly, J.R., de Saint Blanquant, G. (1990). Action of phosphine (PH_3) on production of sterigmatocystin by various fungal strains isolated from foodstuffs. *Food Additives and Contaminants* **7**: 26–28.

Lutsky, I.I., Mor, N. (1981). Alimentary toxic aleukia (septic angina, endemic panmyelotoxicosis, alimentary hemorrhagic aleukia): t-2 toxin-induced intoxication of cats. *American Journal of Pathology* **104**(2): 189–191.

Madsen, A., Hald, B., Mortensen, H.P. (1983). Feeding experiments with ochratoxin A contaminated barley for bacon pigs-3. Detoxification by ammoniation heating +NaOH, or autoclaving. *Acta Agriculturae Scandinavica* **33**: 171–175.

Marasas, W.F.O. (2001). Discovery and occurrence of the fumonisins: A historical perspective. *Environmental Health Perspectives* **109**: 239–243.

Marasas, W.F.O., Jaskiewics, K., Venter, F.S., van Schalkwyk, D.J. (1988). *Fusarium moniliforme* contamination of maize in oesophageal cancer areas in Transkei. *South African Medical Journal* **74**: 110–114.

Marasas, W.F.O., Kellerman, J.S., Pienaar, J.G., Naude, T.W. (1976). Leukoencephalomalacia: A mycotoxicosis of Equidae caused by *Fusarium moniliforme* Sheldon. *Onderstepoort Journal of Veterinary Research* **43**: 113–122.

Marquardt, R.R., Frolich, A.A. (1992). A review of recent advances in understanding ochratoxicosis. *Journal of Animal Science* **70**: 3968–3988.

Martinez, A.J., Monsalve, C. (1989). Aflatoxin occurrence in 1985–86 corn from Venezuela and its destruction by the extrusion process. In: C.E. O'Rear, G.C. Llewellyn (Eds.), *Biodeterioration Research 2*. Plenum Press, New York, pp. 251–259.

Micco, C., Miraglia, M., Onori, R., Libanori, A., Brera, C., Mantovani, A., Macri, C. (1991). Effect of combined exposure to ochratoxin A and penicillic acid on residues and toxicity in broilers. *La Ravista della Societa Italiana di Scienza dell'Allimentazione* **20**: 101–108.

Ministry of Agriculture, Fisheries and Food (MAFF). (1982). The Fertilisers and Feedingstuffs (Amendment) Regulations 1982. SI No. 386.

Missmer, S.A., Suarez, L., Felkner, M., Wang, E., Merrill, A.H. Jr., Rothman, K.J., Hendricks, K.A. (2006). Exposure to fumonisins and the occurrence of neural tube defects along the Texas-Mexico border. *Environmental Health Perspectives* **114**: 237–241.

Muller, H.M. (1983). A survey of methods of decontaminating mycotoxins. I. Physical methods. *Animal Research Development* **18**: 70–96.

Nordic Working Group on Food Toxicology and Risk Evaluation (NNT). (1991). *Nordiske Seminar-og Arbeidsrapporter* (1991:545). Health Evaluation of Ochratoxin A in Food Products, Nordic Council of Ministers, Copenhagen, Denmark, pp. 1–29.

Osborne, B.G., Ibe, F.I., Brown, G.L., Patagine, F., Scudamore, K.A., Banks, J.N., Hetmanski, M.T. (1996). The effects of milling and processing on wheat contaminated with Ochratoxin A. *Food Additives and Contaminants* **13**: 141–153.

Oswald, I.P., Comera, C. (1998). Immunotoxicity of mycotoxins. *Revue de Medecine Veterinaire* **149**: 585–590.

Pelletier, M.J., Reizner, J.R. (1992). Comparison of fluorescence sorting and colour sorting for the removal of aflatoxin from large group of peanuts. *Peanut Science* **19**: 15–20.

Peraica, M., Domijan, A.-M., Jurjevic, Z., Cvjetkovic, B. (2002). Prevention of exposure to mycotoxins from food and feed. *Archives of Industrial Hygiene and Toxicology* **53**: 229–237.

Peraica, M., Radic, B., Lucic, A., Pavlovic, M. (1999). Toxic effects of mycotoxins in humans. *Bulletin of the World Health Organization* **77**(9): 754–766.

Pfohl-Leszkowicz, A., Manderville, R. (2012). An update on direct genotoxicity as a molecular mechanism of ochratoxin A carcinogenicity. *Chemical Research in Toxicology* **25**: 252–262.

Pfohl-Leszkowicz, A., Molinié, A., Tozlovanu, M., Manderville, R.A. (2008). Combined toxic effects of ochratoxin A and citrinin, in vitro and in vivo. In: D.P. Siantar, M.W. Trucksess, P.M. Scott, E.M. Herman (Eds.), *Food Contaminants, Mycotoxins and Food Allergen*. ACS Symposium Series 1001, Chapter 3, pp. 56–80, Washington DC.

Pósa, R., Donkó, T., Bogner, P., Kovács, M., Repa, I., Magyar, T. (2011). Interaction of *Bordetella bronchiseptica, Pasteurella multocida* and fumonisin B1 in the porcine respiratory tract followed up by computed tomography. *Canadian Journal of Veterinary Research* **75**: 176–183.

Pósa, R., Magyar, T., Stoev, S.D., Glávits, R., Donkó, T., Repa, I., Kovács, M. (2013). Use of computed tomography and histopathologic review for lung lesions produced by the interaction between mycoplasma hyopneumoniae and fumonisin mycotoxins in pigs. *Veterinary Pathology* **50**(6): 971–979.

Price, R.L., Jorgensen, K.V. (1985). Effects of processing on aflatoxin levels and on mutagenic potential of tortillas made from naturally contaminated corn. *Journal of Food Science* **50**: 347–349.

Ramasamy, S., Wang, E., Hennig, B., Merrill, A.H. (1995). Fumonisin B_1 alters shingolipid metabolism and disrupts the barrier function of endothelial cells in culture. *Toxicology and Applied Pharmacology* **133**: 343–348.

Ramos, A.J., Fink-Gremmels, J., Hernandes, E. (1996). Prevention of toxic effects of mycotoxins by means of non-nutritive absorbent compounds. *Journal of Food Protection* **59**: 631–641.

Ramos, C.M., Martinez, E.M., Carrasco, A.C., Puente, J.H.L., Quezada, F., Perez, J.T., Oswald, I.P., Elvira, S.M. (2010). Experimental trial of the effect of fumonisin B_1 and the PRRS virus in swine. *Journal of Animal and Veterinary Advances* **9**: 1301–1310.

Reid, L.M., Zhu, X., Ma, B.L. (2001). Crop rotation and nitrogen effects on maize susceptibility to *Gibberella* (*Fusarium graminearum*) ear rot. *Plant Soil* **237**: 1–14.

Rheeder, J.P., Marasas, W.F.O., Thiel, P.G., Sydenham, E.W., Shephard, G.S., van Schalkwyk, D.J. (1992). *Fusarium moniliforme* and fumonisins in corn in relation to human oesophageal cancer in Transkei. *Phytopathology* **82**: 353–357.

Rosner, H. (1998). Mycotoxin regulations: An update. *Revue de Medecine Veterinaire* **149**: 679–680.

Rotter, R.G., Frohlich, A.A., Marquardt, R.R. (1989). Influence of dietary charcoal on ochratoxin A toxicity in Leghorn chicks. *Canadian Journal of Veterinary Research* **53**: 449–453.

Rotter, R.G., Marquardt, R.R., Frohlich, A.A., Abramson, D. (1990). Ensiling as a means of reducing ochratoxin A concentrations in contaminated barley. *Journal of the Science of Food and Agriculture* **50**: 155–156.

Samarajeewa, U. (1991). In situ degradation of mycotoxins by physical methods. In: J.E. Smith, R.S. Henderson (Eds.), *Mycotoxins and Animal Foods*. CRC Press, Boca Raton, FL, pp. 785–796.

Schneider, D.J., Marasas, W.F., Dale Kuys, J.C., Kriek, N.P., Van Schalkwyk, G.C. (1979). A field outbreak of suspected stachybotryotoxicosis in sheep. *Journal of South African Veterinary Association* **50**(2): 73–81.

Schneider, D.J., Miles, C., Garthwaite, I., van Halderen, A., Wessels, J.C., Lategan, H.J. (1996). First report of ergot-alkaloid toxicity in South Africa. *Onderstepoort Journal of Veterinary Research* **63**: 97–108.

Scott, P.M. (1996). Mycotoxins transmitted into beer from contaminated grain during brewing. *Journal of AOAC International* **79**: 875–882.

Scott, P.M. (1998). Industrial and farm detoxification processes for mycotoxins. *Revue de Medecine Veterinaire* **149**: 543–548.

Scott, P.M., Lawrence, G.A. (1994). Analysis of beer for fumonisins. *Journal of Food Protection* **58**: 1379–1382.

Scudamore, K.A., Banks, J.N. (2004). The fate of mycotoxins during cereal processing. In: D. Barug, H. van Egmond, R. López-García, T. van Osenbruggen, A. Visconti (Eds.), *Proceedings of the Second World Mycotoxin Forum: Meeting the Mycotoxin Menace*, February 17–18, 2003, Wageningen Academic Publishers, Nordwijk, the Netherlands, pp. 165–181.

Scudamore, K.A., Nawaz, S., Hetmanski, M.T. (1998). Mycotoxins in ingredients of animal feeding stuffs: II. Determination of mycotoxins in maize and maize products. *Food Additives and Contaminants* **15**: 30–55.

Scudamore, K.A., Patel, S., Breeze, V. (1999). Surveillance of stored grain from the 1997 harvest in the United Kingdom for ochratoxin A. *Food Additives and Contaminants* **16**: 281–290.

Sewram, V., Shephard, G.S., Marasas, W.F.O., Penteado, M., De Castro, M.F. (2003). Improving extraction of Fumonisin mycotoxins from Brazilian corn-based infant foods. *Journal of Food Protection* **66**: 85–859.

Sharma, R.P. (1993). Immunotoxicity of mycotoxins. *Journal of Dairy Science* **76**: 892–897.

Shibamato, T., Bjeldanes, L.F. (1993). Fungal toxins occurring in foods. In: S. Taylor (Ed.), *Introduction to Food Toxicology*. Academic Press Inc., London, U.K., pp. 97–116.

Shier, W.T. (1998). Estrogenic mycotoxins. *Revue de Medecine Veterinaire* **149**: 599–604.

Sibanda, L., Marovatsanga, L.T., Pestka, J.J. (1997). Review of mycotoxin work in sub-Saharan Africa. *Food Control* **8**: 21–29.

Sreemannarayana, O., Frohlich, A.A., Vitti, T.G., Marquardt, R.R., Abramson, D. (1988). Studies of the tolerance and disposition of ochratoxin A in young calves. *Journal of Animal Science* **66**: 1703–1711.

Steiner, W.E., Rieker, R.H., Battaglia, R. (1988). Aflatoxin contamination in dried figs: Distribution and association with fluorescence. *Journal of Agriculture and Food Chemistry* **36**: 88–91.

Stoev, S., Petkova-Bocharova, T.K.A. (1994). A possible role of ochratoxin A in a disease causation in connection with Balkan endemic nephropathy. In: *Proceedings of the Eighth International Congress on Animal Hygiene*, St. Paul, MN, September 12–16, 1994, pp. 61–64.

Stoev, S.D. (1998). The role of Ochratoxin A as a possible cause of Balkan endemic nephropathy and its risk evaluation. *Veterinary and Human Toxicology* **40**: 352–360.

Stoev, S.D. (2007). Food Safety and some foodborne mycotoxicoses, *Veterinary Africa 2007 Congress*, July 27–28, 2007, Johannesburg, South Africa.

Stoev, S.D. (2008). Complex etiology, prophylaxis and hygiene control in mycotoxic nephropathies in farm animals and humans, Special Issue "Mycotoxins: Mechanisms of Toxicological Activity - Treatment and Prevention", Section "Molecular Pathology". *International Journal of Molecular Sciences* **9**: 578–605.

Stoev, S.D. (2010a). Studies on carcinogenic and toxic effects of ochratoxin A in chicks. Special issue "Ochratoxins". *Toxins* **2**: 649–664.

Stoev, S.D. (2010b). Studies on some feed additives and materials giving partial protection against the suppressive effect of ochratoxin A on egg production of laying hens. *Research in Veterinary Science* **88**: 486–491.

Stoev, S.D. (2013a). Food safety and increasing hazard of mycotoxin occurrence in foods and feeds. *Critical Reviews in Food Science and Nutrition* **53**(9): 887–901.

Stoev, S.D. (2013b). The specific multi-Mycotoxic nature of some foodborne mycotoxicoses and the hazard for animals or humans, *Mycotoxins Workshop 2013*: *Mycotoxins Research in Food*: *Challenges and Perspectives*, October 7–8, 2013, Department of Animal Health, North West University, Mafikeng Campus, Mafikeng, South Africa.

Stoev, S.D., Angelov, G., Pavlov, D., Pirovski, L. (1999). Some antidotes and paraclinical investigations in experimental intoxication with ochratoxin A and penicillic acid in chicks. *Veterinarski Arhives* **69**: 179–189.

Stoev, S.D., Anguelov, G., Ivanov, I., Pavlov, D. (2000a). Influence of ochratoxin A and an extract of artichoke on the vaccinal immunity and health in broiler chicks. *Experimental and Toxicologic Pathology* **52**: 43–55.

Stoev, S.D., Daskalov, H., Radic, B., Domijan, A., Peraica, M. (2002a). Spontaneous mycotoxic nephropathy in Bulgarian chickens with unclarified mycotoxin aetiology. *Veterinary Research* **33**: 83–94.

Stoev, S.D., Denev, S. (2013). Porcine/chicken or human nephropathy as the result of joint mycotoxins interaction. Special issue "Recent Advances in Ochratoxins Research". *Toxins* **5**(9): 1503–1530.

Stoev, S.D., Denev, S., Dutton, M.F., Njobeh, P.B., Mosonik, J.S., Steenkamp, P., Petkov, I., (2010b). Complex etiology and pathology of mycotoxic nephropathy in South African pigs. *Mycotoxin Research* **26**: 31–46.

Stoev, S.D., Djuvinov, D., Mirtcheva, T., Pavlov, D., Mantle, P. (2002d). Studies on some feed additives giving partial protection against ochratoxin A toxicity in chicks. *Toxicology Letters* **135**: 33–50.

Stoev, S.D., Dutton, M., Njobeh, P., Mosonik, J., Steenkamp, P. (2010a). Mycotoxic nephropathy in Bulgarian pigs and chickens: Complex aetiology and similarity to Balkan Endemic Nephropathy. *Food Additives and Contaminants Part A* **27**: 72–88.

Stoev, S.D., Goundasheva, D., Mirtcheva, T., Mantle, P.G. (2000b). Susceptibility to secondary bacterial infections in growing pigs as an early response in ochratoxicosis. *Experimental and Toxicologic Pathology* **52**: 287–296.

Stoev, S.D., Grozeva, N., Hald, B. (1998b). Ultrastructural and toxicological investigations in spontaneous cases of porcine nephropathy in Bulgaria. *Veterinarski Arhives* **68**: 39–49.

Stoev, S.D., Gundasheva, D., Zarkov, I., Mircheva, T., Zapryanova, D., Denev, S., Mitev, Y. et al. (2012). Experimental mycotoxic nephropathy in pigs provoked by a mouldy diet containing ochratoxin A and fumonisin B_1. *Experimental and Toxicologic Pathology* **64**: 733–741.

Stoev, S.D., Hald, B., Mantle, P. (1998a). Porcine nephropathy in Bulgaria: A progressive syndrome of complex of uncertain (mycotoxin) etiology. *The Veterinary Record* **142**: 190–194.

Stoev, S.D., Koynarsky, V., Mantle, P.G. (2002b). Clinicomorphological studies in chicks fed ochratoxin A while simultaneously developing coccidiosis. *Veterinary Research Communications* **26**: 189–204.

Stoev, S.D., Paskalev, M., MacDonald, S., Mantle, P.G. (2002c). Experimental one year ochratoxin A toxicosis in pigs. *Experimental and Toxicologic Pathology* **53**: 481–487.

Stoev, S.D., Stefanov, M., Denev, S., Radic, B., Domijan, A.-M., Peraica, M. (2004). Experimental mycotoxicosis in chickens induced by ochratoxin A and penicillic acid and intervention by natural plant extracts. *Veterinary Research Communications* **28**: 727–746.

Stoev, S.D., Stoeva, J., Anguelov, G., Hald, B., Creppy, E.E., Radic, B. (1998c). Haematological, biochemical and toxicological investigations in spontaneous cases with different frequency of porcine nephropathy in Bulgaria. *Journal of Veterinary Medicine Series A* **45**: 229–236.

Stoev, S.D., Vitanov, S., Anguelov, G., Petkova-Bocharova, T., Creppy, E.E. (2001). Experimental mycotoxic nephropathy in pigs provoked by a mouldy diet containing ochratoxin A and penicillic acid. *Veterinary Research Communications* **25**: 205–223.

Stoloff, L., Trucksess, M.W. (1981). Effect of boiling, frying, and baking on recovery of aflatoxin from naturally contaminated corn grits or cornmeal. *Journal of Association of Official Analytical Chemisry* **64**: 678–680.

Sydenham, E.W., Thiel, P.G., Marasas, W.F.O., Shepard, G.S., van Schalkwyk, D.J., Koch, K.R. (1990). Natural occurrence of some *Fusarium* mycotoxins in corn from low and high oesophageal cancer prevalence areas of the Transkei, southern Africa. *Journal of Agricultural and Food Chemistry* **38**: 1900–1903.

Tai, J.H., Pestka, J.J. (1988). Impaired murine resistance to *Salmonella typhimurium* following oral exposure to the trichothecene T-2 toxin. *Food and Chemical Toxicology* **26**: 691–698

Taranu, I., Marin, D.E., Bouhet, S., Pascale, F., Bailly, J.D., Miller, D.J., Pinton, P., Oswald, I.P. (2005). Mycotoxin fumonisin B1 alters the cytokine profile and decreases the vaccinal antibody titer in pigs. *Toxicological Sciences* **84**: 301–307.

Tong, C., Draughon, H. (1985). Inhibition by an antimicrobial food additives of ochratoxin A production by *Aspergillus sulphureus* and *Penicillium viridicatum. Applied Environmental Microbiology* **49**: 1407–1411.

van Egmond, H.P. (1989). Current situation on regulations for mycotoxins: Overview of tolerances and status of standard methods of sampling and analysis. *Food Additives and Contaminants* **6**: 139–188.

van Waes, J.G., Starr, L., Maddox, J., Aleman, F., Voss, K.A., Wilberding, J., Riley, R.T. (2005). Maternal fumonisin exposure and risk for neural tube defects: Mechanisms in an in vivo mouse model. *Birth Defects Research. Part A. Clinical and Molecular Teratology* **73**: 487–497.

WHO. (1991). Evaluation of certain food additives and contaminants. Thirty-seventh report of the joint FAO/WHO Expert Committee on Food Additives (WHO technical report series 806). World Health Organization, Geneva, Switzerland, pp. 29–31.

WHO/FAO. (1995). Report of the Joint FAO/WHO Expert Consultation on Application of Risk Analysis to Food Standards Issues. WHO/FNU/FOS 95.3, Geneva, Switzerland.

Ziprin, R.I., Holt, P.S., Mortensen, R. (1987). T-2 toxin effects on the serum amyloid P-component (SAP) response of *Listeria monocytogenes* and *Salmonella typhimurium* infected mice. *Toxicology Letters* **39**: 177–184.

10 Exposure of Radionuclides in Foods

Major Challenges

Francisco Javier Guillén Gerada

CONTENTS

10.1 INTRODUCTION

Radionuclides are present in foodstuffs as a consequence of their occurrence in the environment. They can be classified according to their origin: anthropogenic radionuclides released due to nuclear accidents, such as Chernobyl and Fukushima, and naturally occurring radionuclides, related mainly to a different geologic composition of the earth's crust. These radionuclides can be transferred into foodstuffs and/or accumulated in foodstuffs, the main pathway for their incorporation in humans. Thus, the protection of population against exposure of radionuclides in foodstuffs follows the general reference level of 1 mSv/year above background for all sources of ionizing radiation (ICRP, 1999). This protection is enforced by international treaties. Consolidated version of Articles 35 and 36 of EURATOM treaty takes the public protection to radioactivity into account (EU, 2010).

> Article 35: Each member state shall establish the facilities necessary to carry out continuous monitoring of the level of radioactivity in air, water, and soil and to ensure compliance with the basic standards. The Commission shall have the right of access to such facilities; it may verify their operation and efficiency.
>
> Article 36: The appropriate authorities shall periodically communicate information on the checks referred to in Article 35 to the Commission so that it is kept informed of the level of radioactivity to which the public is exposed.

To accomplish these articles, member states of EURATOM treaty are obliged to inform the European Commission of the radionuclide content in foodstuffs, and several monitoring programs are carried out (Máté et al., 2015). The quality of the measurements is also considered as a part of these programs by proficiency tests (Merešová and Wätjen, 2014; Wätjen et al., 2012) and reference materials (Wätjen et al., 2014).

The annual effective dose due to foodstuff consumption is influenced by the radiotoxicity of radionuclides present, how much foodstuff is consumed annually, and the activity level of those radionuclides. There are two main scenarios: (1) following a nuclear accident or any other case of radiological emergency and (2) existing or routine exposure. A practical way to achieve the goal of protecting the public is to set reference/guidance levels by international organisms (European Union [EU], IAEA, FAO/WHO, Codex Alimentarius [CA]) to be used in trade and. In this chapter, a critical analysis of these regulations to detect any possibilities for improvement is presented.

10.2 OCCURRENCE OF RADIONUCLIDES IN THE ENVIRONMENT

The occurrence of radionuclides in the environment is usually categorized according to their origin: naturally occurring and anthropogenic. The first group includes those radionuclides belonging to the natural series of uranium, thorium, and actinium, cosmogenic radionuclides, and ^{40}K. The latter constitutes 0.012% of natural potassium. Cosmogenic radionuclides (^{7}Be, ^{10}Be, ^{22}Na, etc.) are formed by the interaction of high-energy cosmic rays with atomic nuclei in the atmosphere. The radionuclides belonging to the natural of uranium (^{238}U, ^{234}U, ^{230}Th, ^{226}Ra, ^{222}Rn, ^{210}Pb, ^{210}Po, etc.), thorium (^{232}Th, ^{228}Th, ^{228}Ra, ^{224}Ra, ^{220}Rn, etc.), and actinium (^{235}U being the head of the series) are distributed worldwide. However, there are certain geological processes and formations that result in a local enhancement of these radionuclides. The industrial use of these deposits as raw materials is known as NORM (naturally occurring radioactive material) industries. Mining industries (uranium, iron, copper, etc.), the production of fertilizers from phosphate rocks, and burning nonnuclear fuels (coal, oil, gas) are some examples of NORM activities (IAEA, 2003). The by-products and residues generated in these industries could have enhanced the concentration of naturally occurring radionuclides due to the chemical processes they were subjected to and/or the way they were managed. In some cases, NORM industries can pose a radiological hazard.

Anthropogenic radionuclides were first introduced into the environment in 1945 as a consequence of the atom bomb blasts during World War II. During the 1950s and 1960s, until its ban in 1963, 520 atmospheric nuclear weapon tests took place, releasing huge amounts of radionuclides into the atmosphere (see Table 10.1). Due to atmospheric air circulation, the released radionuclides became distributed worldwide, which is usually known as global fallout (UNSCEAR, 2000). As the majority of tests took place in the northern hemisphere, the deposition of radionuclides was greater than that in the southern hemisphere. The deposition was also inhomogeneous in latitude because of circulation patterns and airflows in the stratosphere and troposphere, being greater in the temperate regions, with a maximum deposition of ^{90}Sr in the 40°–50° latitude bands, and lower in the equatorial and polar regions (UNSCEAR, 2000).

Anthropogenic radionuclides were also released into the environment as effluents from nuclear fuel reprocessing factories, plutonium fabrication plants, nuclear waste storage sites, and accidents in nuclear reactors such as Windscale, Three Mile Island, Chelyabinsk, Chernobyl, and, more recently, Fukushima. Chernobyl accident (April 26, 1986) was the most serious accident occurred, marking a point of inflexion for radioecology and radioprotection. The most heavily contaminated areas (>3700 kBq/m^2 of ^{137}Cs) were located mostly within the 30 km radius around Chernobyl. However, it also seriously affected other countries such as Finland, Sweden, Norway, Germany, and

TABLE 10.1
Estimation of Activities of ^{131}I, ^{137}Cs, and ^{90}Sr Released in Global Fallout and Chernobyl and Fukushima Daiichi Accidents

Event	Year	^{131}I (PBq)	^{137}Cs (PBq)	^{90}Sr (PBq)	Reference
Global fallout	1950–1960	675,000	948	622	UNSCEAR (2000)
Chernobyl	1986	1,800	85	10	UNSCEAR (2000)
Fukushima Daiichi	2011	100–500	6–20	1	IAEA (2014a)

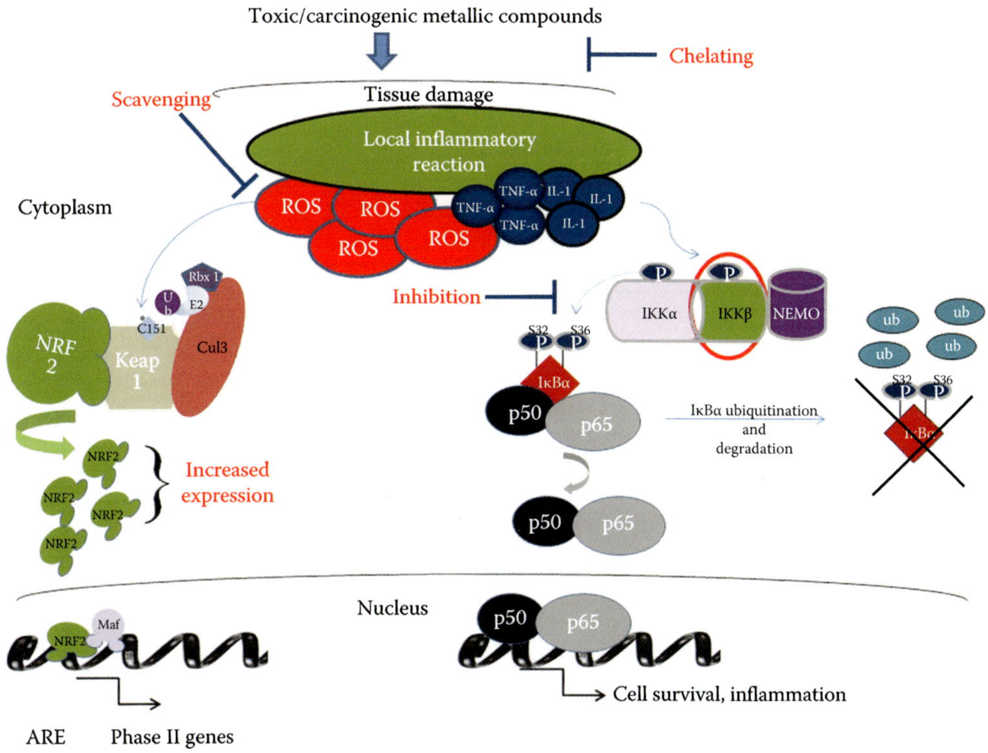

FIGURE 3.1 Multiple protective effects of curcumin related to some toxic food contaminants. The desirable effects of curcumin, indicated by text written in red, include chelating of potentially carcinogenic heavy metals, scavenging of ROS, inhibition of inflammatory signaling, and increased expression of adaptive defense system against oxidative stress.

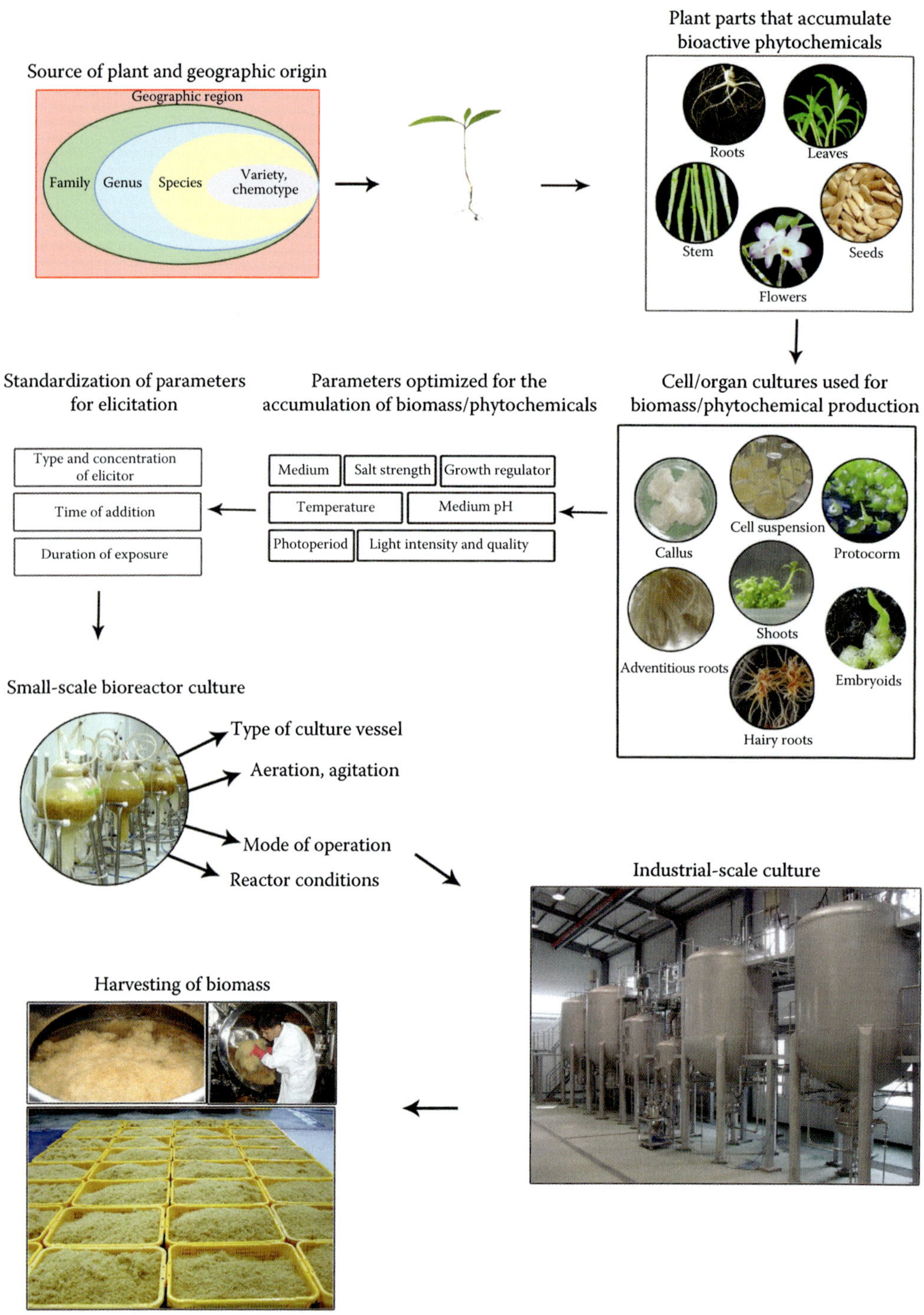

FIGURE 8.1 A schematic depiction of the plant cell, tissue, and organ culture (PCTOC) processes for the production of PCTOC raw material.

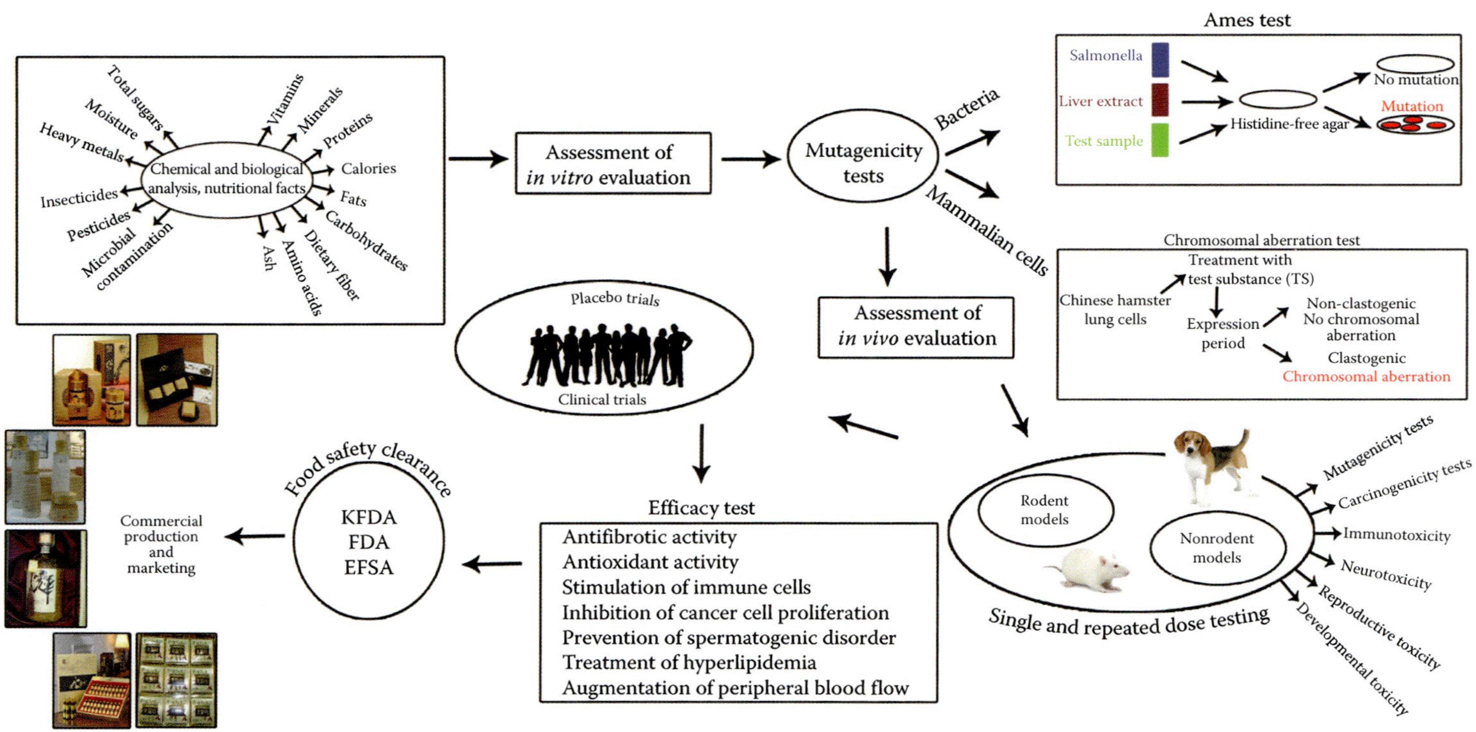

FIGURE 8.3 A schematic depiction of the biosafety evaluation of the plant cell, tissue, and organ culture raw material/product.

FIGURE 9.1 An inappropriate store in rural area of Limpopo Province of South Africa, which would not be able to preserve feed from rain.

FIGURE 9.2 An inappropriate manner for drying or preserving of maize flour in the rural area of Limpopo Province of South Africa.

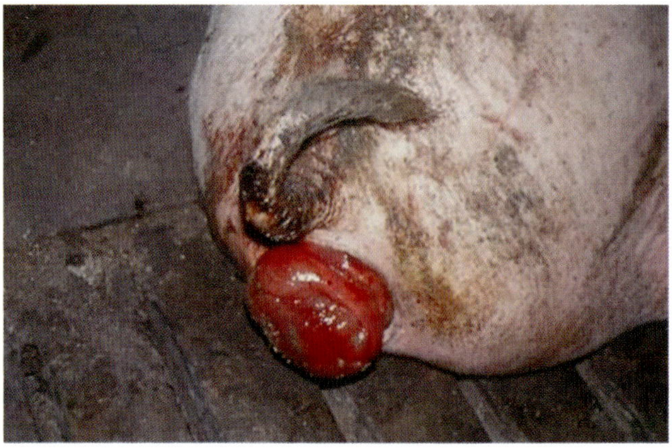

FIGURE 9.3 A rectal prolapse in spontaneous case of fusariotoxicosis in pig from Bulgaria due to the estrogenic effect of mycotoxin zearalenone.

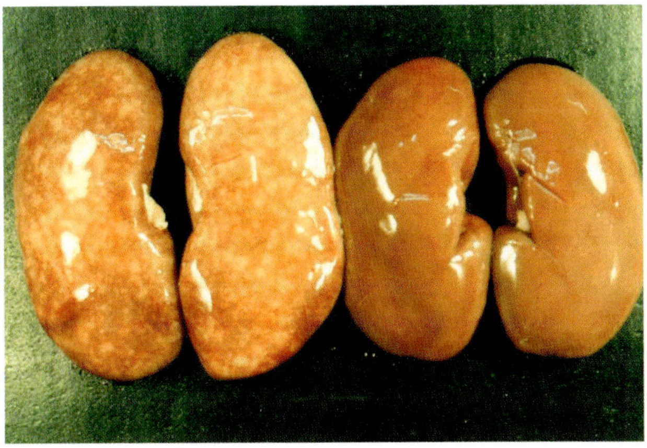

FIGURE 9.4 Spontaneous case of mycotoxic porcine nephropathy identified at the slaughterhouse in Bulgaria. Enlarged and mottled appearance of kidneys from a 6- to 8-month-old pig (left) and normal kidneys in a pig of the same age (right).

FIGURE 9.5 Nervous symptoms such as flexion of legs and sitting posture in chicks fed on moldy diet containing 790 ppb ochratoxin A and 2000–5000 ppb penicillic acid for 10 months.

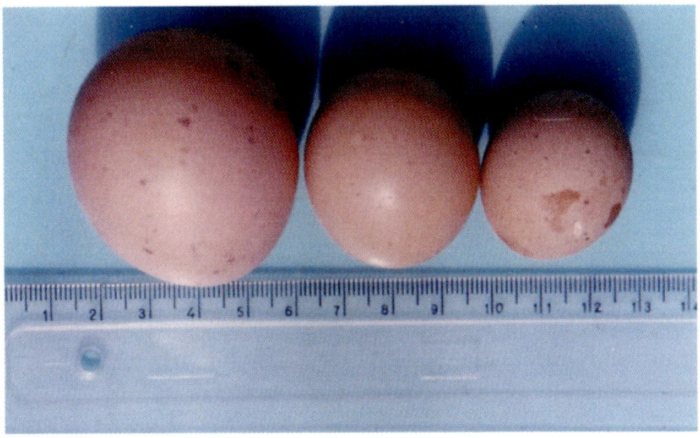

FIGURE 9.6 Small size of eggs from laying hens exposed to 5 ppm OTA in a diet with the weight of the eggs of 15.8 g (center) and 25.8 g (right). Normal size of an egg from a laying hen without OTA treatment (left).

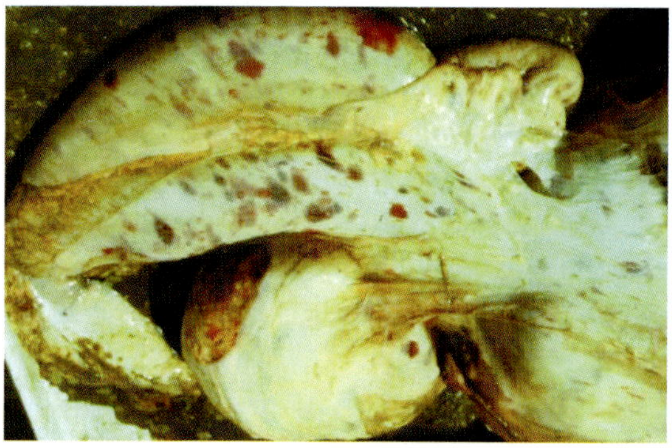

FIGURE 9.7 Spontaneous case of stachybotryotoxicosis in cattle. Strong hemorrhagic diathesis on the serosa of the abomasus and rumen and ulcerative necrotic damages, which are translucent under the serosa.

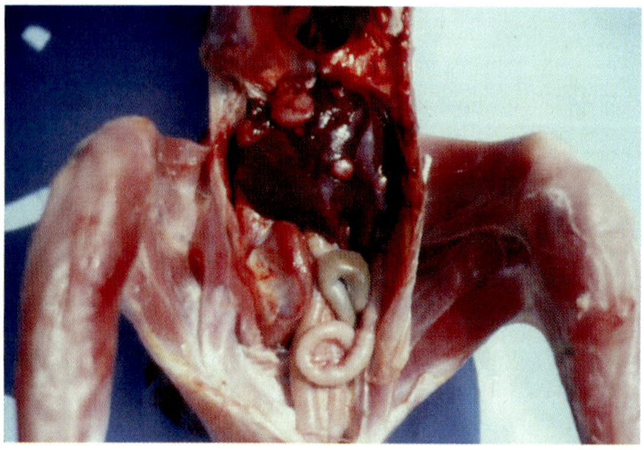

FIGURE 9.8 Adenocarcinoma in the liver of a chick exposed to 5 ppm ochratoxin A for 10 months. Large gray-white neoplastic foci are seen in the liver and protruded significantly above its surface.

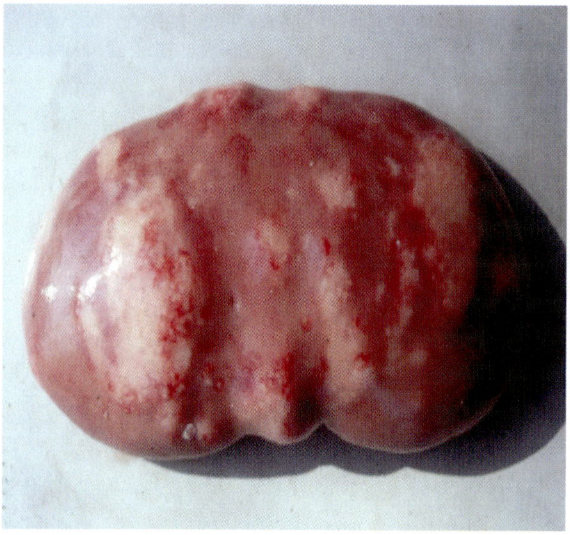

FIGURE 9.9 Kidney with neoplastic tissue (fibroadenoma)—a spontaneous case of MPN in Bulgaria.

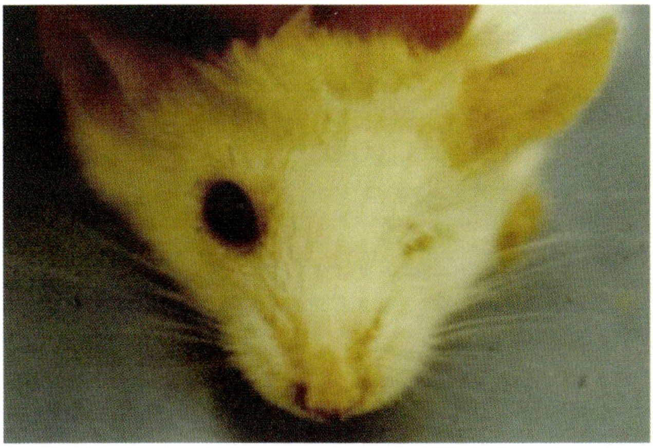

FIGURE 9.10 Malformation (monophthalmia) in mouse, whose mother was exposed to 10 ppm ochratoxin A via the feed up to day eighth of the pregnancy.

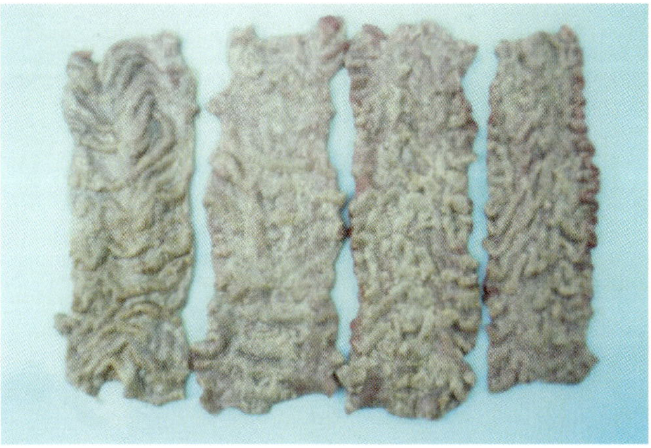

FIGURE 9.11 Diphtheroid accretions on colon mucosa of a pig given 3 ppm ochratoxin A for 17 days and sick by salmonellosis.

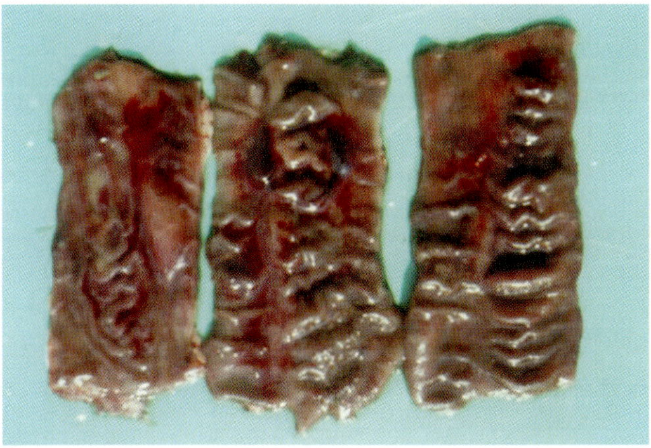

FIGURE 9.12 Hemorrhagic dysentery in a pig 47 days after commencing a diet containing 1 ppm ochratoxin A.

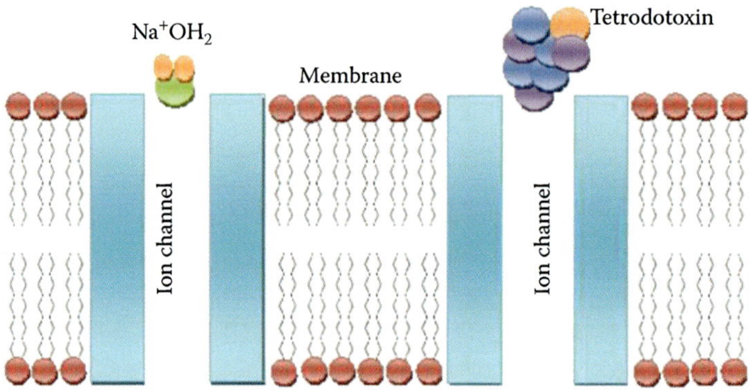

FIGURE 12.1 Plasma membrane with ion channels and hydrated sodium ion and tetrodotoxin.

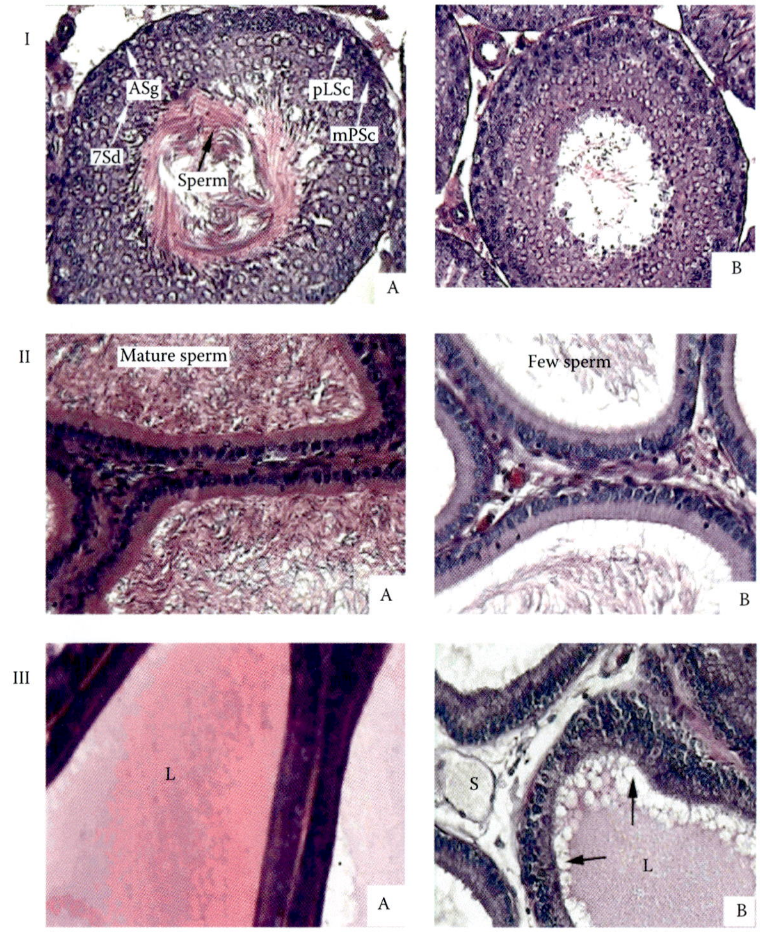

FIGURE 18.2 Photomicrographs of paraffin-embedded hematoxylin and eosin–stained histopathological sections. (IA) Rat testicular section (400×) from control animals. (IB) Testis section (400×) from vanadate-treated animals (0.4 mg/kg, i.p. for 26 days). ASg, spermatogonia A; pLSc, preleptotene spermatocytes; mPSc, mid-pachytene spermatocytes; 7Sd, step 7 spermatid. (IIA) Rat cauda epididymis section (400×) from control animals. (IIB) Cauda epididymis section (400×) from vanadate-treated animals (0.4 mg/kg, i.p. for 26 days). (IIIA) Rat ventral prostate gland section (400×) from control animals. (IIIB) Ventral prostate gland section (400×) from vanadate-treated animals (0.4 mg/kg, i.p. for 26 days). L, lumen; S, stroma. Arrows indicate vacuoles.

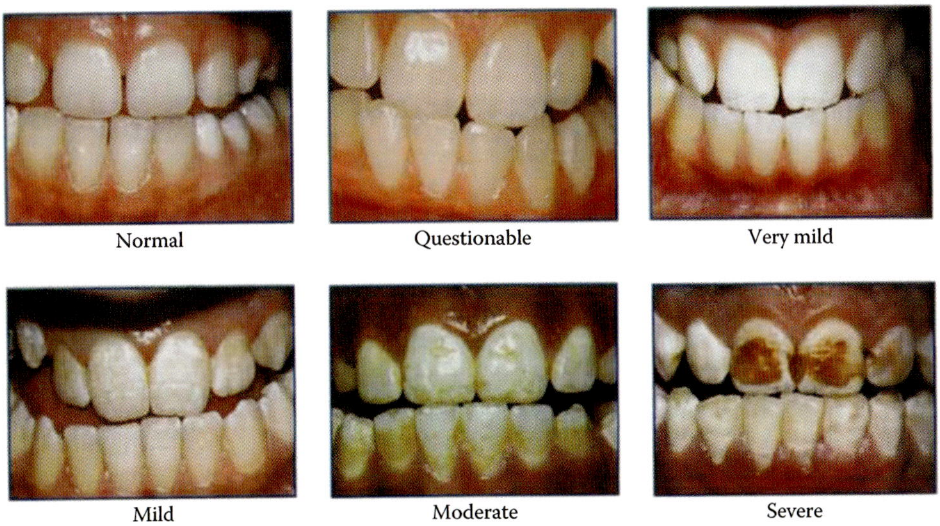

FIGURE 19.4 Dean classification used to categorize dental fluorosis. (Dean et al., 1942.)

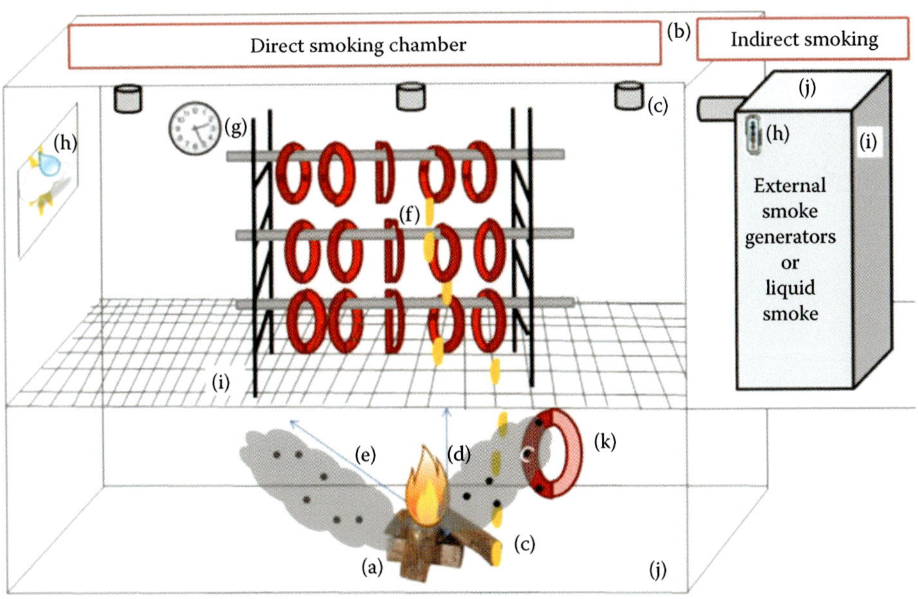

FIGURE 21.6 Smoking chamber representing the process variables mentioned by Codex Alimentarius (CAC/RCP 68-2009) to control the formation and contamination of polycyclic aromatic hydrocarbons. (a) Fuel, (b) direct or indirect (c) smoke generation process (d) distance to product (e) position to product (f) fat content and dripping (g) duration of smoking (h) temperature (i) cleanliness of equipment (j) design of chamber (k) casing (variable not listed in CAC) (•) tar (bio-oil). (From Ledesma, E. et al., *Food Control*, 60, 64, 2016. With permission.)

Austria, where there were areas in which about 40–185 kBq/m^2 of ^{137}Cs was deposited (UNSCEAR, 2000). The Fukushima Daiichi accident (2011) in Japan also released significant quantities of radionuclides into the atmosphere and ocean. Table 10.1 shows the estimated activity level of ^{131}I, ^{137}Cs, and ^{90}Sr released in global fallout and Chernobyl and Fukushima accidents. These radionuclides are the most important ones from a radiological point of view.

10.3 EFFECTIVE DOSE DUE TO INGESTION OF FOODSTUFFS

The effect of ionizing radiation on health is quantified using the concept of dose rate, which takes into account the energy deposited by an α-, β-particle, or γ-ray per unit of mass. As the biological effects of ionizing radiation in different tissues vary, the effective committed dose, which is the tissue-weighted sum of the equivalent doses in all tissues and organs of the body, is used to represent the stochastic health risk associated with the ingestion of radionuclides and an integration time of 50 year for adults. In the case of internal exposure due to ingestion of foodstuffs, it can be expressed by the following expression:

$$D(\text{Sv/year}) = \sum_{i} e_i(g)(\text{Sv/Bq}) \cdot A_i(\text{Bq/kg}) \cdot m(\text{kg/year}) \tag{10.1}$$

where

$e_i(g)$ is the effective committed dose per unit uptake of the ingested radionuclide j for an individual belonging to age group g and an integration time of 50 year for adults and up to 70 year for children (ICRP, 2012, 2013)

A_i is the specific activity of the radionuclide i that the foodstuff contains per fresh mass

m is the fresh mass of the foodstuff ingested per year by a normal individual

The summation extends over all radionuclides present in foodstuffs. In the case of the protection of the public, a reference level for the effective dose of 1 mSv is recommended by the International Commission on Radiological Protection (ICRP), considering all radiation exposure situations (ICRP, 1999). As it can be derived from Equation 10.1, the exposure to the radionuclides present in foodstuffs depends on the following three main variables:

1. Which radionuclide is present?
2. At what concentration?
3. How much food is consumed?

10.4 RADIOTOXICITY OF RADIONUCLIDES DUE TO INGESTION

Thus, the first concern when analyzing the exposure to radionuclides in foodstuffs is to determine which ones are present at the highest content. The dose coefficients, e(g), can be used to establish the radiotoxicity of the radionuclides and hence its potential risk. Table 10.2 shows, as a way of example, the dose coefficients for adults, which take into account the decay schemes and the subsequent emission of ionizing radiations of a given radionuclide. It decreases in the following order:

$$^{210}\text{Po} > {}^{210}\text{Pb}, {}^{228}\text{Ra} > {}^{226}\text{Ra} > {}^{232,230,228}\text{Th} > {}^{224}\text{Ra} > {}^{234,235,238}\text{U} > {}^{90}\text{Sr} > {}^{131}\text{I} > {}^{134,137}\text{Cs} > \cdots$$

An important fact of the effective dose is that it is not an individual specific dose quantity but independent of the individual's body-related parameters. They are referred to a reference man, which is a hypothetical entity defined by anatomical and physiological parameters of reference individuals serving as a basis for biokinetics model describing the fate of the ingested radionuclides (ICRP, 2012).

TABLE 10.2
Dose Coefficients h(g) for Adults (>17 y of Age), Expressed in Sv/Bq (ICRP, 2012, 2013), and Half-Life ($T_{½}$) of Different Anthropogenic and Naturally Occurring Radionuclides

Anthropogenic Radionuclides			Naturally Occurring Radionuclides		
Radionuclide	**$T_{½}$**	**e(g) (Sv/Bq)**	**Radionuclide**	**$T_{½}$**	**e(g) (Sv/Bq)**
$^{3}H^{a}$	12.35 y	$4.2·10^{-11}$	$^{3}H^{a}$	12.35 y	$4.2·10^{-11}$
^{14}C	5730 y	$5.8·10^{-10}$	^{7}Be	53.3 d	$2.8·10^{-11}$
$^{35}S^{b}$	87.44 d	$7.7·10^{-10}$	^{40}K	$1.28·10^{9}$ y	$6.2·10^{-9}$
^{60}Co	5.271 y	$3.4·10^{-9}$	^{210}Po	138.38 d	$1.2·10^{-6}$
^{89}Sr	50.5 d	$2.6·10^{-9}$	^{210}Pb	22.3	$6.9·10^{-7}$
^{90}Sr	29.12 y	$2.8·10^{-8}$	^{224}Ra	3.66 d	$6.5·10^{-8}$
^{99}Tc	$2.13·10^{5}$ y	$6.4·10^{-10}$	^{226}Ra	1600 y	$2.8·10^{-7}$
^{103}Ru	39.28 d	$7.3·10^{-10}$	^{228}Ra	5.75 y	$6.9·10^{-7}$
^{106}Ru	368.2 d	$7.0·10^{-9}$	^{228}Th	1.9131 y	$7.2·10^{-8}$
^{110m}Ag	249.9 d	$2.8·10^{-9}$	^{230}Th	$7.7·10^{4}$ y	$2.1·10^{-7}$
^{125}Sb	2.77 y	$1.1·10^{-9}$	^{232}Th	$1.4·10^{10}$ y	$2.3·10^{-7}$
^{131}I	8.04 d	$2.2·10^{-8}$	^{234}U	$2.4·10^{5}$ y	$4.9·10^{-8}$
^{134}Cs	2.062 y	$1.9·10^{-8}$	^{235}U	$7.0·10^{8}$ y	$4.7·10^{-8}$
^{137}Cs	30.0 y	$1.3·10^{-8}$	^{238}U	$4.4·10^{9}$ y	$4.5·10^{-8}$
^{192}Ir	74.02 d	$1.4·10^{-9}$			
^{238}Pu	87.74 y	$2.3·10^{-7}$			
^{239}Pu	$2.4·10^{4}$ y	$2.5·10^{-7}$			
^{240}Pu	6537 y	$2.5·10^{-7}$			
^{241}Am	432.2 y	$2.0·10^{-7}$			

[a] For organically bound tritium.
[b] For organic sulfur.

10.5 DIET COMPOSITION

The annual food consumption is another key factor that can influence the dose. A number of surveys have been carried out to analyze the nutritional habits of different countries (Byrom et al., 1995; Aarkrog et al., 2000; Suáñez and Robles, 2002; Beresford et al., 2001; MAGRAMA, 2009; Carini and Pellizzoni, 2011; Awudu et al., 2012). Table 10.3 shows average values for some of the main groups of foodstuff in Spain and in the United Kingdom. It can be observed that the diet varies from country to

TABLE 10.3
Average Values of Foodstuff Consumption, Expressed in kg f.w. person/y, for Adults

	Spain		United Kingdom
Foodstuffs	**(Suáñez and Robles, 2002)**	**(MAGRAMA, 2009)**	**(Byrom et al., 1995)**
Dairy	126.1	112.5	115
Cereals	67.5	38	50
Vegetables	101.4	81.7	70
Fruits	76.5	94.9	40.8
Meat	68.8	50	32.5
Fish	35.9	27.4	10.1
Mushrooms	1.3	—	1.0

country due to different age groups and culinary customs and also according to the type of population (urban or rural). Rural populations used to consume a higher percentage of local and/or self-produced vegetables (Carini and Pellizzoni, 2011). Differences between the average values obtained from different surveys were also observed. Regarding minority foodstuffs, such as mushrooms, the preferences of certain types of population also influence the consumption, ranging from 0.06 to 58.4 kg f.w./year worldwide and 0.06 to 25.68 kg f.w./year for the United Kingdom (Guillén and Baeza, 2014).

There are different approaches used by different countries for a sampling diet, Total Diet Studies, to assess the radiation exposure of the population, which are as follows:

- Sampling all food groups, except beverages, in proportion derived from consumption studies, used in the United Kingdom
- Analyzing individual parts of the diet in a "mixed basket" approach, used in the United States (USFDA, 2006)
- Collecting all food consumed by a person during a certain period of time and homogenizing it, used in Spain

Another aspect to be considered in the diet is the effect of global market, that is, the import and export of foodstuffs. In this sense, the CA assumed that 10% of the diet is of imported food for deriving guideline levels (CA, 2011). Other assumptions are usually made in order to minimize the effect of variation in food consumption. CA also assumes that an adult consumes 550 kg of food/year and infant consumes 220 kg (CA, 2011). Consumption of 730 L of water/year for adults is considered for the calculation of the indicative dose (ID) of water intended for human consumption (EU, 2013).

10.6 RADIONUCLIDE CONTENT IN FOODSTUFFS: REFERENCE/GUIDANCE LEVELS

To assure the population protection from ionizing radiation, ICRP considers that public should not be exposed to effective doses higher than 1 mSv/year above the level of background radiation (ICRP, 1999). This reference level comprises all kind of exposures, including external and internal. Therefore, it also sets a reference level to exposure to radionuclides in food, which is considered to be within the level of internal exposure. The accomplishment of this level can be achieved using the definition of the effective dose (Equation 10.1). However, this poses difficulties in practice, because a strict interpretation implies that all foodstuffs should be assayed for all radionuclides, which can be more complicated taking into account that the composition of the diet can vary from country to country and also in time. A solution to these problems is to use reference levels or guidance levels that have been implemented by different international organizations, such as the EU (EU 1989a,b) or CA, with consultations to FAO, WHO, and IAEA (CA, 2011). These levels are based on the fact that the intake of a single radionuclide does not imply an annual effective dose higher than 1 mSv (ICRP, 1999).

The population exposure to radiation sources, in particular to foodstuffs, can be classified into the following two main categories (IAEA, 2014b):

1. Following a nuclear accident or any other case of radiological emergency, defined as "a nonroutine situation that necessitates prompt action, primarily to mitigate a hazard or adverse consequences for human health and safety, quality of life, property, or the environment. (…) It includes situations for which prompt action is warranted to mitigate the effects of a perceived hazard due to (a) the energy resulting from a nuclear chain reaction or from the decay of the products of a chain reaction or (b) radiation exposure."
2. Existing or routine exposure due to (a) contamination of areas by residual radioactive materials deriving from past activities or after an emergency has been declared as ended; (b) commodities, including food, feed, drinking water, and construction materials; (c) natural sources.

As solving a nuclear emergency is top priority, these reference/guidance levels were the first to be set. In fact, the first recommendations by the European Commission date from May 6, 1986 (EC, 1986; Farmer, 2012), just a few days after Chernobyl accident, to April 26, 1986. It regarded milk, milk products, fruit, and vegetables. It has been updated and extended to other foodstuffs and also feeding stuffs (EU, 1989a,b, 1990), being applicable to member states of EU (see Table 10.4). The guidance levels in CA for foodstuffs after a nuclear emergency have a more international scope and are also used in trading (see Table 10.5). After Fukushima accident, Japanese Government set up provisional regulatory values for radioactive materials in foodstuffs (see Table 10.6; MHLW, 2011; Hamada and Ogino, 2012). As it can be observed, these levels are mainly focused on anthropogenic radionuclides, especially radioiodine, radiostrontium, and radiocesium, which are expected to be released in nuclear accidents (see Table 10.1). Comparing the different levels, a lack of uniformity can be observed. This can be observed when considering equivalent foodstuff groups defined by different regulators, foodstuffs consumed by adults, for example, other foodstuffs for EU, foodstuffs other than infant for CA,

TABLE 10.4
Maximum Permitted Levels for Foodstuffs, Expressed as Bq/kg f.w., Following a Nuclear Accident or Any Other Case of Radiological Emergency

	Maximum Permitted Levels for Foodstuffs (Bq/kg f.w.)				
Radionuclides	**Baby Foods**	**Dairy Produce**	**Other Foodstuffs**	**Liquid Foodstuffs**	**Minor Foodstuffs**
Sr (^{90}Sr)	75	125	750	125	7,500
I (^{131}I)	150	500	2000	500	20,000
α-emitters (^{239}Pu, ^{241}Am)	1	20	80	20	800
Other with $T_{½}$ > 10 d (^{134}Cs, ^{137}Cs)[a]	400	1000	1250	1000	12,500

Sources: EU, Council Regulation (EURATOM) Nº 994/89 of 12 April 1989 laying down maximum permitted levels of radioactive contamination in minor foodstuffs following a nuclear accident or any other case of radiological emergency, 1989a; EU, Council Regulation (EURATOM) Nº 2218/89 of 18 July 1989 amending Regulation (Euratom) Nº3954/87 laying down maximum permitted levels of radioactive contamination of foodstuff and feedingstuff following a nuclear accident or any other case of radiological emergency, 1989b.

[a] ^{14}C and ^{40}K excluded.

TABLE 10.5
Guidance Levels for Radionuclides in Foodstuffs Contaminated Following a Nuclear or Radiological Emergency for Use in International Trade

	Guidance Levels CA (Bq/kg f.w.)		
Radionuclides	**Infant Foods**	**Foods Other Than Infant Foods**	**Minor Foodstuffs**
^{238}Pu, ^{239}Pu, ^{240}Pu, ^{241}Am	1	10	100
^{90}Sr, ^{106}Ru, ^{129}I, ^{131}I, ^{235}U	100	100	1,000
^{35}S, ^{60}Co, ^{89}Sr, ^{103}Ru, ^{134}Cs, ^{137}Cs, ^{144}Ce, ^{192}Ir	1000	1,000	10,000
^{3}H, ^{14}C, ^{99}Tc	1000	10,000	100,000

Sources: CA, Codex Alimentarius, Codex General Standards for Contaminants and Toxins in Food and Feed (CODEX STAN 193-1995), 1995; CA, Codex Alimentarius, Fact Sheet on Codex Guideline Levels for Radionulices in Foods Contaminated Following a Nuclear or Radiological Emergency-Prepared by Codex Secretariat, May 2, 2011.

TABLE 10.6
Provisional Regulation Values of Radioactive Materials in Food in Accordance with Food Sanitation Act, Expressed in Bq/kg, by the Japanese Government in 17 March 2011, as a Consequence of Fukushima Daiichi Accident

	Food Groups	
Radionuclide	**Drinking Water/Milk, Dairy Products**	**Vegetables/Grains/Meat, Eggs, Fish, etc.**
Radioactive iodine (^{131}I)	300 (100 infants[a])	2000[b]
Radioactive cesium	200	500
Uranium	20	100
α-Emitting nuclides of Pu and transuranic elements (^{238,239,240}Pu, ^{241}Am, ^{242,243,244}Cm)	1	10

Sources: MHLW, Ministry of Health, Labour and Welfare, Japan, Handling of food contaminated by radioactivity (Relating to the accident at the Fukushima Nuclear Power Plant; http://www.mhlw.go.jp/english/topics/foodsafety/dl/110318-1.pdf, accessed July 21, 2015; Hamada, N. and Ogino, H., *J. Environ. Radioact.*, 111, 83, 2012.

[a] In the case of milk used for baby feeding.

[b] For iodine, this food group includes vegetables (except root vegetables and tubers) and fishery products.

and vegetables, grains, meat, eggs, fish, etc., for Japanese regulations. Reference levels for radioiodine in this group in EU and Japan are 2000 Bq/kg, while only 100 in CA. This difference can be solved partially due to the fact that ^{131}I is a short-lived radionuclide. If the foodstuff can be stored, it takes about 35 days to decay to the level proposed by CA. In the case of radiocesium, there are three different levels for the same foodstuff group, Japanese one being the most restrictive, 500 Bq/kg, while the other two are similar, 1000 and 1250 Bq/kg. Radiostrontium was not considered in Japanese provisional regulatory values due to the characteristics of the accident. Both EU And CA consider it, but with very different values, 750 and 100 Bq/kg, respectively. Finally, Japanese provisional regulatory values take into account the uranium content, 100 Bq/kg. Uranium is not specifically cited by EU levels, but it belongs to the group of α-emitters, 80 Bq/kg. CA only considers ^{235}U, with a guideline level of 100 Bq/kg. These reference levels are very similar for the three regulatory bodies.

Other foodstuff groups, in particular, milk and those destined for infant consumption, have more restrictive levels than those for adults, because the dose coefficients are higher for that age group. The levels for minor foodstuffs, which is considered to be consumed scarcely or at very low rates, are set to 10 times those of major consumption (Tables 10.3 and 10.4). A list of minor foodstuffs can be found in EU (1989a). Feeding stuff is also of great importance since it generally comprised feeding cattle destined for human consumption. Only EU has regulated the maximum content of radionuclides to be present in feeding stuff, setting the maximum levels of ^{134}Cs+^{137}Cs according to the animals to be fed (Pigs 1250 Bq/kg f.w.; poultry, lambs, calves 2500 Bq/kg f.w.; other 5000 Bq/kg f.w.) (EU, 1990).

As it was said earlier, these levels focus mainly on regulating the presence of anthropogenic radionuclides, because nuclear emergencies are mainly associated with accidents in nuclear power plants or industries related to them, such as fuel production or reprocessing. However, there are raw materials and by-products from industries that have an enhanced content of naturally occurring radionuclides (NORM industries), where accidents and releases of radioactive materials can also occur. The CA levels only consider the ^{3}H and ^{14}C, which have a double origin, natural and anthropogenic, and ^{235}U. The Japanese and EU levels do consider in their definition the release of naturally occurring radionuclides. In fact, as the majority of them (^{234,235,238}U, ^{228,230,232}Th, ^{224,226}Ra, ^{210}Po) are α-emitting radionuclides, they can be included in the groups "α-emitters," although the main objective is the detection of ^{238,239,240}Pu, ^{241}Am, and ^{242,243,244}Cm. Other naturally occurring radionuclides

(^{210}Pb, ^{228}Ac, ^{228}Ra) can also be included in the group "other with $T_{½}$ > 10 days" defined by EU. Naturally occurring radionuclides ^{14}C and ^{40}K are excluded from this group due to their low radiotoxicity and since there are some foodstuffs with high levels of stable elements, such as bananas that have a high potassium content and hence ^{40}K.

The existing exposure to radionuclides in food is less regulated. There is a great distinction regarding the emergency exposure, as both naturally occurring and anthropogenic radionuclides are already present in the environment. In emergency scenario, the regulations provide a set of levels for all foodstuffs, including drinking water. While in this scenario, the regulation for radionuclides in drinking water and food is separated. Thus, EU has regulated the water intended for human consumption considering the following parametric values (EU, 2013):

- *Radon*: 100–1000 Bq/L. To be adjusted by each member state within these range. Remedial action is deemed to be justified on radiological protection grounds where radon concentrations exceed 100 Bq/L.
- *Tritium*: 100 Bq/L. If tritium concentration exceeds its parametric value, an analysis of the presence of other artificial radionuclides is required.
- *ID*: 0.1 mSv. ID means the committed effective dose for one year of ingestion resulting from all radionuclides whose presence has been detected in water intended for human consumption, of natural and artificial origin, but excluding tritium, ^{40}K, radon and short-lived radon decay products.

This regulation also sets the monitoring frequency of water intended for human consumption according to the volume of water distributed or produced each day within a supply zone (EU, 2013). The calculation of ID does not imply the determination of all radionuclides present in water but allows the use of reliable screening strategies to be concreted by each member state. These strategies may include the screening for a certain individual radionuclide or radionuclides, or gross alpha or gross beta screening. In the first case, if the activity concentration exceeds 20% of the corresponding derived value (see Table 10.7), an analysis of additional radionuclides will be required, to be defined by each member state, taking into account all relevant information about likely sources

TABLE 10.7
Derived Concentration of Radioactivity in Water Intended for Human Consumption, Calculated for an Annual Dose of 0.1 mSv

Natural Radionuclide	Derived Concentration (Bq/L)	Artificial Radionuclide	Derived Concentration (Bq/L)
^{238}U	3.0	^{14}C	240
^{234}U	2.8	^{90}Sr	4.9
^{226}Ra	0.5	$^{239+240}Pu$	0.6
^{228}Ra	0.2	^{241}Am	0.7
^{210}Pb	0.2	^{60}Co	40
^{210}Po	0.1	^{134}Cs	7.2
^{222}Rn[a]	100–1000[b]	^{137}Cs	11
^{3}H[a]	100	^{131}I	6.2

Source: EC, COUNCIL DIRECTIVE 2013/51/EURATOM of 22 October 2013 laying down requirements for the protection of the health of the general public with regard to radioactive substances in water intended for human consumption. Official Journal of the European Union. 7.11.2013. L 296/12, 2013.

[a] Parametric value.

[b] To be set by each member state.

of radioactivity. The recommended screening values for gross alpha and gross beta indices are 0.1 and 1.0 Bq/L, respectively. The correct use of these indices needs some clarifications. They are not activity levels. They just give information of the total radioactivity present in the form of α- or β-emitters (gross alpha and gross beta, respectively). The sum of all α-emitters determined by other radiochemical procedures is the same order of magnitude than gross alpha index with a precision of 10%–28%, depending on the method used for determining gross alpha (Montaña et al., 2013).

In the case of Spain, where the most important source of anthropogenic radionuclides is global fallout, the presence of radionuclides in water bodies is first and foremost due to naturally occurring radionuclides. The characteristics of natural surface and ground waters are the result of the incorporation into the original rainwater of diverse chemical elements found in the soils through which the water flows. The radionuclides of the natural radioactive series can be present at relatively high concentrations in granitic and metamorphic rocks such as gneisses, black clays, and phosphate rocks (Suarez et al., 2000). The correlation between soils with high levels of natural radioactivity and the activities incorporated into the waters has also been confirmed in different geological environments (Otton, 1994; Baeza et al., 1995; Banks et al., 1998). The connection is of course especially strong for groundwaters (Hess et al., 1985; Orloff et al., 2004). Therefore, there are water supplies that surpass the gross alpha screening levels (Baeza et al., 2014). In this case, further analyses are required to determine which radionuclide is the main contributor. Table 10.7 lists the derived concentration of natural and artificial radionuclides in water. Thorium was not included in the lists mainly because it has very low solubility in water. These derived concentrations are calculated considering the ingestion of a single radionuclide. However, this rarely occurs, as there is a mixture of them present in water, and the calculated ID can surpass the 0.1 mSv/year with lower concentrations (Baeza et al., 2014). There are some remedial options to consider in this situation. One of the options is the modification of the standard procedures in drinking water treatment plants, because they do not significantly remove naturally occurring radionuclides in water, mainly uranium and radium (Baeza et al., 2012, 2014).

Regarding other foodstuffs, the legislation is not so developed in this scenario, as the reference/guidance levels by EU and CA only apply in situations related to nuclear accidents or radiological events and do not apply to routine monitoring purposes. The EU has imposed some restrictions to foodstuff and feeding stuff import from areas affected by Chernobyl and Fukushima accidents. These regulations only take the sum of ^{134}Cs and ^{137}Cs into consideration, whose maximum levels are as follows:

- *Chernobyl-affected areas* (EU, 2008): 370 Bq/kg for milk, dairy produce, and infant food and 600 Bq/kg for the remaining foodstuffs. The validity of these levels has been extended to 2020 (EU, 2009).
- *Fukushima-affected areas* (EU, 2014): These values replace on a provisional basis as shown in Table 10.3 to assure consistency with maximum levels currently applied in Japan
  - *Foodstuffs*: 50 Bq/kg for foods for infant and young children and milk and milk-based drinks, 10 Bq/kg for mineral water and similar drinks and tea brewed from unfermented leaves, and 100 Bq/kg for the remaining foodstuffs.
  - *Feeding stuff*: 100 Bq/kg for cattle and horses, 80 Bq/kg for pigs, 160 Bq/kg for poultry, and 40 Bq/kg for fish, with the exemption of food for ornamental fish.

In this scenario, the reference level of 1 mSv/year for the exposure of public is the only framework valid. Therefore, the question of which radionuclide is to be routinely monitored arises. Table 10.8 lists the effective dose for adults and several natural and anthropogenic radionuclides due to the ingestion of foodstuffs in different countries. The range of variation of effective dose due to natural and anthropogenic radionuclides is the following:

- Natural radionuclides (^{210}Po, ^{210}Pb, ^{40}K, ^{226,228}Ra, ^{228,230,232}Th, U): $8{\cdot}10^{-6}$–6.98 mSv/year
- Anthropogenic radionuclides (^{137}Cs, ^{90}Sr): $6{\cdot}10^{-7}$–0.258 mSv/year

TABLE 10.8
Effective Annual Dose for Adults, Expressed in mSv/Year, Due to the Consumption of Radionuclides Already Present in Foodstuffs (Existing Exposure) from Different Countries

Country	Foodstuff	Radionuclides	Dose (mSv/Year)	Reference
Argentina	Tea, Mate	^{210}Po	0.0063 – 0.080	Colangelo et al. (1992)
	Fish	^{210}Po	$(6–7) \times 10^{-4}$	Colangelo et al. (1994)
Austria	Milk	^{137}Cs	$7 \cdot 10^{-5} – 1.3 \times 10^{-4}$	Landstetter et al. (2013)
		^{90}Sr	$(1.0–4.2) \times 10^{-4}$	
		^{40}K	0.004 – 0.027	
	Meat	^{137}Cs	$6.8 \times 10^{-5} – 1.3 \times 10^{-4}$	
		^{40}K	$5.8 \times 10^{-4} – 0.025$	
	Mushroom	^{137}Cs	1.9×10^{-4}	
		^{40}K	0.0016	
Brazil	Wholediet	^{210}Po	0.0055 – 0.0187	Cunha et al. (2001)
		^{210}Pb	0.0013 – 0.0043	
Cuba	Fish	$^{210}Po/^{137}Cs$	$0.448/6 \times 10^{-5}$	Alonso-Hernández et al. (2000)
	Crustacean	$^{210}Po/^{137}Cs$	$5.6 \times 10^{-4}/5 \times 10^{-7}$	
	Mollusc	$^{210}Po/^{137}Cs$	$6.6 \times 10^{-3}/3 \times 10^{-7}$	
Denmark	Fish	^{90}Sr	7.3×10^{-4}	Aarkrog et al. (2000)
	Crustacean	^{90}Sr	4×10^{-5}	
	Sea mammals	^{90}Sr	5.3×10^{-4}	
	Meat	^{90}Sr	$3.1 \times 10^{-4} – 0.0033$	
	Fish	^{137}Cs	0.0078	
	Crustacean	^{137}Cs	1.2×10^{-4}	
	Sea mammals	^{137}Cs	0.015	
	Meat	^{137}Cs	0.0028 – 0.258	
Egypt	Vegetables	^{210}Po	$3.2 \times 10^{-5} – 0.038$	Din (2011)
	Fruit	^{210}Po	0.0015 – 0.013	
	Cereals	^{210}Po	$8 \times 10^{-6} – 0.026$	
Finland	Cereals	^{210}Po, ^{210}Pb	$3 \times 10^{-4} – 0.022$	Turtiainen et al. (2011)
	Berries	^{210}Po	$8.6 \times 10^{-4} – 0.0051$	Vaarama et al. (2009)
		^{210}Pb	$2.1 \times 10^{-4} – 0.0013$	
		^{137}Cs	$2.6 \times 10^{-4} – 0.0073$	
	Mushroom	^{210}Po	$2.1 \times 10^{-4} – 0.0026$	
		^{210}Pb	$2.7 \times 10^{-5} – 2.6 \times 10^{-4}$	
		^{137}Cs	$(1.4 – 3.7) \times 10^{-5}$	
France	Fish, mollusc	^{210}Po	0.046 – 0.129	Connan et al. (2007)
Ghana	Cereals	$^{226,228}Ra$, ^{228}Th, ^{40}K	0.05 – 1.33	Awudu et al. (2012)
	Potato	$^{226,228}Ra$, ^{228}Th, ^{40}K	0.38	
	Milk	U	3×10^{-5}	Giri et al. (2011)
		^{226}Ra	0.0022	
		^{230}Th	0.0016	
		^{210}Po	0.0085	
	Cereals	^{210}Po	$8 \times 10^{-5} – 0.128$	Kannan et al. (2001)
	Vegetables	^{210}Po	$1.2 \times 10^{-4} – 0.006$	
	Milk	^{210}Po	2.3×10^{-4}	

(Continued)

TABLE 10.8 (*Continued*)
Effective Annual Dose for Adults, Expressed in mSv/Year, Due to the Consumption of Radionuclides Already Present in Foodstuffs (Existing Exposure) from Different Countries

Country	Foodstuff	Radionuclides	Dose (mSv/Year)	Reference
	Meat	^{210}Po	$2.4 \times 10^{-4} - 3.4 \times 10^{-4}$	
India	Fish	^{210}Po	0.093 – 0.594	
	Seafood	^{210}Po	0.61 – 3.4	
	Mollusc	^{210}Po	0.032 – 0.515	
	Crustacean	^{210}Po	0.069 – 0.266	
	Fish	^{210}Po	0.011 – 0.100	Khan et al. (2011)
	Mollusc	^{210}Pb	0.0048 – 0.016	
	Crustacean	^{210}Pb	0.012 – 0.019	
	Fish	^{210}Pb	0.0039 – 0.24	
	Milk	^{210}Po	0.0021	
	Meat	^{210}Po	0.0045	
	Cereals	^{210}Po	0.007	Meli et al. (2014)
Italy	Vegetables	^{210}Po	0.0078	
	Fish	^{210}Po	0.086	
	Wine	^{238}U	$6 \times 10^{-6} - 1.7 \times 10^{-4}$	Desideri et al. (2010)
	Wine	^{210}Po	$2.7 \times 10^{-4} - 0.049$	
	Fish	^{134}Cs	$5.7 \times 10^{-5} - 5 \times 10^{-4}$	Chen (2013)
	Fish	^{137}Cs	$9.1 \times 10^{-5} - 8 \times 10^{-4}$	
	Fish	^{210}Po	0.036 – 0.32	
Japan	Fish	^{228}Ra	0.0012 – 0.011	
	Wholediet	$^{134}Cs + {}^{137}Cs$	0.0013 – 0.099	Harada et al. (2013)
	Mushrooms	^{137}Cs	$(4.0 - 16) \times 10^{-5}$	Muramatsu et al. (1991); Ban-Nai et al. (2004)
Lebanon	Fish	^{210}Po	0.017 – 1.11	Aoun et al. (2015)
		^{210}Pb	$7.2 \times 10^{-3} - 0.45$	
	Fish	^{210}Po	0.044	
	Crustacean	^{210}Po	0.186	Alam et al. (2011)
Malaysia	Mollusc	^{210}Po	0.517	
	Rice	^{40}K	0.029 – 0.058	
	Rice	^{226}Ra	0.026 – 0.101	Asaduzzaman et al. (2015)
	Rice	^{232}Th	0.042 – 0.252	
Montenegro	Fish	^{137}Cs	6×10^{-5}	Antovic et al. (2011)
	Fish	^{226}Ra	4.06×10^{-3}	
Norway	Berries	^{210}Po, ^{210}Pb, ^{40}K	0.05	Gwynn et al. (2013)
	Mushrooms	^{210}Po, ^{210}Pb, ^{40}K	0.50	
Pakistan	Cereal	^{40}K, ^{226}Ra, ^{232}Th	0.217	Tufail et al. (2010)
	Fish	^{210}Po, ^{210}Pb	0.024 – 0.028	
	Seafood	^{210}Po, ^{210}Pb	0.0042 – 0.0075	Nadal et al. (2011)
Spain	Fruit	^{210}Po, ^{210}Pb	$(4.7 - 5.6) \times 10^{-4}$	
	Vegetables	^{210}Po, ^{210}Pb	$(3.9 - 4.9) \times 10^{-4}$	
	Rice	^{210}Po, ^{210}Pb		
	Mushrooms	^{137}Cs	$6 \times 10^{-7} - 1.9 \times 10^{-3}$	Baeza et al. (2004)

(*Continued*)

TABLE 10.8 (*Continued*)
Effective Annual Dose for Adults, Expressed in mSv/Year, Due to the Consumption of Radionuclides Already Present in Foodstuffs (Existing Exposure) from Different Countries

Country	Foodstuff	Radionuclides	Dose (mSv/Year)	Reference
	Milk	^{238}U	3.9×10^{-4}	
	Milk	^{226}Ra	4.7×10^{-4}	
Slovenia	Milk	^{230}Th	2.9×10^{-4}	Štrok et al. (2012)
	Milk	^{210}Po	0.0032	
	Milk	^{210}Pb	0.0046	
Syria	Fish	^{210}Po	0.628	Al-Masri et al. (2000)
	Fish	^{210}Pb	0.341	
Taiwan	Mollusc	^{210}Po	0.041	Lee et al. (2013)
	Mollusc	^{210}Po	1.9 – 4.2	Aközcan et al. (2013)
	Mollusc	^{210}Pb	0.054 – 0.126	
	Vegetables	^{40}K	9.7×10^{-4} – 0.020	
	Vegetables	^{226}Ra	0.0079 – 0.014	
	Vegetables	^{232}Th	3.6×10^{-4} – 0.003	
Turkey	Vegetables	^{137}Cs	$(1.6 – 7.5) \times 10^{-4}$	Canbazoğlu et al. (2013)
	Fruits	^{40}K	6.1×10^{-4} – 0.016	
	Fruits	^{226}Ra	$(2.4 – 9.2) \times 10^{-3}$	
	Fruits	^{232}Th	0.0044 – 0.033	
	Fish	^{226}Ra, ^{232}Th, ^{137}Cs, ^{40}K	0.0057 – 0.019	Görür et al. (2012)
	Fish	^{210}Po	2×10^{-6} – 0.011	Çatal et al. (2012)
	Seafood	^{210}Po	0.04 – 1.22	Dewar et al. (2014)
	Cereal	^{210}Po, ^{210}Pb	0.027	
	Fruit	^{210}Po, ^{210}Pb	0.0051	McDonald et al. (1999)
United Kingdom	Vegetables	^{210}Po, ^{210}Pb	0.0017	
	Whole diet	^{137}Cs	2.6×10^{-4} – 0.027	
	Whole diet	^{90}Sr	0.0011 – 0.0019	Sánchez et al. (1999)
	Whole diet	^{14}C	0.0046 – 0.0098	
	Mushrooms	^{137}Cs	1.47×10^{-7} – 0.018	Barnett et al. (1999, 2001)
USA	Fish	^{210}Po	0.17 – 6.98	Noshkin et al. (1994)
	Fish	^{210}Pb	0.13 – 0.50	

The effective dose due to natural radionuclides in routine scenarios is usually higher than that due to anthropogenic ones. The radionuclides that contributed mostly to the dose are those with higher e(g) coefficients ($^{210}Po > ^{210}Pb$, $^{228}Ra > ^{226}Ra$,…). Regarding the same foodstuff group, it can be observed that the contribution due to ^{210}Po is usually higher than those due to radiocesium (Alonso-Hernández et al., 2002; Chen, 2013). However, there are some locations and "critical groups," which usually consume wild foodstuffs (e.g., wild mushrooms, berries, reindeer, and caribou meat in Arctic environments), in which the radiocesium contribution is significant (Aarkrog et al., 2000). The contribution of ^{210}Po and ^{210}Pb due to the consumption of fish and seafood can be significant to the annual dose. Both ^{210}Pb and ^{210}Po show a high affinity for organic matter and biological activity (phytoplankton and zooplankton), which can also be responsible for the association with particulate matter suspended in seawater, being greater in the case of ^{210}Po (Carvalho, 1995; Masqué et al., 2002). This association is reflected in the marine food chain, decreasing their content in the following order: molluscs > crustaceans > fish (Alam and Mohamed, 2011; Carvalho, 2011).

10.7 CONCLUSIONS

The public exposure to radionuclides in foodstuffs is considered in international regulations and recommendations, in which the reference level is 1 mSv/y for the effective dose (ICPR, 1999; EU, 2010). The following factors influence the effective dose: (1) which radionuclide is ingested, (2) how much foodstuff is annually ingested, and (3) the activity level in it. The radiotoxicity of naturally occurring radionuclides is usually higher than that of anthropogenic ones. The ingestion of different foodstuffs is estimated by different national surveys about nutritional habits. These surveys give different results for different countries, reflecting differences in culinary customs, food availability, etc. Surveys carried out by several organisms and times show variations in the composition of the annual food consumption. Therefore, the most usual way to assure the public protection is the setting of reference/guidance levels for activities in different foodstuffs. These levels are usually referred to scenarios following a nuclear accident or any other case of radiological emergency. As nuclear emergencies usually involve the release of anthropogenic radionuclides, EU reference levels, CA guideline levels, and provisional regulation levels in Japan are focused on them. Comparing similar foodstuff groups defined by each of them, a lack of uniformity can be observed, being the levels very different, as in the case of radioionide, radiocesium, and radiostrontium. Emergencies involving naturally occurring radionuclides are not fully considered in these regulations. CA only considers ^{235}U, while in EU and Japanese provisional regulatory levels, they can be considered partially in the groups "α-emitters" and "others with $T_{1/2} > 10$ days," although the focus in these groups is in anthropogenic radionuclides.

When considering existing exposure, a major difference can be observed, and water intended for human consumption is separated from other foodstuffs, which does not occur in emergencies. It is regulated by a set of parametric values, radon, tritium, and ID, from which derived concentrations for naturally occurring and anthropogenic radionuclides are estimated, since the derived concentrations are derived from these parameters and not the other way round. In areas unaffected by any major nuclear accident, such as in Spain, aquifers with high content of naturally occurring radionuclides, mainly uranium and radium, can surpass the parametric level for ID. Thus, remediation is required, which can be achieved by modifying the processes in drinking water treatment plants to optimize the removal of uranium and radium. These methods have been successfully applied to operating plants, and ID decreased below the parametric level. The reference levels for other foodstuff groups in this scenario are scarce and limited to maximum levels for imports from countries affected by Chernobyl and Fukushima accidents. Reported values for effective dose for different foodstuffs imply that in this scenario the contribution of naturally occurring radionuclides can be even higher than those due to anthropogenic ones.

ACKNOWLEDGMENT

The authors are grateful to the Autonomous Government of Extremadura (Junta de Extremadura) for the financial support to the LARUEX research group (FQM001).

REFERENCES

Aarkrog, A., Dahlgaard, H., and Nielsen, S.P. 2000. Environmental radioactive contamination in Greenland: A 35 years retrospect. *The Science of the Total Environment* 245: 233–248.

Akozcan, S. 2013. Levels of ^{210}Po and ^{210}Pb in mussel and sediments in Candarli Gulf and the related dose assessment to the coastal population. *Marine Pollution Bulletin* 73: 11–15.

Alam, L. and Mohamed, C.A.R. 2011. Natural radionuclide of ^{210}Po in the edible seafood affected by coal-fired power plant industry in Kapar coastal area of Malaysia. *Environmental Health* 10: 43.

Al-Masri, M.S., Mamish, S., Budeir, Y., and Nashwati, A. 2000. ^{210}Po and ^{210}Pb concentrations in fish consumed in Syria. *Journal of Environmental Radioactivity* 49: 334–352.

Alonso-Hernandez, C., Diaz-Asencio, M., Munos-Caravaca, A., Suarez-Morell, E., and Avila-Moreno, R. 2002. ^{137}Cs and ^{210}Po dose assessment from marine food in Cienfuegos Bay (Cuba). *Journal of Environmental Radioactivity* 61: 203–211.

Antovic, I. and Antovic, N.M. 2011. Determination of concentration factors for Cs-137 and Ra-226 in the mullet species *Chelon labrosus* (Mugilidae) from the South Adriatic Sea. *Journal of Environmental Radioactivity* 102: 713–717.

Aoun, M., El Samad, O., Bou Khozam, R., and Lobinski, R. 2015. Assessment of committed effective dose due to the ingestion of ^{210}Po and ^{210}Pb in consumed Lebanese fish affected by a phosphate fertilizer plant. *Journal of Environmental Radioactivity* 140: 25–29.

Asaduzzaman, K., Khandaker, M.U., Amin, Y.M., and Mahat, R. 2015. Uptake and distribution of natural radioactivity in rice from soil in north and west part of peninsular malaysia for the estimation of ingestion dose to man. *Annals of Nuclear Energy* 76: 85–93.

Awudu, A.R., Faanu, A., Darko, E.O., Emi-Reynolds, G., Adukpo, O.K., Kpeglo, D.O. et al. 2012. Preliminary studies on ^{226}Ra, ^{228}Ra, ^{228}Th and ^{40}K concentrations in foodstuffs consumed by inhabitants of Accra metropolitan area, Ghana. *Journal of Radioanalytical and Nuclear Chemistry* 291: 635–641.

Baeza, A., del Río, L.M., Jiménez, A., Miró, C., and Paniagua, J.M. 1995. Factors determining the radioactivity levels of waters in the province of Cáceres (Spain). *Applied Radiation and Isotopes* 46(10): 1053–1059.

Baeza, A., Hernández, S., Guillén, F.J., Moreno, G., Manjón, J.L., and Pascual, R. 2004. Radiocaesium and natural gamma emitters in mushrooms collected in Spain. *Science of the Total Environment* 318: 59–71.

Baeza, A., Salas, A., and Guillén, J. 2012. Adaptation of working conditions of an operating drinking water treatment plant to remove naturally occurring radionuclides. *Water, Air and Soil Pollution* 223: 5057–5069.

Baeza, A., Salas, A., Guillén, J., and Muñoz-Serrano, A. 2014. Association of naturally occurring radionuclides in sludges from Drinking Water Treatment Plants previously optimized for their removal. *Chemosphere* 97: 108–114.

Banks, D., Reimann, C., and Skarphagen, H. 1998. The comparative hydrochemistry of two granitic island aquifers: The Isles of Stilly, UK and the Hvaler Islands, Norway. *Science of the Total Environment* 209: 169–183.

Ban-Nai, T., Muramatsu, Y., and Yoshida, S. 2004. Concentration of ^{137}Cs and ^{40}K in mushrooms consumed in Japan and radiation dose as a result of their dietary intake. *Journal of Radiation Research* 45: 325–332.

Barnett, C.L., Beresford, N.A., Frankland, J.C., Self, P.L., Howard, B.J., and Marriott, J.V.R. 2001. Radiocaesium intake in Great Britain as a consequence of the consumption of wild fungi. *Mycologist* 15(3): 98–104.

Barnett, C.L., Beresford, N.A., Self, P.L., Howard, B.J., Frankland, J.C., Fulker, M.J., Dodd, B.A., and Marriott, J.V.R. 1999. Radiocaesium activity concentrations in the fruit bodies of macrofungi in Great Britain and an assessment of dietary intake habits. *Science of the Total Environment* 231: 67–83.

Beresford, N.A., Voigt, G., Wright, S.M., Howard, B.J., Barnett, C.L., Prister, B. et al. 2001. Self-help countermeasure strategies for populations living within contaminated areas of Belarus, Russia and Ukraine. *Journal of Environmental Radioactivity* 56: 215–239.

Byrom, J., Robinson, C., Simmonds, J.R., Waiters, B., and Taylor, R.R. 1995. Food consumption rates for use in generalized radiological dose assessments. *Journal of Radiological Protection* 15(4): 335–341.

CA, Codex Alimentarius. 1995. *Codex General Standards for Contaminants and Toxins in Food and Feed* (CODEX STAN 193-1995), World Health Organization and Food and Agriculture Organization of the United Nations, Rome, Italy.

CA, Codex Alimentarius. May 2, 2011. *Fact Sheet on Codex Guideline Levels for Radionulices in Foods Contaminated Following a Nuclear or Radiological Emergency*, Prepared by Codex Secretariat, World Health Organization and Food and Agriculture Organization of the United Nations, Rome, Italy.

Canbazoğlu, C. and Doğru, M. 2013. A preliminary study on ^{226}Ra, ^{232}Th, ^{40}K and ^{137}Cs activity concentrations in vegetables and fruits frequently consumed by inhabitants of Elazığ Region, Turkey. *Journal of Radioanalytical and Nuclear Chemistry* 295: 1245–1249.

Carini, F. and Pellizzoni, M. 2011. Attitudes of rural population in emergency exposure situations. *Journal of Environmental Radioactivity* 102: 363–369.

Carvalho, F.P. 1995. ^{210}Pb and ^{210}Po in sediments and suspended matter in the Tagus estuary, Portugal. Local enhancement of natural levels by wastes from phosphate ore processing industry. *Science of the Total Environment* 159: 201–214.

Carvalho, F.P. 2011. Polonium (^{210}Po) and lead (^{210}Pb) in marine organisms and their transfer in marine food chains. *Journal of Environmental Radioactivity* 102: 462–472.

Çatal, E.M., Uğur, A., Ozden, B., and Filizok, I. 2012. ^{210}Po and ^{210}Pb variations in fish species from the Aegean Sea and the contribution of ^{210}Po to the radiation dose. *Marine Pollution Bulletin* 64: 801–806.

Chen, J. 2013. Evaluation of radioactivity concentrations from the Fukushima nuclear accident in fish products and associated risks to fish consumers. *Radiation Protection Dosimetry* 157(1): 1–5.

Colangelo, C.H., Huguet, M.R., and Oliveira A.A. 1994. Levels of ^{210}Po in marine species of the south Atlantic ocean. *Journal of Radioanalytical and Nuclear Chemistry, Letters* 188(1): 65–71.

Colangelo, C.H., Huguet, M.R., Palacios, M.A., and Oliveira, A.A. 1992. Levels of ^{210}Po in some beverages and in tobacco. *Journal of Radioanalytical and Nuclear Chemistry, Letters* 166(3): 195–202.

Connan, O., Germain, P., Solier, L., and Gouret, G. 2007. Variations of ^{210}Po and ^{210}Pb in various marine organisms from Western English Channel: Contribution of ^{210}Po to the radiation dose. *Journal of Environmental Radioactivity* 97: 168–188.

Cunha, I.L., Bueno, L., Fávaro, D.I.T., Maihara, V.A., and Cozzollino, S. 2001. Analysis of ^{210}Pb and ^{210}Po in Brazilian foods and diets. *Journal of Radioanalytical and Nuclear Chemistry* 247(2): 447–450.

Desideri, D., Roselli, C., and Meli, M.A. 2010. Intake of ^{210}Po, ^{234}U and ^{238}U radionuclides with wine in Italy. *Food and Chemical Toxicology* 48: 650–657.

Dewar, A., Camplin, W., Barry, J., and Kennedy, P. 2014. A statistical approach to investigating enhancement of polonium-210 in the Eastern Irish Sea arising from discharges from a former phosphate processing plant. *Journal of Environmental Radioactivity* 138: 289–301.

Din, K.S. 2001. Determination of ^{210}Po in various foodstuffs and its annual effective dose to inhabitants of Qena City, Egypt. *Science of the Total Environment* 409: 5301–5304.

EU. 1989a. Council Regulation (EURATOM) N° 994/89 of 12 April 1989 laying down maximum permitted levels of radioactive contamination in minor foodstuffs following a nuclear accident or any other case of radiological emergency, No. L 101/17–18.

EU. 1989b. Council Regulation (EURATOM) N° 2218/89 of 18 July 1989 amending Regulation (Euratom) N°3954/87 laying down maximum permitted levels of radioactive contamination of foodstuff and feedingstuff following a nuclear accident or any other case of radiological emergency, No. L 211/4–5.

EU. Commission regulation (EURATOM) N°770/90 of 29 March 1990 laying down maximum permitted levels of radioactive contamination of feedingstuff following a nuclear accident or any other case of radiological emergency, No L 83/78–79.

EU. 2008. Reglamento (CE) n° 733/2008 del Consejo de 15 de julio de 2008 relativo a las condiciones de importación de productos agrícolas originarios de terceros países como consecuencia del accidente ocurrido en la central nuclear de Chernobil. Diario Oficial de la Unión Europea, 30.7.2008, L201/1-7. (in Spanish).

EU. 2009. Reglamento (CE) N° 1048/2009 del Consejo de 23 de octubre de 2009 por el que se modifica el Reglamento (CE) no 733/2008 relativo a las condiciones de importación de productos agrícolas originarios de terceros países como consecuencia del accidente ocurrido en la central nuclear de Chernóbil. Diario Oficial de la Unión Europea 6.11.2009. L 290/4. (in Spanish).

EU. 2010. Consolidated version of the treaty establishing the European Atomic Energy Community. Luxembourg, Publications Office of the European Union Publications. Office of the European Union, Luxembourg, 120p.

EU. 2014. Commission Implementing Regulation (EU) No 322/2014 of 28 March 2014 imposing special conditions governing the import of feed and food originating in or consigned from Japan following the accident at the Fukushima nuclear power station. Official Journal of the European Union. 29.3.2014. L 95/1–11.

EC. 1986. 86/156/EEC: Commission Recommendation of 6 May 1986 addressed to the Member States concerning the coordination of national measures taken in respect of agricultural products as a result of radioactive fallout from the Soviet Union. Official Journal L 118, 07/05/1986 P. 0028 – 0028.

EC. 2013. COUNCIL DIRECTIVE 2013/51/EURATOM of 22 October 2013 laying down requirements for the protection of the health of the general public with regard to radioactive substances in water intended for human consumption. Official Journal of the European Union. 7.11.2013. L 296/12.

Farmer, A.M. (Ed.). 2012. *Manual of European Environmental Policy*, 1043pp. Routledge, London, U.K.

Giri, S., Singh, G., Jha, V.N., and Tripathi, R.M. 2011. Risk assessment due to ingestion of natural radionuclides and heavy metals in the milk samples: A case study from a proposed uranium mining area, Jharkhand. *Environmental Monitoring Assessment* 175: 157–166.

Görür, F.K., Keser, R., Akcay, N., and Dizman, S. 2012. Radioactivity and heavy metal concentrations of some commercial fish species consumed in the Black Sea Region of Turkey. *Chemosphere* 87: 356–361.

Guillén, J. and Baeza, A. 2014. Radioactivity in mushrooms: A health hazard? *Food Chemistry* 154: 14–25.

Gwynn, J.P., Nalbandyan, A., and Rudolfsen, G. 2013. ^{210}Po, ^{210}Pb, ^{40}K and ^{137}Cs in edible wild berries and mushrooms and ingestion doses to man from high consumption rates of these wild foods. *Journal of Environmental Radioactivity* 116: 34–41.

Hamada, N. and Ogino, H. 2012. Food safety regulations: What we learned from the Fukushima nuclear accident. *Journal of Environmental Radioactivity* 111: 83–89.

Harada, K.H., Fujii, Y., Adachi, A., Tsukidate, A., Asai, F., and Koizumi, A. 2013. Dietary Intake of Radiocesium in Adult Residents in Fukushima Prefecture and Neighboring Regions after the Fukushima Nuclear Power Plant Accident: 24-h Food-Duplicate Survey in December 2011. *Environmental Science Technology* 47: 2520–2526.

Hess, C.T., Michel, J., Horton, T.R., Prichard, H.M., and Coniglio, W.A. 1985. The occurrence of radioactivity in public water supplies in the United States. *Health Physics* 48: 553–586.

IAEA. 2003. Extent of Environmental Contamination by Naturally Occurring Radioactive Material (NORM) and Technological Options for Mitigation. IAEA Technical reports series n° 419, Vienna, Austria.

IAEA. 2014a. IAEA report on radiation protection after the Fukushima Daiichi accident: Pomoting confidence and understanding. *International Experts Meeting*, Vienna, Austria, February 17–21, 2014.

IAEA. 2014b. Radiation Protection and Safety of Radiation Sources: International Basic Safety Standards. General Safety Requirements. IAEA Safety Standards Series No. GSR Part 3, Vienna, Austria.

ICRP. 1999. Protection of the Public in Situations of Prolonged Radiation Exposure. ICRP Publication 82. *Annals of ICRP* 29(1–2): 1–110.

ICRP. 2012. Compendium of Dose Coefficients based on ICRP Publication 60. ICRP Publication 119. *Annals of ICRP* 41(1): 1–130.

ICRP. August 2013. Corrigenda to ICRP Publication 119: Compendium of Dose Coefficients based on ICRP Publication 60 [Ann. ICRP 41(s) 2012]. *Annals of the ICRP* 42(4): 345–362.

Kannan, V., Iyengarb, M.A.R., and Ramesh, R. 2001. Dose estimates to the public from ^{210}Po ingestion via dietary sources at Kalpakkam (India). *Applied Radiation and Isotopes* 54: 663–674.

Khan, M.F. and Wesley, S.G. 2011. Assessment of health safety from ingestion of natural radionuclides in seafoods from a tropical coast, India. *Marine Pollution Bulletin* 62: 399–404.

Landstetter, C., Zapletal, M., Sinojmeri, M., and Katzlberger, C. 2013. Measurements of natural and artificial radionuclides in food samples and water for human consumption in Austria for the calculation of the ingestion dose. *Journal of Radioanalytical and Nuclear Chemistry* 296: 905–908.

Lee, H. and Wang, J. 2013. Annual dose of Taiwanese from the ingestion of ^{210}Po in oysters. *Applied Radiation and Isotopes* 73: 9–11.

MAGRAMA. 2009. Ministerio de Agricultura, Alimentación y Medio Ambiente. Dossier de Consumo Alimentario en España. Año móvil Julio 2008—Junio 2009. http://www.magrama.gob.es/es/alimentacion/temas/consumo-y-comercializacion-y-distribucion-alimentaria/a%C3%B1o_movil_jul08_jun09_tcm7-7862.pdf, accessed July 21, 2015 (in Spanish).

Masque, P., Cabeza, J.A.S., Bruach, J.M., Palacios, E., and Canals M. 2002. Balance and residence times of ^{210}Pb and ^{210}Po in surface waters of the northwestern Mediterranean Sea. *Continental Shelf Research* 22: 2127.

Máté, B., Sobiech-Matura, K., and Altzitzoglou, T. 2015. Radionuclide monitoring in foodstuff: Overview of the current implementation in the EU countries. *Journal of Radioanalytical and Nuclear Chemistry* 303: 2547–2552.

McDonald, P., Jackson, D., Leonard, D.R.P., and McKay, K. 1999. An assessment of ^{210}Pb and ^{210}Po in terrestrial foodstuffs from regions of England and Wales. *Journal of Environmental Radioactivity* 43: 15–29.

Meli, M.A., Desideri, D., Roselli, C., and Feduzi, L. 2014. Assessment of ^{210}Po in Italian diet. *Food Chemistry* 155: 87–90.

Merešová, J. and Wätjen, U. 2014. Results of an EC laboratory comparison on ^{40}K, ^{90}Sr and ^{137}Cs in dried bilberry powder. *Applied Radiation and Isotopes* 87: 443–446.

MHLW. 2011. Ministry of Health, Labour and Welfare. Japan. Handling of food contaminated by radioactivity (Relating to the accident at the Fukushima Nuclear Power Plant, http://www.mhlw.go.jp/english/topics/foodsafety/dl/110318-1.pdf, July 21, 2015.

Montaña, M., Fons, J., Corbacho, J.A., Camacho, A., Zapata-García, D., Guillén, J. et al. 2013. A comparative experimental study of gross alpha methods in natural waters. *Journal of Environmental Radioactivity* 118: 1–8.

Muramatsu, Y., Yoshida, S., and Sumiya, M. 1991. Concentrations of radiocesium and potassium in basidiomycetes collected in Japan. *Science of the Total Environment* 105: 29–39.

Nadal, M., Casacuberta, N., Garcia-Orellana, J., Ferré-Huguet, N., Masqué, P., Schuhmacher, M., and Domingo, J.L. 2011. Human health risk assessment of environmental and dietary exposure to natural radionuclides in the Catalan stretch of the Ebro River, Spain. *Environmental Monitoring Assessment* 175: 455–468.

Noshkin, V.E., Robison, W.L., and Wong, K.M. 1994. Concentration of ^{210}Po and ^{210}Pb in the diet at the Marshall Islands. *The Science of the Total Environment* 155: 87–104.

Orloff, K.G., Mistry, K., Charp, P., Metcalf, S., Marino, R., Shelly, T., Melaro, E., Donohoe, A.M., and Jones, R.L. 2004. Human exposure to uranium in groundwater. *Environmental Research* 94: 319–326.

Otton, J.K. 1994. Natural Radioactivity in the Environment. Technical report, U.S. Geological Survey, Energy Resource Surveys Program, Reston, VA.

Sanchez, A.L., Walters, C.B., Singleton, D.L., Wood, N.H., and Mondon, K. 1999. Food ingestion doses from artificial radionuclides in Cumbrian diets, ten years post-Chernobyl. *Journal of Environmental Radioactivity* 46: 301–317.

USFDA. November 17, 2006. U.S. Food and Drug Administration. Total Diet Study Statistics on Radionuclide Results. Revision 1, 1999–2005. http://www.fda.gov/downloads/Food/FoodScienceResearch/TotalDietStudy/UCM184305.pdf, accessed July 21, 2015.

UNSCEAR. 2000. United Nations Scientific Committee on the Effects of Atomic Radiation. UNSCEAR 2000 Report to the General Assembly, with scientific annexes, New York.

Sanchez, A.L., Singleton, D., Waiters, B., and Cobb, J. 1997. Radionuclides in whole diets of people living near the Sellafield nuclear complex, UK. *Journal of Radioanalytical and Nuclear Chemistry* 226(1–2): 267–274.

Štrok, M. and Smodiš, B. 2012. Transfer of natural radionuclides from hay and silage to cow's milk in the vicinity of a former uranium mine. *Journal of Environmental Radioactivity* 110: 64–68.

Suáñez, A. and Robles, B. 2002. Estudio comparativo sobre dietas y hábitos alimentarios en la población española. *Radioprotección* IX(31): 27–33 (in Spanish).

Suarez, E., Fernández, J.A., Baeza, A., Moro, M.C., García, D., Moreno, J. et al. 2000. Proyecto Marna. Mapa de radiación gamma natural. Colección de Informes Técnicos 5. Consejo de Seguridad Nuclear. Madrid (in Spanish).

Turtiainen, T., Kostiainen, E., and Hallikainen, A. 2011. ^{210}Pb and ^{210}Po in Finnish cereals. *Journal of Environmental Radioactivity* 102: 438–442.

Tufail, M., Javied, S., Akhtar, N., and Akhter, J. 2010. Assessment of annual effective dose from natural radioactivity intake through wheat grain produced in Faisalabad, Pakistan. *Journal of Radioanalytical and Nuclear Chemistry* 283: 585–590.

Vaaramaa, K., Solatie, D., and Aro, L. 2009. Distribution of ^{210}Pb and ^{210}Po concentrations in wild berries and mushrooms in boreal forest ecosystems. *Science of the Total Environment* 408: 84–91.

Wätjen, U., Altzitzoglou, T., Ceccatelli, A., Dikmen, H., Emteborg, H., Ferreux, L. et al. 2012. Results of an international comparison for the determination of radionuclide activity in bilberry material. *Applied Radiation and Isotopes* 70: 1843–1849.

Wätjen, U., Spasova, Y., Vasile, M., Szántó, Z., Emteborg, H., and Voitsekhovych, O. 2014. Certification of the reference material IRMM-426 for radionuclides in dried bilberries. *Applied Radiation and Isotopes* 87: 475–479.

11 Infant Toxicology*

Overview and Considerations for the Safety Assessment of Products for Infants

Yen-Ching Wu, Jeffrey Fisher, and April Neal-Kluever

CONTENTS

* *Note*: The data and interpretations expressed in this book represent that of the authors and not necessarily that of the US FDA.

11.1 INTRODUCTION

In 1993, the National Academy of Scientists cited qualitative and quantitative differences in the toxicity of pesticides between children and adults and concluded that children cannot be merely considered "little adults" in terms of their biological response to chemicals (National Research Council 1993). This landmark report provided the impetus for the passage of legislation that contributed to stricter safety standards for infants and children, such as the Food Quality Protection Act (1996). In recognition of the existing differences in the biological responses to chemical exposures between children and adults, U.S. federal agencies developed regulatory guidelines that focused on the developing life stages that encompassed the prenatal time period through adolescence. Infants, defined as 12 months of age or younger, represent a subpopulation of interest based on exposure and biological factors. As an example, infants predominantly consume a single source of nutrition (i.e., sole-source consumers) for at least the first 6 months after birth, relying on breast milk and/or infant formula for sustenance, which potentially increases their exposure to chemicals migrating from one packaging type into expressed breast milk or formula. Furthermore, infants consume more food per kilogram body weight per day than adults, which may also raise their exposure to dietary exposure to chemicals. Other factors that alter exposure paradigms for infants compared to other life stages include increased mouthing behavior and elevated respiration rates. Finally, infants undergo rapid biological maturation and exhibit different physiologies from adults that may impact their potential sensitivity to chemical exposures. The aim of this chapter is to introduce current standardized protocols used to assess the safety of products for infants, to provide a critical evaluation of their strengths and weaknesses, and to propose some changes that may bridge existing gaps in these study paradigms. To accomplish these objectives, we will provide an overview of the biological differences that are known between infants and adults, discuss potential sensitivities particular to infants, and review current protocols that provide pertinent information for the safety assessment of infant products when conducted by researchers. We will end with our recommendations to optimize these current guidelines in the context of the safety assessment of products for infants.

11.2 BACKGROUND

Infancy (0–12 months) has emerged as an important life stage for consideration in the safety assessment of food packaging on the basis of both exposure and biological factors. The American Academy of Pediatricians recommends consumption of only formula and/or breast milk until at least the age of 6 months (http://www.aap.org/en-us/advocacy-and-policy/aap-health-initiatives/HALF-Implementation-Guide/Age-Specific-Content/Pages/Infant-Food-and-Feeding.aspx#none). Thus, the first 6 months after birth represent a special scenario in which infants are likely to consume a sole source of nutrition: formula and/or breast milk. Infant foods (formula and breast milk) may contact various food contact materials, such as baby bottles, nipples, storage bags, and cans (infant formula). Chemical components of food contact materials may migrate into food at very low levels. For example, infant formula may contain migrants from commercial packaging, such as the canister or bottle the infant formula is packaged in prior to sale. These migrants are similar or identical to those found in foods packaged in similar materials and consumed by a broader age range. Since the infant diet is fairly limited, this may account for their increased exposures to select chemical migrants contained within certain types of food packaging when compared with older age groups (European Food Safety Authority 2005; Foster et al. 2010). In contrast, older consumers experience very low exposures to any one chemical migrant from food packaging, since they eat a variety of foods packaged in different materials. Currently, FDA considers such exposure elements and their impact on infant safety for food contact materials.

The increased exposure of infants to certain types of chemical migrants can also be accounted by their consumption of higher amounts of food, based on the grams of food consumed per kilogram body weight (Lawrie 1998; National Research Council 1993; Neal-Kluever et al. 2014). During infancy, humans exhibit the highest intake of food than at any other age category, including adolescence (Lawrie 1998). Figure 11.1 exhibits the food intake of infants aged 1 year and younger, excluding breast milk due to difficulty in quantifying its intake, from the National Health and Nutrition Examination Survey (NHANES). However, infants fed breast milk also demonstrate increased food consumption during the first 6 months after birth compared to other age groups (Arcus-Arth et al. 2005), and breast milk may also be stored in food contact materials during storage or bottle feeding. Overall, the lack of a diverse diet and the proportionally higher intake of food intake may cause infants to experience higher exposure levels of particular migrants relative to adults.

In addition to having distinct exposures to chemical migrants, infants undergo rapid changes in developing tissues and organ systems that may alter their biological response to chemicals (Neal-Kluever et al. 2014). Although many physiological systems in infants undergo rapid maturational changes, we have highlighted five systems based on their potential for latent, permanent, or delayed adverse effects from chemical migrant exposure. It is important to note that the differences between infants and other life stages may not always predispose infants to toxic injury. As some examples, infants may not express the enzyme needed to activate a chemical to the toxic species, may not express the target of the toxic species, or may clear the toxic species from the body faster than other life stages. Thus, an important element of the infant safety assessment is to identify key differences in the pharmacokinetic (PK) or pharmacodynamic (PD) profile of a specific chemical and whether these differences would result in an elevated sensitivity in infants. To this end, we have also summarized key considerations regarding infant PK or toxicokinetic parameters in the following section.

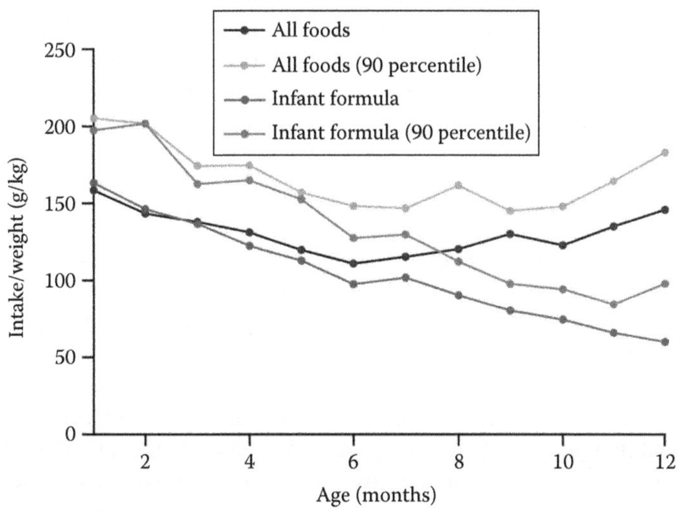

FIGURE 11.1 Mean consumption (grams of food per kilogram of body weight per day) of all foods (infant formula and any other foods) and infant formula (excluding breast milk) by infants during the first 12 months of life. Values are derived from the tabulated two-day National Health and Nutrition Examination Survey data (2005–2010). (From Centers for Disease Control and Prevention, Atlanta, GA; mean and 90th percentile data for age, weight, and daily consumption.)

11.2.1 Nervous System

Several critical developmental processes are ongoing in the nervous system during the immediate postnatal period and do not become complete until years to decades after birth. These include neuronal differentiation, myelination, and migration. Additionally, maturational changes in brain regions during the first postnatal months to years allow for behavioral and motor development. Chemicals that disrupt these maturational processes may have lasting effects on cognition and behavior in children, as identified after exposure to developmental neurotoxicants such as lead, methyl mercury, and ethanol (Choi 1986; Grandjean and Landrigan 2006; Neal and Guilarte 2013). Infants may experience different concentrations of circulating chemicals in the brain due to differences in blood–brain transporter properties, increased brain to body weight ratio, and decreased levels of circulating serum-binding proteins (Saunders et al. 2012, 2013; Watson et al. 2006). Thus, infants may be more susceptible to developmental neurotoxicants on the basis of both toxicokinetic and toxicodynamic factors.

11.2.2 Immune System

The immune system is complex and mediates both innate and adaptive immune functions. Innate immunity is characterized by low specificity and is mediated by soluble factors (e.g., complement) and the actions of leukocytes. Innate immunity is generally considered deficient in infants compared to older age groups and may predispose infants to infection (Kollmann et al. 2012). Adaptive immunity is characterized by high specificity and memory and is mediated largely by the actions of specialized T and B cells. There are two types of adaptive immunity: cell-mediated immunity (CMI: mediated by T cells) and humoral immunity (HI: mediated by B cells). In infants, CMI is considered to be different, but not deficient, based on a well-characterized T-helper type 2 (Th2) cell bias compared to other age groups. Essentially, in infants, the Th2 population is more active and efficient than the Th1 population (Zaghouani et al. 2009). This Th2 bias may predispose infants to higher incidences of allergic reactions and infections compared to other age groups (Kristjansson et al. 2005; Zaghouani et al. 2009). In contrast, HI is considered deficient in infants, as demonstrated by the difficulty in vaccinating very young infants against communicable disease (Siegrist and Aspinall 2009). Infants produce lower levels of antibodies and antibody persistence is decreased for months to years after birth, depending on the type of immunoglobulin (Luebke et al. 2004; Siegrist and Aspinall 2009). Infants further display a limited B-cell response relative to adults or older children (Siegrist and Aspinall 2009). These deficiencies in HI may predispose infants to infection.

11.2.3 Reproductive System

The reproductive organs of human females do not undergo maturational events during infancy. Instead, critical events in the development of female reproductive organs occur prenatally or at adolescence (Beckman and Feuston 2003; Table 11.1). In contrast, the male reproductive system undergoes important maturational events in the developing testes (e.g., Sertoli cells proliferate and mature from birth for about 6 months). The number of Sertoli cells established during the first 6 months directly correlates with the number of spermatogonial cells supported in adulthood and is therefore linked to adult male fertility (Cortes et al. 1987; Lemasters et al. 2000). Furthermore, the developing testes are predisposed to potential toxic injury because the blood–testes barrier may not be fully established until puberty (Camatini et al. 1981; Setchell 2008; Vitale et al. 1973).

As shown in Table 11.1, reproductive organs generally develop *in utero* and are immature at birth in humans, whereas sexual maturation and related events do not occur until puberty (Beckman and Feuston 2003; Lemasters et al. 2000). This presents a challenge for identifying causative or associative agents for reproductive dysfunction, since there can be a prolonged latent period between exposure and presentation of potential adverse outcomes.

TABLE 11.1
Attainment of Developmental Reproductive or Endocrine Milestones in Humans

Reproductive or Endocrine Milestone	Age	
	Male	Female
Folliculogenesis/spermatogonial genesis	Spermatogonia increase from birth through puberty	Folliculogenesis is completed before birth.
Testicular descent	Completed prenatally	N/A.
Preputial separation	Late gestation—3 years	N/A.
Adrenarche	6–8 years (Ibanez et al. 2000)	6–8 years (Ibanez et al. 2000).
Puberty initiation (gonadarche)	12–14 years	8–10 years.
Uterine maturation	N/A	Development begins around the age of 8 and completes at 20 (Holm et al. 1995).
Menarche	N/A	8–14 years
First ovulation	N/A	6 months postmenarche
Spermatogenesis/follicular maturation	Late puberty (mean 13.4 years)	Puberty
Testicular maturation	3 years after puberty onset	N/A
Sexual maturity	Mid–late puberty	11–16 years
Development of HPG axis	Complete at birth	Complete at birth
Development of HPA axis	Complete at birth	Complete at birth

Sources: Neal-Kluever, A. et al., *Food Chem. Toxicol.*, 70C, 68, 2014; Adapted from Marty, M.S. et al., *Birth Defects Res. B, Dev. Reprod. Toxicol.*, 68(2), 125, 2003; Beckman, D.A. and Feuston., M. *Birth Defects Res. B, Dev. Reprod. Toxicol.*, 68(2), 137, 2003 except where references included.

N/A, not applicable.

11.2.4 Endocrine System

Although the actions of a variety of hormonal (endocrine) systems are required for physiological development, three hormonal systems (axes) have particular importance due to the potential for latent, adverse effects. At birth, the hypothalamic–pituitary–gonadal (HPG), hypothalamic–pituitary–adrenal (HPA), and hypothalamic–pituitary–thyroid (HPT) axes are intact and functional. However, after an initial perinatal burst of activity, the HPG and HPA axes both enter a juvenile pause until puberty (Bolt et al. 2002; Chemes 2001; Grumbach 2002). Unlike the HPG and HPA axes, the activity of the HPT axis is sustained after birth and is required for the physiological development of several systems (Raymond and LaFranchi 2010).

Some chemicals have the potential to induce latent effects on the HPA or HPG axes, altering their activity during puberty. In the case of the HPG axis, some evidence suggests that the effects on the HPG by certain chemicals result in altered pubertal timing in humans. Early life exposure to dichlorodiphenyltrichloroethane (DDT) (Ouyang et al. 2005), polybrominated biphenyls (Blanck et al. 2000), and persistent halogenated organic chemicals (Den and Schoeters 2006) has been suggested to either accelerate puberty in girls (Buck Louis et al. 2008) or delay puberty in boys (Buck Louis et al. 2008; Korrick et al. 2011). The potential for delayed adverse effects on reproductive maturation can raise concern for exposure during early life stages. Moreover, adverse effects in HPT signaling during postnatal development may potentially result in abnormal skeletal development, mental retardation, and behavioral deficits (Gilbert et al. 2012; Kim and Mohan 2013; Zimmermann 2011). However, examples of such actions on the HPT axis in humans during postnatal development that result in adverse outcomes are scant (Boas et al. 2012; Goodman et al. 2010).

11.2.5 Skeletal System

Postnatal skeletal development exhibits several milestones that together result in the fusion of ossification centers, increased bone vascularity, as well as longitudinal and diametric growth (Zoetis et al. 2003). Skeletal modeling is particularly active during the early postnatal period when the rate of calcium deposition in infancy exceeds that of adult or even adolescent levels (Abrams 2003; Price et al. 1994). Skeletal modeling allows bones to change shape, gain diameter, and develop the marrow cavity. Disruptions of skeletal modeling can predispose infants to increased fracture risks at later life stages (Cooper et al. 1997; Heaney et al. 2000). Chemicals that may affect the process of skeletal remodeling or growth may thus pose an increased risk for infants due to potential long-term effects.

11.2.6 Pharmacokinetic Processes (Absorption, Distribution, Metabolism, and Excretion)

The PK processes in infants change drastically on a weekly or even daily scale, particularly in the immediate neonatal period (0–4 weeks of age). Furthermore, data regarding the PK behavior of chemicals are limited in infants, except in specific cases of pediatric drugs and a few environmental contaminants. However, PK processes can have an overwhelming impact on how infants respond to exogenous chemicals, resulting in either increased or decreased clearance, metabolism, half-life, bioavailability, distribution, as well as other PK parameters. The following text lists some brief considerations for each of the major PK processes. For further reading, we recommend the excellent reviews by Ginsberg and Alcorn (Alcorn and McNamara 2002, 2003; Ginsberg et al. 2002, 2004a, 2004b). A summary of the maturation of these parameters is shown in Table 11.2.

TABLE 11.2
General Trends in Pharmacokinetic Function in Infants Compared to Adults

PK Parameter	Neonatal: 2 mo	3–5 mo	6–11 mo
Oral absorption rate	~	~	~
GI first-pass effects (Alcorn and McNamara 2003)	↓	↓	↔
GFR (DeWoskin and Thompson 2008)	↓	↓	↔
Maximum tubular secretion rate (DeWoskin and Thompson 2008)	↓	↓	↓
Renal clearance	↓	↔	↔
Biliary clearance (Sussman et al. 1962; Vest 1962)	↓	ND	ND
CYP1A2 activity	↓	↔	↑
CYP2E1 activity	↓	↔	↑
CYP3A (not 3A7) activity	↓	↔	↓
Other CYP activity	↓	↔	↑
Glucuronidation activity (Miyagi and Collier 2007, 2011)	↓	↓	↓
N-acetylation	↓	↓	↓
Sulfation activity (Alcorn and McNamara 2003; Richard et al. 2001)	~	~	~
Epoxide hydrolase activity	↓	↔	↔
Alcohol dehydrogenase activity	↓	↔	↑
Carboxylesterase activity (Pope et al. 2005)	ND	↔	↔

Sources: Neal-Kluever, A. et al., *Food Chem. Toxicol.*, 70C, 68, 2014; Adapted from Ginsberg, G. et al., *Pediatrics*, 113(4), 973, 2004; Ginsberg, G. et al. *Toxicol. Appl. Pharmacol.*, 198(2), 164, 2004b with additions from other sources as noted.

↔ Equivalent to adult; ↑ increased relative to adult; ↓ decreased relative to adult; ~ variable; ND, no data identified.

11.2.6.1 Gastrointestinal Absorption

Infants experience large fluctuations in gastric pH, gastric emptying, gut motility, and intestinal transit time, which all can influence gastrointestinal (GI) absorption. Some classes of chemicals (e.g., metals, organic compounds) exhibit increased total GI absorption in infants compared to adults (Ginsberg et al. 2004b). Factors that affect the rate of GI absorption include the oral bioavailability of chemicals as well as the maturational state of the gut. In the case of the former, the extent of oral bioavailability for most chemicals in infants compared to adults remains largely unknown (Alcorn and McNamara 2003). However, infants have elevated gut permeability in the immediate postnatal period, allowing whole proteins (e.g., immunoglobulins) to be absorbed (Drozdowski et al. 2010; Vukavić 1983, 1984). The extent and molecular weight cutoff for absorption of large chemicals may be influenced by the timing of gut closure, the process of increased selectivity, and lower permeability for GI absorption, but none of these factors are well defined in infants.

11.2.6.2 Distribution

The systemic distribution of a chemical can be influenced by many factors, such as the amount and character of plasma proteins; the relative size of fluid, fat, and tissue compartments of the body; the partition coefficient (solubility) of the compound; and the expression and function of transporters (Neal-Kluever et al. 2014). Depending on physicochemical characteristics of a given chemical, such as its hydrophobicity or hydrophilicity, the volume of distribution and biological half-life may be quite different in infants compared to adults. Furthermore, lower levels of serum-binding proteins may result in increased free drug (or chemical) levels in serum and can be compounded by the presence of increased serum fatty acid and bilirubin levels in the neonate. These endogenous substances can compete with exogenous chemicals for binding sites on albumin and alpha$_1$-acid glycoprotein, thus potentially increasing the amount of unbound toxicant (Alcorn and McNamara 2003).

11.2.6.3 Metabolism

The metabolic profile of infants is significantly different from that of adults. There are large differences in the expression and activities of both phase 1 and phase 2 enzymes (Table 11.2). Overall, comparison of a broad range of metabolic and elimination pathways, for example, cytochrome p450 (CYP) 1A2, CYP2C, CYP3A, miscellaneous CYPs, and glucuronidation, indicates that although metabolic clearance is slower in the immediate postnatal period, it approaches adult levels by 6 months of age (Table 11.2). Although deficient metabolism may be protective if the toxicant requires metabolic activation to produce adverse effects, the potential for adverse events to occur due to deficient metabolic pathways exists. Evidence of this includes adverse outcomes in infants that lack glucuronide conjugation (chloramphenicol: anemia) or reduced CYP activity (valproic acid: hepatotoxicity) (Dreifuss et al. 1987; Ginsberg et al. 2004; Vest 1962). The possibility of an adverse event occurring in infants due to deficient metabolism is dependent on whether sufficient alternative pathways exist in the postnatal period.

11.2.6.4 Excretion

Infants have immature renal and hepatic systems that receive less blood flow and exhibit reduced functional capacity, which together can impede renal and hepatic clearances of chemicals (Alcorn and McNamara 2003; Ginsberg et al. 2004b; Neal-Kluever et al. 2014). In general, the half-life and clearance for many chemicals are prolonged in infants relative to adults until approximately 6 months of age or until renal and hepatic processes mature (Ginsberg et al. 2002).

11.3 SAFETY ASSESSMENT OF PRODUCTS FOR INFANTS

Multiple agencies have developed safety assessments for products that may contact pediatric populations including infants. These agencies include the FDA (food and drugs), the Environmental Protection Agency (EPA: agricultural products), and the Consumer Protection Safety Commission

(CPSC: consumer goods). Section 11.4 will discuss various toxicity testing protocols and strategies that may be informative in infant safety assessments.

11.4 CURRENT PROTOCOLS AND TEST STRATEGIES

11.4.1 Overview

In this section, we review and critique current *d*evelopmental *a*nd *r*eproductive *t*oxicity (DART) studies based on their applicability to infant safety assessment and discuss their strengths and limitations. Design qualities of a study that are pertinent to infant safety assessment include the use of juvenile animals and the incorporation of endpoints sensitive to detect adverse effects in the physiologic systems of interest. It would also be desirable to detect latent effects and differentiate transient events from permanent deleterious effects. Ideally, such studies would also capture relevant information to enable comparison of chemical effects in adults versus offspring. Although no single study design exists to test the points just described, several protocols can be successfully utilized in infant safety assessments provided that any inherent limitations are minimized. Toxicity study protocols that provide useful data for the evaluation of chemical safety within the infant population include the traditional DART studies. These studies typically employ rodents and examine potential chemical adverse effects on adult reproductive systems as well as the developing offspring. While these DART studies are not individually tailored to the conditions of exposure to the human infant and do not have the capability of detecting all possible adverse outcomes, which is biologically, logistically, and economically challenging, many of these protocols do capture relevant endpoints that have utility in safety assessments for this age group (see Table 11.3). Further detailed information on all DART studies can be found within the original documents themselves or collectively in reviews (Parker 2006; Tyl and Marr 2006). Studies contained within "Toxicological Principles for the Safety Assessment of Food Ingredients," otherwise known as Redbook 2000, will be referenced by chapter (e.g., Redbook Chapter IV.C.9.a). Moreover, we will also consider the applicability of other testing approaches such as the nonclinical juvenile toxicity paradigm as well as physiologically based pharmacokinetic (PBPK) modeling.

11.4.2 Screening Studies

Toxicity screening studies (OPPTS 870.3500, OPPTS 870:3650/OECD 422, OPPTS 870.3550/OECD 421) are designed for hazard identification, the first step in risk assessment, and are useful for this purpose for infant safety assessments (OECD 1995, 1996; U.S.E.P.A 1996, 2000a,b). If potential hazards are identified, a screening study can be used to improve subsequent study design, for example, by providing dosage information that can be further applied to more extensive studies or refining the selection of study endpoints. While screening studies require fewer resources than comprehensive toxicity studies (e.g., fewer animals, less test article required, shorter duration, etc.), they are prone to Type I (produce false positives) and Type II statistical errors (produce false negatives) as a consequence of reduced sample size (Holson et al. 2006). Screening studies can vary depending on the treatment periods they require and the types of endpoints they capture. For example, chemical exposure to the pregnant dam in the preliminary developmental toxicity screen (OPPTS 870.3500) is restricted to a specific period of pregnancy, whereas in other screens (e.g., OPPTS 870.3650/OECD 422 and OPPTS 870.3550/OECD 421), it begins prior to pregnancy and ends after birth (refer to Table 11.3). Moreover, while all screening studies capture reproductive and developmental endpoints, only the combined repeated dose toxicity study (OPPTS 870.3560/OECD 422) incorporates neural, immunological, and behavioral endpoints, which potentially increases its likelihood of detecting chemical hazards, although these observations are limited to the adult life stage. Overall, while screening studies are informative, they are not appropriate for making risk assessment decisions and cannot replace comprehensive studies.

TABLE 11.3
Summary of Current Developmental and Reproductive Toxicity (DART) Guidelines

Guideline	Sbj.	Exposure Period	Adverse Effects Assessed	Maternal Endpoints	Offspring Endpoints
OPPTS 870.3500 Preliminary D Tox. Screen	Rat/ms	Portion of organogenesis (i.e., GDs 7–11 for rat and mouse)	Toxic effects to the conceptus after administration to the mother during preg.	bw, CO, gross exam, Tox. signs, preg. data. Moribund/dead: preg. signs and uterine exam, DOD/if they survived until term	Average No. of live/dead at birth and average bw on PNDs 1 and 3, gross anomalies and presence of milk; dead pups: external exam with abnormal gross soft tissue noted
OPPTS 870:3650/ OECD (422): Combined Repeated Dose Tox. Study with R D/D Tox. Screening Test	Rat	♂s: min. 4 wks total (2 wks prior to mating through 2 wks postmating); ♀s: 2 wks prior to mating to PND 4	♂♀: R performance; limited info. on systemic Tox, neurotoxicity, and/or immunotoxicity; offspring: limited detection of postnatal manifestations	bw, clinical BC, CO, FC, FOBs, WC, gross necropsy, mating parameters, No. of imp. sites and corpora lutea, hematological tests, organ histopath. (e.g., liver, kidney, adrenal, thymus, spleen, brain, heart, stomach, intestines, and R organs), DOD/if they survived until term; optional: urinalysis and serum markers of tissue damage	Within 24 hrs. of birth: bw, sex, No. of live pups, still births, and runts, the presence of gross abnormalities; PND 4 live pups: No., bw, sex; CO during lactation (EPA) or abnormal behavior (OECD)
OPPTS 870.3550/ OECD (421): R/D Tox. Screening Test	Rat	♂s: min. 4 wks total (2 wks prior to mating through 2 wks postmating); ♀s: 2 wks prior to mating to PND 4	♂♀: R performance; offspring: limited detection of postnatal manifestations	bw, CO, FC, mating parameters, gross necropsy, No. of imp. sites and corpora lutea, histopath. of R organs, DOD/if they survived until term	Within 24 hrs. of birth: bw, sex, No. of live pups, still births, and runts, the presence of gross abnormalities; repeated for live pups at PND 4; CO during lactation
OECD (414): Prenatal D Tox. Assay/OPPTS 870.3700	Rat/rbt	Imp. to 1 day prior to parturition or fertilization to scheduled C-section	Maternal Tox.; offspring: embryofetal death, altered growth, structural changes	bw, CO, FC, WC, gross necropsy, uterine exam, weight of gravid uterus, No. of corpora lutea, and imp. sites	No. of embryofetal deaths and live fetuses, sex and bw, external changes, skeletal and soft tissue changes; Rbts: internal structures examination in the head (e.g., brain, tongue)
FDA Redbook D Tox. studies (Chapter IV.C.9.b.)	Rat/rbt	Imp. to 1 day prior to parturition or fertilization to scheduled caesarean section	Maternal Tox.; offspring: embryofetal death, altered growth, structural changes, and functional deficiency	bw, CO, FC, WC, gross necropsy, uterine exam, weight of gravid uterus and other organs, No. of corpora lutea and imp. sites	No. of embryofetal deaths and live fetuses, sex and bw, external changes, skeletal and soft tissue changes, fetal crown to rump distance; Rbts: internal structure exam of the head (e.g., brain, tongue)

(Continued)

TABLE 11.3 (*Continued*)
Summary of Current Developmental and Reproductive Toxicity (DART) Guidelines

Guideline	Sbj.	Exposure Period	Adverse Effects Assessed	Maternal Endpoints	Offspring Endpoints
OECD (415:): One-generation R Tox. study	Rat/ms	♂s, 10 wks (rat) or 8 wks (mouse) prior to mating until end of mating; ♀s, 2 wks prior to mating until F1 weaning	♂♀, R performance; offspring: limited detection of postnatal manifestations	bw, CO, FC, gross necropsy, mating parameters, histopath. of R organs if necessary, DOD/if they survived until term.	Morning after birth: bw and sex, No. of stillbirths and live births, gross abnormalities; on PNDs 4, 7, and wkly until study term; No., bw, and CO
OECD (443): Extended One-Generation R Tox. Study	Rat	♂♀s, min. 10 weeks (2 wks prior to mating until of F1 pups); F1 offspring: continuously postweaning until sacrifice	♂♀, R. performance and integrity; offspring: development (physical and functional), DIT and DNT; systemic toxicity in pregnant and lactating ♀s and in young and adult offspring	bw, clinical BC, CO, FC, WC, estrus length and pattern, hematology tests, T4/TSH assays, mating parameters, histopath. and weight of organs (e.g., brain, liver, kidneys, heart, spleen, thymus, pituitary, thyroid, adrenal glands, target organs, and R organs); No. of imp. sites and corpora lutea, quantitative evaluation of primordial follicles, gross necropsy, DOD/if they survived until term	At birth, No. and sex of stillbirths, live births, gross abnormalities; gross necropsy of surplus pups (PND4), AGD: (PNDs 0–4); live pups, bw (on PNDs 0, 1, 4, 7, 21, then wkly until term. including pubertal onset); presence of nipple/areolae in ♂s (PND 12/13); FC and WC postweaning; cohort 1, R D Tox., A) estrus cycle length/pattern, sperm analysis, clinical BC, hematology tests, T4/TSH assays, urinalysis, histopath. and weight of organs (see maternal endpoints, includes lymph nodes, splenic lymphocyte subpop. analysis), No. of corpora lutea and primordial follicles B); weight of R organs and pituitary; cohort 2, DNT, A) FOBs (e.g., following OECD 424 and 426), brain weight and full neurohistopath, B) brain weight and microscopic exam (PND 21/22); cohort 3, DIT, TDAR

(*Continued*)

TABLE 11.3 (*Continued*)
Summary of Current Developmental and Reproductive Toxicity (DART) Guidelines

Guideline	Sbj.	Exposure Period	Adverse Effects Assessed	Maternal Endpoints	Offspring Endpoints
OPPTS 870.3800: R and Fertility Effects/FDA Redbook Guideline for R studies (Chapter IV.C.9.a)/OECD (416): Two-generation R Tox. study	Rat	Two generations: ♀♂s, 10 wks before mating through weaning of F1 offspring; F1, start at weaning, 10 wks before mating, through weaning of F2 offspring	♂♀, R performance and integrity; offspring, growth and development, neonatal morbidity, mortality, lactation, weaning, R capacity, targeted organs, and possible detection of generational cumulative effects	bw, CO, FC, FOBs, WC, gross necropsy, mating parameters, weights and histopath. of organs (e.g., brain, pituitary, liver, kidneys, adrenal glands, spleen, R organs, and targeted organs), estrus lengths/pattern, histopath. of ovary to detect qualitative depletion of primordial follicles, No. of imp. sites	No. and sex of stillbirths, live births, gross abnormalities at birth; on PNDs 1, 4, 7, 21, at pubertal onset, term., No., sex, bw; F1, FC, age of pubertal onset if selected for mating), gross necropsy, histopath. and weights of organs (see maternal endpoints), FOBs (FDA and OECD only); F1♂, sperm analysis; F1♀, estrus cycle length and pattern, quantitative evaluation of primordial follicles, No. of imp. sites; F1/F2 weanlings, weight of the brain, spleen and thymus, gross necropsy; F2, bw, AGD on day 0 if treatment-related effect on sex ratio or sexual maturation of F1; FDA, optional neurotoxicity and immunotoxicity screenings or additional teratology phase
ICH S5(R2): Study 4.1.3	Rat/rbt	Imp. through closure of hard palate	Maternal Tox.; offspring, embryofetal death, altered growth, structural changes	bw, FC, CO, gross evaluation of placenta, No. of corpora lutea and imp. sites, observations that have proved value in other Tox. studies	No. of embryofetal deaths and live fetuses, sex and bw, external changes, skeletal and soft tissue changes
ICH S5(R2): Study 4.1.2	Rat	F0: imp. through end of lactation	Maternal Tox.; offspring, D effects (physical and functional)	bw, CO, FC, FOBs, mating parameters, gross necropsy, No. of imp. sites, observations that have proved value in other Tox. studies	No. of live and dead fetuses and bw at birth, preweaning and postweaning survival and growth/bw, maturation and fertility, FOBs, gross necropsy and possible histopath. of organs with findings

Abbreviations: AGD, anogenital distance; BC, biochemistry; bw, body weight; C-section, caesarean section; CO, clinical observations; D, developmental; DIT, D immunotoxicity test; DNT, D neurotoxicity test; DOD, date of death; FC, feed consumption; ♀, females; FOBs, functional observations (e.g., sensory activity, motor activity, lacrimation, piloerection, unusual respiratory patterns); GD, gestational day; gross necropsy, exam of external surface of the body and all orifices, cavities (cranial, thoracic, and abdominal), and their contents; histopath., histopathology; hrs., hours; Imp., implantation; ♂, male; mating parameters, dates of pairing, insemination date, parturition date, precoital interval, preg. duration; ms, mouse; No., number; PND, postnatal day; preg., pregnancy; pubertal onset, vaginal patency/balanopreputial cleavage; R, reproductive; rbt, rabbit; sbj., subject; subpop., subpopulation; T4, thyroxin; TDAR, T-cell-dependent antibody response assay; term., termination; Tox., toxicity; TSH, thyroid-stimulating hormone; WC, water consumption; wk, week.

11.4.3 Developmental Studies

DART studies vary depending on the time window during which they assess potential adverse chemical effects. The developmental studies (OECD 414: Prenatal Developmental Toxicity Assay/OPPTS 870.3700 and Redbook Chapter IV.C.9.b) focus primarily on possible reproductive toxicities present in the parental generation as well as *in utero* developmental abnormalities induced in the offspring as a result of prenatal exposures (Table 11.3; OECD 2011a; U.S.E.P.A 1998a). There is a distinction between the sensitivities of the developing embryo or fetus and that of the postnatal animal. Therefore, these studies are less informative for infant risk or safety assessment, although they may be helpful in identifying a health risk.

11.4.4 Generational Studies

Generational DART studies (OECD 415, OECD 416, OECD 443, OPPTS 870.3800, and Redbook Chapter IV.C.9.a) are designed to assess fertility and reproductive outcomes after chemical exposure and vary regarding the developmental endpoints that they detect, which in turn establishes their relevance to infant risk or safety assessment (OECD 1983, 2001, 2011b; U.S.E.P.A 1998b). For example, the one-generation reproduction toxicity guideline OECD 415 does not evaluate fertility effects in the offspring (see Table 11.3), making it less informative in the context of infant risk or safety assessment than the two-generational reproduction studies, which comprehensively capture adverse effects in the F_1 offspring during all of its life stages, starting from its conception to the weaning of its own offspring (the F_2 generation). Two-generational studies also include more endpoints that monitor sexual development in the offspring such as balanopreputial separation. For these reasons, the FDA typically recommends two-generational reproduction studies following the Redbook 2000's guidelines, which are similar to the OECD 416 and EPA's OPPTS 870.3800 (see Table 11.3; OECD 2001; U.S.E.P.A 1998b). Using the Redbook study guidelines, potential sensitivity differences in the F_1 offspring versus the parent may be identified, which may address whether juvenile animals are more sensitive to chemical exposure than adults. Such differences may signal the possibility of enhanced risk to the human infant population. Two caveats present in all generational studies are the following: (1) chemical exposures in offspring begin *in utero* and (2) there is a potential for maternally induced developmental adverse events in the pups. The ability to link the onset of adverse events to postnatal chemical exposures can be confounded by *in utero* exposures that occur during testing. Excessive maternal toxicity can give rise to, among other effects, (1) reduced *in utero* nutrient transfer to the offspring, leading to abnormalities in the pups at birth, (2) reduced milk quantity or quality in the mother during nursing, resulting in stunted growth in the pups, and (3) delayed postnatal development of the pups because of maternal neglect. As a result, it is sometimes impossible to identify whether adverse developmental effects are a result of exposure to the test article or due to secondary effects stemming from excessive maternal toxicity (Beyer et al. 2011). Moreover, without additional PK assessments, the amount of chemical that is passed to the offspring by placental and lactational transfer is largely unknown. Despite these potential drawbacks to the two-generational reproduction study, one positive aspect of the study is that direct chemical exposure to F_1 pups postweaning does occur. This is in contrast to the one-generation or teratology studies in which exposure to the offspring occurs only via placental and lactational transfers.

Although the two-generational reproduction studies have an extended period of offspring observation as compared to the one-generation reproduction studies, this test protocol may miss potential latent adverse effects, because sacrifice of the F_1 generation typically occurs immediately after weaning of the F_2 pups. The significance of extending the study period is noted by Holson and colleagues (2006), who cited a study (Snow and Tam 1979) that demonstrated that the phenotypic effects of a chemical can manifest in different ways depending on the time of observation. Deleterious adverse effects from early life exposures that typically manifest into more severe phenotypes later in life may be missed if studies end prematurely. Moreover, the two-generational reproduction study only uses one male or female offspring per litter for further mating and analyses, increasing the likelihood of missing detrimental effects that may occur at lower frequencies (Cooper 2009). To circumvent both issues, Buschmann (2006) suggests recruiting some of the surplus animals from the two-generational reproduction study for a chronic study, which offers the advantage that all life stages would then be examined, would raise the number of offspring per litter that is utilized for both mating and analyses, and would increase the sensitivity of the study. Stand-alone chronic studies (e.g., OPPTS 870.4100/Redbook Chapter IV.C.6) examine latent effects for up to two years in rodents; however, treatment does not begin until after weaning, thereby potentially missing a critical window important for infant safety assessment (U.S.E.P.A 1998c). In addition, although the Redbook 2000 contains a protocol that adds an *in utero* phase to chronic toxicity studies (Redbook Chapter IV.C.8), the endpoints in this study design do not include reproductive endpoints. The National Toxicology Program (NTP) conducted a study on ethinyl estradiol to assess whether it could have cumulative effects on an animal's reproductive system and lead to the development of cancers, by adding a chronic phase to a multigenerational reproductive study. The addition of the chronic study provided latent endpoints such as the presence of uterine polyps that were undetected in the multigenerational study alone, indicating that with this particular chemical, the extension of the study period raised the sensitivity of the study in detecting toxicity (National Toxicology Program 2010). However, this approach is not standard and is logistically and economically challenging; coordination of the starting times for both types of studies might be difficult to achieve due to the increased number of animals in the study and the longer study duration.

The two-generational reproduction study comprehensively captures both reproductive and developmental parameters, but these endpoints may not be the most sensitive for detecting toxicity. For example, in a retrospective study, 33 chemicals were shown to increase toxicity in adult F_1 offspring and subsequent generations versus the parental generation and juvenile F_1 offspring (Piersma et al. 2011). However, only 4 of these 33 chemicals are classified as fertility toxicants by the European Classification, Labeling and Packaging Regulation (EU-CLP) Annex VI. In addition, only two of these chemicals were recognized as developmental toxicants, suggesting that more sensitive parameters exist outside the reproductive and developmental parameters. For example, Hessel et al. (2015) observed that the developing immune system was either more sensitive or equally sensitive than general developmental toxicity parameters in the nine toxicants that they reviewed. Therefore, the range of endpoints captured by a study can affect its overall sensitivity in detecting adverse effects, which, as previously noted, directly impacts the outcome of an infant risk or safety assessment.

The Agricultural Chemical Safety Assessment (ASCA) Technical Committee of the International Life Science Institute Health and Environmental Sciences Institute (ILSI HESI) proposed a modified tiered testing approach, otherwise known as the Extended One Generation Reproductive Toxicity Study (EOGRTS) (OECD 443), which incorporates neural and immune endpoints without a trigger criterion, increases the number of offspring assessed (from one to three per sex per litter), and extends the observational period (OECD 2011b). Instead of using a weight-of-evidence approach to justify the inclusion of specialized cohorts, the EOGRTS recommends a weight-of-evidence approach to *exclude* such cohorts. This potentially results in the default study design including these cohorts, unless there are data to justify their exclusion. Because the EOGRTS incorporates more endpoints beyond reproductive measures, as well as an increase in the use of F_1 offspring, which raises its statistical power, this likely boosts its utility in infant risk or safety assessment. The EOGRTS was designed to have equivalent sensitivity as the two-generational reproduction design in detecting adverse effects but requires fewer resources: it eliminates mating of the F_1 generation unless warranted.

Although the EOGRTS captures neurological endpoints without needing a trigger, the testing paradigm is not as comprehensive as standard developmental neurotoxicity (DNT) testing guidelines

(OPPTS 870.6300, OECD 426) (OECD 2007; U.S.E.P.A 1998d). The EOGRTS assesses animal motor activity only once around postnatal day (PND) 70, which is a study limitation as toxicities can phenotypically manifest differently depending on the period of observation. In contrast, standard DNT test guidelines capture motor activity at multiple time points, which decreases the likelihood of missing an adverse effect. Extending the length of observation may also reveal whether an effect is reversible, but the EOGRTS is also limited in this regard: it stops its neurobehavioral testing and neurohistopathology at young adult stages, requiring studies to be conducted up to PND 75 and PND 90, respectively. The EOGRTS also does not routinely incorporate cognitive testing, such as learning and memory testing, unless prior data suggest such an assessment is needed. This is a detriment to the testing paradigm, as cognitive tests enable researchers to examine multiple systems simultaneously. Learning and memory tasks require the concerted activity of multiple brain regions and the functional involvement of the muscular and skeletal systems. Although it is impossible to pinpoint the mechanistic cause for a failed learning and memory task without additional testing, results from such tests are informative as they indicate whether or not there is a potential for adverse effect as a result of chemical exposure. Overall, the DNT battery that exists within the EOGRTS is limited as compared to the more comprehensive OECD 426 DNT study, which incorporates more endpoints (see Section 5).

Recently, NTP introduced a modified version of the one-generation reproduction study known as the "modified one-generation" (MOG) study design (Foster 2014), which is proposed to be an improvement over the classical multigenerational study design. For example, to raise the sensitivity to detect postnatal developmental effects in comparison to other DART studies, they retain more offspring per litter post weaning for examination (in most DART littering studies, only 1 male or female per litter is retained). The MOG study design can obtain information on the effects of chemicals on prenatal development, postnatal development, and reproduction while requiring fewer animals. Furthermore, the MOG study design divides the offspring of treated rats at PND 21 into different study cohorts: (1) subchronic toxicity cohort (focuses on identifying the targeted organs, (2) teratology cohort, (3) littering cohort, and (4) developmental immunotoxicity (DIT)/DNT cohort, if needed. Although this design permits the evaluation of multiple adverse outcomes at different life stages of a rat simultaneously, identifying the developmental windows negatively affected by chemical exposure is still difficult, due to prenatal contact in the offspring. As discussed earlier, secondary adverse events in offspring due to maternal toxicity may confound the identification of primary, test-article-related adverse events. An improvement to this study in the context of infant risk or safety assessment would be to vary the treatment conditions by having two litter cohorts with one that does not receive any chemical treatments until after the F_1 offspring is weaned, such that the contribution to toxicity by early life exposures can be assessed and any additional effects exerted by direct exposures postnatally can be measured. Moreover, a drawback to the MOG study design is that severe developmental toxicity that leads to reduced pup survival in the F_1 offspring may reduce the number of animals available for all study cohorts.

11.4.5 Specialized Studies

The Redbook 2000 two-generation guideline (Chapter IV.C.9.a) suggests the use of additional DIT or DNT testing based on a weight-of-evidence approach. Similarly, the OECD 416 recommends functional investigations such as sensory function and motor activity measurements (OECD 2001). Although the aim of a weight-of-evidence approach is to reduce upfront costs as well as encourage appropriate follow-up studies, it is unlikely that findings from reproductive endpoints will reveal common DIT changes (Dietert and Zelikoff 2008) that would trigger the need for additional immunotoxicity testing nor would they warrant any further DNT testing. Criteria that could prompt further DNT testing, as designated by the Work Group by the EPA and National Institute of Drug Abuse (NIDA), require prior knowledge of the chemical. Additional information would include whether the chemical is a central nervous system (CNS) behavioral teratogen such as an adult neuropathic/neuroactive agent, or a hormonally active compound and developmental toxicant that does not necessarily produce CNS effects (Claudio et al. 2000). Therefore, without substantial prior knowledge of potential activity of the test article, it is unlikely that DNT testing would occur. A designated section of Redbook 2000 (Chapter IV.C.10) discusses neurotoxicity screening, recommending the use of structure–activity relationships (SARs), published literature and other documentation, as well as empirical testing, for example, conducting separate short-term animal studies with varying chemical dosages and performing histopathology exams and clinical evaluations. In summary, despite the flexibility of the two-generation guideline in expanding its breadth of adverse outcome detection, its weight-of-evidence approach limits the probability that these additional studies actually occur, which potentially reduces its utility in infant risk or safety assessment.

DNT and DIT testing can be important specialized studies to consider for infant risk or safety assessment as the developing immune and nervous systems show distinct properties and susceptibilities from mature systems, as discussed earlier. The general scope of the DNT guideline (OPPTS 870.6300/OECD 426) is the detection of gross neurologic and behavioral deficits throughout postnatal development and adulthood. One key recommendation for this protocol is the incorporation of a positive control, as this helps to establish the sensitivity of a test method (Crofton et al. 2004, 2008). The added value of a DNT study for assessing risk has been reviewed by several groups. For example, a retrospective survey of studies completed by commercial laboratories in the United States, Europe, and Japan on different class of compounds (81% were pharmaceuticals) indicates that neurobehavioral parameters defined the overall *n*o-*o*bserved-*e*ffect *l*evel (NOEL) within the F_1 offspring in 3 out of 113 studies and also had NOELs at doses equivalent to other parameters in 15% of the studies (Middaugh et al. 2003). Moreover, the neurobehavioral parameter was the sole NOEL determinant in two of these studies. Because these parameters defined the overall NOEL in only small number of cases in the retrospective study, this suggests that the DNT study might have little value for risk assessment. However, another recent retrospective analysis revealed that DNT studies that were conducted in accordance with OPPTS 870.6300 or OECD 426 guidelines had a great impact on the risk assessment of pesticides: DNT parameters in the offspring were often the most toxicologically sensitive parameter identified in 69 studies submitted to the EPA Office of Pesticide Programs (OPP) (Raffaele et al. 2010). Of these studies, 23 (33%) deal with adverse developmental neurotoxic effects, which were the sole finding in determining the lowest-observed-effect level (LOEL) (7 for motor, 6 for brain morphometrics, 4 for auditory startle response, and 2 for developmental delays and changes in functional observation battery tasks). These two retrospective studies suggest that the utility of a DNT study can be influenced by the type of chemical that is being assessed. Moreover, as the latter survey reveals, certain DNT endpoints can be more sensitive than others, since clinical signs and neuropathology endpoints were never identified in the effects observed at the LOEL. At the time these studies were conducted, qualitative neuropathological assessments did not require extensive sampling or the use of sensitive stains (Raffaele et al. 2010). NTP has since modified its protocol for neuropathological evaluation in routine toxicity screening of chemicals, drugs, and biological agents by increasing the standard number of required brain slices from three to seven and expanding the anatomical regions evaluated. These modifications offer a more comprehensive examination of the brain and increase the likelihood of detecting a potential neurotoxicant at a larger number of specific sites (Rao et al. 2011, 2014).

Some major challenges for DNT testing include the selection of an appropriate test model (i.e., the toxicokinetic profile of a juvenile animal differs from that of an adult and the toxicokinetic profile of a human also differs from another animal species) and translation of behavioral deficits observed in rodents to human health risks. An additional consideration is that differences between the brain and GI maturation timing may lead to altered absorption of chemicals in a rat pup and decrease its applicability as a model for infant exposures. Although there is no easy solution to these problems, exploring additional models besides rodents for evaluating DNT that more closely resembles humans in physiology may improve the situation. Currently, researchers are exploring both the zebrafish (Selderslaghs et al. 2010; Tierney 2011) and the miniature pig (Bode et al. 2010) as additional test models of neurotoxicity.

While there is an established guideline for DNT testing (OPPTS 870.6300, OECD 426), none exist for evaluating DIT, with the exception of those that are designed to examine adult immunotoxicity (e.g., Redbook Chapter V.C, OPPTS 870.7800 [U.S.E.P.A 1998e], and the ICH(S8) Immunotoxicity Studies for Human Pharmaceuticals, 2006). Direct application of these adult immunotoxicity protocols for assessing DIT is problematic, because some functional endpoints may not be appropriate in developing animals. For example, a 21-day-old rodent that is immunocompetent may exhibit suboptimal T-cell-dependent antigen responses (TDARs) (Collinge et al. 2012), which is also an age that matches the human infantile life stage (Holsapple et al. 2003). However, the value of the TDAR assay to detect DIT in children as young as 18 months of age was showcased in an epidemiological study that used this method to identify an association between exposures to environmental pollutants (i.e., polychlorinated biphenyls) soon after birth and reduced production of antibodies after routine childhood immunizations and therefore reduced immune protection in children (Heilmann et al. 2006). In contrast, a retrospective examination of TDARs conducted in rodents according to the OPPTS 870.7800 guideline, which were exposed to 78 pesticide chemicals from 46 separate chemical classes, indicates that a reduction in TDAR response occurred only in five studies at the high-dose level, suggesting that TDARs are not the most sensitive assay for recognizing DIT (Gehen et al. 2014).

The immunotoxicology testing paradigm that was adopted by the FDA has not been revised since the 1993 "Draft Redbook II," which utilizes a two-level testing strategy approach to animal testing. Level 1 is classified as noninvasive and can be conducted in rodents that are used simultaneously in other toxicity studies (e.g., reproductive studies) and includes hematological, clinical, and pathological screenings. Level 2 is more functionally based and requires concurrent satellite groups of test animals or a separate additional follow-up study (Hinton 1995) that includes assays such as the TDAR. The Redbook II also recommends the use of the maximum tolerated dose (MTD), in addition to an intermediate dose and a control dose, with treatment occurring for only 30 days, whereas the OPPTS 870.7800 stipulates that the highest dose should not produce significant stress, malnutrition, or fatalities but produce some measurable sign of general toxicity. Using the MTD is problematic as it can lead to stress or create immunological effects that are not necessarily relevant to human risk evaluation (Collinge et al. 2012). A major problem with both guidelines is that both focus primarily on the detection of immunosuppression, whereas other possible adverse outcomes may exist, such as hypersensitivity (i.e., allergy development), immune dysfunction, and autoimmunity. Since the 1970s, there has been a steady increase in the incidence of childhood allergy diseases that is partially attributed to environmental factors (e.g., chemicals) (Dietert and Zelikoff 2008). With regard to measuring allergic potential, the FDA Redbook II admits that better animal models aside from the rat are needed for this evaluation (Hinton 1995), but no concrete suggestions are included. Establishing a cohesive DIT guideline is therefore needed and is currently under development (DeWitt et al. 2012).

11.4.6 Nonclinical Juvenile Toxicity Studies

Although the EOGRTS and NTP MOG guideline might have enhanced sensitivity in developmental toxicity detection as compared to two-generational reproduction studies, the prenatal offspring exposure and potential maternal toxicity confounders can reduce their utility in infant safety or risk assessment. Nonclinical juvenile toxicity guidelines address these issues, since the protocol allows targeted dosing of pups prior to weaning. Juvenile toxicity studies are intended to help identify postnatal developmental toxicities that are inadequately assessed in reproductive toxicity assessments and may evaluate differences in sensitivity between adult and immature animals. Both the FDA at the Center for Drug Evaluation and Research (CDER) and the European Medicine Agency (EMEA) have their own guidelines (EMEA 2005; US FDA 2006), which are required for New Drug Applications in North America or Market Authorization Applications in Europe. One major difference between these studies and the DART studies mentioned earlier is that they do not follow a conventional study design and operate on a case-by-case basis, which allows greater flexibility in terms of design. However, there is an intensive planning process, including the (1) age and species of the juvenile animal, (2) sex and sample size, (3) route of administration and dosage formulation, (4) frequency and duration of exposure, and (5) selection of toxicological endpoints and timing of monitoring. Because the subjects of nonclinical juvenile toxicity guidelines are medical compounds that typically have been tested previously in adult animals and where initial PK and PD data likely already exist, the aim of these studies is largely centered on an identified target organ system or organ systems (e.g., reproductive, pulmonary, renal, skeletal, and GI). For example, the particular species of animal and treatment schedules are selected after establishing the comparative developmental status of the test subject and the major organ of concern in humans, which also hinges on pharmacological, PK, and toxicological properties of the medicinal product. Similarly, the selection of toxicological endpoints and timing of observations are chosen to enable the evaluation of drug effects on overall postnatal growth of the animal as well as on postnatal development of specific targeted organ systems.

11.4.7 ICH Guidelines

Additional FDA/CDER guidelines designed for the evaluation of drugs that may be applicable for infant risk or safety assessment include the ICH Harmonized Tripartite Guideline ICH S5(R2) entitled "Detection of Toxicity to Reproduction for Medicinal Products & Toxicity to Male Fertility," which is required when registering new pharmaceuticals*. The guideline is divided into three parts (studies 4.1.1, 4.1.2, and 4.1.3) based on the reproductive stage it covers (refer to Table 11.3). Study 4.1.1 is a teratogenicity study and has little utility with respect to infant safety assessment, since offspring only receive *in utero* chemical exposures and are sacrificed prior to birth. In contrast, study 4.1.2 provides the most relevant information, because the offspring receive direct postnatal chemical exposures and the protocol detects functional deficits (e.g., behavior, maturation, and reproduction), examining both developmental and latent adverse effects until the offspring also reproduces. However, components of study 4.1.2 that reduce its applicability for infant safety assessment are the same as those mentioned for the two-generational reproductive studies. Study 4.1.3 is a developmental toxicity study that is similar to others implemented by other agencies (e.g., FDA: Redbook 2000 Ch.IV.C.9.b, EPA: OPPTS 870.3700, OECD 414) (refer to Table 11.3) and has limited utility in guiding infant safety assessments for the same reasons as study 4.1.1.

11.4.8 DART Study Summary

In this section, we have described the various regulatory protocols that may be useful in an infant risk or safety assessment. By far, the most useful of these are the generational DART studies that incorporate a postnatal exposure period, include a variety of endpoints to detect adverse effects in physiological systems of interest, and have the potential to pick up delayed or latent effects. However, none of the studies include all endpoints relevant to infant safety. For example, reproductive and developmental endpoints contained within a DART study may not be the most sensitive in detecting toxicity, especially in the case of pesticides, which tend to have neurotoxic effects that are captured using neural parameters. Although DART studies monitor the effects of chemicals at multiple life stages and therefore can be used to assess the safety of a range of products to be used in the general population, understanding the risk of exposure to infants requires our focused attention to a specific time period from birth (preweaning) to 6 months of age. Therefore, uncertainty regarding chemical transfer from the mother to the offspring is a large data gap.

---

* Federal Register 59(183). http://www.ich.org/fileadmin/Public_Web_Site/ICH_Products/Guidelines/Safety/S5/Step4/S5_R2__Guideline.pdf. Accessed on September 22, 1994.

11.4.9 PBPK Modeling and Infants

Quantitative chemical risk or safety assessments routinely use an uncertainty factor of 10 to account for sensitive subpopulations and life stages because of a lack of data or gaps in knowledge. These models translate external exposures or intake of chemicals to internal exposures such as blood or organ chemical concentrations. Because the physiology and maturation of organ function and enzymes systems are age dependent, predicting internal exposures to chemicals is a challenge. PBPK models account for age-dependent changes in physiology (e.g., cardiac output, blood flows, and organ growth) and chemical-specific biological determinants (e.g., metabolic rate, urinary excretion, and binding properties). PBPK models have the ability to extrapolate across dose, routes of exposure, and species, while classical PK analyses, which rely on fitting data, are limited in these regard.

A PBPK model is a mathematical representation of the body, which while greatly simplified still accounts for important aspects of physiology and biochemistry. A PBPK model predicts systemic uptake of a chemical by several routes including inhalation and oral ingestion, the movement of a chemical from blood to organs, its metabolism in the liver, and clearance in urine or feces. A variety of simulation software packages are used to perform PBPK model simulations.

A few selected infant and rat pup PBPK models are briefly discussed in the following to demonstrate the utility of early life stage PBPK models. The first series of PBPK models for the nursing infant and rat pup were for a class of chemicals called solvents (Byczkowski and Fisher 1995; Byczkowski et al. 1994; Fisher et al. 1997, 1990). These multiroute solvent exposure models simulated maternal lactational transfer of solvents to the nursing rat pup or infant whereby solvent intake by the nursing rat pup or infant was expressed as mg solvent/kg bw/day. The human maternal/infant simulation models provided estimates of infant exposure to solvents from lactating women occupationally exposed to solvent vapors.

Yoon and colleagues (2011) developed a human nursing neonatal model for manganese (Mn) exposure to the lactating mother and nursing infant. Mn is an essential element and its main source of exposure is from food. Mn is a neurotoxicant at high exposure levels. With underdeveloped homeostatic factors for Mn in neonates, excess exposure of infants to Mn was a health concern. Using a complex model, the authors concluded that the homeostatic set points for the young are different than adults, reflecting the demand for Mn during normal growth and development. Mn is highly regulated in an age-dependent manner by transporter proteins and provides both protection and nutrition to the brain and biliary excretion into the small intestine. Enterocyte turnover during maturation affects the oral uptake of Mn (bioavailability).

Gascon and colleagues (2013) developed a PBPK model based on data from (14 months old, n = 1175), who were exposed to the persistent organic pollutants (POPs), polychlorinated biphenyl (PCB)-153, dichlorodiphenyldichloroethylene (DEE), and hexachlorobenzene (HCB) via breast milk. This PBPK model for POPs was used to assess exposures and accounted for both prenatal and postnatal stages of infancy. The authors concluded that prenatal exposure to PCB-153 was associated with changes in psychomotor development, while postnatal exposure was not.

Building on a series of previous PBPK models for chlorpyrifos, Smith and colleagues (2014) provided theoretical predictions in 6-month-old infants. Concern for infant exposure to chlorpyrifos stems from animal studies that suggested that juvenile animals are more susceptible than adults to chlorpyrifos toxicity. Computational modeling predicted that infants exposed orally to chlorpyrifos at doses of 0.6 mg/kg or higher would have higher serum concentrations of the active metabolite chlorpyrifos-oxon coupled with higher levels of red blood cell cholinesterase inhibition, when compared to adults exposed to an equivalent dose of chlorpyrifos. However, at much lower doses of chlorpyrifos that are more consistent with environmental exposure levels to chlorpyrifos, the reverse was predicted: adults would have slightly higher internal exposures to chlorpyrifos-oxon and greater cholinesterase inhibition compared to 6-month-old infants. In summary, computational models such as the chlorpyrifos model offer valuable insights into dose–response relationships involving pharmacokinetics and pharmacodynamics and ultimately aids in the protection of pediatric populations.

In conclusion, PBPK models for infants or rat pups provide the opportunity to address several questions about early life stages such as evaluating age-dependent toxicity and teasing out PK differences between early and later life stages. The PBPK models can be complex or relatively simple, accounting only for age-dependent information on maturation of body function.

11.5 SUMMARY AND CONCLUSION

U.S. federal agencies have established risk or safety assessment approaches that focus on the developing life stages from the prenatal time period through adulthood. It may be appropriate to perform a separate safety assessment for specific human subpopulations when there are differences in exposure, toxicokinetics, or toxicodynamic parameters compared to the general or adult population (CFSAN Food Advisory Committee 2015). In the context of food packaging, infants are predominantly sole-source consumers for at least the first 6 months after birth, relying on breast milk and/or infant formula for sustenance. Furthermore, infants consume more food per kilogram body weight per day than adults, which may also raise their exposure to chemicals migrating from food packaging. Finally, infants undergo rapid biological maturation and exhibit different physiologies from adults that may impact their potential sensitivity to chemical exposures. To ensure that products are safe for infants, it may be necessary to conduct specialized testing and perform a separate safety assessment for this subpopulation.

This chapter provides additional information regarding the utility of various study protocols for the safety assessment of products for infants, expanding on that given by Neal-Kluever and colleagues (2014). In evaluating chemicals in the context of infant safety, DART studies can often provide useful information, particularly the extended one-generation and two-generation protocols. Although these studies encompass the largest selection of endpoints relevant to infant safety assessments and ensue over the majority of life stages (e.g., pregnancy, fetal development, early postnatal development, young adulthood), data gaps are still present. We will conclude by discussing common data gaps and areas of uncertainty and providing some potential ameliorative approaches.

11.5.1 All Currently Validated DART Protocols Have Inherent Uncertainty Regarding the Internal Exposure of the Pup to the Test Article

Unless a specialized PK/toxicokinetic study is conducted, all the DART study protocols outlined herein rely on maternal transfer of the test article to the fetus or preweaned pup through placental or lactational transfer, respectively. This increases uncertainty in a safety assessment, because it is difficult to know whether the test article is transferred at all, or, if it is transferred, what exposure or dose levels are achieved in the pups.

Potential ameliorative approaches: Direct dosing of the pups in the immediate postnatal period will provide confidence in the exposure level of the pups to the test article. This approach has been successfully used in an extended 90-day subchronic toxicity study design employing bisphenol A as the test article (Delclos et al. 2013); however, oral gavage of the early postnatal pups can be technically challenging. As PBPK models for the infant life stage evolve, computational modeling may be another approach used to estimate the internal dose of preweaned pups to test articles used in regulatory toxicity studies (Patterson et al. 2013; Yang et al. 2013; Yang and Fisher 2014).

11.5.2 Maternal Toxicity Can Confound Developmental Toxicity Studies by Inducing Secondary (Indirect) Adverse Effects on Pup Development

In DART protocols that initiate exposure in the pregnant dam (1-generation, 2-generation, and prenatal toxicity protocols), excessive maternal toxicity can result in indirect effects on the fetal pups that are not a result of direct interaction of the test article (Beyer et al. 2011). In addition to prenatal developmental effects, excessive maternal toxicity can also induce secondary adverse effects in the early postnatal pups. Secondary developmental effects due to excessive maternal toxicity are a common confounder in DART studies and are not a unique problem to safety assessments for infants (Beyer et al. 2011). However, distinguishing primary from secondary effects can increase the understanding of the test article's mode of action.

Potential ameliorative approaches: A dose-range-finding study in pregnant or lactating animals can be conducted to identify appropriate dose levels for a DART study, which should decrease the occurrence of or prevent excessive maternal toxicity. Direct dosing of early postnatal pups may allow isolation of the effects on the dam from that on the pups.

11.5.3 Detection of Delayed or Latent Effects

11.5.3.1 Inclusion of Special Cohorts to Detect Latent or Delayed Effects

As discussed earlier, many traditional toxicity protocols do not include examination of markers for delayed or latent adverse effects. These may include, but are not limited to, developmental neurotoxic effects, developmental immunotoxic effects, as well as endocrine or reproductive effects. The EOGRTS, MOG, and 2-generation protocols allow the inclusion of special cohorts of animals to test endpoints that may detect latent or delayed effects on these physiological systems. However, the need for these cohorts is largely determined by known information on the chemicals prior to the initiation of the study. Thus, for chemicals that lack an extensive dataset, it may be challenging to predict whether special cohorts are needed and, if so, which cohorts and endpoints are appropriate.

Potential ameliorative approaches: Screening studies, including dose-range-finding studies, may provide hazard identification information that can be used to predict whether or not specialized studies or cohorts are needed in a DART study. Additionally, the use of structural alert analysis, such as quantitative structural–activity relationships (QSAR), may provide information regarding the need for specialized studies. In all cases, a dedicated literature search to identify all relevant toxicological information on the test article is essential prior to designing an appropriate study.

11.5.3.2 Coverage of All Relevant Life Stages

Early life exposures can have the potential for lasting effects that may present in later life stages. Some latent effects will present only at specific life stages; for example, the effects on fertility may only present during the reproductive life stage. Additionally, a neurodegenerative effect may present only in elderly animals. Currently, the extended one-generational or two-generational reproduction protocols test the largest array of life stages and exclude only the elderly life stage. The *in utero* addition to a carcinogenicity study (Redbook Ch IV.C.8) may provide information on an elderly life stage but does not include sufficient endpoints to test for DART.

Potential ameliorative approaches: Should a need be identified for testing at later life stages, a chronic phase could be added to a generational protocol. However, this is a potentially onerous addition to an already economically costly protocol. Unfortunately, scientific understanding of early life effects that present in elderly life stages remains at a rudimentary level. As more is learned about this type of latent or delayed effect, additional approaches may be identified.

11.5.4 Identification of the Appropriate Animal Model for Testing

Finding an appropriate animal model that best represents a human infant is challenging due to differences in the developmental milestones of various organ systems across species among many other factors. For example, assigning an exposure regimen in a rodent to match the maturation status of the human infant brain, in the case of a DNT study, does not necessarily mean that the GI system will also correspond developmentally: a crawling rat pup on PNDs 3–12 corresponds to a nine-month old crawling baby (Wood et al. 2003), while a 20-day-old rat has a comparative intestinal barrier maturity as observed in a full-term human baby during the first week of postnatal life (McCracken and Lorenz 2001) or has a gut that resembles that of an early preterm infant at birth (McCracken and Lorenz 2001; Puiman and Stoll 2008).

Potential ameliorative approaches: A series of informative reviews examining the comparative development of individual different organs/systems across species were published in 2003 by *Birth Defects Research Part B* and remain the most authoritative treatise on the subject (Beckman and Feuston 2003; Holsapple et al. 2003; Marty et al. 2003; Wood et al. 2003; Zoetis et al. 2003; Zoetis and Hurtt 2003). These articles can serve as a starting point when designing or interpreting animal studies that model humans. While none of the animal models used in laboratory research will perfectly mimic all of the organ systems of a human infant, it may be possible to optimize, at a minimum, for the primary organ system or systems of interest. Identification of the apical adverse effects of a test article on developing systems is essential to determine the primary organ system of interest. Therefore, consideration of all available toxicological data, including that of screening studies and *in silico* modeling, may be necessary before determining the appropriate test species and age for a DART study.

REFERENCES

Abrams, S. A. 2003. Normal acquisition and loss of bone mass. *Hormone Research* 60 (Suppl 3): 71–76, PM:14671401.

Alcorn, J. and P. J. McNamara. 2002. Ontogeny of hepatic and renal systemic clearance pathways in infants: Part I. *Clinical Pharmacokinetics* 41 (12): 959–998, PM:12222995.

Alcorn, J. and P.J. McNamara. 2003. Pharmacokinetics in the newborn. *Advanced Drug Delivery Reviews* 55 (5): 667–686, PM:12706549.

Arcus-Arth, A., G. Krowech, and L. Zeise. 2005. Breast milk and lipid intake distributions for assessing cumulative exposure and risk. *Journal of Exposure Analysis and Environmental Epidemiology* 15 (4): 357–365, PM:15562290.

Beckman, D. A. and M. Feuston. 2003. Landmarks in the development of the female reproductive system. *Birth Defects Research. Part B, Developmental and Reproductive Toxicology* 68 (2): 137–143, PM:12866705.

Beyer, B. K., N. Chernoff, B. R. Danielsson, K. Davis-Bruno, W. Harrouk, R. D. Hood, G. et al. 2011. ILSI/HESI maternal toxicity workshop summary: Maternal toxicity and its impact on study design and data interpretation. *Birth Defects Research Part B: Developmental and Reproductive Toxicology* 92 (1): 36–51, PM:21312321.

Blanck, H. M., M. Marcus, P. E. Tolbert, C. Rubin, A. K. Henderson, V. S. Hertzberg, R. H. Zhang, and L. Cameron. 2000. Age at menarche and tanner stage in girls exposed in utero and postnatally to polybrominated biphenyl. *Epidemiology (Cambridge, Mass.)* 11 (6): 641–647, PM:11055623.

Boas, M., U. Feldt-Rasmussen, and K. M. Main. 2012. Thyroid effects of endocrine disrupting chemicals. *Molecular and Cellular Endocrinology* 355 (2): 240–248, PM:21939731.

Bode, G., P. Clausing, F. Gervais, J. Loegsted, J. Luft, V. Nogues, and J. Sims. 2010. The utility of the minipig as an animal model in regulatory toxicology. *Journal of Pharmacological and Toxicological Methods* 62 (3): 196–220, PM:20685310.

Bolt, R. J., M. M. van Weissenbruch, H. N. Lafeber, and Delemarre-van de Waal HA. 2002. Development of the hypothalamic-pituitary-adrenal axis in the fetus and preterm infant. *Journal of Pediatric Endocrinology & Metabolism* 15 (6): 759–769, PM:12099385.

Buck Louis, G. M., L. E. Gray, Jr., M. Marcus, S. R. Ojeda, O. H. Pescovitz, S. F. Witchel, W. Sippell et al. 2008. Environmental factors and puberty timing: Expert panel research needs. *Pediatrics* 121 Suppl 3: S192–S207, PM:18245512.

Buschmann, J. 2006. Critical aspects in reproductive and developmental toxicity testing of environmental chemicals. *Reproductive Toxicology* 22 (2): 157–163, PM:16782303.

Byczkowski, J. Z. and J. W. Fisher. 1995. A computer program linking physiologically based pharmacokinetic model with cancer risk assessment for breast-fed infants. *Computer Methods and Programs in Biomedicine* 46 (2): 155–163, PM:7796584.

Byczkowski, J. Z., J. M. Gearhart, and J. W. Fisher. 1994. "Occupational" exposure of infants to toxic chemicals via breast milk. *Nutrition (Burbank, Los Angeles County, Calif.)* 10 (1): 43–48, PM:8199422.

Camatini, M., E. Franchi, and I. DeCurtis. 1981. Differentiation of inter-sertoli junctions in human testis. *Cell Biology International Reports* 5 (2): 109.

CFSAN Food Advisory Committee. 2015. Susceptible Populations held December 16–17, 2014 at White Oak, MD.

Chemes, H. E. 2001. Infancy is not a quiescent period of testicular development. *International Journal of Andrology* 24 (1): 2–7, PM:11168644.

Choi, B. H. 1986. Methylmercury poisoning of the developing nervous system: I. Pattern of neuronal migration in the cerebral cortex. *Neurotoxicology* 7 (2): 591–600, PM:3537862.

Claudio, L., W. C. Kwa, A. L. Russell, and D. Wallinga. 2000. Testing methods for developmental neurotoxicity of environmental chemicals. *Toxicology and Applied Pharmacology* 164 (1): 1–14, PM:10739739.

Collinge, M., L. A. Burns-Naas, G. J. Chellman, T. T. Kawabata, W. J. Komocsar, J. R. Piccotti, J. Shenton, and D. Wierda. 2012. Developmental immunotoxicity (DIT) testing of pharmaceuticals: Current practices, state of the science, knowledge gaps, and recommendations. *Journal of Immunotoxicology* 9 (2): 210–230, PM:22428536.

Cooper, C., C. Fall, P. Egger, R. Hobbs, R. Eastell, and D. Barker. 1997. Growth in infancy and bone mass in later life. *Annals of the Rheumatic Diseases* 56 (1): 17–21, PM:9059135.

Cooper, R. L. 2009. Current developments in reproductive toxicity testing of pesticides. *Reproductive Toxicology* 28 (2): 180–187, http://www.sciencedirect.com/science/article/pii/S0890623809001014.

Cortes, D., J. Muller, and N. E. Skakkebaek. 1987. Proliferation of Sertoli cells during development of the human testis assessed by stereological methods. *International Journal of Andrology* 10 (4): 589–596, PM:3654012.

Crofton, K. M., J. A. Foss, U. Hass, K. F. Jensen, E. D. Levin, and S. P. Parker. 2008. Undertaking positive control studies as part of developmental neurotoxicity testing: A report from the ILSI Research Foundation/Risk Science Institute expert working group on neurodevelopmental endpoints. *Neurotoxicology and Teratology* 30 (4): 266–287, PM:17681747.

Crofton, K. M., S. L. Makris, W. F. Sette, E. Mendez, and K. C. Raffaele. 2004. A qualitative retrospective analysis of positive control data in developmental neurotoxicity studies. *Neurotoxicology and Teratology* 26 (3): 345–352, PM:15113596.

Delclos, K. B., L. Camacho, S. M. Lewis, M. M. Vanlandingham, J. R. Latendresse, G. R. Olson, K. J. Davis et al. 2013. Toxicity evaluation of bisphenol A administered by gavage to Sprague-Dawley rats from gestation day 6 through postnatal day 90. *Toxicological Sciences* 139 (1): 174–197, PM:24496637.

Den, H. E. and G. Schoeters. 2006. Endocrine disrupters and human puberty. *International Journal of Andrology* 29 (1): 264–271, PM:16466548.

DeWitt, J. C., M. M. Peden-Adams, D. E. Keil, and R. R. Dietert. 2012. Current status of developmental immunotoxicity: Early-life patterns and testing. *Toxicologic Pathology* 40 (2): 230–236, PM:22109713.

Dietert, R. R. and J. T. Zelikoff. 2008. Early-life environment, developmental immunotoxicology, and the risk of pediatric allergic disease including asthma. *Birth Defects Research. Part B, Developmental and Reproductive Toxicology* 83 (6): 547–560, PM:19085948.

Dreifuss, F. E., N. Santilli, D. H. Langer, K. P. Sweeney, K. A. Moline, and K. B. Menander. 1987. Valproic acid hepatic fatalities: a retrospective review. *Neurology* 37 (3): 379–385, PM:3102998.

Drozdowski, L. A., T. Clandinin, and A. B. Thomson. 2010. Ontogeny, growth and development of the small intestine: Understanding pediatric gastroenterology. *World Journal of Gastroenterology* 16 (7): 787–799, PM:20143457.

EMEA (2005). Guideline on the need for non-clinical testing in juvenile animals on human pharmaceuticals for pediatric indications. Available at: http://www.ema.europa.eu/docs/en_GB/document_library/Scientific_guideline/2009/09/WC500003306.pdf. Accessed on June 30, 2016.

European Food Safety Authority. 2005. Opinion of the scientific panel on food additives, flavourings, processing aids and materials in contact with food on a request from the commission related to semicarbazide in food. *The EFSA Journal* 219: 1–36.

Fisher, J., D. Mahle, L. Bankston, R. Greene, and J. Gearhart. 1997. Lactational transfer of volatile chemicals in breast milk. *American Industrial Hygiene Association Journal* 58 (6): 425–431, PM:9183837.

Fisher, J. W., T. A. Whittaker, D. H. Taylor, H. J. Clewell, III, and M. E. Andersen. 1990. Physiologically based pharmacokinetic modeling of the lactating rat and nursing pup: A multiroute exposure model for trichloroethylene and its metabolite, trichloroacetic acid. *Toxicology and Applied Pharmacology* 102 (3): 497–513, PM:2315918.

Food and Drug Administration of the Health and Human Services Department. April 13, 2006. Immunotoxicity studies for human pharmaceuticals. *Federal Register* 71: 19193–19194.

Food Quality Protection Act. 1996. Pub. L. no. 104-170, 110 Stat 1489.

Foster, E., J. C. Mathers, and A. J. Adamson. 2010. Packaged food intake by British children aged 0 to 6 years. *Food Additives and Contaminants. Part A, Chemistry, Analysis, Control, Exposure and Risk Assessment* 27 (3): 380–388, PM:20063226.

Foster, P. M. 2014. Regulatory Forum opinion piece: New testing paradigms for reproductive and developmental toxicity—The NTP modified one generation study and OECD 443. *Toxicologic Pathology* 42 (8): 1165–1167, PM:24862797.

Gascon, M., M. A. Verner, M. Guxens, J. O. Grimalt, J. Forns, J. Ibarluzea, N. Lertxundi et al. 2013. Evaluating the neurotoxic effects of lactational exposure to persistent organic pollutants (POPs) in Spanish children. *Neurotoxicology* 34: 9–15, PM:23085522.

Gehen, S. C., A. M. Blacker, D. R. Boverhof, T. R. Hanley, Jr., C. E. Hastings, G. S. Ladics, H. Lu, and F. O. O'Neal. 2014. Retrospective evaluation of the impact of functional immunotoxicity testing on pesticide hazard identification and risk assessment. *Critical Reviews in Toxicology* 44 (5): 407–419, PM:24601769.

Gilbert, M. E., J. Rovet, Z. Chen, and N. Koibuchi. 2012. Developmental thyroid hormone disruption: Prevalence, environmental contaminants and neurodevelopmental consequences. *Neurotoxicology* 33 (4): 842–852, PM:22138353.

Ginsberg, G., D. Hattis, R. Miller, and B. Sonawane. 2004a. Pediatric pharmacokinetic data: Implications for environmental risk assessment for children. *Pediatrics* 113 (4): 973–983.

Ginsberg, G., D. Hattis, and B. Sonawane. 2004b. Incorporating pharmacokinetic differences between children and adults in assessing children's risks to environmental toxicants. *Toxicology and Applied Pharmacology* 198 (2): 164–183, PM:15236952.

Ginsberg, G., D. Hattis, B. Sonawane, A. Russ, P. Banati, M. Kozlak, S. Smolenski, and R. Goble. 2002. Evaluation of child/adult pharmacokinetic differences from a database derived from the therapeutic drug literature. *Toxicological Sciences: An Official Journal of the Society of Toxicology* 66 (2): 185–200, PM:11896285.

Goodman, J. E., L. E. Kerper, C. P. Boyce, R. L. Prueitt, and L. R. Rhomberg. 2010. Weight-of-evidence analysis of human exposures to dioxins and dioxin-like compounds and associations with thyroid hormone levels during early development. *Regulatory Toxicology and Pharmacology* 58 (1): 79–99, PM:20416351.

Grandjean, P. and P. J. Landrigan. 2006. Developmental neurotoxicity of industrial chemicals. *Lancet* 368 (9553): 2167–2178, PM:17174709.

Grumbach, M. M. 2002. The neuroendocrinology of human puberty revisited. *Hormone Research* 57 Suppl 2: 2–14, PM:12065920.

Heaney, R. P., S. Abrams, B. Dawson-Hughes, A. Looker, R. Marcus, V. Matkovic, and C. Weaver. 2000. Peak bone mass. *Osteoporosis International: A Journal Established as Result of Cooperation between the European Foundation for Osteoporosis and the National Osteoporosis Foundation of the USA* 11 (12): 985–1009, PM:11256898.

Heilmann, C., P. Grandjean, P. Weihe, F. Nielsen, and E. Budtz-Jorgensen. 2006. Reduced antibody responses to vaccinations in children exposed to polychlorinated biphenyls. *PLoS Medicine* 3 (8): e311, PM:16942395.

Hessel, E. V., E. C. Tonk, P. M. Bos, Loveren H. van, and A. H. Piersma. 2015. Developmental immunotoxicity of chemicals in rodents and its possible regulatory impact. *Critical Reviews in Toxicology* 45 (1): 68–82, PM:25372701.

Hinton, D. M. 1995. Immunotoxicity testing applied to direct food and colour additives: US FDA 'Redbook II' Guidelines. *Human & Experimental Toxicology* 14 (1): 143–145, PM:7748605.

Holsapple, M. P., L. J. West, and K. S. Landreth. 2003. Species comparison of anatomical and functional immune system development. *Birth Defects Research. Part B, Developmental and Reproductive Toxicology* 68 (4): 321–334, PM:14666995.

Holson, J. F., M. D. Nemec, D. G. Stump, L. E. Kaufman, P. Lindström, and B. J. Varsho. 2006. Significance, reliability, and interpretation of developmental and reproductive toxicity study findings. In R. D. Hood (ed.), *Developmental and Reproductive Toxicology: A Practical Approach*, 2nd edn. Boca Raton, FL: CRC Press, pp. 329–423.

Ibanez, L., J. Dimartino-Nardi, N. Potau, and P. Saenger. 2000. Premature adrenarche-normal variant or forerunner of adult disease? *Endocrine Reviews* 21 (6): 671–696, PM:11133068.

Kim, H.-Y. and S. Mohan. 2013. Role and mechanisms of actinos of thyroid horone on the skeletal development. *Bone Research* 2: 146–161.

Kollmann, T. R., O. Levy, R. R. Montgomery, and S. Goriely. 2012. Innate immune function by Toll-like receptors: Distinct responses in newborns and the elderly. *Immunity* 37 (5): 771–783, PM:23159225.

Korrick, S. A., M. M. Lee, P. L. Williams, O. Sergeyev, J. S. Burns, D. G. Patterson, W. E. Turner et al. 2011. Dioxin exposure and age of pubertal onset among Russian boys. *Environmental Health Perspectives* 119 (9): 1339–1344, PM:21527364.

Kristjansson, S., S. P. Bjarnarson, G. Wennergren, A. H. Palsdottir, T. Arnadottir, A. Haraldsson, and I. Jonsdottir. 2005. Respiratory syncytial virus and other respiratory viruses during the first 3 months of life promote a local TH2-like response. *Journal of Allergy and Clinical Immunology* 116 (4): 805–811, PM:16210054.

Lawrie, C. A. 1998. Different dietary patterns in relation to age and the consequences for intake of food chemicals. *Food Additives and Contaminants* 15 (Suppl): 75–81, PM:9602915.

Lemasters, G. K., S. D. Perreault, B. F. Hales, M. Hatch, A. N. Hirshfield, C. L. Hughes, G. L. Kimmel et al. 2000. Workshop to identify critical windows of exposure for children's health: Reproductive health in children and adolescents work group summary. *Environmental Health Perspectives* 108 Suppl 3: 505–509, PM:10852850.

Luebke, R. W., C. Parks, and M. I. Luster. 2004. Suppression of immune function and susceptibility to infections in humans: Association of immune function with clinical disease. *Journal of Immunotoxicology* 1 (1): 15–24, PM:18958637.

Marty, M. S., R. E. Chapin, L. G. Parks, and B. A. Thorsrud. 2003. Development and maturation of the male reproductive system. *Birth Defects Research. Part B, Developmental and Reproductive Toxicology* 68 (2): 125–136, PM:12866704.

McCracken, V. J. and R. G. Lorenz. 2001. The gastrointestinal ecosystem: A precarious alliance among epithelium, immunity and microbiota. *Cellular Microbiology* 3 (1): 1–11, PM:11207615.

Middaugh, L. D., D. Dow-Edwards, A. A. Li, J. D. Sandler, J. Seed, L. P. Sheets, D. L. Shuey et al. 2003. Neurobehavioral assessment: a survey of use and value in safety assessment studies. *Toxicological Sciences* 76 (2): 250–261, PM:12915715.

National Research Council. 1993. *Pesticides in the Diets of Infants and Children*. Washington, DC: National Academy Press.

National Toxicology Program. 2010. NTP Technical Report on the Toxicology and Carcinogenesis Study of Ethinyl Estradiol in Sprague-Dawley Rats, Research Triangle Park, NC.

Neal, A. P. and T. R. Guilarte. 2013. Mechanisms of lead and manganese neurotoxicity. *Toxicology Research* 2: 99–114.

Neal-Kluever, A., J. Aungst, Y. Gu, K. Hatwell, K. Muldoon-Jacobs, A. Liem, A. Ogungbesan, and M. Shackelford. 2014. Infant toxicology: State of the science and considerations in evaluation of safety. *Food and Chemical Toxicology* 70C: 68–83, PM:24824476.

OECD. 1983. Test No. 415: One-generation reproduction toxicity study. *OECD Guidelines for the Testing of Chemicals* Section 4, /content/book/9789264070844-en.

OECD. 1995. Test No. 421: Reproduction/developmental toxicity screening test. *OECD Guidelines for the Testing of Chemicals* Section 4, /content/book/9789264070967-en.

OECD. 1996. Test No. 422: Combined repeated dose toxicity study with the reproduction/developmental toxicity screening test. *OECD Guidelines for the Testing of Chemicals* Section 4, /content/book/9789264070981-en.

OECD. 2001. Test No. 416: Two-generation reproduction toxicity. *OECD Guidelines for the Testing of Chemicals* Section 4, /content/book/9789264070868-en.

OECD. 2007. Test No. 426: Developmental neurotoxicity study. *OECD Guidelines for the Testing of Chemicals* Section 4, /content/book/9789264067394-en.

OECD. 2011a. Test No. 414: Prenatal development toxicity study. *OECD Guidelines for the Testing of Chemicals* Section 4, http://www.oecd-ilibrary.org/environment/test-no-414-prenatal-development-toxicity-study_9789264070820-en.

OECD. 2011b. Test No. 443: Extended one-generation reproductive toxicity study. *OECD Guidelines for the Testing of Chemicals* Section 4, /content/book/9789264185371-en.

Ouyang, F., M. J. Perry, S. A. Venners, C. Chen, B. Wang, F. Yang, Z. Fang, T. Zang, L. Wang, X. Xu, and X. Wang. 2005. Serum DDT, age at menarche, and abnormal menstrual cycle length. *Occupational and Environmental Medicine* 62 (12): 878–884, PM:16299097.

Parker, R. M. 2006. Testing for reproductive toxicity. In R.D. Hood (ed.), *Developmental and Reproductive Toxicology*, 2nd edn. Boca Raton, FL: CRC Press, pp. 425–487.

Patterson, T. A., N. C. Twaddle, C. S. Roegge, R. J. Callicott, J. W. Fisher, and D. R. Doerge. 2013. Concurrent determination of bisphenol A pharmacokinetics in maternal and fetal rhesus monkeys. *Toxicology and Applied Pharmacology* 267 (1): 41–48, PM:23261975.

Piersma, A. H., E. Rorije, M. E. W. Beekhuijzen, R. Cooper, D. J. Dix, B. Heinrich-Hirsch, M. T. Martin et al. 2011. Combined retrospective analysis of 498 rat multi-generation reproductive toxicity studies: On the impact of parameters related to F1 mating and F2 offspring. *Reproductive Toxicology* 31 (4): 392–401, http://www.sciencedirect.com/science/article/pii/S0890623810003515.

Price, J. S., B. O. Oyajobi, and R. G. Russell. 1994. The cell biology of bone growth. *European Journal of Clinical Nutrition* 48 Suppl 1: S131–S149, PM:7516288.

Puiman, P. and B. Stoll. 2008. Animal models to study neonatal nutrition in humans. *Current Opinion in Clinical Nutrition and Metabolic Care* 11 (5): 601–606, PM:18685456.

Raffaele, K. C., J. Rowland, B. May, S. L. Makris, K. Schumacher, and L. J. Scarano. 2010. The use of developmental neurotoxicity data in pesticide risk assessments. *Neurotoxicology and Teratology* 32 (5): 563–572, PM:20398750.

Rao, D. B., P. B. Little, D. E. Malarkey, R. A. Herbert, and R. C. Sills. 2011. Histopathological evaluation of the nervous system in National Toxicology Program rodent studies: A modified approach. *Toxicologic Pathology* 39 (3): 463–470, PM:21430177.

Rao, D. B., P. B. Little, and R. C. Sills. 2014. Subsite awareness in neuropathology evaluation of National Toxicology Program (NTP) studies: A review of select neuroanatomical structures with their functional significance in rodents. *Toxicologic Pathology* 42 (3): 487–509, PM:24135464.

Raymond, J. and S. H. LaFranchi. 2010. Fetal and neonatal thyroid function: review and summary of significant new findings. *Current Opinion in Endocrinology, Diabetes, and Obesity* 17 (1): 1–7, PM:19901830.

Saunders, N. R., R. Daneman, K. M. Dziegielewska, and S. A. Liddelow. 2013. Transporters of the blood-brain and blood-CSF interfaces in development and in the adult. *Molecular Aspects of Medicine* 34 (2–3): 742–752, PM:23506907.

Saunders, N. R., S. A. Liddelow, and K. M. Dziegielewska. 2012. Barrier mechanisms in the developing brain. *Frontiers in Pharmacology* 3: 46, PM:22479246.

Selderslaghs, I. W., J. Hooyberghs, Coen W. De, and H. E. Witters. 2010. Locomotor activity in zebrafish embryos: A new method to assess developmental neurotoxicity. *Neurotoxicology and Teratology* 32 (4): 460–471, PM:20211722.

Setchell, B. P. 2008. Blood-testis barrier, junctional and transport proteins and spermatogenesis. *Advances in Experimental Medicine and Biology* 636: 212–233, PM:19856170.

Siegrist, C. A. and R. Aspinall. 2009. B-cell responses to vaccination at the extremes of age. *Nature Reviews Immunology* 9 (3): 185–194, PM:19240757.

Smith, J. N., P. M. Hinderliter, C. Timchalk, M. J. Bartels, and T. S. Poet. 2014. A human life-stage physiologically based pharmacokinetic and pharmacodynamic model for chlorpyrifos: Development and validation. *Regulatory Toxicology and Pharmacology* 69 (3): 580–597, PM:24200834.

Snow, M. H., and P. P. Tam. 1979. Is compensatory growth a complicating factor in mouse teratology? *Nature* 279 (5713): 555–557, PM:571964.

Tierney, K. B. 2011. Behavioural assessments of neurotoxic effects and neurodegeneration in zebrafish. *Biochimica et Biophysica Acta* 1812 (3): 381–389, PM:21035547.

Tyl, R. W. and M. C. Marr. 2006. Developmental toxicity test methodology. In R.D. Hood (ed.), *Developmental and Reproductive Toxicology: A Practical Approach*, 2nd edn. Boca Raton, FL: CRC Press, pp. 201–261.

U.S.E.P.A. 1996. OPPTS 870.3500 Preliminary developmental toxicity screen. *Health Effects Test Guidelines* (EPA 712-C-96-205), http://nepis.epa.gov/Exe/ZyNET.exe/P100IRM8.TXT?ZyActionD=ZyDocument&Client=EPA&Index=1995±Thru±1999&Docs=&Query=&Time=&EndTime=&SearchMethod=1&TocRestrict=n&Toc=&TocEntry=&QField=&QFieldYear=&QFieldMonth=&QFieldDay=&IntQFieldOp=0&ExtQFieldOp=0&XmlQuery=&File=D%3A\zyfiles\Index%20Data\95thru99\Txt\00000034\P100IRM8.txt&User=ANONYMOUS&Password=anonymous&SortMethod=hl-&MaximumDocuments=1&FuzzyDegree=0&ImageQuality=r75g8/r75g8/x150y150g16/i425&Display=plf&DefSeekPage=x&SearchBack=ZyActionL&Back=ZyActionS&BackDesc=Results%20page&MaximumPages=1&ZyEntry=1&SeekPage=x&ZyPURL.

U.S.E.P.A. 1998a. OPPTS 870.3700 Prenatal developmental toxicity study. *Health Effects Test Guidelines* EPA 712-C-98-207, http://www.regulations.gov/#!documentDetail;D=EPA-HQ-OPPT-2009-0156-0017.

U.S.E.P.A. 1998b. OPPTS 870.3800 Reproduction and fertility effects. *Health Effects Test Guidelines* (EPA 712-C-98-208), http://www.regulations.gov/#!documentDetail;D=EPA-HQ-OPPT-2009–0156-0018.

U.S.E.P.A. 1998c. OPPTS 870.4100 Chronic toxicity. *Health Effects Test Guidelines* (EPA 712-C-98-210), http://www.regulations.gov/#!documentDetail;D=EPA-HQ-OPPT-2009-0156-0019.

U.S.E.P.A. 1998d. OPPTS 870.6300 Developmental neurotoxicity study. *Health Effects Test Guidelines* (EPA 712-C-98-239), http://www.regulations.gov/#!documentDetail;D=EPA-HQ-OPPT-2009*0156-0042.

U.S.E.P.A. 1998e. OPPTS 870.7800 Immunotoxicity. *Health Effects Test Guidelines* (EPA 712-C-98-351), http://www.regulations.gov/#!documentDetail;D=EPA-HQ-OPPT-2009-0156-0049.

U.S.E.P.A. 2000a. OPPTS 870.3550 Reproduction/developmental toxicity screening test. *Health Effects Test Guidelines* (EPA 712-C-00-367), http://www.regulations.gov/#!documentDetail;D=EPA-HQ-OPPT-2009-0156-0015.

U.S.E.P.A. 2000b. OPPTS 870.3650 Combined repeated dose toxicity study with the reproduction/developmental toxicity screening test. *Health Effects Test Guidelines* (EPA 712-C-00-368), http://www.regulations.gov/#!documentDetail;D=EPA-HQ-OPPT-2009*0156-0016.

US FDA. 2006. Guidance for industry: Nonclinical safety evaluation of pediatric drug products. Available at http://www.fda.gov/downloads/drugs/guidancecomplianceregulatoryinformation/guidances/ucm079247.pdf. Accessed on June 30, 2016.

Vest, M. F. 1962. Conjugation of sulfobromophthalein in newborn infants and children. *The Journal of Clinical Investigation* 41: 1013–1020, PM:13925656.

Vitale, R., D. W. Fawcett, and M. Dym. 1973. The normal development of the blood-testis barrier and the effects of clomiphene and estrogen treatment. *Anatomical Record* 176 (3): 331–344, PM:4716418.

Vukavić, T. 1983. Intestinal absorption of IgA in the newborn. *Journal of Pediatric Gastroenterology and Nutrition* 2 (2): 248–251, PM:6875749.

Vukavić, T. 1984. Timing of the gut closure. *Journal of Pediatric Gastroenterology and Nutrition* 3 (5): 700–703, PM:6502370.

Watson, R. E., J. M. Desesso, M. E. Hurtt, and G. D. Cappon. 2006. Postnatal growth and morphological development of the brain: A species comparison. *Birth Defects Research. Part B, Developmental and Reproductive Toxicology* 77 (5): 471–484, PM:17066419.

Wood, S. L., B. K. Beyer, and G. D. Cappon. 2003. Species comparison of postnatal CNS development: functional measures. *Birth Defects Research Part B: Developmental and Reproductive Toxicology* 68 (5): 391–407, PM:14745989.

Yang, X., D. R. Doerge, and J. W. Fisher. 2013. Prediction and evaluation of route dependent dosimetry of BPA in rats at different life stages using a physiologically based pharmacokinetic model. *Toxicology and Applied Pharmacology* 270 (1): 45–59, PM:23566954.

Yang, X. and J. W. Fisher. 2014. Unraveling bisphenol A pharmacokinetics using physiologically based pharmacokinetic modeling. *Frontiers in Pharmacology* 5: 292, PM:25620931.

Yoon, M., J. D. Schroeter, A. Nong, M. D. Taylor, D. C. Dorman, M. E. Andersen, and H. J. Clewell, III. 2011. Physiologically based pharmacokinetic modeling of fetal and neonatal manganese exposure in humans: describing manganese homeostasis during development. *Toxicological Sciences: An Official Journal of the Society of Toxicology* 122 (2): 297–316, PM:21622944.

Zaghouani, H., C. M. Hoeman, and B. Adkins. 2009. Neonatal immunity: Faulty T-helpers and the shortcomings of dendritic cells. *Trends in Immunology* 30 (12): 585–591, PM:19846341.

Zimmermann, M. B. 2011. The role of iodine in human growth and development. *Seminars in Cell and Developmental Biology* 22: 645–652.

Zoetis, T., and M. E. Hurtt. 2003. Species comparison of anatomical and functional renal development. *Birth Defects Research. Part B, Developmental and Reproductive Toxicology* 68 (2): 111–120, PM:12866702.

Zoetis, T., M. S. Tassinari, C. Bagi, K. Walthall, and M. E. Hurtt. 2003. Species comparison of postnatal bone growth and development. *Birth Defects Research. Part B, Developmental and Reproductive Toxicology* 68 (2): 86–110, PM:12866701.

12 Naturally Occurring Toxicants as Etiologic Agents of Foodborne Disease

Raj K. Keservani, Rajesh K. Kesharwani, and Anil K. Sharma

CONTENTS

12.1 INTRODUCTION

Foodborne diseases (FBDs) and illness are due to the consumption of contaminated food, and the presence of toxins produced by pathogenic bacteria has been of vital concern to public health. More than 250 different FBDs have resulted via direct or indirect food contamination. It has been found that the bacteria alone are the causative agents of two-thirds of FBDs (Olsen 2000). Not only foodborne illnesses affect the health of individuals, but it can also have dramatic economic impact. The economic losses from various factors, such as medical treatment, lost wages and productivity, loss of business, recall and destruction of products, and investigation of the outbreaks, can be very high (Doors 1999). Tropical fishes produce many toxins including ciguatera that acts as poison for other animals and is also produced by salmon raised exclusively in the farms. The fishes having ciguatera are found in the areas of the Pacific and northern Caribbean, and these are classified as high-risk area for fishing. However, imported fish have occasionally caused outbreaks in the United States (DiNubile and Hokama 1995).

Harmful algal blooms (HABs) are increasing worldwide as a result of climate change and global marine traffic. HABs contain high concentrations of algal toxins. Toxin-contaminated shellfish cannot be detected by taste, sight, or smell; the toxins are heat stable and therefore are not destroyed by cooking. Human consumption of toxin-contaminated shellfish leads to illness (Bouchouar et al. 2014). Many seafoods are responsible for causing allergies, and among them, histamine poisoning or scombroid poisoning is a classical example published in 1830 (Minns 2014). The scombroid compound was first discovered in the following fish (i.e., Scombridae): tuna and mackerel.

Many observe pufferfish having an extremely potent neurotoxin, that is, tetrodotoxin (TTX), which is found in the liver and ovaries of all the species of the genus *Takifugu*, although the tissue(s) in which it is localized can differ among species. TTX helps to save pufferfish from predators and is also part of its major defense strategy. TTX also acts as pheromones to attract male during spawning (Itoia et al. 2014). Mushroom poisoning is caused due to ingestion of toxic compound present in a mushroom. This poisoning produces different symptoms, which leads to gastrointestinal problems and sometimes death of the person. Nonidentification of edible and nonedible mushroom is the biggest reason for poisoning as both are very similar.

Mushrooms are the fruiting bodies of a group of higher fungi that have evolved along with higher plants during the course of the evolution (Diaz 2005a); and is belong to Aspergillus Class. Approximately 250 species are known, and among them, many are producing beneficial secondary metabolites used in pharmaceuticals (Brakhage et al. 2008). A famous cholesterol-lowering drug known as lovastatin is produced by *Aspergillus terreus*, and other *Aspergilli* secrete antifungals (griseofulvin), antitumor drugs (terrequinone A), and antibiotics (penicillin and cephalosporin) (Hoffmeister and Keller 2007, Keller et al. 2005).

Many uncharacterized toxin compounds are produced by *Aspergilli* through various metabolic pathways and are known as mycotoxins. These belong to a group of structurally diverse chemical compounds, which show various biological activities and some of them are carcinogenic (Sweeney and Dobson 1998). From the study done by Takao and coworker, it has been proven that consumption of red chili pepper contaminated with compound aflotoxin is one of the causative agents of gallbladder cancer (GBC) in countries like Bolivia and Peru. In countries where aflotoxin is present as a contaminant in red chili pepper, high GBC incidence rates have been documented. The GBC and the association between the two should be confirmed by a case-control study (Takao et al. 2012). Pyrrolizidine alkaloid (PA)–containing plants are found throughout the world and are probably the most common plants that act as causative agents of poisoning of livestock, wildlife, and humans. PAs are potent liver toxins that under some conditions can be carcinogenic (Stegelmeier 2011).

In numerous articles on the antinutritional properties of lectins (Gupta 1987, Kumar et al. 2013, Vasconcelos and Oliveira 2004), the oral toxicity of lectins was reviewed. It was reported that very high concentrations of particular raw lectins, such as that isolated from uncooked red kidney beans (Freed 1999), can cause toxic reactions when exposed as such to humans, characterized by general nausea, bloating, vomiting, and diarrhea. Consumption of Ericaceae family plants (*Pieris*, *Rhododendron*, *Kalmia*, and *Agarista*), leaves, flowers, or secondary products (honey) cause toxic effect due to the presence of diterpene grayanotoxin. It has been reported that the diterpene grayanotoxin–contaminated honey may cause intoxication, specifically characterized by hypotension, dizziness, and atrioventricular block, and also it is responsible for the large number of fatal casualties (Jansen et al. 2012).

12.2 NATURALLY OCCURRING TOXINS

Several toxins are present in nature, but here we discuss about some naturally occurring toxins that are harmful to human and animal; these are described in Table 12.1.

12.2.1 Ciguatoxin

Ciguatoxin is found in the flesh of certain reef fishes and is responsible for causing ciguatera, a type of foodborne illness. Ciguatoxin was first reported in dinoflagellates, such as *Gambierdiscus toxicus*, which is native of tropical and subtropical water environment. These dinoflagellates live together along with algae, corals, and many seaweeds and are eaten by many herbivorous fishes, which in turn are eaten by larger carnivorous fish. In this way, the toxins move up the food chain, resulting in biomagnification (Swift and Swift 1993). Ciguatoxin is the most common toxin that accumulates in tropical fish (Lewis 2001, Pearn 2001). Mortality induced by that toxin is extremely rare as it accounts for ca. 0.1% (Bagnis et al. 1979). Ciguatoxin is stable and cannot be inactivated by treatments such as cooking, freezing, salting, or pickling. Fish accumulating this toxin originating from tropical areas may be consumed irrespective of all seasons and the site of their catch area (Watters 1995).

The mechanisms of ciguatoxin action can be traced directly to excitable nervous and muscular layers: the generation and conduction of action potential. Ciguatoxin is characterized by the ability to bind with voltage-sensitive sodium channels, leading to their opening in the membranes of resting cells. It is manifested by an inflow of Na^+ ions, cell depolarization, and appearance of a spontaneous action potential in the excited cells. The cardiovascular effect, evoked by ciguatoxin activity, is due to the positively inotropic effect on myocardium. By acting on voltage-sensitive Na^+ channels, ciguatoxin evokes an inflow of Na^+ ions into the cell. This, in turn, activates downstream mechanisms; as a result, the cell expels Na^+ ions and absorbs calcium. Although a considerable increase in calcium concentration is buffered by sarcoplasmic reticulum, a local increase in calcium concentrations leads to an increase in the force of myocardium contraction. A similar mechanism of inducing an intracellular inflow of calcium, induced by ciguatoxin, is observed in intestinal endothelial cells. It contributes to the enhanced secretion and is manifested by diarrhea (Lehane and Lewis 2000).

TABLE 12.1
Some Naturally Occurring Toxicants

S. No.	Natural Toxins
1.	Ciguatera poisoning
2.	Shellfish toxins (PSP, DSP, NSP, ASP, AZP)
3.	Scombroid toxin
4.	Tetrodotoxin (pufferfish)
5.	Mushroom toxins
6.	Aflatoxins
7.	Pyrrolizidine alkaloids
8.	Phytohemogglutinin (red kidney bean poisoning)
9.	Grayanotoxin (honey intoxication)

Ciguatoxin intoxication is characterized by gastrointestinal disorders (vomiting, diarrhea, and contractions of abdominal cavity organs), neurological symptoms (headaches and nausea, ataxia, sensibility disorders, muscular and joint pains), itching of the skin of palm, feet and lips, as well as less common cardiovascular symptoms (arrhythmia, bradycardia or tachycardia, reduced blood pressure) (Isbister and Kiernan 2005, Watters 1995). There is no antidote in the case of ciguatoxin poisoning; only adjunct therapy is undertaken. Since ciguatoxin is highly soluble in fats, its absorption from the intestine is rapid and considerable. Nevertheless, early occurrence of vomiting and diarrhea may facilitate the removal of some amount of the toxin before its absorption (Lehane and Lewis 2000). There is some evidence that calcium channel blocker–type drugs, such as nifedipine and verapamil, are effective in treating some of the symptoms that remain after the initial sickness resolves, such as poor circulation and shooting pains through the chest. These symptoms are due to the cramping of arterial walls caused by maitotoxin. Ciguatoxin lowers the threshold for opening voltage-gated sodium channels in the synapses of the nervous system (Davis and Villar 1986).

12.2.2 Shellfish Toxins

Numerous Shellfish poisoning is caused by a group of toxins produced by planktonic algae (dinoflagellates, in most cases) on which shellfish feed. Numerous shellfish toxins have been documented around the world; these include toxins that are currently monitored by the FDA (Deeds 2012).

12.2.2.1 Paralytic Shellfish Poisoning

PSP is caused by water-soluble alkaloid neurotoxins that are collectively referred to as saxitoxins or paralytic shellfish toxins (PSTs). Shellfish absorb active toxins or nontoxins along with food, and the latter are hydrolyzed into the active forms inside the gastrointestinal tract of the shellfish (Isbister and Kiernan 2005, Watters 1995). Saxitoxin and its derivative, that is, gonyautoxin, block sodium channels susceptible to TTX, thus blocking Na^+ influx into the cells: this disturbs the nervous conductance and leads to motor and sensory nervous abnormalities (Isbister and Kiernan 2005). The symptoms include gastrointestinal disorders appearing in the first 30–60 minutes, followed by sensation disorders manifested by pinching and numbness of tongue and lips that, with time, extends to the face, neck, fingers, and toes. In most fatal cases, death occurs within the first 12 hours (Lehane and Olley 2001, Meier and White 1995).

12.2.2.2 Diarrhetic Shellfish Poisoning

DSP is caused by a group of lipid-soluble polyether toxins that includes okadaic acid, dinophysistoxins (DSTs), and a series of fatty acid esters of okadaic acid and DSTs. Okadaic acid is a strong inhibitor of phosphatases, enzymes closely linked with metabolic processes in cells. Diarrhea in humans is thought to be induced by excessive phosphorylation of proteins that control the secretion of sodium through the cells of intestinal epithelium (Hallegraeff et al. 1995).

12.2.2.3 Neurotoxic Shellfish Poisoning

NSP is caused by a group of lipid-soluble polyether toxins called brevetoxins. NSP-causing toxins in shellfish include intact algal brevetoxins and their metabolites (collectively known as NSTs). Poisoning occurs after consumption of shellfish contaminated by brevetoxin, produced by dinoflagellates *Karenia brevis*. Its effect on an organism is reflected in the stimulation of sodium channels in the cell wall (Novak 1998). Symptoms include gastrointestinal disorders as well as sensation disorders in the form of pinching and numbness of the lips, tongue, and throat. Vertigo, muscular pains, and alternating sensation of warmth and coolness are also likely to occur (Hughes and Merson 1976).

12.2.2.4 Amnesic Shellfish Poisoning

ASP is caused by the neurotoxin domoic acid (DA), a water-soluble, nonprotein, excitatory amino acid. Isomers of DA have been reported but are less toxic than DA itself. DA is a water-soluble, thermostable amino acid produced by microscopic alga of the genus *Nitzschia* (Isbister and Kiernan 2005). Its activity is based on the inhibition of adenyl cyclase activity in cellular membranes (Nijjar and Grimmelt 1994) and changes in the structure of myelin sheath of axons (Schumued and Slikker 1999). DA has neurotoxic, immunotoxic, and genotoxic activities (Dizer et al. 2001). Its neurotoxic properties are manifested by neuronal degeneration and necrosis of hippocampus regions and amygdala region.

12.2.2.5 Azaspiracid Shellfish Poisoning

AZP is caused by the lipid-soluble toxin azaspiracid and several derivatives (AZAs). To date, more than 30 AZA analogues have been identified, with three analogues routinely monitored in shellfish. AZP symptoms include nausea, vomiting, severe diarrhea, and contractions of abdominal cavity organs (Satake et al. 1998).

12.2.3 Scombrotoxin

Scombroid food poisoning is a foodborne illness that results from eating spoiled (decayed) fish (Clark et al. 1999). Scombrotoxin is a combination of many chemicals: histamine is the most prominent one. There are three elements that are required for histamine fish poisoning to occur. First, the fish must have high levels of free histidine. Second, certain bacteria that produce the enzyme histidine decarboxylase, particularly *Morganella morganii*, *Klebsiella pneumoniae*, or other Enterobacteriaceae, must be present in the fish (Bjornsdottir et al. 2009). This enzyme is responsible for the conversion of histidine to histamine (Smolinska et al. 2014). Finally, there must have been some form of inappropriate handling or temperate abuse that allows the multiplication of these bacteria and thus the production of histamine. As histamine is heat stable, cooking does not reduce the risk of illness (Smolinska et al. 2014). Scombrotoxin poisoning is closely linked to the accumulation of histamine in these fish (Benner 2012).

Symptom onset is rapid, usually within minutes to a few hours after the consumption of the implicated fish (Heymann 2004). Symptoms may vary for different individuals depending on underlying medical conditions and medications but can include flushing of the face, neck, and upper arms, oral numbness and/or burning, metallic taste, headache, itchy rash, hives, nausea, vomiting, diarrhea, and difficulties in swallowing. The illness is generally self-limiting, and recovery usually occurs within 24 hours. When required, antihistamines are used to treat symptoms (Lehane and Olley 1999).

12.2.4 Tetrodotoxin

Tetrodotoxin, frequently abbreviated as TTX, is a potent neurotoxin. The name derives due to its presence in four fishes, that is, pufferfish, porcupinefish, ocean sunfish or mola, and triggerfish; together they belong to the order Tetraodontiformes. Later on, it was found that other than Tetraodontiformes, TTX was present in several other animals (rough-skinned newt, blue-ringed octopus [Hogan 2008], and Naticidae [Hwang et al. 1991]). TTX is actually the product of some symbiotic bacteria that are living inside the body of these fishes, and these bacteria belong to certain species of *Pseudomonas*, *Pseudoalteromonas*, *Vibrio*, etc. TTX is extremely toxic. The toxin can enter the body by ingestion, injection, or inhalation, or through abraded skin (Patocka and Stredav 2002). The mechanism of toxicity is due to the blockage of fast voltage-gated sodium channels.

TTX does not affect the flow of K^+ ions between the channels, but it very much affects the Na^+ channels and blocks its movement. The hydrated Na^+ ion binds reversibly for nanoseconds as compared to binding of TTX to the channel that is relatively strong (Kd = 10–10 nM). It is proposed that this binding results from the interaction of the positively charged guanidino group on the TTX

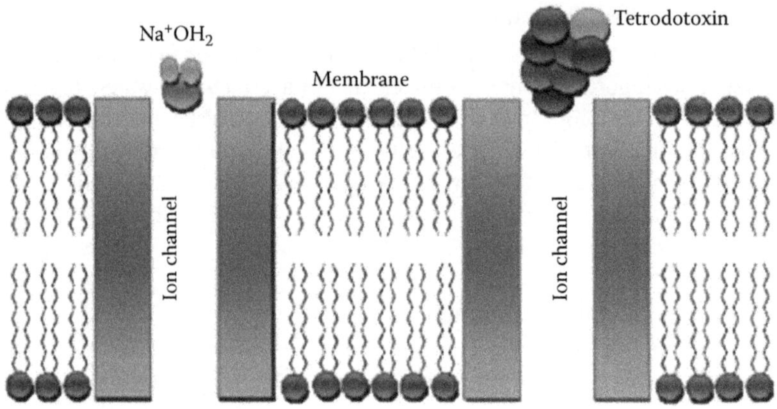

FIGURE 12.1 **(See color insert.)** Plasma membrane with ion channels and hydrated sodium ion and tetrodotoxin.

and negatively charged carboxylate groups on side chains in the mouth of the channel. Saxitoxin, a natural product from dinoflagellates, acts in a similar way and is also a potent nerve poison (Noda et al. 1986) (Figure 12.1).

Symptoms typically develop within 30 minutes of ingestion but may be delayed by up to 4 hours; however, death may occur within 17 minutes of ingestion (Clark et al. 1999). Paresthesia of the lips and tongue is followed by hypersalivation, sweating, headache, weakness, lethargy, incoordination, tremor, paralysis, cyanosis, aphonia, dysphagia, seizures, dyspnea, bronchorrhea, bronchospasm, respiratory failure, coma, and hypotension. Gastroenteric symptoms are often severe and include nausea, vomiting, diarrhea, and abdominal pain. Cardiac arrhythmias may precede complete respiratory failure and cardiovascular collapse.

No antidote has been developed for TTX. Therapy is supportive and based on symptoms, with aggressive early airway management. If ingested, treatment can consist of emptying the stomach, feeding the victim with activated charcoal to bind the toxin, and taking standard life-support measures to keep the victim alive until the ill effects of the poison have worn off. Alpha-adrenergic agonists are recommended in addition to intravenous fluids to combat hypotension. Anticholinesterase agents have been used with mixed success (Clark et al. 1999).

12.2.5 Mushroom Poisoning

Mushroom poisoning (also known as mycetism) is due to harmful effects of ingestion of toxic substances present in the mushroom. These symptoms can vary from slight gastrointestinal disorder to sometimes death. The toxins present are secondary metabolites produced in specific biochemical pathways in the fungal cells. Mushroom poisoning is usually the result of ingestion of wild mushrooms after misidentification of a toxic mushroom as an edible species. The most common reason for this misidentification is close resemblance in terms of color and general morphology of the toxic mushroom species with edible species. There are many folk traditions concerning the defining features of poisonous mushrooms (Hall 2003).

The severity of mushroom poisoning may vary, depending on the geographic location where the mushroom is grown, growth conditions, the amount of toxin delivered, and the genetic characteristics of the mushroom. Boiling, cooking, freezing, or processing may not alter some mushroom's toxicity. Mushroom exposure in children is an infrequent but perennial problem for parents and clinicians. Parental anxiety is generally high because of fears of unknown or untoward effects. The challenges for clinicians are to identify such poisonings, to discern whether poisoning has taken place, to order appropriate diagnostic studies, and to prescribe reasonable therapy. The diverse

TABLE 12.2
List of Mushroom Toxins

Mushroom Toxins	References
Cyclopeptides (Amatoxin)	Diaz (2005a,b), Berger and Guss (2005a,b), Goldfrank (2011)
Gyromitrins (monomethylhydrazine)	Diaz (2005a,b), Berger and Guss (2005a,b), Goldfrank (2011)
Orellanine	Diaz (2005a,b), Berger and Guss (2005a,b), Goldfrank (2011), Judge et al. (2010)
Psilocybin	Diaz (2005a,b), Berger and Guss (2005a,b), Goldfrank (2011)
Muscarine	Diaz (2005a,b), Berger and Guss (2005a,b), Goldfrank (2011)
Muscimol and ibotenic acid	Diaz (2005a,b), Berger and Guss (2005a,b), Goldfrank (2011)
Coprine	Diaz (2005a,b), Berger and Guss (2005a,b), Goldfrank (2011)
Nephrotoxins (norleucine)	Diaz (2005a,b), Berger and Guss (2005a,b), Goldfrank (2011)
Immunoactive toxins	Diaz (2005a,b), Berger and Guss (2005a,b), Goldfrank (2011)
Gastrointestinal irritants	Diaz (2005a,b), Berger and Guss (2005a,b), Goldfrank (2011)
Hemolytic toxins	Diaz (2005a,b), Berger and Guss (2005a,b), Goldfrank (2011)
Mycotoxins	Diaz (2005a,b), Berger and Guss (2005a,b), Goldfrank (2011)

nature of mushroom toxicities, their ubiquitous distribution, and the relative infrequency of the ingestions make these challenges all the more still difficult to overcome (Diaz 2005a).

Each poisonous mushroom species contains one or more toxins, which may be classified on the basis of the mushroom's physiologic and clinical effects on human beings, the target organ's toxicity, and the time of symptom onset (Table 12.2).

12.2.5.1 Cyclopeptides

Cyclopeptides include amatoxins, and *Amanita phalloides* (death cap), *Amanita virosa* (destroying angel), *Amanita verna* (fool's mushroom), *Amanita bisporigera*, *Galerina autumnalis* (autumn skullcap), and *Galerina sulcipes* are the most common mushrooms implicated in liver injury and death among the amatoxin-containing mushrooms.

12.2.5.2 Gyromitrin

Gyromitrin is a volatile hydrazine derivative synthesized by certain species of false morel (*Gyromitra esculenta*) and is easily confused with the early false morel (*Verpa bohemica*). Gyromitrin poisoning typically occurs after ingestion of the toxin-containing mushrooms but may also result from inhalation of the cooking vapors during their preparation.

12.2.5.3 Orellanine

Orellanine is a nephrotoxic compound that is synthesized by several species of *Cortinarius* mushrooms. Orellanine-containing species include *Cortinarius orellanusand* and *Cortinarius speciosissimus*; both are commonly found in Europe and Japan but not in North America. *Cortinarius* species that may contain small amounts of orellanine include *Cortinarius gentilis, Cortinarius rainierensis*, and *Cortinarius splendens henrici*; however, there are very few confirmed cases of *Cortinarius*-induced renal failure in North America (Judge et al. 2010).

12.2.5.4 Psilocybin

Psilocybin and psilocin are found in a number of mushroom genera, including *Psilocybe cubensis*, *Psilocybe semilanceata* (liberty cap), *Panaeolus cyanescens* (previously referred to as *Copelandia* species), *Gymnopilus spectabili* (Big Laughing Jim), *Conocybe cyanopus*, *Psathyrella foenisecii*, and several species of *Pluteus*. Psilocybin and psilocin are serotonin (5-HT2) agonists and, when ingested, cause psychedelic effects similar to those of lysergic acid diethylamide (LSD).

12.2.5.5 Muscarine

Mushrooms that contain muscarine are commonly found in yards, parks, and wooded areas throughout the United States, Europe, and Asia. Species from the genera *Clitocybe* and *Inocybe* (see the following images) are most commonly responsible for muscarinic mushroom poisoning in the United States. *Clitocybe dealbata* (the sweating mushroom) may be confused with the edible fairy ring champignon (*Marasmius oreades*) or sweetbread mushroom (*Clitopilus prunulus*). Other muscarine-containing mushrooms include species belonging to the genera *Boletus*, *Mycena*, and *Omphalotus*. Muscarine stimulates M1 and M2 types of postganglionic cholinergic receptors (muscarinic receptors) present in the autonomic nervous system. This action results in parasympathetic stimulation similar to that caused by the release of endogenous acetylcholine at postganglionic receptors of smooth muscle and exocrine glands.

Muscarine-containing mushrooms typically produce cholinergic symptoms such as sweating, facial flushing, salivation, lacrimation, vomiting, abdominal cramps, diarrhea, urination, and miosis; occasionally, bradycardia, hypotension, and dizziness develop (Benjamin 1995).

There are no specific antidotes available for muscarinic mushroom toxicity. As in this report, patients with severe symptoms and signs may require atropine, hemodynamic monitoring, and aggressive fluid management to reverse the symptoms (Diaz 2005b, Erguven et al. 2007, Goldfrank 2011, Stallard and Edes 1989). George and Hegde studied muscarinic toxicity among family members after consumption of mushrooms and reported that patients with muscarinic mushroom toxicity have early onset of symptoms and they respond well to atropine and symptomatic supportive care (George and Hegde 2013).

12.2.5.6 Muscimol and Ibotenic Acid

Amanita muscaria (fly agaric) and *Amanita pantherina* (panthercap) are the mushrooms that synthesize ibotenic acid and muscimol, both of which are excitatory neurotoxins and may be mildly hallucinogenic.

Ibotenic acid is structurally similar to glutamic acid and acts as an agonist at the glutamic acid receptors (NMDA receptors) in the CNS.

12.2.5.7 Coprine

A few species of mushrooms, including *Coprinopsis atramentaria* (formerly known as *Coprinus atramentarius*), commonly referred to as inky cap or tippler's bane and mistaken for the edible *Coprinus comatus* (shaggy mane), produce coprine, an amino acid that is metabolized to 1-aminocyclopropanol in the human body. This metabolite blocks acetaldehyde dehydrogenase, and in the presence of alcohol, acetaldehyde builds up, resulting in a disulfiram reaction.

12.2.5.8 Norleucine

Other nephrotoxic mushrooms, such as *Amanita smithiana* and *Amanita proxima*, have also been associated with an acute oliguric renal failure. *Amanita smithiana* may be mistaken for the matsutake mushroom (*Tricholoma magnivelare*) by foragers. These mushrooms cause vomiting and diarrhea 1–12 hours after ingestion, followed by a transient elevation of transaminases, and then oliguric renal failure results in 3–6 days.

12.2.5.9 Bronchoalveolar Allergic Syndrome

An immune reaction is believed to be the cause of the bronchoalveolar allergic syndrome seen after the inhalation of spores of some puffball (*Lycoperdon*) mushroom species.

12.2.5.10 Involutin

Ingestion of *Paxillus involutus* may result in the acute onset of abdominal pain, nausea, vomiting, and diarrhea within 30 minutes to 3 hours of ingestion, followed by an immune complex–mediated hemolytic anemia with hemoglobinuria, oliguria, anuria, and acute renal failure.

12.2.5.11 Gastrointestinal Toxins

Hundreds of mushrooms contain toxins that can cause gastrointestinal symptoms (e.g., nausea, vomiting, diarrhea, and abdominal pain) similar to those observed with more dangerous mushrooms. They include *Chlorophyllum molybdites* (green gill), *Boletus piperatus* (pepper bolete), and *Agaricus arvensis* (horse mushroom), among many others.

12.2.5.12 Erythromelalgia Syndrome

Two species of mushrooms, *Clitocybe acromelaga* (in Japan) and *Clitocybe amoenolens* (in Europe), cause a painful burning sensation with reddening of the skin several days after eating them. In Europe, the *C. amoenolens* mushroom has been mistaken for the edible mushroom *Lepista inversa*. The suspected toxin is acromelic acid A.

12.2.6 Aflatoxins

The term *aflatoxin* was coined around 1960 after the discovery that the source of "Turkey 'X' disease" was due to the toxins produced by *Aspergillus flavus* (Wannop 1961). Aflatoxins are naturally occurring mycotoxins that are formed during secondary metabolism in *Aspergillus parasiticus* and *A. flavus* species. Aflatoxins have a strong carcinogenic property, and (Hudler 1998) after ingestion, it reaches to the liver and may be metabolized to epoxide intermediate or be hydroxylated and be converted into the less harmful aflatoxin M_1. The main feed sources of aflatoxins are peanut meal, maize, and cottonseed meal. Many researchers reported that there is a linear relationship between the amount of aflatoxin M1 in milk and aflatoxin B1 in feed consumed by animals (Bakirci 2001).

At least 14 different types of aflatoxin are produced in nature (Boutrif 1998). Aflatoxin B_1 is considered the most toxic and is produced by both *A. flavus* and *A. parasiticus*. Aflatoxins G_1 and G_2 are produced exclusively by *A. parasiticus*. While the presence of *Aspergillus* in food products does not always indicate that harmful levels of aflatoxin are also present, it does imply a significant risk in consumption. Aflatoxins M_1 and M_2 originally were discovered in the milk of cows that fed on moldy grain. In addition to aflatoxins B1 and B2, *A. flavus* also produces many other mycotoxins such as cyclopiazonic acid, kojic acid, beta-nitropropionic acid, aspertoxin, aflatrem, and aspergillic acid (Goto et al. 1996).

The fungus *A. flavus* is a weak and opportunistic plant pathogen, affecting many agricultural crops such as maize (corn), cotton, groundnuts (peanuts), as well as tree nuts such as Brazil nuts, pecans, pistachio nuts, and walnuts. Preharvest contamination of these crops with aflatoxins is common. *A. flavus* also causes the spoilage of postharvest grains during storage. Because *A. flavus* lacks host specificity (St Leger et al. 2000), it can attack the seeds of both monocots and dicots and seeds produced both above the ground (corn) and below the ground (peanuts). Under favorable conditions, *A. flavus* can cause ear rot in maize, resulting in significant economic losses to farmers (Richard and Payne 2003, Robens and Cardwell 2003, 2005). Aflatoxin B1 has been associated with growth impairment in children (Gong et al. 2002, 2004, Shouman et al. 2012), suppressed immune function (Turner et al. 2003), hepatomegaly (Gong et al. 2012), and death due to acute poisoning (Probst et al. 2007).

12.2.6.1 Factors Affecting Aflatoxin Biosynthesis

Many biotic and abiotic environmental factors influence aflatoxin biosynthesis (Bennett and Papa 1988, Demain 1998, Kale et al. 1991, 1996, Yabe et al. 1988) including nutritional factors such as carbon and nitrogen source, environmental factors such as water activity and temperature, and physiological conditions such as pH (Cotty 1988) and bioreactive agents (Guo et al. 2005, Payne and Brown 1988).

12.2.6.1.1 Carbon

The best-known nutritional factors affecting aflatoxin biosynthesis are carbon and nitrogen sources (Adye and Mateles 1964, Bennett et al. 1979, Luchese and Harrigan 1993). The relationship of carbon source and aflatoxin formation has been well established. Simple sugars such as glucose, sucrose, fructose, and maltose support aflatoxin formation, while peptone, sorbose, or lactose does not (Buchanan and Lewis 1984, Payne and Brown 1998). Woloshuk et al. (1997) reported the link between alpha amylase activity and aflatoxin production in *A. flavus*.

12.2.6.1.2 Nitrogen

Nitrogen is closely linked to aflatoxin production (Payne and Brown 1998). Media containing asparagine, aspartate, alanine, ammonium nitrate, ammonium nitrite, ammonium sulfate, glutamate, glutamine, and proline support aflatoxin production, while sodium nitrate– and sodium nitrite–containing media do not (Davis et al. 1967, Reddy et al. 1971, 1979). Nitrate was reported to have a suppressive effect on aflatoxin production, and overexpression of aflR gene by additional copies of aflR overcomes the negative regulatory effect on aflatoxin pathway gene transcription (Chang et al. 1995).

12.2.6.1.3 Temperature

Aflatoxin formation is directly affected by temperature. Optimal aflatoxin production is observed at temperatures near 30°C (28°C–35°C). When temperature increases to above 36°C, aflatoxin production is nearly completely inhibited (Diener and Davis 1970, O'Brian et al. 2007, Roy and Chourasia 1989, Schroeder and Hein 1968). The conditions of high humidity and warm temperatures can give rise to the highest levels of aflatoxins in food (Adams and Moss 2005). They are toxic, immunosuppressive, mutagenic, teratogenic, and carcinogenic to animals and humans (Kocabas and Sekerel 2003, Peraica et al. 1999, Piva et al. 1995).

12.2.6.1.4 pH

Aflatoxin biosynthesis in *A. flavus* occurs in acidic media, but it is inhibited in alkaline media (Cotty 1988).

12.2.6.1.5 Water

Severe aflatoxin outbreaks in corn in hot weather and drought conditions have been documented (Cotty and Jaime-Garcia 2007, Sanders et al. 1984). The possible scenarios may include a combination of these factors: (1) the plant defense mechanism is weakened under water stress; (2) higher insect feeding and associated injuries to plant tissues, thus providing entry opportunities for fungal invasion; and (3) more fungal spores dispersed into the air under dry climate.

12.2.6.1.6 Oxidative Stress Condition

Oxidative stress and aflatoxin biosynthesis are related in *A. parasiticus* (Jayashree et al. 2000a, Kim and Bennett 2006, Mahoney and Molyneux 2004, Reverberi et al. 2006). Oxidative stress induces aflatoxin formation in *A. parasiticus* (Jayashree and Subramanyam 2000b). Treatment of *A. flavus* with tert-butyl hydroperoxide or gallic acid induced significant increases in aflatoxin production (Kim 2006). Similar treatment of *A. parasiticus* also induced aflatoxin production (Reverberi et al. 2005, 2006). Hydrolysable tannins significantly inhibit aflatoxin biosynthesis, with the main antiaflatoxigenic constituent in these tannins being gallic acid (Mahoney and Molyneux 2004).

12.2.6.1.7 Development Stage

Sporulation and sclerotial formation are associated with secondary metabolism (Bennett et al. 1986, Calvo et al. 2002, Chang et al. 2002, Hicks et al. 1997). Spore formation and secondary metabolite formation occur at about the same time (Hicks et al. 1997, Trail et al. 1995). Some mutants that are deficient in sporulation are unable to produce aflatoxins (Bennett and Papa 1988), and some compounds that inhibit sporulation in *A. parasiticus* also inhibit aflatoxin formation (Reib 1982).

12.2.6.1.8 Plant Metabolites

Plant metabolites play some role in aflatoxin formation (Greene-McDowelle et al. 1999, Zeringue et al. 1996, 2002, Zeringue and Bhatnagar 1993). Wright et al. reported that, under certain conditions, n-decyl aldehyde reduces not only fungal growth of *A. parasiticus* but also aflatoxin production by over 95% compared to the control (Wright et al. 2000). The 13(S)-hydroperoxide derivative of linoleic acid, the reaction product of lipoxygenase (encoded by L2 LOX gene from maize), is reported to reduce aflatoxin production (Wilson et al. 2001).

12.2.7 Pyrrolizidine Alkaloids

Pyrrolizidine alkaloids (PAs; sometimes referred to as necine bases) are a group of naturally occurring alkaloids based on the structure of pyrrolizidine. Pyrrolizidine alkaloids are produced by plants as part of a defense mechanism and coping strategy against insect herbivores. More than 660 PAs and PA N-oxides have been identified in over 6000 plants, and about half of them exhibit hepatotoxicity (Radominska-Pandya 2010). The effect of PA poisoning is mainly due to the consumption of plants or plant products (mainly species of *Senecio*) directly and its secondary products like honey. Apart from such type of ingestion, PAs enters into human body through food items like eggs and milk of animal that feed on plants containing PAs and its products.

In honey isolated from *Senecio* species, the total concentration of PAs was 0.3–3.2 μg/kg. PAs could be detected in the concentration range of 30–70 μg/kg in honey from the Alpine foothills of Switzerland. No cases of PA poisoning have been attributed to the ingestion of tainted honey or milk (Roder 1995). Eggs have been implicated as a source of PA toxicity in a case involving the use of unregulated grain feed from *Heliotropium* (Edgar and Smith 2000). The PAs, which have minimal toxicity in their original form, are metabolized in the liver and can become toxic metabolites, depending on the PA and on the liver enzymes (Huxtable and Cooper 2000). Pyrrolizidine alkaloids became an issue for safety concern in relation to herbal medicine about 20 years ago, when an herb enthusiast consumed comfrey (*Symphytum officinale*) on a daily basis and suffered from liver damage (Stickel and Seitz 2000). A new concern arose with the use of another popular herb: coltsfoot (*Tussilago farfara*). Coltsfoot was widely used for coughs in Europe and America and has also been used by the Chinese for the same purpose. There was a report from Switzerland in 1988 regarding liver damage leading to the death of a newborn due to the consumption of coltsfoot tea by the mother during pregnancy (Roulet et al. 1988).

12.2.8 Phytohemagglutinin (Red Kidney Bean Poisoning)

Lectins are widely used reagents in the study of glycoconjugates in solution and on cells and for cell characterization and separation (Sharon 2008). Next to this, lectins are also used as tools for novel techniques such as in a microarray for a high-throughput analysis of glycans and glycoproteins (Hu et al. 2012) and in new data-storing techniques, where carbohydrates are used as hardware for information coding (Gabius et al. 2011). Phytohemagglutinin (PHA, or phytohemagglutinin) has been found in plants as lectins, especially in legumes. The two closely related proteins of PHA are leucoagglutinin PHA-L (leukocytes) and PHA-E (erythrocytes). Phytohemagglutinin shows carbohydrate-binding specificity for a complex oligosaccharide containing galactose, N-acetylglucosamine, and mannose (Hamelryck et al. 1996). This form of food poisoning is a chemical intoxication and is apparently due to the presence of high levels of phytohemagglutinins in beans. Hemagglutinins (lectins) are proteins or glycoproteins, capable of binding to specific carbohydrate residues (Pusztai 1986), which have been detected in a wide variety of leguminous seeds including soybeans (*Glycine max*), lentils (*Lens esculenta*), lima beans (*Phaseolus lunatus*), and red kidney beans (*Phaseolus vulgaris*) (Liener 1974).

Phytohemagglutinin is a sugar-based protein (glycoprotein), which is the main cause of toxin. But some of the highest concentrations of this glycoprotein have been found in red kidney beans.

This toxin is denatured if red kidney beans are cooked at very high temperature. Symptoms are nausea, vomiting, and diarrhea. Abdominal pains appear 2–3 hours after the consumption of kidney beans. However, these appear soon after consumption, but they also disappear quickly after some time. Most symptoms resolve by themselves within a few hours. But there have been symptoms that lead to hospitalization. This depends on the quantity of beans consumed and dehydration. Persistent vomiting or diarrhea can result in the depletion of fluids, and hence, electrolytes need to be given to the patient (http://www.medic8.com/healthguide/food-poisoning/red-kidney-bean-toxins.html). Inactivation of the toxin in beans may be achieved by soaking overnight followed by boiling in freshwater for at least 10 minutes (Bender and Reaidi 1982). Heating beans at a lower temperature is reported to result in an increase in levels of lectin (Bender and Reaidi 1982); hence, there is the observation that beans cooked at 82°C (which were still palatable) were toxic, whereas those cooked at 91°C were not (Coffey et al. 1985).

Zacharko-Siembida et al. studied and demonstrated that red kidney bean lectin increased the serotonin reservoir in the duodenum and thus may be an effective stimulant of the gut maturation in suckling mammals (Zacharko-Siembida et al. 2014). Raw red kidney bean (RRKB) causes bacterial overgrowth of the small intestine due to the lectin phytohemagglutinin (PHA) (Andrews and Jayne-Williams 1974, Banwell et al. 1983, Ramadass et al. 2010).

12.2.9 Grayanotoxins (Honey Intoxication)

Rhododendron mainly constitutes grayanotoxins of the family Ericaceae. These toxins have been found in honey nectar and cause a very rare poisonous reaction called grayanotoxin poisoning, honey intoxication, or rhododendron poisoning (Demircan et al. 2009). More than 25 grayanotoxin isoforms have been isolated from *Rhododendron* (Qiang et al. 2011). Grayanotoxin I (grayanotaxane-3,5,6,10,14,16-hexol 14-acetate) is also known as andromedotoxin, acetylandromedol, rhodotoxin, and asebotoxin (Rahway 1983).

Its chemical structure has been fully elucidated as a diterpene, a polyhydroxylated cyclic hydrocarbon with a 5/7/6/5 ring structure that does not contain nitrogen (Tallent et al. 1957). Grayanotoxin 1 present in *Rhododendron simsii* (Poon et al. 2008) has been reported from a case in Hong Kong while in the honey from Grouse Mountain, BC and Canada, which causes a similar type of poisoning, only grayanotoxins 2 and 3 were found (Yavuz et al. 1991). Currently, grayanotoxins 1 and 3 are thought to be the principal toxic isomers (Gunduz et al. 2008, Scott et al. 1971, Wong et al. 2002, Yavuz et al. 1991). Plants contain numerous compounds that, when beneficial to humans, are categorized as "medicinal" and, when harmful, they are termed *poisonous*. Secondary products derived from plants, such as honey, can contain a number of chemical compounds that, depending on their concentration and application, can also be considered medicinal or poisonous (Viuda-Martos et al. 2008). Clinical signs are evident between several hours and one day following the ingestion of the plant (Casteel and Wagstaff 1989). The observed signs are vomiting, depression, abdominal pain, severe drooling, muscle tremors, tachycardia, weakness, dyspnea, moaning, ataxia, recumbency, and opisthotonos (Armien et al. 1995, Pereira et al. 2008). Gross lesions are nonspecific, and serous bleeding in the intestines, and aspiration pneumonia in two animals were observed. (Armien et al. 1995, Casteel and Wagstaff 1989). It causes a sharp burning sensation in the throat and is thus also referred to as bitter honey (Gunduz et al. 2006).

In general, spontaneous poisoning cases occur in the winter, when the plant is green and when other forages have dried up. Accidental poisoning occurs when prunings or clippings are discarded within the reach of animals (Casteel and Wagstaff 1989). The consumption of "mad honey," or honey, Labrador tea, cigarettes containing grayanotoxin has been associated with human poisoning, and various decoctions are used in alternative medicine (Jansen et al. 2012). The voltage-gated sodium channels of the neurons are most likely a prominent target of grayanotoxins. In animal studies, Onat et al. found that injecting a small dose equivalent to 50 mg of honey intracerebroventricularly in anaesthetized albino rats caused marked bradycardia and respiratory depression

(Onat et al. 1991). Kim et al. reported grayanotoxin intoxication in three patients who ingested either blossom leaves or medicinal preparations made from *Rhododendron* species (Kim et al. 2000). Children who eat the plants are also at risk. A 9-year-old boy from Korea consumed about 10 *Rhododendron schlippenbachii* flowers and, besides the common signs of grayanotoxin intoxication, presented with impaired consciousness and delirium 26 hours later. All symptoms resolved completely after another 17 hours (Kim et al. 2008).

12.3 CONCLUSIONS

Naturally occurring toxins are harmful to human beings; some of them have been described in the preceding texts. Shellfish poisoning occurs after ingestion of organisms contaminated with infectious agents or concentrated toxins. Toxins concentrated in the flesh of shellfish can produce syndromes that include PSP, ASP, diarrheic shellfish poisoning, and neurologic shellfish poisoning. Altogether, seafood toxins are responsible for about 4% of foodborne outbreaks. Grayanotoxins are present in many members of the *Ericaceae* plant family. Medicinal use of grayanotoxin is not well understood, and care should be taken when consuming grayanotoxin-containing herbal preparation. Naturally occurring toxins such as ciguatera poisoning, shellfish toxins (PSP, DSP, NSP, ASP, AZP), scombroid toxin, TTX (pufferfish), mushroom toxins, aflatoxins, pyrrolizidine alkaloids, phytohemogglutinin (red kidney bean poisoning), and grayanotoxin (honey intoxication) are harmful to human health and cause various diseases in humans; hence, before eating these toxin-containing foods, please be aware and take note of their disease-causing effects.

REFERENCES

Adams, M.R. and M.O. Moss. 2005. *Food Microbiology*, New Age International (P) Limited, Publishers, New Dehli, India, pp. 228–244.

Adye, J. and R.I. Mateles. 1964. Incorporation of labelled compounds into aflatoxins. *Biochim. Biophys. Acta.* 86:418–420.

Andrews, A.T. and D.J. Jayne-Williams. 1974. The identification of a phytohaemagglutinin in raw navy beans (*Phaseolus vulgaris* L.) toxic for Japanese quail (*Coturnix coturnix japonica*). *Br. J. Nutr.* 32:181–188.

Armien, A.G. et al. 1995. Intoxicação experimental por *Rhododendron ledifolium* (Ericaceae) em ovinos. *Pesquisa Veterinária Brasileira.* 15(1):1–9.

Bakirci, I. 2001. A study on the occurrence of aflatoxin M1 in milk and milk products produced in Van province of Turkey. *Food Control.* 12(1):47–51.

Banwell, J.G., D.H. Boldt, J. Meyers, and F.L. Jr. Weber. 1983. Phytohemagglutinin derived from red kidney beans (*Phaseolus vulgaris*): A cause for intestinal malabsorption associated with bacterial overgrowth in the rat. *Gastroenterology.* 84:506–515.

Bender, A.E. and G.B. Reaidi. 1982. Toxicity of kidney beans (*Phaseolus vulgaris*) with particular reference to lectins. *J. Plant Foods.* 4:15–22.

Benjamin, D.R. 1995. Muscarine poisoning. In: *Mushrooms: Poisons and Panaceas*, WH Freeman, New York, pp. 340–350.

Benner, R.A. 2012. *Bad Bug Book, Foodborne Pathogenic Microorganisms and Natural Toxins*, 2nd edn. Shellfish toxins (PSP, DSP, NSP, ASP, AZP). Food and Drug Administration, New Hampshire Ave Silver Spring, MD.

Bennett, J.W., P.M. Leong, S.J. Kruger, and D. Keyes. 1986. Sclerotial and low aflatoxigenic morphological variants from haploid and diploid *Aspergillus parasiticus. Experientia.* 42:848–851.

Bennett, J.W. and K.E. Papa. 1988. The aflatoxigenic *Aspergillus* spp. *Adv. Plant Pathol.* 6:265–279.

Bennett, J.W., P.L. Rubin, L.S. Lee, and P.N. Chen. 1979. Influence of trace elements and nitrogen sources on versicolorin production by a mutant strain of *Aspergillus parasiticus. Mycopathologia.* 69:161–166.

Berger, K.J. and D.A. Guss. 2005a. Mycotoxins revisited: Part I. *J. Emerg. Med.* 28(1):53–62.

Berger, K.J. and D.A. Guss. 2005b. Mycotoxins revisited: Part II. *J. Emerg. Med.* 28(2):175–183.

Bjornsdottir, K., G.E. Bolton, P.D. McClellan-Green, L.A. Jaykus, and D.P. Green. 2009. Detection of gram-negative histamine-producing bacteria in fish: A comparative study. *J. Food Prot.* 72(9):1987–1991.

Bouchouar, E., S. Bruzzese, C. Pyles, and K. Stechyshyn. 2014. Shellfish toxins a public health concern for Canadians. *Environ. Health Rev.* 57(1):16–21.

Boutrif, E. 1998. Prevention of aflatoxin in pistachios. *Food Nutr. Agric.* 21:32–38.

Brakhage, A.A., J. Schuemann, S. Bergmann, K. Scherlach, V. Schroeckh, and C. Hertweck. 2008. Activation of fungal silent gene clusters: A new avenue to drug discovery. *Prog. Drug Res.* 66:3–12.

Buchanan, R.L. and D.F. Lewis. 1984. Regulation of aflatoxin biosynthesis: Effect of glucose on activities of various glycolytic enzymes. *Appl. Environ. Microbiol.* 48:306–310.

Calvo, A.M., R.A. Wilson, J.W. Bok, and N.P. Keller. 2002. Relationship between secondary metabolism and fungal development. *Microbiol. Mol. Biol. Rev.* 66:447–459.

Casteel, S. and J. Wagstaff. 1989. Rhododendron macrophyllum poisoning in a group of goats and sheep (Toxic Plant Case Reports). *Vet. Human Toxicol.* 31:176–177.

Chang, P.K., J.W. Bennett, and P.J. Cotty. 2002. Association of aflatoxin biosynthesis and sclerotial development in *Aspergillus parasiticus*. *Mycopathologia.* 153:41–48.

Chang, P.K., K.C. Ehrlich, J. Yu, D. Bhatnagar, and T.E. Cleveland. 1995. Increased expression of *Aspergillus parasiticus* aflR, encoding a sequence-specific DNA-binding protein, relieves nitrate inhibition of aflatoxin biosynthesis. *Appl. Environ. Microbiol.* 61:2372–2377.

Clark, R.F., S.R. Williams, S.P. Nordt, and A.S. Manoguerra. 1999. A review of selected seafood poisonings. *Undersea Hyperb. Med.* 26(3):175–184.

Coffey, D.G., M.A. Uebersax, G.C. Hosfield, and J.R. Brunner. 1985. Evaluation of the hemagglutinating activity of low-temperature cooked beans. *J. Food Sci.* 50:78–87.

Cotty, P. 1988. Aflatoxin and sclerotial production by *Aspergillus flavus*: Influence of pH. *Phytopathology.* 78:1250–1253.

Cotty, P.J. and R. Jaime-Garcia. 2007. Influences of climate on aflatoxin producing fungi and aflatoxin contamination. *Int. J. Food Microbiol.* 119:109–115.

Davis, N.D., U.L. Diener, and V.P. Agnihotri. 1967. Production of aflatoxins B1 and G1 in chemically defined medium. *Mycopathol. Mycol. Appl.* 31:251–256.

Davis, R. and L. Villar. 1986. Symptomatic improvement with amitriptyline in ciguatera fish poisoning. *N. Engl. J. Med.* 315(1):65.

Deeds, J. 2012. *Bad Bug Book, Foodborne Pathogenic Microorganisms and Natural Toxins*, 2nd edn. Shellfish toxins (PSP, DSP, NSP, ASP, AZP). Food and Drug Administration, New Hampshire Ave Silver Spring, MD.

Demain, A.L. 1998. Induction of secondary metabolism. *Int. Microbiol.* 1:259–264.

Demircan, A., A. Keleş, F. Bildik, G. Aygencel, N.O. Dogan, H.F., and Gomez, H.F. 2009. Mad honey sex: Therapeutic misadventures from an ancient biological weapon. *Ann. Emerg. Med.* 54(6):824–829.

Diaz, J.H. 2005a. Evolving global epidemiology, syndromic classification, general management, and prevention of unknown mushroom poisonings. *Crit. Care Med.* 33(2):419–426.

Diaz, J.H. 2005b. Syndromic diagnosis and management of confirmed mushroom poisonings. *Crit. Care Med.* 33(2):427–436.

Diener, U.L. and N.D. Davis. 1970. Limiting temperature and relative humidity for aflatoxin production by *Aspergillus flavus* in stored peanuts. *J. Am. Oil Chem. Soc.* 47:347–351.

DiNubile, M. and Y. Hokama. 1995. The Ciguatera poisoning syndrome from farm-raised Salmon. *Ann. Intern. Med.* 122:2.

Doors, S. 1999. *Food Safety: Current Status and Future Needs*, American Academy of Microbiologists, Washington, DC.

Edgar J.A. and L.W. Smith. 2000. Transfer of pyrrolizidine alkaloids into eggs: Food safety implications. In: Tu, A.T. and Gaffield, W. (eds.), *Natural and Selected Synthetic Toxins: Biological Implications*, American Chemical Society, Washington, DC, pp. 118–128.

Erguven, M. et al. 2007. Mushroom poisoning. *Indian J. Pediatr.* 74:847–852.

Freed, D.L.J. 1999. Do dietary lectins cause disease? *Br. Med. J.* 318:1023–1024.

Gabius, H.J., S. Andre, J. Jiménez-Barbero, A. Romero, and D. Solis. 2011. From lectin structure to functional glycomics: Principles of the sugar code. *Trends Biochem. Sci.* 36:298–313.

George, P. and N. Hegde. 2013. Muscarinic toxicity among family members after consumption of mushrooms. *Toxicol. Int.* 20(1):113–115.

Goldfrank, L.R. 2011. Mushrooms. In: Nelson, L.S., Lewin, N.A., Howland, M.A., Hoffman, R.S., Goldfrank, L.R., and Flomenbaum, N.E., *Goldfrank's Toxicologic Emergencies*, 9th edn. McGraw-Hill, New York, pp. 1522–1536.

Gong, Y.Y., K. Cardwell, A. Hounsa, S. Egal, P.C. Turner, A.J. Hall, and C.P. Wild. 2002. Dietary aflatoxin exposure and impaired growth in young children from Benin and Togo: Cross sectional study. *Br. Med. J.* 325:20–22.

Gong, Y.Y., A. Hounsa, S. Egal, P.C. Turner, A.E. Sutcliffe, K. Cardwell, and C.P. Wild. 2004. Postweaning exposure to aflatoxin results in impaired child growth: A longitudinal study in Benin, West Africa. *Environ. Health Perspect.* 112(13):1334–1338.

Gong, Y.Y., S. Wilson, J.K. Mwatha, M.N. Routledge, J.M. Castelino, B. Zhao, G. Kimani H.C. Kariuki, B.J. Vennervald, D.W. Dunne, and C.P. Wild. 2012. Aflatoxin exposure may contribute to chronic hepatomegaly in Kenyan school children. *Environ. Health Perspect.* 120(6):893–896.

Goto, T., D.T. Wicklow, and Y. Ito. 1996. Aflatoxin and cyclopiazonic acid production by a sclerotium-producing *Aspergillus Tamarii* strain. *Appl. Environ. Microbiol.* 62:4036–4038.

Greene-McDowelle, D.M., B. Ingber, M.S. Wright, H.J.Jr. Zeringue, D. Bhatnagar, and T.E. Cleveland. 1999. The effects of selected cotton-leaf volatiles on growth, development and aflatoxin production of *Aspergillus parasiticus. Toxicon.* 37:883–893.

Gunduz, A., S. Turedi, R.M. Russell, and F. Ayaz. 2008. Clinical review of grayanotoxin/mad honey poisoning past and present. *Clin. Toxicol.* 46:437–442.

Gunduz, A., S. Turedi, H. Uzun, and M. Topbas. 2006. Mad honey poisoning. *Am. J. Emerg. Med.* 24(5):595–598.

Guo, B.Z., C.C. Holbrook, J. Yu, R.D. Lee, and R.E. Lynch. 2005. Application of technology of gene expression in response to drought stress and elimination of preharvest aflatoxin contamination. In: Abbas, H.K. (ed.), *Aflatoxin and Food Safety*, CRC Press, Boca Raton, FL, pp. 313–331.

Gupta, Y. 1987. Anti-nutritional and toxic factors in food legumes: A review. *Plant Foods Hum. Nutr.* 37:201–228.

Hall, I.R. 2003. *Edible and Poisonous Mushrooms of the World*, Timber Press, Portland, OR, p. 103.

Hallegraeff, G.M., D.M. Anderson, and A.D. Cembella (eds.) 1995. *Manual on Harmful Marine Microalgae.* IOC Manuals and Guides, UNESCO, Paris, France, p. 33.

Hamelryck, T.W., M.H. Dao-Thi, F. Poortmans, M.J. Chrispeels, L. Wyns, and R. Loris. 1996. The crystallographic structure of phytohemagglutinin-L. *J. Biol. Chem.* 271(34):20479–20485.

Heymann, D.L. (ed.) 2004. *Control of Communicable Disease Manual*, 18th edn. American Public Health Association, Washington, DC.

Hicks, J.K., J.H. Yu, N.P. Keller, and T.H. Adams. 1997. Aspergillus sporulation and mycotoxin production both require inactivation of the FadA G alpha protein-dependent signaling pathway. *EMBO J.* 16:4916–4923.

Hoffmeister, D. and N.P. Keller. 2007. Natural products of filamentous fungi: Enzymes, genes, and their regulation. *Nat. Prod. Rep.* 24:393–416.

Hogan, C.M. 2008. Rough-Skinned Newt *Taricha granulosa.* GlobalTwitcher.com. Retrieved on April 6, 2009. http://www.medic8.com/healthguide/food-poisoning/red-kidney-bean-toxins.html.

Hu, D., H. Tateno, A. Kuno, R. Yabe, and J. Hirabayashi. 2012. Directed evolution of lectins with sugar-binding specificity for 6-sulfo-galactose. *J. Biol. Chem.* 287:20313–20320.

Hudler, G.W. 1998. *Magical Mushrooms, Mischievous Molds: The Remarkable Story of the Fungus Kingdom and Its Impact on Human Affairs*, Princeton University Press, Princeton, NJ.

Hughes, J.M. and M.W. Merson. 1976. Current concepts, fish and shellfish poisoning. *New Engl. J. Med.* 295:1117–1120.

Huxtable, R.J. and R.A. Cooper. 2000. Pyrrolizidine alkaloids: Physiochemical correlates of metabolism and toxicity. In: Tu, A.T. and Gaffield, W. (eds.), *Natural and Selected Synthetic Toxins: Biological Implications*, American Chemical Society, Washington, DC, pp. 100–117.

Hwang, D.F., Tai, K.P., Chueh, C.H., Lin, L.C., and Jeng, S.S. 1991. Tetrodotoxin and derivatives in several species of the gastropod Naticidae. *Toxicon.* 29(8):1019–1024.

Isbister, G.K. and M.C. Kiernan. 2005. Neurotoxic marine poisoning. *Lancet Neurol.* 4:219–228.

Itoia, S. et al. 2014. Larval pufferfish protected by maternal tetrodotoxin. *Toxicon.* 78:35–40.

Jansen, S.A. et al. 2012. Grayanotoxin poisoning: 'mad honey disease' and beyond. *Cardiovasc. Toxicol.* 12(3):208–215.

Jayashree, T., R.J. Praveen, and C. Subramanyam. 2000a. Regulation of aflatoxin production by Ca(2+)/calmodulin-dependent protein phosphorylation and dephosphorylation. *FEMS Microbiol. Lett.* 183:215–219.

Jayashree, T. and C. Subramanyam. 2000b. Oxidative stress as a prerequisite for aflatoxin production by *Aspergillus parasiticus. Free Radic. Biol. Med.* 29:981–985.

Judge, B.S., J.F. Ammirati, G.H. Lincoff, J.H. Trestrail, and P.B. Matheny. 2010. Ingestion of a newly described North American mushroom species from Michigan resulting in chronic renal failure: *Cortinarius orellanosus. Clin. Toxicol. (Phila).* 48(6):545–549.

Kale, S. and J.W. Bennett. 1991. Strain instability in filamentous fungi. In: Bhatnagar, D., Lillehoj, E.B., and Arora, D.K. (eds.), *Handbook of Applied Mycology, Mycotoxins in Ecological Systems*, Taylor & Francis, New York, Vol. 5, pp. 311–332.

Kale, S.P., J.W. Cary, D. Bhatnagar, and J.W. Bennett. 1996. Characterization of experimentally induced, nonaflatoxigenic variant strains of *Aspergillus parasiticus. Appl. Environ. Microbiol.* 62:3399–3404.

Keller, N.P., G. Turner, and J.W. Bennett. 2005. Fungal secondary metabolism—From biochemistry to genomics. *Nat. Rev. Microbiol.* 3:937–947.

Kim, A.J., J.S. Kim, D.W. Shin, K.J. Baek, S.B. Han, and Y.J. Lee. 2000. Grayanotoxin intoxication-3 case reports. *J. Korean Soci. Emerg. Med.* 11:372–377.

Kim, C., D.S. Kim, H.W. Lee, Y.M. Ahn, and J.H. Uhm. 2008. A case of grayanotoxin intoxication presenting with mental changes and vomiting. *Korean J. Pediatric Gastroent. Nutr.* 11:223–225.

Kim, J.H., B.C. Campbell, R. Molyneux, N. Mahoney, K.L. Chan, J. Yu, J.R. Wilkinson, J. Cary, D. Bhatnagar, and T.E. Cleveland. 2006. Gene targets for fungal and mycotoxin control. *Mycotoxin Res.* 22:3–8.

Kocabas, C.N. and B.E. Sekerel. 2003. Does systemic exposure to aflatoxin B1 cause allergic sensitization? *Allergy.* 58:363.

Kumar, S., A.K. Verma, M. Das, S. Jain, and P.D. Dwivedi. 2013. Clinical complications of kidney bean (*Phaseolus Vulgaris* L.) consumption. *Nutrition.* 29(6):821–827.

Lehane, L. and R.J. Lewis. 2000. Ciguatera: Recent advances but the risk remains. *Int. J. Food Microbiol.* 61:91–125.

Lehane, L. and J. Olley. 1999. Histamine (Scombroid) fish poisoning. In: *National Office of Animal and Plant Health.* Department of Agriculture Fisheries and Forestries, Canberra, Australia.

Lewis, R.J. 2001. The changing face of ciguatera. *Toxicon.* 39:97–106.

Liener, I.E. 1974. Phytohaemagglutinins: Their nutritional significance. *J. Agric. Food. Chem.* 22:17–22.

Luchese, R.H. and W.F. Harrigan. 1993. Biosynthesis of aflatoxin-the role of nutritional factors. *J. Appl. Bacteriol.* 74:5–14.

Mahoney, N. and R.J. Molyneux. 2004. Phytochemical inhibition of aflatoxigenicity in *Aspergillus flavus* by constituents of walnut (*Juglans regia*). *J. Agric. Food Chem.* 52:1882–1889.

Meier, J. and J. White. 1995. *Handbook of Clinical Toxicology of Animal Venoms and Poisons*, 1st edn., CRC Press, Boca Raton, FL, pp. 141–158.

Minns, A. 2014. Scombroid fish poisoning. *The Official Newsletter of the California Poison Control System.* 12(4):1.

Nijjar, M.S. and B. Grimmelt. 1994. Domoic acid inhibits adenylate cyclase activity in rat brain membranes. *Mol. Cell. Biochem.* 136:105–111.

Noda, M., T. Ikeda, T. Kayono, H. Suzuki, H. Takeshima, M. Kurasaki, H. Takahashi, and S. Numa. 1986. Existence of distinct sodium channel messenger RNAs in rat brain. *Nature.* 320:188–192.

Novak, S.M. 1998. Foodborne illness-chemical fish and shellfish poisoning, *Clin Microbiol Newsletter.* 20(3):17–21.

O'Brian, G.R., D.R. Georgianna, J.R. Wilkinson, J. Yu, H.K. Abbas, D. Bhatnagar, T.E. Cleveland, W.C. Nierman, and G.A. Payne. 2007. The effect of elevated temperature on gene transcription and aflatoxin biosynthesis. *Mycologia.* 99:232–239.

Olsen, S.J., L.C. MacKinnon, J.S. Goulding, N.H. Bean, and L. Slutsker. 2000. Surveillance for food borne-disease outbreaks-United Sates 1993–1997. *MMWR CDC Surveill. Summ.* 49(No. SS-1):1–64.

Onat, F.Y., B.C. Yegen, R. Lawrence, A. Oktay, and S. Oktay. 1991. Site of action of grayanotoxins in mad honey in rats. *J. Appl. Toxicol.* 11:199–201.

Patocka, J. and L. Stredav. 2002. Brief review of natural nonprotein neurotoxins. In Price, R. (ed.), *ASA Newsletter*, Applied Science and Analysis Inc., Aberdeen, MD, Vol. 89, pp. 16–23.

Payne, G.A. and M.P. Brown. 1998. Genetics and physiology of aflatoxin biosynthesis. *Annu. Rev. Phytopathol.* 36:329–362.

Pearn, J. 2001. Neurology of Ciguatera. *J. Neurol.* 70:4–8.

Peraica, M., B. Radic, A. Lucic, and M. Pavlovic. 1999. Toxic effects of mycotoxins in humans. *Bull. World Health Org.* 77:754–766.

Pereira, E.C. et al. 2008. Intoxicação por azaléia (Rhododendron spp.) em dois caprinos. In: *Proceedings of the Latin American Veterinary Conference*, Lima, Peru http://www.ivis.org/home.asp. Accessed: July 15, 2013.

Piva, G., F. Galvano, A. Pietri, and A. Piva. 1995. Detoxification methods of aflatoxin. A review. *Nutr. Res.* 15:689–715.

Poon, W.T. et al. 2008. Grayanotoxin poisoning from *Rhododendron simsii* in an infant. *Hong Kong Med. J.*14:405–407.

Probst, C., H. Njapau, and P.J. Cotty. 2007. Outbreak of an acute aflatoxicosis in Kenya in 2004: Identification of the causal agent. *Appl. Environ. Microbiol.* 73:2762–2764.

Pusztai, A. 1986. The role in food poisoning of toxins from plants. In: Robinson, R.K. (ed.), *Developments in Food Microbiology*, Vol. 2. Elsevier Applied Science Publishers, London, U.K., pp. 179–194.

Qiang, Y., B. Zhou, and K. Gao. 2011. Chemical constituents of plants from the genus *Rhododendron*. *Chem. Biodiv.*, 8:792–814.

Radominska-Pandya, A. 2010. Invited speakers. *Drug Metab. Rev.* 42(1):1–2.

Rahway, N.J. 1983. The Merck index. *Merck.* 10:652–653.

Ramadass, B., K. Doklandny, P.L. Moseley, Y.R. Patel, and H.C. Lin. 2010. Sucrose Co-administration reduces the toxic effect of lectin on gut permeability and instestinal bacterial colonization. *Dig. Dis. Sci.* 55:2778–2784.

Reddy, T.V., L. Viswanathan, and T.A. Venkitasubramanian. 1971. High aflatoxin production on a chemically defined medium. *Appl. Microbiol.* 22:393–396.

Reddy, T.V., L. Viswanathan, and T.A. Venkitasubramanian. 1979. Factors affecting aflatoxin production by *Aspergillus parasiticus* in a chemically defined medium. *J. Gen. Microbiol.* 114:409–413.

Reib, J. 1982. Development of *Aspergillus parasiticus* and formation of aflatoxin B1 under the influence of conidiogenesis affecting compounds. *Arch. Microbiol.* 133:236–238.

Reverberi, M., A.A. Fabbri, S. Zjalic, A. Ricelli, F. Punelli, and C. Fanelli. 2005. Antioxidant enzymes stimulation in *Aspergillus parasiticus* by *Lentinula edodes* inhibits aflatoxin production. *Appl. Microbiol. Biotechnol.* 69:207–215.

Reverberi, M., S. Zjalic, A. Racelli, A.A. Fabbri, and C. Fanelli. 2006. Oxidant/antioxidant balance in *Aspergillus parasiticus* affects aflatoxin biosynthesis. *Mycotoxin Res.* 22:39–47.

Richard, J.L. and G.A. Payne. 2003. *Mycotoxins: Risks in Plant, Animal and Human Systems*, Council for Agricultural Science and Technology (CAST): Ames, IA, pp. 101–103.

Robens, J.F. and K. Cardwell. 2003. The costs of mycotoxin management to the USA: Management of aflatoxins in the United States. *J. Toxicol.* 22:139–152.

Robens, J.F. and K. Cardwell. 2005. The cost of mycotoxin management in the United States. In: Abbas, H.K. (ed.), *Aflatoxin and Food Safety*, CRC Press, Boca Raton, FL, pp. 1–12.

Roder, E. 1995. Medicinal plants in Europe containing pyrrolizidine alkaloids. *Pharmazie.* 50:83–98.

Roulet, M. et al. 1988. Hepatic veno-occlusive disease in newborn infant of a woman drinking herbal tea. *J. Pediatr.* 112(3):433–436.

Roy, A.K. and H.K. Chourasia. 1989. Effect of temperature on aflatoxin production in *Mucuna pruriens* seeds. *Appl. Environ. Microbiol.* 55:531–532.

Sanders, T.H., P.D. Blankenship, R.J. Cole, and R.A. Hill. 1984. Effect of soil temperature and drought on peanut pod and stem temperatures relative to *Aspergillus flavus* invasion and aflatoxin contamination. *Mycopathologia.* 86:51–54.

Satake, M., K. Ofuji, H. Naoki, K. James, A. Furey, T. McMahon, J. Silke, and T. Yasumoto. 1998. Azaspiracid, A New Marine Toxin Having Unique Spiro Ring Assemblies, Isolated From Irish Mussels, *Mytilus edulis*. *J. Am. Chem. Soc.* 120:9967–9968.

Schroeder, H.W. and H. Jr. Hein. 1968. Effect of diurnal temperature cycles on the production of aflatoxin. *Appl. Microbiol.* 16:988–990.

Schumued, L. and W. Jr. Slikker. 1999. Black–Gold: A simple, high-resolution histochemical label for normal and pathological myelin in brain tissue sections. *Brain Res.* 837:289–297.

Scott, P.M., B.B. Coldwell, G.S. Wilberg. 1971. Grayanotoxins. Occurrence and analysis in honey and a comparison of toxicities in mice. *Food Cosmet. Toxicol.* 9:179–184.

Sharon, N. 2008. Lectins: Past, present and future. *Biochem. Soc. Trans.* 36:1457–1460.

Shouman, B.O., D.E. Morsi, S. Shabaan, A.H. Abdel-Hamid, and A. Mehrim. 2012. Aflatoxin B1 level in relation to child's feeding and growth. *Indian J. Pediatr.* 79:56–61.

Smolinska, S., M. Jutel, R. Crameri, and L. O'Mahony. 2014. Histamine and gut mucosal immune regulation. *Allergy.* 69(3):273–281.

St Leger, R.J., S.E. Screen, and B. Shams-Pirzadeh. 2000. Lack of host specialization in *Aspergillus flavus*. *Appl. Environ. Microbiol.* 66:320–324.

Stallard, D. and T.E. Edes. 1989. Muscarinic poisoning from medications and mushrooms. A puzzling symptom complex. *Postgrad. Med.* 85:341–345.

Stegelmeier, B.L. 2011. Pyrrolizidine alkaloid-containing toxic plants (Senecio, Crotalaria, Cynoglossum, Amsinckia, Heliotropium, and Echium spp.). *Vet. Clin. North. Am. Food. Anim. Pract.* 27(2):419–428.

Stickel, F. and H.K. Seitz. 2000. The efficacy and safety of comfrey. *Public Health Nutr.* 3(4A):501–508.

Sweeney, M.J. and A.D. Dobson. 1998. Mycotoxin production by *Aspergillus*, *Fusarium* and *Penicillium* species. *Int. J. Food Microbiol.* 43:141–158.

Swift, A. and T. Swift. 1993. Ciguatera. *J. Toxicol. Clin. Toxicol.* 31(1):1–29.

Takao, A. et al. 2012. Aflatoxin contamination of red chili pepper from Bolivia and Peru, countries with high gallbladder cancer incidence rates. *Asian Pacific J. Cancer Prev.* 13(10):5167–5170.

Tallent, W.H., M.L. Riethof, and E.C. Horning. 1957. Studies on the occurrence and structure of acetylandromedol (andromedotoxin). *J. Am. Chem. Soc.* 79:4548–4554.

Trail, F., N. Mahanti, and J. Linz. 1995. Molecular biology of aflatoxin biosynthesis. *Microbiology.* 141:755–765.

Turner, P.C., S.E. Moore, A.J. Hall, A.M. Prentice, and C.P. Wild. 2003. Modification of immune function through exposure to dietary aflatoxin in gambian children. *Environ. Health Perspect.* 111:217.

Vasconcelos, I.M. and J.T.A. Oliveira. 2004. Antinutritional properties of plant lectins. *Toxicon.* 44:385–403.

Viuda-Martos, M., Y. Ruiz-Navajas, J. Ferna´ndez-Lo´pez, and J.A. Pe´rez-A´ lvarez. 2008. Functional properties of honey, propolis, and royal jelly. *J. Food Sci.* 73:117–124.

Wannop, C.C. 1961. The histopathology of Turkey X disease in Great Britain. *Avian Dis.* 5(4):371–381.

Watters, M.R. 1995. Organic neurotoxins in seafood. *Clin. Neurol. Neurosurg.* 97:119–124.

Wilson, R.A., H.W. Gardner, and N.P. Keller. 2001. Differentiation of aflatoxin-producing and non-producing strains of *Aspergillus flavus* group. *Lett. Appl. Microbiol.* 33:291–295.

Woloshuk, C.P., J.R. Cavaletto, and T.E. Cleveland. 1997. Inducers of aflatoxin biosynthesis from colonized maize kernels are generated by an amylase activity from *Aspergillus flavus. Phytopathology.* 87:164–169.

Wong, J., E. Youde, B. Dickinson, and M. Hale. 2002. Report of the Rhododendron feasibility study. School of Agricultural and Forest Sciences. University of Wales, Bangor, U.K., http://www.wildresources.co.uk/reports/ rhodo_eng.pdf.

Wright, M.S., D.M. Greene-McDowelle, H.J. Zeringue, D. Bhatnagar, and T.E. Cleveland. 2000. Effects of volatile aldehydes from aspergillus-resistant varieties of corn on *Aspergillus parasiticus* growth and aflatoxin biosynthesis. *Toxicon.* 38:1215–1223.

Yabe, K., H. Nakamura, Y. Ando, N. Terakado, H. Nakajima, and T. Hamasaki. 1988. Isolation and characterization of *Aspergillus parasiticus* mutants with impaired aflatoxin production by a novel tip culture method. *Appl. Environ. Microbiol.* 54:2096–2100.

Yavuz, H., A. Özel, I. Akkus, and I. Erkul. 1991. Honey poisoning in Turkey. *Lancet.* 337:789–790.

Zacharko-Siembida, A. et al. 2014. Red Kidney Bean (*Phaseolus vulgaris*) lectin stimulation increases the number of enterochromaffin cells in the small intestine of suckling piglets. *Bull. Vet. Inst. Pulawy.* 58(2):289–294.

Zeringue, H.J., Jr. 2002. Effects of methyl jasmonate on phytoalexin production and aflatoxin control in the developing cotton boll. *Biochem. Syst. Ecol.* 30:497–503.

Zeringue, H.J. and D. Bhatnagar. 1993. Neem and control of aflatoxin contamination. In: Singh, R.P., Chari, M.S., Raheja, A.K., and Kraus, W. (eds.), *Neem and Environment*, Science Publishers, Inc., Enfield, NH, Vol. 2, pp. 713–727.

Zeringue, H.J., Jr. R.L. Brown, J.N. Neucere, and T.E. Cleveland. 1996. Relationships between C6-C12 alkanal and alkenal volatile contents and resistance of maize genotypes to *Aspergillus flavus* and aflatoxin production. *J. Agric. Food Chem.* 44:403–407.

13 Natural Toxins and Antinutrients in Plants and Fungi

Ecological Biochemistry of Food

L.B. Borse, S.L. Borse, and N.A. Gujarathi

CONTENTS

13.1 NATURAL TOXINS IN PLANTS AND FUNGI

Living organisms naturally produce chemicals called as natural toxins. These toxins are not harmful to the organisms themselves, but they may be toxic to other creatures, including humans, when consumed (JECFA 2001). Some plants naturally produce their toxic compounds. As for example, under certain conditions, in ocean, microscopic algae (tiny plants) can produce compounds that are toxic to humans but not to shellfish who eat this algae. When humans eat shellfish that contain these toxins, sickness can easily occur. To avoid this, active monitoring program is conducted to ensure that shellfish, which are being sold, do not contain the "shellfish toxins." Some agencies like Department of Fisheries and Oceans, the Canadian Food Inspection Agency, and Environment Canada are involved in joint administration of this active monitoring program (JECFA 2001, Leon et al. 2016).

13.2 NATURAL TOXINS IN FOOD

Food that we consume contains various natural chemicals, including carbohydrates, sugars, proteins, and vitamins, which are essential and beneficial. But some foods may contain potentially harmful natural toxins. Sometimes, food contains toxin in the form of naturally occurring pesticide that is used to prohibit insect attack or to protect the plant when damaged by harsh weather, handling, UV light, or microbes. The adverse reaction and the severity of symptoms manifested in our body depend on our own sensitivity to a natural toxin, the concentration of the toxin in the food, and the amount of the food consumed (Rahman et al. 2014). Here are some examples of commonly eaten foods that may contain natural toxins. The maximum limit is given in the Australia New Zealand Food Standards Code (ANZFA 2001, Lopez et al. 2016).

13.2.1 Cassava and Bamboo Shoots

Naturally occurring cyanogenic glycosides in raw or unprocessed cassava and bamboo shoots may contain toxin hydrogen cyanide. Cassava is also known as yuca, tapioca (a processed product of cassava), gaplek, or manioc. Bamboo shoots, a traditional ingredient of Asian cuisine, are obtained from the underground stems of the bamboo plant. To avoid exposure to the toxins, cassava is converted into sweet cassava before eating. First, it is peeled and sliced and then cooked thoroughly, by baking, boiling, or roasting. Frozen cassava, and frozen peeled cassava, should also be prepared in the same way as mentioned earlier. Fresh bamboo shoots should be sliced in half lengthwise: the outer leaves peeled away and any fibrous tissue at the base trimmed off. The remaining fresh shoots should be thinly sliced into strips and boiled in lightly salted water for 8–10 minutes (ANZFA 2001, Bajaj et al. 2016).

13.2.2 Fish: Escolar and Oilfish

Other types of food poisoning are due to escolar (*Lepidocybium flavobrunneum*) and oilfish (*Ruvettus pretiosus*), which leads to a type of oily diarrhea, called keriorrhea. For example, 45%–67% people were found to be ill due to oily diarrhea after eating these species of fish, which was observed after investigation of some Australian outbreaks. A significant number of illnesses are observed due to the consumption of these fish as the symptoms can be mild and short-lived. The oily diarrhea is produced by indigestible oil present in these fish, which before being excreted out accumulates in the rectum.

Symptoms for mild diarrhea are oily orange or yellow discharge and that of severe diarrhea are nausea and vomiting. Other symptoms may include stomach cramps, loose bowel, and headache. The symptoms can occur suddenly without warning, usually within 2½ hours of fish consumption, but range from 1 to 90 hours later. Symptoms may last for one or two days. Unlike other forms of diarrhea, the oily diarrhea caused by these fish does not result in significant loss of electrolytes from body fluid and is not life-threatening. Pregnant women and patients with a bowel problem or malabsorption are advised not to consume these fish. Initially, it is advised to consume only a small portion of fish. If you experience gastrointestinal symptoms, do not consume it anymore and seek medical advice if symptoms persist (ANZFA 2001).

13.2.3 Fish Oil: Naturally Occurring Mercury

Fish oil is very beneficial for your health, and everyone should eat two to three servings of fish in a week for good health. However, some of the larger species of fish, such as shark, marlin, and swordfish, may contain some levels of naturally occurring mercury, because they are predator and eat smaller fish and they also live for long time absorbing mercury from the ocean. This high level of mercury can affect babies in the womb in terms of attention, learning, and memory in their early years (ANZFA 2001).

13.2.4 Fish: Ciguatera Food Poisoning

Larger and older fish that live in warm ocean water produce ciguatera toxin that causes an unusual type of food poisoning. Poisoning results due to consumption of fish containing the ciguatera toxin. These contaminated fish have eaten smaller fish that eats the algae that shelters a tiny organism responsible for producing a toxin that causes ciguatera. Symptoms emerge within 24 hours, which include numbness around the fingers, toes, and mouth, a burning sensation when in contact with cold, joint pain, nausea, itchiness, and breathing difficulties. Poisoning may also occur with ingestion of coral trout, Spanish mackerel, reef cod, barracuda, emperor, groper, sturgeon, trevally, and kingfish. Any warmwater predatory fish over 6 kg should be treated with suspicion. Under no circumstances should the head, roe, or liver be eaten, and it is wise to eat a small portion of any large fish to test for reactions 24 hours before serving. In case of sickness develops, consumption of this food is not advised (ANZFA 2001, Bene et al. 2016).

13.2.5 Fish: Scombroid (Histamine) Fish Poisoning

Scombroid or histamine poisoning may arise to a person ingested a fish that has undergone inadequate temperature control. It can occur when fish is not chilled to 4°C or below. The result is a high level of histamine building up in the fish flesh (ANZFA 2001, Petrovic et al. 2016).

Symptoms may vary but include burning and tingling of the lips and mouth, flushing of the face, dizziness, headaches, palpitations, sweating, vomiting and diarrhea, blurred vision, and an itchy rash, often on the face, neck, chest, and upper back.

In severe conditions, breathing difficulties can occur mainly in those suffering from asthma. Recent casualty in Bali has focused on scombroid poisoning. The conversation enlisted Bob Mead, associate professor, metabolic biochemist, toxicologist, and chair of the Forensic Biology and Toxicology program (ANZFA 2001).

13.2.6 Fruit Seeds and Pits

Amygdaline is present in the inner part of apple and pear seeds and in the inner stony pit (kernel) of apricots and peaches, which is a cyanogenic glycoside. Amygdaline gets converted into hydrogen cyanide in the stomach causing discomfort or illness. It can produce fatal effects sometimes. A high level of the toxin is present in different types of apricot kernels that can release cyanide into the body when eaten. Adults do not eat more than three apricot kernels a day and children should be advised to avoid it as per Food Standards Australia New Zealand. Occasional eating seed or pipes from these fruits are not a concern but regular habit of eating the seed from these fruits can be dangerous (ANZFA 2001).

13.2.7 Kumara

Kumara is a member of the sweet potato family; in response to injury, insect attack, and other stress, it can produce toxins. The most common toxin, ipomeamarone, makes the kumara taste bitter. The observed reports are avoidable that cattle deaths occur after they have eaten moldy kumara. Toxin levels are usually highest in the damaged organ. It is recommended that before cooking, any damaged parts on kumara should be removed. After cooking, if it tastes bitter, then it is advisable not to eat it (ANZFA 2001).

13.2.8 Parsnip

Commonly, parsnips contain a group of natural toxins known as furocoumarins. These are probably produced for protecting the plant when it has been stressed. The peel of the plant contains the highest concentration of the toxin. But it is also found on the surface layer of the plant or around any damaged areas. One of the furocoumarin toxins can cause stomachache and may also cause a painful skin reaction when with the parsnip plant is when exposed in UV rays of the sun. To reduce the level of toxin, it is important to peel the parsnip before cooking and remove any damaged parts. The toxin levels are reduced when the parsnip is cooked freely by baking, microwaving, or boiling with water, discarding the water after use (ANZFA 2001, Fernandes et al. 2016).

13.2.9 Potatoes

Naturally occurring toxins called glycoalkaloids are found in all potatoes. The level of toxins are usually low, but the sprouts and the peel of potatoes have higher toxin level, which taste bitter. Plants produce these toxins in response to stress, for example, from microorganisms and UV light, and damage such as bruising. The amount of toxin depends on the type of potato and the growing conditions. Symptoms like severe stomach pain and also death due to glycoalkaloid poisoning may be reported, but it is very unusual (McMartinv et al. 2016). It is strictly advised to avoid eating the sprouts and to remove any green or damaged parts before cooking as glycoalkaloids are not destroyed by cooking, and also bitter parts of the potatoes should not be consumed. Green potato crisp should not be eaten. Potatoes are always stored in a dark, cool, and dry place (ANZFA 2001, Guarischi-Sousa et al. 2016).

13.2.10 Kidney Beans

There are many types of beans that contain toxins called lectins; especially, red kidney beans contain the highest concentrations of lectin. Ingestion of four to five raw beans can cause severe stomachache, vomiting, and diarrhea. To destroy the toxins, before cooking, soak the beans for at least five hours and then boil them briskly in freshwater for at least 10 minutes. Cooking at low temperature, for example, in a slow cooker may not destroy the toxin. Improper cooking of beans can be more toxic than raw ones. Tinned beans can be used without further cooking (ANZFA 2001, Zhang et al. 2016).

13.2.11 Rhubarb

Rhubarb contains a natural toxin called oxalic acid. Especially, it is present in highest concentrations in the leaves and these should not be eaten. The age of the plant, the season, the climate, and the type of soil determine the amount of toxin in the plant. Symptoms of toxicity include muscle twitching, vomiting, cramps, headache, breathing difficulties, abnormal and rapid heart rate, pain, convulsions, and coma (ANZFA 2001).

13.2.12 Zucchini (Courgette)

Zucchini contains natural toxins known as cucurbitacins, which has a bitter taste. Bitterness of zucchinis has been known for a long time but is rarely found in commercially grown zucchinis. Symptoms of toxicity include vomiting, stomach cramps, diarrhea, and collapse. It is strictly advised to avoid eating of bitter zucchini (http://www.foodsafety.asn.au/).

13.2.13 Wild Mushrooms

Commercially grown and sold mushrooms are safe. However, the death cap mushroom is deadly and poisonous, contains fungus that is responsible for 90% of all deaths related to mushroom toxicity, and is commonly found in southeastern Australia near established oak trees, and possibly some other trees are harmonious, usually during late summer to early winter after heavy rain or heavy irrigation (ANZFA 2001, Fernandes et al. 2016).

Poison in one mushroom is enough to kill an adult. This poison is not inactivated by cooking or peeling, and all parts of the plant are poisonous. Symptoms of toxicity occur from 6 to 24 hours or more after ingestion of mushrooms. Symptoms may appear for 1–2 days giving a false impression of recovery. Symptoms include violent stomach pains, nausea, vomiting, and diarrhea. The toxin may cause serious liver damage, leading to death from liver failure (ANZFA 2001, Fernandes et al. 2016). If a person after eating poisonous mushrooms gets sick, he or she should seek for immediate medical attention; the sooner the treatment begins, the better the chances of survival. Medical assistance should be sought by calling your local poisons information center or hospital emergency departments. For absolute safety, avoid any wild mushrooms, unless it has been definitely identified as nonpoisonous.

13.3 NATURAL TOXINS IN FOOD PLANTS

Secondary metabolic products or low-molecular-weight endogenous toxins are the toxic food components of plant origin. Products of primary metabolism such as macronutrients and micronutrients are involved in energy metabolism such as photosynthesis, growth, and reproduction. Secondary metabolic products such as plant pigments, flavors, and compounds are species specific and give the plant its particular characteristics and also protect the plants. Some of these products are growth inhibitors, neurotoxins, carcinogens, and teratogens (Omaye 2004, Lopez et al. 2016) that produce toxicity in the individual when taken orally. Plant toxins are classified into alkaloids such as pyrrolizidine alkaloids and glycoalkaloids, cyanogenic glycosides, lectins, saponins, and antinutrients. Some of these are discussed in the following sections.

13.3.1 Alkaloids

It is one of the bitter components of plants found widely in nature and has pharmacological benefits (ANZFA 2001, Lopez at al. 2016). They are secondary plant metabolites of plants. They are often basic nitrogen-containing compounds able to form salts with acid. They are isolated from roots, seeds, leaves, or bark of some species. At least 40% of plant families belong to Amaryllidaceae, Compositae, Leguminosae, Liliaceae, Papaveraceae, and Solanaceae. Normally, alkaloids are found in plants and in common foods.

Pyrrolizidine alkaloid is one of the widely found alkaloids in plants. These are found in about 6000 plant species or 3% of flowering plants. They are widely distributed in all climatic regions of the world (WHO 1988, ANZFA 2001). Plants that contain pyrrolizidine alkaloids are usually ingested through contaminated crops, since these plants grow as weeds along with food crops such as corn or wheat and harvested with the grain. Another source of toxins may be by intake of herbal foods containing these toxins. Avoiding the intake of alkaloid-containing foods and herbal preparations can prevent poisoning (Adekanmi et al. 2009). Some effective measures in agriculture are applied to reduce the contamination of pyrrolizidine-containing plants in food crops. Common food plants contain a natural pesticide against common pests such as the glycoalkaloids. Common examples include alkaloids such as solanine present in potatoes and tomatine present in tomatoes (Maga 1980).

13.3.2 Glycoalkaloids

Glycoalkaloids contain steroidal alkaloid that is coupled to one or more monosaccharides. Generally, Solanaceae families of plants produce glycoalkaloid toxins. In potatoes, the main glycoalkaloids are α-solanine and α-chaconin, which are present in the parenchyma cells of the periderm and the cortex of the tubers. These chemicals are toxic to insects and animals and serve to protect the plant from predators. Glycosides are present in small amounts in potato tubers and are concentrated in the skin of the potato and also in the areas of high metabolic activity such as the eye regions. Green potatoes or blighted potatoes contain the highest amount of toxin, which have very high levels in the sprouts. This toxin cannot be eliminated by washing, soaking, or cooking. Cooked potatoes that contain high amount of solanine have a bitter taste and cause a burning sensation in the throat.

13.3.2.1 Toxicity

The glycoalkaloids found in fresh, undamaged edible plants do not normally cause toxicity. However, toxic effects can be produced at higher doses.

13.3.2.1.1 Acute Toxicity

The major problem for glycoalkaloid toxicity is its acute toxicity. Many cases of human poisonings are reported (sometimes fatal) due to the ingestion of green, damaged, or sprouted potatoes with high levels of glycoalkaloids. Acute toxicity syndromes in humans have been observed at glycoalkaloid levels of more than 2.8 mg/kg body weight (Jadhaw et al. 1981). The onset of symptoms is exhibited from minutes to 2 days after eating of toxic potatoes, with longer incubation periods generally associated with the more severe cases (Maga 1980, Jadhaw et al. 1981). Symptoms are diarrhea, vomiting, and severe abdominal pain. In severe cases, neurological problems may be observed including drowsiness and apathy, confusion, weakness, and vision disturbances followed by unconsciousness and in some cases death. Glycoalkaloids also have saponin-like properties and can disrupt membrane function in the gastrointestinal tract, leading to hemorrhagic damage. This damage can become severe enough to cause death with necrosis.

13.3.2.1.2 Chronic Toxicity

The lack of chronic toxicity data, an adequate number of observed-adverse-effect level for potato glycoalkaloids has not been assessed, and doses are tolerable. Tolerable daily intake for humans has not been determined (WHO 1993, Glover et al. 2015).

13.3.3 Cyanogenic Glycoside

They are found in at least 2000 plant species, of which large number of species are used as food. They are secondary metabolites of plants containing amino acids. There are approximately 25 cyanogenic glycosides known. Different cyanogenic glycosides are found in different cyanogenic food plants, for example, taxiphyllin in bamboo shoots and linamarin in cassava (WHO 1993, Glover et al. 2015).

Important staple foods in some parts of the world (such as cassava and sorghum) contain cyanogenic glycosides. Other edible plants containing cyanogenic glycosides include bamboo shoots, flaxseeds, and seeds of fruits such as apricot and peach, seeds of peas and beans such as lima beans, and shell of soya beans.

13.3.3.1 Toxicity

Cyanogenic glycosides containing plant produce cyanide toxin on ingestion. The plant species has a corresponding hydrolytic enzyme (β-glucosidase). In the presence of water, the nontoxic cyanogenic glycosides are hydrolyzed by the enzyme producing cyanohydrins, which quickly decompose to the toxic hydrogen cyanide. In this way, cyanogenic plants are protected against predators. Cyanogenic glycosides, cyanohydrins, and hydrogen cyanide are collectively known as cyanogens.

13.3.3.1.1 Acute Toxicity

Symptoms of acute cyanide poisoning include rapid shallow breathing, decease in blood pressure, rapid pulse, dizziness, headache, stomach pains, vomiting, diarrhea, mental confusion, reduced sensitivity, cyanosis with twitching, and convulsions followed by terminal coma (WHO 1993, Glover et al. 2015).

When the cyanide level exceeds the limit, death due to cyanide poisoning can occur (Baker et al. 1988). The extent of poisoning of cyanide-containing food is dependent on body weight. For example, a child or person with less body weight is not able to detoxify the cyanide from a meal of inadequately prepared cassava or bamboo shoots. The acute lethal dose of hydrogen cyanide for humans is reported to be 0.5–3.5 mg/kg body weight. In adults, approximately 50–60 mg of free cyanide constitutes a lethal dose.

13.3.3.1.2 Chronic Toxicity

Dietary intake of cyanide can cause chronic diseases. Individuals who have dietary deficiency such as inadequate protein and iodine intake may cause adverse effects. For example, individuals with enlarged thyroid and stunted growth due to iodine deficiency may demonstrate chronic toxicity with continuous dietary cyanide exposure.

Individuals with inadequate dietary intake of protein and iodine manifest neurological diseases such as konzo and tropical ataxic neuropathy (TAN). Konzo is an upper motor neuron disease characterized by irreversible but nonprogressive symmetric spastic paraparesis with an abrupt onset, whereas TAN is a severe neurological syndrome with clinical symptoms of optical atrophy, angular stomatitis, sensory gait ataxia, and neurosensory deafness (Harvey et al. 1985, WHO 1993).

13.4 ANTINUTRIENTS IN PLANTS

The result is a decrease in the body's ability to use the nutrients, even though they are present in the food. Secondary, or unintentional, effects on the plant can occur when genetic material is introduced. With the transfer of one gene, multiple changes in the plant can result. The net result was a lack of significantly increased antinutrient levels in these foods; however, it is important to recognize that these substances could exist and we need to be aware of them.

13.4.1 Rapeseed Oil

Naturally occurring antinutrients in rapeseed oil are glucosinolates and phytic acid.

Glucosinolates toxicity: It causes a significant increase in the thyroid glands, a decrease in thyroxin synthesis, an impairment in metabolism, a decrease in iodine absorption, and a decrease in protein digestion. However, significant differences in the glucosinolate content are observed in the modified form as well as in nontransgenic plants. Genetically modified (GM) lines fit within the established range for glucosinolate.

Phytic acid toxicity: This causes a significant decrease in the bioavailability of calcium, magnesium, iron, zinc, copper, and manganese, as well as a major reduction in protein and starch utilization (Ye et al. 2016).

13.4.2 Maize

Naturally occurring antinutrient in maize is primarily phytic acid. Phytic acid toxicity significantly reduces the bioavailability of calcium, magnesium, iron, zinc, copper, and manganese and reduces primarily protein and starch utilization.

The green sequence found in plant modified to control European corn bit, could encode for protease inhibitor (inhibition of trypsin and chymotrypsin, decreased digestion of protein) (Baumgartner 2000). No protease inhibitor is found in modified maize lines (Nwosu 2010).

13.4.3 Tomato

Naturally occurring antinutrients in tomato are glycoalkaloids and alpha-tomatine (Musundire et al. 2016). Glycoalkaloids' toxicity induces gastrointestinal symptoms, kidney inflammation and nephrotoxicity, metal binding, a reduction in iron absorption, and reduction in enzyme activities. The toxicity of Alpha-tomatine toxicity causes a decrease during maturation, hence red ripe tomatoes loose almost all tomatine when left on plant for 2–3 days.

Shorter ripening times established via genetic modification could lead to an increase in tomatine level. GM tomatoes were investigated for glycoalkaloid levels, which were found to be within levels of non-GM tomato pastes.

13.4.4 Potato

Naturally occurring antinutrients in potato are solanine, protease inhibitors, and phenols (Guarischi-Sousa et al. 2016). Solanine toxicity causes gastrointestinal symptoms, kidney inflammation, metal binding, reduced iron absorption, and inhibition of enzymes. Protease inhibitor toxicity causes an inhibition of trypsin and chymotrypsin and decreased digestion of protein. Phenol toxicity causes reduced bioavailability of trace elements.

13.5 ANTINUTRITIONAL FACTORS FOUND IN PLANTS

13.5.1 Tannins

They are widely used as antinutrients due to the presence of antinutritional factors (ANFs) found in plants. Tannin is a plant bitter in taste and is astringent in nature, and it is a polyphenolic compound that either binds or precipitates proteins, amino acids, and alkaloids. They are mostly found in various trees and shrub foliages, seeds, and agroindustrial by-products (Dube et al. 2001). Tannins have a property of binding to protein to form reversible and irreversible complexes due to the existence of a number of phenolic hydroxyl groups (Patra and Saxena 2010). Tannins are water soluble. They have molecular weight greater than 500. Tannins are divided into two main types: hydrolysable tannins and condensed tannins (CTs). The two types differ in their nutritional and toxic effects. The CTs have more antinutrient effects than hydrolysable tannins. The hydrolysable tannins may produce toxic effects due to hydrolysis in rumen. Tannins are heat stable (Mueller 2001).

13.5.2 Cyanogens

Cyanogens are glycosides of aglycone sugars and aglycone cyanide. On hydrolysis, HCN is released by enzymes that are present in the cytosol. This inflicts damage to the plant. Ruminants are more susceptible to cyanide toxicity than nonruminants, because hydrolysis takes place in the rumen by microbial activity (Patel et al. 2013). After absorption, the HCN is rapidly metabolized in the liver by the enzyme rhodanese that converts toxic cyanide to nontoxic thiocyanate (SCN). Excess cyanide ion inhibits the cytochrome oxidase, which inhibits ATP formation. Because of this, tissues suffer energy deprivation and death follows rapidly. The lethal dose of HCN for cattle and sheep is 2.0–4.0 mg/kg body weight (Robson 2007).

13.5.3 Oxalate

Oxalic acid forms strong bonds with other minerals such as calcium, magnesium, sodium, and potassium. This results in the formation of oxalate salts, which is an antinutrient. When it is digested, it comes in contact with the nutrients in the gastrointestinal tract. Upon release, oxalic acid binds with nutrients, making them inaccessible to the body. If food containing excessive amounts of oxalic acid is consumed regularly, nutritional deficiencies are likely to occur along with severe irritation to the lining of the gut. In ruminants, microbial flora can readily metabolize soluble oxalates; oxalic acid is very insignificant as an antinutritive factor (Habtamu and Negussie 2014). The deficiency of calcium may occur in grazing animals after ingesting some of the various tropical grasses such as buffel grass (*Cenchrus ciliaris*), pangola grass (*Digitaria decumbens*), setaria (*Setaria sphacelata*), and kikuyu grass (*Pennisetum clandestinum*) that contain soluble oxalates in sufficient concentration. Oxalic acid reacts with calcium to produce insoluble calcium oxalate that reduces absorption of calcium. This disturbs the ratio of absorption of calcium/phosphorus, resulting in mobilization of bone mineral leading to hypocalcemia. Secondary hyperparathyroidism or osteodystrophia fibrosa may result due to prolonged mobilization of bone mineral (Rahman and Kawamura 2011).

13.5.4 Nitrates

In 1895, nitrate toxicity was found with cornstalk poisoning. But it was not recognized as the main toxicant during that period. In late 1930s, oat hay poisoning in the high plains region was recognized due to nitrate ingestion (Launchbaugh et al. 2001). Some of the fodder crops such as sudan grass, pearl millet (Andrews and Kumar 1992), and oats (Singh and Bauddh 2011) can accumulate nitrate at potentially toxic levels. Nitrate poisoning is better called as nitrite poisoning. On consumption of forages, nitrate is normally converted in the rumen to nitrite to ammonia to amino acid to protein. When forages have an unusually high concentration of nitrate, the animal cannot complete the conversion and nitrite accumulates, which is absorbed into the bloodstream. This nitrite will convert hemoglobin (the oxygen-carrying molecule) into methemoglobin, which cannot carry oxygen (Benjamin 2006). An animal suffering from nitrate (nitrite) poisoning actually dies from asphyxiation or lack of oxygen.

13.5.5 Protease Inhibitors

Protease inhibitors are widely distributed in the seeds of most cultivated legumes and cereals. They are the most common class of ANFs of plant origin. Protease inhibitors have the ability to inactivate proteolytic enzymes within the gastrointestinal tract of animals. Protease inhibitors due to their protein nature may be easily denatured by heat processing although some residual activity may still remain in the commercially produced products. The antinutrient activity of protease inhibitors is associated with growth inhibition and pancreatic hypertrophy (Chunmei et al. 2010).

13.5.6 Aflatoxins and Gossypol

This group includes aflatoxin B1, B2, G1, and G2 (AFB1, AFB2, AFG1, and AFG2, respectively). In addition, aflatoxin M1 (AFM1) has been identified in the milk of dairy cows consuming AFB1-contaminated feeds. The aflatoxigenic *Aspergilli* are generally regarded as storage fungi, proliferating under conditions of relatively high moisture/humidity and temperature. Aflatoxin contamination is, therefore, almost exclusively confined to tropical feeds such as oilseed by-products derived from groundnuts, cottonseed, and palm kernel. Aflatoxin contamination of maize is also an important problem in warm humid regions where *Aspergillus flavus* may infect the crop prior to harvest and remain viable during storage. Surveillance of animal feeds for aflatoxins is an ongoing issue, owing to their diverse forms of toxicity and also because of legislation in developed countries (D'Mello and Macdonald 1998). Gossypol pigment cottonseed occurs in free and bound forms. In whole seeds, gossypol exists essentially in the free form, but variable amounts may bind with protein during processing to yield inactive forms. Free gossypol is toxic, which causes organ damage, cardiac failure, and death. Cottonseed meal fed to bulls can cause sperm abnormalities and reduce sperm production.

13.6 FUNGAL ANTINUTRIENTS

Fungi are diverse and numerous, with more than a million species of fungi known. Fungi are unicellular or multicellular organisms (plantlike) with threadlike hyphae but without chlorophyll, which plants use in the production of food for themselves. Because they do not have chlorophyll, they do not produce food, they get their food through the absorption of nutrients from others; therefore, nutrition in fungi is heterotrophic and absorptive. They secrete enzymes that break down organic matters and then absorb soluble products (Chang and Miles 1991). Mushrooms grown in the forest are more nutritious and important for medicinal purposes (Manzi et al. 1999). Mushrooms are rich in nutrients containing necessary and important components like protein, sugars, glycogen, lipids, vitamins, amino acids, and fibers and other nutrients necessary for the proper functioning of the body. Several conditions affect the nutritional composition of these mushrooms such as the growing site, the type of substrate, mushroom type, and developing stages. Apart from the nutritional value of some mushrooms, there are a lot of mushrooms that are not edible due to the presence of some ANFs that are detrimental to normal health functioning (Okolie and Gbuji 2002).

13.6.1 Mycotoxins

The word mycotoxin is from the Greek word "mykes," which means fungus, and the Latin word "toxicum," which means poison. Mycotoxins are toxic chemical compounds produced by fungi that can grow in crops in the field or after harvest. They can affect the food such as cereals, nuts, fruit and dried fruit, coffee, cocoa, spices, oilseeds, and milk. There are now more than 300 known mycotoxins having different chemical structures and different modes of action—of which some target the kidney, liver, or immune system and some are carcinogenic. Among these, common mycotoxins are aflatoxins, ochratoxin A, ergot alkaloids, fumonisins, patulin, trichothecenes (such as deoxynivalenol that is also known as vomitoxin), and zearalenone (JECFA 2001, Leon et al. 2016).

Mycotoxins are the secondary metabolites of fungi that impair animal health and productivity (D'Mello and Macdonald 1998). Forages and cereals are frequently contaminated by mycotoxin after contact with infected plants having particular pathogenic fungi. Contamination may also occur during processing and storage of harvested products and feed whenever environmental conditions are appropriate for spoilage fungi. Ambient temperature and moisture are the main factors responsible for fungal colonization and mycotoxin production. Toxic fungi are subdivided into "field" (plant pathogenic) and "storage" (saprophytic/spoilage) organisms. *Claviceps*, *Fusarium*, *Neotyphodium*, and *Alternaria* are classical examples of field fungi, while *Penicillium* and *Aspergillus* exemplify storage organisms. Mycotoxigenic species are differentiated according to specific environmental requirements for growth and secondary metabolism. Thus, *A. flavus*, *A. parasiticus*, and *A. ochraceus* grow readily under warm, humid conditions, while *Penicillium expansum* and *P. verrucosum* under temperate condition, whereas *Penicillium* mycotoxins occur widely in temperate foods, particularly cereal grains.

13.7 CONCLUSION

Protein content is very important for nutritive foods. Thus, degradation of proteins by heat and various harmful chemicals, including pesticides, fungicides, and formaldehyde, should be substantially reduced. Moreover, formaldehyde is carcinogenic, and, in addition, its improper use can lower the absorption of amino acids from the intestine. Another way to protect proteins against excessive degradation in the rumen is the use of tannins, which may form reversible complexes with proteins. CTs have the ability to improve weight gain, wool production, and reproductive efficiency in sheep fed with temperate forages, which reduces the impact of gastrointestinal problems. Effective processing techniques could help to reduce or eliminate the adverse effects of these antinutritive constituents in plant protein sources and improve their nutritive values. Supplementation of selected minerals, amino acids, and vitamins can help to reduce or neutralize the negative effect of ANFs in plant protein sources for livestock nutrition. Studies have shown that saponins have exhibited a variety of activities including antitumor, cholesterol-lowering, immune-enhancing, anticancer, antioxidants, and cardioprotective effects. Potential beneficial effects of protease inhibitors are still being investigated, although lower incidences of pancreatic cancer have been observed in populations where the intake of soybean and its products is high. The animal studies have demonstrated that protease inhibitors can alleviate symptoms of pancreatic cancer, and they may also act as anticarcinogenic agents. Significant research is going on worldwide to increase the beneficial protein content and significantly reduce the antinutritive content of the regular food that is consumed by animals and humans.

REFERENCES

Adekanmi, O.K., O.F. Oluwatooyin, and A. Adebowale. 2009. Yemisi influence of processing techniques on the nutrients and antinutrients of tigernut (*Cyperus esculentus* L.). *World J. Dairy Food Sci.* 4(2): 88–93.

Andrews, D.J. and K.A. Kumar. 1992. Pearl millet for food, feed and forage. In: D.L. Sparks (ed.), *Advances in Agronomy.* Academic Press, Newark, CA, Vol. 48, pp. 90–139.

Australia New Zealand Food Authority. 2001. Pyrrolizidine alkaloids in food. A toxicological review and risk assessment. Technical Report Series No. 2. ANZFA. November. Available from: www.foodstandards.gov.au/_srcfiles/TR2.pdf. Wellington, New Zealand. Accessed on 2016.

Bajaj, J.K., P. Salwan, and S. Salwan. 2016. Various possible toxicants involved in thyroid dysfunction: A review. *J. Clin. Diag. Res.* 10(1): FE01–FE03.

Baker, D.C., R.F. Keeler, and W.P. Garfield. 1988. Mechanism of death in Syrian hamsters gavaged potato sprout material. *Toxicol. Pathol.* 16(3): 333–339.

Bene, C., R. Arthur, H. Norbury, E.H. Allison, M. Beveridge, S. Bush, L. Campling et al. 2016. Contribution of fisheries and aquaculture to food security and poverty reduction: Assessing the current evidence. *World Dev.* 79: 177–196.

Benjamin, D.N. 2006. Effects of fertilizer application and cutting interval on nitrate accumulation in bermudagrass. M.Sc. (Agri.), Thesis, University Tennessee, Martin, TN.

Chang, S.T. and Miles, P.G. 1991. Recent trends in world production of cultivated edible mushrooms. *Mushroom J.* 503: 15–18.

Chunmei, G., P. Hongbin, S. Zewei, and Q. Guixin. 2010. Effect of soybean variety on anti-nutritional factors content, and growth performance and nutrients metabolism in rat. *Int. J. Mol. Sci.* 11: 1048–1056.

D'Mello, J.P.F. and A.M.C. Macdonald. 1998. Fungal toxins as disease elicitors. In J. Rose, (ed.), *Environmental Toxicology: Current Developments*, Gordon and Breach Science Publishers, Amsterdam, the Netherlands, pp. 253–289.

Dube, J.S., J.D. Reed, and L.R. Ndlovu. 2001. Proanthocyanidins and other phenolics in Acacia leaves of Southern Africa. *Anim. Feed Sci. Technol.* 91, 59–67.

Fernandes, A., J.C.M. Barreira, A.L. Antonio, M.B.P.P. Oliveira, A. Martins, and I.C.F.R. Ferreira. 2016. Extended use of gamma irradiation in wild mushrooms conservation: Validation of 2 kGy dose to preserve their chemical characteristics. *Food Sci. Technol.* 67: 99–105.

Glover, R.L., N.J. Connors, C. Stefan, E. Wong, R.S. Hoffman, L.S. Nelson, M. Milstein, S.W. Smith, and M. Swerdlow. 2015. Electromyographic and laboratory findings in acute *Solanum torvum* poisoning. *Clin. Pharmacol.* 61–65.

Guarischi-Sousa, R., M. Puigvert, N.S. Coll, M.I. Siri, M.J. Pianzzola, M. Valls, and J.C. Setubal. 2016. Complete genome sequence of the potato pathogen *Ralstonia solanacearum* UY031. *Stand Genomic Sci.* 11(7): 1–8.

Habtamu, F. and R. Negussie. 2014. Antinutritional factors in plant foods: Potential health benefits and adverse effects. *Int. J. Nutr. Food Sci.* 3(4): 284–289.

Harvey, M.H., B.A. Morris, M. McMillan, and V. Marks. 1985. Measurement of potato steroidal alkaloids in human serum and saliva by radioimmunoassay. *Hum. Toxicol.* 4: 503–512.

Jadhaw, S.J., R.P. Sharma, and J.K. Salunke. 1981. Naturally occurring toxic alkaloids in foods. *CRC Crit. Rev. Toxicol.* 13: 21–103.

JECFA. 2001. Safety evaluation of certain mycotoxins in food. FAO Food and Nutrition Paper 74/WHO, Food Additive Series 47, Geneva, Switzerland, pp. 281–415.

Khare, N. and S. Baumgartner. 2000. Dally-like protein, a new *Drosophila* glypican with expression overlapping with wingless. *Mech. Dev.* 99(1–2): 199–202.

Launchbaugh, K.L., F.D. Provenza, and J.A. Pfister. 2001. Herbivore response to anti-quality factors in forages. *J. Range Manage.* 54: 431–440.

Leon, N., A. Pastor, and V. Yusa. 2016. Target analysis and retrospective screening of veterinary drugs, ergot alkaloids, plant toxins and other undesirable substances in feed using liquid chromatography–high resolution mass spectrometry. *Elsvier Talanta.* 149: 43–52.

Lopez, P., D. Venema, T.D. Rijk, A.D. Kok, J.M. Scholten, H.G.J. Mol, and M.D. Nijs. 2016. Occurrence of *Alternaria* toxins in food products in The Netherlands. *Elsevier Food Control.* 60: 196–204.

Maga, J.A. 1980. Potato alkaloids. *CRC Crit. Rev. Toxicol.* 12: 371–405.

Manzi, P., L. Gambelli, S. Marconi, V. Vivanti, and L. Pizzoferrato. 1999. Nutrients in edible mushrooms: An inter-species comparative study. *Food Chem.* 65: 477–482.

McMartin, K., D. Jacobsen, and K.E. Hovda. 2016. Antidotes for poisoning by alcohols that form toxic metabolites, *Br. J. Clin. Pharmacol.* 1–11.

Mueller, I. 2001. Analysis of hydrolysable tannins. *Anim. Feed. Sci. Technol.* 91: 3–20.

Musundire, R., I.M. Osuga, X. Cheseto, J. Irungu, and B. Torto. 2016. Aflatoxin contamination detected in nutrient and anti-oxidant rich edible stink bug stored in recycled grain containers. 11(1): e0145914.

Nwosu, J.N. 2010. Effect of soaking, blanching and cooking on the anti-nutritional properties of asparagus bean (*Vigna sesquipedis*) flour. *Nat. Sci.* 8(8): 163–167.

Okolie, P.N. and A.D. Gbuji. 2002. Cyanide profiles of some Nigerian edible mushrooms in raw and processed states. *J. Med. Biomed. Res.* 1(2): 60–64.

Omaye, S.T. 2004. *Food and Nutritional Toxicology*, CRC Press, New York.

Patel, P.A.S., S.C. Alagundagi, and S.R. Salakinkop. 2013. The anti-nutritional factors in forages—A review. *Curr. Biotica.* 6(4): 516–526.

Patra, A.K. and J. Saxena. 2011. Exploitation of dietary tannins to improve rumen metabolism and ruminant nutrition. *J. Sci. Food Agric.* 91(1): 24–37.

Petrovic, J., J. Babic, S. Jaksic, B. Kartalovic, D. Ljubojevic, and M. Cirkovic. 2016. Fish product–borne histamine intoxication outbreak and survey of imported fish and fish products in Serbia. *J. Food Prot.* 79(1): 90–94.

Rahman, M.M. and O. Kawamura. (2011). Oxalate accumulation in forage plants: Some agronomic, climatic and genetic aspects. *Asian-Aust. J. Anim. Sci.* 24(3): 439–448.

Rahman, M.R., M.M. Rahman, W.E. Wan Khadijah, and R.B. Abdullah. 2014. Follicle stimulating hormone (FSH) dosage based on body weight enhances ovulatory responses and subsequent embryo production in goats. *Asian Aust. J. Anim. Sci.* 27(9): 1270–1275.

Robson, S. 2007. Prussic acid poisoning in Livestock. www.dpi.nsw.gov.au/primefacts. Accessed on 2016.

Singh, R.P. and K. Bauddh. 2011. Differential toxicity of cadmium to mustard (*Brassica juncia* L.) genotypes under higher metal levels, *J. Environ. Biol.* 32: 355–362.

World Health Organization (WHO). 1988. *Pyrrolizidine Alkaloids.* Environmental Health Criteria 80. WHO, Geneva, Switzerland.

World Health Organization (WHO). 1993. *Solanine and Chaconine.* WHO Food Additives Series 30. WHO, Geneva, Switzerland.

Ye, Y. et al. 2016. The *Saccharomyces cerevisiae* Mre11-Rad50-Xrs2 complex promotes trinucleotide repeat expansions independently of homologous recombination. *DNA Repair* 43: 1–8

Zhang, Z., T. Yang, N. Mi, Y. Wang, G. Li, L. Wang, and Y. Xie. 2016. Antifungal activity of monoterpenes against wood white-rot fungi. *Int. Biodeter. Biodegr.* 106: 157–160.

14 A Review on Mushroom Toxins

Md. Akil Hossain and Seung-Chun Park

CONTENTS

14.1 INTRODUCTION

A mushroom is the fleshy, spore-bearing fruiting body of a fungus. Mushrooms are now gaining superfood status as they are loaded with a number of the most important nutrients on the planet. They have been used as components of human diet and are used extensively in cooking, in many cuisines (notably Chinese, Korean, European, and Japanese). Though they can be classified as neither meat nor vegetable, mushrooms are known as the "meat" of the vegetable world. Nutritional values that can be taken from mushrooms vary to some extent according to different species and different analyses, but most likely, in general, the water content is about 89%, protein 3%, fat 0.4%, carbohydrates 6%, and minerals about 1%. Edible mushrooms are also rich in B vitamins, vitamin K, vitamin D, and sometimes vitamins A and C (Perera and Li 2011). These vitamins are well retained during cooking and in canned or dried mushrooms. Although much of the interest in mushrooms arises from their use as food, their other essential features also attract the collector.

Mushrooms are considered as therapeutic food as some species are widely used in the food and pharmaceutical industry due to the presence of essential nutrients and pharmacologically active substances. These fungal species were described as popular remedies in ancient oriental documents and some of them became ingredients in traditional medicine (Lima et al. 2012). Mushrooms are useful in preventing diseases such as hypercholesterolemia, diabetes, hypertension, and cancer (Bobek and Galbavy 1999; Bobek et al. 1995). Studies have also shown that certain mushroom species have antitumor, antibacterial, antiviral, antiparasitic, anti-inflammatory, antithrombotic, cardiovascular disease inhibition, and immunomodulation properties (Mau et al. 2002). Currently, several extracts (polysaccharide peptide, polysaccharide-K, lentinan) are widely used in China, Japan, and Korea as potential adjuvants to radiation treatments and chemotherapy (Borchers et al. 2008; Sullivan et al. 2006).

The worldwide diversity of mushroom species is roughly estimated at 0.14 million. Among them, 10% are well known and 5% are considered to have varying levels of edibility. More than 2000 species are harmless and 700 are recognized to have substantial pharmacological properties (Jo et al. 2014). A series of species have been considered as harmful to health and defined as toxic species. The common toxic mushroom species, their toxic substances, and toxic effects are listed in Table 14.1. Some mushroom species also contain toxic and beneficial substances simultaneously (Nieminen et al. 2006). The ingestion of toxin-containing mushrooms is accidental and toxicity with mushroom consumption is contributing significantly to increased number of global unintentional food poisoning. There are numerous mushroom poisoning cases in different countries each year. Carcinogenicity, renal failure, alterations in cardiac and respiratory rates, rhabdomyolysis, and other effects were observed in toxicity studies with various species. The accidental poisoning with this species occurs frequently through the misidentification of species, and this event often occurs because of the inadequate accessible information on poisonous mushrooms and their potential toxicity (Jo et al. 2014). Before consuming a mushroom species, it is essential to characterize their toxicological profile.

TABLE 14.1
List of Toxic Mushrooms, Their Toxins, and Toxic Effects

Toxic Species	Toxins	Toxic Effects
Agaricus albolutescens	Gastrointestinal irritants	Gastrointestinal
Agaricus arvensis	Gastrointestinal irritants	Gastrointestinal
Agaricus hondensis	Gastrointestinal irritants	Gastrointestinal
Agaricus placomyces	Gastrointestinal irritants	Gastrointestinal
Agaricus silvicola	Gastrointestinal irritants	Gastrointestinal
Agaricus xanthodermus	Gastrointestinal irritants	Gastrointestinal
Amanita bisporigera	Amanitins (cyclopeptides)	Cytotoxicity (lethal)
Amanita brunnescens	Gastrointestinal irritants, amanitins (cyclopeptides)	Cytotoxicity (lethal)
Amanita capensis	—	Cytotoxicity (lethal)
Amanita chlorinosma	—	Gastrointestinal
Amanita citrina	—	Neurological, hallucinogenic
Amanita cokeri	Ibotenic acid, muscimol	—
Amanita cothurnata	Ibotenic acid, muscimol	Neurological, muscarinic
Amanita flavoconia	—	Neurological, muscarinic
Amanita flavivolvata	—	Neurological, muscarinic
Amanita gemmata	Ibotenic acid, muscimol	Neurological, muscarinic
Amanita mappa	—	Neurological, muscarinic
Amanita muscaria	Ibotenic acid, muscimol	Neurological, muscarinic
Amanita ocreata	Amanitins (cyclopeptides)	—
Amanita pantherina	Ibotenic acid, muscimol	Neurological, muscarinic (very toxic)
Amanita parcivolvata	—	Neurological, muscarinic
Amanita phalloides	Amanitins (cyclopeptides)	Cytotoxic (very lethal)
Amanita porphyria	—	Hallucinogenic
Amanita regalis	—	Neurological, muscarinic
Amanita rubescens	—	Neurological, muscarinic
Amanita solitaria	—	Neurological, muscarinic
Amanita spreta	—	Neurological, muscarinic
Amanita strobiliformis	—	Neurological, muscarinic
Amanita tenuifolia	Amanitins (cyclopeptides)	Cytotoxic (lethal)
Amanita verna	Amanitins (cyclopeptides)	Cytotoxic (lethal)
Amanita virosa	Amanitins (cyclopeptides)	Cytotoxic (lethal)
Amanita smithiana	Orellanine	—
Boletus eastwoodiae	Muscarine, gastrointestinal irritants	Gastrointestinal
Boleltus erythropus	Muscarine	—
Boletus luridus	Gastrointestinal irritants	—
Boletus purpureus	Muscarine	—
Boletus satanas	Muscarine, gastrointestinal irritants	Gastrointestinal
Calvatia gigantea	—	Gastrointestinal
Cantharellus floccosus	—	Gastrointestinal
Chlorophyllum molybditis	Gastrointestinal irritants	—
Citheronia splendens	Orellanine	—
Clavaria formosa	—	Gastrointestinal
Clavaria gelatinosa	—	Gastrointestinal
Clitocybe cerussata	Muscarine	Neurological, muscarinic
Clitocybe claviceps	Disulfiram-like reaction	Neurological, muscarinic
Clitocybe dealbata	Muscarine	Neurological, muscarinic

(*Continued*)

TABLE 14.1 (*Continued*)
List of Toxic Mushrooms, Their Toxins, and Toxic Effects

Toxic Species	Toxins	Toxic Effects
Clitocybe illudens	Muscarine	Neurological, muscarinic
Clitocybe olearia	—	Neurological, muscarinic
Clitocybe rivulosa	Muscarine	Neurological, muscarinic
Clitocybe subilludens	—	Neurological, muscarinic
Clitocybe sudorifica	Muscarine	Neurological, muscarinic
Clitocybe toxica	—	Cytotoxic
Clitocybe trunicicola	—	Neurological, muscarinic
Clitocybe clavipes	Coprine	—
Conocybe filaris	Amanitins (cyclopeptides)	—
Conocybe cyanopus	Psilocybin, psilocin, hallucinogenic indoles	—
Coprinus atramentarius	Coprine, Disulfiram-like reaction	Disulfiram-like effect
Cortinarius orellanus	Orellanine	Cytotoxic
Cortinarius traganus	—	Gastrointestinal
Cortinarius orellanoides	Orellanine	—
Cortinarius rainierensis	Orellanine	—
Cortinarius limonius	Orellanine	—
Cortinarius rubellus	Orellanine	—
Entoloma grande	Gastrointestinal irritants	—
Entoloma lividum	Gastrointestinal irritants	—
Entoloma lividum	—	Gastrointestinal
Entoloma sinuatum	Gastrointestinal irritants	—
Entoloma strictius	Gastrointestinal irritants	—
Galerina autumnalis	Amanitins (cyclopeptides)	Cytotoxic
Galerina marginata	Amanitins (cyclopeptides)	Cytotoxic
Galerina sulciceps	—	Cytotoxic
Galerina venenata	Amanitins (cyclopeptides)	Cytotoxic
Gomphus kauffmanii	—	Neurological, cytotoxic
Gymnopilus decurrens	Psilocybin, psilocin	Hallucinogenic
Gymnopilus spectabilis	Psilocybin, psilocin	Hallucinogenic
Gyromitra brunnea	Gyromitrin, monomethylhydrazine	—
Gyromitra caroliniana	Gyromitrin, monomethylhydrazine	—
Gyromitra esculenta	Gyromitrin, monomethylhydrazine	Gyromitrin poisoning
Gyromitra fastigiata	Gyromitrin, monomethylhydrazine	—
Gyromitra infula	Gyromitrin, monomethylhydrazine	—
Hebeloma crustuliniforme	Gastrointestinal irritants	—
Helvella esculenta	—	Cytotoxic, hemolytic
Helvella gigas	—	Cytotoxic
Helvella infula	—	Cytotoxic
Helvella underwoodii	—	Cytotoxic
Helvella lacunose	Monomethylhydrazine (MMH)	—
Helvella elastica	Monomethylhydrazine (MMH)	—
Inocybe geophylla	Muscarine	—
Inocybe lilacina	Muscarine	—
Inocybe pudica	Muscarine	—
Inocybe godego	—	Neurologic, muscarinic
Inocybe infelix	—	Neurologic, muscarinic

(*Continued*)

TABLE 14.1 (*Continued*)
List of Toxic Mushrooms, Their Toxins, and Toxic Effects

Toxic Species	Toxins	Toxic Effects
Inocybe infida	—	Neurologic, muscarinic
Inocybe lacera	—	Neurologic, muscarinic
Inocybe napipes	—	Neurologic, muscarinic
Inocybe nigrescence	—	Neurologic, muscarinic
Inocybe patouillardii	Muscarine	Neurologic, muscarinic (lethal)
Inocybe picrosma	—	Neurologic, muscarinic
Lactarius chrysorrheus	—	Gastrointestinal
Lactarius glaucescens	—	Gastrointestinal
Lactarius helvus	—	Gastrointestinal
Lactarius lignyotus	—	Gastrointestinal
Lactarius plumbeus	—	Gastrointestinal
Lactarius porninsis	—	Gastrointestinal
Lactarius rufus	Gastrointestinal irritants	Gastrointestinal
Lactarius torminosus	—	Gastrointestinal
Lactarius trivialis	—	Gastrointestinal
Lactarius uvidus	Gastrointestinal irritants	Gastrointestinal
Lactarius vellereus	Gastrointestinal irritants	Gastrointestinal
Lactarius piperatus	Gastrointestinal irritants	—
Lampteromyces japonicus	—	Cytotoxic
Lepiota brunneoincarnata	Amanitins (cyclopeptides)	Cytotoxic
Lepiota fuscovinacea	Amatoxins	Cytotoxic
Lepiota helveola	Amatoxins	Cytotoxic
Lepiota schulzeri	—	Cytotoxic
Lepiota helveola	Amanitins (cyclopeptides)	—
Lycoperdon subincarnatum	—	Gastrointestinal
Marasmius urens	Gastrointestinal irritants	—
Naematoloma fasciculare	Gastrointestinal irritants	Gastrointestinal, cytotoxic
Omphalotus olearius	Gastrointestinal irritants	—
Panaeolus acuminatus	—	Hallucinogenic
Panaeolus retirugis	—	Hallucinogenic
Panaeolus semiovatus	—	Hallucinogenic
Panaeolus subbalteatus	Psilocybin, psilocin	Hallucinogenic
Panaeolus foenisecii	Psilocybin, psilocin	Hallucinogenic
Panus stipticus	—	Gastrointestinal
Paxillus involutus	—	Gastrointestinal, cytotoxic
Pholiota squarrosa	Gastrointestinal irritants	—
Psilocybe baeocystis	Psilocybin, psilocin	Hallucinogenic
Psilocybe cubensis	Psilocybin, psilocin	Hallucinogenic
Psilocybe mexicana	Psilocybin, psilocin	Hallucinogenic
Psilocybe semperviva	Psilocybin, psilocin	Hallucinogenic
Psilocybe cyanescens	Psilocybin, psilocin	Hallucinogenic
Psilocybe semilanceata	Psilocybin, psilocin	Hallucinogenic
Psilocybe silvatica	Psilocybin, psilocin	Hallucinogenic
Rhodophyllus sinuatus	—	Gastrointestinal
Rumaria flavobrunnescens	—	Cytotoxic
Russula emetica	Gastrointestinal irritants	Gastrointestinal

(*Continued*)

TABLE 14.1 (*Continued*)
List of Toxic Mushrooms, Their Toxins, and Toxic Effects

Toxic Species	Toxins	Toxic Effects
Russula foetens	—	Gastrointestinal
Russula fragilis	—	—
Russula nondorbingi	—	Neurological, cerebral
Sarcosphaera crassa	Gyromitrin, monomethylhydrazine	—
Scleroderma aurantium	Gastrointestinal irritants	Neurological
Scleroderma cepa	—	Neurological
Stropharia aeruginosa	Psilocybin, psilocin	—
Tricholoma album	—	Gastrointestinal
Tricholoma flavobrunneum	—	Gastrointestinal
Tricholoma pardinum	Gastrointestinal irritants	Gastrointestinal
Tricholoma pessundatum	—	Gastrointestinal
Tricholoma saponaceum	—	Gastrointestinal, hemolytic
Tricholoma sejunctum	—	Neurological, gastrointestinal
Verpa bohemica	—	Neurological

Sources: Barceloux, D.G., Muscarine-containing mushrooms and muscarine toxicity (clitocybe and inocybe species), in *Medical Toxicology of Natural Substances: Foods, Fungi, Medicinal Herbs, Plants, and Venomous Animals*, John Wiley & Sons, Inc., Hoboken, NJ, 2008; Concon, J.M., *Food Toxicology. Parts A and B*, Marcel Dekker, New York, 1988; Deshpande, S.S., Mushroom toxins, in *Handbook of Food Toxicology*, Marcel Dekker, Inc., New York, 2002; Faulstich, H. et al., *Biochemistry*, 19(14), 3334, 1980.

14.2 CAUSES OF TOXICITY

There are a number of reasons for mushroom poisoning. The most common causes are listed as follows:

1. There are several thousands of mushroom species. A few of them have been documented for their possible poisonous or toxic properties.
2. Misidentification of edible and poisonous mushroom is the vital cause of mushroom poisoning in humans and animals. The reasons for misidentifications are as follows:
   a. Morphological features show similarity among numerous mushroom species. Sometimes, the toxic and nontoxic species of mushrooms can only be differentiated by the microscopic assessment of mushroom tissues and their cell structures.
   b. Environment or growth conditions may change the morphological characteristics of any kind of mushrooms. For instance, the toxic mushroom *Amanita muscaria* can come in different colors (white, yellowish orange, and dark red) with varying intensity of the individual shade/color and depending on growth conditions. The orange-capped poisonous *A. muscaria* may be often confused with the edible *Amanita cesarean*.
   c. Mechanical factors (force of heavy rains, pressure exerted by the growing mushroom on the soil, remnants of volva in the ground at the time of pulling up mushroom, etc.) may modify the typical recognizing features of any mushroom species that may cause confusion and wrong identity. For instance, vital recognizing features of amanitas (remnants of the partial and universal veils, annulus, and volva) can be detached by mechanical means.

d. Sometimes, the victim tries to utilize traditional knowledge gained in one geographic area and apply it in another area. Particularly, *Amanita phalloides* exhibit similarities with *Volvariella volvacea*, a "paddy-straw" mushroom in Asia. At young stage, both mushrooms contain light color and are surrounded by a universal veil.

3. Cases of allergic reactions or other kind of sensitivities to individual species of mushroom even with edible species are not uncommon. Some species may cause sickness in some individuals but no reaction in others.
4. Different people react with varying susceptibilities to different toxic mushrooms.
5. The nature and cause of this toxicity depend on the amount of mushrooms consumed. The level of toxicity also depends on various factors including the health condition and age of victims, the growth conditions of mushrooms, and the amount of toxins present in the mushrooms.
6. The procedure of mushroom cooking or preparation influences their edibility or toxicity. For example, *Paxillus involutus* seems harmless after cooking and three people out of four have died by eating inadequately boiled mushroom of this species (Bschor and Mallach 1963). Ingestion of insufficiently cooked or raw *Leiopotherapon plumbeus, Clitocybe olearia* (*Pleurotus olearius*), *Lactarius porninsis*, and morels (*Verpa bohemica* or *Gyromitra* spp.) is toxic too (Pilat and Usak 1961).
7. Ingestion of some mushroom species with other incompatible food or medicine may lead to toxicity. For example, mushrooms are commonly fried in butter, but the use of too much butter may cause illness. Moreover, *Coprinus* species contain an unusual toxin, namely, "coprime," which inhibits an enzyme (aldehyde dehydrogenase) required to break down alcohol. This toxin is hazardous only when it is ingested within a few days of alcohol ingestion.

14.3 TYPES OF MUSHROOM TOXINS

Mushroom toxins are generally divided into seven categories:

1. Amatoxins (cyclopeptides)
2. Orellanus (*Cortinarius* species)
3. Gyromitrin (monomethylhydrazine)
4. Muscarine
5. Ibotenic acid
6. Psilocybin
7. Coprine

14.4 AMATOXINS, PHALLOTOXINS, AND VIROTOXINS

14.4.1 Physiochemical Properties

The three most important categories of toxic peptides in poisonous mushrooms are amatoxins, virotoxins, and phallotoxins (Wieland 1986). Unusually hydroxylated amino acids and a tryptophan unit linked to sulfur are present in these toxic compounds. Amatoxins, for instance, α-amanitin or β-amanitin, are bicyclic octapeptides. These toxic compounds are present in *Amanita*, *Lepiota*, and *Galerina* and are lethal when mushrooms are ingested (Koppel 1993). Amatoxins are mainly composed of an amino acid ring and bridged with a sulfur atom. The structures of nine amatoxins are presented in Figure 14.1. Most of these amatoxins are present in very small amounts or even trace amounts in mushrooms and are very rarely poisonous. Thus, they are not considered as potent toxic agents of mushroom. Among these, α-amanitin, β-amanitin, and γ-amanitin are the major amatoxins that contribute significantly to hepatorenal toxicity. α-Amanitin, acidic

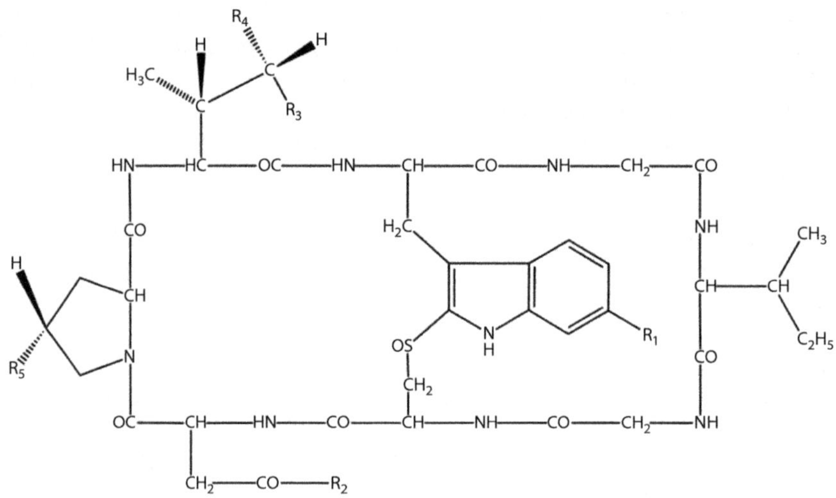

	R_1	R_2	R_3	R_4	R_5
α-Amanitin	OH	NH_2	CH_2OH	OH	OH
β-Amanitin	OH	OH	CH_2OH	OH	OH
γ-Amanitin	OH	NH_2	CH_3	OH	OH
ε-Amanitin	OH	OH	CH_3	OH	OH
Amanin	H	OH	CH_2OH	OH	OH
Amanin amide	H	NH_2	CH_2OH	OH	OH
Amanullin	OH	NH_2	CH_3	H	OH
Amanullinic acid	OH	OH	CH_3	H	OH
Proamanullin	OH	NH_2	CH_3	H	H

FIGURE 14.1 Chemical structure of amatoxins.

β-amanitin, and γ-amanitin each constitute 10% of *A. phalloides* mushrooms. The amatoxin content in these mushrooms is about 3.5 mg/g in dry basis. These compounds are stable and do not degrade at the time of cooking even at high temperatures like 250°C–280°C. They are resistant to freezing at −25°C, and *A. phalloides* produces fatal hepatoxicity when ingested after being in storage for 7–8 months in refrigerator (Himmelmann et al. 2001). Amatoxins degrade gradually when stored in aqueous solutions, when placed in an open place, and when exposed to sunlight or neon light for a long time.

Many cyclic heptapeptides, namely, phallotoxins, also exist in some toxic mushroom species, more commonly in *Amanita* species (*A. phalloides* var. *verna*, *A. phalloides*, *Amanita bisporigera*, and *Amanita virosa*). These kinds of toxins include phalloin, phalloidin, phallisin, phallisacin, and phallacidin (Enjalbert et al. 1992; Faulstich et al. 1973). The chemical structures of phallotoxins are shown in Figure 14.2. Generally, these seven different toxins are considered as phallotoxins. These bicyclic heptapeptides contain threonine, tryptophan, cysteine, hydroxyproline, two alanine residues, and mono-, di-, trihydroxyleucine. These phallotoxins are relatively stable because of the bicyclic structure of those molecules. The geometric features of the basic peptide ring do not change due to chemical modifications in the side chains, and similarly, the toxicities do not diminish as a result of these modifications in general (Wieland and Wieland 1972). However, nontoxic products can be produced by cleaving the ring by heating along with Raney nickel or acid hydrolysis.

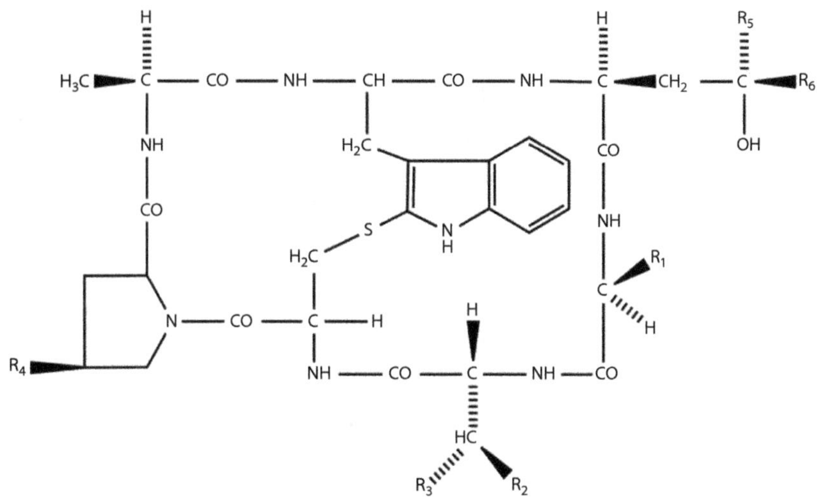

	R_1	R_2	R_3	R_4	R_5	R_6
Phalloin	CH_3	CH_3	OH	OH	CH_3	CH_3
Phalloidin	CH_3	CH_3	OH	OH	CH_2OH	CH_3
Phallisin	CH_3	CH_3	OH	OH	CH_2OH	CH_2OH
Prophalloin	CH_3	CH_3	OH	H	CH_3	CH_3
Phallacin	$CH(CH_3)_2$	OH	COOH	OH	CH_3	CH_3
Phallacidin	$CH(CH_3)_2$	OH	COOH	OH	CH_2OH	CH_3
Phallisacin	$CH(CH_3)_2$	OH	COOH	OH	CH_2OH	CH_2OH

FIGURE 14.2 Chemical structure of phallotoxins.

Phallotoxins are poorly absorbed by the gastrointestinal tract, which possibly restricts the systemic toxicity due to the ingestion of these mushrooms even though they are very much hepatotoxic.

Virotoxins are monocyclic heptapeptides that contain serine (Figure 14.3) and have phalloidin-like activities. This peptide is only present in *A. virosa* and there is a similarity between the amino acid sequence of phallotoxins and virotoxins (Faulstich et al. 1980). Generally, the only dissimilarity is the presence of D-serine in the place of L-cysteine and the absence of 2,3-trans-3,4-dihydroxy-L-proline and 2′-(methylsulfonyl)-L-tryptophan in virotoxins. Viroisin, viroidin, deoxoviroisin, deoxoviroidin, and ala-viroidin are the common virotoxins.

14.4.2 Poisonous Parts

There are approximately 35 amatoxin-containing mushroom species in three genera: *Amanita*, *Galerina*, and *Lepiota*. Probably, most of the *Amanita* mushroom species have amatoxins, even the edible species of mushroom contain trace amounts of these toxins. The most toxic mushroom species having amatoxins is *A. phalloides*. The approximate amount of amatoxin and phallotoxin in diverse samples of *A. phalloides* mushrooms is presented in Table 14.2. Roughly, 2–3 mg/g of amatoxin is present in the dry tissue of *A. phalloides*. Thus, a small amount of fresh mushrooms (around 50 g) possibly has lethal amounts of amatoxin, depending on the concentration of amatoxins. The concentrations of amatoxin in *Amanita* species depend on many factors. Environmental factors, the parts of the plant, toxin type, types of soil, location of cultivation, and species type of *Amanita* are vital factors, among others. Toxic species of *Galerina* such as *Galerina marginata*

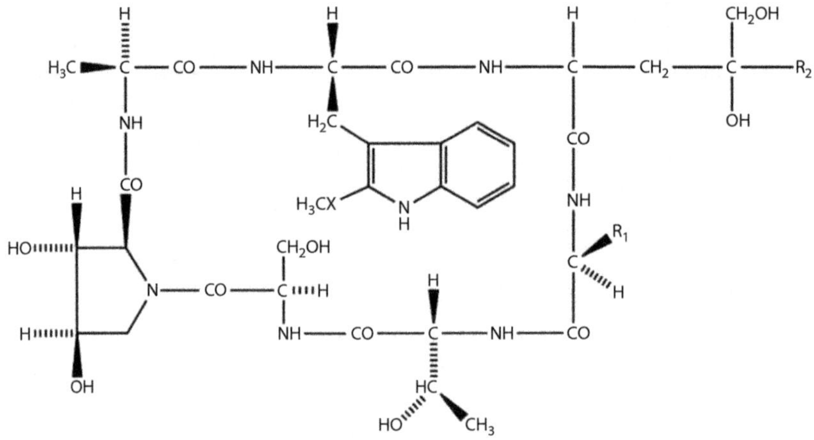

	X	R_1	R_2
Viroidin	SO_2	$CH(CH_3)_2$	CH_3
Deoxoviroidin	SO	$CH(CH_3)_2$	CH_3
Ala[1]-viroidin	SO_2	CH_3	CH_3
Ala[1]-deoxoviroidin	SO	CH_3	CH_3
Viroisin	SO_2	$CH(CH_3)_2$	CH_2OH
Deoxoviroisin	SO	$CH(CH_3)_2$	CH_2OH

FIGURE 14.3 Chemical structure of virotoxins.

comprise about 0.8–1.5 mg/g dry weight of α-amanitin and β-amanitin (Faulstich et al. 1974). The hepatorenal syndrome resulting from amatoxin-containing *Galerina* mushroom ingestion is similar to the ingestion of toxic *Amanita* species. The only difference is that the *Galerina* mushrooms have no phallotoxins that are abundant in toxic species of *Amanita*. Finally, the major toxic ingredients of these poisonous mushrooms are amatoxins instead of virotoxins or phallotoxins.

14.4.3 Symptoms of Poisoning

After 6–24 hours of ingestion, the cyclopeptide comprising toxic mushrooms causes the phalloides syndrome. This syndrome includes acute abdominal pain along with profuse diarrhea, violent bloody vomiting, and nausea. The early stage of severe gastritis symptoms resolves gradually, and the patient enters a symptom-free stage. The duration of this symptom-free phase is inconsistent but is 3–5 days in general. This condition is confusing, and at the same time gradually emerging for kidney and liver injury.

The second phase of this poisoning lasts for 24–36 hours and is characterized by extensive vomiting, nausea, acute abdominal cramps, and watery diarrhea that initiates acute dehydration, acid–base disorders, oliguria, and loss of electrolytes. At this stage, the strong sign of amatoxin poisoning is evident as diarrhea continues. Disturbances in blood coagulation, particularly the decrease of prothrombin time, start in this phase.

Phase III is regarded as an enhancement of clinical signs, whereas raised levels of transaminases reveal the onset of liver necrosis. The duration of this period is about 12–24 hours. The final phase represents encephalopathy, hepatic failure, and commonly severe renal failure. Furthermore, internal bleeding is observed and may cause complications or even death. Mostly, deaths occur within 5–8 days after the ingestion of mushrooms; however, patients may die within 20–25 days (Faulstich and Wieland 1992).

TABLE 14.2
Estimated Amount of Major Amatoxins and Phallotoxins in *Amanita phalloides* and LD_{50} Values of Amatoxins and Phallotoxins

Toxic Compounds	Dry Weight (mg/g)	LD_{50} (mg/kg, White Mouse)
Amatoxins		
α-Amanitin	1.00–2.50	0.30–0.60
β-Amanitin	1.20–3.25	0.50
γ-Amanitin	0.20–0.65	0.20–0.50
ε-Amanitin	—	0.30–0.60
Amanin	—	0.50
Amanin amide	—	0.50
Amanullin	—	>20.00
Amanullinic acid	—	>20.00
Proamanullin	—	>20.00
Phallotoxins		
Phalloin	—	1.50
Phalloidin	0.90–2.20	2.00
Phallisin	—	2.00
Prophalloin	—	>20.00
Phallacin	—	1.50
Phallacidin	2.40–4.30	1.50
Phallisacin	1.70–6.00	4.50

Sources: Antkowiak, W.Z., The chemistry and toxicology of mushroom alkaloids, in M.S. Blum (ed.), *Chemistry and Toxicology of Diverse Classes of Alkaloids*, Alaken, Inc., Fort Collins, CO, 1996; Barceloux, D.G., Muscarine-containing mushrooms and muscarine toxicity (clitocybe and inocybe species), in *Medical Toxicology of Natural Substances: Foods, Fungi, Medicinal Herbs, Plants, and Venomous Animals*, John Wiley & Sons, Inc., Hoboken, NJ, 2008; Concon, J.M., *Food Toxicology. Parts A and B*, Marcel Dekker, New York, 1988; Deshpande, S.S., Mushroom toxins, in *Handbook of Food Toxicology*, Marcel Dekker, Inc., New York, 2002.

The rate of mortality associated with the intoxication by *A. phalloides* is considerably higher in children under 10 years old (about 51%) than that in adults (about 17%) (Floersheim and Bianchi 1984).

14.4.4 Mechanism of Poisoning

The 3D structures of the peptide molecules reveal the main regulators of the biological activity of amatoxins and phallotoxins. Minor changes in the basic structural may abolish the toxicity completely. The structural changes may alter the conformation and, accordingly, confine the capability of these toxins to bind with the receptor. The mechanisms of action of amatoxins and phallotoxins diverge with respect to the cytopathogenic effects along with their primary target tissues. The amatoxins are strongly nephrotoxic and hepatotoxic, while the phallotoxins are only hepatotoxic in general.

Biochemically, amatoxins (specifically α-amanitin) block the catalytic site of the enzyme by binding with ribonucleic acid (RNA) polymerase II. In that way, the toxins inhibit the transcription of deoxyribonucleic acid (DNA), messenger RNA (mRNA) development, and, thus, protein synthesis. The RNA polymerase II inhibition describes some of the biochemical faults noted after poisoning with α-amanitin, specifically, the reduction in the level of blood protein, the reduction in the level of glycogen synthetase in the liver and intermittently in muscles, and resulting glycogen content of these tissues. The reduction in the level of liver nicotinamide-adenine dinucleotide (NAD) and adenosine triphosphate (ATP) and the escape of potassium from liver cells also follow the inhibition of protein synthesis (Faulstich and Wieland 1992).

Amatoxins are potent inhibitors of mammalian DNA–dependent RNA polymerase II at very low concentrations (5–10 ng/mL); consequently, these toxins interfere with protein synthesis through the inhibition of elongation essential for transcription (Zanotti et al. 1992). The necrosis of liver, intestine, and kidney cells develops due to the excessive rates of replication and preeminent close contact with these toxins. The amatoxin poisoning among different species of animal varied substantially. The lower mammals, such as rodents, are comparatively resistant to the effects of amatoxins: the nonmammalian polymerases exhibit a little sensitivity to amanitins. On the other hand, higher mammals such as humans, cats, dogs, and guinea pigs are comparatively hypersensitive to amatoxins (Vetter 1998).

Phallotoxins and virotoxins interact very precisely with the equilibrium system between F-actin and G-actin, which acts a fundamental role in the survival and functions of a cell. By tightly binding to F-actin, they stabilize the structure of the filamentous assembly, averting its dissociation and therefore shifting the equilibrium almost entirely to the F-actin state. Liver is affected almost completely by these toxins. Phalloidins also inhibit the glycogen synthesis; accelerate the autolytic reactions on liver nucleotides and proteins; and release lysosomal enzymes. Parenteral administration of phallotoxins to animals stimulates the polymerization of G-actin and stabilizes the F-actin filaments. The hepatorenal failure in human or other mammals by virotoxins and phallotoxins is limited since both the toxins exhibit lower absorption by the gastrointestinal tract. The LD_{50} values of several amatoxins and phallotoxins are listed in Table 14.2.

14.4.5 Treatment

Some therapeutic techniques have been suggested by Faulstich after diagnosing amatoxin poisoning by serum and urine analysis (Faulstich 1979):

1. During the first 36 hours after the meal, cleansing of stomach and intestines is mandatory. Activated charcoal should be used to bind these toxins in the GI tract, and restoration of water and electrolyte balance is also required.
2. Within 48 hours after the toxin is ingested, it is compulsory to remove amatoxins from blood by using any one of the removal techniques, such as forced diuresis, hemoperfusion by polymeric adsorbents, hemofiltration, and plasmapheresis.
3. The infusion of glucose is needed to prevent hypoglycemia.
4. Drug treatment is also recommended, with 100–300 mg/day of thioctic acid or higher concentrations of penicillin or silymarin.

14.5 ORELLANINES

14.5.1 Physiochemical Properties

Orellanine is a colorless, polar, bipyridyl *N*-oxide compound having blue-fluorescence properties. The Polish physician Grzymala isolated this compound for the first time from *Cortinarius orellanus* (Grzymala 1962). Generally, this toxic substance is present in nephrotoxic mushrooms of the genus *Cortinarius*. The decomposition of this compound is slow in UV light and in high temperature (150°C–180°C) and is subsequently converted to orelline, a bipyridyl compound (Andary et al. 1986; Antkowiak 1985). The chemical structures of orellanine, orellinine, and orelline are presented in Figure 14.4. Orellanine decomposes rapidly at temperatures above 267°C although orellanine is stable in intact mushroom for many years (Antkowiak and Gessner 1979). Orelline is a compound having blue and yellow fluorescence and is not toxic to animals (Antkowiak 1985). Both orelline and orellanine are insoluble in water and organic solvents. However, these compounds are slightly soluble in dimethyl sulfoxide and methanol (Antkowiak and Gessner 1979).

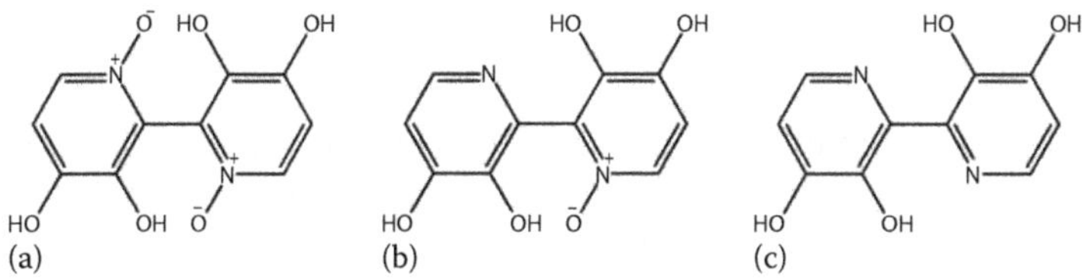

FIGURE 14.4 Chemical structures of (a) orellanine, (b) orellinine, and (c) orelline.

14.5.2 Poisonous Parts

Orellanine is present in *Cortinarius* species including *Cortinarius henricii*, *Cortinarius fluorescens*, *C. orellanus*, *Cortinarius rubellus*, and *Cortinarius rainierensis* (Rapior 1988). The concentration of orellanine was 0.45%–1.4% in *C. henricii* and *C. orellanus* (Oubrahim et al. 1997) as quantified using direct electron spin resonance (Table 14.3). One study showed that the content of orellanine in the caps of mushroom was roughly three times greater than in the stalks (Barceloux, D. G. 2008). The orellanine concentrations in *C. rubellus* (*Cortinarius speciosissimus*, *Cortinarius orellanoides*) and *C. orellanus* were 9 mg/g and 14 mg/g, respectively, from dried mushroom samples (Prast et al. 1988).

14.5.3 Symptoms of Poisoning

Ingestion of orellanine-containing mushroom causes delayed onset of severe tubule-interstitial nephropathy that can progress to chronic renal insufficiency. The first stage of intoxication occurs after 12 hours to 14 days of ingestion and shows mild symptoms (Danel et al. 2001). Apart from general weakness, the symptoms can be a typical form of acute gastritis that includes nausea, vomiting, mild irregular abdominal pain, anorexia, and diarrhea. This prerenal phase also includes chills, intense thirst, burning sensation in the mouth, headache, polyuria, polydipsia, muscular pains, and constipation (Schumacher and Hoiland 1983). In minor cases, these symptoms subside within a few days and renal dysfunction is not developed. In more severe cases, there is a rapid increase in liquid intake, driving the patient to drink several liters per day. Primarily, the urine output consistently increases resulting to renal failure, which ultimately leads to the patient's death. Signs of renal failure include oliguria, albuminuria, anuria, uremia, and the imbalance of electrolytes that lasts for the next 1 or 2 weeks, whereas vomiting of bile, pain in the abdomen, and subicteric state are the indications of hepatic damage. Death may be preceded by drowsiness, coma, and convulsions. Mortalities have been shown to vary from 10% to 20% (Deshpande 2002).

TABLE 14.3
Content of Orellanine in Selected *Cortinarius* Species

Species of *Cortinarius*	Dry Weight (%)
Cortinarius orellanus	1.10
Cortinarius rainierensis	0.85
Cortinarius rubellus	0.20–0.50
Cortinarius henricii	0.45

Source: Oubrahim, H. et al., *J. Chromatogr. A*, 758(1), 145, 1997.
Analytical method of direct electron spin resonance with limit of detection = 5 μg.

14.5.4 Mechanism of Poisoning

The mechanisms of action of orellanines are still not fully elucidated unlike the well-documented histopathological changes and clinical symptoms that arise from the ingestion of orellanine-containing mushrooms. The orellanine and its toxic metabolite (orellinine) possibly affect the epithelium of the renal tubule (Delpech et al. 1990; Oubrahim et al. 1998). The activities of orellanine and two biologically active cationic bipyridines, diquat and paraquat, in physiological system are related to their chemical structures. Paraquat is 4,4′-isomer of diquat, and both these compounds are known for herbicidal attributes. Schumacher and Hoiland proposed a similar pathogenic mechanism of action for all these compounds, assuming their ability to inhibit the intracellular metabolic systems dependent on reduced oxidized nicotinamide-adenine dinucleotide (NADPH) plus H^+, by participation in the electron transport process (Figure 14.5; Schumacher and Hoiland 1983). Moreover, the cytotoxic properties are also exerted by concomitantly produced superoxide, by direct means or by the production of other reactive forms of oxygen, for instance, singlet oxygen. This suggested intracellular toxic mechanism with NADPH plus H^+ reduction would need a long time to barren the cells to the grade of unavoidable necrosis. This is in accordance with the delayed toxic effects observed in this type of mushroom poisoning. Although this assumption may be correct, it has not been validated yet by investigating the electrochemical behavior of orellanine. Moreover, the electrochemical properties of paraquat and diquat are different from orellanine as per the study by Richard et al. (1988).

Consumption of nephrotoxic *Cortinarius* species produces acute interstitial fibrosis, interstitial nephritis, and severe focal tubular damage. The toxicity of this species is due to the metabolites having isoxazolium core (e.g., orellinine), and these intermediate materials are obtained by the photochemical rearrangement of orellanine. These metabolites lead to organ damage by binding with various proteins in the body. It was notified by Rapior that keeping the orellanine in the dark place can purify the compound and display low toxicity *in vivo*, whereas the toxic response of these metabolites may increase in the presence of light (Rapior 1988). These scrutinies indicating the significance of photoexcitation supported proposed phototoxicity mechanism of orellanine (Figure 14.6). However, the exact mechanism of toxicity remains unclear. The LD_{50} values of orellanine are 4.9, 8.3, and 8.0 mg/kg in the cat (per os), mouse, and guinea pig (parenteral), respectively. These values and the symptoms observed in victims indicate the extreme toxicity of this compound (Concon 1988).

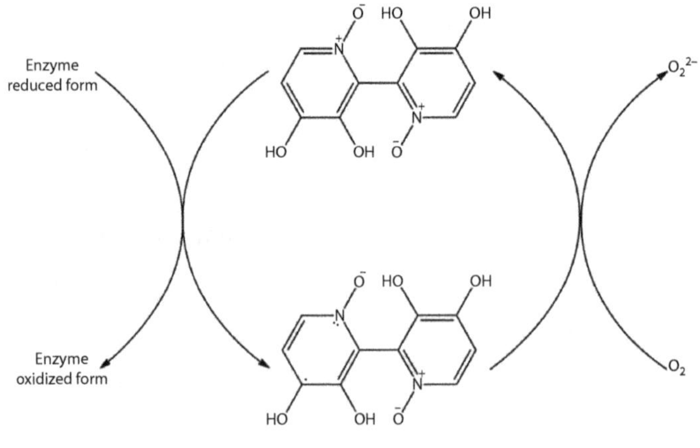

FIGURE 14.5 Intracellular toxic mechanism proposed for orellanine.

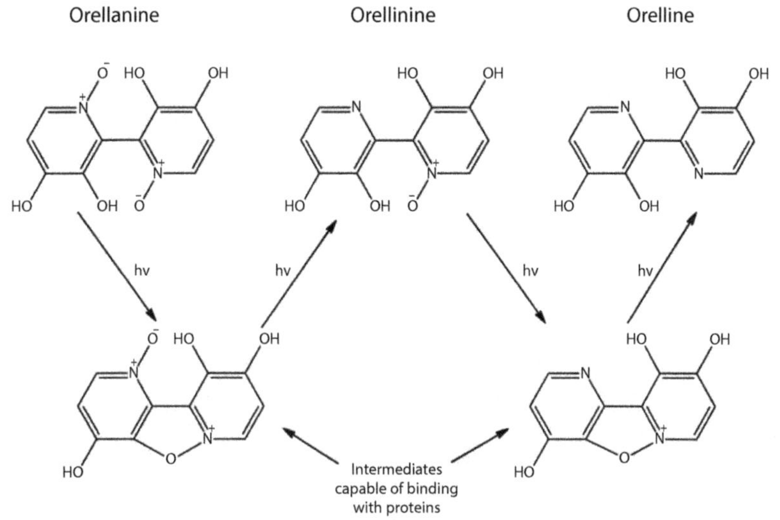

FIGURE 14.6 Photodecomposition of orellanine.

14.5.5 Treatment

Supportive care, proper maintenance of body fluid, and electrolyte balance are the backbone of treatment in orellanine toxicity. Hemodialysis may be required, and continuous observation of the patient for renal function is necessary as there is a possibility of kidney dysfunction. The materials used in hemoperfusion (e.g., activated charcoal) can adsorb orellanine according to some *in vitro* studies (Fulde et al. 1998). However, the efficacy of hemodialysis or hemoperfusion is not confirmed as the clinical and toxicokinetic data are insufficient. Investigational analysis of orellanine in animal revealed that noticeable amounts of orellanine can be excreted by the kidneys within 24 hours and the nephrotoxicity develops within 40 hours of ingestion (Nieminen and Pyy 1976; Prast et al. 1988). The quantity of orellanine is significant clinically but if this is detected after 1–2 weeks of ingestion, it may not be possible anymore to remove the toxins by hemoperfusion or hemodialysis. The use of hemodialysis in this period is desirable depending on factors that support renal function. The antidotes in orellanine poisoning are not available yet as there are still various factors that need to be considered. These factors include the delay of supportive therapy application after poisoning, the amount of orellanine in the body, the preexisting diseases that might cause complications (especially renal disease), and the overall health condition of the patient (Horn et al. 1997). In general, the renal transplantation takes place for more than 10 months afterward the ingestion, as the renal function improves slowly (Danel et al. 2001). It was reported in different case studies that *N*-acetylcysteine and corticosteroids play a role in removing orellanine toxicity (Kilner et al. 1999). However, these treatments have not been clinically proven yet because the extent of renal dysfunction varies after orellanine poisoning and the clinical data are not sufficient enough, and thus, the efficacies of these drugs are still not confirmed.

14.6 GYROMITRIN

14.6.1 Physicochemical Properties

The presence of harmful hydrazine derivatives in *Gyromitra esculenta* and other related species is the main reason for the toxic effects of these mushrooms. Gyromitrin (2-ethylidene-1-formyl-1-methylhydrazine, Figure 14.7a) is the main constituent of this mushroom species and is most toxic among nine hydrazones isolated from *G. esculenta* (Michelot and Toth 1991). Gyromitrin

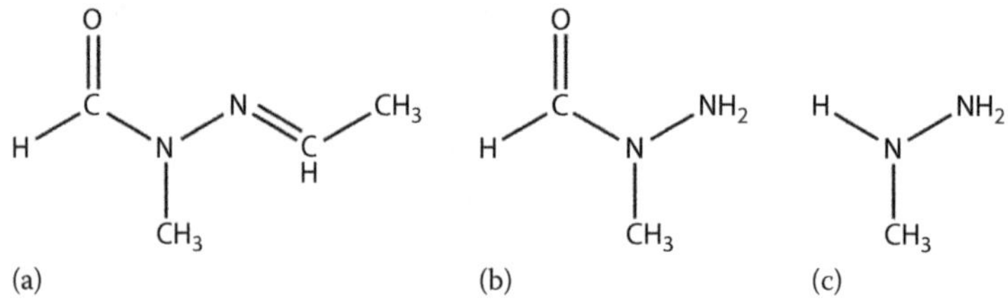

FIGURE 14.7 Chemical structures of (a) gyromitrin, (b) *N,N*-formyl methylhydrazine, and (c) *N*-methylhydrazine.

is an unstable, volatile, colorless, heat-labile, and water-soluble compound that readily undergoes oxidation and hydrolysis even at normal condition. In the gastrointestinal tract, gyromitrin hydrolyzes in the presence of an acid and is converted into 1-formyl-1-methylhydrazine (Figure 14.7b). This intermediate compound later transformed into extremely toxic *N*-methylhydrazine (Figure 14.7c) by losing the formyl residue. Other homologs of gyromitrin also exist, where the acetaldehyde is substituted by other aldehydes. These contain the derivatives of 3-methylbutanal, hexanal, and pentanal. The toxin can be eliminated from *Gyromitra* mushrooms by storing them in dry condition for a long time or by boiling vigorously in water. The edibility of this species cannot be ensured by simple parboiling. It was suspected that gyromitrin is possibly bonded with components having high molecular weight (e.g., glycoside) in the mushroom and is therefore unstable (Deshpande 2002).

14.6.2 Poisonous Parts

Gyromitrin does not exist in all species of *Gyromitra*, and the concentration of this toxin differs largely even within the same species. This difference is due to the variations in climatic conditions and some unrecognized factors. Variable toxicity, which depends on many factors including the physiological conditions of an individual, can occur in the consumption of mushrooms that have gyromitrin. The ingestion of liquid or juice that is extracted from the cooked mushrooms induces toxicity significantly. There are at least eight other hydrazine compounds in the species *G. esculenta* aside from gyromitrin, but it is the most potent toxic compound (Michelot and Toth 1991). Scientific investigations proved that the amount of gyromitrin in various specimens of false morels (*G. esculenta*) varies significantly in the range of 40–730 mg/kg weight in wet basis (Andary et al. 1984). The gyromitrin content in fresh mushrooms is 1.2–1.6 g/kg, while dried mushrooms contain 3 mg/kg of that toxin.

14.6.3 Symptoms of Poisoning

The symptoms of gyromitrin toxicity predominantly comprise gastrointestinal and neurological effects. Generally, mild gastrointestinal symptoms arise after the consumptions of mushrooms that have gyromitrin. There are a number of factors including the content of this toxin in mushroom and the procedures of preparation, which may dominate to appear those symptoms. The symptoms appear suddenly, and the gastrointestinal symptoms arise from 2 to 8 hours up to 2 days after the ingestion of mushrooms or the inhalation of vapor produced at the time of cooking. Possibly, the delay in the onset of GI symptoms is similar to that due to amatoxin toxicity (Koppel 1993). The symptoms are categorized as GI symptoms (nausea, vomiting, abdominal pain) and nonspecific symptoms (headache, general weakness, malaise, enlargement of liver with pain, and abnormal thirst); however, these are only exceptionally accompanied by diarrhea. Neurological symptoms including dizziness, tremor, nystagmus, ataxia, slurred speech, and fatigue may also appear. In severe conditions, hepatic failure, seizures, and coma occur 1–3 days after the ingestion, and subsequently hypovolemia and hypoglycemia develop, which makes the situation more complicated (Giusti and Carnevale 1974). This illness is resolved within 2–6 days in majority of the cases (Leathem and Dorran 2007). But hepatorenal syndrome including hemoglobinuria, liver injury, and anuria may occur after the consumption of enormous amounts of inadequately boiled false morels. Major indications of this stage are nervous maladies, for instance, cramps, difficulties in moving, deliria, and unconsciousness. The patient may die if there he or she has a weak circulatory system (Faulstich and Wieland 1992).

14.6.4 Mechanism of Poisoning

Hydrazine compounds react with pyridoxal-5-phosphate and subsequently inhibit the enzymatic activation of pyridoxine. In this way, the synthesis of GABA decreases and hence the susceptibility to seizures enhances (Lampe 1979). However, it was exhibited in *in vivo* experiments that the level of reduction of GABA synthesis in the brain was low in response to the toxic quantities of monomethylhydrazine (*N*-methyl hydrazine) (Maynert and Kaji 1962). Therefore, the poisonous effects of gyromitrin are not only attributed by monomethylhydrazine. The particular metabolites of gyromitrin that are responsible for the gastrointestinal and neurological toxicity are not yet determined. In animal studies, *N*-methyl-*N*-formyl hydrazine, a metabolite of gyromitrin, depleted cytochrome P-450 within 8–12 hours of the administration of gyromitrin (Braun et al. 1979), whereas the other metabolites of gyromitrin (*N*-methyl hydrazine) did not show similar hepatotoxic effects. Hydrazine compounds show irritating effects to the mucous membranes and also the gastrointestinal tract. Anorexia, inflammation of lung and liver, weight loss, and interstitial nephritis were observed after the inhalation of hydrazine compounds in animal experiments (Comstock et al. 1954). Moreover, the hydrazine compounds are responsible for the depletion of glutathione in erythrocytes and for the destruction of macromolecules in the liver, possibly through the development of free methyl radicals.

14.6.5 Treatment

Commonly occurring complication in severe gyromitrin toxicity is the destruction of fluid and electrolyte balance. Generally, convulsion or coma does not occur infrequently, and the typical supportive care should be given to resolve these clinical conditions. The supportive care in this condition is to apply respiratory support, diazepam or lorazepam, and to evaluate the concentration of blood glucose quickly. The symptoms of complication usually appear in less than 1–2 hours after eating gyromitrin-containing mushrooms so that decontamination may not require. Although there are no clinical data or even reliable *in vivo* animal experimental data regarding the treatment for gyromitrin poisoning, the use of activated charcoal is the possible treatment. Pyridoxine (25 mg/kg or 5 g intravenous dose up to a daily adult dose of 15–20 g) has been recommended for serious neurological symptoms (e.g., coma, seizures) for gyromitrin poisoning based on the similarity between gyromitrin and hydrazine toxicities (Hanrahan and Gordon 1984; Kirklin et al. 1976). However, some animal studies reported that pyridoxine is able to reduce the neurotoxicity that arises from few hydrazine compounds such as *N*-methyl hydrazine and may not be effective to nullify the toxicity due to other hydrazine compounds like *N*-methyl-*N*-formyl hydrazine (Toth and Erickson 1977). The clinical data are not sufficient to prove the effectiveness of pyridoxine in the treatment of neurotoxicity accompanied by the consumption of gyromitrin-containing mushrooms (Zelnick et al. 2003).

According to an animal study, pyridoxine is not effective in the treatment of gyromitrin-associated liver injury (Braun et al. 1979). The same palliative treatment is recommended in gyromitrin-induced hepatic encephalopathy as that used for amatoxin poisoning. In severe conditions, it was suggested to check the complete blood count; levels of creatinine, serum electrolytes, hepatic aminotransferases, glucose, and ammonia; and prothrombin time of the patient. There are some other suggested treatment options with insufficient clinical evidence and folic acid and *N*-acetylcysteine are among them. The daily dose of folinic acid of 20–200 mg has been proposed by some authors, but the effectiveness of this drug in the treatment of gyromitrin poisoning is not yet clinically proven.

14.7 MUSCARINE

14.7.1 Physiochemical Properties

The main toxic component of mushrooms belonging to the genera *Inocybe* and *Clitocybe* is the quaternary trimethylammonium salt of 2-methyl-3-oxy-5-aminomethyl tetrahydrofuran, which is called L-(+)-muscarine (Genest et al. 1968). This compound binds to the acetylcholine muscarinic receptor in the postganglionic and autonomic nerve fibers in the eyes, exocrine glands, gastrointestinal tract, bladder, cardiovascular system, and lungs. Binding of muscarine to these acetylcholine muscarinic receptors leads to depolarization. There are some similarities between the structure of acetylcholine and that of muscarine as shown in Figure 14.8. Other enantiomers of muscarine such as *epi*-muscarine, *allo*-muscarine, and *epi-allo*-muscarine are also present in muscarine-containing mushrooms. These enantiomers are not responsible for the cholinergic effects of muscarine-containing mushrooms as they are inactive compared to L-(+)-muscarine. L-(+)-Muscarine is colorless, water-soluble, and heat stable compound, and the toxicity of these mushrooms having muscarine cannot be diminished by cooking for long time (Barceloux 2008).

14.7.2 Poisonous Parts

The content of muscarine in *Inocybe* species is highly variable and the variation is up to 0.8% of dry weight (Barceloux 2008). The muscarine contents in *Inocybe patouillardii* and *AA. muscaria* are 370 mg/kg and 2–3 mg/kg of dry weight, respectively. Thus, *I. patouillardii* is obviously lethal. Furthermore, the fruiting bodies of *A. phalloides, Amanita pantherina, Boletus luridus, Clitocybe dealbata, Boletus calopus, Clitocybe gibba, Clitocybe hydrogramma, Clitocybe vermicularis, Collybia peronata, Clitopilus intermedius, Hypholoma fasciculare, Hygrocybe nigrescens, Lactarius trivialis, Lactarius rufus, Mycena pura, Mycena pelianthia, Rhodophyllus rhodopolius, P. involutus, Rhodophyllus sinuatus, Russula emetica, Tylopilus felleus*, and *Tricholoma sulphureum* contain ≤20 mg/kg (dry weight) of muscarine and its isomers (Antkowiak 1996).

14.7.3 Symptoms of Poisoning

Clinical features of muscarine toxicity include cholinergic effects, for instance, urgency, salivation, diaphoresis, blurred vision, and diarrhea (Stallard and Edes 1989). Muscarinic symptoms generally appear following the gastrointestinal and autonomic effects few minutes to 2 hours after the ingestion of mushrooms that have muscarine. The autonomic effects associated with the ingestion of muscarine-containing mushrooms include salivation, headache, fasciculations, lacrimation, blurred vision, miosis, and diaphoresis, whereas abdominal pain, diarrhea, nausea,

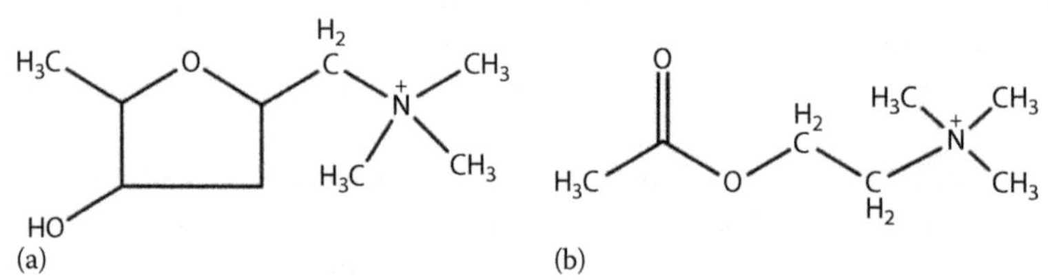

FIGURE 14.8 Chemical structures of (a) muscarine and (b) acetylcholine.

and vomiting are the symptoms of gastrointestinal complications. Dyspnea and wheezing may arise from muscarine-induced bronchoconstriction. Hypotension is not frequently the result, but cardiovascular effects commonly occur along with bradycardia after the consumption of muscarine-containing mushrooms. Usually, symptoms are not severe and patients may recover from this condition within 6 hours.

14.7.4 Mechanism of Poisoning

Through its structural similarity to acetylcholine (Figure 14.8), muscarine binds to the acetylcholine receptor on synapses of the nerve endings of smooth muscles and endocrine glands, causing the well-known parasympathomimetic effects. Muscarine binds strongly with the muscarine type of acetylcholine receptors in smooth muscles of exocrine glands, eye, heart, and gastrointestinal tract; acetylcholine uses an extra "nicotinic action" by affecting ganglia and voluntary muscles. Muscarine opposes esterase activity, as it is not an ester-like acetylcholine. This toxin is not degraded and induces unceasing stimulation of the affected neurons (Faulstich and Wieland 1992). It is more selective and the selective binding affinity of stereoisomers of muscarine strongly depends on the modifications of structure, configuration, and conformation of the rigid muscarine molecule. The hydroxyl group transarranged to the other two substituents plays a key role (Antkowiak 1996). The change of the CH–OH group into carbonyl yields muscarone, which is more potent than muscarine, but is less selective. It shows strong stimulating and blocking effects on neuromuscular junctions and ganglionic synapses as revealed by acetylcholine.

14.7.5 Treatment

The initial phase of patient care includes the assessment of muscarinic effects due to the ingestion of muscarine-containing mushrooms. Usually, the symptoms of poisoning appear at a later time (roughly later than 1–2 hours of ingestion); thus, there is no need of decontamination. Severe poisoning by the ingestion of muscarine-containing mushrooms does not occur frequently because of the low gastrointestinal absorptions and bioavailability of muscarine. Atropine is the drug that is used to antagonize the muscarinic effects. Cardiovascular collapse, bronchospasm, and excessive respiratory secretion are the potential muscarinic symptoms that may be life-threatening. Cardiac monitoring, checking of intravenous fluid level, and measuring oxygen saturation of the patient are needed and, depending on the situation, intubation may be required. The selective antidote for the drying of respiratory secretions is atropine. Starting doses of atropine for different age groups are presented in Table 14.4, and doses are dependent on the severity of the intoxication. Clinical response may dictate the administration of further smaller doses. To determine the correct endpoint

TABLE 14.4
Starting Doses of Atropine to Antagonize Muscarine Poisoning

Age (Years)	Dose (mg)
<2	0.2
2–4	0.3
4–10	0.4
10–14	0.5
>14	0.5–1.0

Source: Deshpande, S.S., Mushroom toxins, in *Handbook of Food Toxicology*, Marcel Dekker, Inc., New York, 2002.

of atropine therapy is very important for preventing atropinization and drying of secretions is the appropriate endpoint. Commonly, the treatment may terminate within 24 hours by ensuring the nullification of muscarinic symptoms. So far, no effective approach has been reported to assure the complete eradication of muscarine.

14.8 MUSCIMOL AND IBOTENIC ACID

14.8.1 Physiochemical Properties

The water-soluble isoxazole compounds, namely, muscimol and ibotenic acid, are the possible primary biologically active compounds in *A. muscaria*, *A. pantherina*, and *Amanita strobiliformis*. Ibotenic acid is α-amino-(4-hydroxy-isoxazole-2)-yl-acetic acid and structurally related to β-oxo-L-glutamine, which may be derived from β-oxo-L-glutamine. Muscimol is produced by the decarboxylation of ibotenic acid. The toxic effects of these mushrooms may also be exerted by muscazone and oxazolone (Michelot and Melendez-Howell 2003). Muscazone is the lactame isomer of muscimol and shows low pharmacological effects in contrast to muscimol (Michelot and Melendez-Howell 2003). Ibotenic acid possibly undergoes a photoinduced rearrangement and subsequently produces muscazone (Goth et al. 1967). The other toxin related to isoxazole is tricholomic acid and is present in *Tricholoma muscarium*. The chemical structures of these toxins are presented in Figure 14.9. At the time of food preparation and digestion in the GI tract, the ibotenic acid undergoes decarboxylation and is converted to muscimol, which is mainly responsible for the psychotropic effects of these mushrooms.

14.8.2 Poisonous Parts

A significant quantity of muscimol and ibotenic acid is found in *A. muscaria*, *A. pantherina*, *Amanita cothurnata*, and *A. strobiliformis*, while a small quantity is found of those in *Amanita gemmata* (Chilton and Ott 1976). The amount of ibotenic acid is significantly more than that of muscimol in the fruiting bodies of these mushroom species. The concentrations of these compounds are variable among those *Amanita* species and even in different samples of same species. The concentrations of isoxazole compounds are comparatively higher in yellow tissue beneath the red cuticles, gills, and white flesh than that in stalks (Eugster 1967; Gore and Jordan 1982). It was shown in a report that the amounts of muscimol and ibotenic acid were higher in caps (380 μg/g muscimol and 990 μg/g ibotenic acid) than that in stalks (80 μg/g muscimol and 230 μg/g ibotenic acid) of fresh samples of *A. muscaria* (Gennaro et al. 1997). *A. muscaria* contains a higher amount of muscarine than that found in certain *Clitocybe* and *Inocybe* species. Only *A. muscaria* contains muscazone when collected during the summer. Moreover, the heavy metals including cadmium, selenium, and vanadium are also present in *A. muscaria* at high concentrations.

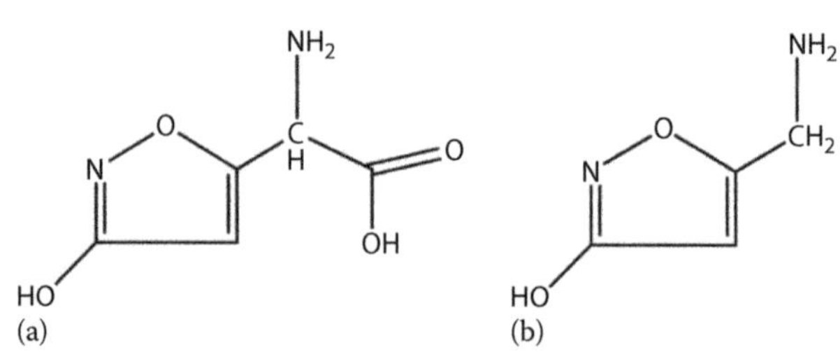

FIGURE 14.9 Chemical structures of (a) ibotenic acid and (b) muscimol.

14.8.3 Symptoms of Poisoning

These toxins of mushroom exhibit poisonous effects on the central nervous system, causing pantherina syndrome. The clinical manifestations of this syndrome are comparable with ethanol intoxication. The initial symptoms of poisoning may occur 15–30 minutes after the ingestion of large amount of these mushrooms (Satora et al. 2005). But in general cases, the symptoms appear after 1–3 hours of toxic mushroom ingestion. The poisonous properties of these poisons initially represent in the central nervous system with spatial distortion, euphoria, anxiety, mydriasis, increased sensitivity to visual and auditory stimuli, depression, ataxia, disorientation, blurred vision, agitation, tremor, convulsions, and fasciculations (Bosman et al. 1965; Satora et al. 2006). This causes increase in body temperature and pulse rate. The ingestion of these toxins with psilocybin and psilocin alters the visual perception without producing defined hallucinations. Prior to complete resolution, a deep sleep with vivid dreams happens frequently. Nonresponsive, extremely dilated pupils may be perceived in severe conditions. Finally, coma along with respiratory depression and cardiovascular collapse may result. Usually, the symptoms are not severe and are resolved automatically, principally for *A. pantherina* and *Amanita regalis*. The resolution occurs within 4–24 hours without any abnormalities (Elonen et al. 1979; Gelfand and Harris 1982). Residual headache may continue in some patients for several days.

14.8.4 Mechanism of Toxicity

The chemical structures of muscimol and ibotenic acid are related with some naturally occurring excitatory amino acids including tricholomic acid, domoic acid, quisqualic acid, and kainic acid, and the neurotransmitting properties of these compounds are due to their structural similarities. Ibotenic acid and muscimol are conformationally restricted derivatives of glutamic acid and gamma-aminobutyric acid (GABA), respectively. In animal models, ibotenic acid binds to glutamic acid receptors, while the muscimol affects GABA receptors by altering resting potential and membrane permeability (DeFeudis 1982; Walker et al. 1971). Particularly, ibotenic acid acts on inhibitory glutamate receptors that are intimately linked to GABA and ionotropic glycine receptors (Cleland 1996). Glutamic acid is the key excitatory neurotransmitter in the central nervous system of mammals. Muscimol is a substrate for the GABA-metabolizing enzyme (GABA transaminase) and an inhibitor of glial and neuronal GABA uptake. Ibotenic acid was approximately five times less toxic than muscimol in rodent models (Scotti de Carolis et al. 1969). According to the investigation of mouse brain homogenates in *in vitro* studies, ibotenic acid converts to muscimol through decarboxylation in the brain (Nielsen et al. 1985). Possibly, the adverse effects of these mushrooms in the central nervous system are caused by the conversion of ibotenic acid to muscimol through decarboxylation (Michelot and Melendez-Howell 2003). The LD_{50} values of muscimol and ibotenic acid by oral and intravenous administration in rats are presented in Table 14.5.

14.8.5 Treatment

Supportive care is the recommended treatment modality. Intubation may be required depending on the quantity seizures present and the severity of respiratory depression. Convulsions associated with the ingestion of these toxins-containing mushrooms can be treated by lorazepam, diazepam, or other benzodiazepines. Generally, decontamination is not necessary. In most cases, close monitoring of the patient is required other than the application of drug therapy such as atropine and physostigmine and others. It is indicated that the behavior of agitated patients should be monitored by placing them in a serene environment. Severely agitated patients should be treated with benzodiazepines (lorazepam, diazepam).

TABLE 14.5
LD_{50} Values of Isoxazole Compounds by Two Different Routes of Administration in Rats

Name of Toxin	Routes of Administration	LD_{50} (mg/kg)
Muscimol	Oral	45
	Intravenous	4.5
Ibotenic acid	Oral	129
	Intravenous	42

Source: Deshpande, S.S., Mushroom toxins, in *Handbook of Food Toxicology*, Marcel Dekker, Inc., New York, 2002.

14.9 PSILOCYBIN AND PSILOCIN

14.9.1 Physiochemical Properties

The structures of psilocybin and related alkaloids are less complicated and have a 3,4-distributed indole ring system like the mycotoxin ergot alkaloids. Psilocybin is a derivative of tryptophan and chemically represented as 4-phosphoryloxy-3-[2-(dimethylamino)-ethyl]-indole or 4-phosphoryloxy-dimethyltryptamine (Figure 14.10a). The psilocin (Figure 14.10b) is the dephosphorylated compound that easily oxidizes and decomposes and is not hallucinogenic in nature. Furthermore, norbaeocystin and baeocystine are the nonmethylated and monomethylated derivatives of psilocybin, respectively.

14.9.2 Poisonous Parts

The compounds psilocybin and psilocin and their derivatives are found in *Psilocybe baeocystis* (Leung and Paul 1967). Generally, *Conocybe cyanopus*, *Panaeolus subbalteatus*, *Pluteus salicinus*, and *Psilocybe semilanceata* have the highest amount (>0.5% of dry weight) of psilocybin (Ohenoja et al. 1987). *P. semilanceata* and *P. subbalteatus* are the only psilocybin-containing mushrooms that are present abundantly in the middle and northern parts of Europe and are toxic (Stijve and Kuyper 1985). Small quantities of psilocin and 0.2%–0.4% of psilocybin are found in the fruiting bodies of *Psilocybe mexicana*.

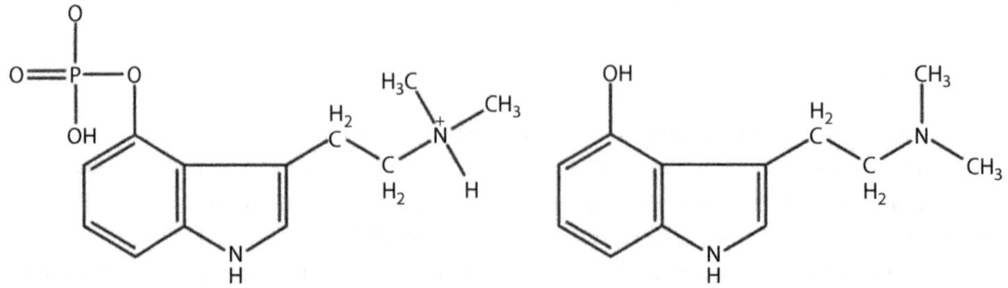

(a) (b)

FIGURE 14.10 Chemical structures of (a) psilocybin and (b) psilocin.

14.9.3 Symptoms of Poisoning

The symptoms of poisoning begin to appear within 30 minutes after the ingestion of roughly 20 g of fresh mushrooms or 4–8 mg of pure psilocybin. The responses against this toxin in different individuals may vary. One patient may feel relaxed, while the other may experience anxiety, dizziness, or tension. In some cases, abdominal discomfort and nausea together with diarrhea and vomiting may occur. After this first preliminary phase, the visual effects such as the objects with closed eyes and insight of brilliant colors usually arise. Panic responses including insanity or the fear of death may occur. The visual effects increase throughout the subsequent periods and disappear later. The stimulation of the sympathetic or central nervous system may lead to the systemic effects and may be accompanied by high blood pressure, rapid heartbeat, and papillary dilatation in association with reduced body temperature and low level of blood sugar (Faulstich and Wieland 1992).

14.9.4 Mechanism of Poisoning

Psilocybin is rapidly converted to psilocin in the body through dephosphorylation. Psilocin is a mixed receptor agonist, and several serotonergic receptors including 5-HT_{1A}, 5-HT_{2A}, and 5-HT_{2C} are partially inhibited by this toxin. It has a high affinity for the 5-HT_{2A} serotonin receptor in the brain, where it mimics the effects of serotonin (5-HT, or 5-hydroxytryptamine). Psilocin binds less strongly with other serotonergic receptors 5-HT_{1A}, 5-HT_{2C}, and 5-HT_{1D}.

14.9.5 Treatment

Therapeutic intervention in adults is rarely required. The poisoning symptoms may diminish naturally; however, the application of phenothiazine or diazepam may be needed in some cases. The LD_{50} values of psilocybin are 275 and 420 mg/kg, respectively, in the case of intravenous and intraperitoneal administrations of the toxins in mice (Antkowiak 1996).

14.10 COPRINE

14.10.1 Physiochemical Properties

Coprine is the major active component of *Coprinus atramentarius* and this mushroom does not have disulfiram or calcium cyanamide (Hatfield and Schaumberg 1975). The administration of this compound to experimental animals does not show any activity and indicates that a metabolite of coprine is mainly responsible for the toxic effects of this mushroom. Coprine, or *N*5-(1-hydroxycyclopropyl)-L-glutamine (Figure 14.11), is a derivative of glutamine that is soluble in water and resistant to heat in neutral solutions. The aldehyde dehydrogenase is not be inhibited by coprime *in vitro* condition.

14.10.2 Poisonous Parts

The average concentration of coprine in fresh samples of *C. atramentarius* was roughly estimated at 160 mg/kg, while the samples of young mushroom contain approximately 50% of the coprine accounted from the samples of mature mushroom (Lindberg et al. 1977). Coprine is also found in *Coprinus quadrifidus*, *Coprinus insignis*, and *Coprinus variegates*. The concentration of coprine cannot be affected considerably by the cooking or boiling of the mushroom (Michelot 1992).

14.10.3 Symptoms of Poisoning

The toxic responses to ethanol may be felt as early as from 15 minutes to 2 hours (typically about 3–6 hours) after the ingestion of *C. atramentarius* mushrooms and may remain up to 48–72 hours. Symptoms experienced by patients due to ethanol poisoning include facial flushing, headache, metallic taste in mouth, paresthesias, vomiting, lightheadedness, palpitations, diaphoresis, chest pain, and tachycardia: these constitute the disulfiram-like reactions that accompany the ingestion of *C. atramentarius* mushrooms. Usually, abdominal pain and diarrhea do not appear. The extents of symptoms are variable and mainly depend on the amount of ethanol and coprine consumed. Other factors that may modulate the intensity of toxic responses are individual sensitivity, the techniques of food preparation, and the duration between ingestion of ethanol and the consumption of *C. atramentarius* mushrooms. Usually, the symptoms of toxicity associated with the ingestion of coprime-containing mushrooms are mild and short lasting and subside quickly without any difficulties. The onset of esophageal rupture and atrial fibrillation are the recorded complications at the time of disulfiram-like reaction, which may resolve within few hours (Caley and Clark 1977; Meyer et al. 1971). The complications associated with disulfiram–ethanol reaction are metabolic acidosis, shock, myocardial infarction, and cardiac dysrhythmias. However, it is not yet clear if these adverse reactions diminish subsequently after the consumption of coprine-containing mushrooms and ethanol. Fatalities related to *C. atramentarius* ingestion are also not well recognized.

14.10.4 Mechanism of Poisoning

Coprine is converted into 1-aminocyclopropanol in the body, which is then further converted into a potent toxin, namely, 1-hydroxy cyclopropyl ammonium ion (Figure 14.11). This toxin blocks an essential sulfhydryl group of acetaldehyde dehydrogenase and reversibly inhibits the low-Km acetaldehyde dehydrogenase in the liver (Wiseman and Abeles 1979). At the time of alcohol metabolism, the acetaldehyde deposition occurs in the blood by the aforementioned mechanism. The poisoning symptoms are mostly responsible for the elevation of blood acetaldehyde. However, the precise mechanism of this toxicity is not clear because in the experimental studies, the elevated concentrations of acetaldehyde individually lead to toxicity as observed from the consumption of coprine-containing mushrooms (Nakano et al. 1974).

14.10.5 Treatment

Supportive care is the treatment modality of coprine toxicity. The most important supportive measures are the treatment of anxiety with lorazepam, diazepam, or other benzodiazepine drugs and the recovery of fluid and electrolyte imbalance. Generally, the clinical representations of coprine intoxication are insignificant. However, severe complications such as cardiac dysrhythmias and hypotension may arise in some cases. The use of pulse oximetry, cardiac monitoring, and insertion of an intravenous line to administer fluids are essential treatment techniques for those patients who have a history of cardiac disease or other critical aberrant signs. The direct-acting vasopressor such as norepinephrine may be administered to the hypotensive patients who are intoxicated by coprine in association with ethanol as the fluid therapy is not responsive in those patients. Generally, decontamination is not recommended for this toxicity since vomiting is recurrent and the absorption of

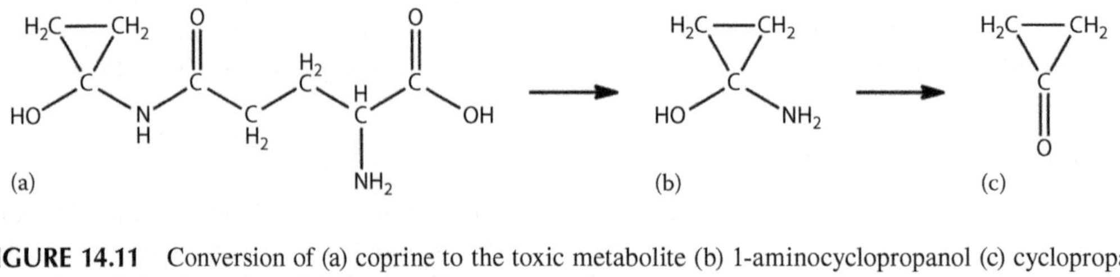

FIGURE 14.11 Conversion of (a) coprine to the toxic metabolite (b) 1-aminocyclopropanol (c) cyclopropanone, through the formation of an intermediate compound.

ethanol is rapid. Coprine or the poisonous metabolites of coprine mostly cannot be eliminated by any antidotes or treatment modalities. People who eat coprine-containing mushrooms are advised to avoid ethanol or ethanol-having foods, drinks, and medicines up to 48–72 hours after the onset of toxic symptoms.

14.11 OTHER TOXINS

Some other toxins of mushroom including bufotenine, bufotenine *N*-oxide, hemolysis-causing agents, phallolysin, lectins, and gastrointestinal irritants are reported in different articles. Bufotenine is 5-substituted tryptamine and is present in *A. muscaria*, *Amanita mappa*, *Amanita porphyria*, *Amanita tormentella*, and *A. pantherina*. The mushrooms having these compounds show hallucinogenic effects in consumers. Other related alkaloidal toxin, namely, bufotenine *N*-oxide, was identified in *Amanita citrina*. Bufotenine is considered as psychomimetic agent. Hemolysis-causing agents were found in different species of Strophariaceae and Hygrophoraceae families; in *Boletaceae*, *Russula*, and *Agaricales* species; and in *Gymnopilus*, *Mycena*, *Amanita*, *Hebeloma*, and *Oudemansiella* genera. Phallolysin is the vital hemolytic causing agent in *A. phalloides*. Several species of Agaricaceae, Cortinariaceae, Russulaceae, Strophariaceae, and Tricholomataceae families contain lectins or hemagglutinins (Seeger and Wiedmann 1972). However, the gastrointestinal symptoms were found in about 40% of all poisoning cases associated with the ingestion of mushrooms (Alder 1960). The reports of gastrointestinal symptoms are very few since the complications resolve within few days and the rates of death due to these symptoms are quite rare.

14.12 CONCLUSION

There are many intoxications reported each year in those countries where the consumption rate of these mushrooms is high. The poisoning due to the ingestion of mushrooms is mostly because of the misidentification of species. Toxic species contain hazardous toxins that are capable of generating diverse syndromes. The number of fatalities recorded due to the ingestion of toxic mushrooms depends on the amount of mushroom consumed. Many effects are frequently observed in different people from a single mushrooms species. Moreover, the chemical compositions of any particular species of mushroom depend on many factors. So far, there are no reliable techniques to determine edible and poisonous mushrooms. However, appropriate documentation is needed to avoid these accidents. Therefore, people should not eat any unfamiliar edible wild species until there is clear-cut and considerable evidence that the particular species is safe.

REFERENCES

Alder, A. E. 1960. Die Pilzvergiftungen in der Schweiz wahrend 40 Jahren. *Schweiz. Z. Pilzkunde* 38:9.

Andary, C., G. Privat, and M. J. Bourrier. 1984. Thin-layer spectrofluorometric microanalysis of monomethylhydrazine in *Gyromitra esculenta*. *Journal of Chromatography* 287 (2):419–424.

Andary, C., S. Rapior, A. Fruchier, and G. Privat. 1986. Cortinaires de la section Orellani: Photod é composition et hypoth è se de la phototoxicit é de l'orellanine. *Cryptogamie Mycologie* 7:12.

Antkowiak, W. Z. 1996. The chemistry and toxicology of mushroom alkaloids. In M. S. Blum (ed.), *Chemistry and Toxicology of Diverse Classes of Alkaloids*. Fort Collins, CO: Alaken, Inc.

Antkowiak, W. Z. and W. P. Gesner. 1979. The structures of orellanine and orelline. *Tetrahedron Letters* 20 (21):4.

Antkowiak, W. Z. and W. P. Gesner. 1985. Photodecomposition of orellanine and orellinine, the fungal toxins of *Cortinarius orellanus* Fries and *Cortinarius speciosissimus*. *Experientia* 41:3.

Barceloux, D. G. 2008. Muscarine-containing mushrooms and muscarine toxicity (clitocybe and inocybe species). In Barceloux, D. G. (ed.), *Medical Toxicology of Natural Substances: Foods, Fungi, Medicinal Herbs, Plants, and Venomous Animals*. Hoboken, NJ: John Wiley & Sons, Inc.

Bobek, P. and S. Galbavy. 1999. Hypocholesterolemic and antiatherogenic effect of oyster mushroom (*Pleurotus ostreatus*) in rabbits. *Die Nahrung* 43 (5):339–342.

Bobek, P., L. Ozdin, and L. Kuniak. 1995. The effect of oyster mushroom (*Pleurotus ostreatus*), its ethanolic extract and extraction residues on cholesterol levels in serum, lipoproteins and liver of rat. *Die Nahrung* 39 (1):98–99.

Borchers, A. T., A. Krishnamurthy, C. L. Keen, F. J. Meyers, and M. E. Gershwin. 2008. The immunobiology of mushrooms. *Experimental Biology and Medicine* 233 (3):259–276.

Bosman, C. K., L. Berman, M. Isaacson, B. Wolfowitz, and J. Parkes. 1965. Mushroom poisoning caused by *Amanita pantherina*. Report of 4 cases. *South African Medical Journal* 39 (39):983–986.

Braun, R., U. Greeff, and K. J. Netter. 1979. Liver injury by the false morel poison gyromitrin. *Toxicology* 12 (2):155–163.

Bschor, F. and H. J. Mallach. 1963. Vergiftungen durch den Kahlen Krempling (*Paxillus involutus*), eine genie\bare Pilzart. *Archives of Toxicology* 20 (2):14.

Caley, M. J. and R. A. Clark. 1977. Cardiac arrhythmia after mushroom ingestion. *British Medical Journal* 2 (6103):1633.

Chilton, W. S. and J. Ott. 1976. Toxic metabolites of *Amanita pantherina*, *A. cothurnata*, *A. muscaria* and other Amanita species. *Lloydia* 39 (2–3):150–157.

Cleland, T. A. 1996. Inhibitory glutamate receptor channels. *Molecular Neurobiology* 13 (2):97–136.

Comstock, C. C., L. H. Lawson, E. A. Greene, and F. W. Oberst. 1954. Inhalation toxicity of hydrazine vapor. *A.M.A. Archives of Industrial Health* 10 (6):476–490.

Concon, J. M. 1988. *Food Toxicology. Parts A and B*. New York: Marcel Dekker.

Danel, V. C., P. F. Saviuc, and D. Garon. 2001. Main features of *Cortinarius* spp. poisoning: A literature review. *Toxicon: Official Journal of the International Society on Toxinology* 39 (7):1053–1060.

DeFeudis, F. V. 1982. Muscimol and GABA-receptors: Basic studies and therapeutic implications. *Reviews in Pure & Applied Pharmacological Sciences* 3 (3):319–379.

Delpech, N., S. Rapior, A. P. Cozette, J. P. Ortiz, P. Donnadieu, C. Andary, and G. Huchard. 1990. Outcome of acute renal failure caused by voluntary ingestion of *Cortinarius orellanus*. *Presse Medicale* 19 (3):122–124.

Deshpande, S. S. 2002. Mushroom toxins. In *Handbook of Food Toxicology*. Marcel Dekker, Inc., New York.

Elonen, E., L. Tarssanen, and M. Harkonen. 1979. Poisoning with brown fly agaric, *Amanita regalis*. *Acta Medica Scandinavica* 205 (1–2):121–123.

Enjalbert, F., C. Gallion, F. Jehl, H. Monteil, and H. Faulstich. 1992. Simultaneous assay for amatoxins and phallotoxins in *Amanita phalloides* Fr. by high-performance liquid chromatography. *Journal of Chromatography* 598 (2):227–236.

Eugster, C. H. 1967. Isolation, structure, and syntheses of central-active compounds from *Amanita muscaria* (L. ex Fr.) hooker. *Psychopharmacology Bulletin* 4 (3):18–19.

Faulstich, H. 1979. New aspects of amanita poisoning. *Klinische Wochenschrift* 57 (21):1143–1152.

Faulstich, H., A. Buku, H. Bodenmuller, and T. Wieland. 1980. Virotoxins: Actin-binding cyclic peptides of *Amanita virosa* mushrooms. *Biochemistry* 19 (14):3334–3343.

Faulstich, H., D. Georgopoulos, and M. Bloching. 1973. Quantitative chromatographic analysis of toxins in single mushrooms of *Amanita phalloides*. *Journal of Chromatography* 79:257–265.

Faulstich, H., D. Georgopoulos, M. Bloching, and T. Wieland. 1974. Analysis of the toxins of amanitin-containing mushrooms. *Zeitschrift fur Naturforschung. Section C: Biosciences* 29 (1):86–88.

Faulstich, H. and Wieland, T. 1992. Mushroom poisons. In A. T. Tu (ed.), *Handbook of Natural Toxins*. New York: Marcel Dekker Inc.

Floersheim, G. L. and L. Bianchi. 1984. Ethanol diminishes the toxicity of the mushroom *Amanita phalloides*. *Experientia* 40 (11):1268–1270.

Fulde, K., J. Bohler, E. Keller, and A. W. Frahm. 1998. Efficiency of haemoperfusion materials at removing the fungal toxin orellanine from human plasma. *Die Pharmazie* 53 (1):58–59.

Gelfand, M. and C. Harris. 1982. Poisoning by *Amanita pantherina*. A description of two cases. *The Central African Journal of Medicine* 28 (7):159–163.

Genest, K., D. W. Hughes, and W. B. Rice. 1968. Muscarine in Clitocybe species. *Journal of Pharmaceutical Sciences* 57 (2):331–333.

Gennaro, M. C., D. Giacosa, E. Gioannini, and S. Angelino. 1997. Hallucinogenic species in *Amanita muscaria*. Determination of muscimol and ibotenic acid by ion-interaction HPLC. *Journal of Liquid Chromatography & Related Technologies* 20 (3):413–424.

Giusti, G. V. and A. Carnevale. 1974. A case of fatal poisoning by *Gyromitra esculenta*. *Archives of Toxicology* 33 (1):49–54.

Gore, M. G. and P. M. Jordan. 1982. Microbore single-column analysis of pharmacologically active alkaloids from the fly agaric mushroom *Amanita muscaria*. *Journal of Chromatography A* 243 (2):323–328.

Goth, H., A. R. Gagneux, C. H. Eugster, and H. Schmid. 1967. 2(3H)-Oxazolone durch Photoumlagerung von 3-Hydroxyisoxazolen. Synthese von Muscazon. *Helvetica Chimica Acta* 50 (1):137–142.

Grzymala, S. 1962. L'isolement de l'orellanine, poison du Cortinarius orellanus Fries et l'etude de ses effects anatomopathologiques. *Bulletin de la Société Mycologique de France* 78:11.

Hanrahan, J. P. and M. A. Gordon. 1984. Mushroom poisoning. Case reports and a review of therapy. *JAMA* 251 (8):1057–1061.

Hatfield, G. M. and J. P. Schaumberg. 1975. Isolation and structural studies of coprine, the disulfiram-like constituent of *Coprinus atramentarius*. *Lloydia* 38 (6):489–496.

Himmelmann, A., G. Mang, and S. Schnorf-Huber. 2001. Lethal ingestion of stored *Amanita phalloides* mushrooms. *Swiss Medical Weekly* 131 (41–42):616–617.

Horn, S., J. H. Horina, G. J. Krejs, H. Holzer, and M. Ratschek. 1997. End-stage renal failure from mushroom poisoning with *Cortinarius orellanus*: Report of four cases and review of the literature. *American Journal of Kidney Diseases: The Official Journal of the National Kidney Foundation* 30 (2):282–286.

Jo, W. S., M. A. Hossain, and S. C. Park. 2014. Toxicological profiles of poisonous, edible, and medicinal mushrooms. *Mycobiology* 42 (3):215–220.

Kilner, R. G., R. J. D'Souza, D. B. Oliveira, I. A. MacPhee, D. R. Turner, and J. B. Eastwood. 1999. Acute renal failure from intoxication by *Cortinarius orellanus*: Recovery using anti-oxidant therapy and steroids. *Nephrology, Dialysis, Transplantation: Official Publication of the European Dialysis and Transplant Association—European Renal Association* 14 (11):2779–2780.

Kirklin, J. K., M. Watson, C. C. Bondoc, and J. F. Burke. 1976. Treatment of hydrazine-induced coma with pyridoxine. *The New England Journal of Medicine* 294 (17):938–939.

Koppel, C. 1993. Clinical symptomatology and management of mushroom poisoning. *Toxicon: Official Journal of the International Society on Toxinology* 31 (12):1513–1540.

Lampe, K. F. 1979. Toxic fungi. *Annual Review of Pharmacology and Toxicology* 19:85–104.

Leathem, A. M. and T. J. Dorran. 2007. Poisoning due to raw *Gyromitra esculenta* (false morels) west of the Rockies. *CJEM* 9 (2):127–130.

Leung, A. Y. and A. G. Paul. 1967. Baeocystin, a mono-methyl analog of psilocybin from *Psilocybe baeocystis* saprophytic culture. *Journal of Pharmaceutical Sciences* 56 (1):146.

Lima, A. D., R. Costa Fortes, M. R. Carvalho Garbi Novaes, and S. Percario. 2012. Poisonous mushrooms: A review of the most common intoxications. *Nutricion Hospitalaria* 27 (2):402–408.

Lindberg, P., R. Bergman, and B. Wickberg. 1977. Isolation and structure of coprine, the in vivo aldehyde dehydrogenase inhibitor in *Coprinus atramentarius*; Syntheses of coprine and related cyclopropa4one derivatives. *Journal of the Chemical Society. Perkin Transactions* 1 (6):684–691.

Mau, J. L., H. C. Lin, and C. C. Chen. 2002. Antioxidant properties of several medicinal mushrooms. *Journal of Agricultural and Food Chemistry* 50 (21):6072–6077.

Maynert, E. W. and H. K. Kaji. 1962. On the relationship of brain gamma-aminobutyric acid to convulsions. *The Journal of Pharmacology and Experimental Therapeutics* 137:114–121.

Meyer, J. H., J. E. Herlocher, and J. Parisian. 1971. Esophageal rupture after mushroom-alcohol ingestion. *The New England Journal of Medicine* 285 (23):1323.

Michelot, D. 1992. Poisoning by *Coprinus atramentarius*. *Natural Toxins* 1 (2):73–80.

Michelot, D. and L. M. Melendez-Howell. 2003. Amanita muscaria: Chemistry, biology, toxicology, and ethnomycology. *Mycological Research* 107 (Pt 2):131–146.

Michelot, D. and B. Toth. 1991. Poisoning by *Gyromitra esculenta*—A review. *Journal of Applied Toxicology* 11 (4):235–243.

Nakano, J., A. C. Gin, and S. K. Nakano. 1974. Effects of disulfiram on cardiovascular responses to acetaldehyde and ethanol in dogs. *Quarterly Journal of Studies on Alcohol* 35 (Pt A):620–634.

Nielsen, E. O., A. Schousboe, S. H. Hansen, and P. Krogsgaard-Larsen. 1985. Excitatory amino acids: Studies on the biochemical and chemical stability of ibotenic acid and related compounds. *Journal of Neurochemistry* 45 (3):725–731.

Nieminen, L. and K. Pyy. 1976. Sex differences in renal damage induced in the rat by the Finnish mushroom, *Cortinarius speciosissimus*. *Acta Pathologica et Microbiologica Scandinavica. Section A, Pathology* 84 (2):222–224.

Nieminen, P., M. Kirsi, and A. M. Mustonen. 2006. Suspected myotoxicity of edible wild mushrooms. *Experimental Biology and Medicine* 231 (2):221–228.

Ohenoja, E., J. Jokiranta, T. Makinen, A. Kaikkonen, and M. M. Airaksinen. 1987. The occurrence of psilocybin and psilocin in Finnish fungi. *Journal of Natural Products* 50 (4):741–744.

Oubrahim, H., J. M. Richard, and D. Cantin-Esnault. 1998. Peroxidase-mediated oxidation, a possible pathway for activation of the fungal nephrotoxin orellanine and related compounds. ESR and spin-trapping studies. *Free Radical Research* 28 (5):497–505.

Oubrahim, H., J. M. Richard, D. Cantin-Esnault, F. Seigle-Murandi, and F. Trecourt. 1997. Novel methods for identification and quantification of the mushroom nephrotoxin orellanine. Thin-layer chromatography and electrophoresis screening of mushrooms with electron spin resonance determination of the toxin. *Journal of Chromatography A* 758 (1):145–157.

Perera, P. K. and Y. Li. 2011. Mushrooms as a functional food mediator in preventing and ameliorating diabetes. *Functional Foods in Health and Disease* 4:11.

Pilat, A. and Usak, O. 1961. *Mushrooms and Other Fungi.* London, U.K.: Peter Nevill Ltd.

Prast, H., E. R. Werner, W. Pfaller, and M. Moser. 1988. Toxic properties of the mushroom *Cortinarius orellanus.* I. Chemical characterization of the main toxin of *Cortinarius orellanus* (Fries) and *Cortinarius speciosissimus* (Kuhn & Romagn) and acute toxicity in mice. *Archives of Toxicology* 62 (1):81–88.

Rapior, S. 1988. Contribution a l'etude de *Cortinarius orellanus* (Fr.) Chimiotaxinomic, culture in vitro, toxicite., University of Montpellier, Montpellier, France.

Richard, J. M., J. Louis, and D. Cantin. 1988. Nephrotoxicity of orellanine, a toxin from the mushroom *Cortinarius orellanus. Archives of Toxicology* 62 (2–3):242–245.

Satora, L., D. Pach, B. Butryn, P. Hydzik, and B. Balicka-Slusarczyk. 2005. Fly agaric (*Amanita muscaria*) poisoning, case report and review. *Toxicon: Official Journal of the International Society on Toxinology* 45 (7):941–943.

Satora, L., D. Pach, K. Ciszowski, and L. Winnik. 2006. Panther cap *Amanita pantherina* poisoning case report and review. *Toxicon: Official Journal of the International Society on Toxinology* 47 (5):605–607.

Schumacher, T. and K. Hoiland. 1983. Mushroom poisoning caused by species of the genus *Cortinarius fries. Archives of Toxicology* 53 (2):87–106.

Scotti de Carolis, A., F. Lipparini, and V. G. Longo. 1969. Neuropharmacological investigations on muscimol, a psychotropic drug extracted from *Amanita muscaria. Psychopharmacologia* 15 (3):186–195.

Seeger, R. and R. Wiedmann. 1972. Zum Vorkommenn von Hamolysinen and Agglutininen in hoheren Pilzen (Basidiomyceten): Untersuchungen an 293 Arten. *Archiv fur Toxikologie* 29 (3):29.

Stallard, D. and T. E. Edes. 1989. Muscarinic poisoning from medications and mushrooms. A puzzling symptom complex. *Postgraduate Medicine* 85 (1):341–345.

Stijve, T. and T. W. Kuyper. 1985. Occurrence of psilocybin in various higher fungi from several European countries. *Planta Medica* 51 (5):385–387.

Sullivan, R., J. E. Smith, and N. J. Rowan. 2006. Medicinal mushrooms and cancer therapy: Translating a traditional practice into Western medicine. *Perspectives in Biology and Medicine* 49 (2):159–170.

Toth, B. and J. Erickson. 1977. Reversal of the toxicity of hydrazine an analogues by pyridoxine hydrochloride. *Toxicology* 7 (1):31–36.

Vetter, J. 1998. Toxins of *Amanita phalloides. Toxicon: Official Journal of the International Society on Toxinology* 36 (1):13–24.

Walker, R. J., G. N. Woodruff, and G. A. Kerkut. 1971. The effect of ibotenic acid and muscimol on single neurons of the snail, *Helix aspersa. Comparative and General Pharmacology* 2 (6):168–174.

Wieland, T. 1986. *Peptides of Poisonous Amanita Mushrooms.* Berlin, Germany: Springer-Verlag.

Wieland, T. and O. Wieland. 1972. The toxic peptides of Amanita species. In A.C. Solomon Cadis and S.J. Ail (eds.), *Fungal Toxins.* New York: Academic Press, Inc.

Wiseman, J. S. and R. H. Abeles. 1979. Mechanism of inhibition of aldehyde dehydrogenase by cyclopropranone hydrate and the mushroom toxin coprine. *Biochemistry* 18 (3):427–435.

Zanotti, G., G. Petersen, and T. Wieland. 1992. Structure-toxicity relationships in the amatoxin series. Structural variations of side chain 3 and inhibition of RNA polymerase II. *International Journal of Peptide and Protein Research* 40 (6):551–558.

Zelnick, S. D., D. R. Mattie, and P. C. Stepaniak. 2003. Occupational exposure to hydrazines: Treatment of acute central nervous system toxicity. *Aviation, Space, and Environmental Medicine* 74 (12):1285–1291.

15 Dioxins and Related Compounds in the Human Food Chain

Sidney J. Stohs

CONTENTS

15.1 INTRODUCTION

Polychlorinated dibenzo-*p*-dioxins (PCDDs), polychlorinated dibenzofurans (PCDFs), and dioxin-like polychlorinated biphenyls (PCBs) are ubiquitous in food of animal origin, and to a much lesser extent, these compounds are also present in fruits and vegetables (Larsen, 2006). 2,3,7,8-Tetrachloro-p-dioxin (TCDD) (see Figure 15.1) is the most potent of the dioxin-like compounds, is prototypical of many halogenated aromatic compounds, and is one of the most potent toxicants known to man. It is believed that up to 99% of human exposure to dioxins occurs through contamination of food, with small amounts being ingested via inhalation (Fuster et al., 2002; Parzelfall, 2002). Isolated cases of industrial contamination of air and animal feed and the now banned use of contaminated polychlorinated pesticides have also led to occasional poisonings of humans and livestock (Parzefall, 2002). Cases of poisonings have decreased with time due to better industrial practices and safety measures.

Dioxins and dioxin-like compounds represent a large group of structurally related chemicals that are environmentally and biologically persistent due to their chemical stability, low aqueous and high lipid solubility, and poor metabolism. They have a common mechanism of action that involves binding to the aryl hydrocarbon receptor (AHR). Prior to industrialization, exceedingly low amounts of dioxins were present in the environment. The low levels that did exist were formed through natural combustion and geological processes (Schecter et al., 2006).

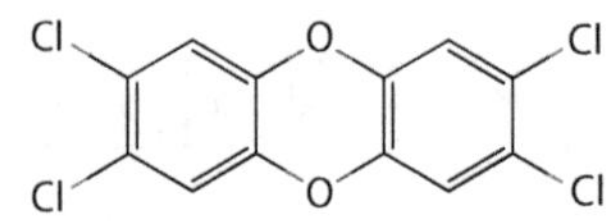

FIGURE 15.1 2,3,7,8-Tetrachlorodibenzo-*p*-dioxin.

The primary sources of dioxin-like compounds are the industrial production of chlorinated organic compounds such as triclosan, polychlorinated biphenyls, pesticides, herbicides, and fungicides, as well as wood preservatives and chemical intermediates such as pentachlorophenol and hexachlorobenzene, the chlorine bleaching of paper and pulp products, and emissions and incomplete combustion from industrial, municipal, hospital, hazardous waste, and sewage-sludge incinerators (Mukerjee, 1998; Zheng et al., 2008). Other sources include emissions from electric furnaces in steel mills, metallurgical industries as copper smelters, and wire reclamation incinerators. Trace amounts are also produced by emissions from motor vehicles using diesel fuel or leaded gasoline, in forest fires and residential burning of wood (Mukerjee, 1998).

These compounds have been of concern because of the clinical and subclinical illnesses and adverse effects that have been attributed to them. The transfer of dioxins into the human food chain, the mechanism of action, the half-lives of these compounds, their relative toxic equivalencies, and demonstrated toxic effects in humans will be reviewed. Finally, the current state of exposures and the degree to which we should be concerned will be discussed.

15.2 CHEMISTRY AND TOXIC EQUIVALENTS

As previously noted, dioxins and structurally related compounds constitute a group of chemicals that induce a common spectrum of effects through a common mechanism of action. Dioxins are composed of two benzene rings connected by two oxygen atoms and are chlorinated in four to eight positions. The most toxic and widely studied of the dioxins is TCDD (Figure 15.1). Other members of the dioxin-like group include dibenzofurans (PCDFs) and planar polychlorinated biphenyls. Figure 15.2 shows the structure of 2,3,4,7,8-dibenzofuran (2,3,4,7,8-PCDF), while the structure of 3,3′,4,4′,5-pentachlorobiphenyl (PCB 126) is presented in Figure 15.3, which are the most toxic of the PCDF and PCB groups of compounds.

Hundreds of congeners of dioxins and dioxin-like compounds are formed during synthetic processes involving industrial and incineration activities. Human exposure results in the uptake and retention of a large number of these compounds that exhibit different binding affinities for AHR, which is discussed in the section about mechanism of action, and therefore differing toxic potentials. As a consequence, various PCDFs, PCDDs, and PCBs have been assigned dioxin toxic equivalency factors (TEFs) based on their relative toxic potencies relative to TCDD, which has been assigned a TEF of 1.0 (van den Berg et al., 2006).

An expert World Health Organization panel has assigned TEF values to 7 PCDDs, 10 PCDFs, 4 non-ortho-substituted PCBs, and 8 mono-ortho-substituted chlorinated biphenyls (van den Berg et al., 2006). 1,2,3,7,8-Pentacholorodibenzo-*p*-dioxin (1,2,3,7,8-pentaCDD) is equitoxic to TCDD and has also been assigned a TEF of 1.0, while other PCDDs are less toxic and have lower TEFs.

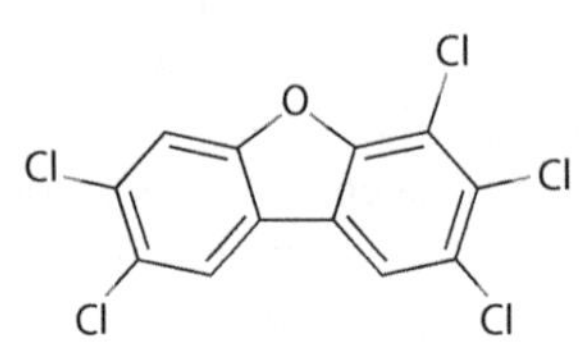

FIGURE 15.2 2,3,4,7,8-Pentachlorodibenzofuran.

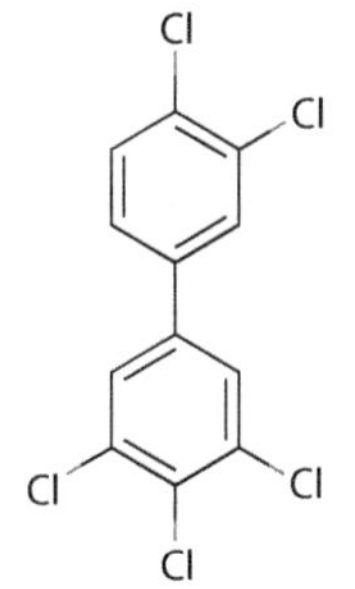

FIGURE 15.3 3,3′,4,4′,5-Pentachlorobiphenyl.

2,3,4,7,8-Pentachloro-dibenzofuran (2,3,4,7,8-pentaCDF) is the most toxic of the PCDFs and has been assigned a TEF of 0.3. 3,3′,4,4′,5-Pentachlorobiphenyl (known as PCB 126) is the most toxic of the PCBs and has a TEF of 0.1. These TEF values are used by simply multiplying the concentration or amount of a dioxin by its TEF, which gives the toxicological equivalence (TEQ) to TCDD. The total toxicological potency or potential of a mixture of dioxins can then be determined by summing these values to give the total TEQ. Thus, the total toxicological potential of a mixture of dioxins in samples such as soil, fat, foods, or milk can be determined (van den Berg et al., 2006; Schecter et al., 2006; Tuomisto, 2012).

15.3 DIOXINS IN THE FOOD CHAIN

15.3.1 General Observations

With the exception of occupational exposure and isolated industrial accidents, virtually all human exposure to dioxins and dioxin-like compounds occurs through the food chain via consumption of contaminated meat and fish or contaminated plant materials. As previously noted, primary sources of dioxins have arisen in association with the production of chlorinated pesticides, halogenated organic chemicals, and the chlorine bleaching of wood pulp and paper products. Lesser amounts are produced by incomplete incineration processes, open burning of wastes, and forest fires (Mukerjee, 1998; Zheng et al., 2008).

Industrial fumes and incinerator products have resulted in the contamination of soils and plants used for forage by animals. For example, a PCB industrial production plant polluted soil and forage of surrounding agricultural fields, resulting in significant contamination of cattle and the subsequent contamination of milk and meat products (Turrio-Baldasarri et al., 2009). The involuntary ingestion of contaminated forage and soil is believed to be the primary cause of the contamination of animal products (Rychen et al., 2014). Because of the highly fat-soluble nature of dioxins, accumulation occurs in adipose tissue associated with meat, liver, and milk, which can then be passed on to human consumers.

15.3.2 Soil Contamination

Various studies have examined the potential for human foodborne exposure to dioxins associated with the recycling of sewage sludge on agricultural lands. In essence, the use of sewage sludge on land use for grazing animals as well as agricultural crops is correlated with increased human exposure to dioxins (Rideout and Teschke, 2004), and the concentrations in foods such as vegetables, meats, and milk are linearly related to the concentrations of the dioxins in the soil (Passuello et al., 2010). However, it has also been shown that large increases of dioxins in soil concentrations from sewage sludge lead to measurable increase in plant contamination due to the lipophilic nature of the dioxins and therefore the ability to be taken up by plants (Rideout and Teschke, 2004). Exposure of crops to air polluted with dioxins results in much greater human exposure (Turrio-Baldasarri et al., 2007) because of the direct deposition of the dioxins on the plants.

15.3.3 Animal Sources

Consumption of chickens that have been exposed to dioxins constitutes another food source of dioxins. Dioxins concentrate in the eggs from chickens and other fowl. Cases of exposure to dioxins via this source have represented isolated cases and are not associated with the general consumption of chickens and eggs. Sources of dioxins in these cases include industrial pollution, the use of sludge on fields, and the use of feeds contaminated with oils or pulp that contain dioxins (Traag et al., 2006; Hoang et al., 2014; Malisch and Kotz, 2014; Torres et al., 2013; Squadrone et al., 2015). Incidents of high dioxin exposure via these sources have declined in recent years, as is generally the case due to greater restrictions on incineration and industrial pollution.

A case of feeding poultry and some pig feed contaminated with PCBs occurred in Belgium (Covaci et al., 2008). An estimated 500 tons of animal feed was contaminated with 50 kg PCBs that also contained small amounts of dioxins. Signs of poultry poisonings were first observed, with more than 2500 poultry and pig farms being affected. A large food-monitoring program for PCB markers was implemented. Due to rapid action, it was deemed unlikely that adverse effects were inflicted on the population.

Another interesting and isolated case of feeding dioxin-contaminated feed to animals occurred in Ireland, with the subsequent recall and destruction of 30,000 tons of pork products and the slaughter of many hogs and some cattle (Marnane, 2012). The feed was contaminated as the result of the use of hot gases from the combustion of contaminated fuel oil for the drying of the feed. This represents a situation that was quickly and efficiently addressed in order to eliminate the source of human exposure (Jacob et al., 2011), and this also demonstrates how broad the range of exposure of dioxins can be in a short period of time.

15.3.4 Marine Sources

In the marine environment, dioxins adhere to sediments that are consumed by zooplankton, which are subsequently consumed by crustaceans and small fish, which are ultimately consumed by the larger fish and sea life that enters the human food chain. Some species of fish such as mackerel appear to concentrate more dioxins than others. In this aquatic food chain, the dioxins are concentrated in the fat of the fish, resulting in higher levels of exposure to human consumers than the original concentrations in the marine sediments (Peltonen et al., 2014; Wang et al., 2015). In areas where wastes have been dumped into lakes and oceans from paper mills and various industrial plants, the consumption of fish constitutes a primary source of dioxin intake (Hsu et al., 2007).

15.3.5 Recent Changes

Various studies have demonstrated that significant decreases in the dioxin contamination of soil, water, and air have occurred over the past 25–30 years. One study has shown that industrial emissions into the air have decreased by a factor of 5 over this period of time (Bruckmann et al., 2013). As a consequence, soil and vegetation contamination has also decreased and so has the exposure of the general population through the food chain (Buchert et al., 2001). Decreases in dioxin exposure have been confirmed by significant decreases in the blood levels of PCDDs, PCDFs, and TEQs, although notably no decreases were observed for PCB 126 (Consonni et al., 2012).

Currently, the greatest risk of exposure to dioxins is associated with the accidental exposure to the food chain through contaminated animal feeds or through industrial accidents (Burchet et al., 2001). Although exposure to dioxins has markedly decreased in recent years, washing of fruits and vegetables, trimming of fat from meats, consuming low-fat dairy products, and simple cooking of food can substantially decrease exposure to dioxin-like compounds as well as other potentially harmful contaminants (Roeder et al., 1998; Zabik and Zabik, 1999).

15.4 MECHANISM OF ACTION

15.4.1 General Mechanism

PCDDs, PCDFs, and PCBs are all believed to exert their effects through a common mechanism of action. An event that is common to all tissues that exhibit toxicological consequences due to exposure to dioxin and dioxin-like compounds is binding to AHR. A tremendous amount of research over the past 35 years has elucidated the cascade of events that occurs following the binding of a ligand such as TCDD to the AHR with subsequent gene regulation and expression, the influence on other signal transduction mechanisms, and other mechanistic aspects that subsequently lead to adverse events such as carcinogenicity and teratogenicity as well as immunological and developmental issues (Pohjanvirta, 2012).

The production of reactive oxygen species (ROS) and the subsequent development of oxidative stress and inflammation occur as the result of the binding of a ligand to the AHR, which leads to oxidative DNA damage, lipid peroxidation, membrane and organelle damage, loss of tissue function, and ultimately cell death (Stohs, 1990). The persistent activation of the AHR by dioxins leads to persistent production of ROS, while preventing or inhibiting the AHR from performing its normal homeostatic functions.

Numerous studies have demonstrated that diverse enzymatic pathways, membranes, organelles, and cells are involved and responsible for the generation of ROS, oxidative stress and inflammation, and the consequences thereof (Stohs and Hassoun, 2012). TCDD is an exceedingly potent inducer of cytochrome P3450 1A1 in both hepatic and extrahepatic tissues as the intestinal mucosa and macrophage and a major mechanism for the production of ROS.

Various lines of investigation support the role of the AHR as the mediator of oxidative stress and inflammation in response to dioxin-like compounds. In congenic mice that respond or are resistant to TCDD and have different AHR loci, an excellent correlation existed between ROS production and relative lethality (Alsharif et al., 1994). Furthermore, the use of AHR knockout mice has clearly shown the role of the AHR in the production of ROS and subsequent oxidative stress (Sugihara et al., 2001). Null, nonresponsive ARH (–/–) mice do not respond to TCDD and its congeners and produce lipid peroxidation.

15.4.2 Oxidative Stress and Toxicity

The relationship between production of oxidative stress and toxicity by dioxins has been demonstrated in various studies, which have been summarized by Stohs and Hassoun (2012). The administration of antioxidants such as vitamin E, butylated hydroxyanisole (BHA), retinol acetate, selenium, vitamin E succinate, resveratrol, ellagic acid, and a tea melanin are all capable of preventing to varying degrees the dioxin-induced production of ROS, lipid peroxidation, DNA damage, and lethality. BHA and resveratrol appear to be most effective with BHA protecting 100% of the animals from a lethal dose of TCDD so long as it was administered on a daily basis, reflecting the difference in the short half-life of BHA and the long half-life of TCDD.

Induction of hepatic cytochrome P4501A1 (CYP1A1) is a sensitive and commonly used indicator of agonistic activity with respect to the AHR. In recent years, research has identified an increasingly large number of compounds that bind and either activate or inhibit the AHR. These compounds include not only the dioxins and dioxin-like compounds but also pesticides, pharmaceuticals, and natural products that are found in the human diet. Most of these compounds that act as agonists of AHR exhibit potencies as inducers of CYP1A1 that are two to three orders of magnitude less than TCDD. Examples of naturally occurring compounds that bind to AHR and exhibit antagonistic activity include curcumin, diosmin, quercetin, chrysin, resveratrol, naringin, myricetin, apigenin, and tangeretin (Safe et al., 2012). These compounds have been associated with tissue protective properties and may therefore modulate the effects of dioxins and dioxin-like compounds.

In summary, the data indicate that the toxic effects of dioxin-like compounds are at least in part due to an altered redox state, leading to the production of ROS, inflammation, and oxidative stress with subsequent tissue damage. Naturally occurring compounds in the diet such as bioflavonoids may offer protection against the toxic effects of dioxins.

15.5 HUMAN TOXICITY

15.5.1 Acute and Chronic Toxic Effects

There are no reports to show that acute toxic effects occur in humans under normal dietary habits and *normal* dietary intake. Although dioxins are highly lipophilic and accumulate in fat tissues, usual background dietary levels are insufficient to result in accumulation that results in the manifestation of acute toxic or chronic symptoms of toxicity. It is estimated that a several hundredfold higher intake of dioxins would be required to reach a body burden sufficient to produce overt signs of toxicity in normal humans (Parzefall, 2002). Cases of acute toxicity to dioxins have been isolated to a number of industrial accidents resulting in the release of large amounts of mixtures of dioxins in confined areas, exposure to high concentrations of pesticides containing dioxin contaminants, or exposure to oils or waste materials containing high concentrations of dioxins (Schecter et al., 2006; White and Birnbaum, 2009; White et al., 2012).

Due to its high lipid solubility and resistance to metabolism, the half-life of TCDD is approximately 9 years, although there appears to be widespread variability between individuals. The half-life of TCDD in humans is approximately 150 times longer than in rats (Geyer et al., 2002; Schecter et al., 2006). Other dioxins and dioxin-like compounds exhibit half-lives that can be either longer or shorter than TCDD, depending on their lipid solubility and resistance to metabolism (Ogura, 2004; Schecter et al., 2006).

Acute and chronic toxic effects of dioxins have been studied extensively in a variety of animal species, and animals were used to characterize the effects of mixtures of various dioxins (Larsen, 2006). It is generally believed that the effects of dioxins observed in animals can occur in humans under conditions of acute exposure or exposure of sufficiently high amounts over a protracted period of time. It should be noted that the responsiveness of humans to dioxins is about one order of magnitude less than the responsiveness of rats (Charnley and Kimbrough, 2006; Connor and Aylward, 2006), and therefore, direct extrapolations from rats to humans cannot be made.

Accidental exposures have provided information regarding the acute effects of dioxins. Long-term epidemiological studies of these exposed populations have provided information on the chronic effects due to storage in fat and the long half-lives. The most widely recognized and most common characteristics associated with acute dioxin exposure involve chloracne and hyperpigmentation (Geusau et al., 2001; Schecter et al., 2006). High-dose exposure may also result in fatigue, weight loss, pruritus, pain in extremities, gastrointestinal symptoms, and general malaise. These manifestations resolve with time as the dioxins are slowly eliminated from the body, assuming no further exposure (Geusau et al., 2002). Increases in serum levels of various hepatic enzymes are common observations.

The human health effects of dioxins and dioxin-like compounds have been summarized by a number of authors (Mukerjee, 1998; Schecter et al., 2006; White and Birnbaum, 2009; Marinkovic et al., 2010; White et al., 2012). Long-term exposures to dioxins may be associated with disruption of the immune, endocrine, nervous, gastrointestinal, hepatic, pulmonary, cardiovascular, reproductive, and dental systems. This broad spectrum of effects has been clearly demonstrated in various animal models but is less clearly demonstrated in humans (Larsen, 2006). The extent to which these effects may occur in humans depends upon many factors such as the amount of initial exposure, duration of exposure, composition of the dioxin mixture, age at which exposure occurred, body weight and amount of body fat, and individual variability. As a consequence, inconsistent effects have been reported (Mandal, 2005; Larsen, 2006).

Hepatic effects as the result of environmental and occupational exposure include increases in transaminase, dehydrogenase, and alkaline phosphatase enzymes (Geusau et al., 2001). Dioxins are well known as potent inducers of the hepatic cytochrome P450 system, particularly CYP4501A1, which is responsible for the metabolism of many drugs (Whitlock, 1999) and for the production of ROS (Stohs, 1990). In addition, CYP1A1 is involved in the oxidative metabolism of a large number of endogenous substances including estrogens, arachidonic acid and other eicosanoids, bilirubin, and melatonin with resultant physiological consequences (Rifkind, 2006). These metabolic effects may relate to the general toxic manifestations that occur.

Nanogram per kilogram mixtures of dioxins have been shown to produce oxidative stress including lipid peroxidation and DNA damage in rat hepatic and brain tissues when given subchronically (13 weeks) and chronically (30 weeks) (Hassoun et al., 2001, 2002). The induction of oxidative stress was dose dependent based on the use of TEQ values. The levels at which the dioxin mixtures were given were considered to be at the high end of human environmental exposure. Whether such effects occur in humans at comparable levels is not known. It was subsequently shown that administration of ellagic acid and vitamin E succinate significantly modulated in a beneficial manner the dioxin mixture effects on the levels of both antioxidant enzymes and glutathione levels, providing protective effects (Hassoun et al., 2006).

Humblet et al. (2008) have reviewed the association between dioxins and cardiovascular disease (CVD). Ten of the 12 cohorts that were included in the review involved occupational exposure. The authors concluded, based on their systematic review, that dioxin exposure resulted in a significant dose-related increase in ischemic heart disease (IHD) with a more modest association with all CVD mortalities.

Age of exposure to dioxins and dioxin-like compounds appears to be very important with respect to toxicity. Prenatal exposure through breast feeding appears to be correlated with a greater incidence of adverse events than exposure at later years in life. Adverse effects in infants due to dioxins have been demonstrated when maternal exposure was based on occupational, accidental, and background environmental intakes (Popadopoulou et al., 2013; Neugebauer et al., 2015). As previously noted, the dioxin intake of breast-fed infants is one to two orders of magnitude higher than formula fed infants due to the concentration and accumulation of dioxins in fat (Geis et al., 2007).

Various studies have indicated that prenatal and early perinatal exposure results in decreased birth weight (Kishi et al., 2013), as well as neurodevelopmental issues including decreased attention performance and behavioral issues (Ten Tusscher et al., 2014; Neugebauer et al., 2015) and decreased immune function including an increase in respiratory infections and allergies (Ten Tusscher and Koppe, 2004). Furthermore, males are more susceptible to the effects of dioxins than females, with shorter gestation times and a greater number of genital abnormalities (Papadopoulou et al., 2013; Vifeiadi et al., 2013, 2014).

As greater measures are undertaken to decrease production and therefore exposure to dioxins, exposure of infants and children as well as adults has resulted in a decrease in the incidence of adverse events (Wilhelm et al., 2008). Charnley and Kimbrough (2006) concluded that because environmental levels of dioxin-like compounds have continued to decline, it is unlikely that children alive today would experience exposures to levels that would produce adverse health effects. Environmental accidents and production of dioxin-like compounds with the release of large amounts of these toxins have been well curtailed in recent years. However, the potential for industrial accidents with the release of a large amount of dioxins and subsequent contamination of defined areas cannot be ruled out.

Exposure to high amounts of dioxins can have long-lasting consequences. Various studies have shown that exposure to the herbicide Agent Orange in Korea (Yi et al., 2014) and dioxin exposure in association with various industrial accidents (Manuwald et al., 2012; Wang et al., 2013; Onozuka et al., 2014) have resulted in an increase in all-cause mortality with the effect appearing to be more pronounced in men than in women.

15.5.2 Carcinogenicity

The carcinogenicity of dioxins and dioxin-like compounds is well documented in various animal studies, and the mechanisms associated therewith have been determined and are beyond the scope of this chapter. It has been much more difficult to demonstrate positive associations of dioxins with human cancers (Boffetta et al., 2011). In 1997, the International Agency for Research on Cancer (IARC, 1997) concluded that there was limited evidence of carcinogenicity of dioxins in humans. This conclusion was based primarily on mechanistic considerations and animal studies. Subsequently, IARC classified the evidence for the carcinogenicity of dioxins in humans as sufficient based on all cancers combined (Baan et al., 2009). These conclusions have been based on exposure to high levels of dioxins as a result of the use of herbicides contaminated with dioxins and various industrial accidents.

The Institute of Medicine (2006) reviewed the evidence regarding carcinogenicity associated with the use of the herbicide Agent Orange in Vietnam and concluded that there was sufficient evidence to link dioxin exposure with soft-tissue sarcomas, chronic lymphocytic leukemia, Hodgkin's disease, and non-Hodgkin's lymphoma. The Institute of Medicine also concluded that there was *limited or suggestive* evidence of an association of dioxin exposure with cancers of the lung, laryngea, trachea, bronchus, and prostate as well as multiple myeloma.

Whether cancers can be caused by the *normal* environmental background levels of dioxin-like compounds is a matter of conjecture, particularly in light of the fact that these levels have declined in recent years due to better industrial practices and fewer industrial accidents. Statistically, it has been easier to show an increase in *all-cancer* risk as opposed to specific neoplasms (Apostoli et al., 2011; Boffetta et al., 2011).

15.6 DISCUSSION AND CONCLUSIONS

Dioxins and dioxin-like compounds are among the most potent and persistent toxicants known to humans, with virtually all exposure occurring through the food chain. Historically, exposure has occurred primarily through industrial exhausts, incineration of waste products, occupational exposure, exposure to contaminated halogenated organic pesticides, and the consumption of contaminated foliage and feedstocks by animals whose products are ultimately consumed by humans. Lesser amounts are produced by open burning of wastes and forest fires.

Much information regarding the toxicological effects and mechanism of action has been derived from animal studies. It is important to keep in mind that various animal species are at least an order of magnitude more sensitive to dioxins than humans, and as a consequence, it has been difficult to demonstrate many of the toxic effects in humans. Virtually all information regarding human toxicity has been derived from industrial accidents, the heavy use of contaminated chlorinated pesticides, and the contamination of animal feeds with high levels of dioxin-like compounds.

The primary source of human exposure to dioxins by the general population involves the consumption of contaminated animal products. Examples include beef, pork, fish, eggs, high fat milk, and chicken. Due to the high lipid solubility of dioxins and dioxin-like compounds and their long half-lives, animal lipids are the prime source of exposure for the general public. Lesser amounts are consumed through vegetables and fruits as well as exposure to contaminated soil. With respect to bodies of water that have been contaminated with dioxins, accumulation usually begins with zooplankton that consume sediments that contain dioxins, and these organisms are consumed by a larger sea life with larger species ultimately being eaten by humans. Thus, the dioxins move up the food chain.

Because much information has been acquired over the past 40 years regarding sources, mechanisms, and the toxicity of dioxins, in most countries, measures have been undertaken to greatly limit production, release, and exposure. As a consequence, levels in ambient air, soils, and human blood levels have significantly declined over the past 20–30 years. As a consequence, there is now little threat of widespread poisonings. The greatest threat of dioxin and dioxin-like compound exposure resides with isolated industrial accidents or human exposure through the inadvertent distribution of contaminated food products arising from feeding dioxin-tainted feeds to animals.

In summary, a global health threat has been greatly curtailed in recent years by a concerted effort involving many governmental, industrial, and educational agencies as well as concerned citizens and foundations. Due to the long half-lives and high lipid solubility, the long-term effects of human exposure to dioxin-like compounds remain to be fully elucidated. Nevertheless, continued vigilance is essential.

REFERENCES

Alsharif, N.Z., Lawson, T., Stohs, S.J. 1994. Oxidative stress induced by 2,3,7,8-tetrachlorodibenzo-*p*-diopxin is mediated by the aryl hydrocarbon (ah) receptor complex. *Toxicology* 92: 39–51.

Apostoli, P., Bergonazi, R., Catalani, S. 2011. Classification as carcinogenic for 2,3,7,8-tetrachloro-para-dioxin: An eventful journey. *G. Ital. Med. Lav. Ergon.* 33: 84–99.

Baan, R., Grosse, Y., Straif, K., Secretan, B., El Ghissassi, F., Bouvard, V., Benbrahim-Tallaa et al. 2009. A review of human carcinogens—Part F: Chemical agents and related occupations. *Lancet Oncol.* 10: 1143–1144.

Boffetta, P., Mundt, K.A., Adami, H.O., Cole, P., Mandel, J.S. 2011. TCDD and cancer: A critical review of epidemiological studies. *Crit. Rev. Toxicol.* 41: 622–636.

Bruckmann, P., Hiester, E., Klees, M., Zetsch, C. 2013. Trends in PCDD/F and PCB concentrations and depositions in ambient air in Northwestern Germany. *Chemosphere* 93: 1471–1478.

Buchert, A., Cederberg, T., Dyke, P., Fiedler, H., Furst, P., Hanberg, A., Hossienpour, J. et al. 2001. Dioxin contamination in food. Bayreuth, Germany, from September 28 to October 1, 2000. *Environ. Sci. Pollut. Res. Int.* 8: 84–88.

Charnley, G., Kimbrough, R.D. 2006. Overview of exposure, toxicity, and risks to children from current levels of 2,3,7,8-tetrachlorodibenzo-*p*-dioxin and related compounds in the USA. *Food Chem. Toxicol.* 44: 601–615.

Connor, K.T., Ayward, L.L. 2006. Human responses to dioxin: Aryl hydrocarbon receptor (AHR) molecular structure, function, and dose-response data for enzyme induction indicate an impaired human AHR. *J. Toxicol. Environ. Health B Crit. Rev.* 9: 147–171.

Consonni, D., Sindaco, R., Bertazzi, P.A. 2012. Blood levels of dioxins, furans, dioxin-like PCBs, and TEQs in general populations: A review, 1989–2010. *Environ. Int.* 44: 151–162.

Covaci, A., Voorspoels, S., Schepens, P., Jorens, P., Blust, R., Neels, H. 2008. The Belgian PCB/dioxin crisis—8 years later: An overview. *Environ. Toxicol. Pharmaccol.* 25: 164–170.

Fuster, G., Schuhmacher, M., Domingo, JL. 2002. Human exposure to dioxins and furans: Application of the substance flow analysis to health risk assessments. *Environ. Sci. Pollut. Res. Int.* 9: 241–249.

Geis, A., Neumeier, G., Rappolder, M., Konietzka, R. 2007. Risk assessment of dioxins and dioxin-like PCBs in food—Comments by the German Federal Environmental Agency. *Chemosphere* 67: S344–S349.

Geusau, A., Abraham, K., Sator, M.O., Stingl, G., Tschachler, E. 2001. Severe 2,3,7,8-tetrachlorodibenzo-*p*-dioxin intoxication: Clinical and laboratory effects. *Environ. Health Perspect.* 1209: 865–869.

Geusau, A., Schmaldienst, S., Derfler, K., Papke, O., Abraham, K. 2002. Severe 2,3,7,8-tetrachlorodibenzo-*p*-dioxin (TCDD) intoxication: Kinetics and trials to enhance elimination in two patients. *Arch. Toxicol.* 76: 316–325.

Geyer, H.J., Schramm, K.W., Feicht, E.A., Behechti, A., Steinberg, C., Bruggemann, R., Poiger, H., Henkelmann, B., Kettrup, A. 2002. Half-lives of tetra-, penta-, hexa-, hepta-, and octachlorodibenzo-*p*-dioxin in rats, monkeys, and humans—A critical review. *Chemosphere* 48: 631–644.

Hassoun, E.A., Li, F., Abushaban, A., Stohs, S.J. 2001. Production of superoxide anion, lipid peroxidation, and DNA damage in the hepatic and brain tissues of rats after subchronic exposure to mixtures of TCDD and its congeners. *J. Appl. Toxicol.* 21: 211–219.

Hassoun, E.A., Vodhanel, J., Holden, B., Abushaban, A. 2006. The effects of ellagic acid and vitamin E succinate on antioxidant enzymes activities and glutathione levels in different brain regions of rats after subchronic exposure to TCDD. *J. Toxicol. Environ. Health A* 69: 381–393.

Hassoun, E.A., Wang, HJ., Abushaban, A., Stohs, S.J. 2002. Induction of oxidative stress in the tissues of rats after chronic exposure to TCDD, 2,3,4,7,8-pentacholrodibenzofuran, and 3,3′,4,4′,5-pentacholorphenol. *J. Toxicol. Environ. Health A* 65: 825–842.

Hoang, T.T., Traag, W.A., Murk, A.J., Hoogenboom, R.L. 2014. Levels of polychlorinated dibenzo-p-dioxins, dibenzofurans (PCDD/Fs) and dioxin-like PCBs in free range eggs from Vietnam, including potential health risks. *Chemosphere* 114: 268–274.

Hsu, M.S., Hsu, K.Y., Wang, S.M., Chou, U., Huang, N.C., Liao, C.Y., Yu, T.P., Ling, Y.C. 2007. A total diet study to estimate PCDD/Fs and dioxin-like PCBs intake from food of Taiwan. *Chemosphere* 676: S65–S70.

Humblet, O., Birnbaum, L., Rimm, E., Mittleman, M.A., Hauser, R. 2008. Dioxins and cardiovascular disease mortality. *Environ. Health Perspect.* 116: 1443–1448.

Institute of Medicine, Committee to Review the Effects in Vietnam Veterans of Exposure to Herbicides. 2006. *Veteran and Agent Orange: Update.* National Academy Press, Washington, DC.

International Agency for Research on Cancer. 1997. Polychlorinated dibenzo-para-dioxins. *IARC Monogr. Eval. Carcinogenic Risks Humans* 69: 33–343.

Jacob, D.J., Lok, C., Morley, K., Powell, D.A. 2011. Government management of two media-facilitated crises involving dioxin contaminated food. *Public Underst. Sci.* 20: 261–269.

Kishi, R., Kobayashi, S., Ikeno, T., Araki, A., Miyahita, C., Itoh, S., Saski, S. et al. 2013. Ten years of progress in the Hokkaido birth cohort study on environment and children's health: Updated 2013. *Environ. Health Prev. Med.* 18: 429–450.

Larsen, J.C. 2006. Risk assessments of polychlorinated dibenzo-*p*-dioxins, polychlorinated dibenzofurans, and dioxin-like polychlorinated biphenyls in food. *Mol. Nutr. Food Res.* 50: 885–896.

Malisch, R., Kotz, A. 2014. Dioxins and PCBs in feed and food—Review from European perspective. *Sci. Total Environ.* 491–492: 2–10.

Mandal, P.K. 2005. Dioxin: A review of its environmental effects and its aryl hydrocarbon receptor biology. *J. Comp. Physiol. B* 175: 221–230.

Manuwald, U., Velasco Garrido, M., Berger, J., Manz, A., Baur, X. 2012. Mortality study of chemical workers exposed to dioxins: Follow-up 23 years after chemical plant closure. *Occup. Environ. Med.* 69: 636–642.

Marinkovic, N., Pasalic, D., Ferencak, G., Grskovic, B., Savljenic Rukavina, A. 2010. Dioxins and human toxicity. *Arh. Hig. Rada Toksikol.* 61: 445–453.

Marnane, I. 2012. Comprehensive environmental review following the pork PCB/dioxin contamination incident in Ireland. *J. Environ. Monit.* 14: 2551–2556.

Mukerjee, D. 1998. Health impact of polychlorinated dibenzo-*p*-dioxins: A critical review. *J. Air Waste Manag. Assoc.* 48: 157–165.

Neugebauer, J., Wittsiepe, J., Kasper-Sonnenberg, M., Schoneck, N., Scholmerich, A., Wilhelm, M. 2015. The influence of low pre- and perinatal exposure to PCDD/Fs, PCBs, and lead on attention performance and attention-related behavior among German school-aged children: Results from the Duisburg birth cohort. *Int. J. Hyg. Environ. Health* 218: 153–162.

Ogura, I. 2004. Half-life of each dioxin and PCB congener in the human body. *Organohalogen Comp.* 66: 3329–3337.

Onozuka, D., Hirata, T., Furue, M. 2014. Net survival after exposure to polychlorinated biphenyls and dioxins: The Yusho study. *Environ. Int.* 73: 28–32.

Papadopoulou, E., Vafeiadi, M., Agramunt, S., Mathianaki, K., Karakosta, P., Spanaki, A., Besselink, H. et al. 2013. Maternal diet, prenatal exposure to dioxins and other persistent organic pollutants and anogenital distance in children. *Sci. Total Environ.* 461–462: 222–229.

Parzefall, W. 2002. Risk assessment of dioxin contamination in human food. *Food Chem. Tox.* 40: 1185–1189.

Passuello, A., Mari, M., Nadal, M., Schuhmacher, M., Domingo, J.L. 2010. POP accumulation in the food chain: Integrated risk model for sewage sludge application in agricultural soils. *Environ. Int.* 36: 577–583.

Peltonen, H., Ruokojarvi, P., Korhonen, M., Kiviranta, H., Flinkman, J., Verta, M. 2014. PCDD/Fs, PCBs and PBDEs in zooplankton in the Baltic Sea—Spatial and temporal shifts in the congener-specific concentrations. *Chemosphere* 114: 172–180.

Pohjanvirta, R. 2012. *The AH Receptor in Biology and Toxicology.* Hoboken, NJ: John Wiley & Sons.

Rideout, K., Teschke, K. 2004. Potential for increased human foodborne exposure to PCDD/F when recycling sewage sludge on agricultural land. *Environ. Health Perspect.* 112: 959–969.

Rifkind, A.B. 2006. CYP1A in TCDD toxicity and in physiology with particular reference to CYP-dependent arachidonic acid metabolism and other endogenous substrates. *Drug Metab. Rev.* 38: 291–335.

Roeder, R.A., Garber, M.J., Schelling, G.T., 1998. Assessment of dioxins in foods from animal origins. *J. Anim. Sci.* 76: 142–151.

Rychen, G., Jurjanz, S., Fournier, A., Toussaint, H., Feidt, C. 2014. Exposure of ruminants to persistent organic pollutants and potential of decontamination. *Environ. Sci. Pollut. Res. Int.* 21: 6440–6447.

Safe, S., Chadalapaka, G., Jutooru, I. 2012. AHR-active compounds in the human diet. In R. Pohjanvirta (ed.), *The AH Receptor in Biology and Toxicology*, Chapter 24. Hoboken, NJ: John Wiley & Sons, pp. 2331–2342.

Schecter, A., Birnbaum, L., Ryan, J.R., Constable, J.D. 2006. Dioxins: An overview. *Environ. Res.* 101: 419–428.

Squadrone, S., Brizio, P., Nespoli, R., Stella, C., Abete, M.C. 2015. Human dietary exposures and levels of polychlorinated dibenzo-p-dioxins (PCDDs), polychlorinated dibenzofurans (PCDFs), dioxin-like polychlorinated biphenyls (PCBs) and non-dioxin-like polychlorinated biphenyls (BDL-PCBs) in free-range eggs close to a secondary aluminum smelter, Northern Italy. *Environ. Pollut.* 206: 429–436.

Stohs, S.J. Oxidative stress induced by 2,3,7,8-tetrachlorodibenzo-p-dioxin (TCDD). *Free Radic. Biol. Med.* 9, 79–90 (1990).

Stohs, S.J., Hassoun, E.A. 2012. Dioxin-activated AHR: Toxic responses and the induction of oxidative stress. In R. Pohjanvirta (ed.), *The AH Receptor in Biology and Toxicology*, Chapter 15. Hoboken, NJ: John Wiley & Sons, pp. 229–244.

Sugihara, K., Kitamura, S., Ohta, S., Yamashita, K., Yasuda, M., Fujii-Kuriyama, Y. 2001. Aryl hydrocarbon receptor (AHR)-mediated induction of xanthine oxidase/xanthine dehydrogenase activity by 2,3,7,8-tetrachlorodibenzo-*p*-dioxin. *Biochem. Biophys. Res. Commun.* 281: 1093–1099.

Ten Tusscher, G.W., Koppe, J.G. 2004. Perinatal dioxin exposure and later effects—A review. *Chemosphere* 54: 1329–1336.

Ten Tusscher, G.W., Leijs, M.M., de Boer, L.C., Legler, J., Olie, K., Sekreijse, H., van Dijk, B.W. et al. 2014. Neurodevelopmental retardation, assessed clinically with magnetoencephalography and electroencephalography, associated with perinatal dioxin exposure. *Sci. Total Environ.* 491–492: 235–239.

Torres, P.J., Leite, C., Krauss, T., Weber, R. 2013. Landfill mining from a deposit of the chlorine/organochlorine industry as source of dioxin contamination of animal feed and assessment of the responsible processes. *Environ. Sci. Pollut. Res. Int.* 20: 1958–1965.

Traag, W.A., Kan, C.A., van der Weg, G., Onstenk, C., Hoogenboom, L.A. 2006. Residues of dioxins (PCDD/Fs and PCBs in eggs, fat and livers of laying hens following consumption of contaminated feed. *Chemosphere* 65: 1518–1525.

Tuomisto, J. 2012. The toxic equivalency principle and its applications in dioxin risk assessment. In R. Pohjanvirta (ed.), *The AH Receptor in Biology and Toxicology*, Chapter 23. Hoboken, NJ: John Wiley & Sons, pp. 317–330.

Turrio-Baldassarri, L., Abate, V., Alivernini, S., Battistelli, C.L., Carasi, S., Casella, M., Iacovella, N. et al. 2007. A study of PCB, PCDD/PCDF industrial contamination in a mixed urban-agricultural area significantly affected food chain and the human exposure. Part 1: Soil and feed. *Chemosphere* 67: 1822–1830.

Turrio-Baldassarri, L., Alivernini, S., Carasi, S., Casellia, M., Fuselli, S., Iacovella, N., Iamiceli, A.L., La Rocca, C., Scarcella, C., Battistelli, C.L. 2009. PCB, PCDD and PCDF contamination of food of animal origin as the effect of soil pollution and the cause of human exposure in Brescia. *Chemosphere* 76: 278–285.

Van den Berg, M., Birnbaum, L.S., Denison, M., de Vito, M., Farland, W., Feeley, M., Fiedler, H. et al. 2006. The 2005 World health organization re-evaluation of human and mammalian toxic equivalency factors for dioxins and dioxin-like compounds. *Toxicol. Sci.* 93: 223–241.

Vifeiadi, M., Agramunt, S., Papadopoulou, E., Besselink, H., Mathianaki, K., Karakosta, P., Spanaki, A. et al. 2013. In utero exposure to dioxins and dioxin-like compounds and anogenital distance in newborns and infants. *Environ. Health Perspect.* 121: 125–130.

Vifeiadi, M., Agramunt, S., Pedersen, M., Besselink, H., Chatzi, L., Fthenou, E., Fleming, S. et al. 2014. In utero exposure to compounds with dioxin-like activity and birth outcomes. *Epidemiology* 25: 215–224.

Wang, L., Weng, S., Wen, S., Shi, T., Sun, G., Zeng, Y., Qi, C., Chen, W. 2013. Polychlorinated dibenzo-*p*-dioxins and dibenzofurans and their association with cancer mortality among workers in one automobile foundry factory. *Sci. Total Environ.* 433: 104–111.

Wang, X., Zhang, H., Zhang, L., Shang, X., Tong, Z., Yu, X., Wu, Y. 2015. Assessment of dioxin-like compounds intake from various marine fish from Zhoushan fishery, China. *Chemosphere* 118: 163–169.

White, S.S., Birnbaum, L.S. 2009. An overview of the effects of dioxins and dioxin-like compounds on vertebrates, as documented in human and ecological epidemiology. *J. Environ. Sci. Health C Environ. Carcinog. Ecotoxicol. Rev.* 27: 197–211.

White, S.S., Fenton, S.E., Birnbaum, L.S. 2012. Adverse health outcomes caused by dioxin-activated AHR in humans. In R. Pohjanvirta (ed.), *The AH Receptor in Biology and Toxicology*, Chapter 22. Hoboken, NJ: John Wiley & Sons, pp. 307–316.

Whitlock, P.J. Jr. 1999 Induction of cytochrome P4501A1. *Ann. Rev. Pharmacol. Toxicol.* 39: 103–125.

Wilhelm, M., Wittsiepe, J., Lemm, F., Ranft, U., Kramer, U., Furst, P., Roseler, S.C. et al. 2008. The Duisburg birth cohort study: Influence of the prenatal exposure to PCDD/Fs and dioxin-like PCBs on thyroid hormone status in newborns and neurodevelopment of infants until the age of 24 months. *Mutat. Res.* 659: 83–92.

Yi, S.W., Ryu, S.Y., Ohrr, H., Hong, J.S. 2014. Agent orange exposure and risk of death in Korean Vietnam veterans: Korean veterans' health study. *Int. J. Epidemiol.* 43: 1825–1834.

Zabik, M.E., Zabik, M.J. 1999. Polychlorinated biphenyls, polybrominated biphenyls, and dioxin reduction during processing/cooking food. *Adv. Exp. Med. Biol.* 459: 213–231.

Zheng, G.J., Leung, A.O., Jiao, L. P., Wong, M.H. 2008. Polychlorinated dibenzo-*p*-dioxins and dibenzofurans pollution in China: Sources, environmental levels and potential human health impacts. *Environ. Int.* 34: 1050–1061.

16 A Review on Mercury Toxicity in Food

Alessia Carocci, Alessia Catalano, Graziantonio Lauria, Maria Stefania Sinicropi, and Giuseppe Genchi

CONTENTS

16.1 INTRODUCTION

The World Health Organization has recognized mercury among the top 10 chemicals of *major public health concern* (World Health Organization 2013). Extensive anthropogenic emission and the use of Hg have caused worldwide mercury contamination in many aquatic and terrestrial ecosystems (Sunderland and Mason 2007; Selin et al. 2010).

After being introduced into the environment, mercury undergoes a series of complicated transport and transformation processes during its biogeochemical cycle that is closely associated with chemical forms of mercury present in different phases of the environment (Liu et al. 2012). A principal form of mercury to which general populations are exposed is methylmercury. The major adverse effects of methylmercury exposure in human beings are neurological effects (WHO's Methylmercury-Environmental Health Criteria 101 1990; Harada 1995; Clarkson et al. 2003; Carocci et al. 2014) followed by renal (Li et al. 2010b), cardiovascular (Chan and Egeland 2004), reproductive (Itai et al. 2004), and immune (Gardner et al. 2010) system effects.

Ingestion of contaminated food represents one of the major pathways of human mercury exposure; thus, the issue of mercury in food, particularly the consumption of contaminated fish such as tuna, swordfish, shark, and marine mammals such as whales, has been receiving a considerable attention (Clemens et al. 2012). The accumulation of heavy metals in fish can occur due to metals that are naturally present in the aquatic environment, but it can also be exacerbated by anthropogenic activities such as industrial activity and pollution (Järup 2003).

The total mercury (THg) content of fish can consist of a combination of several mercury species: methylmercury, ethylmercury, and inorganic mercury (Morel et al. 1998). Human exposure to chemical contaminants can be characterized by examining biomarkers. Total Hg in hair (THHg) and total Hg in blood (TBHg) are both validated biomarkers of MeHg intake correlated with seafood consumption in general human populations (Berglund et al. 2005). Blood and hair Hg concentrations (Clarkson and Magos 2006) are considered as good biomarkers of MeHg exposure, while with respect to prenatal exposure, both cord blood and maternal hair have been used (Cernichiari et al. 1995). Generally, Hg in hair is 200–300 times more concentrated than that in blood (Miklavcic et al. 2011). Hair protein (α-keratin) is very rich in sulfhydryl groups (—SH), and mercury and methylmercury compounds have high tendency to bind to sulfur. However, in many circumstances, the benefits of fish consumption, which lowers the risk of suboptimal neurodevelopment in infants, children and adolescents have been evaluated with respect to fish people who don't eat fish.

Many authors state that the benefits of fish consumption generally outweigh the negative effects associated with it, for example, methylmercury exposure from eating top predatory species with high mercury levels in their tissues (Mozaffarian and Rimm 2006). In general, health authors encourage fish and other seafood as a part of a balanced diet since seafood is an important source of nutrients, including proteins and omega-3 polyunsaturated fatty acids (PUFAs). To prevent elevated exposure to methylmercury and other toxic contaminants, avoidance and restricted consumption of certain fish species (tuna, swordfish, shark, pike, brown trout, and arctic char) are recommended. In addition, to protect our health from the risk induced by mercury, it is best to limit the consumption of fish to 2–3 serving per week, by varying the choice and preference to fish of smaller sizes with lowest levels of mercury in the tissues.

Actually, there are currently global efforts underway to establish food consumption guidelines in order to reduce the risk of mercury exposure and institute mercury management practices in reducing mercury emissions and to promote the benefits humans get of fish consumption. Even though fish ingestion represents the primary human exposure pathway for methylmercury, it has been found recently that methylmercury exposure also occurs through rice ingestion (Sakamoto et al. 2007; Li et al. 2010a). This exposure can pose a variety of health risks, mainly neurological damages, especially in fetal brain.

In some villages in China, in mercury mining regions, rice ingestion is an important methylmercury exposure pathway (Rothenberg et al. 2013). Rice is a staple food for more than half of the world's population, mostly in developing countries. But in any case, we have not forgotten that, compared with fish and sea mammals that are highly nutritious food, rice is of poor nutritional quality and lacks specific beneficial nutrients such as omega-3 PUFAs, selenium, and essential amino acids. Other sources of methylmercury are exposure to fossil fuel emission and various commercial products like creams, germicidal soaps, teething powder, analgesics, vaccinations, batteries, and incandescent lights (Goldman and Shannon 2001). The current sources of human exposure to elemental mercury include dental amalgams, barometers, thermometers, and sphygmomanometers (Guzzi et al. 2006). In addition, another source of ethyl mercury is thimerosal (sodium ethyl mercury thiosalicylate), which is used as a topical antiseptic and preservative in vaccines (Dórea et al. 2011).

16.2 CHEMICAL FORMS OF MERCURY

Mercury (hydrargyrum) is a heavy metal (having an atomic weight of 200.59, density 13.59 g/cm^3, melting point –39°C, and boiling point 359°C) of well-known toxicity. It is responsible for the disasters that occurred in two Japanese villages, near Minamata Bay in 1953 and along the Agano River in Niigata in 1967 (Tsubaki and Irukajama 1997), and in Iraq in 1971–1972 (Bakir et al. 1973; Skerfving and Copplestone 1976). Mercury is an extremely rare element in the Earth's crust usually in the form of mineral cinnabar (mercury sulfide, HgS), having an average mass abundance of only 0.09 mg/kg (Sinicropi et al. 2010). It exists in several forms: inorganic mercury, which includes liquid metallic mercury and mercury vapor (Hg^0) and mercurous (Hg_2^{++}) and mercuric (Hg^{++}) salts, and organic mercury, with methylmercury (CH_3Hg, MeHg), ethylmercury (C_2H_5Hg, EtHg), and phenylmercury (C_6H_5Hg, PhHg).

In the atmosphere, elemental mercury (Hg^0) constitutes the majority of Hg (>90%) and is the predominant form in the gaseous phase, which facilitates the long-range transport of Hg at a global scale (Pirrone and Mahaffey 2005). It is derived from volcanic and geothermal outgassing activities and from Earth's movements. It is released by chlor-alkali plants (Biester et al. 2002) and coal-fired power plants (Novoa-Munoz et al. 2008) and then settles in oceans, rivers, and lakes.

In aquatic and soil environments, mercury is primarily present in its mercuric form, including inorganic (e.g., mercuric hydroxide) and organic mercuric compounds, and secondarily as Hg^0 (Ullrich et al. 2001; Fitzgerald et al. 2007). Although only a minor fraction of total mercury is present as MeHg (typically less than 10% and 3% in water and soil/sediment, respectively), the formation of this compound is an important step in mercury cycling (Ullrich et al. 2001). This is because MeHg can be bioaccumulated along the food web and reach high concentrations in organisms, particularly in aquatic environments. MeHg results from the biomethylation of inorganic mercury primarily by sulfate-reducing bacteria present in organisms living in aqueous environment and phytoplankton in the ocean (Morel et al. 1998).

In fishes and wildlife that prey on fish, MeHg can be the dominant form of mercury species owing to bioaccumulation and biomagnification (Fitzgerald et al. 2007). MeHg may also result from the methylation of inorganic mercury by microorganisms in the mouth, when mercury vapor is released from amalgam dental filling (Guzzi et al. 2006), and by nonenzymatic methylation, when vitamin B12 in the form of methylcobalamin donates a methyl group to mercury (Bertilsson and Neujahr 1971). The different mercury forms are interconvertible *in vivo*; for example, inhaled elemental mercury vapor is absorbed through the mucous membrane of the lungs and rapidly oxidized to other forms. MeHg is easily absorbed through the gut and deposits in many tissues and crosses the blood–brain barrier. Inorganic mercury is absorbed through the intestines; it also crosses the blood–brain barrier, even though it is less toxic than MeHg.

16.3 MERCURY TOXICITY

Mercury in all forms poisons cellular function by altering the tertiary and quaternary structures of proteins and by binding with sulfhydryl and selenohydryl groups; thus, it can potentially impair the function of any organ or any subcellular structure. Human toxicity varies with the form of mercury, the dose, and the rate of exposure (Bernhoft 2012).

The chief target organ for inhaled Hg^0 vapor is the brain, but peripheral nerve function, renal function, immune function, endocrine and muscle function, and several types of dermatitis have also been described. In the inorganic form, mercury is absorbed from the gastrointestinal tract and acts to produce inflammatory reactions in the kidneys and gastrointestinal apparatus, with abdominal pain, vomiting, and diarrhea. In the cardiovascular system, acute inorganic mercury exposition *in vivo* promotes the reduction of myocardial force development (Azevedo et al. 2012). Chronic poisoning with mercury salts is rare, usually also involving concomitant occupational exposure to mercury vapor. Inorganic mercury accumulates in the human breast and is secreted in breast milk, which can damage the developing infant's CNS and pulmonary and nephrotic systems (Counter and Buchanan 2004). Inorganic mercury exposure can also induce Kawasaki disease (Mutter and Yeter 2008), which results from the impairment of the immune system. As regards organic mercury compounds, MeHg is the most toxic form of mercury and its principal target tissues in humans are those in the central nervous system. The rate at which MeHg is absorbed in the gastrointestinal tract is high (about 90%–95%), which contributes to its high toxicity and degree of bioavailability (Nielsen and Andersen 1992). Once organic mercury compounds enter the body, they are rapidly distributed to tissues, mainly those in the liver (approximately 50%), CNS, and kidneys. The methylmercury easily crosses the blood–placenta barrier and accumulates more in the fetus than in the mother. In addition, umbilical cord blood levels of mercury tend to be higher than maternal blood levels of mercury (Vather et al. 2000). Mercury poisoning can result in mental retardation, neurological deficits, blindness, loss of hearing, dysarthria, developmental deficits, and death (Guzzi and La Porta 2008).

Mercury and methylmercury induce mitochondrial dysfunction, which reduces ATP synthesis and increases lipid, protein, and DNA peroxidation. Breakage of mitochondria, increased lipid peroxidation, and oxidation of proteins and nucleic acids in the brain appear to be important factors in conditions such as amyotrophic lateral sclerosis and Alzheimer's disease (Qiu et al. 2007; Johnson and Atchison 2009; Bento-Abreu et al. 2010; Carocci et al. 2014). There are good evidences linking mercury with hemolytic anemia and aplastic anemia, because mercury is thought to compete with iron for binding to hemoglobin, which results in impaired hemoglobin formation (Pyszel et al. 2005).

16.4 FOOD SOURCES OF MERCURY HUMAN EXPOSURE

Human mercury exposure particularly occurs through the ingestion of methyl, dimethyl, and ethylmercury from contaminated seafood. The consumption of shellfish and fish is by far the most significant source of ingestion-related mercury exposure in humans and animals (Carvalho et al. 2005; Chen et al. 2012). Hg and MeHg are present in very small concentrations in oceans, lakes, and rivers. However, at the start of the food chain, they are absorbed, usually as methylmercury, by algae, which in turn are then eaten by fishes and other organisms at higher levels of the food chain. Fishes efficiently absorb methylmercury that is accumulated in the muscle tissue. This entails the bioaccumulation of mercury in the adipose tissues of the successive trophic levels: zooplankton, small nekton, larger fish, and so on in the food chain.

It is well established that the majority of MeHg exposure in humans originates from eating fish and other types of seafood (Mergler et al. 2007). More than 90% of total Hg in the fish muscle is present as MeHg; this is found in the highest concentrations in predatory fish such as swordfish, shark, and tuna. Concentrations of methylmercury in aquatic species depend on the levels of environmental contamination and on the predatory nature and life span of the species. A positive relationship between fish's size/age and THg and MeHg concentrations has been found for numerous fish species investigated. Larger, longer-living predators (e.g., swordfish, shark) have higher tissue mercury concentrations, while smaller or shorter-lived species (e.g., shellfish, salmon) have very low mercury concentrations (Mozaffarian and Rimm 2006).

Species on the food chain can amass body concentrations of mercury up to ten times higher than the species they consume. MeHg is readily bioavailable and biomagnifies in the food chain so that fish, at higher trophic levels, regularly have Hg concentrations about 10^5–10^6 fold higher than the concentration in the water in which they live (Stein et al. 1996). Even though MeHg contamination of fish is a major public health concern, many people worldwide rely on fish as their major protein source, and fish consumption is still encouraged among the general population because of the health benefits associated with it, for example, fish is rich source of omega-3 PUFAs and essential amino acids. Fish consumption is to be considered the primary pathway of methylmercury exposure in the world for those people who live along the coast.

However, in the inland regions of China, most of the residents eat little amount of fish, but they live in areas where a significant amount of mercury is present in the environment. Recent studies have reported high levels of MeHg in rice from Hg-polluted areas (Li et al. 2010a). A research by Barrett (2010) from China demonstrates that rice, a staple food for a billion people, can be a primary source of methylmercury in areas where there is substantial inorganic mercury pollution. Zhang et al. (2010) performed a study in Guizhou Province, known as "mercury capital" of China, a typical environment in inland China with severe Hg pollution essentially due to Hg mining and smelting. They found that rice consumption is by far the predominant pathway of MeHg exposure to the general population of Guizhou. In a study of 98 persons from the Wanshan Hg mining area, hair MeHg levels indicated MeHg exposure. Rice, the staple food of the local inhabitants, also showed high total Hg (THg) and MeHg levels. Pork meat, vegetable, and drinking water samples collected in Wanshan Hg mining area contained highly elevated THg, but very low levels of MeHg.

The relationships between the estimated rice MeHg intake and hair MeHg levels ($r = 0.65$, $p < 0.001$) confirmed that rice with high MeHg levels indeed was the main route of MeHg exposure for the local residents in the Wanshan Hg mining area (Feng et al. 2008). Rothenberg and coauthors (2013) have quantified in China total mercury (THg) and methylmercury (MeHg) in rice grain grown in three sites using water-saving rice cultivation methods and in one Hg-contaminated site, where rice was grown under flooded conditions. The study suggested that a portion of Hg species in rice grain was derived from the soil and translocation of Hg species from soil to rice grain was independent of irrigation practices and Hg levels, although other factors can also be important.

In another study (Meng et al. 2014), 155 rice plants were collected from 10 mining areas in three provinces of China (Hunan, Guizhou, and Guangdong), where most of mercury mining takes place. Although the degree of Hg contamination varied significantly among different mining areas, rice seed showed the highest ability for methylmercury (MeHg) accumulation. Both concentrations of total mercury (THg) and MeHg in rice plants were significantly correlated with Hg levels in soil, indicating that soil is still an important source for both inorganic mercury (IHg) and MeHg in rice plants. The obvious discrepancy between the distribution patterns of THg and MeHg reflected different pathways of inorganic Hg and MeHg accumulation. Water-soluble Hg plays an important role in MeHg accumulation in rice plants.

Qiu et al. (2008) report that methylmercury in rice (*Oryza sativa* L.), grown in abandoned mercury mining areas, contained mercury levels in its edible portion 10–100 times higher than other crop plants. The results demonstrate that rice is a MeHg bioaccumulative plant and the main MeHg source for human exposure in the study areas. The elevated MeHg level in rice may be likely linked with its high methylation ability in paddy soils. Seasonal irrigation of paddy field produces anaerobic conditions in superficial soil, which favors *in situ* methylation of inorganic Hg (Gilmour and Henry 1991; Porvari and Verta 1995). However, the mechanism of the uptake of MeHg in rice tissue is still unclear. Rothenberg et al. (2013) demonstrated that more alkaline conditions at the highly contaminated site possibly reduced the bioavailability of Hg species, preventing MeHg uptake in rice grain.

16.5 EFFECTS OF NUTRIENTS ON METHYLMERCURY TOXICITY

Population is primarily exposed to MeHg through consumption of contaminated fish and marine mammals; however, recent studies have reported high levels of MeHg in rice and confirmed that in China, the main human exposure to MeHg is related to frequent rice consumption in mercury-polluted areas. Mercury can counteract the positive effects of omega-3 PUFAs in fish (Khan and Wang 2009). Compared with fish, rice is of poor nutritional quality and lacks specific nutrients identified as having health benefits (e.g., selenium, *n*-3 LCPUFAs, and essential amino acids); however, the effects of these nutrients on the toxicity of MeHg deserve special attention (Li et al. 2010b).

In contrast to mercury, selenium is an essential trace element, and it is known to play a protective role on genetic damage and possibly on cancer. As constituent of selenoproteins, it has structural and enzymatic roles, and it is known as an antioxidant and catalyst for the production of active thyroid hormones (Rayman 2000). Selenium is also reported to ameliorate the toxic effects of methylmercury (Berry and Ralston 2008). MeHg toxicity may be due in part to impaired selenium-dependent enzyme synthesis or activity because MeHg covalently binds selenium in the active sites of selenoenzymes (particularly glutathione peroxidase and thioredoxin reductase), thereby inhibiting their activity (Seppanen et al. 2004). This counteracting effect may explain oxidative damage attributable to methylmercury.

Methylmercury in fish and marine mammals tissues is bound to the —SH and —SeH groups of the cysteine and selenocysteine residues in proteins (Harris et al. 2003; Carocci et al. 2014), and this compound is not destroyed by any cooking or cleaning process. MeHg is a "soft" Lewis acid and therefore has strong thermodynamic affinity for thiolic groups (—SH) that are "soft" Lewis bases (Amirbahman et al. 2002). Thiolic groups are present in cysteine, tripeptide glutathione, and cysteine residues of proteins and enzymes and are known to play important structural and functional roles in the chemistry of proteins, because they are located within the active sites of many enzymes and directly involved in catalysis. With their Hg and MeHg complexes, these thiolic groups are the primary sites of Hg binding in biological systems. Since selenols (—SeH, in selenocysteine) chemically resemble thiols (—SH) and can readily replace thiols in cysteines, Hg and MeHg complexes with selenol-containing biomolecules widely occur in biological systems. However, there is one major difference, considering the lower pK_a value and the stronger nucleophilicity of selenol (pK_a 5.7) with respect to thiol (pK_a 8.4): selenols are in general more powerful nucleophiles than thiols and more reactive (Khan et al. 2009; Carocci et al. 2014).

Therefore, Hg and MeHg complexes with selenols tend to be thermodynamically even more stable. Selenoproteins play important roles in protecting against Hg toxicity; in fact, they may bind this metal through their highly reactive selenol group, and their antioxidative properties help to eliminate the reactive oxygen species induced *in vivo* by Hg. Chen et al. (2006) found that Hg exposure increased both selenoprotein and glutathione peroxidase (GSH-Px) in Hg miners compared with the control group. The protective effect of Se against MeHg does not appear to involve Hg absorption in the intestine or excretion of Hg in the urine or feces; it also does not appear to affect the rate of demethylation of MeHg, but Se modified the form of MeHg, altering its distribution by freeing MeHg from blood proteins (Chapman and Chan 2000).

Selenium is known to bioaccumulate in fish, so exposure to MeHg from fish consumption may be associated with exposure to increased levels of Se. Recent studies report that the beneficial effects of *n*-3 long-chain polyunsaturated fatty acids (*n*-3 LCPUFAs) can obscure the determination of adverse effects of prenatal MeHg exposure in longitudinal observational studies (Strain et al. 2008). Both *n*-6 and *n*-3 PUFAs are well-known essential nutrients in the human diet. Fish and seafood, meat, and eggs are three main food sources that contribute to *n*-3 LCPUFA intake, with fish and seafood being the major contributor (Meyer at al. 2003). The eicosapentaenoic acid (EPA; C20:5 *n*-3) and docosahexaenoic acid (DHA; C22:6 *n*-3) have been increasingly identified as having health benefits (Ginsberg and Toal 2009). DHA has been associated with a number of beneficial effects on neurocognitive and ocular functions. There is increasing evidence from human studies confirming the protective effects of *n*-3 LCPUFA on the intoxication of MeHg from fish consumption. While it is important to reduce the amount of MeHg exposure from the consumption of MeHg-contaminated fish, it is equally important to consume fish in order to get enough *n*-3 LCPUFA. Balancing the risks and benefits of fish consumption has become an increasingly important goal of fish consumption advisories (Li et al. 2010b).

Rice contains very low concentrations of EPA and DHA, and cereal-based products and dishes only contributed 2.55% of total source. MeHg exposure from rice consumption and disparities in *n*-3 LCPUFA for the rice-eating population may mean serious harm due to MeHg exposure in the same exposure level (Meyer et al. 2003). Furthermore, the type and amount of proteins consumed might affect the uptake and distribution of Hg in the body. Generally, the content of proteins is 15%–25% in fish and human body easily digests it (Wang et al. 2007). Fish proteins are of very high quality and contain sufficient amounts of all the essential amino acids required by the body for the growth and maintenance of lean muscle tissue.

The contents of protein were about 8% in rice (Yang et al. 2002). Rice proteins are of low quality, because they do not contain all the essential amino acids. Moreover, rice proteins have fiber membrane around, and they are more difficult to be absorbed by human body compared with animal proteins. Therefore, the effects of protein on the toxicity of MeHg in rice should be of more concern. Few studies have examined the effects of malnutrition on the metabolism and toxicity of Hg, having generally a deleterious effect. Studies regarding the influence of dietary protein levels on the fate of MeHg in mice demonstrated that low protein intakes have been associated with increased Hg in the brain of the mouse (Li et al. 2010a). The results indicated that low protein diet feeding might decrease urinary Hg excretion by increasing the retention of MeHg metabolite(s) in renal cells and suggested that the dietary protein status, which could modulate the metabolism of thiol compounds, played an important role in determining the fate of MeHg.

Some results suggested the presence in brain capillaries of a transport system capable of selectively mediating MeHg uptake across the brain capillary endothelial cell membrane (Aschner and Clarkson 1988). Cysteine can act as a carrier for MeHg across the blood–brain barrier and thus alter Hg distribution by increasing Hg levels in the brain and increasing neurotoxicity; however, cysteine may detoxify MeHg by preventing the inhibition of enzymes such as carnitine acyltransferase. Adequate protein intake prolongs survival after oral doses of MeHg. Few studies have examined the effects of malnutrition on the metabolism and toxicity of Hg, but malnutrition in general has a deleterious effect. Inadequate protein increased MeHg-induced mortality in mice (Adachi et al. 1994), and methionine deficiency during MeHgCl exposure caused an increase in serum prostaglandins.

16.6 PHYTOREMEDIATION

Metals are naturally present in the Earth's crust at various concentrations. Although many metals, such as Cu, Ni, Co, Mn, Fe, and Zn, are essential for cells, all metals are toxic at high concentrations. Any metal and metalloid species may be considered a contaminant if they are in a form or concentration that causes detrimental human or environmental effects (McIntyre 2003). Excessive levels of many metals in soil can result in crop yield reduction and poor quality and quantity of agricultural products and pose significant hazards to human, animal, and ecosystem health (Blaylock and Huang 2000).

Two different strategies and technologies have been employed for soil decontamination. The first one consists in the application of different agronomical practices to minimize the bioavailability of heavy metals in soil that include pH modification, organic matter and fertilizer managements, and choice of suitable vegetables for a particular soil. The second method is the use of phytoremediation techniques in which metal-accumulating plants are used to transport and concentrate metals from soil into the aboveground shoots, which are then harvested. Phytoremediation represents a low-cost alternative to traditional techniques such as soil removal.

In 1885, Baumann identified plants capable of accumulating high concentrations of Ni. Minguzzi and Vergnano (1948) identified plants capable of accumulating up to 1% Ni in shoots. After these first identifications of accumulating plants, a great deal of research has been conducted to find and study other plants able to extract and accumulate heavy metals from contaminated soil. Plant species expressing the bacterial mercuric ion reductase gene, *merA*, convert ionic mercury Hg^{++} from growth substrates to the less toxic volatile metallic mercury Hg^0. This activity removes mercury from the plant through volatilization. The goal is to develop plants that intercept and remove Hg^{++} from polluted aquatic systems before it can undergo bacterial methylation to the toxic methylmercury. Heaton and coauthors (2003) introduced this bacterial mercuric reductase gene, *merA*, into rice *Oryza sativa* L. by particle gun bombardment to produce rice with less toxic metallic mercury, Hg^0. This activity confers mercury resistance to plants and removes mercury through volatilization. The *merA*-expressing rice germinated and grew on semisolid growth medium spiked with sufficient Hg^{++} to kill the nonengineered (wild-type) controls. Meagher and Heaton (2005) have engineered plants with *merA* gene that remove MeHg from media and convert it to its volatile form. In this form of transgenic phytoremediation, *Arabidopsis* plants were engineered with *merA* and *merB* genes, from a bacterium that is resistant to mercury (Rugh et al. 1996). Also eastern cottonwood (*Populus deltoides*), a better candidate tree for phytoremediation in riparian ecosystem, was transformed with the *merA* gene (Che et al. 2003). Transgenic cotton wood shoots had normal growth on medium containing 25 mM Hg^{++}, a mercury concentration that killed the wild-type shoots.

Many examples of these phytoremediation techniques that have been successfully used are known, including the restoration of abandoned Hg and Au mine workings. The level of mercury bioaccumulation in wild plants; the distribution of bioavailable Hg, elemental Hg, and total Hg in soil; and the concentration of total gaseous Hg in ambient air was studied at three different mining sites in the Wanshan mercury mining district of China. Results of the study of Wang and collaborators (2011) indicated that three plant species (*Macleaya cordata* L., *Achillea millefolium* L., and *Pteris vittata* L.) showed enhanced accumulation of Hg and therefore may have potential for use in the phytoremediation of soils of the Wanshan mining area.

Tangahu et al. (2011) in their review reported that accumulated Hg in *Brassica juncea Czern* is much higher than in other plant species. Muddarisna and coauthors (2013) in their studies demonstrated that *Lindernia crustacea* (L.) *F.*, *Paspalum conjugatum* L., and *Cyperus kyllingia Endl.* are potentially useful for the phytoremediation of mercury-contaminated soil. Addition of ammonium thiosulfate to the soil contaminated for the presence of Hg increased Hg accumulation in plant. Growth and biomass production in maize grown in remediated soil increased 74% and 67%, especially after phytoremediation with *Paspalum conjugatum*.

The cost of the phytoremediation is lower than that of traditional processes both *in situ* and *ex situ*, and there is the possibility of the recovery and reuse of precious metals by companies specializing in phytomining. The phytoremediation method maintains the biological properties and physical structure of the soil. This technique is environmentally friendly and potentially cheap and offers the possibility of biorecovery of the heavy and radioactive metals.

REFERENCES

Adachi, T., Yasutake, A., Hirayama, K. 1994. Influence of dietary levels of protein and sulfur amino acids on the fate of methylmercury in mice. *Toxicology* 93:225–234.

Amirbahman, A., Reid, A.L., Haines, T.A., Kahl, J.S., Arnold, C. 2002. Association of methylmercury with dissolved humic acids. *Environmental Science and Technology* 36:690–695.

Aschner, M., Clarkson, T.W. 1988. Uptake of methylmercury in the rat brain, effects of amino acids. *Brain Research* 462:31–39.

Azevedo, B.F., Furieri, L.B., Peçanha, F.M., Wiggers, G.A., Vassallo, P.F., Simões, M.R., Fiorim, J. et al. 2012. Toxic effects of mercury on the cardiovascular and central nervous systems. *Journal of Biomedicine and Biotechnology*. Article ID 949048, 11pp.

Bakir, F., Damluji, S.F., Amin-Zaki I. 1973. Methylmercury poisoning in Iraq: An interuniversity report. *Science* 181:230–241.

Barrett, J.R. 2010. Rice is a significant source of methylmercury: Research in China assesses exposures. *Environmental Health Perspectives* 118:A398.

Baumann, A. 1885. Das Verhalten von Zinksalzen gegen Pflanzen und im Boden. *Landwirtsch Vers* 3:1–53.

Bento-Abreu, A., Van Damme, P., Van Den Bosch, L., Robberecht, W. 2010. The neurobiology of amyotrophic lateral sclerosis. *European Journal of Neuroscience* 31:2247–2265.

Berglund, M., Lind, B., Björnberg, K.A., Palm, B., Einarsson, O., Vahter, M. 2005. Inter-individual variations of human mercury exposure biomarkers: A cross-sectional assessment. *Environmental Health* 4:20.

Bernhoft, R.A. 2012. Mercury toxicity and treatment: A review of the literature. *Journal of Environmental and Public Health*. Article ID 460508, 10pp.

Berry, M.J., Ralston, N.V.C. 2008. Mercury toxicity and the mitigating role of selenium. *EcoHealth* 5:456–459.

Bertilsson, L., Neujahr, H.Y. 1971. Methylation of mercury compounds by methylcobalamin. *Biochemistry* 10:2805–2808.

Biester, H., Müller, G., Schöler, H.F. 2002. Binding and mobility of mercury in soils contaminated by emissions from chlor-alkali plants. *Science of the Total Environment* 284:191–193.

Blaylock, M.J., Huang, J.W. 2000. Phytoextraction of metals. In: Raskin, I., Ensley, B.D. (eds.), *Phytoremediation of Toxic Metals: Using Plants to Clean-Up the Environment*. New York: Wiley, pp. 53–70.

Carocci, A., Rovito, N., Sinicropi, M.S., Genchi, G. 2014. Mercury toxicity and neurodegenerative effects. *Review of Environmental Contamination and Toxicology* 229:1–18.

Carvalho, M.L., Santiago, S., Nunes, M.L. 2005. Assessment of the essential element and heavy metal content of edible fish muscle. *Analytical and Bioanalytical Chemistry* 382:426–432.

Cernichiari, E., Brewer, R., Myers, G.J., Marsh, D.O., Lapham, L.W., Cox, C., Shamlaye, C.F., Berlin, M., Davidson, P.W., Clarkson, T.W. 1995. Monitoring methylmercury during pregnancy: Maternal hair predicts fetal brain exposure. *Neurotoxicology* 16:711–716.

Chan, H.M., Egeland, G.M. 2004. Fish consumption, mercury exposure, and heart diseases. *Nutrition Review* 62:68–72.

Chapman, L., Chan, H.M. 2000. The influence of nutrition on methyl mercury intoxication. *Environmental Health Perspective* 108:29–56.

Che, D., Meagher, R.B., Heaton, A.C., Lima, A., Rugh, C.L., Merkle, S.A. 2003. Expression of mercuric ion reductase in Eastern cottonwood (*Populus deltoides*) confers mercuric ion reduction. *Plant Biotechnology Journal* 1:311–319.

Chen, C., Yu, H., Zhao, J., Li, B., Qu, L., Liu, S., Zhang, P., Chai, Z. 2006. The roles of serum selenium and selenoproteins on mercury toxicity in environmental and occupational exposure. *Environmental Health Perspective* 114:297–301.

Chen, C.Y., Driscoll, C.T., Lambert, K.F., Mason, R.P., Rardin, LR., Serrell, N., Sunderland, E.M. 2012. Marine mercury fate: From sources to seafood consumers. *Environmental Research* 119:1–2.

Clarkson, T.W., Magos, L. 2006. The toxicology of mercury and its chemical compounds. *Critical Review of Toxicology* 36:609–662.

Clarkson, T.W., Magos, L., Myers, G.J. 2003. The toxicology of mercury-current exposures and clinical manifestations. *The New England Journal of Medicine* 349:1731–1737.

Clemens, S., Monperrus, M., Donard, O.F.X., Amouroux, D., Guerin, T. 2012. Mercury speciation in seafood using isotope dilution analysis: A review. *Talanta* 89:12–20.

Counter, S.A., Buchanan, L.H. 2004. Mercury exposure in children: A review. *Toxicology and Applied Pharmacology* 198:209–230.

Dórea, J.G., Bezerra, V.L.V.A., Fajon, V., Horvat, M. 2011. Speciation of methyl- and ethyl-mercury in hair of breastfed infants acutely exposed to thimerosal-containing vaccines. *Clinica Chimica Acta* 17:1563–1566.

Feng, X., Li, P., Qiu, G.L., Wang, S.F., Li, G.H., Shang, L.H., Meng, B. et al. 2008. Human exposure to methylmercury through rice intake in mercury mining areas, Guizhou province, China. *Environmental Science and Technology* 42:326–332.

Fitzgerald, W.F., Lamborg, C.H., Hammerschmidt, C.R. 2007. Marine biogeochemical cycling of mercury. *Chemical Reviews* 107:641–662.

Gardner, R.M., Nyland, J.F., Silbergeld, E.K. 2010. Differential immunotoxic effects of inorganic and organic mercury species in vitro. *Toxicology Letters* 198:182–190.

Gilmour, C.C., Henry, EA. 1991. Mercury methylation in aquatic systems affected by acid deposition. *Environmental Pollution* 71:131–169.

Ginsberg, G., Toal, B. 2009. Quantitative approach for incorporating methylmercury risks and omega-3 fatty acid benefits in developing species-specific fish consumption advice. *Environmental Health Perspective* 117:267–275.

Goldman, L.R., Shannon, M.W. 2001. American Academy of Pediatrics: Committee on Environmental Health. Technical report: Mercury in the environment: Implications for pediatricians. *Pediatrics* 108:197–205.

Guzzi, G., La Porta, C.A. 2008. Molecular mechanisms triggered by mercury. *Toxicology* 244:1–12.

Guzzi, G., Minoia, C., Pigatto, P.D., Severi, G. 2006. Methylmercury, amalgams, and children's health. *Environmental Health Perspective* 114:A149.

Harada, M. 1995 Minamata disease: Methylmercury poisoning in Japan caused by environmental pollution. *Critical Review of Toxicology* 25:1–24.

Harris, H.H., Pickering, I.J., George, G.N. 2003. The chemical form of mercury in fish. *Science* 301:1203.

Heaton, A.C., Rugh, C.L., Kim, T., Wang, N.J., Meagher, R.B. 2003. Toward detoxifying mercury-polluted aquatic sediments with rice genetically engineered for mercury resistance. *Environmental Toxicological Chemistry* 22:2940–2947.

Itai, Y., Fujino, T., Ueno, K., Motomatsu, Y. 2004 An epidemiological study of the incidence of abnormal pregnancy in areas heavily contaminated with methylmercury. *Environmental Science* 11:83–97.

Järup, L. 2003 Hazards of heavy metal contamination. *British Medical Bulletin* 68:167–182.

Johnson, F.O., Atchison, W.D. 2009 The role of environmental mercury, lead and pesticide exposure in development of amyotrophic lateral sclerosis. *Neurotoxicology* 30:761–765.

Khan, M.A., Asaduzzaman, A.M., Schreckenbach, G., Wang, F. 2009 Synthesis, characterization and structures of methylmercury complexes with selenoamino acids. *Dalton Transactions* 29:5766–5772.

Khan, M.A., Wang, F. 2009. Mercury–selenium compounds and their toxicological significance: Toward a molecular understanding of the mercury–selenium antagonism. *Environmental Toxicological Chemistry* 28:1567–1577.

Li, P., Feng, X., Qiu, G. 2010a. Methylmercury exposure and health effects from rice and fish consumption: A review. *International Journal of Environmental Research and Public Health* 7:2666–2691.

Li, S.J., Zhang, S.H., Chen, H.P., Zeng, C.H., Zheng, C.X., Li, L.S., Liu, Z.H. 2010b. Mercury-induced membranous nephropathy: Clinical and pathological features. *Clinical Journal of American Society of Nephrology* 5:439–444.

Liu, B., Yan, H., Wang, C., Li, Q., Guédron, S., Spangenberg, J.E., Feng, X., Dominik, J. 2012. Insights into low fish mercury bioaccumulation in a mercury-contaminated reservoir, Guizhou, China. *Environmental Pollution* 160:109–117.

McIntyre, T. 2003. Phytoremediation of heavy metals from soils. *Advances in Biochemical Engineering/ Biotechnology* 78:97–123.

Meagher, R.B., Heaton, A.C.P. 2005. Strategies for the engineered phytoremediation of toxic element pollution: Mercury and arsenic. *Journal of Industrial Microbiology and Biotechnology* 32:502–513.

Meng, M., Li, B., Shao, J.J., Wang, T., He, B., Shi, J.B., Ye, Z., Jiang G.B. 2014. Accumulation of total mercury and methylmercury in rice plants collected from different mining areas in China. *Environmental Pollution* 184:179–186.

Mergler, D., Anderson, H.E., Chan, L.H.M., Mahaffey, K.R., Murray, M., Sakamoto, M., Stern, A.H. 2007. Methylmercury exposure and health effects in humans: A worldwide concern AMBIO. *A Journal of the Human Environment* 36:3–11.

Meyer, B.J., Mann, N.J., Lewis, J.L., Milligan, G.C., Sinclair, A.J., Howe, P.R. 2003. Dietary intakes and food sources of omega-6 and omega-3 polyunsaturated fatty acids. *Lipids* 38:391–398.

Miklavcic, A., Cuderman, P., Mazej, D., SnojTratnik, J., Krsnik, M., Planinsek, P., Osredkar, J., Horvat, M. 2011. Biomarkers of low-level mercury exposure through fish consumption in pregnant and lactating Slovenian women. *Environmental Research* 111:1201–1207.

Minguzzi, C., Vergnano, O. 1948. Il contenuto di nichel nelle ceneri di *Alyssum bertlonii* Desv. *Atti della Società Toscana di Scienze Nataturali, Memorie Serie A* 55:49–77.

Morel, F.M.M., Kraepiel, A.M.L., Amyot, M. 1998. The chemical cycle and bioaccumulation of mercury. *Annual Review of Ecology and Systematics* 29:543–566.

Mozaffarian, D., Rimm, E.B. 2006. Fish intake, contaminants, and human health: Evaluating the risks and the benefits. *The Journal of the American Medical Association* 296:1885–1899.

Muddarisna, N., Krisnayanti, B.D., Utami, S.R., Handayanto, E. 2013. Phytoremediation of mercury-contaminated soil using three wild plant species and its effect on maize growth. *Applied Ecology and Environmental Sciences* 1:27–32.

Mutter, J., Yeter, D. 2008 Kawasaki's disease, acrodynia, and mercury. *Current Medicinal Chemistry* 15:3000–3010.

Novoa-Munoz, J.C., Pontevedra-Pombal, X., Martinez-Cortizas, A., Garcia-Rodeja Gayoso, E. 2008. Mercury accumulation in upland acid forest ecosystems nearby a coal fired power-plant in Southwest Europe (Galicia, NW Spain). *Science Total Environmental* 394:303–312.

Nielsen, J.B., Andersen, O. 1992. The toxicokinetics of mercury in mice offspring after maternal exposure to methylmercury effect of selenomethionine. *Toxicology* 74:233–241.

Pirrone, N., Mahaffey, K.R. 2005. *Dynamics of Mercury Pollution on Regional and Global Scales: Atmospheric Processes and Human Exposures around the World.* New York: Springer-Verlag, 748pp.

Porvari, P. and Verta, M. 1995. Methylmercury production in flooded soils: A laboratory study. *Water, Air, and Soil Pollution* 80:765–773.

Pyszel, A., Wrobel, T., Szuba, A., Andrzejak, R. 2005. Effect of metals, benzene, pesticides and ethylene oxide on the haematopoietic system. *Medycyna Pracy* 56:249–255.

Qiu, C., De Ronchi, D., Fratiglioni, L. 2007. The epidemiology of the dementias: An update. *Current Opinion in Psychiatry* 20:380–385.

Qiu, G., Feng, X., Li, P., Wang, S., Li, G., Shang, L., et al. 2008. Methylmercury accumulation in rice *(Oryza sativa* L*)* grown at abandoned mercury mines in Guizhou, China. *Journal of Agricultural and Food Chemistry* 56:2465–2468.

Rayman, M.P. 2000. The importance of selenium to human health. *Lancet.* 356:233–241.

Rothenberg, S.E., Yu, X., Zhang, Y. 2013. Prenatal methylmercury exposure through maternal rice ingestion: Insights from a feasibility pilot in Guizhou Province, China. *Environmental Pollution* 180:291–298.

Rugh, C.L., Wilde, H.D., Stack, N.M., Thompson, D.M., Summers, A.O., Meagher, R.B. 1996. Mercuric ion reduction and resistance in transgenic *Arabidopsis thaliana* plants expressing a modified bacterial *merA* gene. *Proceedings of the National Academy of Sciences* 93:3182–3187.

Sakamoto, M., Kaneoka, T., Murata, K., Nakai, K., Satoh, H., Akagi, H. 2007. Correlations between mercury concentrations in umbilical cord tissue and other biomarkers of fetal exposure to methylmercury in the Japanese population. *Environmental Research* 103:106–111.

Selin, N.E., Sunderland, E.M., Knightes, C.D., Mason, R.P. 2010. Sources of mercury exposure for U.S. seafood consumers: implications for policy. *Environmental Health Perspectives* 118:137–143.

Seppanen, K., Soininen, P., Salonen, J.T., Lotjonen, S., Laatikainen, R. 2004. Does mercury promote lipid peroxidation? An in vitro study concerning mercury, copper, and iron in peroxidation of low-density lipoprotein. *Biological Trace Element Research* 101:117–132.

Skerfving, S.B., Copplestone, J.F. 1976. Poisoning caused by the consumption of organomercury dressed seed in Iraq. *Bulletin of the World Health Organization* 54:101–112.

Sinicropi, M.S., Amantea, D., Caruso, A., Saturnino, C. 2010. Chemical and biological properties of toxic metals and use of chelating agents for the pharmacological treatment of metal poisoning. *Archives of Toxicology* 84:501–520.

Stein, E.D., Cohen, Y., Winer, A.M. 1996. Environmental distribution and transformation of mercury compounds. *Critical Reviews in Environmental Science and Technology* 26:1–43.

Strain, J.J., Davidson, P., Bonham, M., Duffy, E., Stokes-Riner, A., Thurston, S., Wallace, J. et al. 2008. Associations of maternal long-chain polyunsaturated fatty acids, methylmercury, and infant development in the Seychelles Child Development Nutrition Study. *Neurotoxicology* 29:776–782.

Sunderland, E.M., Mason, R.P. 2007. Human impacts on open ocean mercury concentrations. *Global Biogeochemical Cycles* 21(4):GB4022.

Tangahu, B.V., Abdullah, S.R.S., Basri, A., Idris, M., Anuar, N., Mukhlsin, M. 2011. A review on heavy metals (As, Pb and Hg) uptake by plants through phytoremediaton. *International Journal of Chemical Engineering* Article ID 939161, 31pp.

Tsubaki, T., Irukajama, K. 1997. Minamata disease. In: Tsubaki, T., Irukajama, K. (eds.), *Methylmercury Poisoning in Minamata and Niigata*. Tokyo, Japan: Kodansha, p. 317.

Ullrich, S.M., Tanton, T.W., Abdrashitova, S.A. 2001. Mercury in the aquatic environment: A review of factors affecting methylation. *Critical Reviews in Environmental Science and Technology* 31:241–293.

Vather, M., Akesson, A., Lind, B., Bjors, U., Schutz, A., Berglung, M. 2000. Longitudinal study of methylmercury and inorganic mercury in blood and urine of pregnant and lactating women, as well as in umbilical cord blood. *Environmental Research* 84:186–194.

Wang, J., Feng, X., Anderson, C.W., Zhu, W., Yin, R., Wang, H. 2011. Mercury distribution in the soil-plant-air system at the Wanshan mercury mining district in Guizhou, Southwest China. *Environmental Toxicology and Chemistry* 30:2725–2731.

Wang, R., Sun, W.W., Li, P., Xiao, Z., Nie, G.X. 2007. Analysis and comparison on nutritional compositions of ten species edible fishes in Xinxiang city. *Journal of Xinxiang Medical College* 24:596–598.

World Health Organization. 1990. *Methylmercury*. Environmental Health Criteria 101. Geneva, Switzerland: WHO.

World Health Organization. 2013. *Mercury and Health* (Fact sheet No. 361). Geneva, Switzerland: WHO. http://www.who.int/mediacentre/factsheets/fs361/en/ (accessed 02 September 2015).

Yang, Y.X., Wang, G.Y., Pan, X.C. 2002. *China Food Composition*, 1st edn. Beijing, China: Peking Medical University Press, pp. 10–20.

Zhang, H., Feng, X., Larssen, T., Qiu, G., Vogt, R.D. 2010. In inland China, rice, rather than fish, is the major pathway for methylmercury exposure. *Environmental Health Perspectives* 118:1183–1188.

17 Tea Beverages and Extracts as a Source of Exposure to Lead and Other Heavy Metals

Sidney J. Stohs and Sidhartha D. Ray

CONTENTS

17.1 INTRODUCTION

17.1.1 LEAD TOXICITY

Lead is the most widely scattered, ubiquitous toxic metal in the world (Chen and Hu, 2009; Garcia-Leston et al., 2010), and lead poisoning is an important environmental disease with long-term adverse health effects (Meyer et al., 2008; Filippelli and Laidlaw, 2010). Lead is classified as a possible human carcinogen and is extremely toxic to the brain, kidneys, gastrointestinal tract, liver, bones, and the immune and reproductive systems.

Lead introduced into the blood stream has a half-life of about 30 days. It binds to red blood cells and is distributed throughout the body with primary accumulation in bone and to a lesser extent in liver. The half-life of lead deposited in bones is estimated to be 20–30 years. Toxicologically, lead can be a substitute for calcium and zinc to a lesser extent. In various tissues, the disruption of calcium in mitochondria results in the formation of reactive oxygen species, mitochondrial self-destruction, and cellular apoptosis (programmed cell death) (Mason et al., 2014).

In the brain, lead interferes with neurotransmitter release and can result in lead encephalopathy that may occur in children at lower exposure levels than in adults. High lead exposure can result in mental dullness, headache, irritability, memory loss, tremor, and hallucinations, which may lead to convulsions, paralysis, coma, and death if left untreated (Mason et al., 2014).

17.1.2 Sources of Lead Exposure

The primary route of lead uptake into the body is via oral ingestion whereby lead is absorbed in the intestine and distributed throughout the body. Inhalation may be a primary route of exposure in areas where lead-containing wastes are discharged into the air. The portion of ingested lead that is absorbed is age dependent, with adults typically absorbing less than 5%, whereas absorption in children may be as high as 50% (Filippelli and Laidlaw, 2010). Because of the high absorption of lead in children and their rapid growth and tissue differentiation, children are highly vulnerable to permanent effects of lead poisoning.

Although the ban on leaded gasoline and lead-based paints as well as lead-based solders has reduced lead exposure, current sources of lead include historic contamination via lead-based paints and plumbing in older buildings, lead-containing wastes discharged into the atmosphere, manufacturing of new products, recycling of old lead-containing products such as batteries, improper manufacture of toys, improper disposal of lead-contaminated products, and lead incorporation into plant food products in the form of contaminated groundwater and soil (Meyer et al., 2008; Mason et al., 2014).

An examination of 193 ayurvedic medicines sold via the Internet demonstrated that approximately 20% of these medicines contained detectable lead, mercury, and arsenic with a median lead concentration in some samples as high as 11.5 μg/g (Saper et al., 2008). A review of atypical sources of childhood lead poisoning in the United States during 1966–2006 identified Tibetan herbal supplements, imported candies, leaded cosmetics, assorted metallic jewelry and trinket items, lead-glazed dinnerware, and ingested lead pellets as the most common causes of exposure (Gorospe and Gerstenberger, 2008).

A retrospective clinical investigation has reported a lead poisoning outbreak from the consumption of an Ayurvedic medicine in Durban, South Africa (Mathee et al., 2015). Samples of the implicated medicines were analyzed to determine lead levels. During the acute phase, blood lead levels ranged from 34 to 116 μg/dL, while two years later, it ranged from 13 to 34 μg/dL. The medication capsules had a lead content of 125,235 μg/g. A majority of the study subjects suffered from a wide range of nonspecific ill health symptoms despite a significant delay in the diagnosis of lead poisoning. Findings of this study suggest the potential for lead poisoning from exposure to Ayurvedic medicines. The possibility of an aggravated situation may very likely arise when excessive tea drinkers use these medications.

Questions have been raised regarding the amount of lead exposure from tea (*Camellia sinensis*) infusions (brewed teas) and tea extracts. With the exception of water, tea is the most widely consumed beverage in the world. Due to the widespread consumption of tea infusions and tea extracts present in many dietary supplements and herbal medicines, questions have been raised regarding the possible exposure to significant amounts of lead in these preparations. This chapter examines the uptake of lead by tea (*Camellia sinensis*) and other plants and the amount of lead in tea preparations and the potential for lead exposure from these sources.

17.2 SOIL AS A SOURCE OF LEAD

Lead is naturally present in all soils (Anon, 2004; Chen and Hu, 2009), and according to the University of Massachusetts Department of Plant and Soil Sciences, soils generally contain lead in the range of 15–40 parts per million (ppm), which is also frequently expressed as milligrams per kilogram (mg/kg) or micrograms per gram (μg/g). As a consequence, the relationship between the soil content of lead and other minerals and the lead and mineral content of various plants with emphasis on tea has been examined. China and India produce over 50% of the world's tea supply as well as large amounts of tea extracts that are exported for use in herbal medicines and dietary supplements (Natesan and Ranganathan, 1990).

Tea plants require damp acid soil for growth and temperatures ranging from 10°C to 30°C. Several studies have assessed the soil content of lead in various tea-growing regions of China. In a study by Jin et al. (2005b), the soil lead levels ranged from 32.2 to 36.4 mg/kg, well below the maximum permissible level of 60 mg/kg limit provided for organic tea gardens (plantations) in China. In another investigation, lead levels averaged 42.7 mg/kg in the soils associated with various tea plantations in China (Guo et al., 2008). In this study, the investigators also measured and reported the endogenous levels of cadmium, mercury, arsenic, nickel, and chromium in the soil.

Various studies have shown a positive correlation between soil content of lead and lead in tea leaves. The higher the soil content of lead, the higher the lead content in tea plants (Jin et al., 2005a,b; Han et al., 2006; Guo et al., 2008; Chen et al., 2010). Several studies along these lines have also examined the relationship between the lead content of tea plants grown hydroponically and the lead levels to which the plants are exposed. It appears that the lead content found in tea plants was directly proportional to the concentration of lead used in the nutrient solutions (Kang et al., 2004; Xia et al., 2010). Plant growth was inhibited in solutions containing greater than 100 mg lead/L. Similar concentration-dependent results have been shown when examining lead adsorption by tea root cell walls (Xu et al., 2014). The chlorophyll content of the plants decreased with increasing lead concentration in the solution, while conversely, the malondialdehyde content of tea leaves was enhanced, indicating lead-induced (oxidative) stress (Xia et al., 2010).

Since lead occurs naturally in all soils (Anon, 2004), lead will occur in essentially all plants. For example, lead content was measured in 110 samples of 7 different plant species in Bulgaria (Arpadjan et al., 2008) and was shown to range from 0.2 to 8.6 μg/g. In a study of 130 plant samples from 14 different species in Brazil (Caldas and Machado, 2004), horse chestnut appeared to concentrate lead more than other species, with amounts of lead being as high as 22 μg/g or below in most samples and several samples being as high as approximately 150 μg/g. The contents of heavy metals in 23 herbs from northern China were determined, with a mean lead content of 1.78 ± 1.98 μg/g (Li et al., 2006). The contents of cadmium, mercury, and arsenic were also determined in these samples.

Plants accumulate the elements that are present in the soil. For example, 11 elements including manganese, potassium, sodium, zinc, copper, chromium, iron, cobalt, nickel, cadmium, and lead were measured in tea samples of green and black tea (Soomro et al., 2008. Of the elements tested, potassium was the most abundant, followed by sodium, manganese, and iron. Lead (0.37 μg/g) and cadmium (0.12 μg/g) exhibited the lowest contents in the tea samples.

Soil pH is another important factor with respect to lead uptake by tea plants. The more acidic the soil, the more readily plant extractable the lead becomes (Jin et al., 2005b; Han et al., 2007; Chen et al., 2010). Jin et al. (2005b) have described how nonedaphic factors contribute to the lead accumulation in tea plants and ultimately drinkable tea leaves. Lead adsorption by tea root cell walls has also been shown to be pH dependent (Xu et al., 2014). Using both pot and field experiments, liming with calcium carbonate was shown to increase soil pH by up to 1 pH unit and decrease soil extractable lead by up to 32% (Han et al., 2007).

17.3 LEAD IN TEA

17.3.1 Lead Content

Because China is one of the major sources of tea and tea products exported to the United States and other countries, as well as the fact that tea is the world's most widely consumed beverage after water, the levels of lead in tea (leaves) have been extensively studied in China (Jin et al., 2005a, 2014; Han et al., 2006; Qin and Chen, 2007; Chen et al., 2010; Ning et al., 2011). In a study involving 57 randomly acquired tea samples (green, black, oolong, and scented teas), the mean lead concentration was 1.32 μg/g and ranged from 0.2 to 6.3 μg/g (Qin and Chen, 2007). Lead concentrations in 1225 tea samples collected nationally in China have been assessed, with 32% of the samples exhibiting lead levels that were greater than 2.0 μg/g (Han et al., 2006). An increase in tea lead concentrations was noted from 1989 compared to 2000.

In a study of lead in tea from the Zhejiang Province of China, the lead content was below 5.0 μg/g in all samples tested, the permissible limit allowed by the Chinese Ministry of Agriculture (Jin et al., 2005a). A study involving eight varieties of tea noted that the lead content of young leaves ranged from approximately 0.6 to 6.0 μg/g depending on the variety (Chen et al., 2010). Jin et al. (2014) reviewed lead exposure from all sources including tea and concluded that the average dietary intake in China was 1.232 μg/kg body weight per day.

Lead, cadmium, arsenic, mercury, and copper levels in Chinese Yunnan Pu'er tea were determined (Ning et al., 2011). A total of 30 different brands of the tea involving 150 commercial samples were assayed. Lead content ranged from 0.26 to 3.2 μg/g. Mercury was not detected in 17 of the 30 brands. Copper content was 12–22 μg/g, arsenic content ranged was 0.035–0.24 μg/g, and cadmium levels were very low at 0.0059–0.085 μg/g. The authors concluded that based on the low levels of heavy metals detected in the samples, the teas were safe for consumption.

The lead content of teas has also been assessed in various other countries in addition to China. For example, the lead contents of teas grown or consumed in India (Seth et al., 1973; Natesan and Ranganathan, 2006), Japan (Tsushida and Takao, 2006), Saudi Arabia (Ashraf and Mian, 2008; Al-Othman et al., 2012), Iran (Salahinejad and Aflaki, 2010; Falahi and Hedaiati, 2013), Ethiopia (Ashenef, 2014), Thailand (Nookabkaew et al., 2006), Pakistan (Soomro et al., 2008; Inayat et al., 2010), Taiwan (Shen and Chen, 2008), Germany (Feldheim and Stelte, 1975), Slovak Republic (Haban et al., 2008), Romania (Kekedy-Nagy and Ionescu, 2009), and Nigeria (Mohammed and Sulaiman, 2009; Orisakwe et al., 2015) have been determined.

In a survey of 25 teas from different countries, the concentration of lead was found to be up to 3.1 μg/g (Feldheim and Stelte, 1975). The lead content of 139 green tea samples from Japan ranged from 0.2 to 1.9 μg/g (Tsushida and Takao, 2006), while the lead concentrations from various tea samples in the Slovak Republic ranged from 0.34 to 1.56 μg/g (Haban et al., 2008). In Saudi Arabia, lead concentrations in tea varied from 0.3 to 2.2 μg/kg. In four Pakistani branded tea samples, heavy metals including lead were within international standards, while two unbranded samples exhibited lead levels as high as 15.8 μg/g (Inayat et al., 2010). The results indicate that lead occurs in all teas and that the amount of lead in tea leaf preparations is generally but not always below 3 μg/g.

17.3.2 Factors Affecting the Amount of Lead in Teas

Various studies have shown that the amount of lead present in teas is dependent upon a number of factors including the endogenous amount of lead present in the soil (Jin et al., 2005a,b; Han et al., 2006; Chen et al., 2010; Guo et al., 2008), the organic matter content of the soil, the solubility and extractability of the lead (Jin et al., 2005b; Han et al., 2006; Salahinijad and Aflaki, 2010), the soil pH (Jin et al., 2005b; Han et al., 2007; Chen et al., 2010), the variety of the tea (Han et al., 2006; Shen and Chen, 2008; Chen et al., 2010; Xu et al., 2011), seasonal variations (Han et al., 2007), proximity to industrial sites and atmospheric pollution (Jin et al., 2005a,b; Han et al., 2006), the lead content of the water used in the tea gardens and to process the tea leaves (Borah et al., 2010), the age of the collected tea leaves (Chen et al., 2010), and the manner in which the leaves are processed (Jin et al., 2005a).

Tea plants growing closest to industrial areas and roads were shown to have higher lead content than those growing farther away, presumably due to lead-containing industrial and automobile exhaust (Jin et al., 2005a; Han et al., 2006). The addition of zeolite and apatite to lead-containing soils has been suggested as a practical means of decreasing water-soluble and extractable lead by tea plants (Wang and Zhang, 2008).

With respect to lead uptake, a progression was shown to exist where the lead content of: soil > fine roots > coarse roots > stems > mature leaves > young leaves/shoots, which are used for making commercial tea > seeds (Chen et al., 2010). Similar results have been reported for black tea in India (Natesan and Ranganathan, 1990).

Different varieties of tea take up lead differentially. For example, the Yingshuang variety of tea was shown to be more resistant to lead uptake than the Longjing 43 variety under identical conditions (Xu et al., 2011). Furthermore, differences were observed with respect to the extractability of the lead from the roots of the two varieties of tea. Lead was most readily extractable with acetic acid solution from Longjing 43 roots, while a salt solution was most effective in extracting lead from the Yingshuang roots, indicating a difference in the chemical binding of the lead.

A wide variety of elements are present in tea. Nookabkaew et al. (2006) measured and showed the presence of 19 elements in commercial samples of three herbal teas products, *Camellia sinensis*, *Morus alba*, and *Gynostemma pentaphyllum*. The average lead contents of these three herbs were 3.9, 0.40, and 9.3 μg/g, respectively, demonstrating the differential uptake of lead by various plant species. In addition to most of the elements assayed by Soomro et al. (2008), calcium, magnesium, aluminum, vanadium, mercury, zinc, selenium, barium, and arsenic were also determined. Calcium and magnesium were present in highest concentrations in all samples (not measured in the previous study by Soomro et al., 2008).

Sixteen elements were measured in 14 commercial tea samples in Romania with the lead content of the teas averaging 0.86 μg/g (Kekedy-Nagy and Ionesco, 2009). Seenivasan et al. (2008) examined the content of five heavy metals including lead in 100 black tea samples from seven tea districts in southern India. The mean level of lead in these samples was 0.81 ± 0.32 μg/g.

Twelve elements were assessed in black tea (Salahinejad and Aflaki, 2010). Calcium, magnesium, aluminum, and manganese were present at mg/g levels, while lead, chromium, iron, nickel, copper, and zinc were present at μg/g levels. Lead content ranged from 1.34 to 1.91 μg/g in tea samples from Iran, India, and Ceylon. The contents of lead, mercury, cadmium, chromium, arsenic, and nickel have been determined in soil samples in tea plantations (Guo et al., 2008), and the concentrations of these and other elements are reflected in plants grown on this soil.

Shen and Chen (2008) examined the content of 10 elements in 48 commercial green, oolong, and black tea products in Taiwan. The mean lead content levels for the three tea types were 0.2, 0.4, and 2.1 μg/g, respectively. These results indicate that the method of processing tea for commercial marketing and use influence the lead content of the final tea product.

17.4 TEA EXTRACTS AND INFUSIONS

The extractability of lead from tea leaves and herbal tea products during the preparation of tea decoctions (infusions, brewed teas) has been examined by a number of investigators (Feldheim and Stelte, 1975; Natesan and Ranganathan, 1990; Nookabkaew et al., 2006; Shen and Chen, 2008; Soomro et al., 2008; Schwalfenberg et al., 2013). In general, lead has been shown to have a low solubility and extractability as compared to most other elements. For example, only 7.1% of the total lead was released into the infusion using a standardized procedure whereby 2 g of tea leaves were treated with 100 mL boiling water for 10 minutes (Nookabkaew et al., 2006). Shen and Chen (2008) reported that only 7.1% of lead was released from green tea, while approximately 50% and 59% of lead was released from oolong and black teas, respectively.

Schwalfenberg et al. (2013) examined the lead extractability from 30 different teas sold in tea bags. They analyzed the lead content of both the tea leaves and the tea steeped for 3–4 minutes and 15–17 minutes. All brewed tea and tea leaves contained detectable levels of lead with an increase in lead in brewed tea as the steeping time increased. Chinese oolong teas had the highest levels of lead, followed by green tea and black tea, with white tea having the lowest lead levels..Lead levels ranged from 0.10 to 4.39 μg/L of brewed tea.

Of the lead extracted when brewing black tea, 68% of the total solubilized lead occurred within the first minute (Natasan and Raganathan, 1990). They reported an extractability of about 33%–42% for lead from various black teas. The mean levels of lead in black tea infusion samples at 5, 15, and 60 minutes were approximately 0.80, 0.99, and 1.27 mg/kg of dry tea weight, respectively (Shekoohiyan et al., 2012). Thus, the extractability of lead varies depending upon the type of tea and how it has been processed (Feldheim and Stelte, 1975; Natesan and Ranganathan, 1990; Shen and Chen, 2008; Salahinejad and Afliki, 2010; Shekoohiyan et al., 2012).

The source of lead in brewed tea is not always derived from the tea itself. Tea brewed in traditional metallic teapots has been shown to be a significant source of lead, nickel, and other metals as aluminum, copper, iron, and zinc (Bolle et al., 2011). Teapots were shown to produce a high level of lead in leachates. An additional source of lead is derived from the solder, which may be used to connect the spout to the teapot (Petit et al., 2013).

17.5 LEAD BINDING BY TEA AND CHEMOPROTECTION

The high lead-binding ability of tea plant constituents has been demonstrated in a number of studies that correlates with the low extractability of lead from tea products. Spent (waste) black and green tea leaves serve as excellent lead absorbents (Lavecchia et al., 2010; Zuorro and Lavecchia, 2010). Lead removal efficiencies from aqueous solutions of over 95% were achieved using spent tea leaves as the lead absorbent with a maximum absorbent capacity of approximately 100 mg/g. With respect to lead removal, the following order of efficiency was reported: black tea > green tea > Fuller's earth > activated charcoal.

In a more recent study, Cheraghi et al. (2015) have shown that the lead-binding kinetics of waste black tea leaves is dependent upon multitude parameters including pH, temperature, and the amount of metal concentration and adsorbent. This investigation determined that the maximum monolayer absorption capacity of the spent tea leaves was approximately 167 mg/g at 25°C. Neither this study nor the two previous studies determined what tea constituents were responsible for the lead binding.

Green tea may also exhibit protective effects with respect to ingested lead. Various studies have shown that the effects of lead administered in toxic amounts to rats can be ameliorated by the treatment of the animals with a green tea extract (Hamed et al., 2011; Meki et al., 2011; Khalaf et al., 2012; Mehana et al., 2012; Sha'bani et al., 2015; Winiaraska-Mieczan, 2015). Lead disrupts tissue antioxidant defenses and results in lipid peroxidation and oxidative tissue damage, producing physiological dysfunction. These studies have shown that the administration of a green tea extract protects against lead-induced damage to tissues in the brain, liver, and blood by increasing the antioxidant defenses such as reduced glutathione levels and antioxidant enzymes. Green tea extract decreased lipid peroxidation and DNA damage (Khalaf et al., 2012) and prevented the antifertility effects of lead (Sha'bani et al., 2015).

Green tea infusions were shown to lower lead levels in the brain, liver, and kidneys of lead-treated rats (Winiaraska-Mieczan, 2015). The protective effects of green tea extracts and infusions against lead toxicity have also been confirmed based on histopathological studies (Meki et al., 2011; Mehana et al., 2012). The protective effects of green, black, red, and white tea infusions against the adverse effects of lead and cadmium were compared. The results indicated that the beneficial effects of black, red, and white teas were comparable to green tea.

A study in a $HepG_2$ cell culture system has shown that tea catechins (polyphenolic compounds) protect in a concentration-dependent manner against the toxic effects of lead including decreased cell viability, lipid peroxidation, and altered membrane fluidity (Chen et al., 2002). Galloylated catechins were more protective against lead toxicity than nongalloylated catechins. The results provide a mechanistic explanation for the lead binding and protective effects of tea preparations. Based on these studies, tea preparations may protect against the toxic effects of not only the small amounts of lead inherently present in the extracts but also exogenously consumed lead.

17.6 LEAD INTAKE STANDARDS

Governments throughout the world have established appropriate standards for lead content of teas as well as herbal medicines. The World Health Organization allows 10 μg lead/g (ppm), while China, Germany, and Australia allow 5 μg/g. Japan allows 20 ppm total metals in herbal medicines.

The U.S. Pharmacopeia has not established standards for herbs but allows 3 μg lead/g for drugs. The limits for lead in dietary supplements are 15 μg/day established by the U.S. Environmental Protection Agency and California Proposition 65 and 21 μg/day established by the European Union and World Health Organization (Schwalfenberg et al., 2013).

The daily intake of lead in China has been measured at between 102 and 130 μg (Gao et al., 2006; Li et al., 2006). In the United States, the estimated intake of lead per day from fruits and vegetables is believed to be 15–25 μg (Schreck, 2010), while exposure to air, water, and industrial sources can result in up to a daily lead consumption of 200 μg. Exposure to drugs, herbal medicines, dietary supplements, and candies as chocolate can add up to 15 μg of lead ingested per day (Schreck, 2010).

17.7 SUMMARY AND CONCLUSIONS

Lead is a ubiquitous toxic heavy metal that has been classified as a possible human carcinogen, and contaminated water, industrial wastes, lead-based paints, and contaminated soils represent common sources of exposure. Lead is present in all soil and is absorbed by plants, with the lead uptake being proportional to lead in the soil. Various other factors can also influence the concentration of lead in plants, including soil pH, amount of organic matter in the soil, proximity to highways and industrial areas, variety and species of the plants, age of the plants at collection, and amount of lead in the water used to nourish the plants.

The amount of lead in teas from around the world appears to be relatively low and is generally below 3 μg/g. The amount of lead from tea and tea extracts as well as other herbal extracts is small and usually represents less than 5% of the total daily lead intake. It should be borne in mind that children are much more susceptive to all sources than adults. In adults, at these levels, the probability that significant and health-impacting intake of lead can occur by the ingestion of tea and dietary supplements containing tea and other extracts is low, particularly in light of the fact that lead has a low extractability from tea leaves with tea catechins having a high binding affinity for lead. Furthermore, tea extracts have been shown to protect against lead consumed from other sources. Exposure to lead from air, water, and industrial sources represents much more serious health threats. However, it is important to continue to monitor all sources of exposure to lead and other heavy metals due to its high toxicity and long-term effects.

Conflict of Interest
The authors received no financial support for this chapter and have no conflicts of interest to declare.

REFERENCES

Al-Othman ZA, Yilmaz E, Sumayli HM, Soylak M. 2012. Evaluation of trace metals in tea samples from Jeddah and Jazan, Saudi Arabia by atomic absorption spectrometry. *Bull Environ Contam Toxicol* 89: 1216–1219.

Anon. 2004. Soil lead levels. Interpretations and recommendations. http://umass.edu/plsoils/soilstest/lead1.htm. Accessed on December 20, 2010.

Arpadjan ED, Celik G, Gucer S. 2008. Arsenic, cadmium and lead in medicinal herbs and their fractionation. *Food Chem Toxicol* 46: 2871–2875.

Ashenef A. 2014. Essential and toxic metals in tea (*Camellia sinensis*) imported and produced in Ethiopia. *Food Addit Contam Part B Surveill* 7: 30–36.

Ashraf W, Mian AA. 2008. Levels of selected heavy metals in black tea varieties consumed in Saudi Arabia. *J Bull Environ Contam* 81: 101–104.

Bolle F, Brain W, Petit D, Boutakhit K, Feraille G, Van Loco J. 2011. Tea brewed in traditional metallic teapots as a significant source of lead, nickel and other chemical elements. *Food Addit Contam Part A Chem Anal Control Expo Risk Assess* 28: 1287–1293.

Borah KK, Bhuyan B, Sarma HP. 2010. Lead, arsenic, fluoride and iron contamination of drinking water in the tea garden belt of Darrang district, Assam, India. *Environ Monit Assess* 169: 347–352.

Caldas ED, Machado LL. 2004. Cadmium, mercury and lead in medicinal herbs in Brazil. *Food Chem Toxicol* 42: 599–603.
Chen H, Hu Y. 2009. Lead (Pb) isotopic fingerprinting and its applications in lead pollution studies in China: A review. *Environ Poll* 158: 1134–1146.
Chen L, Yang X, Jiao H, Zhao B. 2002. Tea catechins protect against lead-induced cytotoxicity, lipid peroxidation, and membrane fluidity in HepG2 cells. *Toxicol Sci* 69: 149–156.
Chen Y, Xu J, Yu M, Chen X, Shi J. 2010. Lead contamination in different varieties of tea plant (*Camellia sinensis* L.) and factors affecting lead bioavailability. *J Sci Food Agric* 90: 1501–1507.
Cheraghi M, Sobhanardakani S, Zandipak R, Lorestani B, Merrikhpour H. 2015. Removal of Pb (II) from aqueous solutions using tea leaves. *Iran J Toxicol* 9: 1247–1253.
Falahi E, Hedaiati R. 2013. Heavy metal content of black teas consumed in Iran. *Food Addit Contam Part B Surveill* 6: 123–126.
Feldheim W, Stelte W. 1975. Studies on tea. I. Contents of mercury, lead, and cadmium in tea leaves and tea decoctions. *Z Lebensm Unters Forsch* 159: 293–296. (In German).
Filippelli GM, Laidlaw MAS. 2010. The elephant in the playground: Confronting lead-contaminated soils as an important source of lead burdens in urban populations. *Persp Biol Med* 53: doi:10.1353/pbm.0.0136.
Gao JQ, Li XW, Zhao JL. 2006. 2000 Chinese total diet study—Dietary lead and cadmium intakes. *Wei Sheng Yan Jui* 35: 750–754. (In Chinese).
Garcia-Leston J, Mendez J, Pasaro E, Laffon B. 2010. Genotoxic effects of lead: An updated review. *Environ Int* 36: 623–636.
Gorospe EC, Gerstenberger SL. 2008. Atypical sources of childhood lead poisoning in the United States: A systematic review form 1966–2006. *Clin Toxicol* 46: 728–737.
Guo HY, Zhou WJ, Zhang YZ, Huang YX, Zhou Q, Yan XC, Zhang JX, Chen QC, Mao CM. 2008. Content of soil heavy metals and characteristics of environmental quality in tea plantations of Changsha Baili tea zone. *Huan Jing Ke Xue* 29: 2320–2326. (In Chinese).
Haban M, Habanova M, Otepka, P, Lukac N, Massanyi P. 2008. Concentration of heavy metals in various children's herbal tea types and their correlations. *J Environ Sci Health B* 43: 533–538.
Hamed EA, Meki ARMA, El-Mottaleb NAA. 2011. Protective effect of green tea on lead-induced oxidative damage in rat blood and brain tissue homogenates. *J Physiol Biochem* 66: 143–151.
Han WY, Shi YZ, Ma LF, Ruan JY, Zhao FJ. 2007. Effect of timing and seasonal variation on lead concentration of tea plant (*Camellia sinensis* (L.) O. Kuntze). *Chemosphere* 66: 84–90.
Han WY, Zhao FJ, Shi YZ, Ma LLF, Ruan JY. 2006. Scale and causes of lead contamination in Chinese tea. *Environ Pollut* 139: 125–132.
Inayat A, Khan SR, Chowdhry MN, Waheed A. 2010. Analysis of caffeine and heavy metal content in branded and unbranded tea available in Pakistan. *Pak J Sci Indust Res* 53: 20–24.
Jin CW, He YF, Zhang K, Zhou GD, Zheng SJ. 2005a. Lead contamination in tea leaves and non-edaphic factors affecting it. *Chemosphere* 61: 726–732.
Jin CW, Zheng SJ, He YF, Zhou GD, Zhou ZX. 2005b. Lead contamination in tea garden soils and factors affecting its bioavailability. *Chemosphere* 59: 1151–1159.
Jin Y, Liu P, Wu Y, Min J, Wang C, Sun J, Zhang Y. 2014. A systematic review on food lead concentration and dietary lead exposure in China. *Chin Med J* 127: 2844–2849.
Kang M, Luo Y, Shi Y, Ma L, Han W. 2004. A study on properties of uptake and accumulation of lead by tea plants. *J Tea*. doi: cnki:ISSN:0577-8921.0.2004-02-007. (In Chinese).
Kekedy-Nagy L, Ionesco A. 2009. Characterization and classification of tea herbs based on their metal content. *Acta Univ Sapien Agric Environ* 1: 11–19.
Khalaf AA, Moselhy WA, Abdel-Hamed MI. 2012. The protective effect of green tea extract on lead induced oxidative and DNA damage on rat brain. *Neurotoxicology* 33: 280–289.
Lavecchia R, Pugliese A, Zuorro A. 2010. Removal of lead from aqueous solutions by spent tea leaves. *Chem Engin Transact* 19: 73–78.
Li XW, Gao JQ, Chen JM. 2006. Study on the baseline contents and reference maximum limit standard of heavy metals and harmful elements of 23 Chinese herbs in northern China. *Wei Sheng Yan Jui* 35: 459–463. (In Chinese).
Mason LH, Harp JP, Han DY. 2014. Pb neurotoxicity: Neurophysiological effects of lead toxicity. *Biomed Res.* doi:10.1155/2014/840547.
Mathee A, Naicker N, Teare J. 2015. Retrospective investigation of a lead poisoning outbreak from the consumption of an Ayurvedic medicine: Durban, South Africa. *Int J Environ Res Public Health* 12: 7804–7813. http://www.mdpi.com/1660-4601/12/7/7804/htm.

Mehana EE, Meki AR, Fazili KM. 2012. Ameliorated effects of green tea extract on lead induced liver toxicity in rats. *Exp Toxicol Path* 64: 291–295.

Meki AR, Alghasham A, El-Deep ES. 2011. Effect of green tea extract on lead toxicity in different organs of rats. *Int J Health Sci* 5(Suppl 1): 12–15.

Meyer PA, Brown MJ, Falk H. 2008. Global approach to reducing lead exposure and poisoning. *Mut Res/Rev Mut Res* 659: 166–175.

Mohammed MI, Sulaiman MA. 2009. Analysis of some metals in some brands of tea sold in Kano, Nigeria. *Bayero J Pure Appl Sci* 2: 34–39.

Natesan S, Ranganathan V. 2006. Content of various elements in different parts of the tea plant and in infusions of black tea from Southern India. *J Sci Food Agric* 51: 125–139.

Ning P, Gong C, Zhang Y, Guo K, Bai J. 2011. Lead, cadmium, arsenic, mercury and copper levels in Chinese Yunnan Pu'er tea. *Food Addit Contam Part B Surveill* 4: 28–33.

Nookabkaew S, Rangkadilok N, Satayavivad J. 2006. Determination of trace elements in herbal tea products and their infusions consumed in Thailand. *J Agric Food Chem* 54: 6939–6944.

Orisakwe OE, Mbagwu HQ, Ukpai P, Udowelle NA. 2015. Survey of polycyclic aromatic hydrocarbons and lead in Chinese teas sold in Nigeria: Levels and health implications. *Rocz Panstw Zaki Hig* 66: 225–232.

Petit D, El Houari W, Jacobs K, Baeyens W, Leemakers M. 2013. Trace element content in tea brewed in traditional metallic and stainless steel teapots. *Environ Monit Assess* 185: 8957–8966.

Qin F, Chen W. 2007. Lead and copper levels in tea samples marketed in Beijing, China. *Bull Environ Contam Toxicol* 78: 128–131.

Salahinejad M, Aflaki F. 2010. Toxic and essential mineral elements content of black tea leaves and their tea infusions consumed in Iran. *Biol Trace Element Res* 134: 109–117.

Saper RB, Phillips RS, Sehgal A, Khouri N, Davis RB, Paquin J, Thuppil V, Kales SN. 2008. Lead, mercury, and arsenic in US- and Indian-manufactured ayurvedic medicines sold via the internet. *J Am Med Assoc* 300: 915–923.

Schreck JH. 2010. Lead in herbs and herbal medicines. http://www.drshen.com/lead-in-herbs.htm. Accessed on December 20, 2010.

Schwalfenberg G, Genius SJ, Rodushkin I. 2013. The benefits and risks of consuming brewed tea: Beware of toxic element contamination. *J Toxicol*. doi: 10.1155/2013/370460.

Seenivasan S, Manikandan N, Muraleedharan NN, Selvasundaram R. 2008. Heavy metal content of black teas from south India. *Food Control* 19: 746–749.

Seth TD, Hasan MZ, Sircar S. 1973. Lead content of Indian tea. *Bull Environ Contam Toxicol* 9: 124–128.

Sha'bani N, Miraj S, Rafiejan-Kohpayei M, Namjoo AR. 2015. Survey of the detoxification effect of green tea extract on the reproductive system in rats exposed to lead acetate. *Adv Biomed Res* 4: 155. doi: 10.4103/2277-9175.161582.

Shekoohiyan S, Ghoochani M, Mohagheghian A, Mahvi AH, Yunesian M, Nazmara S. 2012. Determination of lead, cadmium and arsenic in infusion tea cultivated north of Iran. *Iran J Environ Health Sci Eng* 9: 37. doi. 10.1186/1735-2746-9-37.

Shen FM, Chen HW. 2008. Element composition of tea leaves and tea infusions and its impact on health. *Bull Environ Contam Toxicol* 80: 300–304.

Soomro MT, Zahir E, Mohuiddin S, Khan AN, Naqvi II. 2008. Quantitative assessment of metals in local brands of tea in Pakistan. *Pak J Biol Sci* 11: 285–289.

Tsushida T, Takao T. 2006. Zinc, copper, lead and cadmium content in green tea. *J Sci Food Agric* 28: 255–258.

Wang H, Zhang MK. 2008. Effect of natural minerals on immobilization of lead in soils of tea gardens. *J Tea Sci*. doi: CNKI:SUN:CYKK.0.2008-02-016. (In Chinese).

Winiarska-Mieczan A. 2015. The potential protective effect of green, black, red and white tea infusions against adverse effect of cadmium and lead during chronic exposure—A rat model study. *Regul Toxicol Pharmacol* 73: 521–529.

Xia JG, Lan HX, Wu DY. 2010. Lead stress on growth of tea trees and physiological index in leaves of tea. *J Agro-Environ Sci*. doi: CNKI:SUN:NHBH.0.2010-01-010. (In Chinese).

Xu J, Bao JQ, Yu MG, Chen YX. 2014. Lead adsorption by root cell wall of tea plants. *Ying Yong Shen Tai Xue Bao* 25: 427–432. (In Chinese).

Xu J, Yu MG, Chen YX, Fu XP, Duan DC. 2011. Characteristics of distribution and chemical forms of Pb in tea plant varieties. *Ying Yong Sheng Tai Xue Bao* 22: 891–896. (In Chinese).

Zuorro A, Lavecchia R. 2010. Absorption of Pb(II) on spent leaves of green and black tea. *Am J Appl Sci* 7: 153–159.

18 Dual Effects of Vanadium
Toxicity Analysis in Developing Therapeutic Lead-Ups

Rituparna Ghosh and Samudra Prosad Banik

CONTENTS

18.1 INTRODUCTION

Vanadium is a member of group VB of the periodic table. The term *vanadium* originated from the name "Vanadis," the goddess of beauty and fertility in Scandinavian mythology, since the element forms beautiful multicolored chemical compounds (Mukherjee et al., 2004). Andres Manuel Del Rio was the first chemist to provide the idea of this new element in 1801. However, it was discovered by Nils Sefstrom, a Swedish chemist, in 1830. Due to its hardness and its ability to form alloys, ferrovanadium became a popular component of hard steel alloys used in machines and tools. In 1911, Dutton first reported the development of dry cough and irritation of eyes following occupational exposure to the fumes and dust of vanadium oxides (Dutton, 1911). Vanadium has become the subject of interest among nutritionists since the discovery that various marine species have this metal as an essential element (Nriagu, 1998). Other features that favor the status of vanadium as an essential nutrient include rapid excretion, ubiquity in the biosphere, homeostatic regulation with accumulation, and its ability to catalyze the chelation of biologically active compounds.

18.2 CHEMICAL AND PHYSICAL PROPERTIES

Pure vanadium is a bright silver white, soft, and ductile metal and is the 22nd most abundant element in the earth's crust, having atomic number 23. The chemistry of vanadium is complex because this element can exist in oxidation states from −5 to +5 and forms polymers frequently (Nechay, 1984). Predominant forms of vanadium in extracellular fluids and inside the cell are vanadate (VO_3^-, V^{+5}) and vanadyl (VO^{+2}, V^{+4}), respectively. The balance between these states is dependent on pH, ligand availability, and susceptibility to aerial oxidation. Inside, the cell vanadium is reduced by glutathione and other agents to vanadyl species (VO^{+2}) and stabilized as such by various ligands (Nechay, 1984). Vanadyl binds readily to proteins, amino acids, nucleic acids, phosphates, phospholipids, glutathione, citrate, oxalate, lactate, ascorbate, etc. (Barceloux, 1999). Bioavailability and exposure of vanadium to tissues depend upon its conjugation to these various cellular molecules.

18.3 SOURCE OF EXPOSURE

18.3.1 Air

Exposure of the general population to vanadium in air results primarily from the combustion of petroleum, coal, and heavy oils during the generation of electricity and heat. Very little vanadium is present in natural gas. Additional industrial sources of vanadium in the metallurgy industry include discharges from furnaces roasting vanadium slag and vanadium pentoxide smelting furnaces and crucibles that melt ferrovanadium alloys. Anthropogenic sources account for about two-thirds of the atmospheric vanadium (ATSDR, 1992). Continental dust, marine aerosols, and, to a lesser extent, volcanic emission account for the rest of the vanadium in the atmosphere. Very high level of vanadium results from boiler-cleaning operations as a result of the high concentration of vanadium oxides in the dust. Population in areas with high levels of residual fuel oil consumption may also be exposed to high background levels of vanadium, both from increased particulate deposition upon food crops and soil in the vicinity of power plants and higher ambient air levels (WHO, 1988).

18.3.2 Soil

Vanadium is a widely distributed trace metal in the earth's crust. The soil types, which contain the highest concentration of vanadium in India, are Tundra podzols and clays (ATSDR, 1992). Natural releases of vanadium to soil result from weathering of rock bearing vanadium minerals, precipitation of vanadium particulate from the atmosphere, deposition of suspended particulate from water, and plant and animal wastes. Anthropogenic releases of vanadium to soil include the use of certain fertilizers containing materials with a high vanadium content such as rock phosphate, superphosphate, and basic slag as well as disposal of industrial wastes such as slag heaps and mine tailings. Additional release to the environment results from the disposal of vanadium-containing wastes in landfills and from wet and dry deposition of airborne particulate, particularly in areas with high levels of residual fuel oil combustion (ATSDR, 1992).

18.3.3 Water

Drinking water is not an important source of exposure to vanadium for the general population (WHO, 1988). Natural sources of vanadium release to water include wet and dry deposition, soil erosion, and leaching from rocks and soils. The largest amount of vanadium release occurs naturally through water erosion of land surfaces. Anthropogenic releases to water may include leaching from the residue of ores and clays, vanadium-enriched slags, urban sewage sludge, and certain fertilizers, all of which are subjected to rain and groundwater drainage, as well as leachate from ash ponds and coal preparation wastes (WHO, 1988).

18.3.4 Food

Food is the major source of exposure to vanadium for the general population. The most likely pathway of vanadium exposure for the general population is the ingestion of food-bearing soil residue (Byrne and Kosta, 1978). The daily requirement for vanadium probably is small (<10 μg/day) with the highest source of vanadium present in black pepper, dill seeds, mushrooms, parsley, shell fish, spinach, and some prepared foods. In general, seafood contains higher concentrations of vanadium compared to terrestrial animal sources of foods. Beverages, fresh fruit and vegetables, cereals, liver, fats, and oil contain smaller amounts. The processing of foods tends to raise the concentration of vanadium. High concentrations of vanadium exist in tobacco (Barceloux, 1999).

18.4 ENVIRONMENTAL FATE

The global recycling of vanadium involves the release of vanadium from natural and anthropogenic sources to the air, water and soil, the transport of particles of vanadium in water and air, wet and dry deposition, absorption, and complexing (ATSDR, 1992).

18.4.1 Air

Vanadium enters in air as an aerosol of simple or complex vanadium oxides from anthropogenic sources and as mineral particles of less soluble trivalent forms from natural sources.

18.4.2 Soil

Compared with other metals, vanadium is moderately mobile in neutral and alkaline soils, but the mobility of vanadium in acidic soil is relatively small. The weathering of rock may release vanadium to air or water and also precipitate vanadium with polyvalent cations or organic complexing agents.

18.4.3 Water

A relatively small amount of vanadium is transported via water compared with the sedimentation process. In freshwater, vanadium exists as the vanadyl ion (VO_2^+, VO $[OH]^+$) in reducing environments and as the pentavalent vanadate form under oxidizing conditions. Transport of vanadium in water occurs primarily in suspension and, to a much lesser extent, in solution. Consequently, most of the vanadium in freshwater enters seawater and then absorbed to colloids or in suspended form. Removal of vanadium in seawater results from precipitation as hydrated ferric trioxide or by biochemical processes in organic matter. Marine plants and invertebrates accumulate much higher concentrations of vanadium compared with terrestrial plants and animals (ATSDR, 1992).

18.5 TOXICOKINETICS OF VANADIUM

18.5.1 Absorption

The absorption of vanadium compounds depends on their solubility and the route of entry. Inhalation of dusts and fumes is the major route of exposure of workers to vanadium. Although the rates of absorption are not well defined, vanadium compounds including V_2O_5 are well absorbed following instillation into lungs. In contrast, vanadium salts are poorly absorbed from the gastrointestinal tract and animal studies suggest that the amount of vanadium absorbed by the gastrointestinal tract is <1%–2% (Roshchin et al., 1980). Dietary vanadium occurs either as vanadyl (+4) or as vanadate (+5), with the latter being absorbed about three times more effectively by the gastrointestinal tract than the former. The skin probably is a minor route of exposure, especially because of the low solubility of metallic vanadium. The stomach converts most forms of vanadium into vanadyl (VO^{+2}) and vanadium remains in the vanadyl form in the alkaline media of the duodenum (ATSDR, 1992).

18.5.2 Distribution

Following transport in plasma, the distribution of vanadium from the blood is rapid, which might be due to the brisk transport of vanadium from blood to the tissues. Upon supplementation, vanadium has been reported to be incorporated in various organs and tissues including liver, kidney, brain, heart, muscles, and bones (Figure 18.1). The total amount of vanadium in the human body is small with the total body burden averaging about 100 μg V (Byrne and Kosta, 1978).

18.5.3 Metabolism

Vanadium is an element and, as such, is not metabolized. However, in the body, there occurs an interconversion of two oxidation states of vanadium, the tetravalent form, vanadyl (V^{+4}), and the pentavalent form, vanadate (V^{+5}). Vanadium can reversibly bind to transferrin protein in the blood and then be taken up into erythrocytes. Vanadate is considered more toxic than vanadyl, because vanadate is reactive with a number of enzymes and is a potent inhibitor of the Na^+/K^+ ATPase of plasma membranes. There is a slower uptake of vanadyl into erythrocytes compared to the vanadate form. This difference in uptake is due to the time required for the vanadyl form to be oxidized to vanadate. When V^{+4} or V^{+5} is administered intravenously, a balance is reached in which vanadium moves in and out of the cells at a rate that is comparable to the rate of vanadium removal from the blood (ATSDR, 1992). A decrease in glutathione, NADPH, and NADH occurs within an hour after intraperitoneal injection of sodium vanadate in mice. It is believed that vanadate requires these cytochrome P-450 components for oxidation to the vanadyl form (ATSDR, 1992).

18.5.4 Intracellular Transport of Vanadium

Vanadyl compounds have been found to penetrate the erythrocyte membranes through passive diffusion. Their interaction with membranes stabilizes the complexes and stimulates conformational changes of its proteins, although the process of spontaneous oxidation of cationic to anionic ions is also observed (Yang et al., 2003). Vanadate compounds penetrate the cells through anionic channels, where they are reduced to the vanadyl form. Further experiments show that vanadium in the cationic and anionic forms penetrates the intracellular space by means of passive diffusion, thus causing

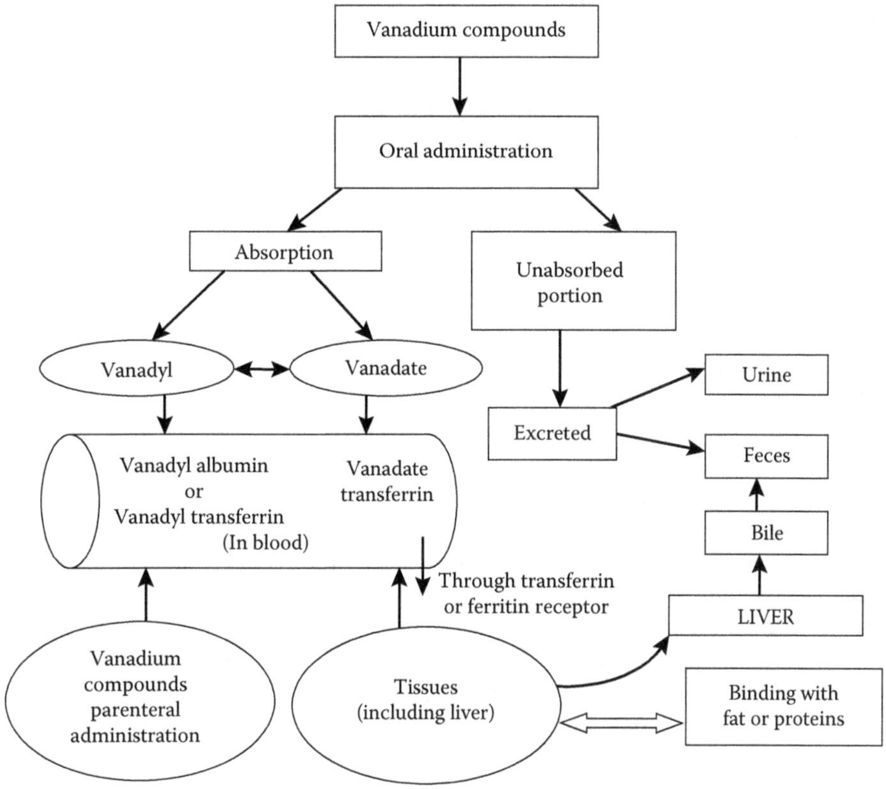

FIGURE 18.1 Vanadium cycle within a human body.

changes in the cellular microvilli and the structure of the actinic filament network (Yang et al., 2004). However, intracellular distributions of vanadium mainly occur in the nucleus and mitochondria. After penetration, vanadate ions undergo the oxidation–reduction processes, with participation of reduced glutathione, catechol, NADPH, NADH, cysteine, ascorbic acid, and glutathione reductase. The form in which vanadium is found inside the cell depends on the quantitative ratio of the entire pool of reducers to metabolic oxidizers and also on the concentration of vanadium itself (Goc, 2006).

18.5.5 Elimination/Excretion

The effects of vanadium administration persist even after vanadium has been withdrawn for several days. Unabsorbed vanadium is excreted in feces. When vanadium was administered through parenteral route, 10% of the vanadium was found in the feces of humans and rats (Barceloux, 1999). Vanadium has been reported to be excreted through bile and urine. It is, thus, the bile route through which a significant amount of vanadium may be ultimately excreted through feces (Figure 18.1).

18.6 PHYSIOLOGICAL RELEVANCE

It is well accepted that vanadium is required in the body as a trace mineral (French and Jones, 1993; Harland and Harden-Williams, 1994) although the precise physiologic actions of vanadium are yet to be established. Studies conducted in animals deprived of vanadium in diet have revealed a host of abnormalities including reduced fertility, higher rates of spontaneous abortion, and decreased milk production. Alongside, there were leg deformations, foot inflammation, and generalized musculoskeletal pain (Anke et al., 1986). Other biochemical alterations in vanadium-deficient goats included decreased levels of isocitrate dehydrogenase, lactate dehydrogenase, serum creatinine, and beta-lipoproteins and elevated serum glucose. Vanadium deficiency affects thyroid metabolism, decreased thyroid weight and thyroid weight–body weight ratio. Vanadium compounds modulate thyroid hormone levels in blood. Plasma triiodothyronine was higher in vanadium-supplemented rats. Vanadium compounds are also known to affect iron, glucose, and lipid metabolism. It is found to be essential for a number of species such as chicken and rats; deficiency symptoms in such species include retarded growth, impairment of reproduction, disturbance of lipid metabolism, and inhibition of Na^+/K^+ ATPase activity in the kidney, brain, and heart (Nriagu, 1998). However, these ill effects are rarely noted in human since a daily dietary intake of approximately 10 μg in humans is sufficient to satisfy the physiological requirements of vanadium. Entry of vanadium through food sources is limited to 40 ng/g of body weight through consumption of fishes, seafoods, and mushrooms (Ikebe and Tanaka, 1979; Myron et al., 1977), which is considered to have no side effects. Under normal physiological conditions, less than 5% of the ingested vanadium in its amount is taken into the blood (Nielsen, 1995a,b). However, bioabsorption can be increased by an additional 5% through synthesized vanadium salts especially those used as insulin-mimetic agent (Bodgen et al., 1982; Wiegmann et al., 1982). Inside the acidic environment of the vascular system, most vanadium is transformed to the vanadyl (IV) state (Chasteen et al., 1986). Residual vanadium in the +V state as vanadate is taken up by the intestinal epithelial cells through a poorly understood phosphate-based anionic transporter system (Cantley et al., 1978). Vanadium circulates as vanadyl in conjunction with transferrin and ferritin (IV only) (Chasteen et al., 1986) and is returned to bone as the primary reservoir for vanadium within the body. The element is excreted chiefly through urine.

18.7 THERAPEUTIC AND PHARMACOLOGICAL APPLICATIONS OF VANADIUM

Vanadium has been associated with a number of pharmacological uses, which include lowering of cholesterol, triglycerides, and glucose levels, diuretic and natriuretic effects, anticarcinogenic effect, contraction of blood vessels, and enhancement of oxygen affinity of hemoglobin and myoglobin (Mukherjee et al., 2004; Poucheret et al., 1998; Rehder, 1992; Thompson et al., 1993; Table 18.1). Vanadyl sulfate caused a significant and sustained decrease in blood pressure in spontaneously hypertensive rats. Bis (maltolato) oxovanadium (IV) was found to decrease appetite and body weight by decreasing hypothalamic neuropeptide mRNA expression and the compound was claimed to be a possible therapeutic agent in obesity (Wang et al., 2001; Wilsky et al., 2001).

Interest in vanadium pharmacology got a shot in the arm with the discovery of vanadium-containing enzymes (Badmaev et al., 1999). These enzymes are capable of replicating the activities of nitrogenase, haloperoxidase, and bromoperoxidase and are testimonial to the indispensible physiological roles of the element. Vanadium sulfate is used as muscle mass builder and used to improve performance in weight training athletes (Fawcett et al., 1997). Vanadium compounds have been reported to induce calcium signaling and to activate Ca^{2+} release–activated channel in Jurkat T lymphocytes and rat basophilic leukemia-2H3 mast cells (Mukherjee et al., 2004). Peroxovanadium compounds have been found to induce NO synthase in the liver of mice and enhance circulating nitrate level in the blood. The control of progression of leishmaniasis by vanadium compounds is believed to be an NO-dependent process. Vanadyl treatment reduced the incidences of the occurrence of urinary stones in nondiabetic rats, and overall, it reduced the mortality rate and did not show the development of renal and testicular tumors (Alimonti et al., 2000; Shi et al., 1996).

Many studies have shown successful exploitation of the toxic effects of vanadium against tumor cells. This stems from the cytotoxic activity of the metal coupled with alteration of the levels of protein tyrosine kinases (PTKs) (Sabbioni et al., 1991; Stem et al., 1993). Vanadium complexes have been reported to have antitumor activity against L-1210 murine leukemia and antineoplastic activity against rat liver

TABLE 18.1
Physiological Effects of Vanadium Uptake

Anticoagulant effect	Strong anticoagulant effect on cell lines.	Stankov et al. (2007)
Antidiabetic agent	Various vanadium compounds, including organovanadium compounds, such as bis(maltolato)oxovanadium(IV), can mimic insulin and improve glycemic control.	Kovel'man et al. (2002)
Antihypertensive activity	In hypertensive rats on a high-sucrose diet, vanadium can check increases in angiotensin II, possibly through an increase in endothelin-1 levels.	Shi et al. (1997)
Antilipemic agent	Pharmacologic amounts of vanadium (approximately 10–60 mcg daily, based on a food intake of 500 g dry weight) may affect cholesterol and triglyceride metabolism.	Harland and Harden-Williams (1994)
Antineoplastic activity	In animals, supplementation with vanadium, in the form of ammonium metavanadate, has been found to exert an antihepatocarcinogenic effect.	Chakraborty et al. (2007)
Antiviral effects	Potent inhibitory effects on cellular targets and bacterial and viral DNA and RNA polymerases, especially on HIV.	Stankov et al. (2007)
Cardiovascular changes	Environmental exposure to vanadium may be with an increased risk of cardiopulmonary disease.	Zhang et al. (2009)
Cytochrome P450 substrate	Vanadium may exert effects on CYP1A1 enzyme activity.	Anwar-Mohamed et al. (2009)
Immunostimulant properties	Vanadium can regulate immune response.	Jaremin (1985)
Immunosuppressant properties		French and Jones (1993)
Nephrotoxic effects	Elevated levels of vanadium inside the body may accumulate in kidney tissues and may be associated with an increased incidence of renal stone disease, distal renal tubular acidosis, and hypokalemic periodic paralysis.	French and Jones (1993)
Neurologic changes	Raised plasma levels of vanadium have been reported in patients with manic depression.	Naylor (1984)

tumors, fluid and solid Ehrlich ascites tumor, and TA3Ha murine mammary adenocarcinoma; they inhibit the growth of human tumor colony formation of HEP-2 epidermoid carcinoma cells (Murthy et al., 2000). Vanadium was found to modulate phases I and II hepatic metabolizing enzymes and improve antioxidant status during the development of carcinogenesis (Mukherjee et al., 2004). More recently, *N,N'*-ethylenebis (pyridoxylideneiminato) vanadium (IV) complex [Pyr2enV(IV)] has been responsible for inducing 93% and 57% of cell mortality in A375 (human melanoma) and A549 (human lung carcinoma) cells, respectively (Strianese et al., 2013). Thus, vanadium compounds offer protection against chemical carcinogenesis, either by inhibition of active carcinogenic metabolites (Evangelou, 2002) or through blocking of the tyrosine phosphatases (Strianese et al., 2013).

Among all prospective therapeutic applications of vanadium, its insulin-mimetic effect is probably the best understood and the most promising one (Cam et al., 2000). Almost all metabolic effects of insulin can be reproduced using only vanadium species. The basis of insulin action involves binding to a specific tyrosine kinase receptor that initiates a cascade of downstream signaling events involving protein kinases. In contrast, vanadium acts independently of the insulin receptor (Fantus et al., 1994; Mooney et al., 1989; Strout et al., 1989). The blood glucose–lowering effect of vanadium in the form of sodium orthovanadate, sodium metavanadate, and vanadyl sulfate has been associated with the modulation of several enzymes including 6-phosphofructokinase, glucokinase, and fructose 2, 6-bisphosphatase. Vanadate was reported to stimulate glucose metabolism (Barceloux, 1999) by the activation and autophosphorylation of solubilized insulin receptor, stimulating the tyrosine kinase activity of the insulin receptor β-subunit (Li et al., 1996), and by the inhibition of phosphoprotein tyrosine phosphatases (Fantus et al., 1995). Recently, more light has been shed on the role of vanadium and its hypoglycemic potential in diabetes therapy (Nielsen, 1998; Nielsen and Uthus, 1990). The collective potential of vanadium to increase glucose uptake, promote glycogen synthesis, and inhibit lipolysis has led to serious consideration of its prospective therapeutic use in the treatment of both type 1 and type 2 diabetes mellitus and the associated cardiological complications. However, studies need to be carried out to confirm any potential cross-reactivity of vanadium with conventional hypoglycemic medications before they are available as formulations on the market shelf.

18.8 SAFETY OF VANADIUM IN THERAPEUTIC AND NUTRACEUTICAL APPLICATION: TOXICITY EVALUATION

In spite of the multitude of health-promoting effects of vanadium verified on animal models, effective human trials are still in their early stages (Eisenberg et al., 1998). Safe use of vanadium in novel therapeutics has been compounded by the toxicity associated with this metal at various body targets.

18.8.1 Metabolic Changes

The general toxic effect of vanadium observed in rabbits was weight loss along with decrease in albumin levels and increase in globulins levels with resultant two-fold decrease of the albumin–globulin ratio, decrease in whole blood cholinesterase levels and total serum protein levels, increase in nucleic acid levels in the blood, and a considerable increase in the blood chloride concentration. Vanadium increases triglyceride concentrations in the liver and blood, decreases the serum cholesterol level, and increases glutamate–oxaloacetate transaminase and glutamate–pyruvate transaminase activity. Vanadium also inhibits the activity of monoamine oxidase, resulting in the accumulation of serotonin in the central nervous system (CNS), leading to functional disturbances, such as bronchospasm, diarrhea, and urinary incontinence, and it is also responsible for the dystrophic and necrotic processes in the kidneys and the high permeability of the blood vessels. Vanadium exposure increases the catabolism of cystine and cysteine and reduces the content of coenzyme A in rat liver and the synthesis of cholesterol (Roshchin et al., 1980).

18.8.2 Hematological Changes

Vanadium reduces the deformability of erythrocytes, causing depressed erythrocyte count, hemoglobin level, and hematocrit index, and also produces peroxidative changes in the erythrocyte membrane, thus leading to hemolysis and anemia (Ścibior et al., 2006; Zaporowska and Wasilewski, 1992; Zaporowska et al., 1993). The reduction in RBC and hemoglobin may suggest that vanadate is toxic to bone marrow. The percentage of reticulocytes increased in the peripheral blood. Following exposure to vanadium, some changes were also noted in white blood cell system. The elevation in white blood cell counts in the rats treated with vanadium indicates a systemic inflammation (al-Bayati et al., 1990). An increase in platelet count as well as the presence of megaplatelets was observed following vanadium exposure. Vanadium induces thrombocytosis and it might correlate with some thromboembolic diseases (González-Villalva et al., 2006).

18.8.3 Respiratory Changes

Exposures to vanadium and its compounds result primarily in adverse effects to the respiratory system (Ress et al., 2003; Sax, 1984; Worle-Knirsch et al., 2007). Animals exposed to vanadium pentoxide dust particles showed marked tracheitis, pulmonary edema, bronchopneumonia, enteritis, marked irritation of the respiratory mucosa, and expiratory difficulty. Pathological changes observed included chronic rhinitis, emphysema, patches of lung atelectasis, bronchopneumonia, and pyelonephritis (Roshchin et al., 1980). In cases of severe poisoning, respiratory failure was followed by death. In the lungs, there was capillary congestion, perivascular and focal edema, bronchitis, and focal interstitial pneumonia.

The effect of vanadium dust in upper respiratory tract is characterized by irritation, rhinitis, wheezing, nasal hemorrhage, cough, sore throat, and chest pain. These symptoms are prominent in boilermakers who are exposed to high concentrations of V_2O_5 fumes, hence the name "boilermakers" bronchitis. An acute bronchitis with dyspnea and expiratory wheezes may develop. Asthma may develop after heavy exposure to vanadium compounds (ATSDR, 1992). Plant workers previously free of lung disease and exposed to vanadium pentoxide developed bronchial hyperreactivity and symptoms of asthma (Irsigler et al., 1999). Vanadate acts directly on human bronchial smooth muscle promoting the release of Ca^{2+} from intracellular store via the production of inositol phosphate second messengers and inhibition of Ca^{2+} ATPase and causing spasms (Cortijo et al., 1997). Occasionally, pulmonary edema and/or pneumonia may result with fatal consequences. Exposure to vanadium may cause a metal fume fever-like syndrome associated with neutrophilic alveolitis associated with marked neutrophilia (Vandenplas et al., 2002), lung inflammation, and apoptosis (Dill et al., 2004; Fortoul et al., 2002; Wang et al., 2003).

18.8.4 Cardiovascular Changes

Severe exposure of animals to vanadium oxides and salts produced cardiovascular changes, namely, occurrence of arrhythmias and extrasystole, prolongation of the Q-RST interval, decrease in the height of the P and T waves of the ECG, perivascular swelling, and fatty changes in the myocardium (Roshchin et al., 1980). Intense vasoconstriction in the spleen, kidney, and intestines and increase in systolic blood pressure had been reported following intravenous injections of sodium ortho- and metavanadate (ATSDR, 1992).

18.8.5 Hepatic Changes

Vanadium has been reported to cause fatty changes with partial cell necrosis in the liver of rat and rabbits. A reduction in the liver tissue respiration and a decrease in the albumin/globulin ratio in the serum were also observed along with vanadium accumulation in the tissue (ATSDR, 1992). Vanadium is a potent inhibitor of many enzymes, while it stimulates adenylate cyclase. It has been shown to inhibit cholesterol biosynthesis and lower plasma cholesterol levels. Lipid peroxidation of rat liver microsomes and mitochondria was induced by vanadium compounds (WHO, 2000). Chronically, histopathological changes observed in the liver following the higher level of inhalation exposure included central vein congestion with scattered small hemorrhages and granular degeneration of hepatocytes (WHO, 2000). Vanadium pentoxide exposure stimulates the secretion of cytokines and chemokines by hepatocytes and causes associated hepatotoxicity. Significant amounts of vanadate accumulate in the intermembrane space of the liver mitochondria and the detoxification mechanism of reduction of vanadate is inadequate enough to prevent damaging actions on hepatic mitochondria (Zychlinski and Byczkowski, 1990).

18.8.6 Renal Changes

Glomerular hyperemia, necrosis of convoluted tubules, albuminuria, fatty changes in the kidney of the rat and rabbit, and granular degeneration of the epithelial cells of the convoluted tubules were reported to be related to acute vanadium exposure (Roshchin et al., 1980). Vanadate has diuretic and natriuretic effects on the kidney in the rat and increases the urinary excretion of calcium, phosphate, bicarbonate, and chloride by the rat kidney. These diuretic and natriuretic effects are thought to be due to the inhibition of Na^+/K^+ ATPase, which causes inhibition of the tubular reabsorption. Chronic exposure to increasing dietary concentrations of sodium metavanadate resulted in vanadium accumulation in the kidneys (Rattner et al., 2006). When renal brush border membrane vesicles are exposed to vanadium pentoxide, there is a time-dependent inhibition of citrate uptake in the membrane, possibly contributing to nephrotoxicity (Sato et al., 2002). Chronically in experimental animals, histopathological changes observed in the kidneys following the higher level of inhalation exposure included marked granular degeneration of the epithelium of the convoluted tubules and corticomedullary microhemorrhagic foci in the kidneys. Data on the effects of vanadium on the human kidney are lacking.

18.8.7 Immunological Changes

Vanadium may affect the normal functioning of the immune system. Exposure of mice to vanadium resulted in a decrease in antibody-forming cells in the spleens challenged with sheep erythrocytes; serum immunoglobulins were not affected, and there was a dose-related increase in resistance to *Escherichia coli* endotoxin lethality and a decrease in resistance to *Listeria* lethality. Peritoneal macrophage activity also decreased but without any effects on viability. There was enlargement of the liver and spleen with enhanced formation of splenic megakaryocytes and red blood cell precursors (ATSDR, 1992).

18.8.8 Neurological Changes

Vanadium has been found to decrease the level of noradrenaline, while it appears to increase those of dopamine and 5-hydroxytryptamine. Vanadium has a selective effect on adrenergic pathways and inhibits protein synthesis in neuroblastoma cells. Neurophysiological effects including disturbances of the CNS, impaired conditioned reflexes and neuromuscular excitability, and decrease in chronaxie ratio of antagonistic muscles have been reported following exposure of dogs and rabbits to vanadium oxides and salts (ATSDR, 1992). Nonspecific signs and symptoms in workers exposed to vanadium compounds when cleaning oil-fired boiler include headache, weakness, nausea, vomiting, and tinnitus. Disorders of nervous system include severe neurotic states and tremor of the fingers and hands (WHO, 1988). Severe acute exposure to vanadium pentoxide has major pathophysiological manifestations on the nervous system (WHO, 2000). Inhalation thereof produces a time-dependent loss of dendritic spines, necrotic-like cell death, and considerable alterations of the hippocampus CA1 neurophile, all of which is associated with spatial memory impairment (Avila-Costa et al., 2006). Additionally, there is a decrease in the number of tyrosine hydroxylase immunoreactive neurons in the pars compacta of the substantia nigra (Avila-Costa et al., 2004). Within the ependymal epithelium, cilia loss, cell sloughing, and cell layer detachment occur after vanadium pentoxide inhalation (Avila-Costa et al., 2005). The damage results in disrupted permeability of the epithelium and promotes access of inflammatory mediators to the underlying neuronal tissue causing injury and neuronal death (Avila-Costa et al., 2005). The Na^+/K^+ ATPase in rat brain is inhibited by vanadate (VO^{3-}) or the vanadyl ion (VO^{+2}). In humans, severe chronic exposure results in nervous disturbances and neurasthenic or vegetative symptoms (WHO, 2000).

18.8.9 Genotoxic and Mutagenic Changes

Vanadium compounds have been shown to inhibit or stimulate the activity of many DNA or RNA enzymes inducing several genotoxic and mutagenic effects (Stemmler and Burrows, 2001). Altamirano-Lozano et al. (1996) showed single-stranded DNA breaks in individual testis cells of mice following vanadium pentoxide exposure. Vanadate has been shown to inhibit the expression of the gene for phosphoryl pyruvate carboxylase in rat hepatoma cells and stimulate DNA synthesis using cultures of quiescent human fibroblasts and in Swiss mouse 3T3 and 3T6 cells. Further, Ivancsits et al. (2002) found that exposure of human fibroblasts to vanadate effectively causes DNA strand breaks. Vanadium(IV)-mediated free radical generation causes 2′-deoxyguanosine hydroxylation, which leads to DNA strand breaks. Vanadyl sulfate can damage DNA in human lymphocytes and HeLa cell (Wozniak and Blasiak, 2004). Vanadium is capable of inducing cytotoxic and cytostatic effects, chromosomal damage, and the induction of apoptosis (Altamirano-Lozano, 1998; Rodriguez-Mercado et al., 2003). In a National Toxicology Program (NTP) study, it was also reported to be carcinogenic in rats and mice (Ress et al., 2003), although the interpretation of the results has recently been disputed (Duffus, 2007).

18.9 CONCERNS ON THE USE OF VANADIUM AS AN ANTIDIABETIC AGENT

Despite the impressive spectrum of the antidiabetic properties of vanadium, vanadium salts have been closely studied for their toxic side effects (Domingo, 1996; Srivastava, 2000). These include gastrointestinal toxicity, diarrhea, dehydration via reduced water intake, appetite suppression, and reduced body weight gain (Blondel et al., 1989; Brichard et al., 1998; Dai et al., 1994; Domingo et al., 1994; Gil et al., 1987; Heyliger et al., 1985; Ramanadham et al., 1989; Sakurai et al., 1990). In fact, the severity of this toxicity has resulted in vanadate-induced death in diabetic animals (Clark et al., 2012; Heyliger et al., 1985). Pilot studies with type 2 diabetic patients have achieved improved cholesterol levels and insulin sensitivity.

18.10 UNDERSTANDING VANADIUM CHEMISTRY TO COMBAT TOXICITY

Over recent years, most of the efforts to minimize vanadium-associated toxicity have been devoted toward modifying the chemical structure of vanadium. Organovanadium complexes have been synthesized to increase uptake in the intestinal lining, as vanadium salts are poorly ingested (McNeill et al., 1995). Bis(maltolato) oxovanadium (IV) had been one such success story to achieve antidiabetic effects at one-half of the dosage required of vanadium salts, with minimal toxicity. Another compound synthesized is naglivan, which showed antidiabetic effects without any significant toxic side effects (Cam et al., 1993). Three new organic complexes were synthesized for vanadyl sulfate administration (Reul et al., 1999). All three compounds exerted better glycemic control and restored glycolytic enzyme function as compared to the vanadium salt therapy alone. The combination of vanadium compounds with hydrogen peroxide produces potent insulin-mimetic agents known as peroxovanadium (pV) complexes. These novel class of compounds unlike vanadate acts on the insulin receptor and its cognate ligands (Bevan et al., 1995; Posner et al., 1994) and prevent their dephosphorylation to constitutively turn on the insulin response. Alternative phytochemical-based strategies for the administration of vanadium such as that in conjunction with *Trigonella* seed powder, a hypoglycemic material itself, have also gained much popularity in recent years in terms of restoring glycemic control in mice model.

18.10.1 Effect on Reproductive Health

Vanadium is a globally recognized threat to reproductive health. Vanadium derivatives act as generator of free radicals, which impairs testicular functions, spermatogenesis, and sperm maturation in rats. Exposure to vanadium (IV) has been found to result in profuse ultrastructural changes in male germ cells with concomitant induction of apoptosis (Aragon and Altamirano-Lozano, 2001; Aragon et al., 2005). The *in vivo* results indicate that an imbalance between the generation of reactive oxygen species and oxygen system protectors occurs in the testis and epididymis of vanadium-treated rats. This was reflected in decreased testicular and epididymal antioxidant enzymes activities such as SOD, catalase, glutathione peroxidase, glucose-6-phosphate dehydrogenase, glutathione-S-transferase, and glutathione reductase. Unlike other antioxidant enzymes, the testicular and epididymal γ-glutamyl transpeptidase activities were elevated after vanadium exposure as part of the adaptation of cells to oxidative stress. The levels of various nonenzymatic antioxidants such as GSH, vitamin C, and vitamin E were also found to have decreased due to vanadium exposure. Moreover, there is a marked elevation in lipid peroxidation level and hydroxyl radical generation that may indirectly lead to an increase in oxidative stress, causing ROS-induced damages to macromolecules such as DNA, proteins, and key enzymes involved in testicular functions (Chandra et al., 2007a–c). Vanadium exposure caused a significant reduction in epididymal sperm number, which indicates an interference with spermatogenesis or may be attributed to the impairment of sperm maturation and secretory functions of epididymal cells. Histopathology of testes (Figure 18.2) is characterized by the suppression of spermatocyte populations and abolition

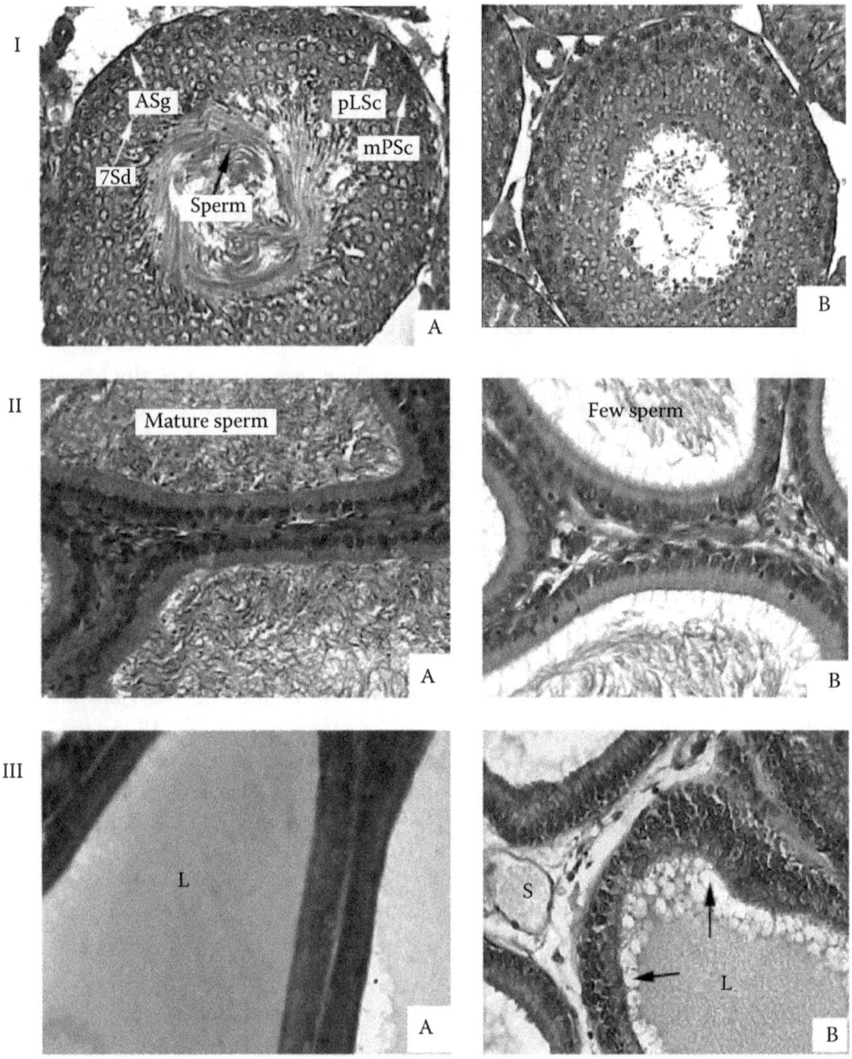

FIGURE 18.2 **(See color insert.)** Photomicrographs of paraffin-embedded hematoxylin and eosin–stained histopathological sections. (IA) Rat testicular section (400×) from control animals. (IB) Testis section (400×) from vanadate-treated animals (0.4 mg/kg, i.p. for 26 days). ASg, spermatogonia A; pLSc, preleptotene spermatocytes; mPSc, mid-pachytene spermatocytes; 7Sd, step 7 spermatid. (IIA) Rat cauda epididymis section (400×) from control animals. (IIB) Cauda epididymis section (400×) from vanadate-treated animals (0.4 mg/kg, i.p. for 26 days). (IIIA) Rat ventral prostate gland section (400×) from control animals. (IIIB) Ventral prostate gland section (400×) from vanadate-treated animals (0.4 mg/kg, i.p. for 26 days). L, lumen; S, stroma. Arrows indicate vacuoles.

of round spermatids beyond stage VII, with the number of earlier germ cells being severely reduced. This might be due to an increase in the number of germ cells apoptosis, as a consequence of adverse effects on Sertoli cells and germ cell interactions. Thus, it may be speculated that an increase in free radical formation relative to loss of antioxidant defense system during vanadium exposure may be implicated in vanadium-induced testicular, epididymal, and spermatogenic toxicities. Aside from testicular modifications, regressive and degenerative changes were also noted in cauda epididymis, ventral prostate gland, coagulating gland, and seminal vesicles. Epididymal sperm number in vanadium-treated rats has significantly decreased (46.3%), showing abnormal sperm morphology with abnormal hook, banana-shaped head, coiled flagellar end, microcephaly, and macrocephaly (Chandra et al., 2010). Female fertility was also severely affected by vanadium exposure. Administration of 200 ppm of sodium metavanadate in adult female rats caused significant reduction in the number of implantation sites and viable fetuses (Morgan and El-Tawii, 2003). Ultrastructural studies indicated associated fibrosis in stroma, disorganized uterine glands, and disruption of columnar epithelium (Shrivastava et al., 2007).

The physiological complications associated with vanadium are currently too many to prescribe it safely in drug formulations. Therefore, in spite of proven requirement as a trace element in our body, this flip side of vanadium has raised serious eyebrows in the medical fraternity.

18.11 CONCLUDING REMARKS

The role of vanadium in promoting human health especially in treating chronic ailments such as type 2 diabetes has been unequivocally established. Vanadium is safe in adults if consumption limit does not exceed 1.8 mg/day as set by the National Institute of Medicine. At higher doses, such as those used to treat diabetes, vanadium frequently causes unwanted side effects including abdominal discomfort, diarrhea, nausea, loss of energy, and CNS disturbances. The most persistent difficulty in designing vanadium-based drugs has been the unequal bioavailability of the metal in various tissue parts owing to the conjugation of vanadium to ferritin or transferring receptor inside the body. Apparently, the amount of elemental vanadium also varies in different commonly available vanadium salts through which dietary supplementation is going to occur, for example, vanadyl sulfate contains 31% elemental vanadium, sodium metavanadate contains 42% elemental vanadium, and sodium orthovanadate contains 28% elemental vanadium. The biggest challenge therefore for the pharmaceutical industry is to find the optimal tissue-specific dosage of vanadium, which will be minimally interfering with the normal physiology and functioning of other regions. Hopefully, with a fast-advancing medical technology, the future is not too far where robust state-of-the-art treatment procedures such as personalized medicine will be able to address these issues successfully to ensure a better disease-free life for the human society.

ACKNOWLEDGMENT

The authors acknowledge Prof. Amar K. Chandra, Dept. of Physiology, University of Calcutta, for his mentorship in some of the studies reported and also the University Grants Commission, Govt. of India, for providing the funds to carry out the studies.

REFERENCES

al-Bayati, M.A., Giri, S.N., and Raabe, O.G. 1990. Time and dose-response study of the effects of vanadate in rats: Changes in blood cells, serum enzymes, protein, cholesterol, glucose, calcium, and inorganic phosphate. *J Environ Pathol Toxicol Oncol* 10:206–213.

Alimonti, A., Petrucci, F., Krachler, M., Bocca, B., and Caroll, S. 2000. Reference values for chromium, nickel and vanadium in urine of youngsters form the urban area of Rome. *J Environ Monit* 2:351–354.

Altamirano-Lozano, M. 1998. Genotoxic effects of vanadium compounds. *Invest Clin* 39:39–47.

Altamirano-Lozano, M., Alvarez-Barrera, L., Basurto-Alcantara, F., Valverde, M., and Rojas, E. 1996. Reprotoxic and genotoxic studies of vanadium pentoxide in male mice. *Teratog Carcinog Mutagen* 16:7–17.

Anke, M., Groppel, B., Gruhn, K., Kosla, T., and Szilagyi, M. 1986. In *Spurenelement-Symposium New Trace Elements*, eds. Anke, M., Baumann, W., Braunlich, H., Bruckner, C., and Groppel, B., pp. 1266–1275, Friedrich-Schiller-Universitat Jena, Jena, Germany.

Anwar-Mohamed, A., Elbekai, R.H., and El-Kadi, A.O. 2009. Regulation of CYP1A1 by heavy metals and consequences for drug metabolism. *Expert Opin Drug Metab Toxicol* 5(5):501–521.

Aragon, M.A. and Altamirano-Lozano, M. 2001. Sperm and testicular modifications induced by subchronic treatments with vanadium (IV) in CD-1 mice. *Reprod Toxicol* 15:145–151.

Aragon, M.A., Ayala, M.E., Fortoul, T.I., Bizarro, P., and Altamirano-Lozano, M. 2005. Vanadium induced ultrastructural changes and apoptosis in male germ cells. *Reprod Toxicol* 20:127–134.

ATSDR (Agency for Toxic Substances and Disease Registry). 1992. *Toxicological Profile for Vanadium,* U.S. Department of Health and Human Services, Public Health Service, Atlanta, GA.

Avila-Costa, M.R., Colín-Barenque, L., Zepeda-Rodriquez, A., Antuna, S.B., Saldivar, O.L., Espejel-Maya, G., Mussali-Galante, P. et al. 2005. Ependymal epithelium disruption after vanadium pentoxide inhalation. A mice experimental model. *Neurosci Lett* 381:21–25.

Avila-Costa, M.R., Fortoul, T.I., Nino-Cabrera, G., Colin-Barenque, L., Bizarro-Nevares, P., Gutierrez-Valdez, A.L., Ordonez-Librado, J.L. et al. 2006. Hippocampal cell alterations induced by the inhalation of vanadium pentoxide (V_2O_5) promote memory deterioration. *Neurotoxicity* 27:1007–1112.

Avila-Costa, M.R., Montiel Flores, E., Colin-Barenque, L., Ordonez, J.L., Gutierrez, A.L., Nino-Cabrera, H.G., Mussali-Galante, P., and Fortoul, T.I. 2004. Nigrostriatal modifications after vanadium inhalation: An immunocytochemical and cytological approach. *Neurochem Res* 29:1365–1369.

Badmaev, V., Prakash, S., and Majeed, M. 1999. Vanadium: A review of its potential role in the fight against diabetes. *J Altern Complement Med* 5:273–291.

Barceloux, D.G. 1999. Vanadium. *Clin Toxicol* 37:265–278.

Bevan, A.P., Drake, P.G., Yale, J.F., Shaver, A., and Posner, B.I. 1995. Peroxovanadium compounds: Biological actions and mechanism of insulin-mimesis. *Mol Cell Biochem* 153(1–2):49–58.

Blondel, O., Bailbe, D., and Portha, B. 1989. In vivo insulin resistance in streptozotocin diabetic rats: Evidence for reversal following oral vanadate treatment. *Diabetologia* 32:185–190.

Boden, G., Chen, Z., Ruiz, J., Van Rossum, G.D.V., and Turco, S. 1996. Effects of vanadyl sulfate on carbohydrate and lipid metabolism in patients with non-insulin-dependent diabetes mellitus. *Metabolism* 45:1130–1135.

Bodgen, J.D., Higashino, H., Lavenhar, M.A., Bauman, J.W. Jr., Kemp, F.W., and Aviv, A. 1982. Balance and tissue distribution of vanadium after short-term ingestion of vanadate. *J Nutr* 10(2):233–245.

Brichard, S.M., Okitolonda, W., and Henquin, J.C. 1998. Long-term improvement of glucose homeostasis by vanadate treatment in diabetic rats. *Endocrinology* 123:2048–2053.

Byrne, A.R. and Kosta, L. 1978. Vanadium in foods and in human body fluids and tissues. *Sci Total Environ* 10:17–30.

Cam, M.C., Brownsey, R.W., and McNeill, J.H. 2000. Mechanisms of vanadium action: Insulin-mimetic and insulin-enhancing agents. *Can J Physiol Pharmacol* 78:829–847.

Cam, M.C., Cros, G.H., Serrano, J.J., Lazaro, R., and McNeill, J.H. 1993. In vivo antidiabetic actions of naglivan, an organic vanadyl compound in streptozotocin-induced diabetes. *Diabetes Res Clin Pract* 29(2):111–121.

Cantley, L.C. Jr., Resh, M.D., and Guidotti, G. 1978. Vanadate inhibits the red cell (Na^+/K^+) ATPase from the cytoplasmic side. *Nature* 272(5653):552–554.

Chakraborty, T., Chatterjee, A., Rana, A., Rana, B., Palanisamy, A., Madhappan, R., and Chatterjee, M. 2007. Suppression of early stages of neoplastic transformation in a two-stage chemical hepatocarcinogenesis model: Supplementation of vanadium, a dietary micronutrient, limits cell proliferation and inhibits the formations of 8-hydroxy-2′-deoxyguanosines and DNA strand-breaks in the liver of sprague-dawley rats. *Nutr Cancer* 59(2):228–247.

Chandra, A.K., Ghosh, R., Chatterjee, A., and Sarkar, M. 2007a. Effects of vanadate on male rat reproductive tract histology, oxidative stress markers and androgenic enzyme activities. *J Inorg Biochem* 101:944–956.

Chandra, A.K., Ghosh, R., Chatterjee, A., and Sarkar, M. 2007b. Vanadium -induced testicular toxicity and its prevention by oral supplementation of zinc sulphate. *Toxicol Mech Method* 17:175–187.

Chandra, A.K., Ghosh, R., Chatterjee, A., and Sarkar, M. 2007c. Amelioration of vanadium induced testicular toxicity and adrenocortical hyperactivity by vitamin E acetate in rats. *Mol Cell Biochem* 306:189–200.

Chandra, A.K., Ghosh, R., Chatterjee, A., and Sarkar, M. 2010. Protection against vanadium-induced testicular toxicity by testosterone propionate in rats. *Toxicol Mech Method* 20(6):306–315.

Chasteen, N.D., Lord, E.M., and Thompson, H.F. 1986. In *Frontiers in Bioinorganic Chemistry*, ed. Xavier, A.V., pp. 133–141, VCH Verlagsgesellschaft, Weinhem, Germany.

Chasteen, N.D., Lord, E.M., Thompson, H.J., and Grady, J.K. 1986. Vanadium complexes of transferrin and ferritin in the rat. *Biochem Biophys Acta* 884(1):84–92.

Clark, T.A., Heyliger, C.E., Kopilas, M., Edel, A.L., Junaid, A., Aguilar, F., Smyth, D.D. et al. 2012. A tea/vanadate decoction delivered orally over 14 months to diabetic rats induces long-term glycemic stability without organ toxicity. *Metabolism* 61(5):742–753.

Cohen, N., Halberstam, M., Shlimovich, P., Chang, C.J., Shamoon, H., and Rossetti, L. 1995. Oral vanadyl sulfate improves hepatic and peripheral insulin sensitivity in patients with non-insulin-dependent diabetes mellitus. *J Clin Invest* 95:2501–2509.

Cortijo, J., Villagrasa, V., Marti-Cabrera, M., Villar, V., Moreau, J., Advenier, C., Morcillo, E.J., and Small, R.C. 1997. The spasmogenic effects of vanadate in human isolated bronchus. *Br J Pharmacol* 121:1339–1349.

D'Onofrio, F., Le, M.Q., Chiasson, J.L., and Srivastava, A.K. 1994. Activation of mitogen activated protein (MAP) kinases by vanadate is independent of insulin receptor autophosphorylation. *FEBS Lett* 340(3):269–275.

Dai, S., Thompson, K.H., Vera, E., and McNeill, J.H. 1994. Toxicity studies on one-year treatment of non-diabetic and streptozotocin-diabetic rats with vanadyl sulphate. *Pharmacol Toxicol* 75(5):265–273.

Dill, J.A., Lee, K.M., Mellinger, K.H., Bates, D.J., Burka, L.T., and Roycroft, J.H. 2004. Lung deposition and clearance of inhaled vanadium pentoxide in chronically exposed F344 rats and B6C3F1 mice. *Toxicol Sci* 77:6–18.

Domingo, J.L. 1996. Vanadium: A review of the reproductive and developmental toxicity. *Reprod Toxicol* 10(3):175–182.

Domingo, J.L., Gomez, M., Sanchez, D.J., Llobet, J.M., and Keen, C.L. 1994. Relationship between reduction in food intake and amelioration of hyperglycemia by oral vanadate in STZ-induced diabetic rats. *Diabetes* 43(10):1267, 1269–1270.

Duffus, J.H. 2007. Carcinogenicity classification of vanadium pentoxide and inorganic vanadium compounds, the NTP study of carcinogenicity of inhaled vanadium pentoxide, and vanadium chemistry. *Regul Toxicol Pharmacol* 47:110–114.

Dutton, W.F. 1911 Vanadiumism. *JAMA* 56:1648–1652.

Eisenberg, D.M., Davis, R.B., Ettner, S.L., Appel, S., Wilkey, S., Van Rompay, M., and Kessler, R.C. 1998. Trends in alternative medicine use in the United States, 1990–1997. *JAMA* 280:1569–1575.

Evangelou, A.M. 2002. Vanadium in cancer treatment. *Crit Rev Oncol Hemat* 42:249–265.

Fantus, I.G., Ahmad, F., and Deragon, G. 1994. Vanadate augments insulin-stimulated insulin receptor Kinase activity and prolongs insulin action in rat adipocytes. Evidence for transduction of amplitude of signaling into duration of response. *Diabetes* 43(3):375–383.

Fantus, I.G., Deragon, G., Lai, R., and Tang, S. 1995. Modulation of insulin action by vanadate: Evidence of a role for phosphotyrosine phosphatase activity to alter cellular signaling. *Mol Cell Biochem* 153:103–112.

Fawcett, J.P., Farquhar, S.J., Thou, T., and Shand, B.I. 1997. Oral vanadyl sulphate does not affect blood cells, viscosity or biochemistry in humans. *Pharmacol Toxicol* 80:202–206.

Fortoul, T.I., Quan-Torres, A., Sanchez, I., Lopez, I.E., Bizarro, P., Mendoza, M.L., Osorio, L.S. et al. 2002. Vanadium in ambient air: Concentrations in lung tissue from autopsies of Mexico City residents in the 1960s and 1990s. *Arch Environ Health* 57:446–449.

French, R.J. and Jones, P.J. 1993. Role of vanadium in nutrition: Metabolism, essentiality and dietary considerations. *Life Sci* 52(4):339–346.

Gil, J., Miralpeix, M., Carreras, J., and Bartrons, R. 1987. Insulin like effects of vanadate on glucokinase activity and fructose-2,6-bisphosphate levels in the liver of diabetic rats. *J Biol Chem* 262:6658–6662.

Goc, A. 2006. Biological activity of vanadium compounds. *Cent Eur J Biol* 1:314–332.

Goldfine, A.B., Simonson, D.C., Folli, F., Patti, M.E., and Kahn, C.R. 1995. Metabolic effects of sodium metavanadate in humans with insulin-dependent and noninsulin-dependent diabetes mellitus in vivo and in vitro studies. *J Clin Endocrinol Metab* 80:3311–3320.

Gonzalez-Villalva, A., Fortoul, T.I., Avila-Costa, M.R., Pinon-Zarate, G., Rodriguez-Laraa, V., Martínez-Levy, G., Rojas-Lemus, M. et al. 2006. Thrombocytosis induced in mice after subacute and subchronic V_2O_5 inhalation. *Toxicol Ind Health* 22:113–116.

Harland, B.F. and Harden-Williams, B.A. 1994. Is vanadium of human nutritional importance yet? *J Am Diet Assoc* 94(8):891–894.

Heyliger, C.E., Tahiliani, A.G., and McNeill, J.H. 1985. Effect of vanadate on elevated blood glucose and depressed cardiac performance of diabetic rats. *Science* 227:1474–1477.

Ikebe, K. and Tanaka, R. 1979. Determination of vanadium and nickel in marine samples by flameless and flame atomic absorption spectrophotometry. *Bull Environ Contam Toxicol* 21:526–532.

Irsigler, G.B., Visser, P.J., and Spangenberg, P.A. 1999. Asthma and chemical bronchitis in vanadium plant workers. 1999. *Am J Ind Med* 35:366–374.

Ivancsits, S., Pilger, A., Diem, E., Schaffer, A., and Rudiger, H.W. 2002. Vanadate induces DNA strand breaks in cultured human fibroblasts at doses relevant to occupational exposure. *Mut Res* 519:25–35.

Jaremin, B. 1985. Immunoregulatory properties of some heavy metals in physiology and pathology conditions. Part II. *Bull Inst Marit Trop Med Gdynia* 36(1–4):81–88.

Kovel'man, I.R., Tochilkin, A.I., Beliaeva, N.F., Gorodetskiĭ, V.K., Kniazhev, V.A., Tkachenko, S.B., Semenov, L.L., and Semenova, N.V. 2002. Modern approaches to pharmacotherapy of diabetes mellitus type II. *Vopr Med Khim* 48(4):337–352.

Li, J., Elberg, G., Crans, D.C., and Shechter, Y. 1996. Evidences for the distinct vanadyl (+4)—Dependent activating system for manifesting insulin-like effects. *Biochemistry* 35:8314–8328.

McNeill, J.H., Yuen, V.G., Dai, S., and Orvig, C. 1995. Increased potency of vanadium using organic ligands. *Mol Cell Biochem* 153(1–2):175–180.

Mooney, R.A., Bordwell, K.L., Luhowskyj, S., and Casnellie, J.E. 1989. The insulin-like effect of vanadate on lipolysis in rat adipocytes is not accompanied by an insulin-like effect on tyrosine phosphorylation. *Endocrinology* 124(1):422–429.

Morgan, A.M. and El-Tawil, O.S. 2003. Effects of ammonium metavanadate on fertility and reproductive performance of adult male and female rats. *Pharmacol Res* 47:75–85.

Mukherjee, B., Patra, B., Mahapatra, S., Banerjee, P., Tiwari, A., and Chatterjee, M. 2004. Vanadium—An element of biological significance. *Toxicol Lett* 150:135–143.

Murthy, M.S., Rao, L.N., Kuo, L.Y., Toney, J.H., and Marks, T.J. 2000. Anti-tumor and toxicologic properties of the organometallic anti-cancer agent vanadocene dichloride. *Inorg Chem Acta* 152:117–124.

Myron, D.R., Givand, S.H., and Nielsen, F.H. 1977. Vanadium content of selected foods as determined by flameless atomic absorption spectroscopy. *J Agric Food Chem* 25(2):297–300.

Naylor, G.J. 1984. Vanadium and manic depressive psychosis. *Nutr Health* 3(1–2):79–85.

Nechay, B.R. 1984. Mechanism of action of vanadium. *Annu Rev Pharmacol Toxicol* 24:501–524.

Nielsen, F. and Uthus, E. 1990. The essentiality and metabolism of vanadium. In *Vanadium in Biological Systems*, ed. Chasteen, N.D., pp. 51–62, Kluwer, Dordrecht, the Netherlands.

Nielsen, F.H. 1995a. Vanadium and its role in life. In *Metal Ions and Biological Systems*, eds. Sigel, H. and Sigel, A., vol. 31, pp. 543–573, Marcel Dekkar, New York.

Nielsen, F.H. 1995b. *Metal-Ligand Interactions in Biological Fluids, Bioinorganic Medicine*, vol. 1, pp. 425–427, Marcel Dekkar, New York.

Nielsen, F.H. 1998. The nutritional essentiality and physiological metabolism of vanadium in higher animals. In *Vanadium Compounds*, pp. 297–307, American Chemical Society, Washington, DC.

Nriagu, J.O. 1998. History, occurrence, and uses of vanadium. In *Vanadium in the Environment*, ed. Nriagu, J.O., pp. 1–24, Wiley, New York.

Posner, B.I., Faure, R., Burgess, J.W., Bevan, A.P., Lachance, D., Zhang-Sun, G., Fantus, I.G., Ng, J.B., Hall, D.A., and Lum, B.S. 1994. Peroxovanadium compounds. A new class of potent phosphotyrosine phosphatase inhibitors which are insulin mimetics. *J Biol Chem* 269(6):4596–4604.

Poucheret, P., Verma, S., Grynpas, M.D., and McNeill, J.H. 1998.Vanadium and diabetes. *Mol Cell Biochem* 188:73–80.

Ramanadham, S., Mongold, J.J., Brownsey, R.W., Cros, G.H., and McNeill, J.H. 1989. Oral vanadyl sulphate in treatment of diabetes mellitus in rats. *Am J Physiol* 257:H904–H911.

Rattner, B.A., McKernan, M.A., Eisenreich, K.M., Link, W.A., Olsen, G.H., Hoffman, D.J., Knowles, K.A., and McGowan, P.C. 2006. Toxicity and hazard of vanadium to mallard ducks (*Anas platyrhynchos*) and Canada geese (*Branta Canadensis*). *Toxicol Environ Health A* 69:331–351.

Rehder, D. 1992. Structure and function of vanadium compounds in living organisms. *Biometal* 5:3–12.

Ress, N.B., Chou, B.J., Renne, R.A., Dill, J.A., Miller, R.A., Roycroft, J.H., Hailey, J.R., Haseman, J.K., and Bucher, J.R. 2003. Carcinogenicity of inhaled vanadium pentoxide in F344/N rats and B6C3F1 mice. *Toxicol Sci* 74:287–296.

Reul, B.A., Amin, S.S., Buchet, J.P., Ongemba, L.N., Crans, D.C., and Brichard, S.M. 1999. Effects of vanadium complexes with organic ligands on glucose metabolism; a comparison study in diabetic rats. *Br J Pharmacol* 126(2):467–477.

Rodriguez-Mercado, J.J., Roldan-Reyes, E., and Altamirano-Lozano, M. 2003. Genotoxic effects of vanadium(IV) in human peripheral blood cells. *Toxicol Lett* 144:359–369.

Roshchin, A.V., Ordzhonikidze, E.K., and Shalganova, I.V. 1980. Vanadium–toxicity, metabolism, carrier state. *J Hyg Epidemiol Microbiol Immunol* 24:377–383.

Sabbioni, E., Pozzi, G., Devos, S., Pintar, A., Casella, L., and Fischbach, M. 1993. The intensity of vanadium(V)-induced cytotoxicity and morphological transformation in BALB/3T3 cells is dependent on glutathione-mediated bioreduction to vanadium(IV). *Carcinogenesis* 14(12):2565–2568.

Sabbioni, E., Pozzi, G., Pintar, A., Casella, L., and Garattini, S. 1991. Cellular retention, cytotoxicity and morphological transformation by vanadium(IV) and vanadium (V) in BALB/3T3 cell lines. *Carcinogenesis* 12(1):47–52.

Sakurai, H., Tsuchiya, K., Nakatsuka, M., Sofue, M., and Kawada, J. 1990. Insulin-like effects of vanadyl ion in streptozotocin induced diabetic rats. *J Endocrinol* 126:451–459.

Sato, K., Kusaka, Y., Akino, H., Kanamaru, H., and Okada, K. 2002. Direct effects of vanadium on citrate uptake by rat renal brush border membrane vesicles (BBMV). *Ind Health* 40:278–281.

Sax, N.I. 1984. *Dangerous Properties of Industrial Materials,* 6th edn., pp. 2717–2720, Van Nostrand Reinhold Company, New York.

Scibior, A., Zaporowska, H., and Ostrowski, J. 2006. Selected haematological and biochemical parameters of blood in rats after subchronic administration of vanadium and/or magnesium in drinking water. *Arch Environ Contam Toxicol* 51 (2006) 287–295.

Shi, S.J., Preuss, H.G., Abernethy, D.R., Li, X., Jarrell, S.T., and Andrawis, N.S. 1997. Elevated blood pressure in spontaneously hypertensive rats consuming a high sucrose diet is associated with elevated angiotensin II and is reversed by vanadium. *J Hypertens* 15(8):857–862.

Shi, X., Jiang, H., Mao, Y., Ye, J., and Saffiotti, U. 1996. Vanadium(IV) mediated free radical generation and related 2′-deoxyguanosine hydroxylation and DNA damage. *Toxicology* 106:27–38.

Shrivastava, S., Jadon, A., and Shukla, S. 2007. Effect of Tiron and its combination with nutritional supplements against vanadium intoxication in female albino rats. *J Toxicol Sci* 32:185–192.

Srivastava, A.K. 2000. Anti-diabetic and toxic effects of vanadium compounds. *Mol Cell Biochem* 206(1–2):177–182.

Stankov, J.M., Uskoković-Marković, S., Holclajtner-Antunović, I., Todorović, M., and Djurdjević, P. 2007. Compounds of Mo, V and W in biochemistry and their biomedical activity. *J Trace Elem Med Biol* 21(1):8–16.

Stemmler, A.J. and Burrows, C.J. 2001.Guanine versus deoxyribose damage in DNA oxidation mediated by vanadium(IV) and vanadium(V) complexes. *J Biol Inorg Chem* 6:100–106.

Stern, A., Yin, X., Tsang, S.S., Davison, A., and Moon, J. 1993. Vanadium as a modulator of cellular regulatory cascades and oncogene expression. *Biochem Cell Biol* 71(3–4):103–112.

Strianese, M., Basile, A., Mazzone, A., Morello, S., Turco, M.C., and Pellecchia, C.J. 2013. Therapeutic potential of a pyridoxal-based vanadium(IV) complex showing selective cytotoxicity for cancer versus healthy cells. *Cell Physiol* 228:2202–2209.

Strout, H.V., Vicario, P.P., Saperstein, R., and Slater, E.E. 1989. The insulin-mimetic effect of vanadate is not correlated with insulin receptor tyrosine kinase activity nor phosphorylation in mouse diaphragm in vivo. *Endocrinology* 124(4):1918–1924.

Thompson, K.H., Leichter, J., and McNeil, J.H, 1993. Studies of vanadyl sulfate as a glucose-lowering agent in STZ diabetic rats. *Biochem Biophys Res Commun* 97:1549–1555.

Vandenplas, O., Binard-Van Caugh, F., Gregoire, J., Brumagne, A., and Larbanois, A. 2002. Fever and neutrophilic alveolitis caused by a vanadium based catalyst. *Occup Environ Med* 59:785–787.

Wang, J., Yuen, V.G., and McNeill, J.H. 2001. Effect of vanadium on insulin sensitivity and appetite. *Metabolism* 50:667–673.

Wang, L., Medan, D., Mercer, R., Overmiller, D., Leornard, S., Castranova, V., Shi, X., Ding, M., Huang, C., and Rojanasakul, Y. 2003. Vanadium-induced apoptosis and pulmonary inflammation in mice: Role of reactive oxygen species. *J Cell Physiol* 195:99–107.

Wang, X., Dyson, M.T., Jo, Y., and Stocco, D.M. 2003. Inhibition of cyclooxygenase-2 activity enhances steroidogenesis and steroidogenic acute regulatory gene expression in MA-10 mouse Leydig cells. *Endocrinology* 144:3368–3375.

WHO (World Health Organization). 1988. Environmental health criteria No. 81: Vanadium, World Health Organization, Geneva, Switzerland.

WHO (World Health Organization). 2000. Vanadium, Chapter 6.12. WHO Regional Office for Europe, Copenhagen, Denmark.

WHO. 1988. Environmental health criteria No. 81: Vanadium, World Health Organization, Geneva, Switzerland.

Wiegmann, T.B., Day, H.D., and Patak, R.V. 1982. Intestinal absorption and secretion of radioactive vanadium *48VO3 in rats and effect of $Al(OH)_3$. *J Toxicol Environ Health* 10(2):233–245.

Wilsky, G.R., Goldfine, A.B., Kostyniak, P.J., McNeil, J.H., Yang, L.O., Khan, H.R., and Crans, D.C. 2001. Effects of vanadium(IV) compounds in the treatment of diabetes: In vitro and in vivo studies with vanadyl sulfate and bis (maltolato) oxovanadium(IV). *J Inorg Biochem* 85:33–42.

Worle-Knirsch, J.M., Kern, K., Schleh, C., Adelhelm, C., Feldmann, C., and Krug, H.F. 2007. Nanoparticulate vanadium oxide potentiated vanadium toxicity in human lung cells. *Environ Sci Technol* 41:331–336.

Wozniak, K. and Blasiak, J. 2004. Vanadyl sulfate can differentially damage DNA in human lymphocytes and HeLa cells. *Arch Toxicol* 78:7–15.

Yang, X.G., Wang, K., Lu, J., and Crans, D.C. 2003. Membrane transport of vanadium compounds and interaction with the erythrocyte membrane. *Coordin Chem Rev* 273:103–111.

Yang, X.G., Yang, X.D., Yuan, L., Wang, K., and Crans, D.C. 2004. The permeability and cytotoxicity of insulin-mimetic vanadium compounds. *Pharmaceut Res* 21:1026–1033.

Zaporowska, H. and Wasilewski, W. 1992. Haematological effects of vanadium on living organisms. *Comp Biochem Physiol C* 102:223–231.

Zaporowska, H., Wasilewski, W., and Słotwińska, M. 1993. Effect of chronic vanadium administration in drinking water to rats. *Biometals* 6:3–10.

Zhang, Z., Chau, P.Y., Lai, H.K., and Wong, C.M. 2009. A review of effects of particulate matter-associated nickel and vanadium species on cardiovascular and respiratory systems. *Int J Environ Health Res* 19(3):175–185.

Zychlinski, L. and Byczkowski, J.Z. 1990. Inhibitory effects of vanadium pentoxide on respiration of rat liver mitochondria. *Arch Environ Contam Toxicol* 19:138–142.

19 Fluoride Toxicity
A Major Health Hazard

*Abbas Ali Mahdi, Ayushi Jain,
Maheep Bhatnagar, and Rekha Bhatnagar*

CONTENTS

19.1 INTRODUCTION

Fluoride is probably the first inorganic ion that drew attention of scientific world for its toxic effects on development, and now the fluoride toxicity through drinking water is well recognized as a global health problem. World Health Organization (WHO, 2010) recommended an optimal level of fluoride in drinking water, for the prevention of dental caries and normal growth of dentition. When fluoride in drinking water is added under fluoridation program in many countries, and also through other dietary sources, foodstuffs, and beverages, the optimal daily uptake for individuals is often exceeded. This results in uncontrolled and unpredictable consumption of fluoride that may often exceed its therapeutic value, thereby inducing negative health impacts within the population as a whole. Fluoride efficiency and deficiency exert an inimical impact on the sustainability of the organism (Mohapatra et al., 2009).

Fluorine (F) is a univalent poisonous gaseous halogen that occurs under standard conditions (i.e., room temperature and normal vapor pressure) as a light yellow-green, pungent, acrid gas. The fluorine molecular weight is 18.998 g/mol, the boiling point is −187°C, and the melting point is −223°C. The fluorine atom is found in nature as a single isotope containing 9 protons, 9 electrons, and 10 neutrons, with an atomic number of 19. It is the most chemically reactive of all elements. Fluorine readily forms compounds with most other elements, including the noble gases krypton, xenon, and radon. The high reactivity is due to its high electronegativity, its unusually low dissociation energy, and the great bond strength of its compounds. Fluorine compounds are characterized by an oxidation state −1 (Preedy, 2015). In soil, water, plants, food, and even in air, fluorine is present in its ionic form. During weathering, fluorine can be leached out and dissolved into water, where its natural concentration depends on the geological, chemical, and physical characteristics of the reservoir (ADA, 2005). It accounts for about 0.3 g/kg of the earth's crust and exists in the form of fluorides in a number of minerals, of which fluorspar, cryolite, and fluorapatite are the most common. In 1990, the Joint FAO/IAEA/WHO Expert Committee regarded it as the third category element that is potentially toxic but may be necessary for certain functions of the human body at low doses (WHO, 1998). Its main physiological function is to prevent dental caries and participate in bone mineralization. However, excessive accumulation of fluoride in the body can cause chronic fluorosis, which can lead to disorders of bone metabolism, dental fluorosis, and skeletal fluorosis (Ayoob and Gupta, 2006; Ismail and Hasson, 2008). In 2010, the WHO estimated tens of millions of patients with dental fluorosis and skeletal fluorosis worldwide (WHO, 2010). Of all the fluorosis regions, China and India are the countries having more serious conditions of fluorosis than others.

Fluoride is an inorganic anion of fluorine, which is naturally found in the biosphere with ubiquitous presence in the environment and is the most reactive anion. This feature also explains why fluorine is mainly found as inorganic compounds. Application of fluoride as systemic (water, salt, sugar, milk, food supplements) and topical (toothpastes, gels, mouth rinses, varnishes) agent has increased its concentration in nature (Almeida et al., 2007). The benefits of fluoride include both caries control and bone formation (Dhar and Bhatnagar, 2009). However, it can negatively affect the quality of the developing mineralized tissues, causing dental and skeletal fluorosis. The variable responses seem to be determined by the dose and time of exposure (Everett, 2011). Thus, the knowledge of fluoride is crucial not only to understand its biological effects but also to optimize fluoride-driven preventive and therapeutic strategies.

19.2 OCCURRENCE, DISTRIBUTION, AND GEOCHEMISTRY

Fluorine is well distributed throughout the earth's crust (0.06%–0.09%) and exists only in combination with other elements as fluoride compounds. The average crustal abundance of fluoride is 300 mg/kg (Tebutt, 1983) that are constituents of minerals. Fluoride is commonly associated with volcanic activity and fumarolic gases. Thermal waters, especially those of high pH, are also rich in fluoride. Fluoride is also found in all natural waters at some concentration. Seawater typically contains about 1 mg/L, while rivers and lakes generally exhibit concentrations of less than 0.5 mg/L. In groundwater, low or high concentrations of fluoride can occur, depending on the nature of the rocks and the occurrence of fluoride-bearing minerals. The occurrence, distribution, and mobility of fluorine in aqueous reservoirs also depend on the interplay of a number of geochemical processes that determine its removal or release into the solution. Environmental conditions, such as water pH, the chemical composition of the system, and the reaction kinetics, strongly influence all these processes (Berger et al., 2012). The natural fluoride concentration varies strongly and depends on the local geological structure and also on local/regional human activities.

Fluorides frequently occur in igneous as well as in metamorphic rocks, especially in alkali rocks like granite, basalt, shale, clays, and calcium phosphate. The average fluorine concentration in crystalline rocks can range from 100 mg/kg up to over 1000 mg/kg, because these contain a large proportion of F-bearing minerals. Minerals having the greatest effect on the hydrogeochemistry of fluoride are fluorite, apatite, mica, amphiboles, fluorspar, rock phosphate, cryolite, certain clays, and villiamite (WHO, 2006; Bernal et al, 2014). Fluorite (CaF_2) is a very common fluoride mineral of low solubility occurring in both igneous and sedimentary rocks. Fluoride compounds like sodium fluoride are soluble in water; aluminum, calcium, and magnesium fluorides are only sparingly soluble and the solubility differs with respective compositions (Borgnino et al., 2013). Minerals of commercial importance include cryolite and rock phosphates. The fluoride salt cryolite is used for the production of aluminum and also as an important pesticide. Rock phosphates are converted into phosphate fertilizers by the removal of up to 4.2% fluoride; the removed and purified fluoride (as fluorosilicates) is a major source of fluoride (Carpenter, 1969; Loganathan et al., 2003).

19.3 ENVIRONMENTAL EXPOSURE

Today, people are exposed to fluoride in two major ways. Systemic fluoride exposure is via water, foods, beverages, and artificial fluoride supplements, and topical fluoride exposure is through dental hygiene products: toothpastes, mouth rinses, professional fluoride gels, foams, and varnishes (Dhar and Bhatnagar, 2009). The WHO and Indian Council of Medical Research (ICMR) guideline value for desirable limit of fluoride in drinking water is 1.0 mg/L and permissible limit is 1.5 mg/L (ICMR, 2005; WHO, 2010). Fluoride is most commonly found in drinking water as a natural contaminant or as an additive intended to provide public health protection (artificial water fluoridation). Fluorosilicic acid, sodium hexafluorosilicate, and sodium fluoride are commonly used in municipal water fluoridation schemes under safe water for drinking (Chae et al., 2006). Exposure to fluoride in human beings goes practically unnoticed. Some proportion of fluoride in the body comes via the increased use of fluoride-containing compounds as additives in toothpaste (1000–1500 ppm), mouthwash (230–900 ppm of fluoride), and in dietary supplements. However, drinking water is the most leading source of intake of this element in humans and daily intakes of fluoride vary widely according to the various sources of exposures taken in totality (Dhar and Bhatnagar, 2009).

19.3.1 WATER

Fluoride concentration in natural water depends on various factors such as temperature, pH, solubility of fluoride-bearing minerals, anion exchange capacity of aquifer materials (OH for F), nature of geological formation, climate, and contact time of water with particular formation. Uncontrolled mining, untreated industrial waste, and the municipal solid waste have polluted surface and groundwater, thus raising the level of fluoride in the groundwater (WHO, 2006). Moreover, agricultural practices like the use of phosphate fertilizer and industrial activities like the use of clays (used in ceramic industries) or burning of coals contribute to high fluoride concentrations in the groundwater. The groundwater fluoride concentration depends majorly on the reaction time with the aquifer minerals and thus the high fluoride concentration in groundwater results due to the long contact time with the aquifers (Levy, 2003).

In areas with relatively high fluoride concentrations in groundwater, drinking water becomes increasingly important as a source of fluoride. Daily exposure in volcanic areas (e.g., the United Republic of Tanzania) with high fluoride levels in drinking water may be up to 20–30 mg for adults. The water containing the highest concentration of fluorides corresponds to water resources located in mountainous regions or areas with geological deposits of marine origin, such as in Southeast Asia and Northeast Africa (WHO, 2006; Doull et al., 2006). Studies conducted in the last decade show that a significant number of people in towns exposed to fluoridated drinking water suffer from health issues and varying degrees of fluorosis. Studies have also reported that the fluoride content in the water supplies is affected by climatic variations such that the fluoride content increases after a drought and decreases after heavy rains. Dental, nonskeletal, and skeletal fluorosis is prevalent and severe in areas that have naturally occurring high levels of fluoride in groundwater, which is used for drinking (Levy, 2003; Susheela, 2003).

19.3.2 Food

Fluoride content in plants and animal food depends on its concentration in soil, water used for irrigation, and air deposition. All vegetation contains some fluoride, which is absorbed from soil and water (Arnesen, 1997). The highest levels in field-grown vegetables are found in curly kale (up to 40 mg/kg fresh weight) and endive (0.3–2.8 mg/kg fresh weight) (Cerklewski, 1997; Chen and Whitford, 1999). Other foods containing high levels include fish (0.1–30 mg/kg) and tea (Cook, 1969). High concentration of fluoride in tea can be caused by high natural concentrations in tea plants or by the use of additives during growth or fermentation (Whyte et al., 2005). Levels in dry tea can be 3–300 mg/kg (average 100 mg/kg), so 3–4 cups of tea may contain approximately 0.4–0.8 mg of fluoride (Cao et al., 2004). In areas where water with high fluoride content is used to prepare tea, the intake via tea can be several times greater. Seafood items are also reported to be rich in fluoride. Diet consisting of fish and aquatic invertebrates is especially rich in fluoride, albeit freshwater fish usually show higher F levels than marine species (EFSA, 2008; Dhar and Bhatnagar, 2009; WHO, 2010). Fish, such as sardines, may contribute to higher dietary fluoride intake if the bones are ingested. A large percentage of fluoride in human body may come from canned foods, drinks processed with fluoridated water, medicines, insecticides, and fertilizers (Ozsvath, 2009).

Some investigations have suggested that the increased consumption of bottled water, soft drinks, and juices (cola, grapefruit, pineapple, apple, nectar, and prune flavors) prepared with fluoridated water may be a significant source of systemic fluoride, and thus, these drinks have been implicated as risk factors for dental fluorosis in young children (Hakkinen et al., 1999). Variability in fluoride concentration among beverages is primarily determined by the fluoride content of the water used during manufacturing and by the fruit used.

The average daily dietary intake of fluoride (expressed on a body weight basis) by children residing in optimally fluoridated (1 ppm) communities is 0.05 mg/kg/day as compared to the communities without optimally fluoridated water where the average intake for children is about 50% lower. Dietary fluoride intake by adults in optimally fluoridated (1 ppm) areas averages 1.4–3.4 mg/day and in nonfluoridated areas averages 0.3–1.0 mg/day. Levels of daily exposure to fluoride also depend on the geographical area. For example, in the Netherlands, the total daily intake is calculated to be 1.4–6.0 mg of fluoride. Food seems to be the source of 80%–85% of fluoride intake; intake from drinking water is 0.03–0.68 mg/day and from toothpaste 0.2–0.3 mg/day (Bartlett et al., 2005; Browne et al., 2005; WHO, 2010).

19.3.3 Air

The natural concentrations of fluoride in air are of the order of 0.5 ng/m^3. If anthropogenic emissions are included, worldwide background concentrations are of the order of 3 ng/m^3. In the Netherlands, concentrations in areas without sources are 30–40 ng/m^3, rising to 70 ng/m^3 in areas with many sources (WHO, 2006; Preedy, 2015). In a survey of fluoride in the air, of some communities in the United States and Canada, concentrations were in the range of 0.02–2.0 μg/m^3. In some parts of India and China, fluoride concentrations in indoor air ranged from 16 to 46 μg/m^3 owing to the indoor combustion of high-fluoride coal (WHO, 2010).

19.3.4 Dental Products

The use of dental products is the direct means of exposure to fluoride. For dental purposes, fluoride preparations may contain low (0.25–1 mg/tablet; 1000–1500 mg of fluorine per kg of toothpaste) or high concentrations (liquids containing 10,000 mg/L and gels containing 4,000–6,000 mg/kg used for local applications) (Dhar and Bhatnagar, 2009). Mouth rinse, ulcer gels, and some dental medicines also contain varying amount of fluorides (WHO, 1996).

19.4 GLOBAL SCENARIO OF FLUORIDE ENDEMICITY

Waters with high fluoride concentrations occur in large and extensive geographical belts associated with (1) sediments of marine origin in mountainous areas, (2) volcanic rocks, and (3) granitic and gneissic rocks. A typical example of the first extends from Iraq and Iran through Syria and Turkey to the Mediterranean region and hence from Algeria to Morocco. Other important examples come from the southern parts of the United States, southern Europe, and the southern parts of the former USSR. The most well-known and documented area associated with volcanic activity follows the East African Rift system from the Jordan valley down through Sudan, Ethiopia, Uganda, Kenya, and the United Republic of Tanzania (WHO, 1996). Groundwater with high fluoride concentrations occurs in many areas of the world including large parts of Africa, China, the Eastern Mediterranean, and southern Asia (India, Sri Lanka) (Figure 19.1). One of the best-known high-fluoride belts on land extends along the East African Rift from Eritrea to Malawi. The Americas and Japan have similar belts (Susheela, 1993; WHO, 2006; Hussain et al., 2010).

It is estimated that around 200 million people, from among 25 nations the world over, are under the dreadful fate of fluorosis. India and China, the two most populous countries of the world, are the worst affected (Susheela, 2003). Worldwide, an estimated 370 million people receive artificially fluoridated water, and 50 million drink water that is naturally fluoridated at or near the optimal level. Overall, some 30 countries practice water fluoridation, and the percentage of populations receiving artificially fluoridated water varies greatly. Countries where fluoridation is practiced (and the percentage of their populations receiving fluoridated water) include Argentina (21%), Australia (61%), Brazil (41%), Canada (43%), Chile (40%), Colombia (80%), Israel (75%), Malaysia (70%), New Zealand (61%), Singapore (100%), the Republic of Ireland (73%), Spain (10%), and the United Kingdom (10%). Most other western European countries have ceased, or never practiced, water fluoridation. In several Latin American countries, where centralized water supplies are often lacking, fluoridated salt is the chosen method of providing dental protection across disparate communities (Susheela, 1993; UNICEF, 1999; WHO, 2010).

The fluoridation of drinking water often generates both strong support and opposition within communities. The practice is recommended by the U.S. Department of Health and Human Services (HHS) to prevent tooth decay. According to the Centers for Disease Control and Prevention (CDC), in 2010, 73.9% of the people in the United States who receive their water from a public water system received fluoridated water (roughly 204.3 million people). The Department of Health and Human Services (HHS), United States, had long recommended an optimal fluoridation level in the range of 0.7–1.2 mg/L to prevent tooth decay (Levy 2003; WHO, 2006).

This practice is controversial because fluoride has been found to have beneficial effects at low levels and is intentionally added to many public water supplies; however, at higher concentrations, it is known to have toxic effects. The Environmental Protection Agency (EPA) regulates the amount of fluoride that may be present in public water supplies to protect against fluoride's adverse

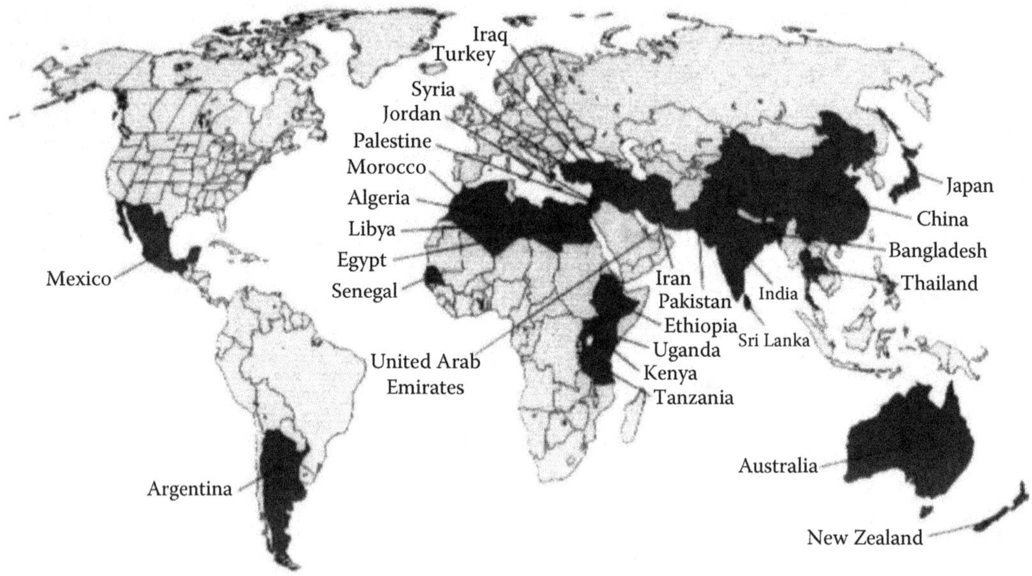

FIGURE 19.1 Global scenario of fluoride endemicity.

health effects. In 1986, EPA issued a drinking water regulation for fluoride that includes an enforceable standard—a maximum contaminant level (MCL)—and an MCL goal (MCLG) of 4 mg/L to protect against adverse effects on bone structure. To address this concern, EPA included in the regulation a secondary (advisory) standard of 2 mg/L to protect children against dental fluorosis and adverse health effects (USEPA, 1996; WHO, 2008).

19.5 INDIAN SCENARIO OF FLUORIDE ENDEMICITY

India is among the 25 nations around the globe where health problems occur due to excess ingestion of fluoride (>1.5 mg/L) by drinking water. Endemic fluorosis remains a challenging and extensively studied national health problem in India. In 1991, 13 out of India's 32 states and territories were reported to have naturally high concentrations of fluoride in water (Susheela, 1993), but this had risen to 17 by 1999 (UNICEF, 1999), and according to a recent report, the number has increased to 21 states (Hussain et al., 2010).

In Indian continent, the higher concentration of fluoride in groundwater is associated with igneous and metamorphic rocks and about 66.6 million people, of which an estimated 6 million are children below the age of 14 years (Ayoob and Gupta, 2006; Vikas et al., 2013) are at risk of developing fluorosis from drinking high-fluoride water. High concentrations of fluoride (>1.5 mg/L) have been reported in the states of Haryana, Delhi, Rajasthan, Karnataka, Uttar Pradesh, Maharashtra, Gujarat, Madhya Pradesh, Andhra Pradesh, Tamil Nadu, Kerala, Jammu and Kashmir, Punjab, Orissa, Himachal Pradesh, and Bihar. The most seriously affected states are Andhra Pradesh, Punjab, Haryana, Rajasthan, Gujarat, Tamil Nadu, and Uttar Pradesh (Teotia et al., 1984; Saini et al., 2013; Hussain et al., 2010).

The highest concentration observed to date in India is 48 mg/L in Rewari District of Haryana (UNICEF, 1999). The high concentrations in groundwater are a result of dissolution of fluorite, apatite, and topaz from the local bedrock. Efforts to address the problem of fluoride in rural water supplies have been led by the Rajiv Gandhi National Drinking Water Mission, with considerable support from external agencies, particularly UNICEF (Susheela, 2003). Further, there have been no comprehensive health surveys for dental fluorosis from which the overall extent of the problem could be assessed. Nevertheless, in the most affected states listed earlier, half or more of the districts have some villages with groundwater supplies having high fluoride concentrations. In these states, 10%–35% of the rural population has been estimated to be at risk, and perhaps an environmental occurrence of total of 66.6 million people in India as a whole may be considered to be affected (WHO, 2010). The prevalence and severity of fluorosis in India increased with increasing fluoride concentration over the past decade. In a number of studies, it has been reported that the prevalence and severity of fluorosis were highest in subjects belonging to the economically poor community (Susheela, 2003; Hussain et al., 2010).

19.6 FLUORIDE PHARMACOKINETICS

The main route for the incorporation of fluoride into the human body is the digestive tract; 90% of the fluoride ingested is absorbed in the stomach. In adults, some 10% of it is deposited in the bones, whereas in children, up to 50% is fixed to the bone tissue. The fluoride levels reach a peak concentration within 20–60 min after ingestion and the concentration declines, usually within 3–6 h following the peak levels, due to the uptake of fluoride by hard tissue and efficient removal by the kidneys. In newborn children, 80%–90% of the fluoride absorbed is retained in the skeletal system. This affinity decreases with age and stabilizes over a period of time (Dhar and Bhatnagar, 2009; Blaszczyk et al., 2012).

19.6.1 Absorption

Approximately 70%–80% of ingested fluoride is absorbed. In an acidic stomach, fluoride is converted into hydrogen fluoride (HF) and about 40% is rapidly absorbed from the stomach. Fluoride not absorbed in the stomach is absorbed in the intestine and is unaffected by pH at this site (Whitford, 1996). Relative to the amount of fluoride ingested, high concentrations of cations that form insoluble complexes with fluoride (e.g., calcium, magnesium, and aluminum) can markedly decrease gastrointestinal fluoride absorption (Whitford, 1996). Fasting dramatically increases the rate of fluoride absorption to nearly 100%, from a 70% to 80% when taken with food. Once absorbed, fluoride rapidly binds with calcium ions and may lead to hypocalcemia. Fluoride also has direct cytotoxic effects and interferes with a number of enzyme systems: it disrupts oxidative phosphorylation, glycolysis, coagulation, and neurotransmission (by binding calcium). Fluoride also inhibits Na^+/K^+-ATPase channels, which may lead to hyperkalemia by extracellular release of potassium. The NRC report, 2001, stated that "many of the untoward effects of fluoride are due to the formation of AlFx [aluminum fluoride] complexes, which are more toxic as compared to hydrofluoric acid" (Villa et al., 2010).

19.6.2 Distribution

Once absorbed into the blood, fluoride readily distributes throughout the body, with approximately 99% of the body burden of fluoride retained in calcium-rich areas such as bone and teeth (dentine and enamel) where it is incorporated into the crystal lattice. Fluoride can cross the placenta and is also found in mother's milk at low levels essentially equal to those in blood (WHO, 1996; Dhar and Bhatnagar, 2009). Under certain conditions, plasma fluoride levels provide an indication of the level of fluoride in the drinking water consumed. Studies have reported that fasting plasma fluoride concentrations of healthy young or middle-aged adults expressed in micromoles per liter are roughly equal to the fluoride concentrations in drinking water expressed as milligrams per liter (Susheela, 2003). Levels of fluoride that are found in the bone vary with the part of the bone examined and with the age and sex of the individual. Bone fluoride is considered to be a reflection of long-term exposure to fluoride (WHO, 2006).

19.6.3 Excretion

Renal excretion is the predominant route for the removal of inorganic fluoride from the body, primarily via urine. Urinary fluoride clearance increases with urine pH due to a decrease in the concentration of hydrofluoric acid. Numerous factors (e.g., diet and drugs) can affect urine pH and thus affect fluoride clearance and retention. Thus, a useful epidemiological tool for monitoring fluoride ingestion is urinary fluoride excretion (Whitford et al., 1976). Approximately 50% of absorbed fluoride is excreted by kidney within 24 h period. The remainder can be retained in the oral cavity and lower digestive tract. Fluoride is also excreted via feces and sweat.

19.7 FLUORIDE TOXICITY AND MECHANISM

Toxicity is the characteristic of an individual organism that responds to the toxic substances at a particular concentration and dose for a specific period of time. "The dose makes the poison" is a principle of toxicology that was expressed by Paracelsus, and it has precluded the chemical toxicity of most of the toxicant molecules. The term *bioaccumulation* refers to the net accumulation over time of metal or nonmetal or any other persistent substances within an organism from both biotic and abiotic (soil, air, and water) sources. The bioaccumulation of fluoride is also known as the residual fluoride toxicity (Preedy, 2015). It is dependent on several other factors such as the dose (how much), the duration (how long), the timing (when in the life cycle), the route of exposure (injection, inhalation, and ingestion), and individual characterization and life styles (Dhar and Bhatnagar, 2009). The potential biological and ecological significance of F accumulation is not easy to evaluate as all the chemical species of fluoride pollutant (HF, F_2, SiF_9, NaF, and H_2S) are very toxic and reactive, having a tendency for accumulation in tissues (WHO, 1996).

19.7.1 Toxicity Range

Soluble fluoride salts, of which sodium fluoride is the most common, are only mildly toxic, although they have resulted in both accidental and self-inflicted deaths from acute poisoning. The lethal dose for most adult humans is estimated at 5–10 g (which is equivalent to 32–64 mg/kg elemental fluoride/kg body weight) (Baselt, 2008). For sodium fluorosilicate (Na_2SiF_6), the median lethal dose (LD_{50}) orally in rats is 0.125 g/kg, corresponding to 12.5 g for a 100 kg adult human being (Merck, 1976). The fatal period ranges from 5 min to 12 h. Micromolar fluoride concentrations promote cell proliferation and growth, whereas a millimolar concentration suppresses cell proliferation and induces apoptosis in hard tissues (Susheela and Bhatnagar, 2002; Levy, 2003; Bai et al., 2010). Due to the high affinity of fluoride to strongly mineralized tissues, fluoride concentrations reach the highest levels in the bones and teeth, and so these materials are frequently used in the bioindication of long-term exposure to fluoride and its toxicity (Ismail and Hasson, 2008; Dhar and Bhatnagar, 2009; Levy, 2014).

19.7.2 Mechanism

Fluoride toxicity in humans has been an area of intense research in the past decade (WHO, 2006). With the availability of ample data demonstrating its effect on the central nervous system (CNS) and immune system (Preedy, 2015), it is increasingly important to understand the molecular events leading to its toxicity. The mechanism of fluoride toxicity is mediated mainly through the generation of oxidative stress, excitotoxicity, and alteration of metabolic pathways (Figure 19.2).

19.7.2.1 Generation of Oxidative Stress

The hypothesis suggesting oxidative stress as a key mechanism in fluoride-induced toxicity has been confirmed by a large number of studies in fluorosis patients, animal models of fluorosis, or cultured cells treated with varying concentration of fluoride. There have been a number of

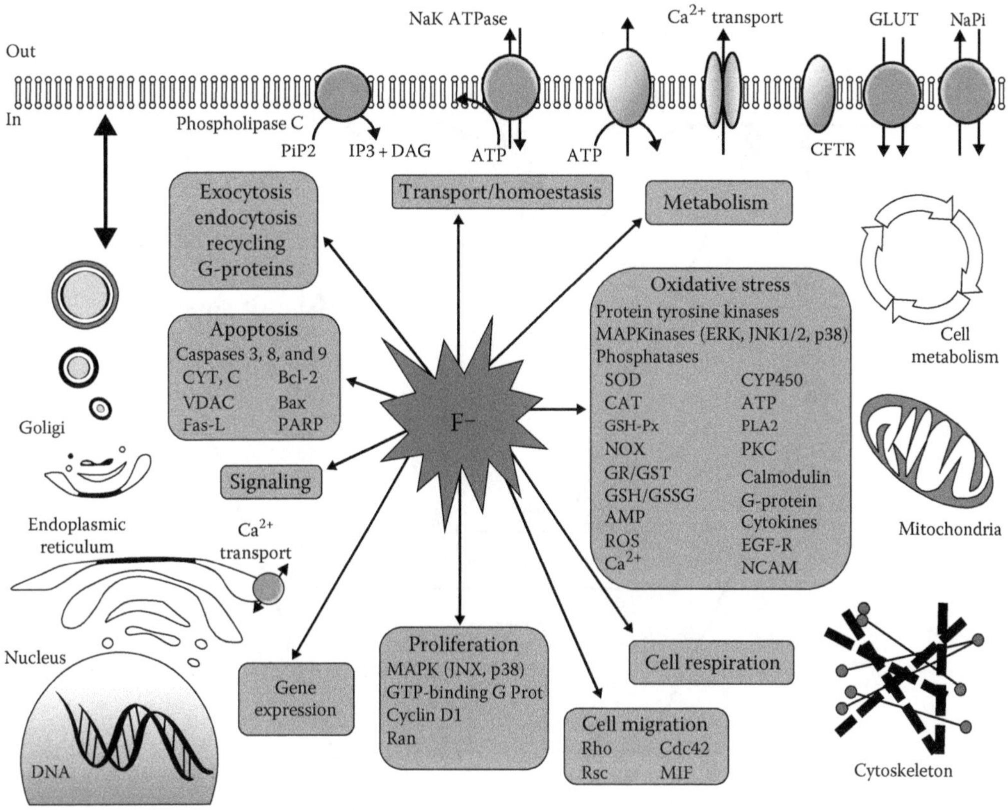

FIGURE 19.2 Possible mechanism of fluoride toxicity in a cell.

reports suggesting downregulation of antioxidant capacity of cellular defense mechanism due to fluoride exposure. These findings have shown reduced antioxidant status, apoptosis, and altered neurological and cardiorespiratory variables (Blaylock, 2004; Dhar and Bhatnagar, 2009; Fina et al., 2012).

Fluoride readily forms strong hydrogen bonds with —OH and —NH species and thus contributes to the induction of free radical generation, thereby increasing oxidative stress (Jain et al., 2015). High level of fluoride elevates the concentration of superoxide anion (O^-), and reduction of superoxide anion results in one of the most reactive moieties, hydrogen peroxide (H_2O_2), which participate in Fenton reaction to generate far more damaging molecules like hydroxyl radical (Preedy, 2015). Fluoride is found to reduce the activity of antioxidant enzymes such as glutathione (GSH), catalase, glutathione peroxidase, and superoxide dismutase (SOD) (Sharma et al., 2014). Also, an elevation in level of malondialdehyde (MDA) was observed in fluorotic patients, which indicates lipid peroxidation, an ultimate reason for oxidative stress (Choi, 2012). Moreover, fluoride can alter glutathione levels, often resulting in the excessive production of ROS at the mitochondrial level, leading to the damage to cellular components.

19.7.2.2 Excitotoxicity

Russell Blaylock was first to suggest excitotoxicity as the central mechanism in fluoride-induced neurotoxicity (Blaylock, 2004) (Figure 19.3). He proposed that microglial activation leads to excitotoxicity that plays a central role in the neurotoxicity of fluoride. Earlier reports have also suggested that fluoride intoxication causes the derangement of glutamate neurotransmission (Blaylock, 2004). An increase in the level of glutamate causes the activation of microglia (immune cells of the

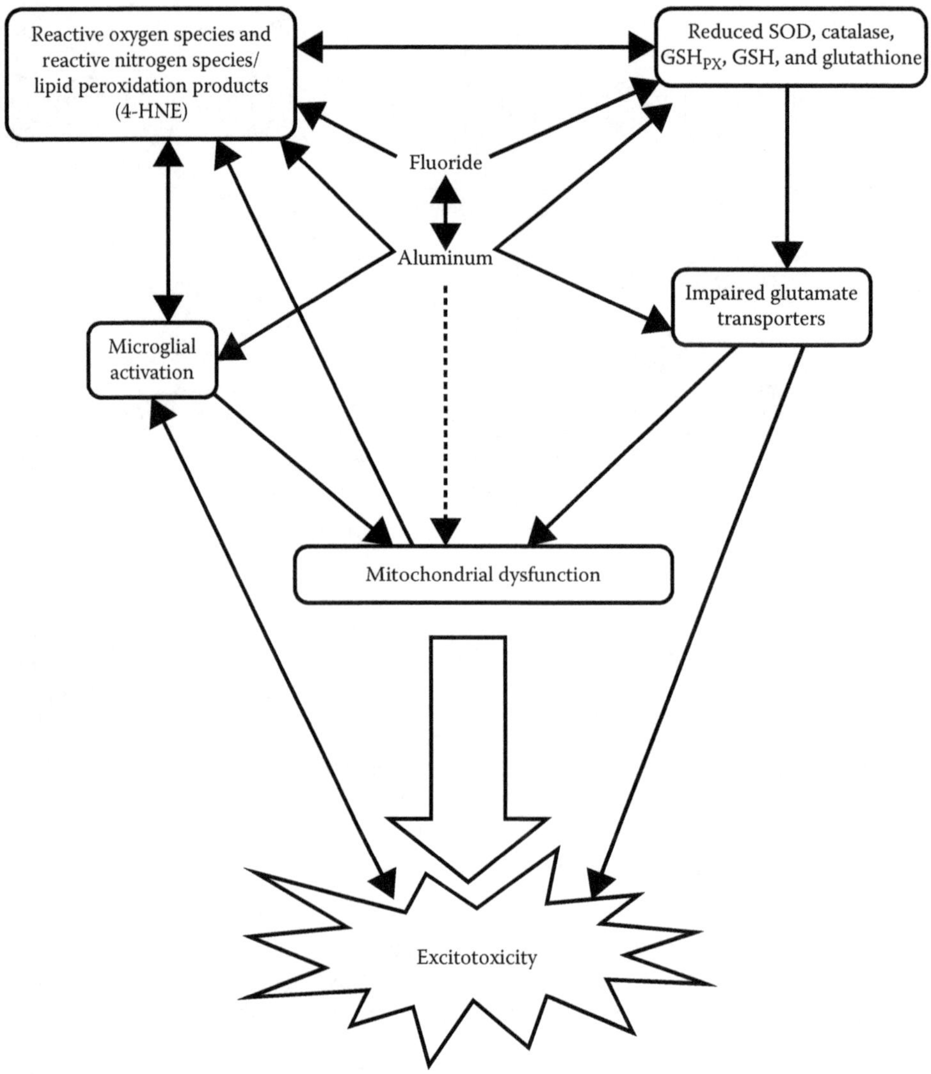

FIGURE 19.3 Possible mechanism of fluoride excitotoxicity in neurons. (From Blaylock, R.L., *Fluoride*, 37, 301, 2004.)

brain) that release cytokines, ultimately leading to neurodegeneration. An accumulation of fluoride is observed in the brain regions rich in glutamate receptors, especially in the hippocampus, dentate gyrus, amygdala, cortex and cerebellum, which further suggest that glutamate plays a key role in fluoride toxicity (Sharma et al., 2014). Another consequence of neurotoxic effect of fluoride is elevation in neurotransmitter nitric oxide (NO); this is because glutamate receptors (NMDAR), which are concentrated at synapse, promote the overproduction of NO (Blaylock, 2004), thereby showing that excitotoxicity is one of the main causative factors behind fluoride toxicity.

19.7.2.3 Alteration of Metabolic Pathways

Affinity of fluoride with Ca^+ and Mg^+ ions inhibits the activity of many enzymes, thereby affecting the signaling process and metabolic pathways. Alteration in Ca^+ homeostasis modifies signaling pathways as calcium simultaneously represents an integrative signal and a central convergence point of many distinct signaling pathways (Teotia et al., 1998; Chen and Whitford, 1999). The biotoxicity of fluoride restrains synthesis of many enzymes including enolase, ATPases, cholinesterases, arginase, esterases, isocitrate dehydrogenase, phosphatases, and aconitase (WHO, 2010; Preedy, 2015). Reports have also suggested that fluoride intoxication alters the activity of neuronal enzymes like LDH, SDH, ALT, AST, CK, ATPases, and acetylcholinesterase (Susheela, 2003; Sharma et al., 2014; Jain et al., 2015). As all these enzymes are key factors in the process of glycolysis, the citric acid cycle, and the electron transport chain, it is likely that fluoride also affects energy generation, the membrane function, the amino acid metabolism, and the nerve impulse transmission in the brain (Preedy, 2015).

Fluoride is a well-known G-protein activator. Activation of heterotrimeric GTP-binding proteins by fluoride requires trace amounts of Al^{3+} or Be^{2+} ions. AlF_x mimics a γ-phosphate at its transition state in a $G\alpha$ protein and is therefore able to inhibit its GTPase activity. AlF_x also forms complexes with small GTP-binding proteins in the presence of their GTPase-activating proteins (GAPs). As phosphate analogs, AlF_x or BeF_x affects the activity of a variety of phosphoryl transfer enzymes. Most of these enzymes are fundamentally important in cell signal transduction or energy metabolism (Susheela, 2003; Dhar and Bhatnagar, 2009; Levy et al., 2014).

19.8 HEALTH HAZARDS

Intensely developing *in vivo* and *in vitro* research provides a growing body of evidence pointing to the negative effects of excessive fluoride on mammals, from the subcellular to the ecological levels (Ayoob and Gupta, 2006; Bai et al., 2010). A number of studies have reported on the acute effects of fluoride exposure following fluoridation overdosing. However, the effects of long-term exposure and assessment of human health to naturally occurring fluoride from drinking water and other environmental sources are the major concern with regard to human health. A large number of epidemiological studies have been conducted in many countries concerning the effects of long-term exposure to fluoride. Excessive exposure to fluoride can give rise to a number of adverse effects, ranging from mild dental fluorosis to crippling skeletal fluorosis.

Acute fluoride toxicity occurred from accidental ingestion of pesticides in the past. Today, accidental overdoses from fluoride-containing substances are a risk especially for children and ingestion of an acute fatal dose of fluoride is very rare (Susheela, 2003). The amount of fluoride considered lethal when taken orally is 35–70 mg fluoride/kg body weight. This is equivalent to 5–10 g sodium fluoride for a 70 kg adult and 1–2 g sodium fluoride for a 15kg child. Depending on the amount ingested and absorbed, symptoms of vomiting, diffuse abdominal pain, diarrhea, hypersalivation, polydypsia, cramping of the extremities, bronchospasm, hyperkalemia, hypocalcemia, ventricular fibrillation, and cardiac arrest can be observed and even in certain cases death may follow (Rabinowitch, 1945; Levy, 2003; Li, 2003). Oral ingestion of 5–8 mg/kg body weight fluoride (NaF) is accepted as a possible toxic dose, which could cause serious life-threatening systemic signs and symptoms requiring immediate hospitalization (Buzalaf and Levy, 2011). Overdoses from fluoride dietary supplements, mouth rinses, toothpastes, medicines, and professional gels can be possible risk sources without proper supervision. Acute poisoning can also terminate in death due to blocking cell metabolism since fluorides inhibit enzymatic processes, particularly metalloenzymes responsible for important vital processes (Li, 2003; Dhar and Bhatnagar, 2009; Preedy, 2015).

Chronic toxicity via excess but nonlethal fluoride consumption, over extended periods, may exhibit dental, skeletal, and nonskeletal manifestations. Chronic fluoride exposures have also been shown to cause cytotoxic effects and interfere with many enzyme systems by disrupting oxidative phosphorylation, glycolysis, coagulation, cellular transmission, and inhibition of Na^+/K^+-ATPase channel, leading to hyperkalemia and cholinergic signs (Li, 2003; Blaylock, 2004; Buzalaf and Levy, 2011; Preedy, 2015). Due to chronic toxicity, bone density slowly increases; the joints stiffen and become painful (Schulz et al., 1986).

19.8.1 Dental Effects

Dental caries are considered a significant dental public health problem in many parts of the world. While fluoride is accepted as an effective method to prevent caries, the excessive consumption of fluoride can put teeth at risk of developing dental fluorosis. Dental fluorosis (enamel fluorosis) is defined as hypocalcified enamel, caused by excess fluoride intake during dental formation that varies from mild opaque lines to pitting and staining of enamel. Dental fluorosis can also be described as a diffuse symmetric hypomineralization disorder of ameloblasts (Susheela, 2003; Dhar and Bhatnagar, 2009). The community fluorosis index is a way of measuring the burden of dental fluorosis in a population. It takes into account the severity and prevalence of dental fluorosis. Multiple mechanisms for dental fluorosis initiation have been proposed including direct fluoride-related effects on ameloblasts (secretory and maturation phases), indirect fluoride-related effects on the forming matrix (nucleation, crystal growth in all stages of enamel formation), and calcium homeostasis, depending on the dose and duration of fluoride exposure (Bronckers et al., 2009; Buzalaf and Levy, 2011).

In clinical terms, dental fluorosis may result in varying degrees of structural damage, superficial porosity, and loss of continuity of the dental enamel layer. Characteristic clinical features of dental fluorosis are discoloration of enamel surface, yellow brown-black strips seen along gum, mottling, loss of glistening, loosening of teeth, and loss of enamel, which gives the teeth a corroded appearance (Baelum et al., 1987) (Figure 19.4). Mottling is best seen on the incisors of the upper jaw. In severe cases and long-term exposures in endemic zones, people lose their teeth at an early age and may become edentate. It is irreversible and only occurs with exposure to fluoride when enamel is developing. The process is of slow progression and instead of being a normal creamy-white translucent color, fluorosed enamel is porous (objectionable secondary staining often occurs) and opaque; teeth can resemble a ghastly white chalk color (light refractivity is greatly reduced because the enamel prism structure is defective) (WHO, 2006; Dhar and Bhatnagar, 2009). Cloudy striated (lines of demarcation) enamel, white specks or blotches, snow capping, and brown pits on teeth are all characteristics of fluorosis. In its more severe form, fluorosed enamel is brittle and prone to erosion and breakage, especially when drilled and pilled. Even in the milder forms, there is an increased enamel attrition. Fluorotic lesions are not just calcified to enamel but can be seen by microscope in dentin as well (Bardsen, 1999).

Because fluoride is a powerful bone- and tooth-seeking element, it also deposits bone or bone-like material externally on the roots of teeth and internally in the tooth's pulp chamber; the calcified material narrows the pulp chamber and thereby interferes with tooth nutrition. To prevent fluorosis from occurring in the most prominent and/or most susceptible teeth, the most critical time to avoid fluoride exposure is the first 3–6 years of a child's life (Bardsen, 1999; Aoba and Fejerskov, 2002; Dhar and Bhatnagar, 2009).

Dental fluorosis may be more than a cosmetic defect if enough fluorotic enamel is fractured and lost, causing pain, adversely affecting food choices, compromising chewing efficiency, and requiring complex dental treatment (Bartlett et al., 2005). Dental fluorosis in nonendemic areas has been associated with the consumption of fluoridated drinks and foods and the misuse of toothpastes, mouth rinses, gels, drops, and tablets. The margin between the beneficial effects of fluoride and the occurrence of dental fluorosis is small and public health programs seek to retain a suitable balance between the two (UNICEF, 1999). Dental fluorosis increases as fluoride concentration increases. At a fluoride concentration of 1 mg/L, about 20% of children have evidence of dental fluorosis, but this fluorosis is of a mild degree of severity and would not be cosmetically obvious to the children or their parents (Dean et al., 1942). Fluoride concentration above 2 mg/L in drinking water causes dental fluorosis in 60% of the affected population, whereas concentration above 6 mg/L affects 100% of the population.

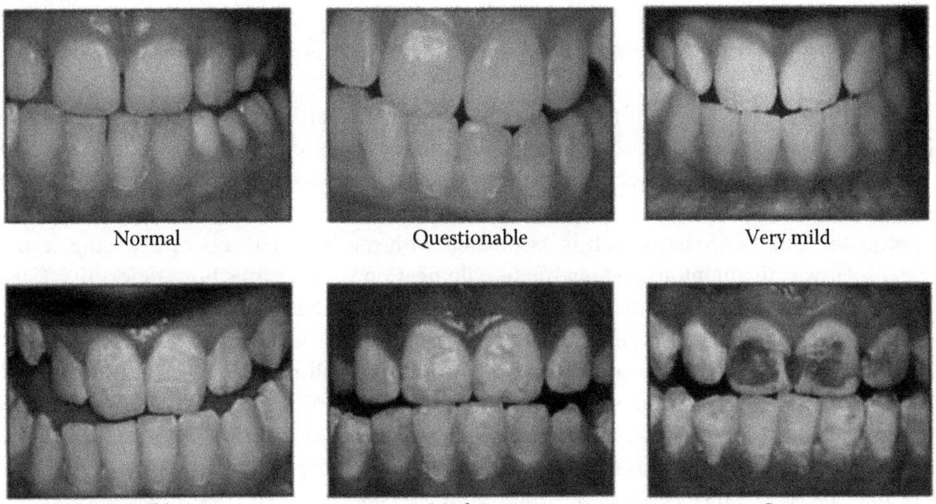

FIGURE 19.4 **(See color insert.)** Dean classification used to categorize dental fluorosis. (Dean et al., 1942.)

19.8.2 Skeletal Effects

Excess nonlethal fluoride consumption over extended periods may result in a serious health problem called skeletal fluorosis. Skeletal fluorosis is a variant of chronic fluoride toxicity that goes usually undetected without radiography (BIS, 2003). Endemic skeletal fluorosis is a geochemistry disease, which occurs in particular geographical environment with high fluoride concentration in drinking water under natural conditions. It is defined as a chronic metabolic bone disease caused by excessive intake of fluoride and with major clinical manifestations of pain in the neck, waist and large joints in limbs, motor dysfunction, and typical abnormal signs of bone x-ray (Susheela, 1993; Li, 2003; Bartlett et al., 2005). Symptoms are increased bone mass in milder cases with pain and calcification of spinal ligaments in more severe stages (Dhar and Bhatnagar, 2009). Irreversible changes of skeletal fluorosis include the increase of bone mass (osteosclerosis), the increase of bone-like tissue (osteomalacia), osteoporosis, enhanced activity of osteoblast, activation of osteoclasts, and acceleration of bone turnover. Among them, the enhanced activity of osteoblasts is an early and dominant link in skeletal fluorosis. Early stages in skeletal fluorosis are not clinically obvious and may be misdiagnosed as (seronegative) rheumatoid arthritis, osteoporosis, or ankylosing spondylitis (Bernstein et al., 1966; Briancon and Meunier, 1981; Schulz et al., 1986; Li, 2003; Levy et al., 2014).

Fluoride increases the stability of the crystal lattice in bone but makes the bone more brittle. The total quantity of fluoride ingested is the single most important factor in determining the clinical course of skeletal fluorosis; the severity of symptoms correlates directly with the level and duration of exposure. Bone changes observed in human skeletal fluorosis are structural and functional, with a combination of osteosclerosis, osteomalacia, osteoporosis, exostosis formation, and secondary hyperparathyroidism in a proportion of patients (Emami et al., 2007; Everett, 2011).

Excessive accumulation of fluoride in the body and calcium malnutrition may promote the occurrence and development of fluorosis, which can aggravate the toxicity of fluoride to many bone structures. An important study done by Ren and Li (1999) reported the effect of fluoride on human bone diseases and the relationship with calcium (most commonly given to fluorosis patients). Teotia et al. (1998) summarized the surveys for more than 30 years from 1963 to 1997. This study collected 45,725 children with skeletal fluorosis. All of them lived in the fluorosis areas after birth, and fluoride contents in drinking water ranged from 1.5 to 25 mg/L. The study found that the children with calcium-deficient diet (<300 mg/day) showed more serious and complicated fluorosis symptoms compared to the children with calcium-adequate diet (>800 mg/day), under the same fluoride dose intake. The incidence rate of metabolic bone diseases, such as rickets, osteoporosis, parathyroid hormone–related bone disease, and leg deformities in the children with a calcium-deficient diet, was over 90%. The incidence rate of metabolic bone diseases in the children with calcium-adequate diet was less than 25%.

Crippling skeletal fluorosis, which is associated with the higher levels of exposure, can result from osteosclerosis, ligamentous and tendinous calcification, and extreme bone deformity (Susheela AK, 2003). This may lead to permanent disability. X-ray examination of the bones reveals thickening and high density of bones (Figure 19.5). Specific symptoms of crippling fluorosis are immobile knee, pelvic and shoulder joints, rigidity of joints, kyphosis, scoliosis, restricted flexion, paraplegia, and quadriplegia. The ossification of vertebrae and constriction of vertebral canal and intervertebral foramen may lead to paralysis (Li, 2003; Susheela, 2003; Levy et al., 2014; Preedy, 2015).

One possible feature of fluorosis is bone fracture, although some studies have reported a protective effect of fluoride on fracture (Jolly et al., 1968; Simonen and Laitinen, 1985). A few investigations on skeletal fluorosis and the risk of fractures include quantitative estimates of the dose–response relationship. Long-term exposure to fluoride in drinking water stimulates osteoblastic bone formation, particularly in cancellous bone, and at low levels (1.00–1.06 ppm), this decreases the risk of overall fractures. At higher levels, however (≥4.32 ppm), fluoride can decrease cortical bone mineral density and increase skeletal fragility, leading to increased fracture risk (Susheela, 1993; WHO, 2010). Toxic levels of fluoride have also been associated with a weakening of bones and an increase in hip and wrist fractures. The U.S. National Research Council reported that fractures with fluoride levels 1–4 mg/L follow a dose–response relationship, but the data are inadequate for drawing firm conclusions about the risk or safety of exposures at 2 mg/L (Fawell et al., 2006; WHO, 2010). A new form of fluorosis characterized by genu valgum and genu varum, osteoporosis of the lower limbs (Figure 19.6), has been reported in recent years in some districts of Andhra Pradesh, Rajasthan, Uttar Pradesh, and Tamil Nadu. The syndrome was observed among people whose staple was sorghum (jowar) (Shivashankara et al., 2000). Children in endemic fluorosis areas in many parts of India often have deaf mutism, knock-knee, and bowlegs (Susheela, 2003).

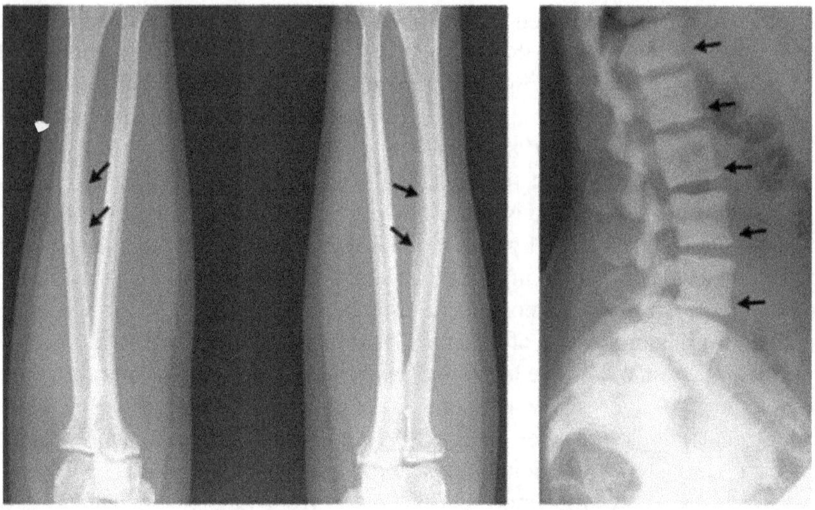

(a) (b)

FIGURE 19.5 Abnormal growth of bone. An x-ray showing (a) calcifications on ligaments (left) and areas of (b) dense bone on the spinal vertebrae (right). Credit: New England Journal of Medicine Case report © 2013.

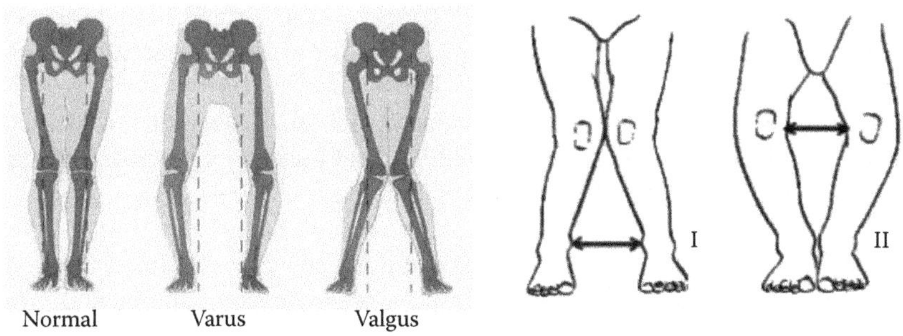

FIGURE 19.6 Genu varum and genu valgum types of skeletal fluorosis zones.

19.8.3 Nonskeletal Effects

A number of epidemiological studies have been carried out to examine other possible adverse outcomes as a consequence of exposure to fluoride, either from drinking water or as a consequence of occupation.

19.8.3.1 Neurotoxicity

One of the most important areas of research is the evaluation of the effect of fluoride on the nervous system and endocrine glands of mammals, including the pineal gland (Choi et al., 2012; Sharma et al., 2014). It has been shown experimentally that fluoride has neurotoxic properties and adversely affects the functioning of the brain, even at small doses. It contributes to the induction of oxidative stress, free radicals, and lipid peroxidation in the brain and inhibits the production of antioxidant enzymes, mitochondrial enzymes of energy, and glutamate transporters. A large number of studies have shown impaired memory and learning ability in animals and humans (Blaylock, 2004; Barbier et al., 2010; Choi et al., 2012; Jain et al., 2015).

Neurodevelopmental disabilities, including autism, attention-deficit hyperactivity disorder, dyslexia, and other cognitive impairments, affect millions of children worldwide, and some diagnoses seem to be increasing in frequency. Chemical toxicants that injure the developing brain are among the known causes for this rise in prevalence. Since 2006, epidemiological studies have documented six additional developmental neurotoxicants—manganese, fluoride, chlorpyrifos, dichlorodiphenyltrichloroethane, tetrachloroethylene, and the polybrominated diphenyl ethers (Grandjean and Landrigan, 2014). Numerous Indian and Chinese studies suggested a possible effect of higher drinking water fluoride levels on learning, memory, and intelligence of children (Doull et al., 2006, Sharma et al., 2014, Jain et al., 2015). A meta-analysis of 27 cross-sectional studies of children exposed to fluoride in drinking water, mainly from China, suggests an average IQ decrement of about seven points in children exposed to raised fluoride concentrations (Choi et al., 2012).

Fluorine is capable of crossing the blood–brain barrier, which can cause biochemical and functional changes in the nervous system. The levels of neurotransmitters such as norepinephrine, 5-hydroxytryptamine, acetylcholine, NO, and their receptors have been found to be altered in fluorosis. Fluoride has also been reported to have a specific effect on protein synthesis in the brain, entailing degenerative changes in the neurons, varying degrees of loss of gray matter, and changes in Purkinje cells in the cerebellar cortex (Blaylock, 2004; Sharma et al., 2008). Moreover, it causes changes in mitochondrial enzymes, granular endoplasmic reticulum, chromatin clumping, damage to the nuclear membrane, and a decrease in the number of synapses, microtubules, and synaptic vesicles. These changes indicate that fluoride can alter the neuronal growth and division in the cortex and may reduce efficiency in neuronal connections, giving rise to abnormal synaptic functioning and impact cognitive development (Blaylock, 2004). Animal studies have also yielded information about the direct toxic effects of fluoride on CNS, including decreased number receptors, lower lipid content, neurodegeneration in the hippocampus and cortex, increased formation of β-amyloid plaques (classic cerebral anomaly in patients presenting Alzheimer's disease), exacerbation of lesions caused by iodine deficit, and accumulation of fluoride in the pineal gland (Li, 2003; Susheela, 2003; Doull et al., 2006; Fina et al., 2012; Sharma et al, 2014).

Neurological manifestations in skeletal fluorosis are nervousness, headache, insomnia, depression, tingling sensation in fingers and toes, excessive thirst (polydypsia), tendency to urinate frequently (polyuria), and seizures, which may result from both hypomagnesemia and hypocalcemia. Different theories on fluoride influencing the brain have been proposed (Levy, 2003; Blaylock, 2004; Ozsvath, 2009), but further research with strong methodology on this critical claim is necessary to draw substantial conclusions (Doull et al., 2006; Fina et al., 2012).

19.8.3.2 Nephrotoxicity

The kidneys play the major role in the removal of fluoride from the body. Normally, kidneys are very efficient and excrete fluoride very rapidly. Fluoride is concentrated at high levels within kidney tubules, so kidney has higher concentration of fluoride than plasma. Fluoride is freely filtered through the glomerular capillaries and reabsorption takes place due to diffusion of HF. Among all the halogens, the renal clearance of fluoride is unusually high (Schmidt, 1986; Dhar and Bhatnagar, 2009; Garcia et al., 2012). It is known that persons suffering from certain forms of renal impairment have a lower margin of safety for the effects of fluoride than the average person. The database "Fluoride in Drinking-Water" available on this subject is, however, too limited to allow a quantitative evaluation of the increased sensitivity to fluoride toxicity of such persons (US EPA, 1985; Fawell et al., 2006).

The WHO (2010) reported that the present data were too limited to permit any quantitative evaluation of possible increased sensitivity due to impaired kidney function. Within the recommended dose, no effects are expected, but chronic ingestion in excess of 12 mg/day is expected to cause adverse effects.

The kidney injury is characterized by failure to concentrate urine, leading to polyuria, and subsequent dehydration with hypernatremia and hyperosmolarity. Inorganic fluoride inhibits adenylate cyclase, activity required for antidiuretic hormone effect on the distal convoluted tubule of the kidney. Fluoride also stimulates intrarenal vasodilation, leading to increased medullary blood flow, which interferes with the countercurrent mechanism in the kidney required for the concentration of urine (Li, 2003; Lupo et al., 2011; Preedy, 2015).

19.8.3.3 Hepatotoxicity

A very few studies have linked fluoride toxicity with hepatic cell structure and functions. Chinoy et al. (2004) and Guo et al. (2003) studied liver functions in fluoride toxicity model system and observed a decrease in organosomatic index, cytoplasmic degranulation and vacuolation of hepatocytes, leukocytic infiltration, nuclear pyknosis, increased number of Kupffer cells, and disorganized architecture of the hepatic cords. Zonal necrosis was also reported (Preedy, 2015).

19.8.3.4 Carcinogenicity

Controversy of alleged risk of fluoridation to increase cancer development is not new. But considering the time gap between factors causing cancer, its clinical onset and different types of cancers onset warrant more research to form a strong hypothesis. The relation of kidney, bladder, thyroid and especially bone cancer to fluoride in the literature is inconclusive to establish a link in humans (Ozsvath, 2009). The WHO also stated that the data are not convincing to support the hypothesis that fluoride causes cancer in humans. Many of the findings suggesting a correlation suffer from methodological flaws (WHO, 2006).

Studies of occupationally exposed populations, primarily from aluminum smelting, pesticide industries, have reported an increased incidence of fluoride toxicity and mortality from lung and bladder cancers. However, the data are inconsistent and in a number of studies the results can be more readily attributed to exposure to other substances than fluoride. There is no consistent evidence to demonstrate any association between the consumption of high fluoridated drinking water and either morbidity or mortality from cancer (US EPA, 1996; WHO, 2010).

19.8.3.5 Reproductive Effects

The majority of the studies investigating the effects of fluoride on the reproductive system have been conducted with high doses and short durations on animals. Human studies are very limited on this subject. An important case control study in human subjects showed fluoride toxicity via biochemical changes in seminal plasma (Chauhan et al., 2014). Numerous detrimental effects of ingesting high fluoride on males have been reported: sperm morphology/mobility, testosterone levels, and follicle-stimulating hormones (Ozsvath, 2009; Chauhan et al., 2013). On the other hand, a systematic review conducted by York University on the systemic effects of public water fluoridation reported no evidence on the reproductive toxicity of fluoride (WHO, 2010).

19.8.3.6 Thyroid

As it is already known that fluoride is more electronegative than iodine, it easily displaces iodine within the body, thereby affecting the functioning of the thyroid gland. Fluoride's suppressive effect on the thyroid is more severe when iodine is deficient and shows deranged thyroid hormonal level within the body. The enzymes that catalyze deiodination are called iodothyronine deiodinases and fluoride is known to alter the activity of the deiodinases. The TSH output from the pituitary gland is also inhibited by fluoride, thus reducing the thyroid output from the thyroid glands. Fluoride competes for the receptor sites on the thyroid gland, which respond to TSH (Levy, 2003; Susheela, 2003; Preedy, 2015).

19.8.3.7 Birth Defects

Fluoride can cross the placenta and incorporate into the tissues of the fetus (Susheela, 2003). Placenta fluoride level is 60% of the maternal blood plasma. The relation of fluoride with human developmental disorders like spina bifida occulta, sudden infant death syndrome, and Down's syndrome has been investigated in the literature, but the results were inconclusive (Susheela, 2003; Doull et al., 2006). High fluoride levels in water did not result in developmental defects in human and animal studies with the exception of ossification problems (Grandjean and Landrigan, 2014). The results of several epidemiological studies on the possible adverse effects of fluoride in drinking water on pregnancy outcome indicate that there is no apparent relationship between the rates of Down's syndrome or congenital malformation and the consumption of fluoridated drinking water (Susheela, 1993; US EPA, 1996).

19.8.3.8 Other Health Effects

Epidemiological studies have also suggested fluoride intake to increase the susceptibility of infertility (Susheela, 2003; Chauhan et al., 2014) and premature aging (Preedy, 2015). A decrease in insulin sensitivity and insulin signaling has also been reported in an animal model of chronic fluoride toxicity (Lupo et al., 2011). Severe fluoride toxicity may result in multiorgan failure. Central vasomotor depression and direct cardiotoxicity may occur. Death usually results from respiratory paralysis, dysrhythmia, or cardiac failure. Fluoride is reported to affect endocrine function in many ways: decreased thyroid function, increased calcitonin activity, secondary hyperparathyroidism, and impaired glucose tolerance whose level/extent varies from person to person (Ozsvath, 2009; Preedy, 2015). Hydrofluoric acid is irritating to the gastric mucosa, especially at toxic dose level of fluoride, and can cause abdominal pain, vomiting, gastritis, constipation, and irritable bowel syndrome (Nopakun et al., 1989; Li, 2003; Doull et al., 2006).

19.9 CONCLUSIONS

Fluoride exposure from numerous sources leads to a wide and variable range of effects, some of which are well documented and some not. The majority of the information on the health effects of fluoride is regarding its efficacy against dental caries. Besides its protective potential, fluoride is a substance that should be handled with caution because in excess it is highly toxic. There is a strong evidence that excessive intake of fluoride has toxic effects on different organ systems, depending on the dose administered, age, and exposure time. Fluorosis severely affects the quality of life, ultimately causing several systemic, metabolic, and degenerative disorders. Its alarming upsurge in the recent years is a matter of deep concern. The body of knowledge on the adverse effects of fluoride on soft tissues and different organs is limited, warranting further in-depth research.

REFERENCES

ADA., 2005. *Fluoridation Facts.* Chicago, IL: American Dental Association.

Almeida, B. S., da Silva Cardoso, V. E., and Buzalaf, M. A., 2007. Fluoride ingestion from toothpaste and diet in 1- to 3-year-old Brazilian children. *Community Dentistry and Oral Epidemiology.* **35**(1): 53–63.

Aoba, T. and Fejerskov, O., 2002. Dental fluorosis: Chemistry and biology. *Critical Reviews in Oral Biology and Medicine.* **13**: 155–170.

Arnesen, A. K. M., 1997. Availability of fluoride to plants grown in contaminated soils. *Plant and Soil.* **191**: 13–25.

Ayoob, S. and Gupta, A. K., 2006. Fluoride in drinking water: A review on the status and stress effects. *Critical Reviews in Environmental Science and Technology.* **36**: 433–487.

Baelum, V., Fejerskov, O., Manji, F., and Larsen, M. J., 1987. Daily dose of fluoride and dental fluorosis. *Tandlaegebladet.* **91**(10): 452–456.

Bai, C., Chen, T., Cui, Y., Gong, T., Peng, X., and Cui, H. M., 2010. Effect of high fluorine on the cell cycle and apoptosis of renal cells in chickens. *Biological Trace Element Research.* **138**: 173–180.

Bardsen, A., 1999. "Risk periods" associated with the development of dental fluorosis in maxillary permanent central incisors: A *meta*-analysis. *Acta Odontologica Scandinavica.* **57**(5): 247–256.

Bartlett, J. D., Dwyer, S. E., Beniash, E., Skobe, Z., and Payne-Ferreira, T. L., 2005. Fluorosis: A new model and new insights. *Journal of Dental Research.* **84**(9): 832–836.

Baselt, R. C., 2008. *Disposition of Toxic Drugs and Chemicals in Man.* Foster City, CA: Biomedical Publications, pp. 636–640.

Berger, T., Peltola, P., Drake, H., and Astrom, M., 2012. Impact of a fluorine- rich granite intrusion on levels and distribution of fluoride in a small boreal catchment. *Aquatic Geochemistry.* **18**: 77–94.

Bernal, N. F., Gleeson, S. A., Dean, A. S., Liu, X.-M., and Hoskin, P., 2014. The source of halogens in geothermal fluids from the Taupo Volcanic Zone, North Island, New Zealand. *Geochimica et Cosmochimica Acta.* **126**: 265–283.

Bernstein, D. S., Sadowsky, N., Hegsted, D. M., Guri, C. D., and Stare, F. J., 1966. Prevalence of osteoporosis in high and low fluoride areas in North Dakota. *JAMA: The Journal of the American Medical Association.* **198**: 499–504.

Blaszczyk, I., Birkner, E., Gutowska, I., Romuk, E., and Chlubek, D., 2012. Influence of methionine and vitamin E on fluoride concentration in bones and teeth of rats exposed to sodium fluoride in drinking water. *Biological Trace Element Research.* **146**: 335–339.

Blaylock, R. L., 2004. Excitotoxicity: A possible central mechanism in fluoride neurotoxicity *Fluoride* **37**: 301–314

Borgnino, L., Garcia, M. G., Bia, G., Stupar, Y., Le Coustumer, P. H., and Depetris, P. J., 2013. Mechanisms of fluoride release in sediments of Argentina's central region. *Science of the Total Environment.* **443**: 245–255.

Briancon, D. and Meunier, P. J., 1981. Treatment of osteoporosis with fluoride, calcium and vitamin. *The Orthopedic Clinics of North America.* **12**: 629–648.

Browne, D., Whelton, H., and O'Mullane D., 2005. Fluoride metabolism and fluorosis. *Journal of Dentistry.* **33**: 177–186.

Bronckers, A. L., Lyaruu, D. M., and DenBesten, P. K., 2009. The impact of fluoride on ameloblasts and the mechanisms of enamel fluorosis. *Journal of Dental Research.* **88**(10): 877–893.

Bureau of Indian Standards (BIS), 2003. Drinking water specification, IS:10500. New Delhi, India: Bureau of Indian Standards.

Buzalaf, M. A. and Levy, S. M., 2011. Fluoride intake of children: Considerations for dental caries and dental fluorosis. *Monographs in Oral Science*. **22**: 1–19.

Buzalaf, M. A. and Whitford, G. M., 2011. Fluoride metabolism. *Monographs in Oral Sciences*. **22**: 20–36.

Carpenter, R., 1969. Factors controlling the marine geochemistry of fluorine. *Geochimica et Cosmochimica Acta*. **33**: 1156–1167.

Cao, J., Luo, S. F., Liu, J. W., and Li, Y., 2004. Safety evaluation on fluoride content in black tea. *Food Chemistry*. **88**: 233–236.

Cerklewski, F. L., 1997. Fluoride bioavailability—Nutritional and clinical aspects. *Nutrition Research*. **17**: 907–929.

Chae, G.-T., Yun, S.-T., Kwon, M.-J., Kim, Y.-S., and Mayer, B., 2006. Batch dissolution of granite and biotite in water: Implication for fluorine geochemistry in groundwater. *Geochemical Journal*. **40**: 95–102.

Chauhan, D. S., Singh, V. P., Tripathi, S., Tomar, A., Tiwari, M., and Tomar, S., 2013. Influence of fluoride exposure on hypothalamic pituitary gonadal axis hormones and semen quality. *Asian Journal of Biological Life Science*. **3**: 201–206.

Chauhan, D. S., Tripathi, S., Singh, V. K., Tomar, S., Tomar, S., Tiwari, M., and Mahdi A. A., 2014. Excessive fluoride delineating biochemical changes in seminal plasma: A case control clinical study. *European Academic Research*. **7**: 9023–9041.

Chen, X. and Whitford, G. M., 1999. Effects of caffeine on fluoride, calcium and phosphorus metabolism and calcified tissues in the rat. *Archives of Oral Biology*. **44**: 33–39.

Choi, A. L., Sun, G., Zhang, Y., and Grandjean, P., 2012. Developmental fluoride neurotoxicity: A systematic review and meta-analysis. *Environmental Health Perspectives*. **120**: 1362–1368.

Chinoy, N. J., Sharma, A. K., Patel, T. N., Memon, R., and Jhala, D. D., 2004. Recovery from fluoride and aluminium induced free radical liver toxicity in mice. *Fluoride*. **37**: 257–263.

Cook, H. A., 1969. Fluoride and tea. *Lancet*. **2**: 329.

Dean, H. T., Arnold, F. A., and Elvolve, E., 1942. Additional studies of the relation of fluoride domestic waters to dental caries experience in 4,425 white children aged 12–14 years in 13 cities in 4 states. *Public Health Reports*. 57: 1155–1179.

Dhar, V. and Bhatnagar, M., 2009. Physiology and toxicity of fluoride. *Indian Journal of Dental Research*. **20**: 350–355.

Doull, J., Boekelheide, K., Farishian, B., Isaacson, R., Klotz, J., and Kumar, J. E. A., 2006. *Fluoride in Drinking Water: A Scientific Review of EPA's Standards*. Washington, DC: Committee on Fluoride in Drinking Water, Board on Environmental Studies and Toxicology, Division on Earth and Life Sciences, National Research Council of the National Academies.

Dowd, F. J., 1999. Saliva and dental caries. *Dental Clinics of North America*. **43**: 579–597.

EFSA (European Food Safety Authority), 2008. Sodium monofluorophosphate as a source of fluoride added for nutritional purposes to food supplements. *The EFSA Journal*. **886**: 1–18.

Emami, M. J., Ghahramani, M. H., Abdinejad, F., and Namazi, H., 2007. Q-angle: An invaluable parameter for evaluation of anterior knee pain. *Archives of Iranian Medicine*. **10**: 24–26.

Everett, E. T., 2011. Fluoride's effects on the formation of teeth and bones, and the influence of genetics. *Journal of Dental Research*. **90**: 552–560.

Fawell, J., Bailey, K., Chilton, J., and Dahi, E., 2006. *Fluoride in Drinking-Water*. Cornwall, U.K.: World Health Organization, pp. 144–264.

Fina, B., Brance, M. L., Brun, L. R., and Rigalli, A., 2012. Fluoride inhibition of oxygen consumption and increased oxidative stress in rats. *Fluoride*. **45**: 343–348.

Garcia, M. G., Lecomte, K. L., Stupar, Y., Formica, S. M., Barrionuevo, M., Vesco, M., Gallara, R., and Ponce, R., 2012. Geochemistry and health aspects of F-rich montainous streams and groundwaters from Sierras Chicas de Cordoba, Argentina. *Environmental Earth Sciences*. **65**: 535–545.

Grandjean, P. and Landrigan, P. J., 2014. Neurobehavioural effects of developmental toxicity. *The Lancet Neurology*. **3**: 330–338.

Guo, X., Sun, G., and Sun, Y., 2003. Oxidative stress from fluoride induced hepatotoxicity in rats. *Fluoride*. **36**: 25–29.

Hakkinen, S., Heinonen, M., Karenlampi, S., Mykkanen, H., Ruuskanen, J., and Torronen, R., 1999. Screening of selected flavonoids and phenolic acids in 19 berries. *Food Research International*. **32**: 345–353.

Hussain, J., Hussain, I., and Sharma K. C., 2010. Fluoride and health hazards: Community perception in a fluorotic area of central Rajasthan (India): An arid environment. *Environmental Monitoring and Assessment*. **162**: 1–14.

ICMR. 2005. Twenty years of RMRCT Jabalpur. A biannual newsletter of regional medical research centre for tribals, Jabalpur. *RMRCT Update*. **204**: 1–2.

Ismail, A. I. and Hasson, H., 2008. Fluoride supplements, dental caries and fluorosis: A systematic review. *Journal of the American Dental Association.* **139**(11): 1457–1468.

Jain, A., Mehta, V., Chittora, R., Mahdi, A. A., and Bhatnagar, M., 2015. Melatonin ameliorates fluoride induced neurotoxicity in young rats: An in *vivo* evidence. *Asian Journal of Pharma Clinical Research.* **4**: 32–40.

Jolly, S. S., Singh, B. M., Mathur, C., and Malhotra, K. C., 1968. Epidemiological, clinical and biochemical Study of endemic dental and skeletal fluorosis in Punjab. *British Medical Journal.* **4**, 427–429

Levy, S. M., 2003. An update on fluorides and fluorosis. *Journal of the Canadian Dental Association.* **69**(5): 286–291.

Levy, S. M., Warren, J. J., Phipps, K., Letuchy, E., Broffitt, B., Eichenberger-Gilmore, J., Burns, T. L. et al. 2014. Effects of life-long fluoride intake on bone measures of adolescents: A prospective cohort study. *Journal of Dental Research.* **93**: 353–359.

Li, L., 2003. The biochemistry and physiology of metallic fluoride: Action, mechanism, and implications. *Critical Reviews in Oral Biology & Medicine.* **14**: 100–114.

Li, G. and Ren, L., 1997. Effects of excess fluoride on bone turnover under conditions of diet with different calcium contents. *Zhonghua Bing Li Xue Za Zhi.* **26**: 277–280.

Loganathan, P., Hedley, M. J., Grace, N. D., Lee, J., Cronin, S. J., Bola, N. S., and Zanders, J. M., 2003. Fertiliser contaminants in New Zealand grazed pasture with special reference to cadmium and fluorine: A review. *Australian Journal of Soil Research.* **41**: 501–532.

Lupo, M., Buzalaf, M. A., and Rigalli, A., 2011. Effect of fluoridated water on plasma insulin levels and glucose homeostasis in rats with renal deficiency. *Biological Trace Element Research.* **140**: 198–207.

Mohapatra, M., Anand, S., Mishra, B. K., Dion, E., Giles, P., and Singh, A., 2009. Review of fluoride removal from drinking water. *Journal of Environmental Management.* **91**: 67–77

Nopakun, J., Messer, H. H., and Voller, V., 1989. Fluoride absorption from the gastrointestinal tract of rats. *Journal of Nutrition.* **119**: 1411–1417.

Ozsvath, D. L., 2009. Fluoride and environmental health: A review. *Reviews in Environmental Science and Biotechnology.* **8**: 59–79.

Preedy, V. R. 2015. *Fluorine Chemistry, Analysis, Function and Effects: Food and Nutritional Components in Focus.* Cambridge, U.K.: The Royal Society of Chemistry.

Rabinowitch, I. M., 1945. Acute fluoride poisoning. *Canadian Medical Association Journal.* **52**(4): 345–349.

Saini, P., Khan, S., Baunthiyal, M., and Sharma, V., 2013. Mapping of fluoride endemic area and assessment of F–1 accumulation in soil and vegetation. *Environmental Monitoring Assessment.* **185**: 2001–2008.

Schmidt, C. W. and Leuschke, W., 1986. Metabolism of fluorine in terminal renal failure. *Zeitschrift fuer Urologie und Nephrologie.* **79**: 481–484.

Schulz, E. E., Engstrom, H., Sauser, D. D., and Baylink, D. J., 1986. Osteoporosis: Radiographic detection of fluoride-induced extra-axial bone formation. *Radiology.* **159**: 457–462.

Sharma, C., Suhalka, P., Sukhwal, P., Jaiswal, N., and Bhatnagar, M., 2014 Curcumin attenuates neurotoxicity induced by fluoride: An in vivo evidence. *Pharmacognosy Magazine.* **10** (37): 61–65.

Sharma, R., Tsuchiya, M., and Bartlett, J. D., 2008. Fluoride induces endoplasmic reticulum stress and inhibits protein synthesis and secretion. *Environmental Health Perspectives.* **116**(9): 1142–1146.

Shivashankara, A. R., Shivarajashankara Y. M., Rao, S. H., Bhat, G. P., 2000. A clinical and biochemical study of chronic fluoride toxicity in children of kheru thanda of gulbarga district, karnataka, India. *Fluoride.* **33**(2): 66–73.

Simonen, O. and Laitinen, O., 1985. Does fluoridation of drinking water prevent bone fragility and osteoporosis? *Lancet.* **2**: 432–433.

Susheela, A. K. (ed.), 1993. *Prevention and Control of Fluoride in India (Health Aspects).* New Delhi, India: Rajiv Gandhi National Drinking Water Mission, Vol. 1, pp. 20–22.

Susheela, A. K., 2003. *Treatise on Fluorosis*, 2nd edn. New Delhi, India: Fluorosis Research & Rural Development Foundation.

Susheela, A. K. and Bhatnagar, M., 2002. Reversal of fluoride induced cell injury through elimination of fluoride and consumption of diet rich in essential nutrients and antioxidants. *Molecular and Cellular Biochemistry.* **234/235**: 335–340.

Susheela, A. K., Kumar, A., Bhatnagar, M., and Bahadur, R., 1993. Prevalence of endemic fluorosis with gastro-intestinal manifestations in people living in some North-Indian villages. *Fluoride.* **26**: 97–104.

Tebutt, T. H. Y., 1983. Relationship between natural water quality and health. United Nations Educational, Scientific and Cultural Organization, Paris, France.

Teotia, S. P. S., Teotia, M., Singh, D. P., Rathour, R. S., Singh, C. V., Tomar, N. P. S., Nath, M., and Singh, N. P., 1984. Endemic fluorosis: Change to deeper bore wells as a practical community-acceptable approach to its eradication. *Fluoride*. **17**: 48–52.

Teotia, M., Teotia, S. P. S., and Singh, K. P., 1998. Endemic chronic fluoride toxicity and dietary calcium deficiency interaction syndromes of metabolic bone disease and deformities in India: Year 2000. *Indian Journal of Pediatrics*. **05**: 371–381.

UNICEF, 1999. State of the art report on the extent of fluoride in drinking water and the resulting endemicity in India. Report by Fluorosis Research & Rural Development Foundation for UNICEF, New Delhi, India.

US EPA, 1996. R.E.D. FACTS, Cryolite, EPA-738-F-96-016, United States Environmental Protection Agency.

Vikas, C., Kushwaha, R., Ahmad, W., Prasannakumar, V., and Reghunath, R., 2013. Genesis and geochemistry of high fluoride bearing groundwater from a semi-arid terrain of NW India. *Environmental Earth Sciences*. **68**: 289–305.

Villa, A., Anabalon, M., Zohouri, V., Maguire, A., Franco, A. M., and Rugg- Gunn, A., 2010. Relationships between fluoride intake, urinary fluoride excretion and fluoride retention in children and adults: An analysis of available data. *Caries Research*. **44**: 60–68.

Whitford, G. M., 1996. The metabolism and toxicity of fluoride. *Monographs in Oral Science*. **16**: 1–153.

Whitford, G. M., Pashley, D. H., and Stringer, G. I., 1976. Fluoride renal clearance: A pH-dependent event. *American Journal of Physiology*. **230**: 527–532.

Windholz, M., ed. 1976. *The Merck Index*, 9th edn. Rahway, NJ: Merck & Co., Inc., p. 211.

World Health Organization (WHO), 1996. *Guidelines for Drinking-Water Quality*, 2nd edn., Vol. 2, Health Criteria and Other Supporting Information. Geneva, Switzerland: World Health Organization.

World Health Organization (WHO), 1998. *Guidelines for Drinking-water Quality*, 2nd edn., Vol. 3, Addendum. Surveillance and Control of Community Supplies. Geneva, Switzerland: World Health Organization.

World Health Organization (WHO), 2008. *Guidelines for Drinking–Water Quality*, 3rd edn. Incorporating the First and Second Addenda. Geneva, Switzerland: WHO Library Cataloguing-in-Publication Data.

World Health Organization (WHO), 2010. *Exposure to Inadequate or Excess Fluoride: A Major Public Health Concern*. Public Health and Environment. Geneva, Switzerland: World Health Organization.

Whyte, M. P., Essmyer, K., Gannon, F. H., and Reinus, W. R., 2005. Skeletal fluorosis and instant tea. *American Journal of Medicine*. **118**(1): 78–82.

20 Antioxidant Additives in Food Preservation and Human Health

Piya Das Ghatak and Chandan K. Sen

CONTENTS

20.1 A BRIEF HISTORY OF FOOD STORAGE

Abundance of agricultural products is often season specific, and they need to be transported to geographically remote areas for human consumption. Irrespective of the form, raw or cooked, there is a pressing need to preserve and store food. Since the dawn of civilization, storing meat on ice and drying food in sun were very primitive practices. Middle Eastern and Oriental civilizations bear evidence of food preservation as early as 12,000 BC. Sailors, pilgrims, transporters, and merchants preserved foods for inevitable reasons of traveling long distances. Drying, smoking, pickling, salting, and honeying were introduced as popular preservation methods. These approaches not only preserved but added flavor to food. The late 1700s witnessed major advancements in food preservation under Napoleon Bonaparte who observed that "an army travels on its stomach." Soldiers in his troop lived off salt-preserved meat. Such meat suffered from compromised nutritional value, resulting in healthcare issues in his army. This prompted the search for better mechanisms of food preservation. Napoleon's French government announced a prize of 12,000 francs for effective innovators in food preservation. Nicolas Appert, later recognized as the "father of food preservation," won the prize. He discovered that boiled food stored in an airtight container may retain its nutritional quality and would not spoil as long as the seal is not broken (Featherstone 2012). Fifty years later, another Frenchman, Louis Pasteur, reported the relation between food spoilage and microorganisms. Pasteurization relies on high-temperature heating aimed at keeping food from spoiling. By 2018, the food preservative market is expected to be worth $2.7 billion (Marketsandmarkets 2015).

20.2 PROBLEMS WITH FOOD STORAGE: PACKAGING

Packaging of processed food is as important as food processing itself. Primarily, a package provides protection and tampering resistance as well as in almost all cases bears a "nutritional facts" label compliant with local regulatory guidelines. Proper packaging is aimed at preserving the food from spoilage by means of direct physical contact or microbial contamination and making the food visually appealing to the consumer (Robertson 2012). Metal cans represent early solutions in food packaging. Almost around the same time, in the early 1800s, paperboard and carton were introduced as food packaging containers, followed by glass and plastic containers. Because improper packaging poses the threat of serious infection to food material, and may damage commerce by limiting consumer appeal, this is an extensive field for research and development.

In recent years, the development of novel food packaging techniques such as modified atmosphere, active packaging, and intelligent packaging has increased the safety, quality, and shelf life of foods (Dainelli et al. 2008). Modified atmosphere packaging (MAP) rests on the principle of air replacement in the package with a more inert gas mixture (Parry 2012). For most food products, different combinations of oxygen (O_2), nitrogen (N_2), and carbon dioxide (CO_2) gases are used. Oxygen scavengers or oxygen absorbers are added to help remove and/or decrease the level of oxygen in the package. These agents are added to foods, which are prone to rancidification. Often, the scavenger mixture of iron powder and sodium chloride is enclosed in a porous sachet or packet, but it can also be part of packaging structures. Intelligent packaging material is an emerging area in food nanotechnology. It is a mode of packaging capable of carrying intelligent functions such as detecting, sensing, and tracing, which help in decision-making to extend shelf life and have the potential to warn about possible problems. pH-sensitive food spoilage indicator is an example of intelligent packaging that uses metabolite information to warn the consumer (Nopwinyuwong et al. 2010).

20.3 FOOD ADDITIVES

By definition, food additives are any ingredients added to food in small amounts. Generally, the purpose is to preserve flavor, maintain the safety and freshness of food, and enhance its taste and appearance. Additives are added to food for a precise purpose during processing to

- Add nutrients: in order to replace vitamins, minerals, and other micronutrients lost during processing or of supplementary dietary value. Examples include riboflavin (vitamin B2) added in cereal and energy bars.
- Help process or prepare food such as herbs and spices.
- Fight infection-related complications. For example, nitrate and nitrite compounds are added to meat products such as sausages and ham to protect against the botulism-causing bacteria *Clostridium botulinum.*
- Make the food more appealing to end consumers. An example is the use of xanthan gum in salad dressings, bakery stuffing, and puddings.

Usually, the labeling on the food container bears the details of direct additives on them. Indirect food additives, on the other hand, are not used or placed in the food on purpose but become part of the food in small amounts during packaging, storing, or other handling-associated processes (USFDA 2004). For example, slight amounts of packaging materials might be present in food during its storage. Food packaging manufacturers are therefore required to demonstrate to the U.S. Food and Drug Administration (FDA) that all such additives are safe for human consumption before the process is approved for commercial use.

20.3.1 Color

Color additive or food coloring is any dye, pigment, or element that changes the original color when added to food or drink, with the objective to improve human consumption (Barrows et al. 2003). Food colorings are used in food for many purposes:

- Compensate color loss caused by exposure to light, air, extreme temperature, humidity, and storage conditions.
- Correct the usual variations in color.
- Enrich the natural color.
- Provide desirable color to colorless foods. For example, without color additives, cola beverages would not be brown.
- Make food more attractive and appetizing.
- Allow consumers to recognize products on sight. For example, strawberry flavored food products are better received if their color resembles a strawberry shade.

FDA-permitted colors are categorized into certified and certification exempt colors.

Certified dyes are synthetically produced, that is, human made. These are broadly used because they are often effective in imparting an intense, even color at low cost. As listed in the following texts, there are nine certified colors approved in the United States to be added to food, the last two of them being of limited use. They are named as "FD&C" (Food, Drugs and Cosmetics) followed by a numeric value (Table 20.1).

The synthesis process is well developed to a point where the coloration does not introduce any undesirable odor or flavor.

TABLE 20.1
Certified Color Additives Approved for Use in Human Food, Subject to Batch Certification

Name	Common Names	Restrictions in Color/Uses
FD&C Blue No. 1	Brilliant blue FCF	Bright blue: used generally
FD&C Blue No. 2	Indigotine	Royal blue: used generally
FD&C Green No. 3	Fast green FCF	Sea green: used generally
FD&C Red No. 3	Erythrosine	Cherry red: used generally
FD&C Red No. 40	Allura red AC	Orange-red: used generally
FD&C Yellow No. 5	Tartrazine	Lemon-yellow: used generally
FD&C Yellow No. 6	Sunset yellow FCF	Orange: used generally
Orange B		Red shade: limited use
Citrus Red No. 2		Orange shade: limited use

Source: Adapted from USFDA, Summary of color additives for use in the United States in foods, drugs, cosmetics, and medical devices, 2015c, last modified November 8, 2015, http://www.fda.gov/ForIndustry/ColorAdditives/ColorAdditiveInventories/ucm115641.htm, accessed August 15, 2015.

Certification exempt colors include natural pigments, that is, those derived from natural sources such as vegetables and minerals. Generally, because of higher production cost from naturally sourced materials, these are more expensive than synthetic certified colors and may add flavors to food.

Example includes caramel color, made from caramelized sugar.

Unlike synthetic certified colors, exempted color additives are not essential to be stated by name on food labels but may be mentioned merely as food colorings or color added (USFDA 2004).

20.3.1.1 Food Coloring Agent

Color is added to replace color lost during preparation or to make the food more attractive. For example, brilliant blue color is often used in ice cream, packaged soups, bottled food colorings, icing, and dairy products to impart a blue color.

20.3.1.2 Color Retention Agent

These are used to preserve the natural color of food or beverage and resist discoloration. For example, ascorbic acid is added to peaches to help retain its bright color during canning (Choi 2002).

20.3.2 Essence

Food essences are mostly liquids derived from plants. Flowers and such other natural sources that do have potent smells and tastes are of culinary value. Often they are described as extracts such as vanilla, cinnamon, lemon, nutmeg, peppermint, and rose.

20.3.3 Condiment

Derived from the Latin *condimentum*, meaning "spice, seasoning," a condiment is a seasoning, sauce, or additional food preparation that is added to food to impart or enhance flavor. Examples include different spices, mayonnaise, soy sauce, tomato ketchup, and vinegar.

20.3.4 Preservatives

Serving as an additive, preservative is a substance that is added to food to prevent the decomposition or spoilage of food by any microbial growth or any undesirable chemical changes (Aversano 1984). Thus, preservatives are additives added to food to protect it from becoming inconsumable.

For example, benzoic acid and sodium benzoates are used in jams, salad dressings, and juices (USFDA 2010). A food preservative or preserving agent is added to food with the objective of inhibiting or retarding the formation of microorganisms such as yeasts and molds or any undesirable chemical change, thus increasing its shelf life (Russell and Gould 2003, Fazlara and Ekhtelat 2012). There are physical and chemical methods of food preservation, and most of the time, these two methods are combined. As the name suggests, chemical preservation takes place by adding chemical compounds to the food, whereas the physical methods include physical treatment of the food, such as UV radiation, freezing, and drying. Chemical preservatives can be broadly classified into two groups: synthetic or artificial and nonsynthetic or natural. Synthetic chemical preservatives can be classified into two classes—antimicrobial and antioxidants. Lactic acid is the most commonly used antimicrobial preservative (Daeschel 1989). Antioxidant additives inhibit the oxidation process of food, especially those that are high in fat content, thus helping in preventing rancidity.

It is important that the amount of preservative used may not exceed the approved safe levels. Even if the preservative amount is within limits, it may not affect food flavor undesirably (Kooistra and Troller 1968, Aversano 1984). Sodium benzoate, sodium nitrite, and potassium sorbate are widely used in the food industry. They are used to prevent biodeterioration of margarine, fresh juices, sweets, cakes, syrups, fish, and meat products by microorganisms. These preservatives exert an antimicrobial effect. Antifungal action requires higher concentrations than that required for antibacterial action (Stanojevic et al. 2009). They may destroy microbial cell membranes, cell walls, or protein structure and enzyme systems, thus acting as microbicidal or microbiostatic (Er et al. 2014). Again, the preservatives added to food are harmful for human health when consumed in high amounts (Walker 1990, Noorafshan et al. 2014). Therefore, studies are ongoing to design new preservative compounds that can be effective against biofilm-forming bacteria at a dose that is not harmful for human consumption (Er et al. 2014).

20.3.4.1 Acidity Regulators

Acidity regulators or pH control agents are food additives aimed at changing or maintaining pH (acidity or basicity) of the food product. They may be organic or mineral acids, bases, neutralizing agents, or buffering agents. Acidity regulators are important for processing, taste, and food safety. Inadequate control of food pH may result in the growth of undesirable bacteria causing potential health hazards. Examples include citric, acetic, lactic acid, and sodium bicarbonate.

20.3.4.2 Anticaking Agents

Anticaking agents were developed to keep food ingredients, mostly powdered and granulated materials, from clumping together after being packaged. Because moisture poses a substantial threat, anticaking agents either act to absorb moisture or act as a sealant, preventing exposure to water or oil. Examples include sodium ferrocyanide and hygroscopic calcium carbonates. Sodium aluminosilicate is present in many commercial table salts as well as dried milk, egg mixes, sugar products, and flours. Some anticaking agents are soluble in water, while others are soluble in alcohol or other organic solvents.

20.3.4.3 Natural Preservatives

Some natural compounds possess antimicrobial properties, which are helpful in preventing bacterial and fungal spoilage as well as the growth of microbial pathogens. There is a growing interest to substitute synthetic or artificial preservatives with natural preservatives. Examples of natural preservatives include salt, lemon juice, sugar, pickling, rosemary extract, honey, oil, vinegar, oregano, clove, and thyme.

20.3.4.4 Artificial or Chemical Preservatives

According to the FDA, the term *chemical preservative* means any chemical that, when added to food, tends to prevent or retard deterioration but does not include common salt, sugar, vinegar,

TABLE 20.2
Few Commonly Used Food Additives

Additive Type	Function	Examples of Uses	Names as They Appear on Product Labels
Preservatives	• Prevent microbial growth (antimicrobials) • Slow or prevent changes in color, flavor, texture, and delay rancidity (antioxidants) • Maintain freshness	Beverages, baked goods, chips, oils, salad dressings, cheese, fruits, vegetables	• Ascorbic acid • Citric acid • Sodium benzoate • Potassium sorbate • BHT • BHA • EDTA • Tocopherols • Sodium nitrite
Sweeteners	• Add sweetness	Beverages, baked goods, confections, table-top sugar, processed foods	• Sucrose • Glucose • Fructose • Sorbitol • Mannitol • Aspartame
Color additives	• Offset color loss due to exposure to light, air, extreme temperatures, moisture, and storage conditions • Add color to colorless foods • Correct natural variation in color	Processed foods—candies, jams, jellies, soft drinks, puddings, pie fillings	• FD&C Blue Nos. 1 and 2, FD&C Green No. 3, FD&C Red Nos. 3 and 40, beta-carotene, caramel color
Flavors and spices	• Enhance the flavor already present in the food	Processed foods	• Monosodium glutamate (MSG) • Hydrolyzed soy protein • Autolyzed yeast extract
Nutrients	• Replace vitamins and minerals lost during processing (enhancement) • Add nutrients that may be lacking in the diet (fortification)	Flour, breads, cereals, rice, macaroni, margarine, salt, milk, energy bars	• Thiamine hydrochloride • Riboflavin (vitamin B2) • Niacin • Folic acid • Ascorbic acid • Vitamin D • Amino acids • Potassium iodide

Source: Adapted from USFDA, Overview of food ingredients, additives & colors, 2010, http://www.fda.gov/Food/IngredientsPackagingLabeling/FoodAdditivesIngredients/ucm094211.htm-foodadd.

spice, or oil extracted from spices, substances added to food by direct exposure to wood smoke, or chemicals applied for their insecticidal or herbicidal properties.

Example includes sodium benzoate (USFDA 2014a) (Table 20.2).

20.4 ANTIOXIDANT

By definition, an antioxidant is a substance that prevents the oxidation of other molecules (Sies 2015). In 1989, Halliwell and Gutteridge defined antioxidant as "any substance that, when present at low concentrations compared with that of an oxidizable substrate, significantly delays or inhibits oxidation of that substrate." In other words, an "antioxidant" is "a molecule capable of slowing or preventing the oxidation of other molecules by being oxidized itself" (Halliwell and Gutteridge 1990).

TABLE 20.3
Few Commonly Used Antioxidants

Antioxidant Name	Use
Ascorbic acid	Beers, cut fruits, dried potatoes, jams
Citric acid	Biscuits, jams, tinned fruits, alcoholic drinks, cheese, and dried soup
Tocopherols	Meat pies and oils
Sodium ascorbate	Dairy-based drink, milk powder, whey products, fruit concentrate
Propyl gallate	Chewing gum, butter oil, ghee, dairy-based desserts, cereals, herbs, spices, seasonings, condiments
Erythorbic acid	Fresh meat, milk, poultry, fish products
Sodium erythorbate	Fermented vegetables, fish products, crustaceans
Butylated hydroxyanisole (BHA)	Cocoa, chocolate products, frozen meat and fish, edible ice, sorbet, milk fat, butter oil
Butylated hydroxytoluene (BHT)	Food supplements, bakery, fish oil, other animal fats
Tertiary butylhydroquinone (TBHQ)	Beverages, bread crumbs, soups, broths
Calcium ascorbate	Dried vegetables, breakfast cereals, egg-based desserts, vinegars

Antioxidants are used in food industry to keep food safe from deterioration caused by oxidation. Unlike microbial (bacterial and fungal) spoilage, oxidation reactions keep occurring even in a refrigerated environment. The broad spectrum of antioxidant preservatives encompasses natural antioxidants such as tocopherols and ascorbic acid, as well as synthetic antioxidants such as propyl gallate (PG), tertiary butylhydroquinone (TBHQ), butylated hydroxyanisole (BHA), and butylated hydroxytoluene (BHT). Primary antioxidants such as BHA or BHT are substituted phenolic-type compounds and act as free radical scavengers to terminate the propagation of lipid peroxidation chain reactions implicated in acceleration of food oxidation. On the other hand, synergists or secondary antioxidants such as thiodipropionates break down hydroperoxides, which normally decompose to form additional free radicals into relatively stable products (Shamberger et al. 1972; Table 20.3).

20.4.1 Principles of Redox Chemistry

The word "redox" originated from the combination of "reduction" and "oxidation," which occur simultaneously (Roth et al. 2001). Thus, redox reactions are such that one reactant is oxidized or loses electron(s), while the other is reduced or gains electron(s) simultaneously (Török et al. 2000). Redox reactions are common and vital to all living organisms (Halliwell 2006).

20.4.2 Oxidation as a Threat to Food Preservation

Oxygen is essential for life but can be toxic at high doses. Apples and other produce (e.g., bananas, peaches, and potatoes) contain an enzyme called polyphenol oxidase (PPO) that reacts with oxygen and iron-containing phenols present in fruits. When the cut or bruised fruit is exposed to ambient oxygen, PPO oxidizes the phenolic compounds to colorless o-quinones, which then produces a brown coloring as a result of its reaction with other amino acids, a very common example of Maillard reaction. One simple way of stopping apples from turning brown is to add a small amount of ascorbic acid–rich lemon juice (Pizzocaro et al. 1993).

20.4.3 Antioxidant Additives

Food antioxidant additives were first introduced to the market over six decades ago (Matill 1947). Because of the health benefits and food preservation aspects of antioxidants, this class of additives is under active development. A key consideration here is to make sure that these additives do not have any distinctive taste of their own. Furthermore, there is a growing interest to identify plant-derived natural antioxidant additives as opposed to reliance on synthetic chemistry. Such preference is primarily based on safety and toxicity concerns that may be posed by synthetic antioxidants (Williams et al. 1999).

20.4.3.1 Food Colors

Several food colors, apart from adding the desired color to the food, also possess antioxidant properties. Most natural food color additives are rich in pigments, which are natural sources of antioxidants (McBride 1996). Some examples are discussed in the following texts.

20.4.3.1.1 Lutein

Lutein adds yellow color to foods (Latin *luteus* meaning "yellow"). Lutein is a xanthophyll occurring in plants, especially in leaves and flowers. The marigold flower is commonly used to extract lutein. Other leafy vegetables such as kale and spinach are also viable sources of lutein. The main coloring components of lutein are its fatty acid esters. Lutein is capable of scavenging superoxide radicals and hydroxyl radicals and inhibiting lipid peroxidation (Sindhu et al. 2010, Canon 2013).

20.4.3.1.2 Carotene

Carotenes are among the most widely occurring pigments in nature. It is used in the food industry as a coloring agent, adding a shade of yellow to orange-red in food products such as nonalcoholic flavored drinks, confectionary, desserts, yogurts, juice, butter, margarine, sauces, and soups. Also, because of its importance as a nutritionally valuable antioxidant, it is commonly used as a dietary supplement and nutrient in the United States (USFDA 2014b).

20.4.3.1.3 Aronia

Aronia fruits produce a bright red color and are a rich source of anthocyanins. Aronia berries are used in juice blends for color and marketed for their antioxidant properties. The antioxidant properties of anthocyanins are recognized in a wide range of health outcome studies (Oszmiański and Wojdylo 2005).

20.4.3.1.4 Anthocyanin

Anthocyanins are water-soluble pigments, which cause red, blue, and purple color in fruits, flowers, and leaves. The word "anthocyanin" is derived from the Greek "anthos" meaning flower and "kyanos" meaning blue. It adds color to foods, which include pickles, soft drinks, and fruit desserts (Lila 2004). Apart from acting as a color additive to foods, anthocyanin is popularly known for its antioxidant property. There is evidence from both *in vivo* and *in vitro* studies that anthocyanin is a potent antioxidant (Tsuda et al. 1996).

20.4.3.2 Condiments

Condiments are substances that add flavor to food. Spices, herbs, and sauces are commonly found in the "condiments" aisle of the grocery shop. The use of condiments represents a safe and reliable approach to add distinct flavor to food without adding the burden of extra calories. Apart from adding distinct aroma, most spices possess antioxidant properties as well (Benzie et al. 2011). Interestingly, it is claimed that spices and herbs include the most antioxidant-rich products among other food categories such as berries and berry products, fruit and fruit juices, nuts and seeds, breakfast cereals, chocolate and sweets, beverages, and vegetables and vegetable products (Carlsen et al. 2010). Clove, oregano, ginger, cinnamon, turmeric, basil, mustard, paprika, and chilies are a few of the most antioxidant-rich spices (Srinivasan 2014). These culinary herbs draw interest over synthetic antioxidants with questionable human health safety records (Hinneburg et al. 2006).

20.4.3.3 Preservatives

20.4.3.3.1 Citric Acid (Food Acid and Acid Regulator)

Citric acid acts both as a food preservative and as an antioxidant additive. Citric acid regulates food pH, thus preventing spoilage. It also helps to retain the color of the food (Ahvenainen 1996). Foods that commonly use citric acid as an additive include ice cream and sorbets, caramel, soda, cider, beer and wine, cheese, many canned and jarred foods, baked goods and cake mixes, frozen fish (particularly herring, shrimp and crab), many processed sweets, and precut and packaged fruits and vegetables (Canon 2013).

20.4.3.3.2 Sulfur Dioxide

Sulfur dioxide is a chemical preservative extensively used in the wine industry. It is also used for other food products such as dehydrated fruits and vegetables and sulfured molasses (USFDA 2014a,b). Sulfur dioxide maintains the colorful appearance of the fruit, prevents rotting by virtue of its antimicrobial activity, and preserves food flavor during processing and subsequent storage (Joslyn and Braverman 1954). The preservation of food sulfur dioxide involves both its antioxidant and antimicrobial effects in most cases.

20.4.3.3.3 Sodium Sulfite

Sodium sulfite is an antioxidant preservative (USFDA 2014a). It is commonly used as a preservative for dried fruits, preserved radish, and flour treatment (DATABASE 2015b).

20.4.3.3.4 Sodium Bisulfite (aka Sodium Hydrogen Sulfite)

Sodium bisulfite is used as an antioxidant to prevent the browning of foods and to kill microbes (DATABASE 2015a).

20.4.3.3.5 Sodium Metabisulfite

Sodium metabisulfite is commonly used as a preservative in commercial wines and also as a flour treatment agent (DATABASE 2015a).

20.4.3.3.6 Vinegar

Vinegar is a commonly used food additive. It is generally used as a preservative, texturizer, and acidifier. Because of its acetic acid content, spoilage-causing microorganisms cannot survive in a vinegar environment. For the same reason, it is commonly used as a meat tenderizer where the acetic acid denatures meat protein. Furthermore, as an acidifier, vinegar helps to maintain a desired pH balance in food. Vinegar has considerable antioxidant properties (Dávalos et al. 2005).

20.4.3.3.7 Tartaric Acid

Tartaric acid acts as an emulsifier and an antioxidant additive. Tartaric acid emulsification is commonly used in baking. In winemaking, it is used to adjust wine acidity level to make the taste more palatable. This acid is used as a natural preservative to preserve jams, soft drinks, and candies as well (Papadopoulos et al. 2001).

20.4.3.3.8 Malic Acid

Malic acid, a food acid regulator and flavoring agent, is added to fruit-flavored drinks, candy, lemon-flavored ice-tea mix, ice cream, and preserves (Papadopoulos et al. 2001).

20.4.3.3.9 Lactic Acid

Lactic acid is a food additive that acts as a neutralizing and flavoring agent. It is commonly used in bakery products, cheese, frozen desserts, fruit butters, and jellies. Lactic acid increases the shelf life of meat (Lin and Yen 1999, Naveena et al. 2006).

20.4.3.4 Nutrients

There are several essential nutrient additives that have considerable antioxidant properties (Machlin and Bendich 1987).

20.4.3.4.1 Vitamin C

Ascorbic acid, commonly known as vitamin C, is a nutrient additive widely used in beer, cut fruits, dried potatoes, and jams with the purpose of preventing discoloration caused by oxidation. Thus, it is also a food preservative. This food additive also compensates for the loss of vitamin C during processing of foods. Dietary ascorbic acid is indispensible to sustain healthy living (Meister 1994).

20.4.3.4.2 Vitamin E

Tocopherols and tocotrienols represent the natural vitamin E family (Sen et al. 2006). These are extensively used in the preservation of edible oil such as sunflower oil, soybean oil, fish oil, and olive oil. They are also commonly used in meat pies and oils to reduce the oxidation of fatty acids and vitamins (Frankel 1996). Palm oil is a rich source of tocotrienols and tocopherols. Along with its antioxidant properties, α-tocotrienol offers major advantage over tocopherols for disorders of the brain and central nervous system (Sen et al. 2004).

20.4.3.5 Chemical/Synthetic Antioxidants

20.4.3.5.1 Gallates

Certain esters of gallic acids, usually termed as gallates are used as antioxidant food additives. Propyl gallate is an ester formed by the condensation of gallic acid and propanol. It is used as a food additive for antioxidant and other properties (USFDA 2013a). Propyl gallate protects fats and oils in food from oxidation (Medina et al. 2013). Examples of foods commonly using propyl gallate include vegetable oils, meat products, soup base, chewing gum, and cereals. Octyl gallate is an ester of 1-octanol and gallic acid. It is used both as an antioxidant and as a food preservative. Dodecyl gallate, or lauryl gallate, is an ester of dodecanol and gallic acid. It is also used both as an antioxidant and as a preservative. The antioxidant properties of dodecyl gallate in fatty products prevent rancidity. Food products that use this additive are snack foods, cereals, margarines, and oils (Noshly 2014).

20.4.3.5.2 Erythorbic Acid and Sodium Erythorbate

Erythorbic acid (D-isoascorbic acid), a stereoisomer of L-ascorbic acid, and its sodium salt, sodium erythorbate, demonstrate antioxidant properties and are used for this purpose in a number of food products (USFDA 2015b). They are also used as a preservative in cured meats, wine, beer, soft drinks, and frozen vegetables (Hui 2006).

20.4.3.5.3 Butylated Hydroxyanisole (BHA)

BHA is a potent lipid-phase antioxidant additive, which is present in a large number of fat and oil products such as potato chips, lard, butter, and margarine. Chemically, it is a phenolic compound and essentially is a combination of two isomeric organic compounds, 2-*tert*-butyl-4-hydroxyanisole and 3-*tert*-butyl-4-hydroxyanisole. It prevents the rancidification of fats by terminating lipid peroxidation, resulting in improved food taste and longer shelf life. Apart from fats and oils, BHA is also used in other food products such as beer, baked goods, beverages, and dry fruits. BHA has been extensively used in the food industry since 1947 for its potential to keep fat and oils safe and odor-free by preventing oxidation (Lam et al. 1979).

20.4.3.5.4 Butylated Hydroxytoluene (BHT)

BHT is an additive very similar to BHA in nature and properties. This is also a phenolic antioxidant, which is lipophilic in nature. BHT is used in the food industry with a similar purpose as served by

BHA, which is fighting lipid peroxidation to save oils, chips, and other fat-containing products from being rancid and to sustain their freshness for longer period of time. Apart from fats and oils, BHT is also added to cereals, breads, crackers, and snack foods (USFDA 2014a).

20.4.4 Antioxidants in Packaging

Food packaging is a critical component of the vast food industry that has a major influence on marketing and overall food commerce. The three most important purposes met by packaging are efficient food storage, enabling long distance shipping, and evoking customer appeal. Thus, food packaging is subject to rigorous research and development. The development of interactive, smart, or intelligent packaging is in demand. Adding antioxidants to the food being packaged is an innovation introduced in recent years with the objective of minimizing food oxidation. Lipid oxidation is the second leading trigger of food putrefaction after microbial growth. Oxidation not only threatens the food content but also is directly implicated in the thermal disgrace of polymers used for food packaging. Synthetic antioxidants such as BPA, some polyphenols, and thioester (such as dilauryl thiodipropionate) compounds are conventionally used to minimize such disgrace. However, these polymer antioxidants are potentially toxic if they were to leech out into the food content. Because of such threat, the use of polymer polyphenols and thioesters is being minimized. A solution to this problem is provided by the addition of food additive antioxidants such as BHA, BHT, and PG to polymer packaging material (Brüggemann et al. 2004, Gómez-Estaca et al. 2014). Thus, edible food antioxidants may be useful to protect packaging material and would pose no threat to human health (Torres-Arreola et al. 2007). Natural antioxidant film coating is commonly used on food to prevent oxidation. Coating food with natural antioxidants such as vitamin E, oregano extract, essential oils, grapefruit seed extract, and green tea is effective in active food packaging (Sanches-Silva et al. 2014). Another approach that is commonly adopted to safeguard food from oxidation involves the slow release of antioxidants from the packaging material into the food over a period of time (Gemili et al. 2010). Along with the original purpose of protecting the food, such an approach also adds nutritional value to the food. Recent research in the area of antioxidant active packaging has led to the development of a film based on ethylene vinyl alcohol copolymer (EVOH) and green tea extract. Food, both aqueous and fat soluble, is coated with this film (López de Dicastillo et al. 2011).

20.5 TOXICOLOGY OF FOOD ADDITIVES

20.5.1 Additive Toxicity

Food additives were introduced to keep food aesthetically appealing over a period of time and to increase the shelf life of food without compromising human health. Ancient Romans and Egyptians used food additives. In those times, addition of natural ingredients such as salt, vinegar, and sugar served the purpose. Later, artificial or synthetic additives were introduced. As safety concerns came to the focus, several food additives were found to be unsafe. Examples include low-calorie sweeteners or sugar substitutes such as saccharin, aspartame, and acesulfame-K that are often used in "diet" sodas. They are also used with tea or coffee. Saccharin, after its accidental discovery in 1869, was banned because of its debatable effects on human health. The FDA banned it in 1977, after it was proven to have carcinogenic effects on rat bladder (Price et al. 1970, Munro et al. 1975). Roosevelt, who ate saccharin regularly, observed "anybody who says saccharin is injurious to health is an idiot" (Smallwood 2014). The ban was repealed in 2000, but it is required that food labels declare the presence of saccharin in it. Recent studies report potential genotoxic effects of saccharin (Bandyopadhyay et al. 2008). Aspartame, another accidental discovery, is widely used as an artificial nonsaccharide sweetener. Safety issues related to aspartame remained a concern for a long period of time. It got approved by the FDA in several stages in 1981, 1983, and finally in 1996.

Although the FDA says, "Aspartame is one of the most exhaustively studied substances in the human food supply, with more than 100 studies supporting its safety," there are studies providing compelling evidences that it possesses serious threat as a carcinogen (USFDA 2003, 2015a, Rencüzogullari et al. 2004, Soffritti et al. 2014).

Another broadly used additive is monosodium glutamate (MSG), the sodium salt of glutamic acid. It is used commercially in a wide variety of foods, which includes meats, poultry, and seafood and in home as a flavor enhancer, in the form of the popular "ajinomoto." MSG is found to induce adverse effects on liver and kidney, as well as in the retina (Manal Said and Nawal 2012, Meister 2013)

Toxicity of some antioxidant additives is addressed in the following texts:

20.5.1.1 Citric Acid

Citric acid, a widely used antioxidant additive, possesses a clastogenic effect as evident in a study conducted with human peripheral lymphocytes. Food additive citric acid has been reported to be able to induce chromosomal aberrations (Yılmaz et al. 2008).

20.5.1.2 Sulfites

Different sulfite salts of sodium, such as sodium sulfite, sodium bisulfite, and sodium metabisulfite, are used as additives in a wide variety of food products as antioxidant preservatives. Due to their potential toxicity, all of these salts are considered as GRAS (generally recognized as safe) by the FDA (USFDA 2013b). Studies have shown that sulfites have adverse effects on the brain, kidney, and liver (Kencebay et al. 2013, Noorafshan et al. 2013, Mahmoud et al. 2015).

20.5.1.3 Gallates

All gallates additives, dodecyl, propyl, and octyl, show some general toxicological activities, including dermatological allergies (Organization 1973, Van der Heijden et al. 1986,). Propyl gallate may cause cancer. A 2009 study shows it is an estrogen antagonist (Amadasi et al. 2008). PG can cause allergic reactions in the form of an asthma attack in some people. It can also cause stomach and skin irritation. The FDA considers PG as GRAS. Recent studies have shown that PG has genotoxic and cytotoxic abilities. It is shown that cells treated with PG lead to early apoptosis and DNA damage in cells (Hamishehkar et al. 2013). Studies conducted in rat hepatocytes also indicate that the hepatotoxic effect of propyl and other gallates is due to mitochondrial dysfunction (Nakagawa and Tayama 1995). Dodecyl gallate has shown to be a strong sensitizer (Meeren 1987).

20.5.1.4 Butylated Hydroxyanisole (BHA)

BHA is deemed by the FDA as GRAS and by the NIH as "reasonably anticipated to be a human carcinogen" (National Institute of Health 1991, USFDA 2014a). There are several studies over last two decades, in various cell lines and in a dose and time-dependent manner, that BHA is capable of inducing cytotoxic effects, such as DNA damage and mitochondrial permeability transition causing apoptosis of cells (Weber 2014). Although many consider BHA as safe when consumed in a low concentration, unfortunately there are studies that have proved its cytotoxic effects even in lower than the recommended doses. BHA is considered as a carcinogen in the state of California (Scorecard 2011). Several *in vivo* mouse studies have shown evidences of BHA acting as a carcinogen (Vandghanooni et al. 2013).

20.5.1.5 Butylated Hydroxytoluene (BHT)

BHT shares safety concerns with BHA when it comes to human consumption. Although the FDA recognizes it as GRAS in the United States, BHA is banned in Japan, Romania, Sweden, Australia, and many European countries. Several studies have been conducted over a period of time to test the activity of BHA from a toxicological perspective in a dose-dependent manner. It has shown to have hepatotoxic effect on rats (Panicker et al. 2014). In the "Final report on safety assessment of BHT," it was reported that it is neither genotoxic nor carcinogenic but it has some dose- and time-dependent cytotoxic effects (Lucas et al. 2000, Lanigan and Yamarik 2001).

20.5.2 Additives and Sensitivity: Allergy and Intolerance

Food additives may cause food allergy. Several synthetic food additives cause allergic reactions when consumed. Sulfites, aspartame, parabens, tartrazine, MSG, nitrates, BHT, and benzoates are few of such additives that are known to cause allergies (Gultekin and Doguc 2013). BHT and BHA may aggravate chronic urticaria (Goodman et al. 1990). The artificial color additive sunset yellow FCF (for coloring food) triggers chronic urticaria (Ibero et al. 1981). Natural color additives are generally safe for human health (Lucas et al. 2000). Sulfites may induce asthma (Stevenson and Simon 1984, Ban et al. 2014). Most of these food allergies are mild in nature and typically resolve within a day or two.

20.6 EFFECTS OF FOOD PRESERVATIVES

20.6.1.1 Children

Nutrition plays a critical role in child development. Current studies support the notion that food additives may contribute to child hyperactivity (Swanson and Kinsbourne 1980, McCann et al. 2007). Furthermore, food additives may cause attention deficit hyperactive disorder (ADHD) in children (Boris and Mandel 1994). ADHD may have a profound influence on child development.

20.6.1.2 Women: Pregnancy

Although statutory warnings caution pregnant women against the consumption of tobacco and alcohol on the packages, no such caution is aimed at the consumption of food additives. While the topic is open for investigation, it is reasonable to hypothesize that significant consumption of food containing food additives may adversely influence child health. For example, orally administered MSG at a late stage of pregnancy is known to cross the placental barrier and cause a series of behavioral abnormalities in a mouse study (Yu et al. 1997).

20.7 FUTURE PERSPECTIVE

With developing countries leading the pack, the food preservative market is rapidly expanding. During storage, packaged foods experience physical and chemical changes that compromise freshness and customer appeal. This represents a massive threat to overall food commerce. Much of this damage is caused by oxidation reactions resulting in major emphasis on the antioxidant food additive market. While on the one hand, antioxidant additives may enhance the nutritional value of food, on the other hand, they prevent food from spoilage. Spoilage is a component of the overall food-waste problem, which is currently estimated to be $165 billion per year in the United States. Solving the problem of food spoilage alone is estimated to feed 1 billion more people by 2050 (RockefellerFoundation 2014).

GLOSSARY

Clastogenic: Giving rise to or inducing disruption or breakages, as of chromosomes.

Sensitizer: A sensitizer is defined by the Occupational Safety and Health Administration (OSHA) as "a chemical that causes a substantial proportion of exposed people or animals to develop an allergic reaction in normal tissue after repeated exposure to the chemical." The condition of being sensitized to a chemical is also called chemical hypersensitivity.

ABBREVIATIONS

ADHD: Attention deficit hyperactive disorder
BHA: Butylated hydroxyanisole
BHT: Butylated hydroxytoluene
FCF: For coloring food
FDA: Food and Drug Administration
FD&C: Food, Drug, and Cosmetic
GRAS: Generally recognized as safe
MAP: Modified atmosphere packaging
MSG: Monosodium glutamate
NIH: National Institutes of Health
OSHA: Occupational Safety and Health Administration
PG: Propyl gallate

REFERENCES

Ahvenainen, R. 1996. New approaches in improving the shelf life of minimally processed fruit and vegetables. *Trends in Food Science & Technology* 7(6):179–187.

Amadasi, A., Mozzarelli, A., Meda, C., Maggi, A., and Cozzini, P. 2008. Identification of xenoestrogens in food additives by an integrated in silico and in vitro approach. *Chemical Research in Toxicology* 22(1):52–63.

Aversano, R. W. 1984. Food preservative composition. Google Patents. US 4476112 A.

Ban, G. Y., Kim, M., Yoo, H. S., Ye, Y. M., and Park, H. S. 2014. Letter to the editor: Two major phenotypes of sulfite hypersensitivity: Asthma and urticaria. *Yonsei Medical Journal* 55(2):542–544.

Bandyopadhyay, A., Ghoshal, S., and Mukherjee, A. 2008. Genotoxicity testing of low-calorie sweeteners: Aspartame, acesulfame-K, and saccharin. *Drug and Chemical Toxicology* 31(4):447–457.

Barrows, J. N., Lipman, A. L., and Bailey, C. J. 2003. Color additives: FDA's regulatory process and historical perspectives. *Food Safety Magazine*, October/November 2003.

Barrows, J. N., Lipman, A. L., Bailey, C. J., and Cianci, S. 2003. Color additives: FDA's regulatory process and historical perspectives [Internet]. The Target Group; 2003 (cited January 10, 2010).

Benzie et al. (2011): Paur, I., Carlsen, M. H., Halvorsen, B. L., et al. Antioxidants in Herbs and Spices: Roles in Oxidative Stress and Redox Signaling. In: Benzie, I. F. F., Wachtel-Galor S., eds. *Herbal Medicine: Biomolecular and Clinical Aspects*, 2nd edn. Boca Raton, FL: CRC Press.

Boris, M. and Mandel, F. S. 1994. Foods and additives are common causes of the attention deficit hyperactive disorder in children. *Annals of Allergy* 72(5):462–467.

Brüggemann, O., Visnjevski, A., Burch, R., and Patel, P. 2004. Selective extraction of antioxidants with molecularly imprinted polymers. *Analytica Chimica Acta* 504(1):81–88.

Canon, E. 2013. What you need to know about citric acid. Accessed August 15. http://ethicalfoods.com/citric-acid/.

Carlsen, M. H., Halvorsen, B. L., Holte, K., Bøhn, S. K., Dragland, S., Sampson, L., Willey, C. et al. 2010. The total antioxidant content of more than 3100 foods, beverages, spices, herbs, and supplements used worldwide. *Nutrition Journal* 9(3):2–11.

Choi, M.H. 2002. Effects of ascorbic acid retention on juice color and pigment stability in blood orange (*Citrus sinensis*) juice during refrigerated storage. *Food Research International* 35(8):753–759.

Daeschel, M. A. 1989. Antimicrobial substances from lactic acid bacteria for use as food preservatives. *Food Technology* (*USA*).

Dainelli, D., Gontard, N., Spyropoulos, D., Zondervan-van den Beuken, E., and Tobback, P. 2008. Active and intelligent food packaging: Legal aspects and safety concerns. *Trends in Food Science & Technology* 19:S103–S112.

DATABASE, Codex General Standard for Food Additives (GSFA) Online. 2015a. Food additive details sodium metabisulfite (223). Last modified 2015, accessed August 15. http://www.codexalimentarius.net/gsfaonline/additives/details.html?id=225.

DATABASE, Codex General Standard for Food Additives (GSFA) Online. 2015b. Food additive details sodium sulfite (221). Last modified 2015, accessed August 15. http://www.codexalimentarius.net/gsfaonline/additives/details.html?id=227.

Dávalos, A., Bartolomé, B., and Gómez-Cordovés, C. 2005. Antioxidant properties of commercial grape juices and vinegars. *Food Chemistry* 93(2):325–330. doi: 10.1016/j.foodchem.2004.09.030.

Er, B., Demirhan, B., Onurdağ, F. K., Özgacar, S. Ö., and Öktem, A. B. 2014. Antimicrobial and antibiofilm effects of selected food preservatives against Salmonella spp. isolated from chicken samples. *Poultry Science* 93(3):695–701.

Fazlara, A. and Ekhtelat, M. 2012. The disinfectant effects of benzalkonium chloride on some important foodborne pathogens. *American-Eurasian Journal of Agricultural & Environmental Science* 12(1):23–29.

Featherstone, S. 2012. A review of development in and challenges of thermal processing over the past 200 years—A tribute to Nicolas Appert. *Food Research International* 47(2):156–160.

Frankel, E. N. 1996. Antioxidants in lipid foods and their impact on food quality. *Food Chemistry* 57(1):51–55.

Gemili, S., Yemenicioğlu, A., and Altınkaya, S. A. 2010. Development of antioxidant food packaging materials with controlled release properties. *Journal of Food Engineering* 96(3):325–332.

Gómez-Estaca, J., López-de-Dicastillo, C., Hernández-Muñoz, P., Catalá, R., and Gavara, R. 2014. Advances in antioxidant active food packaging. *Trends in Food Science & Technology* 35(1):42–51.

Goodman, D. L., McDonnel, J. T., Nelson, H. S., Vaughan, T. R., and Weber, R. W. 1990. Chronic urticaria exacerbated by the antioxidant food preservatives, butylated hydroxyanisole (BHA) and butylated hydroxytoluene (BHT). *Journal of Allergy and Clinical Immunology* 86(4):570–575.

Gultekin, F. and Doguc, D. K. 2013. Allergic and immunologic reactions to food additives. *Clinical Reviews in Allergy & Immunology* 45(1):6–29.

Halliwell, B. 2006. Reactive species and antioxidants. Redox biology is a fundamental theme of aerobic life. *Plant Physiology* 141(2):312–322.

Halliwell, B. and Gutteridge, J. M. C. 1990. The antioxidants of human extracellular fluids. *Archives of Biochemistry and Biophysics* 280(1):1–8.

Hamishehkar, H., Khani, S., Kashanian, S., Ezzati Nazhad Dolatabadi, J., and Eskandani, M. 2013. Geno- and cytotoxicity of propyl gallate food additive. *Drug and Chemical Toxicology* 37(3):241–246.

Hinneburg, I., Dorman, H. J. D., and Hiltunen, R. 2006. Antioxidant activities of extracts from selected culinary herbs and spices. *Food Chemistry* 97(1):122–129.

Hui, Y. H. 2006. *Handbook of Food Science, Technology, and Engineering*. Vol. 149, CRC Press, Boca Raton, FL.

Ibero, M., Eseverri, J. L., Barroso, C., and Botey, J. 1981. Dyes, preservatives and salicylates in the induction of food intolerance and/or hypersensitivity in children. *Allergologia et Immunopathologia* 10(4):263–268.

Joslyn, M. A. and Braverman, J. B. S. 1954. The chemistry and technology of the pre-treatment and preservation of fruit and vegetable products with sulfur dioxide and sulfites. *Advances in Food Research* 5:97.

Kencebay, C., Derin, N., Ozsoy, O., Kipmen-Korgun, D., Tanriover, G., Ozturk, N., Basaranlar, G., Yargicoglu-Akkiraz, P., Sozen, B., and Agar, A. 2013. Merit of quinacrine in the decrease of ingested sulfite-induced toxic action in rat brain. *Food and Chemical Toxicology* 52:129–136.

Kooistra, J. A. and Troller, J. A. 1968. Food preservative compositions and method for inhibiting microbial growth in food products. Google Patents. US 3404987 A.

Lam, L. K. T., Pai, R. P., and Wattenberg, L. W. 1979. Synthesis and chemical carcinogen inhibitory activity of 2-tert-butyl-4-hydroxyanisole. *Journal of Medicinal Chemistry* 22(5):569–571.

Lanigan, R. S. and Yamarik, T. A. 2001. Final report on the safety assessment of BHT (1). *International Journal of Toxicology* 21:19–94.

Lila, M. A. 2004. Anthocyanins and human health: An in vitro investigative approach. *BioMed Research International* 2004(5):306–313.

Lin, M. Y. and Yen, C. L. 1999. Antioxidative Ability of Lactic Acid Bacteria. *Journal of Agricultural and Food Chemistry* 47(4):1460–1466. doi: 10.1021/jf981149l.

López de Dicastillo, C., Nerin, C., Alfaro, P., Catalá, R., Gavara, R., and Hernández-Muñoz, P. 2011. Development of new antioxidant active packaging films based on ethylene vinyl alcohol copolymer (EVOH) and green tea extract. *Journal of Agricultural and Food Chemistry* 59(14):7832–7840.

Lucas, C. D., Hallagan, J. B., and Taylor, S. L. 2000. The role of natural color additives in food allergy. *Advances in Food and Nutrition Research* 43:195–216.

Machlin, L. J. and Bendich, A. 1987. "Free radical tissue damage: Protective role of antioxidant nutrients. *The FASEB Journal* 1(6):441–445.

Mahmoud, A. A. T., Hassan, G. M., Hassan, A. M. S., Latif, A. K. M. A., and Ramadan, M. F. 2015. Demonstrating adverse effects of a common food additive (sodium sulfite) on biochemical, cytological and histopathological parameters in tissues of albino Wister rats. *European Journal of Integrative Medicine* 7(3): 234–242.

Marketsandmarkets. 2015. Food preservative market by type (natural, chemical), function (antimicrobial, antioxidant), application (oil & fat, bakery, dairy, snack, meat, poultry & seafood, confectionery, beverage), & geography—Global trend & forecast to 2018. Accessed August 15. http://www.marketsandmarkets.com/Market-Reports/food-preservatives-market-420.html.

Matill, H.A. 1947. Antioxidants. *Annual Review of Biochemistry* 16:177–192.

McBride, J. 1996. Plant pigments paint a rainbow of antioxidants. United States Department of Agriculture, Accessed August 15. http://agresearchmag.ars.usda.gov/1996/nov/plant.

McCann, D., Barrett, A., Cooper, A., Crumpler, D., Dalen, L., Grimshaw, K., Kitchin, E., Lok, K., Porteous, L., and Prince, E. 2007. Food additives and hyperactive behaviour in 3-year-old and 8/9-year-old children in the community: A randomised, double-blinded, placebo-controlled trial. *The Lancet* 370(9598):1560–1567.

Medina, M. E., Iuga, C., and Alvarez-Idaboy, J. R. 2013. Antioxidant activity of propyl gallate in aqueous and lipid media: A theoretical study. *Physical Chemistry Chemical Physics* 15(31):13137–13146.

Meeren, H. L. M. 1987. Dodecyl gallate, permitted in food, is a strong sensitizer. *Contact Dermatitis* 16(5):260–262.

Meister, A. 1994. Glutathione-ascorbic acid antioxidant system in animals. *Journal of Biological Chemistry-Paper Edition* 269(13):9397–9400.

Meister, B. 2013. Monosodium glutamate (MSG) lesions. *Transplantation, Lesions, and Surgical Techniques: Methods in Neurosciences* 7:70–89.

Munro, I.C., Moodie, C.A., Krewski, D., and Grice, H.C. 1975. A carcinogenicity study of commercial saccharin in the rat. *Toxicology and Applied Pharmacology* 32(3):513–526.

Nakagawa, Y. and Tayama, S. 1995. Cytotoxicity of propyl gallate and related compounds in rat hepatocytes. *Archives of Toxicology* 69(3):204–208.

National Institutes of Health. 1991. Butylated hydroxyanisole. In *Sixth Annual Report on carcinogens (1991) Summary.* U.S. Dept. of Health and Human Services, Public Health Service, pp. 111–112.

Naveena, B.M., Muthukumar, M., Sen, A.R., Babji, Y., and Murthy, T.R.K. 2006. Improvement of shelf-life of buffalo meat using lactic acid, clove oil and vitamin C during retail display. *Meat Science* 74(2):409–415. doi: doi:10.1016/j.meatsci.2006.04.020.

Noorafshan, A., Asadi-Golshan, R., Karbalay-Doust, S., Abdollahifar, M. A., and Rashidiani-Rashidabadi, A. 2013. Curcumin, the main part of turmeric, prevents learning and memory changes induced by sodium metabisulfite, a preservative agent, in rats. *Experimental Neurobiology* 22(1):23–30.

Noorafshan, A., Erfanizadeh, M., and Karbalay-Doust, S. 2014. Sodium benzoate, a food preservative, induces anxiety and motor impairment in rats. *Neurosciences* (*Riyadh*) 19:24–28.

Nopwinyuwong, A., Trevanich, S., and Suppakul, P. 2010. Development of a novel colorimetric indicator label for monitoring freshness of intermediate-moisture dessert spoilage. *Talanta* 81(3):1126–1132. doi: 10.1016/j.talanta.2010.02.008.

Noshly. 2014. Octyl gallate 311 or E311, accessed August 15. http://noshly.com/additive/e311/antioxidant/311/ -. Vcjuj3iRPzJ.

NTP (National Toxicology Program). 2014. *Report on Carcinogens*, 13th edn. Research Triangle Park, NC: U.S. Department of Health and Human Services, Public Health Service. http://ntp.niehs.nih.gov/pubhealth/roc/roc13/.

Organization World Health. 1973. Toxicological evaluation of some food additives including anticaking agents, antimicrobials, antioxidants, emulsifiers and thickening agents, accessed August 15. http://www.inchem.org/documents/jecfa/jecmono/v05je26.htm.

Oszmiański, J. and Wojdylo, A. 2005. Aronia melanocarpa phenolics and their antioxidant activity. *European Food Research and Technology* 221(6):809–813.

Panicker et al. (2014): Panicker, V. P., George, S., and Krishna, D. 2014. Toxicity study of butylated hydroxyl toluene (BHT) in rats. *International Journal of Pharmacy and Pharmaceutical Sciences* 3:758–763.

Papadopoulos, K., Triantis, T., Dimotikali, D., and Nikokavouras, J. 2001. Evaluation of food antioxidant activity by photostorage chemiluminescence. *Analytica Chimica Acta* 433(2):263–268.

Parry, R.T. 2012. *Principles and Applications of Modified Atmosphere Packaging of Foods.* Springer, New York.

Pizzocaro, F., Torreggiani, D., and Gilardi, G. 1993. Inhibition of apple polyphenoloxidase (PPO) by ascorbic acid, citric acid and sodium chloride. *Journal of Food Processing and Preservation* 17:21–30.

Price, J. M., Biava, C. G., Oser, B. L., Vogin, E. E., Steinfeld, J., and Ley, H. L. 1970. Bladder tumors in rats fed cyclohexylamine or high doses of a mixture of cyclamate and saccharin. *Science* 167(3921):1131–1132.

Rencüzogullari, E., Tüylü, B. A., Topaktas, M., Ila, H. B., Kayraldiz, A., Arslan, M., and Diler, S. B. 2004. Genotoxicity of aspartame. *Drug and Chemical Toxicology* 27(3):257–268.

Robertson, G. L. 2012. *Food Packaging: Principles and Practice.* CRC Press, Boca Raton, FL.

Rockefeller Foundation (2015): Food waste and spoilage. https://www.rockefellerfoundation.org/app/uploads/Food-Waste-and-Spoilage.pdf. Accessed July 7, 2016.

Roth, J. P., Yoder, J. C., Won, T. J., and Mayer, J. M. 2001. Application of the Marcus cross relation to hydrogen atom transfer reactions. *Science* 294(5551):2524–2526. doi: 10.1126/science.1066130.

Russell, N. J. and Gould, G. W. 2003. *Food Preservatives*. Springer, New York.

Sanches-Silva, A., Costa, D., Albuquerque, T. G., Buonocore, G. G., Ramos, F., Castilho, M. C., Machado, A. V., and Costa, H. S. 2014. Trends in the use of natural antioxidants in active food packaging: A review. *Food Additives & Contaminants: Part A* 31(3):374–395.

Scorecard. 2011. California state regulatory program lists, accessed August 15. http://scorecard.goodguide.com/chemical-groups/one-list.tcl?short_list_name=p65.

Sen, C. K., Khanna, S., and Roy, S. 2004. Tocotrienol: The natural vitamin E to defend the nervous system? *Annals of the New York Academy of Sciences* 1031(1):127–142.

Sen, C. K., Khanna, S., and Roy, S. 2006. Tocotrienols: Vitamin E beyond tocopherols. *Life Sciences* 78(18):2088–2098.

Shamberger, R. J., Tytko, S., and Willis, C. E. 1972. Antioxidants in cereals and in food preservatives and declining gastric cancer mortality. *Cleveland Clinic Quarterly* 39(2):119–124.

Sies, H. 2015. Oxidative stress: A concept in redox biology and medicine. *Redox Biology* 4:180–183.

Sindhu, E. R., Preethi, K. C., and Kuttan, R. 2010. Antioxidant activity of carotenoid lutein in vitro and in vivo. *Indian Journal of Experimental Biology* 48(8):843–848.

Smallwood, K. 2014. The accidental discovery of saccharin, and the truth about whether saccharin is bad for you, accessed July 6, 2016. http://www.todayifoundout.com/index.php/2014/05/saccharin-discovered-accident/.

Soffritti, M., Padovani, M., Tibaldi, E., Falcioni, L., Manservisi, F., and Belpoggi, F. 2014. The carcinogenic effects of aspartame: The urgent need for regulatory re-evaluation. *American Journal of Industrial Medicine* 57(4):383–397.

Srinivasan, K. 2014. Antioxidant potential of spices and their active constituents. *Critical Reviews in Food Science and Nutrition* 54(3):352–372.

Stanojevic, D., Comic, L., Stefanovic, O., and Solujic-Sukdolak, S. l. 2009. Antimicrobial effects of sodium benzoate, sodium nitrite and potassium sorbate and their synergistic action in vitro. *Bulgarian Journal of Agricultural Science* 15(4):307–311.

Stevenson, D. D. and Simon, R. A. 1984. Sulfites and asthma. *Journal of Allergy and Clinical Immunology* 74(4):469–472.

Swanson, J. M. and Kinsbourne, M. 1980. Food dyes impair performance of hyperactive children on a laboratory learning test. *Science* 207(4438):1485–1487.

Tawfik, M. S. and Al-Badr, N. 2012. Adverse effects of monosodium glutamate on liver and kidney functions in adult rats and potential protective effect of vitamins C and E. *Food and Nutrition Sciences* 3(5):651.

Török, B., Kiricsi, I., Molnár, Á., and Olah, G. A. 2000. Acidity and catalytic activity of a Nafion-H/silica nanocomposite catalyst compared with a silica-supported Nafion sample. *Journal of Catalysis* 193(1):132–138.

Torres-Arreola, W., Soto-Valdez, H., Peralta, E., Cárdenas-López, J. L., and Ezquerra-Brauer, J. M. 2007. Effect of a low-density polyethylene film containing butylated hydroxytoluene on lipid oxidation and protein quality of Sierra fish (*Scomberomorus sierra*) muscle during frozen storage. *Journal of Agricultural and Food Chemistry* 55(15):6140–6146.

Tsuda, T., Shiga, K., Ohshima, K., Kawakishi, S., and Osawa, T. 1996. Inhibition of lipid peroxidation and the active oxygen radical scavenging effect of anthocyanin pigments isolated from Phaseolus vulgaris L. *Biochemical Pharmacology* 52(7):1033–1039.

USFDA. 2003. Reported aspartame toxicity effects, accessed August 15. http://www.fda.gov/ohrms/dockets/dailys/03/jan03/012203/02p-0317_emc-000199.txt.

USFDA. 2004. Overview of food ingredients, additives & colors. Last modified april 2010, accessed August 15. http://www.fda.gov/Food/IngredientsPackagingLabeling/FoodAdditivesIngredients/ucm094211.htm - foodadd.

USFDA. 2010. Overview of food ingredients, additives & colors, accessed August 15, 2015. http://www.fda.gov/Food/IngredientsPackagingLabeling/FoodAdditivesIngredients/ucm094211.htm-foodadd.

USFDA. 2013a. Select committee on GRAS substances (SCOGS) opinion: Propyl gallate. Last modified 08/14/2015, accessed August 15. http://www.fda.gov/Food/IngredientsPackagingLabeling/GRAS/SCOGS/ucm261032.htm.

USFDA. 2013b. Select committee on GRAS substances (SCOGS) opinion: Sodium sulfite. Last modified 04/23/2013, accessed August 15. http://www.fda.gov/Food/IngredientsPackagingLabeling/GRAS/SCOGS/ucm261417.htm.

USFDA. 2014a. CFR—Code of Federal Regulations Title 21. Last modified April 1, 2014, accessed August 15. http://www.accessdata.fda.gov/scripts/cdrh/cfdocs/cfcfr/cfrsearch.cfm?fr=101.22.

USFDA. 2014b. Food additive status list. Last modified 12/16/2014, accessed August 15. http://www.fda.gov/food/ingredientspackaginglabeling/foodadditivesingredients/ucm091048.htm - ftnC.

USFDA. 2015a. Additional information about high-intensity sweeteners permitted for use in food in the United States. Last modified 5/26/15, accessed August 15. http://www.fda.gov/Food/IngredientsPackagingLabeling/FoodAdditivesIngredients/ucm397725.htm.

USFDA. 2015b. Select committee on GRAS substances (SCOGS) opinion: Erythorbic acid (D-isoascorbic acid). Last modified 06/29/2015, accessed August 15. http://www.fda.gov/Food/IngredientsPackagingLabeling/GRAS/SCOGS/ucm261286.htm.

USFDA. 2015c. Summary of color additives for use in the United States in foods, drugs, cosmetics, and medical devices. Last modified 08/11/2015, accessed August 15. http://www.fda.gov/ForIndustry/ColorAdditives/ColorAdditiveInventories/ucm115641.htm.

Van der Heijden, C. A., Janssen, P. J. C. M., and Strik, J. J. T. W. A. 1986. Toxicology of gallates: A review and evaluation. *Food and Chemical Toxicology* 24(10):1067–1070.

Vandghanooni, S., Forouharmehr, A., Eskandani, M., Barzegari, A., Kafil, V., Kashanian, S., and Ezzati Nazhad Dolatabadi, J. 2013. Cytotoxicity and DNA fragmentation properties of butylated hydroxyanisole. *DNA and Cell Biology* 32(3):98–103.

Walker, R. 1990. Toxicology of sorbic acid and sorbates. *Food Additives & Contaminants* 7(5):671–676.

Weber, R. W. 2014. Adverse reactions to the antioxidants butylated hydroxyanisole and butylated hydroxytoluene. *Food Allergy: Adverse Reactions to Foods and Food Additives,* 5th edn., pp. 393–401.

Williams, G. M., Iatropoulos, M. J., and Whysner, J. 1999. Safety assessment of butylated hydroxyanisole and butylated hydroxytoluene as antioxidant food additives. *Food and Chemical Toxicology* 37(9):1027–1038.

Yılmaz, S., Ünal, F., Yüzbaşıoğlu, D., and Aksoy, H. 2008. Clastogenic effects of food additive citric acid in human peripheral lymphocytes. *Cytotechnology* 56(2):137–144.

Yu, T., Zhao, Y., Shi, W., Ma, R., and Yu, L. 1997. Effects of maternal oral administration of monosodium glutamate at a late stage of pregnancy on developing mouse fetal brain. *Brain Research* 747(2):195–206.

21 Industrial Food Processing Contaminants

K.V. Ragavan, Navin K. Rastogi, and Alok Kumar Srivastava

CONTENTS

21.1 INTRODUCTION

Eating habits of humans have changed a lot in the past three to four decades with increased consumption of processed foods driven by industrialization, adoption of western culture, and other social factors (Cordain et al. 2005). Prior to industrialization, food was cooked and consumed at small scale at one's own kitchen, but with the advent of modern machinery, food processing has taken a giant leap and plays a major role in human food consumption. Food processing involves various operations/steps to convert raw food materials from farm to processed food in plate. Food processing encompasses unit operations such as cleaning, sorting, blanching, thermal treatment, fermentation, curing, packaging, and other processes (van Boekel et al. 2010). Novel processing methods such as irradiation (Nayak et al. 2006), high-pressure processing (Rastogi et al. 2007), infrared processing (Rastogi 2012), and UV treatment (Rastogi 2011) were developed and adopted for improving the current processing techniques to produce safe food. A gamut of additives ranging from coloring agents, acidity regulators, flour enhancers, and others are added to food products to enhance their sensory and nutritional attributes (Saltmarsh and Barlow 2013). Packaging is an important part of food processing industry, which is essential for storing and transporting processed foods for long duration and distant places without compromising the quality of food products (Robertson 2012). Different processes make food processing a highly complex process in the present day to ensure safe and nutritious food to consumers. During food processing operations, certain undesired compounds are formed as by-products, which are categorized as food contaminants. There are numerous such compounds formed in various processes mainly due to ingredients, processing conditions, packaging, etc. They affect the food quality and also have an impact on consumers' health by causing ill effects. From the beginning of the twenty-first century, major food processing contaminants are identified, and possible mitigation strategies are developed to reduce or remove the contaminants (Lineback et al. 2012; Stadler et al. 2002). In this chapter, major food contaminants are discussed along with the mechanism of formation and mitigation strategies (Table 21.1).

Food contaminants can be classified on the basis of process/operation involved in the food processing and are discussed in the following sections.

TABLE 21.1
Precursors, Reaction Conditions, and Occurrence in Possible Foods Are Summarized and Tabulated for Food Processing Contaminants

Food Processing Contaminant	Precursors/Starting Materials	Reaction Conditions	Possible Foods	References
Thermal processing contaminants				
Acrolein	Cooking oil, methionine, threonine, ascorbic acid, degradation of carbohydrates, lipid peroxidation	Temperature 180°C, acidic pH	Deep fried fat foods	Abraham et al. (2011); Guth et al. (2013)
Acrylamide	Asparagine, reducing sugars	Temperature above 120°C	Potato chips, bakery foods, coffee, cereal products, etc.	Lineback et al. (2012); Stadler et al. (2002); Zyzak et al. (2003)
AGEs	Lysine, arginine, reducing sugars	Dry heat, a_w 0.4–0.7, basic pH	Fried foods, coffee, milk products, and bakery products	Henle (2005); Poulsen et al. (2013)
Chloropropanol and glycidol	Oils and fats, chlorine, diacylglycerol	High temperature	Hydrolyzed vegetable proteins, edible fats and oils, and infant and baby foods	Velísek et al. (1980); IARC (1995); Zuzana et al. (2009)
Furan	Glucose, fructose, serine, cysteine, ascorbic acid, carotene	Temperature above 90°C	Fruit juices, coffee, instant foods, milk products, etc.	Moro et al. (2012); Nie et al. (2013); Owczarek-Fendor et al. (2012)
HAAs	Amino acids, proteins, nitrogenous bases, glucose	Temperature above 150°C	Meat, meat products and fish	Oz and Kaya (2011a); Szterk (2015)
HMF	Reducing sugars	Acidic conditions	Coffee, bakery foods, malt and malt-based foods, dry fruits	Capuano and Fogliano (2011); Van Putten et al. (2013)
PAHs	Incomplete combustion of organic compounds, oils, and wood	Temperature above 750°C	Roasted, fried, grilled, barbequed, and smoked meat and meat products	Dost and Deli (2012); Purcaro et al. (2013); Tfouni et al. (2014)
Fermentation contaminants				
Acetaldehyde	Autoxidation of phenolic compounds	Fermentative action of *Saccharomyces* yeast, acetic acid bacteria, and LAB	Alcoholic beverages	Lachenmeier et al. (2009a,b); Lachenmeier and Sohnius (2008)
Acrolein	Glycerol and 3-hydroxypropionaldehyde	*LAB* metabolism, storage in acidic conditions, distillation	Distilled spirits and alcoholic beverages	Bartowsky (2012)

(*Continued*)

TABLE 21.1 (*Continued*)
Precursors, Reaction Conditions, and Occurrence in Possible Foods Are Summarized and Tabulated for Food Processing Contaminants

Food Processing Contaminant	Precursors/Starting Materials	Reaction Conditions	Possible Foods	References
BAs	Amino acids	Enzymatic decarboxylation	Cheese, vinegar, wine, alcoholic beverages, and fermented foods	De Mey et al. (2014); Arrieta and Prats-Moya, (2012); Ordóñez et al. (2013); Loizzo et al. (2013)
Ethyl carbamate	Carbamyl phosphate, citrulline, cyanogen, diethyl pyrocarbonate, and urea	Fermentation and distillation	Alcoholic beverages (beer, whisky, etc.), bread, soy sauce, fermented milk, yoghurt, cheese, sugarcane spirits, and cachaça	Sarah Hasnip et al. (2007); Riachi et al. (2014); Weber and Sharypov (2009)
Preservatives and additives				
Benzene	Benzoates, ascorbic acid, and erythorbic acid	Acidic conditions, presence of Cu(II) and Fe(III) ions	Carbonated beverages and roasted and smoked foods	Gardner and Lawrence (1993); Vinci et al. (2011)
	β-carotene, phenylalanine, and terpenes	Canning above 100°C, duration 30 min	Canned infant carrot juice	Lachenmeier et al. (2008, 2010)
MEIs	Caramels and sulfates	Caramelization	Cola drinks	Moon and Shibamoto (2011)
Nitrosamines and analogs	Nitrates, ascorbic acids, and heme	Curing process, acidic condition	Cured meat products	Skibsted (2011)
SEM	Azodicarbonamide	Baking process	Baked products	Becalski et al. (2004); Noonan et al. (2008)
Packaging contaminants				
Bisphenols	PC bottles and lacquer coating in cans	Leaching, depolymerization, repeated usage	Milk and canned foods	Maia et al. (2009)
Mineral oil hydrocarbons	Paperboards, recycled paper, and cardboard	Migration and storage conditions	Packed foods	Biedermann and Grob (2012); Lorenzini et al. (2010, 2013)
Nonylphenol	Epoxy resins and plastics	Leaching and migration	Packed foods	Grün and Blumberg (2009)
Perfluorinated compounds	Food contact paper, microwave popcorn paper bags, pizza box liners, candy wrappers, fast-food wrappers, and PTFE nonstick cookware	Saturated fat and high temperature during microwaving	Packed foods, microwave cooked popcorn	Begley et al. (2005)

(*Continued*)

TABLE 21.1 (*Continued*)
Precursors, Reaction Conditions, and Occurrence in Possible Foods Are Summarized and Tabulated for Food Processing Contaminants

Food Processing Contaminant	Precursors/Starting Materials	Reaction Conditions	Possible Foods	References
Phthalates	PET bottles and plastic lids of glass jars	Volatilization and leaching	Bottled water, beverages and fruit drinks, infant food, and packed foods	Cao (2010)
SEM	Azodicarbonamide used in PVC liners in gaskets and in lids of glass bottles and jars	Leaching	Foods packed in glass jars and packed foods	Nestmann et al. (2005); Stadler et al. (2004)
Disinfectant residues				
HAA and THM	Chlorine and natural organic matter	Chlorination of water	Domestic water	Giller et al. (1997); Villanueva et al. (2003)
SEM	Hypochlorite, arginine, creatinine, creatine, and urea	Thermal processing conditions	Thermal processed foods	Hoenicke et al. (2004)
Novel food processing contaminants				
Alkylcyclobutanones	Lipids and proteins	X-rays and γ-rays	Irradiated food products	Letellier and Nawar (1972); Yu et al. (2012)
Furans	Fructose and HFCS	UV-C radiation	UV-processed foods	Fan and Geveke (2007)

21.2 THERMAL PROCESSES

Thermal processing of food plays a crucial role in imparting taste, flavor, aroma, texture, and color of various foods. It also renders the food safe with a long shelf life by reducing the microbial count. Thermal processing includes baking, heating, frying, roasting, smoking, grilling, barbequing, and popping to bring specific and desired flavor, aroma, and texture to a wide variety of foods (van Boekel et al. 2010). However, as a result of heat and interaction between the components of food, certain foods at particular thermal processing conditions tend to form unwanted and toxic compounds. Many of such compounds are carcinogenic to humans and pose threat to human health. Contaminants formed during heat processing are discussed next.

21.2.1 Acrolein

Food such as fruits and vegetables naturally contains α,β-unsaturated aliphatic carbonyl compounds and their level increases during thermal processing. Acrolein (2-propenal) is the simplest α,β-unsaturated aldehyde formed in foods due to high-temperature degradation of triglycerides, amino acids (threonine and methionine), and carbohydrates (Abraham et al. 2011; Guth et al. 2013). Carbohydrates in food undergo C–C bond breakage during the thermal processing to form intermediates in multiple step reaction to form acrolein. 2-Hydroxypropanal being an important intermediate gets converted into acrolein by enolization at high temperature and acidic pH (Figure 21.1a)

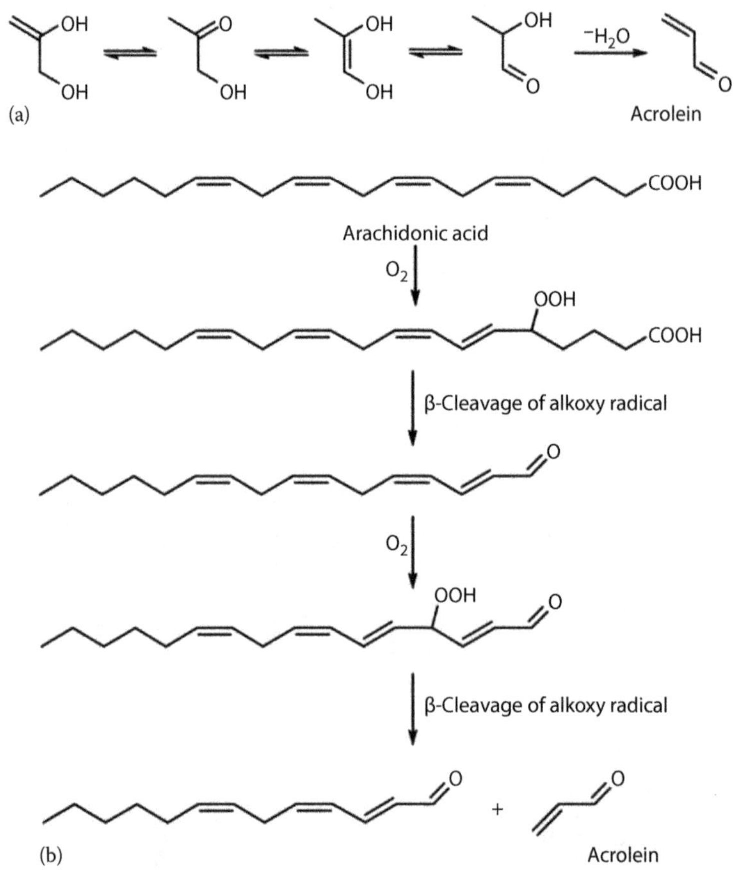

FIGURE 21.1 (a) Acrolein formation from carbohydrate and (b) acrolein formation from lipids. (From Stevens, J.F. and Maier, C.S., *Mol. Nutr. Food Res.*, 52, 7, 2008. With permission.)

(Stevens and Maier 2008). Methionine undergoes Strecker degradation to form ammonia and methional, which further gets converted into acrolein and methanethiol via retro-Michael cleavage reaction (Ballance 1961).

Formation of acrolein in food depends upon the composition of oil, temperature, and heating duration and is found at a concentration of 0.2–1.4 mg/kg in deep fried foods (Abraham et al. 2011; Guth et al. 2013). Cooking oils undergo degradation once it is heated above 180°C, leading to the formation of acrolein. Several pathways are elucidated by various research groups (Stevens and Maier 2008), out of which the one proposed by Esterbauer and coworkers is more practical (Esterbauer et al. 1991). Acrolein formation takes place in two steps: First, lipid hydroperoxides break the C–C bond of lipids, that is, β-cleavage of alkoxy radical into aldehyde with an alkyl fragment; second, hock fragmentation leads to the formation of acrolein and shorter lipid with an aldehyde group and alkyl terminal (Esterbauer et al. 1991; tevens and Maier 2008) (Figure 21.1b).

Acrolein, being highly electrophilic in nature, reacts readily with biomolecules such as amino acids, proteins, lipids, glutathione, and nucleic acids by forming adducts and conjugates. Acrolein inhibits the function of cellular proteins by disturbing their structural conformation due to cross-linking between the amino acids and degradation of sulfhydryl groups (Hua et al. 2010). Acrolein conjugates covalently with glutathione; inhibits its antioxidant property, thereby depleting the antioxidant defense; and also prevents further generation of glutathione in the cellular system (Shi et al. 2011). Acrolein is classified as a pulmonary hazard since it causes mutations in tumor suppressor p53 DNA by forming adducts with guanine residues and leads to chronic pulmonary diseases, lung cancer, asthma, irritation in lungs, and acute lung injury (Bein and Leikauf 2011). It also inhibits DNA repair mechanism in cells and causes bladder cancer in rat models (Tang et al. 2011). IARC has classified acrolein as a class 3 carcinogen (not classifiable as to its carcinogenicity to humans) (IARC Monographs on the Evaluation of Carcinogenic Risks to Humans 1995).

21.2.1.1 Mitigation

Addition of antioxidants such as ascorbyl palmitate and tocopherol concentrate leads to reduction in acrolein formation in foods. Acrolein readily reacts with ascorbate through the Michael addition reaction to form a conjugate, thereby reducing the concentration of acrolein (Stevens and Maier 2008).

21.2.2 Acrylamide

Acrylamide is a well-known contaminant formed during thermal processing of food containing carbohydrates and amino acids through the Maillard reaction above 120°C. Generally, acrylamide is not present in raw food ingredients; however, during thermal processing, it is formed and found in high quantities in carbohydrate-rich plant foods (potato snacks, bakery products, coffee, roasted almonds, cereal products) than in protein-rich animal foods (Lineback et al. 2012; Stadler et al. 2002). Processed plant products tend to contain large amounts of acrylamide, due to the presence of precursors, for the formation of acrylamide such as fructose and sucrose along with the amino acid asparagine (Vinci et al. 2011; Zyzak et al. 2003). Acrylamide is formed as a result of the Maillard reaction in thermal processed foods, which serves as the main pathway for its formation. Initially, the amino acid asparagine undergoes condensation reaction with reducing sugars or carbonyl compounds of the pathway followed by dehydration to Schiff base compounds. It decarboxylates to unstable 5-oxazolidinone intermediates to form stable azomethine ylides that undergo series of proton transfer and β-elimination to form acrylamide (Figure 21.2). Alternate pathways are proposed for the formation of acrylamide as well (Lineback et al. 2012; Stadler et al. 2002; Zyzak et al. 2003). Acrylamide content in various food sources tends to vary from batch to batch in the same food, and hence, the overall human intake of acrylamide varies greatly.

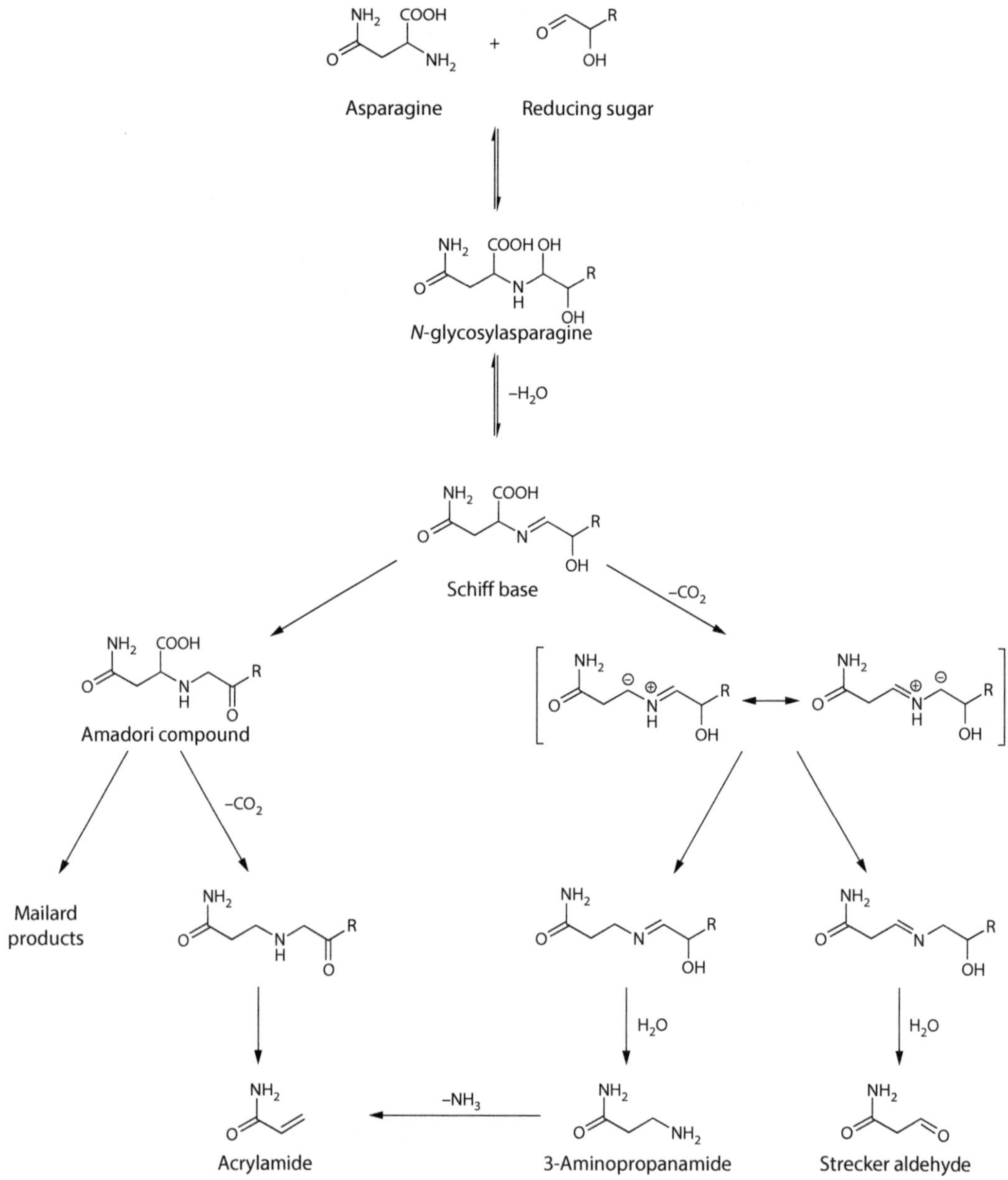

FIGURE 21.2 Formation of acrylamide in thermally processed foods. (Reprinted with permission from Vinci, R.M., Andjelkovic, M., Canfyn, M., Van Overmeire, I., and Van Loco, J., Factors influencing benzene formation from the decarboxylation of benzoate in liquid model systems, *J. Agric. Food Chem.*, 59, 12975–12981. Copyright 2011, American Chemical Society.)

Acrylamide is reactive due to the presence of electrophilic double bond, which can covalently bond with sulfhydryl groups of various biomolecules and hinder their action. Acrylamide is a classified group 2A (probably carcinogenic to humans) genotoxic carcinogen, neurotoxic to animals and humans, and reproductive toxicant (Capuano and Fogliano 2011; IARC 1994). Lots of research has been conducted to remove and reduce acrylamide content in the processed food since its discovery in 2002 (Vinci et al. 2011).

21.2.2.1 Mitigation

Mitigation of acrylamide in different food matrices is discussed briefly in the following section and explained in detail elsewhere (CAC/RCP 67-2009 2009; Food Drink Europe Acrylamide Toolbox 2011 2011).

- *Selection of raw materials*: Selection of potatoes and wheat based on their free asparagine and reducing sugar content, along with pre- and postharvest conditioning to reduce acrylamide formation during processing (Lineback et al. 2012; Zhang et al. 2009).
- *Blanching*: Blanching of potatoes before processing leads to softening of tissues and removal of free asparagine based on solutes used during the process. It also helps in better diffusion of asparaginase during the enzymatic removal of asparagine (Hendriksen et al. 2009). Blanching combined with NaCl treatment reduced 67% of acrylamide in potato chips (Pedreschi et al. 2010). Blanching acts as a means to enhance the transfer of possible acrylamide reducing compounds/molecules into the food matrix.
- *Use of asparaginase (EC 3.5.1.1)*: With the enzyme action, free asparagine gets converted into aspartic acid and ammonia, thereby limiting the formation of acrylamide (Hendriksen et al. 2009). Asparaginase-based acrylamide removal is being followed in laboratory and industrial scale with an efficiency of 85%–90% in food products (Capuano and Fogliano 2011).
- Addition of additives
    - Substitution of leavening and raising agents containing ammonia with sodium and potassium helps in the reduction of acrylamide in baked food products.
    - Replacing reducing sugars with sucrose greatly reduces acrylamide formation (CAC/RCP 67-2009 2009).
    - Addition of acidity regulators such as citric acid resulted in the reduction of acrylamide mainly due to unfavorable pH for acrylamide formation.
- *Fermentation*: *Saccharomyces cerevisiae* (baker's yeast) removes up to 70% of acrylamide during a 48 h fermentation at 30°C in an instant coffee (Akillioglu and Gökmen 2014). Filamentous fungus (*Aspergillus oryzae*) is used for the degradation of acrylamide in food and beverage industries (Wakaizumi et al. 2009).
- *Vacuum treatment*: Dry food products such as biscuits and potato chips were hydrated and subjected to the following conditions (6.67 Pa, 60°C, 5–15 min), which resulted in 43% reduction in the acrylamide content (Anese et al. 2010).

21.2.3 Advanced Glycation End Products

Advanced glycation end product (AGEs) are final products of carbohydrate and protein conjugation resulting from various nonenzymatic pathways such as Maillard, Amadori, and Schiff base reactions during thermal processing of foods. Foods processed by dry thermal process (frying, roasting, grilling, and baking) are found to contain more amount of AGEs than moist thermal processing (boiling, stewing, and steaming). Amino acids (lysine and arginine) and non-protein molecules with free amino group can readily react with molecules containing carbonyl groups such as carbohydrates (reducing sugars), oxidized lipids and ascorbic acid to form AGEs in three different stages. Initially, a Schiff base is formed due to condensation of amine and carbonyl moieties. In the second stage, Schiff base compounds transform into colorless ketoamines, which are highly unsaturated molecules, by Amadori rearrangement. In the last stage, all the intermediates and Amadori products degrade into advanced products (Figure 21.3; Henle 2005; Poulsen et al. 2013). Processing parameters such as high temperature, dry heat, water activity (a_w) in the range of 0.4–0.7, basic pH, and ingredients rich in amines and carbonyl-containing molecules promote the formation of AGEs. AGEs are also endogenously formed in the human body due to a hyperglycemic condition in

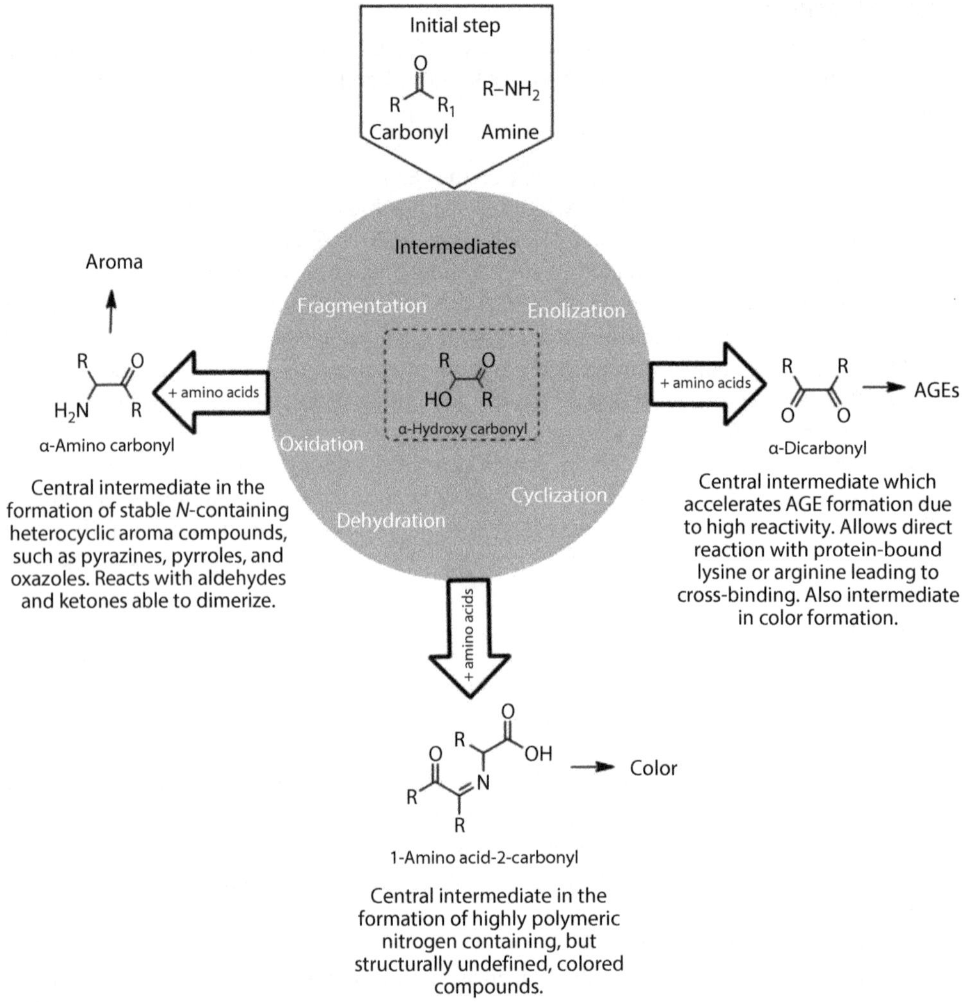

FIGURE 21.3 Formation of advanced glycation end products in thermally processed foods and the important intermediates resulting from the Maillard reaction. (From Poulsen, M.W. et al., *Food Chem. Toxicol. Int. J. Publ. Br. Ind. Biol. Res. Assoc.*, 60, 10, 2013. With permission.)

diabetic patients and accumulate in cells. They are linked in exacerbating some pathological conditions such as renal failure, cardiovascular diseases, rheumatoid arthritis, and Alzheimer's disease (Stirban et al. 2014).

21.2.3.1 Mitigation

AGEs formation in thermally processed foods can be minimized by employing one or more of the following strategies (Uribarri et al. 2010; Yaylayan et al. 2005):

- Thermal processing with high humidity (higher a_w)
- Replacing the main reactants such as lysine and arginine and reducing sugars with substitutes
- Decreasing the pH of the cooking medium by the addition of lemon juice, vinegar, or acidity regulators
- Reducing the formation of Amadori products and subsequently AGEs by the addition of antioxidants such as aminoguanidine and pyridoxine

21.2.4 Chloropropanols and Glycidol

Chloropropanols are a group of chlorine-containing alcohols formed during the thermal processing of fats and oils (Velisek et al. 2003). 3-Chloro-1,2-propanediol or 3-monochloropropane-1,2-diol (3-MCPD) is the predominant among them and is present with glycidol (2,3-epoxy-1-propanol), an organic compound with alcohol and epoxide groups. Dehalogenation of MCPD can also lead to the formation of glycidol. Chloropropanols and glycidol are mostly found in ester form with fats and lipids than their natural form in food products such as hydrolyzed vegetable proteins (Velísek et al. 1980), edible fats and oils (Zelinková et al. 2006), and infant and baby foods (IARC Monographs on the Evaluation of Carcinogenic Risks to Humans 1995; Zuzana et al. 2009). 3-MCPD is also found in minor quantities in neutral lipid fraction of goat milk (Cerbulis et al. 1984). Deodorization of oils is a crucial step employed in industries for the removal of volatile compounds from oil using steam at high temperature and low pressure. The presence of chlorine and chloride-containing molecules, level of diacylglycerol, and processing conditions favor the formation of chloropropanols and glycidyl esters (Andres et al. 2013).

3-MCPD is classified as a nongenotoxic carcinogen. It is found to be toxic to the kidney and cause tumors. It is also linked in affecting male sterility in rats and mice (Andres et al. 2013; Bakhiya et al. 2011). Glycidol acts readily as an alkylating agent and forms DNA adducts in cells. It causes multisite tumors in rodents by acting as a mutagen and classified as group 2A (probably carcinogenic to humans) carcinogen with genotoxic effects by IARC (IARC 2000). It is advised to maintain the concentration of 3-MCPD and glycidol *as low as reasonably achievable* (ALARA) in processed foods.

21.2.4.1 Mitigation

- *Glutathione*: Addition of glutathione reduced 3-MCPD content by 80% in glycerol–NaCl systems (Velisek et al. 2003).
- *Acidic steam*: Deodorization with acidic solutions of formic acid reduced the formation of glycidyl esters by 35% (Matthäus et al. 2011).
- *Washing of raw materials*: Washing raw material with water or 75% ethanol solution reduced 3-MCPD esters by 20%–25% (Matthäus et al. 2011).
- *Alkaline treatment*: 3-MCPD is stable in acidic conditions of protein hydrolysates; however, they form glycidol and glycerol in subsequent steps in alkaline condition. So bringing pH of protein hydrolysates to alkaline conditions is a common procedure to remove 3-MCPD (Hamlet et al. 2011).

21.2.5 Furan

Furan is one of the most studied highly toxic food contaminants formed during thermal processing of foods. It is found in a wide variety of foods such as baby and infant foods, bread and bakery products, coffee, canned foods, and sauces. Furan (C_4H_4O) is a small volatile heterocyclic aromatic compound that is lipophilic in nature (Crews and Castle 2007). The European Food Safety Authority (EFSA) declared that "furan is clearly carcinogenic in rats and mice," which arises mainly due to its ability to form DNA-reactive genotoxic metabolite (genotoxic) (Scientific Report of EFSA 2010). Furan causes dose-dependent hepatocellular adenomas and carcinoma in mice and rats (National Toxicology Program 1993). IARC classified furan as a group 2B carcinogen (possibly carcinogenic to humans) (IARC Monographs on the Evaluation of Carcinogenic Risks to Humans 1995). Studies have deciphered the formation of furan in foods containing carbohydrates in interaction with proteins, lipids, and ascorbic acid (Nie et al. 2013; Owczarek-Fendor et al. 2012).

Carbohydrate-rich (glucose and fructose) foods undergo Millard reactions to form aldotetrose derivatives by reacting with different amino acids (serine and cysteine) by aldol condensation. Aldotetrose derivatives break into intermediates in multiple step reaction to form furan by cyclization in its final step (Yaylayan 2006). Ascorbic acids form furan in a reaction similar to reducing sugars in the presence and absence of air. Ascorbic acid gets converted to dehydroascorbic acid by oxidation and then to 2,3-diketogluconic acid by hydration reaction. In subsequent reaction, it leads to the formation of aldotetrose derivatives (2-deoxyaldotetrose) and then to furan in thermal processed foods (Perez Locas and Yaylayan 2004). Polyunsaturated fatty acids (PUFAs) such as linoleic and linolenic acids undergo series of free radical lipid autoxidation and formation of lipid peroxides. Further, the lipids are broken down by hemolytic cleavage to form highly toxic aldehyde containing alkyl compounds and subsequent cyclization to form furan (Becalski and Seaman 2005; Moro et al. 2012). Furan levels are found to be directly proportional to the degree of unsaturation in the fatty acids and triacylglycerols. For instance, linolenic acid degrades to more amount of furan than linoleic acid (Märk et al. 2006).

Furans are shown to form in thermal processed foods such as fruit and vegetable juice, baby foods, and bakery foods. Carotene in carrot and carbohydrates in bakery products are the major ingredients responsible for the formation of furans (Wegener and López-Sánchez 2010).

21.2.5.1 Mitigation

Furan has a boiling point of 31°C, which evaporates when foods are moderately heated and leaves behind very little amount of residual furan in food matrix; however, in case of coffee, it may lead to loss in other volatile compounds responsible for its aroma (Crews and Castle 2007). In an interesting observation for domestic mitigation strategy, stirring coffee brew while consuming leads to 75%–80% reduction in furan levels due to high temperature of coffee, which is enough for furan to escape due to its volatility (Mesías and Morales 2014).

21.2.5.1.1 Removal or Substitution of Additives

Using nonreducing sugars (sucrose) instead of reducing sugars such as glucose or fructose is a viable solution to reduce furan formation during thermal processing at industrial scale without much affecting the sweetness of the food product (Figure 21.4; Nie et al. 2013). Other ingredients such as ascorbic acid and PUFA, which are responsible for furan formation, are difficult to replace with other

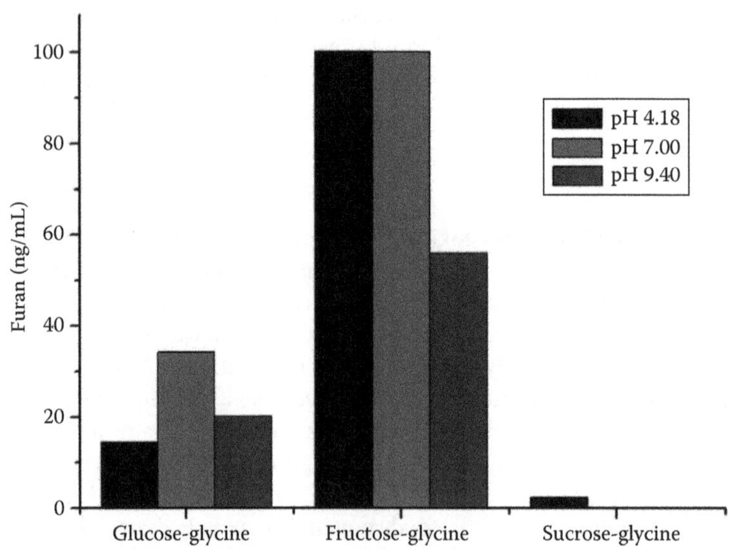

FIGURE 21.4 Furan formation in model systems with various ingredients in simulated sterilization conditions (120°C for 30 min). (From Nie, S. et al., *Food Sci. Hum. Wellness*, 2, 87, Copyright 2013, Creative Commons. With permission.)

substitutes because of their nutritional benefits. Hence, the addition of free radical scavengers and maintaining inert atmosphere during thermal processing will help in reducing furan formation and also retaining most of their health benefits in processed food (Märk et al. 2006; Shinoda et al. 2005).

21.2.5.1.2 Optimization of Processing Condition

Certain changes during processing have proven to reduce the furan content to a large extent and are listed as follows:

- *Processing in open vessel*: Furan evaporates due to its highly volatile nature (Hasnip et al. 2006; Kim et al. 2009).
- *Food irradiation*: Irradiating food with a dose of γ-rays (kGy) leads to 35%–50% reduction in the furan content in biscuits and fat sauce, respectively (Anese and Suman 2013).

Optimization of heating regime during thermal processing to retain food-specific volatiles and reducing furan concentration would be beneficial for adaptation in food processing industry.

21.2.5.1.3 Postprocessing Methods

High-moisture foods such as meat sauce when subjected to vacuum treatment (12 kPa for 10 min) resulted in 67% reduction in furan concentrations without compromising the sensory attributes. However, the same mitigation method was not good at reducing furan contents in low-moisture foods such as biscuits (Anese et al. 2014). Modification of storage conditions such as filling inert N_2 gas instead of O_2 resulted in 30% reduction in furan content in processed foods (Märk et al. 2006). Mitigation strategies for furan are reviewed in detail elsewhere (Anese and Suman 2013; Crews and Castle 2007).

21.2.6 Heterocyclic Aromatic Amines

Heterocyclic aromatic amines (HAAs) are a group of aromatic compounds containing nitrogen in its structure, mainly found in meat, meat products, fish, and other animal sources. Currently, there are more than 25 such amines found in various food products, and they are potent mutagens due to their carcinogenic property. HAA formation in food products depends upon the availability of moisture, free amino acids, proteins, glucose, lipid content, nitrogenous bases, and nucleosides in the ingredients (Oz and Kaya 2011b; Szterk 2015). Generally, HAA has 3 (sometimes 2–5) condensed aromatic rings with one or more nitrogen atoms present in the ring structure. HAAs are classified into two groups of compounds: thermic HAAs, formed at a temperature range of 100°C–300°C, and pyrolytic HAAs, formed above 300°C. Thermic HAAs are chemically aminoimidazoazarene derivatives that are polar in nature, whereas pyrolytic HAAs are mostly nonpolar in nature (Alaejos and Afonso 2011; Oz and Kaya 2011b; Rahman et al. 2014). HAAs readily react with DNA base (deoxyguanosine) and form adducts that are carcinogenic in nature (Turesky and Vouros 2004). In rodent and monkeys, HAA was shown to cause tumors in the breast, colon, prostate, liver, and other organs (Sugimura et al. 2004). The carcinogenic activity of HAA in humans depends on the amount, type, and cooking method of the food product, and hence, maintaining ALARA of HAA is advised. HAA in food products can be reduced and mitigated using the following strategies.

21.2.6.1 Mitigation

- *Antioxidants and spices*: Addition of spices such as turmeric, lemongrass, torch ginger, and curry leaves during grilling of meat has resulted in reducing the formation of HAAs in the range of 40–85 ng/100 g of the food sample (Jinap et al. 2015). Addition of antioxidants of bamboo leaves, tea polyphenols, liquorice extract, phytic acid, and sodium isoascorbate resulted in 53%–77% reduction of different HAAs in pork patties (Zhang et al. 2013). Apart from aforementioned sources, other antioxidant-rich sources such as black pepper, pomegranate seed extract, grape seed, sage, thyme, garlic, and rosemary are used to reduce HAA formation in various meat products (Keşkekoğlu and Üren 2014; Oz and Kaya 2011a). Antioxidant and spices neutralize the free radicals formed during the cooking process, thereby preventing the formation of HAAs.

- *Water-soluble vitamins*: Addition of water-soluble vitamins such as pyridoxamine, pyridoxine, nicotinic acid, biotin, thiamine, and L-ascorbic acid reduced HAA formation above 40% in beef. Vitamins act as antioxidant molecules and also prevent the formation of phenylacetaldehyde, an intermediate compound in the formation of HAAs (Wong et al. 2012).
- *Cooking temperature and time*: Cooking at low temperature for less duration reduces the formation of HAA in meat and meat products. Microwave-based cooking is also an efficient way to reduce HAA formation, which can be adopted at industrial scale for the production of HAA free or lower HAA-containing meat products (Alaejos and Afonso 2011; Haskaraca et al. 2014).
- *Sous-vide cooking*: Cooking food products sealed in airtight bags and exposing to heat in temperature-controlled water bath or steam resulted in reduced concentration of HAAs compared to normal cooking methods (Oz and Zikirov 2015).

21.2.7 5-Hydroxy Methyl Furfuryl

5-Hydroxy methyl furfuryl (HMF) is a heterocyclic furan derivative found in thermally processed carbohydrate-rich foods. It is an intermediate compound formed during Maillard reaction and also from direct dehydration of reducing sugars in acidic conditions. HMF is formed mainly from two pathways: in acidic conditions, formation and concentration of HMF in thermally processed foods directly depend on temperature and duration of processing; however, small concentration of HMF can be formed at low temperatures as well (Capuano and Fogliano 2011). Reducing sugars such as glucose and fructose undergo dehydration and 1,2-enolization in acidic conditions and high temperature to form 3-oxyosone, a key intermediate in the formation of HMF. Further dehydration and cyclization of 3-oxyosone lead to the formation of HMF (Lee and Nagy 1990). In case of dry and high-temperature (pyrolysis) conditions, fructose and sucrose react together to form highly reactive fructofuranosyl cation that directly breaks down and cyclizes to HMF (Figure 21.5; Perez Locas and Yaylayan 2008). HMF was found in a wide variety of processed foods such as coffee, bakery foods, malt and malt-based foods, dry fruits, and vinegar in the range mg/kg–g/kg of the food product. It is one of the most commonly found thermal processing contaminants in large quantities and is used to monitor thermal processing of cereal and cereal-based foods (Capuano and Fogliano 2011; Van Putten et al. 2013). HMF can be converted to sulfooxymethyl furfuryl (SMF) by sulfotransferase enzymes present in humans and rodents. SMF is a highly reactive molecule that has the ability to bind with DNA causing mutagenic effects and cancer (Svendsen et al. 2009). HMF is classified as a weak carcinogen and found to cause colon cancer in rats and also induces mutagenic and genotoxic reactions in humans and bacterial cells (National Toxicology Program 2010).

21.2.7.1 Mitigation

HMF can be either prevented or removed in processed foods by various methods as follows:

- *Vacuum treatment*: Vacuum treatment is an effective method to remove HMF from high-moisture foods (meat sauce); however, there was some loss in odor components. But it is not an effective method for the removal of HMF in low-moisture foods (biscuits, cereal-based products, and roasted coffee) due to high viscosity and entrapment in solid food matrix (Anese et al. 2014).

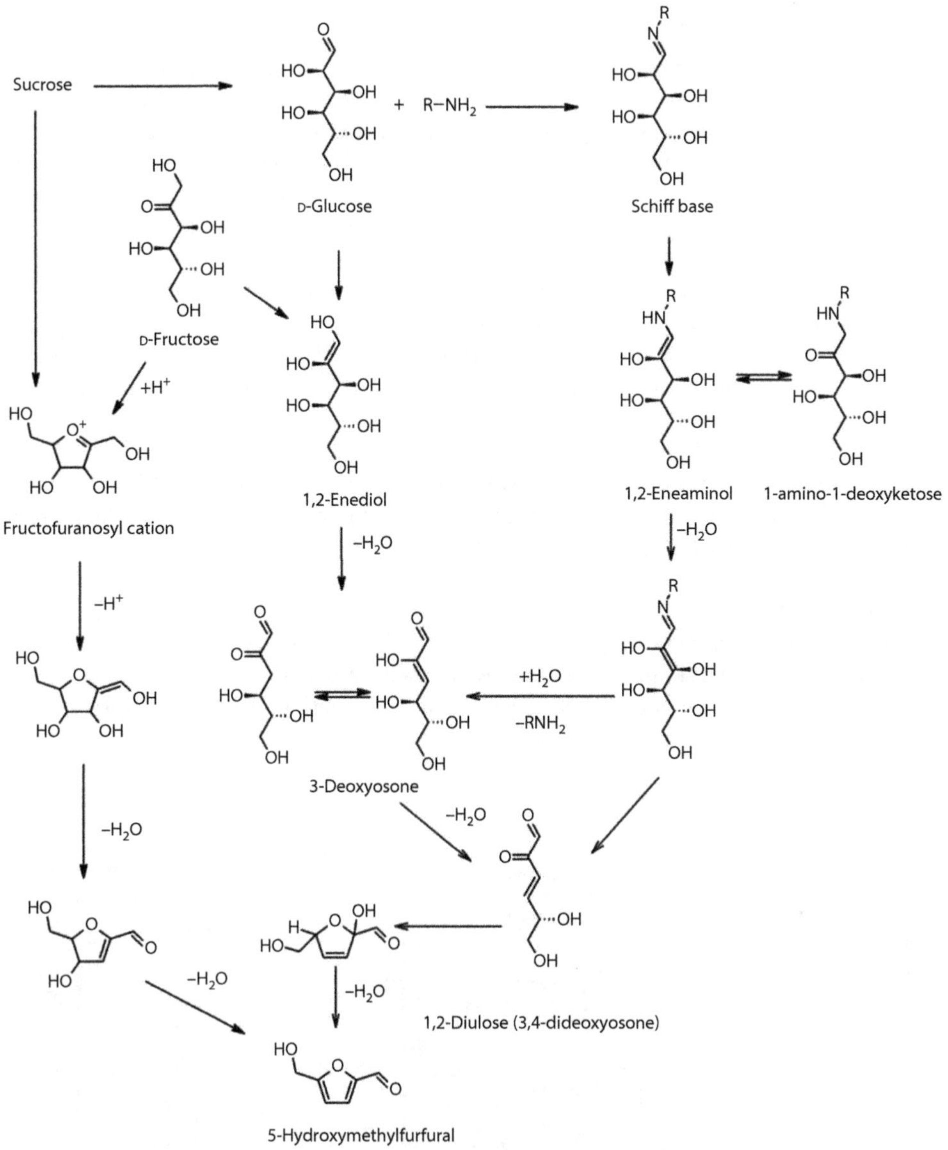

FIGURE 21.5 Proposed reaction mechanism for the formation of 5-hydroxy methyl furfuryl. (From Capuano, E. and Fogliano, V., *LWT Food Sci. Technol.*, 44, 793, 2011. With permission.)

- *Yeast fermentation*: *S. cerevisiae* (baker's yeast) removes up to 99.2% HMF during a 24-hour fermentation at 30°C in an instant coffee supplemented with 10% sucrose (Akillioglu and Gökmen 2014).
- *High-pressure processing*: High-pressure processing of apple juice was found to contain 100-fold less concentration of HMF than pasteurized juice (Saldo et al. 2009).
- Other mitigation strategies such as substitution of ingredients, combined heating with vacuum, and irradiation techniques are followed to remove HMF, which are discussed elsewhere in detail (Anese and Suman 2013; Mogol and Gökmen 2014).

21.2.8 Polycyclic Aromatic Hydrocarbon

Smoking is an ancient food processing technique, in which meat and meat products are exposed to smoke and heat from wood or other fuel to increase the shelf life, enhance the taste, and infuse specific aroma, flavor, and color. Polycyclic aromatic hydrocarbons (PAHs) are tertiary tar products formed due to incomplete combustion of organic compounds, oils, and wood above a temperature of 750°C through condensation of aromatic compounds, and their formation is higher with increasing temperature (Ledesma et al. 2016). Food products contain PAH formed due to contamination from environment and during thermal processing methods such as roasting, frying, grilling, barbequing, and smoking, contaminated packaging, and mineral oil contamination in edible oils (Dost and Deli 2012; Purcaro et al. 2013; Tfouni et al. 2014). Smoking and grilling are the major processing techniques that contribute to the formation and infusion of PAHs in meat and meat products. PAHs generally contain 2–6 rings, and larger molecules with more rings are quite stable compared to their smaller counterparts and are also found to be carcinogenic in nature. PAHs being lipophilic in nature tend to accumulate in tissues and have the potential to cause mutagenic effects (CAC/RCP 68-2009 2015; European commission 2002; WHO Technical Report Series 2015). PAH formation and infusion in foods can be effectively managed by following the 10 processing variables specified by Codex Alimentarius (CAC/RCP 68-2009 2015).

21.2.8.1 Mitigation

- Implementing the processing parameters specified by Codex Alimentarius during food processing operations (Figure 21.6; CAC/RCP 68-2009 2015).
- Smoking temperature less than 600°C prevents the formation of PAH in smoke and food products.

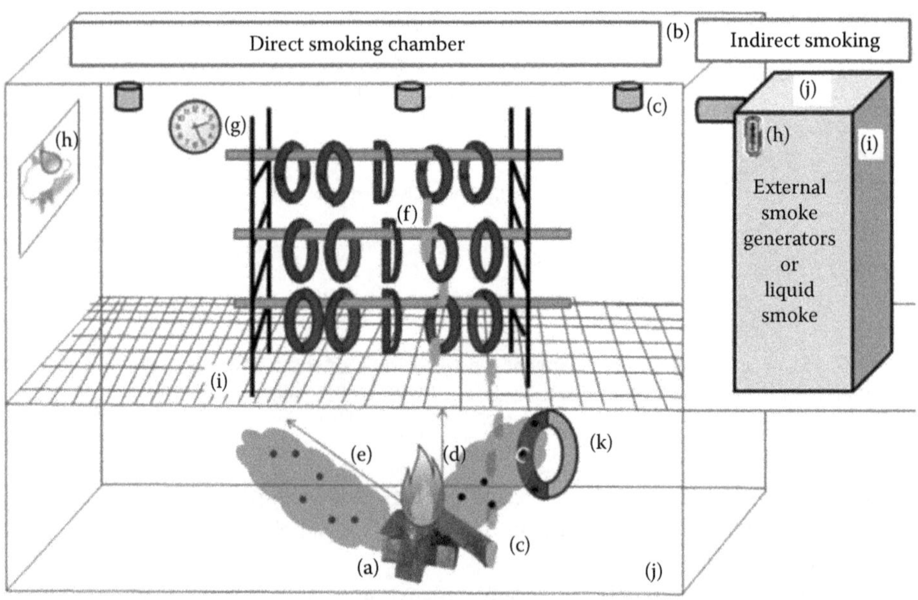

FIGURE 21.6 (See color insert.) Smoking chamber representing the process variables mentioned by Codex Alimentarius (CAC/RCP 68-2009) to control the formation and contamination of polycyclic aromatic hydrocarbons. (a) Fuel, (b) direct or indirect (c) smoke generation process (d) distance to product (e) position to product (f) fat content and dripping (g) duration of smoking (h) temperature (i) cleanliness of equipment (j) design of chamber (k) casing (variable not listed in CAC) (•) tar (bio-oil). (From Ledesma, E. et al., *Food Control*, 60, 64, 2016. With permission.)

- Indirect and novel smoking methods such as liquid smoke (condensed form of smoke) can highly reduce PAH in food systems.
- Controlling other processing parameters such as smoking duration, source of smoke, fat content in foods, and synthetic casing help reducing PAH content in food products (Ledesma et al. 2016).

21.3 FERMENTATION PROCESSES

These classes of contaminants could be encountered during the fermentation of food and beverages, ripening of the products, and possible microbial fermentation by-products.

21.3.1 Acetaldehyde

Acetaldehyde is a volatile compound having fruity aroma at low concentrations and is found naturally in various alcoholic beverages. Acetaldehyde is classified as group 2B carcinogen (possible carcinogen to humans) by IARC, owing to its ability to cause malignant esophageal tumors in experimental animals and aldehyde dehydrogenase–deficient humans (Anon 1999; Baan et al. 2007; Seitz and Stickel 2007). Acetaldehyde acts as a mutagen, causes point mutation, DNA adducts, and sister chromatid exchanges in mammalian cells, and interferes in DNA repair mechanism (Druesne-Pecollo et al. 2009; Seitz and Stickel 2007; Wang et al. 2000). Acetaldehyde is naturally present in alcoholic beverages due to one or the other factors such as the fermentative action of *Saccharomyces* yeast, acetic acid bacteria, lactic acid bacteria (LAB), and autoxidation of phenolic compounds present in alcoholic beverages (Lachenmeier et al. 2009a,b; Lachenmeier and Sohnius 2008). Average acetaldehyde content in alcoholic beverages such as beer (9 mg/L), wine (34 mg/L) and spirits (66 mg/L). However, the maximum concentration varied between 63–1159 mg/L. (Lachenmeier and Sohnius 2008; Lachenmeier et al. 2012). Alcoholic beverages get metabolized into acetaldehyde in human body and get accumulated, which depends on the dose, type, and amount of the alcoholic beverage consumed (Lachenmeier et al. 2009a).

21.3.1.1 Mitigation

Acetaldehyde can be removed by distillation process; however, absolute removal is not a feasible process.

21.3.2 Acrolein

Acrolein is formed in fermented beverages mainly due to the following two factors:

1. *LAB metabolism*: Glycerol dehydratase enzyme present in *LAB* converts glycerol to 3-hydroxypropionaldehyde and spontaneous formation of acrolein due to storage in acidic conditions and heat treatment (distillation process). Acrolein readily reacts with antioxidant (anthocyanin) molecules to form conjugates that contribute to bitterness in red wine (Bartowsky 2012).
2. *Distillation*: During distillation process, glycerol and 3-hydroxypropionaldehyde get dehydrated to acrolein.

21.3.2.1 Mitigation

Membrane-based separation may be a possible solution to selectively remove glycerol and acrolein from alcoholic beverages to prevent the formation of acrolein during distillation process. LAB growth and fermentation of glycerol to acrolein can be reduced by maintaining low pH and higher SO_2 concentrations in red wine during storage (Bartowsky 2012).

21.3.3 Biogenic Amines

Biogenic amines (BAs) are class of low-molecular-weight nitrogen containing compounds found in fermented and processed foods due to microbial or cellular (plant and animal) metabolism (Bodmer et al. 1999). BAs encompass histamine, tyramine, putrescine, cadaverine, phenylethylamine, and their analogs found in various food products including cheese (Loizzo et al. 2013), vinegar (Ordóñez et al. 2013), wine (Arrieta and Prats-Moya 2012), and other fermented foods (De Mey et al. 2014). These compounds are known to have physiological and toxic effects on human health depending on the compound and their concentration. BAs are basically breakdown products of amino acids by enzymatic decarboxylation (decarboxylases) (Alvarez and Moreno-Arribas 2014). BAs are involved in the regulation and control of various physiological processes in humans such as body temperature, digestion process (stomach pH and gastric juice secretion), immune response, cell growth, and differentiation (Ladero et al. 2010). BAs may lead to the formation of carcinogenic N-nitrosamines in foods containing nitrites (Honikel 2008). BAs also involve in the synthesis of nitrogen-containing biomolecules such as nucleic acids, proteins, hormones, alkaloids, and aromatic volatile compounds (Santos 1996). At the contrary, they also cause headache, nausea, blood pressure disorders, carcinogenic effects, and others (Ladero et al. 2010).

21.3.3.1 Mitigation

Fermentation is one of the widely used methods to remove BAs in food by cultivating microorganisms with BA-degrading capacity. Both gram-positive and gram-negative bacteria are capable of producing enzymes that can degrade these amines (Alvarez and Moreno-Arribas 2014). Recently, Giancane and his group have reported paramagnetic iron oxide nanoparticle–based removal of BAs from wine samples by utilizing the magnetic property of the nanoparticles. The proposed method had the potential of large-scale application in industries with advantages such as low cost and easy mitigation process and does not involve any chemical or biological reactions or their by-products, (Bettini et al. 2015).

21.3.4 Ethyl Carbamate

Ethyl carbamate, chemically an ethyl ester of carbamic acid, also called urethane, is a toxic chemical formed during fermentation and distillation process of food products. Ethyl carbamate is a classified group 2A carcinogen (probably carcinogenic to humans) by IARC in 2007 and also genotoxic in various experimental animals (Baan et al. 2007; Cravedi et al. 2015; Weber and Sharypov 2009). Ethyl carbamate is found in various fermented foods such as alcoholic beverages (beer, whisky, etc.), bread, soy sauce, fermented milk, yoghurt, and cheese (Hasnip et al. 2007; Riachi et al. 2014; Weber and Sharypov 2009). Ethyl carbamate is found in quantities ranging from ng to mg per kg/L depending on the food and the processing involved. Different mechanisms of ethyl carbamate formation in fermented foods have been suggested based on the precursors (carbamyl phosphate, citrulline, cyanogen, diethyl pyrocarbonate, and urea). Among them, urea is generated from the metabolism of arginine in yeast cells by arginine degradation pathway in yeast cells. Urea being toxic to yeast cells, it is exported out of the cell, which possibly reacts with the ethanol and forms ethyl carbamate (Figure 21.7; Zhao et al. 2013). In case of grape wine fermentation, citrulline and carbamyl phosphate are formed from the conversion of arginine by arginine deaminase pathway in LAB, which reacts with ethanol separately to form ethyl carbamate outside the bacterial cells (Azevedo et al. 2002).

Cyanide is one unusual precursor for the formation of ethyl carbamate in stone fruit spirits. Cyanogenic glycosides present in stone fruits are converted into cyanide by enzymatic action and thermal decomposition rather than any microbial metabolism. Cyanide then spontaneously oxidizes to cyanate in the presence of ethanol, which further gets catalyzed into ethyl carbamate in the presence of copper ions. The food additive diethyl pyrocarbonate that is used to inhibit the growth of undesired microbes in the preparation of grape wine reacts with the ammonia in alcoholic broth to

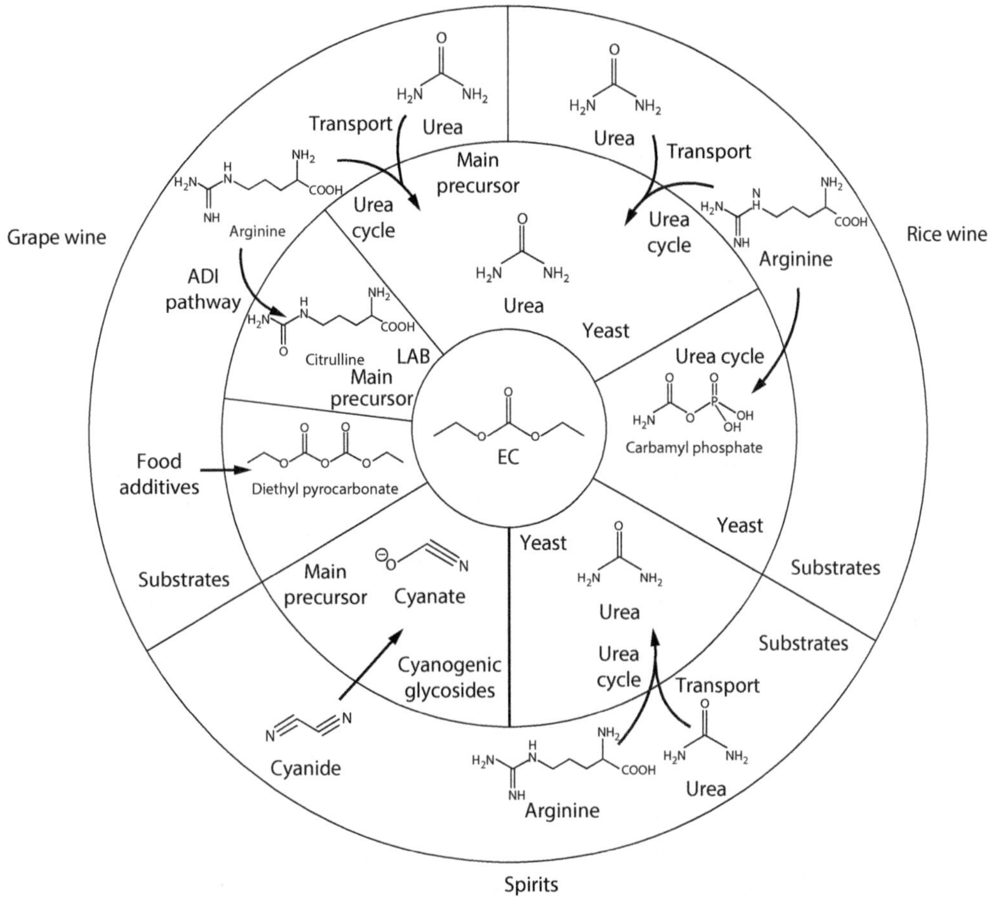

FIGURE 21.7 Sources of precursors and their pathway to formation of ethyl carbamate in fermented foods. (From Zhao, X. et al., *Trends Food Sci. Technol.*, 32, 97, 2013. With permission.)

form ethyl carbamate (Weber and Sharypov 2009). Multiple precursors and pathways are responsible for the formation of ethyl carbamate in alcoholic beverages, which may need one or more mitigation strategies to make them safe.

Cupric ions catalyze the formation of ethyl carbamate from cyanide in stone fruit spirits (Riachi et al. 2014).

21.3.4.1 Mitigation

Various mitigation strategies that are available for the removal/reduction of ethyl carbamate content in fermented foods, which includes physical, chemical, enzymatic, and metabolic methods, are listed as follows:

- Decreasing the precursor content (arginine, citrulline, carbamyl phosphate, and urea)
- Using low-urea-producing yeast strains for fermentation
- Process optimization (low-temperature fermentation process reduces ethyl carbamate formation)
- Filtration and adsorption of ethyl carbamate by charcoal filtration
- Addition of acidic urease enzyme to fermented food to degrade urea
- Removal of cyanide in food material to be fermented
- Removal of Cu(II) ions from fermented foods that act as a catalyst

Based on the feasibility, one or combination of mitigation strategies may be employed for the removal/reduction of ethyl carbamate in food products (Jiao et al. 2014; Park et al. 1993; Weber and Sharypov 2009; Zhao and Johnson 2007).

21.4 PRESERVATIVES AND ADDITIVES

Preservatives and additives are primarily added to improve the shelf life and overall attributes of the food product. In few circumstances, they react with other food ingredients/compounds and form compounds that are antinutritional and toxic and pose various health issues at very low concentration.

21.4.1 Benzene

Benzoic acid (E210) and its salts (sodium benzoate, E211; potassium benzoate, E212; calcium benzoate, E213) are often added as a preservative along with acidity regulators such as ascorbic acid (E300) and erythorbic acid (E315) to soft drinks and other liquid foods to improve the shelf life (Gardner and Lawrence 1993; Thakur and Ragavan 2013). Unfortunately, these food additives react to form benzene in the food system. Benzene is a well-known group 1 carcinogen to humans due to its obvious carcinogenicity for causing nonlymphocytic leukemia (Anon 2014). Benzene may be introduced in the food matrix through other routes such as leaching from packaging materials as a contaminant along with carbon dioxide in carbonated beverages, roasting, and smoking process of food (Vinci et al. 2011).

In case of liquid foods, benzoic acid is decarboxylated by ascorbic acid and transition metal ions [Cu(II) and Fe(III)], where the later acts as a catalyst in the formation of benzene (Gardner and Lawrence 1993). Benzene formation by decarboxylation of benzoate is influenced by various factors such as hydroxyl radicals, buffer, pH of the liquid food, transition metal ions, and the presence of antioxidants in liquid food models. Benzene formation was found to be high in liquid food matrix with a pH range of 2–5, containing ascorbic acid and copper ions (Cu^{2+}). Presence of copper ions in the system significantly influences benzene formation at highest levels (1400 μg/kg) (Vinci et al. 2011).

Canned infant carrot juice is another food product found to contain benzene, but the formation is due to heat treatment. β-Carotene, phenylalanine, and terpenes present in carrot juice may act as a precursor for the formation of benzene due to the thermal treatment involved in the canning of carrot juice to avoid microbial contamination. Canning and heat treatment at a temperature of 100°C and above for at least 30 min lead to significant production of benzene in heat-treated foods (Lachenmeier et al. 2008, 2010).

21.4.1.1 Mitigation

Presence of antioxidants with metal chelators completely prevent the formation of benzene in liquid food samples (Vinci et al. 2011). In case of canned infant carrot juice, optimization of process parameters may help in preventing the formation of benzene (Lachenmeier et al. 2008). Replacement of the additives and catalysts with analog molecules may help in overcoming the formation of benzene in liquid foods without affecting the food quality.

21.4.2 Methylimidazoles

Caramels, a group of food additive (E150 a–d) produced by caramelization process, are added in food to impart dark brown to black color in dark beer, beverages, bakery products, cooked meat, and sauces (Karim and Smith 2015; Sengar and Sharma 2012). Caramels contain a concoction of organic compounds, among which methylimidazole (MEI) and its derivatives were found to have carcinogenic effects and are classified as carcinogens (National Toxicology Program 2007).

MEIs are by-products formed due to the reaction between ammonia and reduction sugars during the high-temperature caramelization process (Karim and Smith 2015). 4(5)-MEI is the common MEI derivative formed during the caramelization process, which is a nitrogen-containing heterocyclic compound. MEI inhibits the functions of cytochrome P450 enzyme and acts as a neurotoxin and convulsion agent in cattle and animals (Karangwa et al. 1990; Lee et al. 2013). MEIs are also formed as a by-product during the Maillard reactions due to the presence of common ingredients (reducing sugars and ammonia/amine groups) and heat treatment as in caramel production. Commercial cola drinks contained MEI in the range of 0.3–0.36 μg/mL much lesser than the model system, where MEI formation was 0.49–0.71 mg/mL (Moon and Shibamoto 2011).

21.4.2.1 Mitigation

Sulfates play a major role in the production of caramel; hence, they are used as an ingredient in the process (Lee et al. 2013). Lee and his group did an interesting study in which they analyzed the role of different additives on the formation of MEI. Among the additives, iron sulfates reduced the 4-MEI formation by 80% during the production of caramel (Seo et al. 2014). The aforementioned study highlights the mitigation of MEI in food systems by the use of food additives.

21.4.3 Nitrosamines and Analogs

Curing is a food processing technique in which a food material especially meat is preserved with the help of nitrite and nitrate salts believed to be followed as early as 3000 BC (American Meat Institute 2015; Sindelar and Milkowski 2011). Curing is also done to impart specific flavor and pink color and to prevent lipid oxidation in meat products (Skibsted 2011). Sodium nitrite (E 250) is a commonly used preservative and curing agent in animal and poultry meat; however, sodium nitrate (E 251) also serves the same purpose in specific foods. Nitrites are significant compounds of public interest in preventing the growth of toxic microorganisms in meat and meat products such as *Clostridium botulinum* and *Listeria monocytogenes*. No other salts are good at preventing the growth of these microorganisms. Nitrite salt breaks down to nitric oxide in meat and reacts with the heme (cofactor in hemoglobin) component of the blood and forms nitrosyl heme (Figure 21.8), which brings the characteristic pinkish red color to the raw cured meat. Nitrates and nitrites are often added in excess to replenish the loss of color to various deterioration pathways during storage. Residual nitrate present in excess leads to the formation of carcinogenic nitrosamines, which is further accelerated by cooking of meat especially deep frying and panfrying methods (Li et al. 2012). Apart from meat and meat products, nitrosamines are found in malted and brewed products due to nitrosation of nitrogen trioxide and nitrogen tetroxide to nitrosamines generally during the kilning

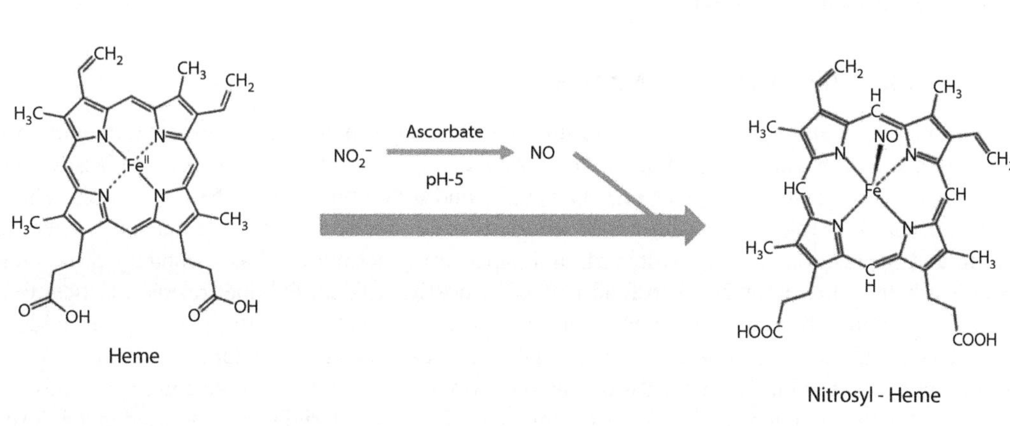

FIGURE 21.8 Formation of nitrosyl heme during curing of meat and meat products, which is responsible for the red color of meat.

process with hot air (Wainright 1986). Nitrosamines are classified as carcinogens in various animal systems (Bogovski and Bogovski 1981; IARC 1978; Lijinsky 1999), but further studies are required to establish a direct link between nitrates in food and its carcinogenicity (Jakszyn and Gonzalez 2006; Michaud et al. 2009).

21.4.3.1 Mitigation

Formation of nitrosamines can be prevented by adding various food additives such as ascorbic acid, erythorbic acid, ascorbyl palmitate, and vitamin E to the meat products. Most of such food additives are derivatives of vitamin C, which acts as a good antioxidant in preventing the oxidation of lipids in meat, thereby reducing the formation of nitrosamines (Herrmann et al. 2015; Tannenbaum et al. 1991). Cooking of meat at lower temperature during frying may also reduce the formation of nitrosamines without compromising the taste of meat.

21.4.4 Semicarbazide

Azodicarbonamide with E number 927a is a flour treatment agent used to improve baking functionality such as bread volume and color during baking process. It is also used as a flour bleaching agent to increase the whiteness of wheat flour due to the formation of gluten. It improves elasticity and gas retention capacity of dough, which aids in holding a large amount of CO_2 during the baking process (Joiner et al. 1963). Azodicarbonamide is stable in dry flours, but in moist flour it forms biurea, a nontoxic intermediate before breaking down to semicarbazide (SEM) and ethyl carbamate during thermal processing. SEM belongs to hydrazine group of chemicals, known for their strong reducing capacity in chemical reactions. SEM was found to be a weak carcinogen (group 3) and genotoxic in experimental animals, and the use of azodicarbonamide is banned in Australia and European Union (Semicarbazide in Food 2005). It is permitted as a food additive in the United States, Canada, and other countries up to 45 ppm (Becalski et al. 2004; Noonan et al. 2008).

SEM found in meat and meat products may also be formed as a breakdown product of nitrofurazone antibiotics used for the treatment of bacterial and protozoan infections in cattle, swine, poultry, and fish. The use of nitrofurazone is prohibited for its carcinogenic activity, and SEM acts as a marker in screening the illegal use of nitrofurazone in meat and meat products (Cooper et al. 2008). SEM was found in minor quantities in natural unprocessed foods such as algae, seaweed, shrimp, and eggs (Hoenicke et al. 2004).

21.4.4.1 Mitigation

Replacing azodicarbonamide with different flour treatment agents will overcome SEM formation in thermally processed dough products.

21.5 PACKAGING CONTAMINANTS

Food packaging is an important process to improve the shelf life and safe transportation and contains important information regarding the food products. Packaging materials include glass, wood, paper, cardboard, plastics (polymer resins), metals, and a combination of these materials. Glass is an excellent inert material toward food matrix; however, due to bulkiness, it finds only specific and limited application. Wood, cardboard, and paper are generally used as secondary packaging materials; hence, direct contact with food matrix is mostly avoided. Polymer resins (plastics) find major application in food packaging owing to their properties such as lightweight, ease of molding to any design and shape, low cost and choice of printing, heat sealing, and color (Kirwan et al. 2011). Depending on the food matrix polarity, owing to depolymerization, breakdown and reactiveness of food constituents, monomers, and residues from the packaging material leach into food matrix over time. A large group of such contaminants have been studied in past few decades, which are briefly discussed next.

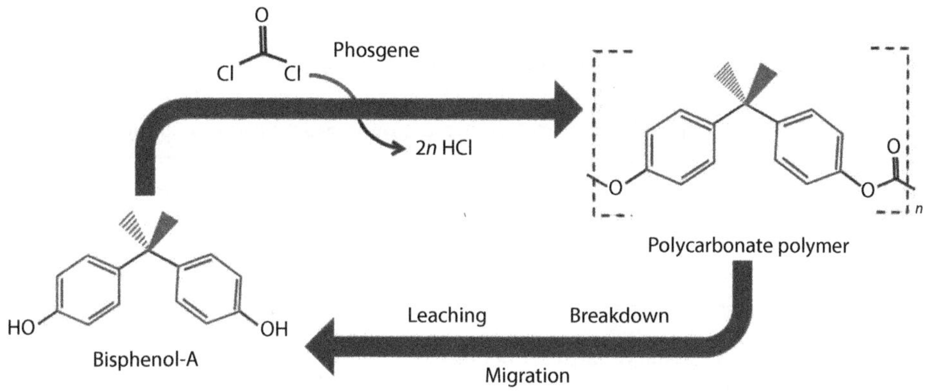

FIGURE 21.9 Formation of polycarbonate polymer from bisphenol A and its breakdown and migration to food matrix. (From Ragavan, K.V. et al., *TrAC Trends Anal. Chem.*, 52, 24, 2013. With permission.)

21.5.1 Bisphenols

Bisphenols are a class of compounds used as monomers in the production of polycarbonate (PC) resins and used for internal coatings in tin cans for food and beverage packaging applications. PC bottles were used extensively as infant feeding bottles and water bottles as a replacement for glass bottles that are fragile in nature. Incomplete polymerization and breakdown lead to leaching of bisphenols to food matrix, and the migration depends on the polarity of food matrix, temperature, pH, and solutes (Figure 21.9; Maia et al. 2009; Ragavan et al. 2013). Bisphenols have similar dihydroxydiphenyl as core chemical structure with different side chains, which is structurally analogous to endocrine hormones estradiol and diethylstilbestrol. Bisphenols possess a weak estrogen-binding capacity, hence classified as endocrine-disrupting compounds. Bisphenols cause a gamut of health problems to humans especially infants and children ranging from metabolic, cardiovascular, reproduction, and developmental disorders and alter immune system (Geens et al. 2011; Liao et al. 2012; Ragavan et al. 2013). PC feeding bottles are replaced with polypropylene bottles in most of the countries.

21.5.2 Brominated Flame Retardants

Brominated flame retardants are a class of organic compounds containing bromine, which are used as additive flame retardants in plastics, electronics, textiles, and other materials. They are detected in food products such as meat, fish, and milk. These molecules tend to bio-accumulate in human tissues and being toxic, they interfere with the functions of thyroid, liver and kidney. However, detailed toxicological studies are required to ascertain their role on human health (Birnbaum and Staskal 2004; Kim et al. 2014; Schecter et al. 2008).

21.5.3 Mineral Oil Hydrocarbons

Mineral oils are a mixture of aromatic and saturated hydrocarbons used as a solvent in printing inks, paperboards, and recycled paper and cardboard. Mineral oil aromatic hydrocarbons are found to be genotoxic carcinogen, whereas saturated hydrocarbons tend to aggregate in human tissues and pose serious liver problems. Migration of mineral oil hydrocarbons was significant in case of foods stored in a recycled paperboard box meant for transportation for prolonged duration. Migration was dependent on temperature, type of mineral oil hydrocarbons, and duration of storage (Biedermann and Grob 2012; Lorenzini et al. 2010, 2013). Migration of mineral oils can be prevented by using an additional barrier material as an internal coating to paperboard or outer coating to food materials.

21.5.4 Nonylphenol

Nonylphenols are a group of amphiphilic alkylphenols consisting of a phenol ring attached to alkyl chain, which are used as a catalyst in the production of epoxy resins and plastics. They are highly stable in nature due to high half-life in years and migrate into food matrix during direct contact with food. Nonylphenols and its esters are classified as xenoestrogens, and they are found to cause breast cancer and obesity and have the potential to disturb endocrine activity by mimicking estrogen activity (Grün and Blumberg 2009; Soares et al. 2008).

21.5.5 Perfluorinated Compounds

Perfluorooctanoic acid (PFOA) and perfluorooctane sulfonate (PFOS) are the major perfluorinated compounds used in the production of food contact paper, microwave popcorn paper bags, pizza box liners, candy wrappers, fast-food wrappers, and polytetrafluoroethylene (PTFE) nonstick cookware. These molecules are lipophilic in nature, and they leach into food by migration of residue from packaging material in μg/kg and mg/kg levels. The presence of saturated fat and high temperature during microwaving of popcorns enhance their migration. PFOA and PFOS are nongenotoxic compounds capable of inducing tumors in livers and pancreas in rodents (Begley et al. 2005).

21.5.6 Phthalates

Phthalates are a group of organic compounds in the form of esters that are primarily used to manufacture Polyethylene terephthalate (PET) bottles, used as plasticizers to soften poly-vinyl chloride (PVC) and as solvents to hold color and aroma in plastic containers meant for food-handling purposes. Phthalates migrate into food matrix majorly due to volatilization and leaching from packaging material such as PVC tubings, PET bottles, and plastic lids of glass jars. Phthalates are found in bottled water, beverages and fruit drinks, infant food, and packed foods and pose serious health problems. Some of the phthalate esters are found to be rodent liver carcinogen and few are teratogenic in nature (Cao 2010).

21.5.7 Semicarbazide

Azodicarbonamide is largely used as a blowing agent in the manufacture of foamed plastic materials such as PVC liners in gasket and in lids of glass bottles and jars. It decomposes to SEM in ppm concentrations and leaches to food matrix (Nestmann et al. 2005; Stadler et al. 2004). European Union has banned the use of plastic lines and gaskets made from azodicarbonamide for use in direct contact with food (Semicarbazide in Food 2005).

21.6 NANOMATERIALS

Nanomaterials are materials with one of its dimension in the range 1–100 nm. Nanomaterials include metal nanoparticles, nanocomposites, carbon nanotubes, graphene nanosheets, and nanoemulsions, displaying fascinating and interesting properties at nanoscale that are different from the bulk state. In the past decade, they are explored and applied in food especially in food packaging (Silvestre et al. 2011; Sozer and Kokini 2009), food safety (Duncan 2011), sensors (Ragavan et al. 2013; Thakur and Ragavan 2013), emulsions, and nutraceuticals (Cushen et al. 2012; Neethirajan and Jayas 2011). Silver nanoparticles and nanocomposites are widely used as an antimicrobial agent in food packaging materials to prevent microbial growth and keep the food material safe (Damm et al. 2007). Zinc nanocrystals (Vermeiren et al. 2002) and carbon nanotubes, (Kang et al. 2007; Warheit et al. 2004) incorporated in packaging materials were found to have antimicrobial, antifungal, and antibiotic properties. TiO_2 nanoparticles can act as an antimicrobial agent in the

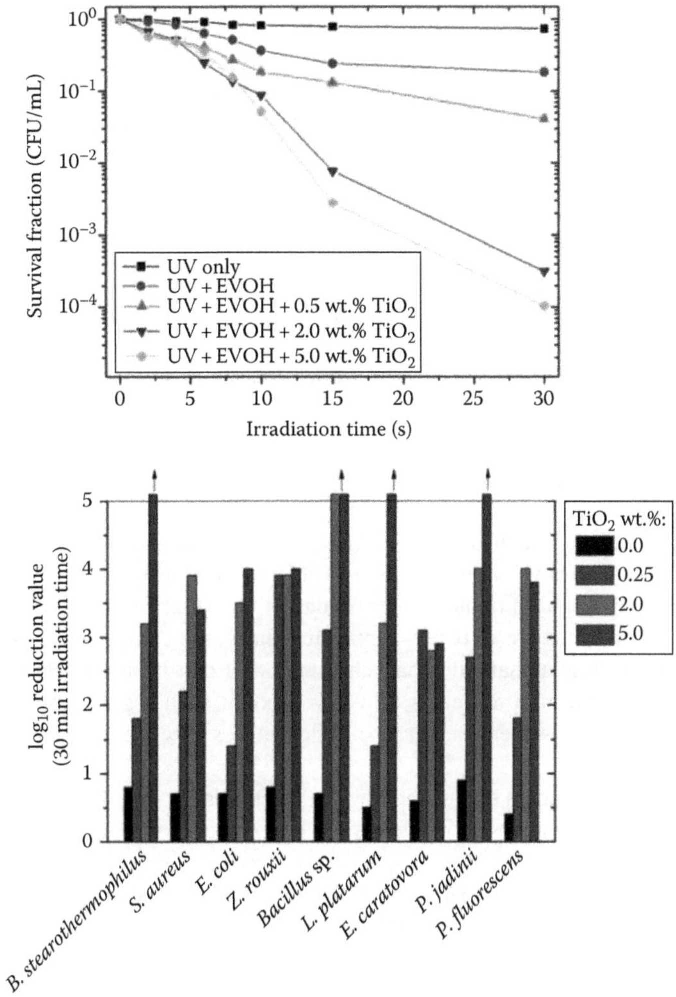

FIGURE 21.10 Antibacterial effect of TiO_2-incorporated ethylene vinyl alcohol films irradiated with UV. (From Duncan, T.V., *J. Colloid Interface Sci.*, 363, 1, 2011. With permission.)

presence of UV light. They are incorporated into ethylene vinyl alcohol films for food packaging applications. They are quite effective against common foodborne pathogens such as *Bacillus cereus*, *Escherichia coli*, and other microorganisms (Figure 21.10; Cerrada et al. 2008). However, they leach into food matrix and enter into consumers' body. Nanomaterials are found to induce inflammation and oxidative stress in cells, promote reactive oxygen species and tumor formation, and are cytotoxic in nature (Carlson et al. 2008). Consumer attitude toward the application of nanotechnology is moderately positive (Frewer et al. 2014). Fate of nanomaterials in humans and toxicological effects of nanomaterials need to be completely assessed to understand their role for safer food applications.

21.7 DISINFECTANT RESIDUES

Disinfectants are used to clean the food processing equipment and to maintain sanitation in food plants. In few cases, disinfectant residues enter the food being processed and have some effect on food quality and safety. Similarly, solvent residues are also found in some food products.

21.7.1 Haloacetic Acid and Trihalomethanes

Chlorination of water is an effective disinfectant process used to kill pathogenic microbes from domestic water. Ideally, an active chlorine concentration of 0.3–0.5 mg/L is advised for chlorination of domestic water. Residual active chlorine present in water systems prevents microbial growth and keeps water safe for a prolonged time. However, it reacts with natural organic matter present in water systems and leads to the formation of carcinogenic chlorinated compounds belonging to the following two main classes:

1. *Haloacetic acid*: HAAs comprise nine compounds in which hydrogen atom of acetic acid is replaced by halogens, especially chlorine, and are mostly weak carcinogens except monochloroacetic acid (Giller et al. 1997).
2. *Trihalomethanes*: THM group consists of four different compounds that are formed by reaction between chlorine and organic and inorganic matter present in water. THMs are classified as group B carcinogens (Villanueva et al. 2003).

21.7.1.1 Mitigation

Ozonation: Disinfection of water with ozone gas is one of the alternative strategies followed to overcome HAA and THM. Ozonation is cost-effective and also has very short life; hence, its residues are not found in treated water (Baker et al. 2002; Guzel-Seydim et al. 2004).

Chloramines: Chloramines are nitrogen-containing inorganic compounds with lower oxidation potential and higher half-life than chlorines and can be used effectively for water disinfection. Monochloramine reacts to a lower extent with organic matter in water and forms lesser THMs. However, the usage of chloramines is related to increased lead exposure (Miranda et al. 2007).

UV treatment: UV treatment is one of the physical methods for disinfection of water without leaving any chemical residues. It is effective in killing bacteria and removing micropollutants (Yang et al. 2014).

Adsorbents: Al-MCM-41 (a mesostructured aluminosilicate) has been found effective in complete removal of HAA in water (Bruzzoniti et al. 2012).

21.7.2 Semicarbazide

Bleaching and disinfection of food ingredients with hypochlorite (12% active Cl) solutions leads to the formation of SEM at ppm–ppb levels. Foods containing arginine, creatinine, creatine, and urea may contain high amount of SEM during thermal processing. It was shown that processing with lower concentrations of active chlorine reduced substantial amount of SEM formation in food ingredients (Hoenicke et al. 2004). It is advised to use disinfectants other than hypochlorite or in a combination to reduce and prevent SEM formation in processed foods.

21.8 NOVEL FOOD PROCESSING

Novel food processing operations are followed by the food processors due to the merits such as quicker processing time, lower costs, low-temperature processing, and better food quality. However, these processes also have some by-products that are considered as contaminants.

21.8.1 Alkylcyclobutanones and Free Radicals

Food irradiation is a physical processing technique in which a food material in prepacked or bulk state is exposed to ionizing radiation, which include electron beam (β-rays) or high-energy photons (x-rays and γ-rays) to reduce microbial count, sterilize, inhibit sprouting of vegetable, and delay ripening of fruits. Food irradiation does not involve any heat source, and it is familiarly known as "cold sterilization technique." Another marked advantage of food irradiation is "residue-free" processing of food, which overcomes the use of water, chemicals, and additives. Food irradiation is largely applied for processing spices and dry fruits, inhibition of sprouting in onions, delay in ripening of fruits, and sterilization of meat and meat products (Calado et al. 2014). As a result of irradiation, reactive species, free radicals, and 2-alkylcyclobutanones (2-ACBs) are formed in the food. Free radicals and reactive species react with various biomolecules such as lipids and proteins and lead to oxidation and rancidity of food products. Free radicals present in the food matrix may damage cells when consumed and cause various health effects (Letellier and Nawar 1972; Yu et al. 2012). Gamma rays–irradiated apple and orange juice are found to contain furan (discussed earlier) dependent on radiation dose (Fan 2005).

2-ACBs are unique radiolytic compounds produced when food containing fatty acids are irradiated. 2-Dodecylcyclobutanone (2-DCB) and 2-tetradecyclobutanone are formed from palmitic acid and stearic acid, respectively, which serve as chemical markers of food irradiation (Zhao et al. 2012). 2-DCB has been shown to cause oxidative stress in mitochondrial cells, finally leading to apoptosis (Yu et al. 2012). 2-ACBs have been shown to be present in commercial, fresh, and nonirradiated cashew nut and nutmeg samples (Variyar et al. 2008). Hence, there is a debate regarding the toxicity of ACBs in irradiated and nonirradiated food samples.

21.8.1.1 Mitigation

Mitigation strategies for removal and reduction of ACBs in irradiated food samples are yet to be explored. Hurdle technology may be one of the possible solutions to reduce the formation of ACBs in fat-containing foods in which low dose of radiation can be effectively combined with other processing methods.

21.8.2 Furan (UV Processing)

UV processing of foods is another novel technique adopted for the processing of plant and animal proteins, for sterilizing fruit juices and sugar syrups, and disinfecting drinking water and soft drinks (Bule et al. 2010; Fan and Geveke 2007; Kuan et al. 2013). According to U.S. FDA, a 5-log reduction in microbial count is necessary during UV-C processing of fruit juices at a wavelength of 254 nm, which corresponds to uniform exposure of the fruit juice with 400 J/m^2 (Guerrero-Beltrán and Barbosa-Cánovas 2004). UV-C sterilization of fruit juices containing fructose leads to the formation of furan (Fan and Geveke 2007). Fructose in fruit juices and sweet drinks mainly comes from high-fructose corn syrup (HFCS) that is used as sweetener. Recently, UV-C treatment of food containing linoleic and linolenic acids (PUFAs) also shown to form furan (Fan 2015).

21.8.2.1 Mitigation

Processing fruit juices in basic pH (8–9) helps in reducing the formation of furan from fructose. Along with that, ascorbic acid has shown to reduce the formation of furan in foods subjected to UV-C processing (Bule et al. 2010). Avoiding or finding different substitutes for α-tocopherol and $FeSO_4$ will help in preventing the formation of furan in fat-containing foods processed in UV-C. Processing at acidic pH (3–4) helps in reducing the formation of furan in foods containing linoleic acid (Fan 2015). UV-C treatment combined with mild heat treatment may help in processing foods at low dose of UV-C to achieve 5-log reduction in microbial count.

21.9 PERSPECTIVE

Overall dietary intake of the contaminants may come from domestic cooking and outdoor eating habits along with industry produced foods. Therefore, it is worth advising and educating both caterers and consumers to follow simple mitigation strategies in their cooking habits to overcome the ill effects of contaminants arising from food processing (CAC/RCP 67-2009 2009; CAC/RCP 68-2009 2015). Regulatory agencies can prepare code of practice for the mitigation of all possible food processing contaminants and strictly implement in food processing industries to produce safe food without contaminants. Most of the food processing contaminants pose serious threat to human health; hence, it is advised to reduce the contaminants to ALARA in processed foods. The presence of more than a single contaminant in a food product is nowadays common; hence, multiple mitigation steps may be needed to remove or reduce the amount of contaminant. Mitigation strategies for contaminants need to be addressed with caution to overcome the secondary ill effects arising due to the formation of new contaminants from existing ones or introduction of new contaminant. At the same time, care is required to retain the specific volatiles and sensory attributes of the food system in which mitigation steps are performed.

Analytical approaches play a major role in screening, detection, and quantification of contaminants and identification of new compounds in processed foods from the complex food matrix. Advancement in analytical techniques will help in detecting new and unidentified compounds in processed foods that may be probable contaminants (Farré and Barceló 2013). Nanotechnology application in food processing and packaging requires complete understanding of the role of nanoparticles in food and their fate after consumption in human system. Detail toxicological evaluation is necessary to arrive to a conclusion about the health benefits and safety claims of nanoparticles in food systems (Rollin et al. 2011; Siegrist et al. 2007).

21.10 CONCLUSIONS

At present, industrial food processing is an indispensable process to fulfill the food demand of ever-growing global population. Combination of factors such as ingredients in raw materials, additives, and processing conditions may lead to the formation and accumulation of undesirable compounds in processed food products collectively called food processing contaminants. Food processing contaminants can be classified on the basis of processing step during which it is formed. They are broadly grouped into thermal processing contaminants, fermentation contaminants, preservatives and additives, packaging contaminants, nanomaterials, disinfectant residues, and novel food processing contaminants. Thermal processing and food packaging contribute to a lot of contaminants in processed and packed foods. Many of the contaminants are carcinogens and endocrine-disrupting compound that cause health problems in humans. However, the formation and accumulation of contaminants in processed foods can be controlled by implementing mitigation strategies at industrial scale. Advanced and novel analytical methods for the detection of contaminants and new mitigation strategies for the removal of contaminants are need of time. Overall, an integrated approach is required for producing safe food, free from processing contaminants, which can win consumer acceptance of novel food processing technologies and keep them healthy.

REFERENCES

Abraham, K., S. Andres, R. Palavinskas, K. Berg, K.E. Appel, and A. Lampen. 2011. Toxicology and risk assessment of acrolein in food. *Molecular Nutrition and Food Research* 55 (9): 1277–1290. doi:10.1002/mnfr.201100481.

Akillioglu, H.G. and V. Gökmen. 2014. Mitigation of acrylamide and hydroxymethyl furfural in instant coffee by yeast fermentation. *Food Research International* 61: 252–256. doi:10.1016/j.foodres.2013.07.057.

Alaejos, M.S. and M.A. Ana 2011. Factors that affect the content of heterocyclic aromatic amines in foods. *Comprehensive Reviews in Food Science and Food Safety* 10 (2): 52–108. doi:10.1111/j.1541-4337.2010.00141.x.

Alvarez, M.A. and M. Victoria Moreno-Arribas. 2014. The problem of biogenic amines in fermented foods and the use of potential biogenic amine-degrading microorganisms as a solution. *Trends in Food Science & Technology* 39 (2): 146–155. doi:10.1016/j.tifs.2014.07.007.

American Meat Institute. 2015. Sodium nitrate: The facts. Accessed August 17. https://www.meatinstitute.org/index.php?ht=a/GetDocumentAction/i/44170.

Andres, S., K.E. Appel, and A. Lampen. 2013. Toxicology, occurrence and risk characterisation of the chloropropanols in food: 2-monochloro-1,3-propanediol, 1,3-dichloro-2-propanol and 2,3-dichloro-1-propanol. *Food and Chemical Toxicology: An International Journal Published for the British Industrial Biological Research Association* 58 (August): 467–478. doi:10.1016/j.fct.2013.05.024.

Anese, M., F. Bot, and M. Suman. 2014. Furan and 5-hydroxymethylfurfural removal from high- and low-moisture foods. *LWT—Food Science and Technology* 56 (2): 529–532. doi:10.1016/j.lwt.2013.12.030.

Anese, M. and M. Suman. 2013. Mitigation strategies of furan and 5-hydroxymethylfurfural in food. *Food Research International* 51 (1): 257–264. doi:10.1016/j.foodres.2012.12.024.

Anese, M., M. Suman, and M. Cristina Nicoli. 2010. Acrylamide removal from heated foods. *Food Chemistry* 119 (2): 791–794. doi:10.1016/j.foodchem.2009.06.043.

Anon. 1999. Acetaldehyde. IARC monographs on the evaluation of carcinogenic risks to humans. http://monographs.iarc.fr/ENG/Monographs/vol71/mono71-11.pdf. Accessed on August 17, 2015.

Anon. 2014. IARC monographs on the evaluation of carcinogenic risks to humans, Lyon, France. http://monographs.iarc.fr/ENG/Publications/internrep/14-002.pdf. Accessed on August 17, 2015.

Arrieta, M.P. and M.S. Prats-Moya. 2012. Free amino acids and biogenic amines in alicante monastrell wines. *Food Chemistry* 135 (3): 1511–1519. doi:10.1016/j.foodchem.2012.06.008.

Azevedo, Z., J.A. Couto, and T. Hogg. 2002. Citrulline as the main precursor of ethyl carbamate in model fortified wines inoculated with *Lactobacillus hilgardii*: A marker of the levels in a spoiled fortified wine. *Letters in Applied Microbiology* 34 (1): 32–36.

Baan, R.t, K. Straif, Y. Grosse, B. Secretan, F. El Ghissassi, V. Bouvard, A. Altieri, and V. Cogliano. 2007. Carcinogenicity of alcoholic beverages. *The Lancet Oncology* 8 (4): 292–293. doi:10.1016/S1470-2045(07)70099-2.

Baker, K.H., J.P. Hegarty, B. Redmond, N.A. Reed, and D.S. Herson. 2002. Effect of oxidizing disinfectants (chlorine, monochloramine, and ozone) on *Helicobacter pylori*. *Applied and Environmental Microbiology* 68 (2): 981–984. doi:10.1128/AEM.68.2.981-984.2002.

Bakhiya, N., K. Abraham, R. Gürtler, K.E. Appel, and A. Lampen. 2011. Toxicological assessment of 3-chloropropane-1,2-diol and glycidol fatty acid esters in food. *Molecular Nutrition and Food Research* 55 (4): 509–521. doi:10.1002/mnfr.201000550.

Ballance, P.E. 1961. Production of volatile compounds related to the flavour of foods from the strecker degradation of DL-methionine. *Journal of the Science of Food and Agriculture* 12 (7): 532–536. doi:10.1002/jsfa.2740120706.

Bartowsky, E. 2012. Lactic acid bacteria (LAB) in grape fermentations—An example of LAB as contaminants in food processing. In *Lactic Acid Bacteria: Microbiological and Functional Aspects*, 4th edn., pp. 343–359. Boca Raton, FL: CRC Press/Taylor & Francis Group.

Becalski, A., B.P.Y. Lau, D. Lewis, and S.W. Seaman. 2004. Semicarbazide formation in azodicarbonamide-treated flour: A model study. *Journal of Agricultural and Food Chemistry* 52 (18): 5730–5734. doi:10.1021/jf0495385.

Becalski, A. and S. Seaman. 2005. Furan precursors in food: A model study and development of a simple headspace method for determination of furan. *Journal of AOAC International* 88 (1): 102–106.

Begley, T.H., K. White, P. Honigfort, M.L. Twaroski, R. Neches, and R.A. Walker. 2005. Perfluorochemicals: Potential sources of and migration from food packaging. *Food Additives & Contaminants* 22 (10): 1023–1031. doi:10.1080/02652030500183474.

Bein, K. and G.D. Leikauf. 2011. Acrolein—A pulmonary hazard. *Molecular Nutrition and Food Research* 55 (9): 1342–1360. doi:10.1002/mnfr.201100279.

Bettini, S., A. Santino, L. Valli, and G. Giancane. 2015. A smart method for the fast and low-cost removal of biogenic amines from beverages by means of iron oxide nanoparticles. *RSC Advances* 5 (23): 18167–18171. doi:10.1039/C5RA01699A.

Biedermann, M. and K. Grob. 2012. On-line coupled high performance liquid chromatography-gas chromatography for the analysis of contamination by mineral oil. Part 2: Migration from paperboard into dry foods: Interpretation of chromatograms. *Journal of Chromatography A* 1255: 76–99. doi:10.1016/j.chroma.2012.05.096.

Birnbaum, L.S. and D.F. Staskal. 2004. Brominated flame retardants: Cause for concern? *Environmental Health Perspectives* 112 (1): 9–17.

Bodmer, S., C. Imark, and M. Kneubühl. 1999. Biogenic amines in foods: Histamine and food processing. *Inflammation Research* 48 (6): 296–300. doi:10.1007/s000110050463.

Bogovski, P. and S. Bogovski. 1981. Animal species in which N-nitroso compounds induce cancer. *International Journal of Cancer* 27 (4): 471–474.

Bruzzoniti, M.C., R.M. De Carlo, C. Sarzanini, D. Caldarola, and B. Onida. 2012. Novel insights in Al-MCM-41 precursor as adsorbent for regulated haloacetic acids and nitrate from water. *Environmental Science and Pollution Research* 19 (9): 4176–4183. doi:10.1007/s11356-012-0900-6.

Bule, M.V., K.M. Desai, B. Parisi, S.J. Parulekar, P. Slade, R.S. Singhal, and A. Rodriguez. 2010. Furan formation during UV-treatment of fruit juices. *Food Chemistry* 122 (4): 937–942. doi:10.1016/j.foodchem.2010.03.116.

CAC/RCP 67-2009. 2009. Code of practice for the reduction of acrylamide in foods. http://www.codexalimentarius.org/standards/list-standards/en/?no_cache=1. Accessed on August 17, 2015.

CAC/RCP 68-2009. 2015. Code of practice for the reduction of contamination of food with polycyclic aromatic hydrocarbons (PAH) from smoking and direct drying processes. Accessed August 17. http://www.codexalimentarius.org/download/standards/11257/CXP_068e.pdf. Accessed on August 17, 2015.

Calado, T., A. Venâncio, and L. Abrunhosa. 2014. Irradiation for mold and mycotoxin control: A review. *Comprehensive Reviews in Food Science and Food Safety* 13 (5): 1049–1061. doi:10.1111/1541-4337.12095.

Cao, X.L. 2010. Phthalate esters in foods: Sources, occurrence, and analytical methods. *Comprehensive Reviews in Food Science and Food Safety* 9 (1): 21–43. doi:10.1111/j.1541-4337.2009.00093.x.

Capuano, E. and V. Fogliano. 2011. Acrylamide and 5-hydroxymethylfurfural (HMF): A review on metabolism, toxicity, occurrence in food and mitigation strategies. *LWT—Food Science and Technology* 44 (4): 793–810. doi:10.1016/j.lwt.2010.11.002.

Carlson, C., S.M. Hussain, A.M. Schrand, L.K. Braydich-Stolle, K.L. Hess, R.L. Jones, and J.J. Schlager. 2008. Unique cellular interaction of silver nanoparticles: Size-dependent generation of reactive oxygen species. *The Journal of Physical Chemistry B* 112 (43): 13608–13619. doi:10.1021/jp712087m.

Cerbulis, J., O.W. Parks, R.H. Liu, E.G. Piotrowski, and H.M. Farrell. 1984. Occurrence of diesters of 3-chloro-1, 2-propanediol in the neutral lipid fraction of goats' milk. *Journal of Agricultural and Food Chemistry* 32 (3): 474–476. doi:10.1021/jf00123a013.

Cerrada, M.L., C. Serrano, M. Sánchez-Chaves, M. Fernández-García, F. Fernández-Martín, A. de Andrés, R.J. Jiménez Riobóo, A. Kubacka, M. Ferrer, and M. Fernández-García. 2008. Self-sterilized EVOH-TiO_2 nanocomposites: Interface effects on biocidal properties. *Advanced Functional Materials* 18 (13): 1949–1960. doi:10.1002/adfm.200701068.

Cooper, K.M., J. Le, C. Kane, and D.G. Kennedy. 2008. Kinetics of semicarbazide and nitrofurazone in chicken eggs and egg powders. *Food Additives & Contaminants. Part A, Chemistry, Analysis, Control, Exposure & Risk Assessment* 25 (6): 684–692. doi:10.1080/02652030701822060.

Cordain, L., S.B. Eaton, A. Sebastian, N. Mann, S. Lindeberg, B.A. Watkins, J.H. O'Keefe, and J. Brand-Miller. 2005. Origins and evolution of the western diet: Health implications for the 21st century. *The American Journal of Clinical Nutrition* 81 (2): 341–354.

Cravedi, E.D., A. Di Domenico, M.L. Férnandez-Cruz, P. Fürst, J. Fink-Gremmels, C.L. Galli, P. Grandjean et al. 2015. Ethyl carbamate and hydrocyanic acid in food and beverages scientific opinion of the panel on contaminants. Accessed August 6. http://www.researchgate.net/profile/Maria_Fernandez-Cruz/publication/242588214_Ethyl_carbamate_and_hydrocyanic_acid_in_food_and_beverages_1_Scientific_Opinion_of_the_Panel_on_Contaminants/links/02e7e5344f1693dc1e000000.pdf.

Crews, C. and L. Castle. 2007. A review of the occurrence, formation and analysis of furan in heat-processed foods. *Trends in Food Science and Technology* 18 (7): 365–372. doi:10.1016/j.tifs.2007.03.006.

Cushen, M., J. Kerry, M. Morris, M. Cruz-Romero, and E. Cummins. 2012. Nanotechnologies in the food industry—Recent developments, risks and regulation. *Trends in Food Science and Technology* 24 (1): 30–46. doi:10.1016/j.tifs.2011.10.006.

Damm, C., H. Münstedt, and A. Rösch. 2007. Long-term antimicrobial polyamide 6/silver-nanocomposites. *Journal of Materials Science* 42 (15): 6067–6073. doi:10.1007/s10853-006-1158-5.

De Mey, E., K. De Klerck, H. De Maere, L. Dewulf, G. Derdelinckx, M.C. Peeters, I. Fraeye, Y.V. Heyden, and H. Paelinck. 2014. The occurrence of N-nitrosamines, residual nitrite and biogenic amines in commercial dry fermented sausages and evaluation of their occasional relation. *Meat Science* 96 (1): 821–828. doi:10.1016/j.meatsci.2013.09.010.

Dost, K. and C. Deli. 2012. Determination of polycyclic aromatic hydrocarbons in edible oils and barbecued food by HPLC/UV-vis detection. *Food Chemistry* 133 (1): 193–199. doi:10.1016/j.foodchem.2012.01.001.

Druesne-Pecollo, N., B. Tehard, Y. Mallet, M. Gerber, T. Norat, S. Hercberg, and P. Latino-Martel. 2009. Alcohol and genetic polymorphisms: Effect on risk of alcohol-related cancer. *The Lancet Oncology* 10 (2): 173–180. doi:10.1016/S1470-2045(09)70019-1.

Duncan, T.V. 2011. Applications of nanotechnology in food packaging and food safety: Barrier materials, antimicrobials and sensors. *Journal of Colloid and Interface Science* 363 (1): 1–24. doi:10.1016/j.jcis.2011.07.017.

Esterbauer, H., R.J. Schaur, and H. Zollner. 1991. Chemistry and biochemistry of 4-hydroxynonenal, malonaldehyde and related aldehydes. *Free Radical Biology & Medicine* 11 (1): 81–128.

European Commission. 2002. Polycyclic aromatic hydrocarbons—Occurrence in foods, dietary exposure and health effects. http://ec.europa.eu/food/fs/sc/scf/out154_en.pdf. Accessed on October 6, 2015.

Fan, X. 2005. Impact of ionizing radiation and thermal treatments on furan levels in fruit juice. *Journal of Food Science* 70 (7): e409–e414. doi:10.1111/j.1365-2621.2005.tb11468.x.

Fan, X. 2015. Furan formation from fatty acids as a result of storage, gamma irradiation, UV-C and heat treatments. *Food Chemistry* 175 (May): 439–444. doi:10.1016/j.foodchem.2014.12.002.

Fan, X. and D.J. Geveke. 2007. Furan formation in sugar solution and apple cider upon ultraviolet treatment. *Journal of Agricultural and Food Chemistry* 55 (19): 7816–7821. doi:10.1021/jf071366z.

Farré, M. and D. Barceló. 2013. Analysis of emerging contaminants in food. *TrAC Trends in Analytical Chemistry* 43: 240–253. doi:10.1016/j.trac.2012.12.003.

Food Drink Europe. 2011. Food drink Europe acrylamide toolbox 2011. http://www.fooddrinkeurope.eu/uploads/publications_documents/Toolboxfinal260911.pdf. Accessed on August 17, 2015.

Frewer, L.J., N. Gupta, S. George, A.R.H. Fischer, E.L. Giles, and D. Coles. 2014. Consumer attitudes towards nanotechnologies applied to food production. *Trends in Food Science & Technology.* 40 (2): 211–225.

Gardner, L.K. and G.D. Lawrence. 1993. Benzene production from decarboxylation of benzoic acid in the presence of ascorbic acid and a transition-metal catalyst. *Journal of Agricultural and Food Chemistry* 41 (5): 693–695. doi:10.1021/jf00029a001.

Geens, T., L. Goeyens, and A. Covaci. 2011. Are potential sources for human exposure to bisphenol-A overlooked? *International Journal of Hygiene and Environmental Health* 214 (5): 339–347. doi:10.1016/j.ijheh.2011.04.005.

Giller, S., F. Le Curieux, F. Erb, and D. Marzin. 1997. Comparative genotoxicity of halogenated acetic acids found in drinking water. *Mutagenesis* 12 (5): 321–328.

Grün, F. and B. Blumberg. 2009. Endocrine disrupters as obesogens. *Molecular and Cellular Endocrinology* 304 (1–2): 19–29. doi:10.1016/j.mce.2009.02.018.

Guerrero-Beltrán, N.J.A. and G.V. Barbosa-Cánovas. 2004. Advantages and limitations on processing foods by UV light. *Food Science and Technology International* 10 (3): 137–147. doi:10.1177/1082013204044359.

Guth, S., M. Habermeyer, M. Baum, P. Steinberg, A. Lampen, and G. Eisenbrand. 2013. Thermally induced process-related contaminants: The example of acrolein and the comparison with acrylamide. *Molecular Nutrition & Food Research* 57 (12): 2269–2282. doi:10.1002/mnfr.201300418.

Guzel-Seydim, Z.B., A.K. Greene, and A.C. Seydim. 2004. Use of ozone in the food industry. *LWT—Food Science and Technology* 37 (4): 453–460. doi:10.1016/j.lwt.2003.10.014.

Hamlet, C.G., L. Asuncion, J. Velíšek, M. Doležal, Z. Zelinková, and C. Crews. 2011. Formation and occurrence of esters of 3-chloropropane-1,2-diol (3-CPD) in foods: What we know and what we assume. *European Journal of Lipid Science and Technology* 113 (3): 279–303. doi:10.1002/ejlt.201000480.

Haskaraca, G., E. Demirok, N. Kolsarıcı, F. Öz, and N. Özsaraç. 2014. Effect of green tea extract and microwave pre-cooking on the formation of heterocyclic aromatic amines in fried chicken meat products. *Food Research International* 63: 373–381.

Hasnip, S., C. Crews, and L. Castle. 2006. Some factors affecting the formation of furan in heated foods. *Food Additives and Contaminants* 23 (3): 219–227. doi:10.1080/02652030500539766.

Hasnip, S., C. Crews, N. Potter, J. Christy, D. Chan, T. Bondu, W. Matthews, B. Walters, and K. Patel. 2007. Survey of ethyl carbamate in fermented foods sold in the United Kingdom in 2004. *Journal of Agricultural and Food Chemistry* 55 (7): 2755–2759. doi:10.1021/jf063121c.

Hendriksen, H.V., B.A. Kornbrust, P.R. Ostergaard, and M.A. Stringer. 2009. Evaluating the potential for enzymatic acrylamide mitigation in a range of food products using an asparaginase from *Aspergillus oryzae*. *Journal of Agricultural and Food Chemistry* 57 (10): 4168–4176. doi:10.1021/jf900174q.

Henle, T. 2005. Protein-bound advanced glycation endproducts (AGEs) as bioactive amino acid derivatives in foods. *Amino Acids* 29 (4): 313–322. doi:10.1007/s00726-005-0200-2.

Herrmann, S.S., K. Granby, and L. Duedahl-Olesen. 2015. Formation and mitigation of N-nitrosamines in nitrite preserved cooked sausages. *Food Chemistry* 174: 516–526. doi:10.1016/j.foodchem.2014.11.101.

Hoenicke, K., R. Gatermann, L. Hartig, M. Mandix, and S. Otte. 2004. Formation of semicarbazide (SEM) in food by hypochlorite treatment: Is SEM a specific marker for nitrofurazone abuse? *Food Additives and Contaminants* 21 (6): 526–537. doi:10.1080/02652030410001712484.

Honikel, K.O. 2008. The use and control of nitrate and nitrite for the processing of meat products. *Meat Science, Symposium on Meat Safety: From Abattoir to Consumer,* 78 (1–2): 68–76. doi:10.1016/j.meatsci.2007.05.030.

Hua, Y., W. Wu, and X. Wu. 2010. Structural modification of soy protein by the lipid peroxidation product acrolein. *LWT—Food Science and Technology* 43 (1): 133–140. doi:10.1016/j.lwt.2009.05.006.

IARC. 1978. *Some N-Nitroso Compounds*, Vol. 17. IARC Monographs on the Evaluation of Carcinogenic Risks to Humans IARC Monographs. Lyon, France: IARC.

IARC. 1994. *Acrylamide*, Vol. 60. IARC Monographs on the Evaluation of Carcinogenic Risks to Humans. Lyon, France: IARC, WHO.

IARC. 1995. IARC Monographs on the Evaluation of Carcinogenic Risks to Humans. Dry Cleaning, Some Chlorinated Solvents and Other Industrial Chemicals. Lyon, France: IARC.

IARC. 2000. Some Industrial Chemicals. IARC Monographs on the Evaluation of Carcinogenic Risk of Chemicals to Humans, Vol. 77. Lyon, France: IARC. http://monographs.iarc.fr/ENG/Monographs/vol77/mono77.pdf.

Jiao, Z., Y. Dong, and Q. Chen. 2014. Ethyl carbamate in fermented beverages: Presence, analytical chemistry, formation mechanism, and mitigation proposals. *Comprehensive Reviews in Food Science and Food Safety* 13 (4): 611–626. doi:10.1111/1541-4337.12084.

Jinap, S., S.Z. Iqbal, and R.M.P. Selvam. 2015. Effect of selected local spices marinades on the reduction of heterocyclic amines in grilled beef (satay). *LWT—Food Science and Technology* 63 (2): 919–926. doi:10.1016/j.lwt.2015.04.047.

Joiner, R.R., F.D. Vidal, and H.C. Marks. 1963. A new powdered agent for flour maturing. *Cereal Chemistry* 40 (5): 539.

Kang, S., M. Pinault, L.D. Pfefferle, and M. Elimelech. 2007. Single-walled carbon nanotubes exhibit strong antimicrobial activity. *Langmuir: The ACS Journal of Surfaces and Colloids* 23 (17): 8670–8673. doi:10.1021/la701067r.

Karangwa, E., G.E. Mitchell, and R.E. Tucker. 1990. High-performance liquid chromatographic determination of 4-methylimidazole in sheep plasma and in ammoniated tall Fescue Hay. *Journal of Chromatography B: Biomedical Sciences and Applications* 532 (January): 105–113. doi:10.1016/S0378-4347(00)83756-1.

Karim, F. and J.S. Smith. 2015. Detection and quantification of 4(5)-methylimidazole in cooked meat. *Journal of Food Science* 80 (2): T465–T471. doi:10.1111/1750-3841.12748.

Keşkekoğlu, H. and A. Üren. 2014. Inhibitory effects of pomegranate seed extract on the formation of heterocyclic aromatic amines in beef and chicken meatballs after cooking by four different methods. *Meat Science* 96 (4): 1446–1451. doi:10.1016/j.meatsci.2013.12.004.

Kim, T.K., Y.K. Lee, Y.S. Park, and K.G. Lee. 2009. Effect of cooking or handling conditions on the furan levels of processed foods. *Food Additives & Contaminants. Part A, Chemistry, Analysis, Control, Exposure & Risk Assessment* 26 (6): 767–775. doi:10.1080/02652030902774656.

Kim, Y.R., F.A. Harden, L.M.L. Toms, and R.E. Norman. 2014. Health consequences of exposure to brominated flame retardants: A systematic review. *Chemosphere* 106 (July): 1–19. doi:10.1016/j.chemosphere.2013.12.064.

Kirwan, M.J., S. Plant, and J.W. Strawbridge. 2011. Plastics in food packaging. In *Food and Beverage Packaging Technology*, eds. Richard, C. and R.K. Kirwan, pp. 157–212. Oxford, U.K.: Wiley-Blackwell. http://onlinelibrary.wiley.com/doi/10.1002/9781444392180.ch7/summary.

Kuan, Y.H., R. Bhat, A. Patras, and A.A. Karim. 2013. Radiation processing of food proteins—A review on the recent developments. *Trends in Food Science & Technology* 30 (2): 105–120. doi:10.1016/j.tifs.2012.12.002.

Lachenmeier, D.W., F. Kanteres, and J. Rehm. 2009a. Carcinogenicity of acetaldehyde in alcoholic beverages: Risk assessment outside ethanol metabolism. *Addiction* 104 (4): 533–550. doi:10.1111/j.1360-0443.2009.02516.x.

Lachenmeier, D.W., T. Kuballa, H. Reusch, C. Sproll, M. Kersting, and U. Alexy. 2010. Benzene in infant carrot juice: Further insight into formation mechanism and risk assessment including cconsumption data from the DONALD study. *Food and Chemical Toxicology* 48 (1): 291–297. doi:10.1016/j.fct.2009.10.012.

Lachenmeier, D.W., M.C. Przybylski, and J. Rehm. 2012. Comparative risk assessment of carcinogens in alcoholic beverages using the margin of exposure approach. *International Journal of Cancer* 131 (6): 995–1003. doi:10.1002/ijc.27553.

Lachenmeier, D.W., H. Reusch, C. Sproll, K. Schoeberl, and T. Kuballa. 2008. Occurrence of benzene as a heat-induced contaminant of carrot juice for babies in a general survey of beverages. *Food Additives & Contaminants. Part A, Chemistry, Analysis, Control, Exposure & Risk Assessment* 25 (10): 1216–1224. doi:10.1080/02652030802036230.

Lachenmeier, D.W., B. Sarsh, and J. Rehm. 2009b. The composition of alcohol products from markets in lithuania and hungary, and potential health consequences: A pilot study. *Alcohol and Alcoholism* 44 (1): 93–102. doi:10.1093/alcalc/agn095.

Lachenmeier, D.W. and E.M. Sohnius. 2008. The role of acetaldehyde outside ethanol metabolism in the carcinogenicity of alcoholic beverages: Evidence from a large chemical survey. *Food and Chemical Toxicology* 46 (8): 2903–2911. doi:10.1016/j.fct.2008.05.034.

Ladero, V., M. Calles-Enriquez, M. Fernández, and M.A. Alvarez. 2010. Toxicological effects of dietary biogenic Aamines. *Current Nutrition & Food Science* 6 (2): 145–156.

Ledesma, E., M. Rendueles, and M. Díaz. 2016. Contamination of meat products during smoking by polycyclic aromatic hydrocarbons: Processes and prevention. *Food Control* 60 (February): 64–87. doi:10.1016/j.foodcont.2015.07.016.

Lee, H.S. and S. Nagy. 1990. Relative reactivities of sugars in the formation of 5-hydroxymethylfurfural in sugar-catalyst model systems 1. *Journal of Food Processing and Preservation* 14 (3): 171–178. doi:10.1111/j.1745-4549.1990.tb00126.x.

Lee, K.G., H. Jang, and T. Shibamoto. 2013. Formation of carcinogenic 4 (5) -methylimidazole in caramel model systems : A role of sulphite. *Food Chemistry* 136 (3–4): 1165–1168. doi:10.1016/j.foodchem.2012.09.025.

Letellier, P.R. and W.W. Nawar. 1972. 2-alkylcyclobutanones from the radiolysis of triglycerides. *Lipids* 7 (1): 75–76.

Liao, C., F. Liu, Y. Guo, H.B. Moon, H. Nakata, Q. Wu, and K. Kannan. 2012. Occurrence of eight bisphenol analogues in indoor dust from the United States and several Asian countries: Implications for human exposure. *Environmental Science & Technology* 46 (16): 9138–9145. doi:10.1021/es302004w.

Lijinsky, W. 1999. N-Nitroso compounds in the diet. *Mutation Research* 443 (1–2): 129–138.

Li, L., P. Wang, X. Xu, and G. Zhou. 2012. Influence of various cooking methods on the concentrations of volatile N-Nitrosamines and biogenic aamines in dry-cured sausages. *Journal of Food Science* 77 (5). doi:10.1111/j.1750-3841.2012.02667.x.

Lineback, D.R., J.R. Coughlin, and R.H. Stadler. 2012. Acrylamide in foods: A review of the science and future considerations. *Annual Review of Food Science and Technology* 3 (1): 15–35. doi:10.1146/annurev-food-022811-101114.

Loizzo, M.R., F. Menichini, N. Picci, F. Puoci, U.G. Spizzirri, and D. Restuccia. 2013. Technological aspects and analytical determination of biogenic amines in cheese. *Trends in Food Science and Technology* 30 (1): 38–55. doi:10.1016/j.tifs.2012.11.005.

Lorenzini, R., M. Biedermann, K. Grob, D. Garbini, M. Barbanera, and I. Braschi. 2013. Migration kinetics of mineral oil hydrocarbons from recycled paperboard to dry food: Monitoring of two real cases. *Food Additives & Contaminants. Part A, Chemistry, Analysis, Control, Exposure & Risk Assessment* 30 (4): 760–770. doi:10.1080/19440049.2013.766765.

Lorenzini, R., K. Fiselier, M. Biedermann, M. Barbanera, I. Braschi, and K. Grob. 2010. Saturated and aromatic mineral oil hydrocarbons from paperboard food packaging: Estimation of long-term migration from contents in the paperboard and data on boxes from the market. *Food Additives & Contaminants. Part A, Chemistry, Analysis, Control, Exposure & Risk Assessment* 27 (12): 1765–1774. doi:10.1080/19440049.2010.517568.

Maia, J., J.M. Cruz, R. Sendón, J. Bustos, J.J. Sanchez, and P. Paseiro. 2009. Effect of detergents in the release of bisphenol a from polycarbonate baby bottles. *Food Research International* 42 (10): 1410–1414. doi:10.1016/j.foodres.2009.07.003.

Märk, J., P. Pollien, C. Lindinger, I. Blank, and T. Märk. 2006. Quantitation of furan and methylfuran formed in different precursor systems by proton transfer reaction mass spectrometry. *Journal of Agricultural and Food Chemistry* 54 (7): 2786–2793. doi:10.1021/jf052937v.

Matthäus, B., F. Pudel, P. Fehling, K. Vosmann, and A. Freudenstein. 2011. Strategies for the reduction of 3-MCPD esters and related compounds in vegetable oils. *European Journal of Lipid Science and Technology* 113 (3): 380–386. doi:10.1002/ejlt.201000300.

Mesías, M. and F.J. Morales. 2014. Reliable estimation of dietary exposure to furan from coffee: An automatic vending machine as a case study. *Food Research International* 61: 257–263. doi:10.1016/j.foodres.2013.07.054.

Michaud, D.S., C.N. Holick, T.T. Batchelor, E. Giovannucci, and D.J. Hunter. 2009. Prospective study of meat intake and dietary nitrates, nitrites, and nitrosamines and risk of adult glioma. *American Journal of Clinical Nutrition* 90 (3): 570–577. doi:10.3945/ajcn.2008.27199.

Miranda, M.L., D. Kim, A.P. Hull, C.J. Paul, and M. Alicia Overstreet Galeano. 2007. Changes in blood lead levels associated with use of chloramines in water treatment systems. *Environmental Health Perspectives* 115 (2): 221–225. doi:10.1289/ehp.9432.

Mogol, B.A. and V. Gökmen. 2014. Mitigation of acrylamide and hydroxymethylfurfural in biscuits using a combined partial conventional baking and vacuum post-baking process: Preliminary study at the lab scale. *Innovative Food Science & Emerging Technologies* 26: 265–270. doi:10.1016/j.ifset.2014.05.002.

Moon, J.K. and T. Shibamoto. 2011. Formation of carcinogenic 4(5)-methylimidazole in maillard reaction systems. *Journal of Agricultural and Food Chemistry* 59 (2): 615–618. doi:10.1021/jf104098a.

Moro, S., J.K. Chipman, J.-W. Wegener, C. Hamberger, W. Dekant, and A. Mally. 2012. Furan in heat-treated foods: Formation, exposure, toxicity, and aspects of risk assessment. *Molecular Nutrition & Food Research* 56 (8): 1197–1211. doi:10.1002/mnfr.201200093.

National Toxicology Program. 1993. Toxicology and carcinogenesis studies of furan (CAS No. 110-00-9) in F344 Rats and B6C3F1 mice(gavage studies). *National Toxicology Program Technical Report Series* 402 (January): 1–286.

National Toxicology Program. 2007. Toxicology and carcinogenesis studies of 4-Methylimidazole (Cas No. 822-36-6) in F344/N Rats and B6C3F1 mice (feed studies). *National Toxicology Program Technical Report Series* 535 (January): 1–274.

National Toxicology Program. 2010. NTP toxicology and carcinogenesis studies of 5-(hydroxymethyl)-2-furfural (CAS No. 67-47-0) in F344/N rats and B6C3F1 mice (gavage studies). *National Toxicology Program Technical Report Series* 554 (June): 7–13, 15–19, 21–31 passim.

Nayak, C.A., S. Chethana, N.K. Rastogi, and K.S.M.S. Raghavarao. 2006. Enhanced mass transfer during solid–liquid extraction of gamma-irradiated red beetroot. *Radiation Physics and Chemistry* 75 (1): 173–178. doi:10.1016/j.radphyschem.2005.03.015.

Neethirajan, S. and D.S. Jayas. 2011. Nanotechnology for the food and bioprocessing industries. *Food and Bioprocess Technology* 4 (1): 39–47. doi:10.1007/s11947-010-0328-2.

Nestmann, E.R., B.S. Lynch, K. Musa-Veloso, G.H. Goodfellow, E. Cheng, L.a. Haighton, and V.M. Lee-Brotherton. 2005. Safety assessment and risk-benefit analysis of the use of azodicarbonamide in baby food jar closure technology: Putting trace levels of semicarbazide exposure into perspective—A review. *Food Additives and Contaminants* 22 (9): 875–891. doi:10.1080/02652030500195312.

Nie, S., J. Huang, J. Hu, Y. Zhang, S. Wang, C. Li, and M. Xie. 2013. Effect of pH, temperature and heating time on the formation of furan in sugar—Glycine model systems. *Food Science and Human Wellness* 2 (2): 87–92. doi:10.1016/j.fshw.2013.05.001.

Noonan, G.O., T.H. Begley, and G.W. Diachenko. 2008. Semicarbazide formation in flour and bread. *Journal of Agricultural and Food Chemistry* 56 (6): 2064–2067. doi:10.1021/jf073198g.

Ordóñez, J.L., R.M. Callejón, M.L. Morales, and M.C. García-Parrilla. 2013. A survey of biogenic amines in vinegars. *Food Chemistry* 141 (3): 2713–2719. doi:10.1016/j.foodchem.2013.05.087.

Owczarek-Fendor, A., B. De Meulenaer, G. Scholl, A. Adams, F. Van Lancker, G. Eppe, E. De Pauw, M.L. Scippo, and N. De Kimpe. 2012. Furan formation in starch-based model systems containing carbohydrates in combination with proteins, ascorbic acid and lipids. *Food Chemistry* 133 (3): 816–821. doi:10.1016/j.foodchem.2012.01.098.

Oz, F. and M. Kaya. 2011a. The inhibitory effect of black pepper on formation of heterocyclic aromatic amines in high-fat meatball. *Food Control* 22 (3–4): 596–600. doi:10.1016/j.foodcont.2010.10.010.

Oz, F. and M. Kaya. 2011b. Heterocyclic aromatic amines in meat: HCAS and meat. *Journal of Food Processing and Preservation* 35 (6): 739–753. doi:10.1111/j.1745-4549.2011.00524.x.

Oz, F. and E. Zikirov. 2015. The effects of sous-vide cooking method on the formation of heterocyclic aromatic amines in beef chops. *LWT—Food Science and Technology* 64 (1): 120–125. doi:10.1016/j.lwt.2015.05.050.

Park, K.K., A. Liem, B.C. Stewart, and J.A. Miller. 1993. Vinyl carbamate epoxide, a major strong electrophilic, mutagenic and carcinogenic metabolite of vinyl carbamate and ethyl carbamate (urethane). *Carcinogenesis* 14 (3): 441–450.

Pedreschi, F., K. Granby, and J. Risum. 2010. Acrylamide mitigation in potato chips by using NaCl. *Food and Bioprocess Technology* 3 (6): 917–921. doi:10.1007/s11947-010-0349-x.

Perez Locas, C. and V.A. Yaylayan. 2004. Origin and mechanistic pathways of formation of the parent furan a food toxicant. *Journal of Agricultural and Food Chemistry* 52 (22): 6830–6836. doi:10.1021/jf0490403.

Perez Locas, C. and V.A. Yaylayan. 2008. Isotope labeling studies on the formation of 5-(hydroxymethyl)-2-furaldehyde (HMF) from sucrose by pyrolysis-GC/MS. *Journal of Agricultural and Food Chemistry* 56 (15): 6717–6723. doi:10.1021/jf8010245.

Poulsen, M.W., R.V. Hedegaard, J.M. Andersen, B. de Courten, S. Bügel, J. Nielsen, L.H. Skibsted, and L.O. Dragsted. 2013. Advanced glycation endproducts in food and their effects on health. *Food and Chemical Toxicology: An International Journal Published for the British Industrial Biological Research Association* 60 (October): 10–37. doi:10.1016/j.fct.2013.06.052.

Purcaro, G., S. Moret, and L.S. Conte. 2013. Overview on polycyclic aromatic hydrocarbons: Occurrence, legislation and innovative determination in foods. *Talanta* 105: 292–305. doi:10.1016/j.talanta.2012.10.041.

Ragavan, K.V., N.K. Rastogi, and M.S. Thakur. 2013. Sensors and biosensors for analysis of bisphenol-A. *TrAC Trends in Analytical Chemistry* 52: 248–260.

Rahman, U.ur, A. Sahar, M. Issa Khan, and M. Nadeem. 2014. Production of heterocyclic aromatic amines in meat: Chemistry, health risks and inhibition. A review. *LWT—Food Science and Technology* 59 (1): 229–233. doi:10.1016/j.lwt.2014.06.005.

Rastogi, N.K. 2011. Opportunities and challenges in application of ultrasound in food processing. *Critical Reviews in Food Science and Nutrition* 51 (8): 705–722. doi:10.1080/10408391003770583.

Rastogi, N.K. 2012. Recent trends and developments in infrared heating in food processing. *Critical Reviews in Food Science and Nutrition* 52 (9): 737–760. doi:10.1080/10408398.2010.508138.

Rastogi, N.K., K.S.M.S. Raghavarao, V.M. Balasubramaniam, K. Niranjan, and D. Knorr. 2007. Opportunities and challenges in high pressure processing of foods. *Critical Reviews in Food Science and Nutrition* 47 (1): 69–112. doi:10.1080/10408390600626420.

Riachi, L.G., A. Santos, R.F a Moreira, and C.a B De Maria. 2014. A review of ethyl carbamate and polycyclic aromatic hydrocarbon contamination risk in cachaça and other brazilian sugarcane spirits. *Food Chemistry* 149: 159–169. doi:10.1016/j.foodchem.2013.10.088.

Robertson, G.L. 2012. *Food Packaging: Principles and Practice*, 3rd edn. Boca Raton, FL: CRC Press.

Rollin, F., J. Kennedy, and J. Wills. 2011. Consumers and new food technologies. *Trends in Food Science and Technology* 22 (2–3): 99–111. doi:10.1016/j.tifs.2010.09.001.

Saldo, J., Á. Suárez-Jacobo, R. Gervilla, B. Guamis, and A.X. Roig-Sagués. 2009. Use of ultra-high-pressure homogenization to preserve apple juice without heat damage. *High Pressure Research* 29 (1): 52–56. doi:10.1080/08957950802715112.

Saltmarsh, M. and S. Barlow. 2013. *Essential Guide to Food Additives*. Cambridge, U.K.: Royal Society of Chemistry.

Santos, M.H.S. 1996. Biogenic amines: Their importance in foods. *International Journal of Food Microbiology* 29 (2–3): 213–231. doi:10.1016/0168-1605(95)00032-1.

Schecter, A., T.R. Harris, N. Shah, A. Musumba, and O. Päpke. 2008. Brominated flame retardants in US food. *Molecular Nutrition & Food Research* 52 (2): 266–272. doi:10.1002/mnfr.200700166.

Scientific Report of EFSA. 2010. Update of results on the monitoring of furan levels in food. Parma, Italy. *EFSA Journal* 8 (7): 1702.

Seitz, H.K. and F. Stickel. 2007. Molecular mechanisms of alcohol-mediated carcinogenesis. *Nature Reviews Cancer* 7 (8): 599–612. doi:10.1038/nrc2191.

Semicarbazide in Food. 2005. Opinion of the scientific panel on food additives, flavourings, processing aids and materials in contact with food on a request from the commission related to semicarbazide in food. http://www.efsa.europa.eu/sites/default/files/scientific_output/files/main_documents/219.pdf. Accessed on October 15, 2015.

Sengar, G. and H.K. Sharma. 2012. Food caramels: A review. *Journal of Food Science and Technology* 51 (September): 1–11. doi:10.1007/s13197–012–0633-z.

Seo, S., M. Hyun Ka, and K.G. Lee. 2014. Reduction of carcinogenic 4(5)-methylimidazole in a caramel model system: Influence of food additives. *Journal of Agricultural and Food Chemistry* 62 (27): 6481–6486. doi:10.1021/jf502008q.

Shi, R., T. Rickett, and W. Sun. 2011. Acrolein-mediated injury in nervous system trauma and diseases. *Molecular Nutrition and Food Research* 55 (9): 1320–1331. doi:10.1002/mnfr.201100217.

Shinoda, Y., H. Komura, S. Homma, and M. Murata. 2005. Browning of model orange juice solution: Factors affecting the formation of decomposition products. *Bioscience, Biotechnology, and Biochemistry* 69 (11): 2129–2137. doi:10.1271/bbb.69.2129.

Siegrist, M., M.-E. Cousin, H. Kastenholz, and A. Wiek. 2007. Public acceptance of nanotechnology foods and food packaging: The influence of affect and trust. *Appetite* 49 (2): 459–466. doi:10.1016/j.appet.2007.03.002.

Silvestre, C., D. Duraccio, and S. Cimmino. 2011. Food packaging based on polymer nanomaterials. *Progress in Polymer Science (Oxford)* 36 (12): 1766–1782. doi:10.1016/j.progpolymsci.2011.02.003.

Sindelar, J.J. and A.L. Milkowski. 2011. Sodium nitrite in processed meat and poultry meats: A review of curing and examining the risk/benefit of its use. *American Meat Science Association White Paper Series* 3: 1–14.

Skibsted, L.H. 2011. Nitric oxide and quality and safety of muscle based foods. *Nitric Oxide: Biology and Chemistry/Official Journal of the Nitric Oxide Society* 24 (4): 176–183. doi:10.1016/j.niox.2011.03.307.

Soares, A., B. Guieysse, B. Jefferson, E. Cartmell, and J.N. Lester. 2008. Nonylphenol in the environment: A critical review on occurrence, fate, toxicity and treatment in wastewaters. *Environment International* 34 (7): 1033–1049. doi:10.1016/j.envint.2008.01.004.

Sozer, N. and J.L. Kokini. 2009. Nanotechnology and its applications in the food sector. *Trends in Biotechnology* 27 (2): 82–89. doi:10.1016/j.tibtech.2008.10.010.

Stadler, R.H, I. Blank, N. Varga, F. Robert, J. Hau, P. a Guy, M.-Cl. Robert, and S. Riediker. 2002. Acrylamide from maillard reaction products. *Nature* 419 (6906): 449–450. doi:10.1038/419449a.

Stadler, R.H., P. Mottier, P. Guy, E. Gremaud, N. Varga, S. Lalljie, R. Whitaker et al. 2004. Semicarbazide is a minor thermal decomposition product of azodicarbonamide used in the gaskets of certain food jars. *The Analyst* 129 (3): 276–281. doi:10.1039/b314206j.

Stevens, J.F. and C.S. Maier. 2008. Acrolein: Sources, metabolism, and biomolecular interactions relevant to human health and disease. *Molecular Nutrition and Food Research* 52 (1): 7–25. doi:10.1002/mnfr.200700412.

Stirban, A., T. Gawlowski, and M. Roden. 2014. Vascular effects of advanced glycation endproducts: Clinical effects and molecular mechanisms. *Molecular Metabolism* 3 (2): 94–108. doi:10.1016/j.molmet.2013.11.006.

Sugimura, T., K. Wakabayashi, H. Nakagama, and M. Nagao. 2004. Heterocyclic amines: Mutagens/carcinogens produced during cooking of meat and fish. *Cancer Science* 95 (4): 290–299.

Svendsen, C., T. Husøy, H. Glatt, J.E. Paulsen, and J. Alexander. 2009. 5-hydroxymethylfurfural and 5-sulfooxymethylfurfural increase adenoma and flat ACF number in the intestine of min/+ mice. *Anticancer Research* 29 (6): 1921–1926.

Szterk, A. 2015. Heterocyclic aromatic amines in grilled beef: The influence of free amino acids, nitrogenous bases, nucleosides, protein and glucose on HAAs content. *Journal of Food Composition and Analysis* 40 (June): 39–46. doi:10.1016/j.jfca.2014.12.011.

Tang, M.-S., H.-T. Wang, Y. Hu, W.-S. Chen, M. Akao, Z. Feng, and W. Hu. 2011. Acrolein induced DNA damage, mutagenicity and effect on DNA repair. *Molecular Nutrition & Food Research* 55 (9): 1291–1300. doi:10.1002/mnfr.201100148.

Tannenbaum, R., S. Wishnok, and C. Leaf. 1991. Inhibition of nitrosamine by ascorbic acid. *The American Journal of Clinical Nutrition* 53: 247S–250S.

Tfouni, S.a.V., G.R. Padovani, R.M. Reis, R.P.Z. Furlani, and M.C.R. Camargo. 2014. Incidence of polycyclic aromatic hydrocarbons in vegetable oil blends. *Food Control* 46: 539–543. doi:10.1016/j.foodcont.2014.06.028.

Thakur, M.S. and K.V. Ragavan. 2013. Biosensors in food processing. *Journal of Food Science and Technology* 50 (4): 625–641.

Turesky, R.J. and P. Vouros. 2004. Formation and analysis of heterocyclic aromatic amine–DNA adducts in vitro and in vivo. *Journal of Chromatography B, Analytical and Biological Aspects of Heterocyclic Aromatic Amines in Foods* 802 (1): 155–166. doi:10.1016/j.jchromb.2003.10.053.

Uribarri, J., S. Woodruff, S. Goodman, W. Cai, X. Chen, R. Pyzik, A. Yong, G.E. Striker, and H. Vlassara. 2010. Advanced glycation end products in foods and a practical guide to their reduction in the diet. *Journal of the American Dietetic Association* 110 (6): 911–916.e12. doi:10.1016/j.jada.2010.03.018.

van Boekel, M., V. Fogliano, N. Pellegrini, C. Stanton, G. Scholz, S. Lalljie, V. Somoza, D. Knorr, P. Rao Jasti, and G. Eisenbrand. 2010. a review on the beneficial aspects of food processing. *Molecular Nutrition and Food Research* 54 (9): 1215–1247. doi:10.1002/mnfr.200900608.

Van Putten, R.J., J.C. Van Der Waal, E. De Jong, C.B. Rasrendra, H.J. Heeres, and J.G. De Vries. 2013. Hydroxymethylfurfural, a versatile platform chemical made from renewable resources. *Chemical Reviews* 113 (3): 1499–1597. doi:10.1021/cr300182k.

Variyar, P.S., S. Chatterjee, M.G. Sajilata, R.S. Singhal, and A. Sharma. 2008. Natural existence of 2-alkylcyclobutanones. *Journal of Agricultural and Food Chemistry* 56 (24): 11817–11823. doi:10.1021/jf802391z.

Velisek, J., P. Calta, C. Crews, S. Hasnio, and M. Dolezal. 2003. 3-chloropropane-1, 2-diol in models simulating processed foods: Precursors and agents causing its decomposition. *Czech Journal of Food Sciences* 21 (5): 153–161.

Velísek, J., J. Davídek, V. Kubelka, G. Janícek, Z. Svobodová, and Z. Simicová. 1980. New chlorine-containing organic compounds in protein hydrolysates. *Journal of Agricultural and Food Chemistry* 28 (6): 1142–1144.

Vermeiren, L., F. Devlieghere, and J. Debevere. 2002. Effectiveness of some recent antimicrobial packaging concepts. *Food Additives and Contaminants* 19 Suppl: 163–171.

Villanueva, C., F. Fernandez, N. Malats, J. Grimalt, and M. Kogevinas. 2003. Meta-analysis of studies on individual consumption of chlorinated drinking water and bladder cancer. *Journal of Epidemiology and Community Health* 57 (3): 166–173. doi:10.1136/jech.57.3.166.

Vinci, R.M., M. Andjelkovic, M. Canfyn, I. Van Overmeire, and J. Van Loco. 2011. Factors influencing benzene formation from the decarboxylation of benzoate in liquid model systems. *Journal of Agricultural and Food Chemistry* 59 (24): 12975–12981. doi:10.1021/jf203233s.

Wainright, T. 1986. The Chemistry of nitrosamine formation: Relevance to malting and brewing. *Journal of the Institute of Brewing* 92 (1): 49–64. doi:10.1002/j.2050-0416.1986.tb04373.x.

Wakaizumi, M., H. Yamamoto, N. Fujimoto, and K. Ozeki. 2009. Acrylamide degradation by filamentous fungi used in food and beverage industries. *Journal of Bioscience and Bioengineering* 108 (5): 391–393. doi:10.1016/j.jbiosc.2009.05.004.

Wang, M., E.J. McIntee, G. Cheng, Y. Shi, P.W. Villalta, and S.S. Hecht. 2000. Identification of DNA adducts of acetaldehyde. *Chemical Research in Toxicology* 13 (11): 1149–1157.

Warheit, D.B., B.R. Laurence, K.L. Reed, D.H. Roach, G.a.M. Reynolds, and T.R. Webb. 2004. Comparative pulmonary toxicity assessment of single-wall carbon nanotubes in rats. *Toxicological Sciences: An Official Journal of the Society of Toxicology* 77 (1): 117–125. doi:10.1093/toxsci/kfg228.

Weber, J.V. and V.I. Sharypov. 2009. Ethyl carbamate in foods and beverages: A review. *Environmental Chemistry Letters* 7 (3): 233–247. doi:10.1007/s10311-008-0168-8.

Wegener, J.W. and P. López-Sánchez. 2010. Furan levels in fruit and vegetables juices, nutrition drinks and bakery products. *Analytica Chimica Acta* 672 (1–2): 55–60. doi:10.1016/j.aca.2010.04.042.

WHO Technical Report Series. 2015. Evaluation of certain food contaminants. Accessed August 17, 2015. http://apps.who.int/iris/bitstream/10665/43258/1/WHO_TRS_930_eng.pdf.

Wong, D., K.-W. Cheng, and M. Wang. 2012. Inhibition of heterocyclic amine formation by water-soluble vitamins in maillard reaction model systems and beef patties. *Food Chemistry* 133 (3): 760–766. doi:10.1016/j.foodchem.2012.01.089.

Yang, W., H. Zhou, and N. Cicek. 2014. Treatment of organic micropollutants in water and wastewater by uv-based processes: A literature review. *Critical Reviews in Environmental Science and Technology* 44 (13): 1443–1476. doi:10.1080/10643389.2013.790745.

Yaylayan, V.A. 2006. Precursors, formation and determination of furan in food. *Journal Für Verbraucherschutz Und Lebensmittelsicherheit* 1 (1): 5–9. doi:10.1007/s00003-006-0003-8.

Yaylayan, V.A., L. Haffenden, F.L. Chu, and A. Wnorowski. 2005. Oxidative pyrolysis and postpyrolytic derivatization techniques for the total analysis of maillard model systems: Investigation of control parameters of maillard reaction pathways. *Annals of the New York Academy of Sciences* 1043 (June): 41–54. doi:10.1196/annals.1333.005.

Yu, D.-Y., Q.-L. Zhao, M. Furuta, S. Todoriki, K. Izumi, K. Yamakage, K. Matsumoto, T. Nomura, and T. Kondo. 2012. Molecular mechanisms of apoptosis induction by 2-dodecylcyclobutanone, a radiolytic product of palmitic acid, in human lymphoma U937 cells. *Apoptosis* 17 (6): 636–645. doi:10.1007/s10495-012-0698-1.

Zelinková, Z., M. Doležal, and J. Velíšek. 2009. Occurrence of 3-chloropropane-1,2-diol fatty acid esters in infant and baby foods. *European Food Research and Technology* 228 (4): 571–578. doi:10.1007/s00217-008-0965-0.

Zelinková, Z., B. Svejkovská, J. Velísek, and M. Dolezal. 2006. Fatty acid esters of 3-chloropropane-1,2-diol in edible oils. *Food Additives and Contaminants* 23 (12): 1290–1298. doi:10.1080/02652030600887628.

Zhang, Y., Y. Ren, and Y. Zhang. 2009. New research developments on acrylamide: Analytical chemistry, formation mechanism, and mitigation recipes. *Chemical Reviews* 109 (9): 4375–4397. doi:10.1021/cr800318s.

Zhang, Y., C. Yu, J. Mei, and S. Wang. 2013. Formation and mitigation of heterocyclic aromatic amines in fried pork. *Food Additives & Contaminants: Part A* 30 (9): 1501–1507. doi:10.1080/19440049.2013.809627.

Zhao, X., G. Du, H. Zou, J. Fu, J. Zhou, and J. Chen. 2013. Progress in preventing the accumulation of ethyl carbamate in alcoholic beverages. *Trends in Food Science and Technology* 32 (2): 97–107. doi:10.1016/j.tifs.2013.05.009.

Zhao, X. and J. Karl Johnson. 2007. Simulation of adsorption of DNA on carbon nanotubes. *Journal of the American Chemical Society* 129 (34): 10438–10445. doi:10.1021/ja071844m.

Zhao, Y., F. Wang, W. Li, A. Li, and Y. Ha. 2012. 2-dodecylcyclobutanone as an irradiation dose indicator to identify and estimate original dose in irradiated ground beef. *European Food Research and Technology* 235 (5): 901–906. doi:10.1007/s00217–012–1818–4.

Zyzak, D.V., R.A. Sanders, M. Stojanovic, D.H. Tallmadge, B.L. Eberhart, D.K. Ewald, D.C. Gruber et al. 2003. Acrylamide formation mechanism in heated foods. *Journal of Agricultural and Food Chemistry* 51 (16): 4782–4787. doi:10.1021/jf034180i.

22 Targeting Cancer Stem Cells by Functional Foods and Their Constituents

Prasad R. Dandawate, Dharmalingam Subramaniam, and Shrikant Anant

CONTENTS

22.1 INTRODUCTION

Despite extensive efforts in cancer research for the identification of therapeutic target, cancer cellular pathways, and drug development for the treatment of various cancer types, cancer is still the second leading cause of death in the United States.[1] Cancer cells use numerous pathways for growth, development, metastasis, and survival under stressful conditions.[2] Multidrug resistance and recurrence of cancer after radiotherapy and surgery are the major problems in overcoming cancer.[3,4] In the present time, cancer stem cells (CSCs) and their signaling pathways are gaining importance because they provide novel approaches to develop new anticancer therapies.[5,6] Tumor is composed of heterogeneous types of cells and a small population of cells, which is supposed to possess high tumor-initiating capacity called "cancer stem cells" or "cancer-initiating cells" (CICs). CSCs/CICs are defined as a small population of cancer cells that have self-renewal and differentiation ability.[7–9] CSCs/CICs have been reported to exhibit resistance to currently available chemotherapy, radiotherapy, and certain molecular targeting therapy[10]; hence, a study of the molecular mechanisms of the maintenance or generation of CSCs/CICs might be useful for targeting CSC/CIC. Current anticancer chemotherapy is unable to eliminate CSCs; hence, novel treatment options targeting CSCs are needed.

Nature is a rich source of food material that we consume daily. In recent decade, food and their active constituents have gained importance because of their ability to target multiple pathways in cancer, CSCs, CSCs signaling pathways, and their wide safety profile.[11] Functional foods are dietary components that possess capacity to induce health-promoting effects and reduce the risk of diseases. These functional foods include food or food ingredients that are used by human being in daily diet such as turmeric, soy, broccoli, and green tea. Several studies identified the active constituents from these food materials and studied them for anticancer activities *in vitro* in animals and in human clinical trials to establish their mechanism of action. Hence, functional foods and their active constituents are important for the treatment of cancer and will serve as lead molecules for future drug development. In this chapter, we are summarizing recent reports on colon, breast, pancreatic, ovarian, and bladder cancer dealing with the presence of CSCs and their signaling pathways, CSC markers, chemoresistance to current drug therapy, and functional foods targeting CSCs and their signaling pathways.

22.1.1 Colon Cancer

Colorectal cancer is the second leading cause of morbidity in the United States among all cancers[12] affecting one in four women and one in three men worldwide.[13] American Chemical Society (ACS) estimated that 102,500 new cases (50,850 men and 51,650 women) would be diagnosed with colon cancer in 2015[1] and also estimated 49,700 deaths (26,100 men and 23,600 women) would occur due to colon cancer in 2015.[1] Colonoscopy is generally used to detect colon cancer, and resection of the early-stage polyps can prevent cancer progression. Persistent development of polyps in the colon may develop into cancer, and hence after completing 50 years of age, frequent colonoscopy is recommended in the United States.[14] Recurrence is a common problem with an estimated 40% of cases within 3–5 years of treatment. The current chemotherapeutic regimen for the treatment of colorectal cancer is 5-fluorouracil, oxaliplatin, and irinotecan or drug combinations FOLFOX or FOLFIRI. The existence of CSCs in colon cancer was established, and CD133, CD166, CD24, CD44, ALDH1, LGR5, and doublecortin and CaM kinase-like-1 (DCLK1), etc., are proposed as colon CSC markers.[15–17] Nakanishi et al. have shown that DCLK1 differentiates between tumor and normal stem cells in the intestine, while specific ablation of DCLK1+ stem cells was resulted in a marked regression of polyps without apparent damage to the normal intestine.[18,19] Recently, long-lived and quiescent intestinal DCLK1 tuft cell populations were reported to play a role in intestinal regeneration and homeostasis. These cells maintain quiescence even after oncogenic mutation, but tissue injury induces their activation and it can serve as colon cancer–initiating cells.[20]

22.1.2 Pancreatic Cancer

Pancreatic cancer is the fourth leading cause of cancer-related deaths in the United States with the highest mortality rate among all cancers. ACS estimated that new cases of 48,960 Americans (24,840 men and 24,120 women) would be diagnosed with pancreatic cancer in 2015. ACS also estimated that 40,560 Americans (20,710 men and 19,850 women) would die of pancreatic cancer in 2015.[1] It is hypothesized that pancreatic cancer will be the second leading cause of cancer deaths in the United States by 2030.[21] The major problems in the treatment of pancreatic cancer are drug resistance.[22,23] Gemcitabine (GEM) is the current treatment option for metastatic pancreatic cancer, while a combination of GEM and 5-fluorouracil and oxaliplatin (5-FU + Ox) could be used for early-stage cancer patients.[24] However, the response rate to current chemotherapy is only about 10%.[25] Recently, CSCs have been proposed to play an important role in drug resistance as well as metastasis in pancreatic cancer.[26] CD44+CD24+ESA+ (epithelial specific antigen), CD133+,[27] and ALDH1 were reported as potential pancreatic CSC markers.[28] Moreover, we identified the expression of DCLK1 in a small proportion of cells in pancreatic cancer.[29] Recently, DCLK1 found to mark a morphologically distinct cell subpopulation with stem cell characteristics in preinvasive pancreatic cancer.[30]

22.1.3 Breast Cancer

Breast cancer is commonly diagnosed in women worldwide.[31] Early detection of breast cancer helped in the reduction of the mortality rate in the United States. ACS estimated that 231,840 (29%) new breast cancer cases will be identified in the United States in 2015 with an estimated death rate of 40,730 (15%).[1] The gene expression profiling of breast cancer suggested heterogeneity with at least six distinct subtypes depending on the presence or absence of estrogen or progesterone receptor expression or the amplification/overexpression of the HER2/ERBB2 locus.[32] It is commonly distinguished into luminal A (estrogen/progesterone positive), luminal B, HER2+, and triple-negative subtypes,[33] while each subtype shows differential drug response. The current standard treatment for breast cancer includes chemotherapy (cisplatin (CIS), paclitaxel, carboplatin, bevacizumab, doxorubicin, cyclophosphamide, docetaxel, and epirubicin), surgery, and radiation.[34] The presence of breast CSCs is reported, and CD44+, CD24-/low, ESA+,[35–37] CD44+/CD24−,[38] CD133,[39] and ALDH[40] were proposed as CSC markers.

22.1.4 Ovarian Cancer

Ovarian cancer is the deadliest gynecologic cancer in the world, with an estimated 21,290 new cases and 14,180 deaths that were projected to occur in the United States in 2015.[1] The overall survival is only little changed despite the advancement in radical surgery and chemotherapy over last 30 years.[41] The majority of ovarian cancers (>90%) originated from epithelial surface.[42] Depending on histological subtypes, epithelial ovarian cancers are serous (68%–71%), clear cell (12%–13%), endometrioid (9%–11%), mucinous (3%), transitional (1%), and mixed (6%).[43] The standard care for ovarian cancer includes surgery, radiation, and chemotherapy consisting of CIS, paclitaxel, or their combinations.[44] Aldehyde dehydrogenase (ALDH) and CD133 have been reported as major CSC markers of ovarian cancer, and high expression of these markers was contributed in poor prognosis disease.[45,46] Furthermore, the peritoneal disseminated metastasis of epithelial ovarian cancer is initiated by the CD44v6-positive subpopulation, and CD44v6 expression serves as a biomarker for the clinical outcome of advanced ovarian cancer patients.[47]

22.1.5 Bladder Cancer

Bladder cancer is the fourth most abundant cancer in males and the eighth in females. ACS estimated that new cases of 74,000 Americans (56,320 men and 17,680 women) would be diagnosed with bladder cancer in 2015. ACS also estimated that 16,000 Americans (11,510 men and 4,490 women) would die of bladder cancer in 2015.[1] The current standard of therapy for bladder cancer includes surgical resection, radiation, and first-line chemotherapy using CIS as the chemotherapeutic agent (e.g., GEM plus CIS [GC] or methotrexate, vinblastine, doxorubicin and CIS [MVAC], dose-dense MVAC). Most patients will progress after a median time of about 8 months and the optimal subsequent systemic treatment remains unsettled.[48] Recent study demonstrated that ALDH1 expression plays an important role in bladder cancer development and prognosis. In addition, cells expressing ALDH1A1 along with CD44 can be potential therapeutic targets in bladder carcinomas.[49]

22.2 CSC SIGNALING PATHWAYS

22.2.1 Wnt Signaling

Dysregulation of Wnt/β-catenin signaling is observed in numerous cancers including colorectal and breast cancer.[50,51] The Wnt signaling pathway was reported to have a role in the regulation of cell proliferation, differentiation, migration, cell polarity, and asymmetric cell division.[52] In patients with familial adenomatous polyposis coli (APC), malfunction of APC led to the accumulation of β-catenin in the cytoplasm. β-catenin usually exists in bound form with membranous E-cadherin. β-catenin enters into the nucleus and triggers the cell cycle, and it precipitates the loss of membranous E-cadherin and decreases its expression.[53] This process led to reduction in cell-to-cell contact hence allowing cells to migrate from epithelium to mesenchyme. This event is called as "epithelial to mesenchymal transition" (EMT).[54] The loss of E-cadherin expression in epithelial cells leads to reduction in cell-to-cell contact, which results in cell motility and migration. EMT plays an important role in certain physiological conditions like embryogenesis and wound healing. The highest concentration of nuclear β-catenin has been reported at the advancing margin in free tumor cells that have lost E-cadherin expression especially in colorectal cancer.[54] The existence of these cells has been associated with metastasis as well as poor survival. Lower β-catenin expression has been reported in adenomas and within the central tumor. Nuclear β-catenin expression enhances survivin expression, which in turn increases cell proliferation and apoptosis resistance.[55] It was also reported that colorectal CSCs are situated at the tumor margin and attain mobility through their high expression of nuclear β-catenin and subsequent reduction of adhesion molecules.[56] These cells form a "front" that functions as a "germinal layer," which can break away to metastasize to generate new colonies of cancer cells at different regions. The mutations in the Wnt/β-catenin pathway are reported to be linked with a number of cancers[57] and associated with controlling self-renewal abilities of CSCs.[58]

22.2.2 Hedgehog Signaling

Aberrantly activated hedgehog (HH) signaling has been identified in the malignant phenotype of several types of human cancers. There is an evidence that HH signaling pathway gene expression is altered in colon neoplasia,[59,60] and also the rate of Ptch and Smo expression is gradually increased in agreement with tumor progression[60] as well as in metastasis.[61,62] Canonical HH signaling genes are expressed in primary colon cancers, metastatic disease, human colon carcinoma cell lines,[63] and tumor xenografts.[64] HH signaling is also critical in the regulation of cellular proliferation, stemness, cell fate determination, and cellular survival.[65] Oncogene-driven signaling pathways converge downstream of Smo on the Gli transcription factors providing noncanonical regulation of HH signaling pathway.[66] Such noncanonical Gli protein activation can hence circumvent the inhibition of Smo resulting in reduced efficacy of or resistance to Smo inhibitors. The Gli genes comprise a family of transcription factors that transcriptionally regulate downstream targets in HH-dependent survival. Gli2 seems to be the primary activator of HH signaling, with Gli1 as a transcriptional target of Gli2, which may amplify HH-induced target gene expression[67,68] Gli1 and Gli2 induce transcription of overlapping and distinct sets of target genes.[67,69] GANT61 was identified in a cell-based screen for small molecule inhibitors of GLI1-mediated transcription and abrogated GLI function in the nucleus, blocked both GLI1- and GLI2-mediated transcription, and inhibited GLI1–DNA binding.[70] The first phytochemical that was observed to inhibit the HH pathway was cyclopamine,[71,72] a naturally occurring compound found in the plant *Veratrum californicum*, commonly called the corn lily. Cyclopamine specifically inhibited Smo activation by targeting sonic hedgehog (Shh) pathway.[71]

22.2.3 Notch Signaling

Notch signaling plays a critical role in the differentiation and maintenance of stem cells. Abnormal activation of the notch signaling has been linked with the development of numerous cancers.[73] Notch pathway comprises the family of mammalian transmembrane receptors (Notch-1–4) for ligands (JAG1, JAG2, delta-like 1–4). Notch receptors upon binding undergo cleavage releasing a notch intracellular domain, which migrates to the nucleus, where it targets genes such as Hes-1 and Hey1,[74] cyclin D,[75] $p21^{CIP1}$,[76] Nuclear Factor-Kappa B (NF-κB),[77] and c-myc.[78] γ-secretase is a multiprotein intramembrane-cleaving protease with an emerging list of protein substrates, including the notch receptors. The γ-secretase complex has four components including presenilin, nicastrin, Pen2, and Aph1 that are all thought to be crucial for the activity.[79] The catalytic domain resides within presenilin, and nicastrin has been suggested to be critical for substrate recognition.[80,81]

22.2.4 Hippo Signaling

The Hippo signaling is crucial in regulating tissue homeostasis, organ size, and tumorigenesis. There are two groups of central kinases, Mst1/2 and Lats1/2, which are modulated by Sav and Mob. When the pathway is active, phosphorylation of the Yes-associated protein 1 (YAP1) or transcriptional coactivator with PDZ-binding motif (TAZ) transcriptional coactivators by Lats1/2 results in cytoplasmic sequestration and degradation, while pathway inactivation allows unphosphorylated YAP/TAZ to enter the nucleus and along with one of four TEA domain (TEAD) family members activate transcription. The porphyrin molecule verteporfin disrupts the formation of the YAP1–TEAD complex by binding to YAP1 and changing its conformation.[82] Dysregulation of the pathway is implicated in many types of cancers; for instance, YAP1/TAZ and TEAD are upregulated in numerous human tumors by mechanisms including gene amplification and silencing of upstream components of the pathway. FAT4 is mutated in 5% of patients with gastric cancer, while SAV1 gene is mutated in colon and renal cancers.[83] Furthermore, reduced MST1/2 and LATS1 expression was found in gastric cancer, and tumors with lymph node metastasis showed minimal levels of LATS1.[84] Moreover, YAP1 is overexpressed at high frequency in several human cancers and can drive cancer development in mouse models.[85,86]

Studies showed that the increased nuclear activity of YAP and TAZ could drive cell transformation and tumorigenesis.[87] By using gene expression profiling technique, it was found that high CSC content in breast cancer tissues overlaps with YAP/TAZ-induced gene expression, suggesting the importance of YAP/TAZ in CSCs.[88] Such oncogenic activity is correlated to an acquisition of mesenchymal properties, like EMT in epithelial tissues. The combined abilities of self-renewal and invasive properties are suggested as the driving force to develop aggressive cancers.[89,90] Furthermore, knockdown of TAZ or YAP alone inhibits the self-renewal ability of CSCs in breast cancer[88] and overexpression of TAZ promotes mammosphere formation and CSC marker expression in non-CSC cancer cell populations.[88] TAZ is also the leading factor to sustain CSC-like properties, and the expression of a nuclear localized TAZ-S89A mutant is sufficient to confer the self-renewal process of breast cancer populations.[88]

22.2.5 Other Stem Cell–Related Signaling Pathways

22.2.5.1 JAK–STAT Signaling

The Janus kinase (JAK) and signal transducer and activator of transcription (STAT) pathway exhibits a crucial role in the signaling of cytokines and growth factors affecting numerous cellular activities such as proliferation, growth, and immune response. The dysregulation or mutations in

JAK–STAT pathways have been found in cancer.[91] JAK–STAT pathway plays an important role in self-renewal and germ line stem cell population maintenance.[92] Hernandez-Vargas and group have demonstrated that sorting of the cells with a putative CSC phenotype (CD44+/CD24 low) showed a constitutive JAK–STAT pathway activation. The JAK–STAT pathway activation may represent one of the characteristics of putative breast CSCs.[93] In our group, we have shown that prolactin treatment to colon cancer cells induced the phosphorylation of JAK2, STAT3, and ERK1/2 as well as enhanced the expression of a Notch-1 receptor ligand, Jagged 1. Notch signaling regulates CRC stem cell population and also stem cell marker genes DCLK1, LGR5, ALDH1, and CD44.[94]

22.2.5.2 PI3K/Akt/mTOR Signaling

The phosphatidylinositol-3-kinase (PI3K)/Akt and the mammalian target of rapamycin (mTOR) signaling pathways are crucial to many physiological and pathological conditions, such as cell proliferation, angiogenesis, metabolism, differentiation, and survival.[95] This signaling pathway can be considered as a master regulator for cancer. It plays a central role in growth, proliferation, motility, survival, and angiogenesis in tumor cells.[96,97] mTOR pathway activation in breast cancer stem cells (BCSCs) is essential for colony-formation capacity *in vitro* and for exhibiting tumorigenic property *in vivo*.[98] A recent study showed that mTOR suppression could decrease ALDH1 activity in colorectal cancer cells.[99]

22.2.5.3 MAPK–ERK Signaling

The activation of mitogen-activated protein kinase (MAPK) pathways led to several cellular responses such as inflammation, growth, differentiation, and cell death. These serine/threonine protein kinases are phosphorylated and activated by MAPK-kinases (MAPKK), which are phosphorylated by MAPKK-kinases, and are in turn activated by interaction with the small GTPase family proteins ultimately connecting the MAPK component to cell surface receptors. The MAPK signaling activation has appeared to play a role in inducing resistance to transforming growth factor-β (TGF-β)-induced apoptosis in CD133+ cells as compared to CD133− cells,[99] whereas enhanced MAPK signaling especially in CD133+ CSCs resulted in the activation of ERK1/2. Chai and group recently demonstrated that metformin enhanced the sensitivity to GEM in pancreatic cancer cells in part due to decreased proliferation of CD133(+) cells and suppressed P70S6K signaling activation via inhibition of phosphorylation of ERK.[100] Interestingly, MAPK-2, another member of the MAPK/ERK pathway, was highly expressed in the CD133-positive prostate progenitor cells.[100]

22.3 FUNCTIONAL FOODS AFFECTING CSCs AND THEIR SIGNALING PATHWAYS

Functional foods and their active constituents are extensively studied for various biological activities because of its common use as food material and as a wide range of safety and therapeutic activity against various diseases and disorders. In cancer, functional foods and their constituents are reported to exert beneficial effects against multiple cancers by targeting multiple cellular pathways and CSCs. Hence, they are proposed to be helpful in elimination of cancer and to control metastasis and recurrence. Here, we are summarizing important constituents found in functional foods. The chemical structures of active ingredients of functional foods are represented in Figure 22.1.

22.3.1 Turmeric

Curcumin is an extensively studied polyphenolic compound derived from turmeric (*Curcuma longa*). It is reported for targeting multiple cellular pathways in cancer and studied for the treatment of a wide range of cancers.[11,101–103] Turmeric is a commonly used ingredient of many food recipes for giving color and flavor. Apart from giving aesthetic appeal to food, turmeric and its

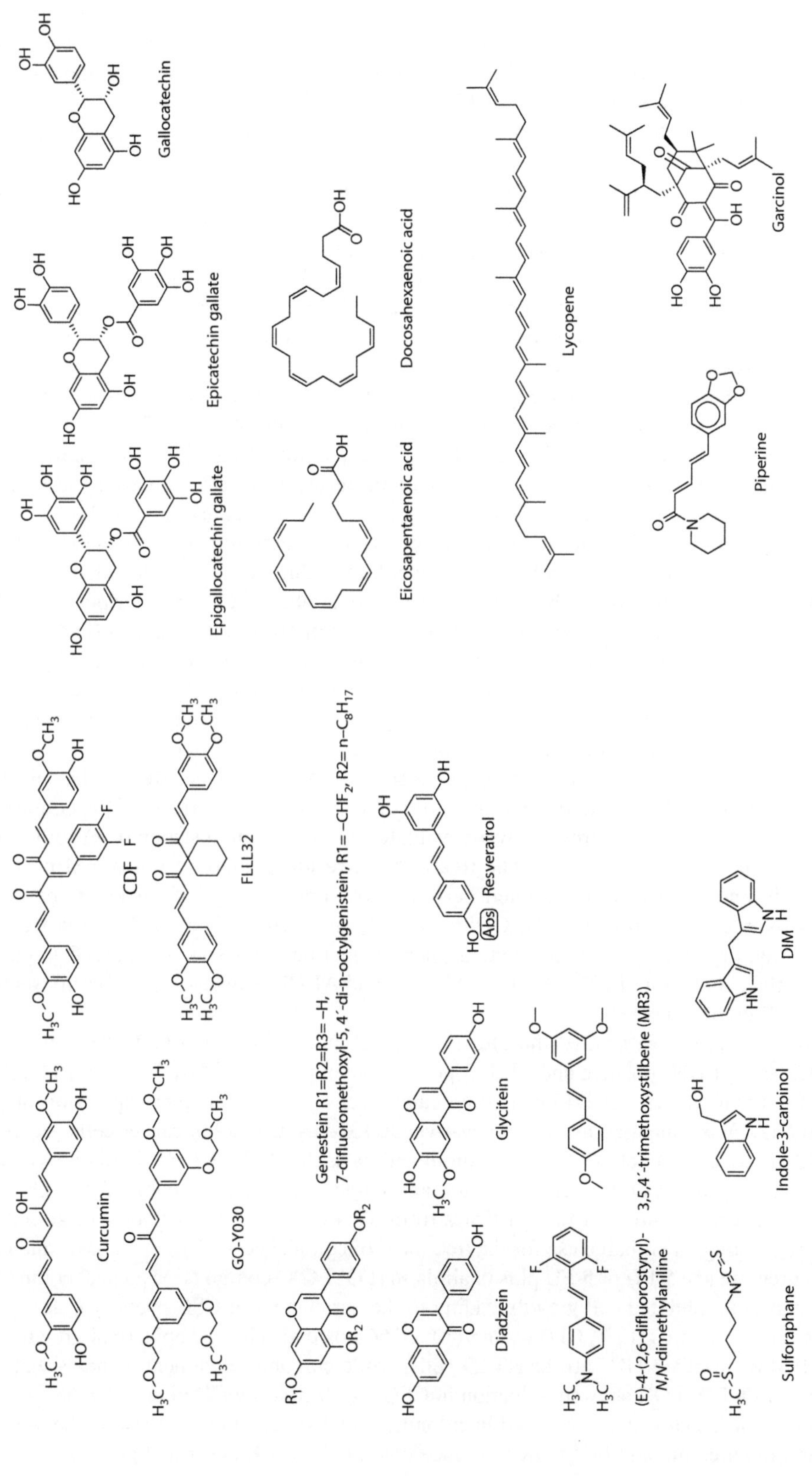

FIGURE 22.1 Chemical structures of active constituents of functional foods.

active constituent curcumin exert several biological effects against disease and disorders including antioxidant,[104] anti-inflammatory,[105] anticancer,[106] antidiabetic,[107] antibacterial,[108] and antiviral[109] activities. Hence, it is one of the most important functional foods in the present time. The anticancer ability of curcumin is widely established, and it is reported to inhibit CSCs and their signaling pathways including notch, Wnt, HH, and STAT3.

Curcumin was found to inhibit HH pathway in many cancers including pancreatic, colon, and breast cancer. We have previously shown that curcumin inhibits notch signaling in esophageal cancer.[106] We found that curcumin exerted dose- and time-dependent inhibition of cell growth and colony formation in esophageal cancer cells by inducing apoptosis through the activation of caspase-3 and increase in Bax to Bcl-2 ratio. It also caused cell cycle arrest by downregulating cyclin D1 levels. Curcumin treatment also decreased the number and sizes of esophagospheres. Further, we have shown that curcumin decreased Notch-1 activation, Jagged-1 and Hes-1 expression, which was mediated via downregulation of presenilin-1 and nicastrin. Curcumin also reduced the expression of miR-21 and miR-34a and enhanced the expression of tumor suppressor let-7a miRNA.[106] Shakibaei et al. have evaluated the effects of 5-FU and curcumin on the status of DNA mismatch repair (MMR) and CSC activity in 3D cultures of HCT116, HCT116+ch3 (complemented with chromosome 3), and their corresponding isogenic 5-FU-chemoresistant derivative clones (HCT116R, HCT116+ch3R). Curcumin pretreatment increased the effectiveness of 5-FU against HCT116R and HCR116+ch3R cells as evidenced by increased colonosphere disintegration, apoptosis induction, and reduced cell growth. Curcumin and/or 5-FU also affected MMR-deficient colon cancer cells in high-density cultures. Curcumin-mediated chemosensitization to 5-FU was suggested by its ability to decrease the number of CSC marker positive cells.[110] Lin et al. have studied the role of STAT3 in colon cancer–initiating cells and the effect of STAT3 inhibitor FLLL32 (curcumin analog) against colon cancer stem cells characterized by ALDH-positive and CD133-positive subpopulation (ALDH(+)/CD133(+)). FLLL32 inhibited the phosphorylation of STAT3 along with its downstream target genes involved in colon cancer stem cells including Bcl-XL, survivin, and Notch-1, -3, and -4. FLLL32 further reduced cell viability and spheroid formation and induced the cleavage of caspase-3 in colon cancer stem cells, suggesting the therapeutic role of FLLL32 in the treatment of colon cancer.[111] The effects of the combination of Src inhibitor dasatinib and curcumin were evaluated against chemoresistant (CR) colon cancer cells enriched in CSC subpopulation. The treatment of residual tumors from ApcMin+/– mice with dasatinib and/or curcumin exhibited between 80% and 90% reduction in the expression of the CSC markers including ALDH, CD133, CD44, and CD166. The combination treatment reduced cellular growth, invasion, and colonosphere formation. It also decreased CSC population by reducing the expression of CD133, CD44, CD166, and ALDH, suggesting the future treatment options for CR colon cancer.[112]

Lin and coworkers have studied the effects of novel curcumin analog GO-Y030 and STAT3 on colon CSCs ([ALDH]-positive and CD133-positive subpopulations [ALDH(+)/CD133(+]). The results showed that ALDH(+)/CD133(+) colon cancer cells possessed higher expression of phosphorylated STAT3 as compared to ALDH-negative/CD133-negative colon cancer cells, suggesting that STAT3 is activated in colon CSCs. Curcumin and its analog GO-Y030 inhibited STAT3 phosphorylation, cell viability, and sphere formation and induced apoptosis in colon CSCs. GO-Y030 was found to reduce the tumor growth of CSCs from SW480 and HCT116 colon cancer cell lines in mouse, suggesting its probable use for the treatment of colon cancer.[113] Yu et al. have studied the effects of curcumin and 5-FU or 5-FU plus oxaliplatin (FOLFOX) combination against colon CSCs. FOLFOX exposure inhibited cell growth leading to the enrichment of CSC phenotype as assessed by increased amount of CD133-, CD44-, and/or CD166-positive cells and epidermal growth factor receptor (EGFR) levels in HCT116 or HT-29 cells, while curcumin treatment alone as well as in combination with FOLFOX showed reduction in CSCs by decreasing CD44 and CD166 as well as EGFR expression. Treatment also resulted in colonosphere disintegration, suggesting the use of the combination of curcumin and FOLFOX to reduce colon CSCs.[114] Kanwar and group have studied the combination of difluorinated curcumin analog (CDF) with 5-FU + Ox against colon CSCs. CDF in combination with 5-FU + Ox exerted potent activity than curcumin in decreasing CD44 and CD166 in CR colon cancer cells. This combination also reduced growth, induced apoptosis, and enhanced disintegration of colonospheres, mediated in part through ABCG2 downregulation, inhibition of EGFR, insulin-like growth factor 1 receptor (IGF-1R), and NF-κB signaling and suppression of β-catenin, cyclooxygenase-2 (COX-2), and c-Myc. In addition, there occurs inactivation of Bcl-xL and an induction of Bax proteins. This study therefore suggested that CDF could act as a chemotherapeutic agent for colon cancer by eliminating CSCs.[115]

The effects of curcumin (10, 20, and 30 μMol/mL) against TGF-β1-induced formation of EMT were also studied in pancreatic cancer cells. Curcumin treatment reduced TGF-β1-induced cell proliferation, cell invasion, expression of Shh, GLI1 and vimentin, and increased expression of E-cadherin, while the treatment also induced apoptosis in Panc-1 cells.[116] Wang and colleagues have shown that curcumin treatment reduced the cell growth and induced apoptosis in pancreatic cancer cells by downregulating the expression of Notch-1, Hes-1, and Bcl-XL proteins. Furthermore, downregulation of Notch-1 by small-interfering RNA before curcumin treatment led to increased cell growth inhibition and apoptosis.[117] Bao and coworkers have shown that CDF decreased the sphere-forming ability of pancreatic cancer cells and increased the disintegration of pancospheres. These effects were found to be associated with reduction in CSC markers such as CD44 and EpCAM, especially in GEM-resistant MIAPaCa-2 cells that consist of high number of CSCs having increased miR-21 and decreased miR-200. CDF treatment resulted in the reduction of tumor growth along with a decrease in NF-κB DNA-binding activity, expression of COX-2 as well as miR-21, and enhanced expression of phosphatase and tensin homolog (PTEN) and miR-200 in tumors, suggesting that the anticancer potential of CDF against pancreatic cancer is mediated by targeting the CSC pathway.[118] Further studies have shown that CDF treatment reduced cell survival, clonogenicity, pancospheres formation, invasive cell migration, and CSC function in pancreatic cancer cells. These effects were found to be mediated by the reduced expression of EZH2 and increased expression of tumor suppressor microRNA like let-7a, b, c, d, miR-26a, miR-101, miR-146a, and miR-200b, c. CDF treatment also inhibited the growth of pancreatic cancer in an orthotopic xenograft model by reducing the expression of EZH2, Notch-1, CD44, EpCAM, and Nanog and increased the expression of let-7, miR-26a, and miR-101.[119] Li et al. have found that CDF treatment decreased the signal transduction in the AR/TMPRSS2-ERG/Wnt signaling network that led to the inactivation of Wnt signaling pathway, suggesting the use of novel curcumin analog for the treatment of pancreatic cancer.[120]

Mohapatra and colleagues have shown that the combination of curcumin and resveratrol exerted synergistic anticancer activity *in vitro* (cigarette smoke condensate mediated transformed breast epithelial cell, MCF-10A-Tr) and *in vivo* (tumor xenograft mice) model system. Resveratrol treatment led to enhanced uptake of curcumin in cells and induced apoptosis in MCF-10A-Tr by increasing the ratio of Bax to Bcl-xL, releasing of cytochrome C, and cleaving of poly-ADP-ribose polymerase (PARP) and caspase-3 in cells. The combination reduced HH–Gli pathway intermediates including Shh, Gli-1, Smo, c-Myc, and cyclin D1 and enhanced $p21^{Waf/Cip1}$ levels *in vitro* and *in vivo*. The combined treatment also resulted in pronounced reduction in Gli-promoter activity, while the unchanged expression of proteins mentioned earlier and Gli-promoter activity in $p21^{Waf/Cip1}$ knockout cells suggested that the anticancer effects of the combination were mediated through $p21^{Waf/Cip1}$-mediated inhibition of HH–Gli pathway.[121] The anti-CSC effects of the combined treatment of curcumin and epigallocatechin gallate (EGCG) were studied in MDA-MB-231 cells and MCF-7 cells transfected with HER in tumor-sphere formation and wound-healing assays. The combined treatment decreased CD44-positive cells and also inhibited STAT3 phosphorylation, while STAT3-NF-κB interaction was retained.[122] Mukherjee and group have found that curcumin inhibits the translocation of β-catenin and hence impeding Slug transactivation. It consequently restored the expression of E-cadherin leading to increased formation of E-cadherin/β-catenin complex and cytosolic retention of β-catenin to inhibit EMT and breast CSCs' migration.[123]

22.3.2 Soy Isoflavones

Isoflavones are the compounds that are present in foods such as soy, beans, and chickpeas. These foods contain significant major amounts of isoflavones such as genistein, daidzein, and glycitein. Numerous studies revealed that soy-containing foods exerted protection against prostate cancer.[124] Soy milk is a commercially available food product that was found to reduce the prostate cancer risk.[125] Soy-containing foods are commonly consumed in various food recipes, especially in salads. Soy isoflavones are reported to inhibit the growth of several cancers[126–128] and CSCs.[129,130]

Iftikhar and Rashid have performed a comparative molecular docking study to predict the binding mode of compounds to β-catenin. Known inhibitors including genistein, isorhamnetin, fisetin, and silibinin exhibited binding energy between −5.68 and −4.98 kcal/mol, whereas newly studied compounds such as catechin, luteolin, coumestrol, and β-naphthoflavone possessed binding energies of −6.50 to −5.22 kcal/mol. These flavonoids and T-cell factors 4 (TCF4) shared common binding residues and therefore compete for binding with β-catenin, providing a model to study novel compounds as β-catenin inhibitors.[131] However, while they compete for β-catenin binding to TCF4, the possibility exists that the compounds may not affect other protein–protein interactions involving β-catenin. Nevertheless, genistein was found to control Wnt genes response during carcinogen response mediated by the methylation of DNA, modifications of histone, and their interactions with the gene's regulatory region in colon carcinogenesis.[132] Hirata et al. have studied the effects of genistein on Wnt signaling cascade in renal cell carcinoma (RCC). Genistein induces apoptosis and decreases proliferation and invasion of RCC cells, mediated by the downregulation of miR-1260b expression.[133]

Genistein treatment has been shown to reduce cell growth and proliferation in colon cancer cell lines (SW480 and HCT15 cells) by increasing the expression of DKK1 via induction of histone acetylation at the DKK1 promoter region.[134] Zhang and group have also studied the effects of genistein on Wnt/β-catenin signaling cascade in azoxymethane (AOM)-induced colon cancer in rats. Genistein treatment reduced AOM-induced aberrant nuclear accumulation of β-catenin, reduced the number of aberrant crypt foci (ACF), and suppressed the expression of Wnt target genes c-Myc and cyclin D1 as well as Wnt signaling genes Wnt5a, Sfrp1, Sfrp2, and Sfrp5.[135] Genistein treatment (75 μM/L, 4 days) decreased cell viability and increased apoptosis in colon cancer cells. The treatment led to reduction in nuclear β-catenin and increase in phospho-β-catenin accumulation along with the increase in Sfrp2 (antagonist of Wnt pathway) mRNA expression. The study suggested that genistein-induced inhibition of β-catenin-mediated Wnt signaling has been mediated via increasing the expression of Sfrp2 gene by demethylating its silenced promoter in DLD-1 cells.[136] Yu et al. have collected CD44 (positive) expressing cells, which showed the properties like CSCs and found to form sphere colonies. Genistein treatment to these cells led to reduction in the expression of Gli1 and CD44 mRNA and blocked the high cell migration capacity in gastric cancer cells.[130] Genistein treatment led to growth inhibition and apoptosis induction in pancreatic cancer cells. The treatment also resulted in the upregulation of miR-34a and the downregulation of Notch-1 expression.[137]

Bao and group have shown that genistein treatment attenuated the forkhead box M1 (FoxM1) overexpression-induced EMT and CSC phenotype in pancreatic cancer cells. Genistein also reduced cell growth, migration, invasion, clonogenicity, pancospheres formation, and EMT phenotype and decreased the expression of CD44 and EpCAM.[138] Genistein treatment resulted in the inhibition of cell growth, apoptosis induction, G2/M phase cell cycle arrest, and NF-κB activity via Notch-1 signaling cascade in MDA-MB-231 cells. Genistein treatment has also downregulated the expression of cyclin B1, Bcl-2, and Bcl-xL in MDA-MB-231 cells.[139] Fan et al. have evaluated the mechanisms behind the inhibitory effects of genistein on BCSCs in MCF-7 breast cancer cell line and their implants in nude mice. Genistein decreased cell growth and induced apoptosis of MCF-7 cells as well as decreased breast CSCs by the downregulation of HH–Gli1 signaling cascade.[140] Postweaning treatment with genistein reduced the mammary adiposity and enhanced the expression of mammary tumor suppressor PTEN and E-cadherin in female mice as compared to control mice on a casein diet. Further, genistein showed a decrease in mature adipocyte numbers and an accumulation of triglyceride as well as transcription levels of PPAR-γ and fatty acid synthase in SV40-immortalized mouse mammary stromal fibroblast-like cells during differentiation into adipocytes. The treatment also increased the gene expression of estrogen receptor-β (ER-β (Esr2)). The anchorage-independent sphere formation was diminished in conditioned medium from genistein-treated adipocytes.[129] Ning et al. have studied the ability of 7-difluoromethoxyl-5, 4′-di-*n*-octylgenistein (DFOG) to inhibit the self-renewal capacity of ovarian cancer stem cells (OCSLCs). It was found that DFOG reduced proliferation, self-renewal ability, and expression of stem cell markers CD133, CD44, and ALDH1 in the spheroid forming cells in the SKOV3 cells. Furthermore, DFOG treatment downregulated the expression of transcription factor FoxM1.[141,142]

22.3.3 Grapes

Grapes and berries are popular fruits in all parts of the world and were found to exert health-improving effects. Grapes and berries are popular ingredients of various salads and fruit recipes. It is consumed as such as well as in the form of juice. Resveratrol is the major compound present in the skin of grapes and berries, which is responsible for beneficial effects of these fruits.[143] Resveratrol is widely explored for its biological activity including antioxidant,[144] anti-inflammatory,[145] antidiabetic,[146] and anticancer[147–152] activities.

Liu et al. have studied the growth inhibitory and apoptosis-inducing effects of resveratrol in HCT116 colon cancer cells. Resveratrol was found to upregulate PTEN expression and reduce Akt1/2 phosphorylation. It also exerted dose-dependent decrease in protein as well as mRNA expression of β-catenin, suggesting that resveratrol exerts antiproliferative effects through the regulation of PTEN/PI3K/Akt and Wnt/β-catenin signaling cascade.[153] Ji and group have used MALAT1 lentiviral shRNA and overexpression constructs in LoVo and HCT116 cells and demonstrated that the anticancer effects of resveratrol were mediated by the inhibition of Wnt/β-catenin signaling cascade and c-Myc, MMP-7, and MALAT1 expression. Resveratrol-induced downregulation of MALAT1 resulted in the reduction of nuclear localization of β-catenin and resulted in the inhibition of Wnt/β-catenin signaling leading to CRC invasion and metastasis inhibition.[154] Zhang et al. have synthesized and tested a series of fluorinated *N,N*-dialkylaminostilbenes and found that (E)-4-(2,6-difluorostyryl)-*N,N*-dimethylaniline inhibits Wnt signaling at nanomolar concentration. It also reduced the growth of human colon cancer cell xenografts in athymic nude mice at a dose of 20 mg/kg.[155] Gao et al. have studied the HH signaling inhibitory effects of resveratrol and EMT in gastric cancer SGC-7901 cells *in vitro*. Resveratrol treatment showed reduction in the expression of Gli-1, Snail, and N-cadherin as well as an increase in the expression of E-cadherin, suggesting the inhibition of HH pathway and EMT. Resveratrol also exerted dose-dependent decrease in cell viability, cell invasion, and metastasis.[156]

Resveratrol treatment also dose dependently inhibited cell proliferation and colony formation of MiaPaCa-2 cells. Moreover, there were reduced levels of Ihh, Ptch, and Smo, suggesting the involvement of HH signaling pathway in the mechanism of action of resveratrol.[157] Mo et al. have demonstrated that resveratrol inhibited cell growth and G0/G1 cell cycle arrest and induced apoptosis in pancreatic cancer cells. Resveratrol treatment reduced the expression of Gli1, Ptc1, and Smo, while similar results were obtained in Gli1 knockdown Panc-1 cells or Gli1-overexpressed BxPC-3 cells, suggesting that resveratrol may not act on the Gli1 promoter directly.[158] Shankar and colleagues have examined the inhibitory effects of resveratrol on stem cell characteristics of pancreatic CSCs (CD133(+)CD44(+)CD24(+)ESA(+)) derived from human primary tumors and Kras (G12D) transgenic mice. Resveratrol treatment also reduced pancreatic cancer growth and development of PanIN lesions in Kras (G12D) mice. In addition, there was a decreased self-renewal capacity of pancreatic CSCs and increased apoptosis by activating caspase 3/7 system as well as reducing Bcl-2 and XIAP expression in CSCs. Moreover, resveratrol treatment inhibited CSCs, migration and invasion as well as reduced the expression of Nanog, Sox-2, c-Myc, Oct-4, and ABCG2 and EMT markers (Zeb-1, Slug, and Snail) in CSCs.[159]

Resveratrol treatment inhibited the proliferation and percentage of BCSCs isolated from MCF-7 and SUM159 cells. Resveratrol treatment decreased the size and number of mammospheres. Resveratrol injection (100 mg/kg/day) reduced tumor xenograft growth in non-obese diabetic/severe combined immunodeficiency (NOD/SCID) mice as well as tumor BCSC population within the tumor. Resveratrol was found to induce autophagy in BCSCs as assessed by transmission electron microscopy, GFP-LC3-II puncta-formation assay, and Western blot of LC3-II, Beclin-1, and Atg-7, while it also suppressed Wnt/β-catenin signaling cascade indicating that resveratrol reduced the percentage of BCSCs in tumor by inducing autophagy and inhibiting Wnt/β-catenin signaling cascade.[160] Tsai et al. have found that 3,5,4′-trimethoxystilbene (MR-3) enhanced the expression of E-cadherin and triggered a cobblestone-like morphology of MCF-7 cells. It also decreased the mesenchymal markers' expression, including snail, slug, and vimentin. MR-3 treatment reduced the invasion and migration of MCF-7 as well as decreased the expression and nuclear translocation of β-catenin. MR-3 restored the activity of glycogen synthase kinase-3β by reducing Akt phosphorylation.[161]

Seino et al. have studied the effects of resveratrol on A2780 human ovarian cancer cells in terms of the viability and self-renewal capacity of CSCs. Resveratrol treatment induced death in ovarian CSCs and exerted reactive oxygen species–dependent inhibition of self-renewal capacity of ovarian CSCs.[162] Zhong et al. have shown the inhibitory effect of resveratrol on two human ovarian cancer *cell* lines, OVCAR-3 and CAOV-3, mediated by the accumulation of G1-phase, apoptosis induction, and downregulation of Wnt, notch, and STAT3 signaling cascade.[163]

22.3.4 Broccoli (Cruciferous Vegetables)

Cruciferous vegetables contain another important phytochemical sulforaphane that contains isothiocyanate group. It is present in vegetables like broccoli, brussels, sprouts, or cabbages, which are consumed as salads as well as in cooked form. Sulforaphane is found to exert cytotoxic activity against several cancers.[164–168]

Khor et al. have studied the effects of sulforaphane on gene expression profile on polyps of small intestine in ApcMin/+ mice by Affymetrix microarray technique. Sulforaphane treatment induced phase II drug–metabolizing enzymes and genes involved in apoptosis, growth, and maintenance. Sulforaphane treatment has also upregulated proapoptotic genes including MBD4, TNFR-7, and TNF (ligand)-11 and downregulated those of prosurvival genes cyclin-D2, integrin-β1, and Wnt-9A. The treatment also decreased COX-2 and increased 15-LOX genes, which are involved in colon carcinogenesis.[169] Hu and group have studied the chemopreventive potential of diet supplemented with sulforaphane (300 and 600 ppm, 3 weeks) in ApcMin/+ mice. The results showed that sulforaphane-supplemented ApcMin/+ mice developed significantly smaller and less number of polyps as well as induced apoptosis and decreased proliferation in small intestine. Immunohistochemical staining showed reduced expression of p-JNK, p-ERK, and p-Akt in adenomas after sulforaphane treatment in ApcMin/+ mice. Sulforaphane treatment did not affect biomarkers of the Wnt signaling pathway, β-catenin and cyclin-D1.[170]

Kallifatidis has shown that CIS, GEM, doxorubicin, 5-fluorouracil, or sulforaphane induced apoptosis and reduced viability; however, combination of a drug and sulforaphane further enhanced the toxicity in pancreatic CSCs. Combination treatments targeted CSCs as indicated by reduction in clonogenicity, spheroid formation, and ALDH1 activity along with a decrease in the expression of Notch-1 and c-Rel. Furthermore, combination treatment was more effective in abolishing potential of tumor initiation and growth of CSC xenografts without causing any side effects in normal cells.[171] Rodova has studied molecular mechanisms of sulforaphane on self-renewing capacity of pancreatic CSCs. Sulforaphane was found to induce apoptosis by downregulating Bcl-2 and caspase activation. The treatment also reduced CSC-derived spheres suggesting its anticlonogenic capacity. Further, it was found to inhibit sonic hedgehog (Shh) signaling pathway and transcriptional activity of Gli by reducing the expression of pluripotency-maintaining factors including Nanog and Oct-4 as well as PDGFRα and Cyclin D1.[172] Li et al. have shown that sulforaphane treatment reduced tumor growth in orthotopically injected pancreatic CSCs in the pancreas of NOD/SCID/IL2Rγ mice, and these effects were mediated through the alteration of Shh–GLI signaling pathway. Sulforaphane treatment (20 mg/kg) exerted 45% reduction in tumor growth of pancreatic cancer and decreased the expression of Smo, Gli 1, and Gli 2 in mouse tissues as well as pluripotency-maintaining transcription factors (Nanog and Oct-4) and angiogenesis markers VEGF and PDGFRα. Sulforaphane treatment also reduced EMT markers such as Zeb-1, enhanced E-cadherin expression, and induced apoptosis by downregulating Bcl-2 and XIAP expression.[173]

Li et al. have found that sulforaphane can target basal-like ductal carcinoma *in situ* (DCIS) stem-like cells, decrease the expression of ALDH1, and reduce mammospheres and colony formation. The miR-140, miR-29a, and miR-21 were found differentially expressed in exosomes from DCIS stem cells, while sulforaphane treatment reprograms DCIS stem cells as suggested by significant changes in exosomal secretion resembling to nonstem cancer cells.[174] Li et al. have shown that significant reduced expression of miR-140 in cancer stem cells and SOX9 and ALDH1 (stem cell factors in DCIS stem cells) are direct targets of miR140. In a model of ER-α-negative/basal-like DCIS, sulforaphane treatment decreased SOX9 and ALDH1 expression and reduced tumor growth *in vivo*.[175] Li et al. have evaluated the potential of sulforaphane (1–5 μM/L) to inhibit breast CSCs and found 65%–80% decrease in ALDH-positive cell population in human breast cancer cells. The treatment also decreased the size (8- to 125-fold) and number of mammospheres (45%–75%). Sulforaphane treatment (50 mg/kg, 2 weeks) has shown reduction in ALDH-positive cells in nonobese diabetic/severe combined immune-deficient xenograft tumors. It eliminated breast CSCs *in vivo* and hence reduced tumor growth after reimplantation of primary tumor cells into the secondary mice. The effects were mediated via the downregulation of Wnt/β-catenin pathway.[176] Li and group have examined the effects of sulforaphane on the mesenchymal stem cell (MSC)-differentiated adipocytes on breast cancer cell migration and found that sulforaphane promoted the self-renewal of MSC and inhibited the differentiation of adipocytes. The sulforaphane-treated adipocytes exhibited hindered cytokine communication with breast cancer cells resulting in reduced migration and tumor formation in breast cancer cells.[177] Forster et al. have shown that sulforaphane increased the levels of Cx43 and E-cadherin, reduced the CSC markers c-Met and CD133, improved the functional morphology of gap junctions, and enhanced gap junctional intercellular communication.[178]

Indole-3-carbinol (I3C) is converted to its dimeric product 3,3′-diindolylmethane (DIM) from glucosinolates, which is present in cruciferous plants including broccoli.[179] DIM is a major derivative of condensation product of I3C under acidic conditions of the stomach.[180] Epidemiological studies have shown that consumption of cruciferous vegetables containing these indoles could reduce the risk of cancer. Leem et al. have studied the effects of DIM treatment on gene expression in colon cancer by using genome-wide transcriptome analysis. The results suggested that DIM treatment upregulated 692 genes while downregulating 731 genes. DIM treatment downregulated β-catenin, Myc, and FOS genes, while it also suppressed c-Myc, β-catenin, and cyclin D1 expression in DLD-1 and HCT116 cells.[181] Semov and group have shown DIM as a selective and potent CSC inhibitor. DIM reduced the sphere formation and overcome CSCs' chemoresistance to paclitaxel, doxorubicin, and SN-38. DIM pretreatment of tumor spheres before implantation in mice reduced the growth of primary tumors as compared to control.[182]

22.3.5 Green Tea Polyphenols

Tea is a very popular drink and consumed worldwide. In recent years, green tea and their components have gained tremendous attention because of their antioxidant,[183] anti-inflammatory,[184] antidiabetic,[185] and anticancer[186–191] activities. Green tea contains numerous polyphenols such as epicatechin, epigallocatechin, epicatechin-3-gallate, and EGCG, which are responsible for the biological activity of green tea.

Umeda et al. have shown that EGCG-induced inhibition of cell growth and reduction in myosin regulatory light chain phosphorylation (at Thr-18/Ser-19) were mediated by 67kDa laminin receptor (67LR). It was found that Caco-2 cells exhibited higher 67LR expression and EGCG (1 μM) accumulated cells in G2/M phase. EGCG treatment also reduced the phosphorylation of myosin phosphatase targeting subunit 1 at Thr-696.[192] EGCG treatment to HT29 cells resulted in the inhibition of GSK3-α and GSK3-β as well as reduced β-catenin phosphorylation.[193] Bose et al. have studied the effects of (–)-EGCG and fish oil combination on intestinal carcinogenesis in ApcMin/+ mice fed a high-fat diet. The combination treatment (9 weeks) exerted better antitumor activity with reduction in tumor number by 53% as compared to controls and compounds alone. The antitumor effects were mediated by apoptosis induction, modulation of Wnt signaling, reduced levels of prostaglandin E2 (PGE2), and decreased Akt phosphorylation as compared to controls, suggesting that low doses of EGCG and in combination with fish oil can reduce tumor multiplicity in ApcMin/+ mice.[194]

The treatment with EGCG led to induction of HMG-box transcription factor 1 (HBP1) transcriptional repressor levels via increase in HBP1 mRNA stability, while HBP1 knockdown lines resulted in reduced sensitivity to EGCG-induced reduction in Wnt signaling and c-MYC. Kim et al. have shown that EGCG treatment decreased breast cancer cell proliferation and invasiveness in HBP1-dependent manner.[195] The effects of EGCG on SUM-149 and SUM-190 inflammatory breast cancer (IBC) cell lines were evaluated. EGCG treatment reduced the growth, invasive properties, and survival of IBC cells and also decreased lymphangiogenesis-promoting genes, including VEGF-D. EGCG inhibited the sphere formation in SUM-149 cells, while the treatment reduced growth and induced apoptosis in the stem SUM-149 cells. EGCG treatment reduced the growth of preexisting tumors derived from ALDH-positive stem SUM-149 cells and their VEGF-D expression in an orthotopic mouse model and hence inhibiting overall aggressiveness of IBC.[196]

Tang et al. have demonstrated that EGCG reduced the expression of Nanog, c-Myc, and Oct-4 and inhibited pancreatic CSCs' self-renewal capacity, while shRNA-induced inhibition of Nanog increased the EGCG effects on self-renewal capacity of pancreatic CSCs. EGCG treatment reduced proliferation and induced apoptosis by reducing Bcl-2 and XIAP expression and caspase-3 activation. The sonic hedgehog (Shh) pathway and its components including smoothened, patched, Gli1 and Gli2 as well as transcriptional activity of Gli was also inhibited by EGCG treatment. EGCG also inhibited EMT by reducing the expression of Snail, Slug, and ZEB1 and TCF/LEF transcriptional activity. Further studies revealed that combination of quercetin with EGCG resulted in synergistic activity on CSCs' self-renewal capacity via the inhibition of TCF/LEF and Gli activities.[197] Appari et al. have studied the efficacy of the combination of catechins with sulforaphane and quercetin on stem cell features and miRNA signaling in pancreatic ductal adenocarcinoma cell line and nonmalignant pancreatic ductal cells. Green tea catechins' (GTCs) mixture reduced cell viability, migration, MMP-2 and MMP-9 expression, ALDH1 activity and induced apoptosis as well as colony and spheroid formation, while GTCs, sulforaphane, or quercetin exhibited superior activity. The treatment also induced the expression of miR-let7-a in cancer cells and not normal cells, and it was found to be associated with inhibition of KRAS.[198] The literature reports suggested that green tea polyphenols especially EGCG can be effectively used as a CSC-eliminating agent for the treatment of cancer.

22.3.6 Polyunsaturated Fatty Acids

Polyunsaturated fatty acids (PUFAs) are major components of physiology, and mainly α-linolenic acid (found in plant oils), eicosapentaenoic acid (EPA), and docosahexaenoic acid (DHA) are important PUFAs. The primary sources of omega-3-fatty acids are marine algae and phytoplankton, while several plant oils such as walnut, flaxseed oil, and hemp oil also contain omega-3-fatty acids. Fish oil, egg oil, and squid oils are animal sources of omega-3-fatty acids. Hence, a large part of population use plant and animal sources for their dietary consumption, and omega-3-fatty acids and their health-promoting effects became a major topic of the study. Omega-3-fatty acids are known to exert antioxidant,[199] neuroprotective,[200] cardioprotective,[201] anti-inflammatory,[202] antidiabetic,[203] and anticancer[204] activities.

Piazzi et al. have evaluated the effectiveness of eicosapentaenoic acid–free fatty acid (EPA-FFA) in preventing inflammation and cancer in the AOM–dextran sodium sulphate (DSS) model of colitis-associated colorectal cancer. The EPA–FFA diet resulted in reduced tumor multiplicity, incidence, and maximum size of tumors. EPA–FFA, especially in initiation arm, decreased cell proliferation and expression of nuclear β-catenin as well as induced apoptosis. In initiation as well as promotion arms, EPA–FFA treatment resulted in enhanced membrane switch from ω-6 to ω-3 PUFAs and decreased the production of PGE2. The treatment with EPA–FFA restored the loss of notch signaling cascade as compared to control and also led to the enrichment of gut microbiota enriched in *Lactobacillus* species.[205] Vasudevan et al. have studied the effects of EPA in combination with FuOx (5-FU + oxaliplatin) against FuOx-resistant (CR) colon cancer, which are significantly enriched with CSCs/CSLCs. EPA treatment alone was effective, but combination treatment exerted superior activity in inhibiting cell growth, sphere formation, enhanced sphere disintegration, decreased proinflammatory metabolites, and reduced the growth of xenografts of CR colon cancer cells in SCID mice. The combination treatment resulted in decreased CSC/CSLC population, enhanced apoptosis by PARP cleavage, increased pPTEN expression, decreased pAkt expression, and normalization of β-catenin expression.[206] De Carlo et al. have demonstrated that EPA was actively incorporated into the membrane lipid of colon cancer (COLO 320 DM) cells, while 25 μM of EPA reduced the cancer cell number without affecting CD133(+) CSLCs. EPA treatment induced the downregulation of CD133 expression and the upregulation of colonic epithelium differentiation markers, such as cytokeratin 20 (CK20) and mucin 2 (MUC2). The treatment sensitized COLO 320 DM cells to 5-FU + Ox, while EPA enhanced CD133(+) CSLCs sensitivity to only 5-FU.[207] Algamas-Dimantov and group have studied the diabesity-induced intestinal ontogenesis in obese, in insulin-resistant db/db mice, in fat-1 transgene coding for desaturation of (n-6) PUFA into (n-3) PUFA, in db/db crossed with fat-1 mice, and in control mice. Diabesity led to enhanced proliferation in colon and epithelial colonocytes and goblet cell dedifferentiation along with enhanced β-catenin and hepatocyte nuclear factor (HNF)-4α transcriptional activities accompanied by enhancement in HNF-4α-bound (n-6) PUFA in the colon. In fat-1 mice, colonic proliferation was restrained, followed by crypt stem cell differentiation into epithelial colonocytes and goblet cells, coupled with decrease in transcriptional activities of β-catenin and HNF-4α and a concomitant increase in HNF-4α-bound (n-3) PUFA at the expense of (n-6) PUFA. These effects including colonic proliferation and differentiation, β-catenin and HNF-4α-responsive genes profile, and the composition of HNF-4α-bound PUFA of db/db mice were reversed to normal wild type by introducing fat-1 gene into the db/db mice, suggesting that (n-3) PUFA may be useful to reduce diabesity-induced intestinal ontogenesis as a preventive therapy for colorectal cancer.[208]

Xue et al. have shown that DHA treatment reduced cell growth and induced G1 cell cycle arrest in 4T1 mouse breast cells and MCF-7 human breast cells. The treatment with DHA to 4T1 mouse breast cells led to a decrease in the expression of β-catenin and T-cell factor/lymphoid-enhancing factor reporter activity as well as downregulation of c-Myc and cyclin D1. Studies in Balb/c mice bearing breast cancer fed with a 5% fish oil–supplemented diet for 30 days showed a reduction in tumor growth through the inhibition of growth and induction of apoptosis. The treatment also decreased β-catenin expression in tumor tissue with enhanced apoptosis and reduced metastasis to lung, suggesting the use of fish oil for prevention and treatment of breast cancer.[209] Mandal et al. have found that a fish oil diet containing DHA and EPA suppresses cancer cell metastasis to bone in a mouse model of MDA-MB-231 human breast cancer cell. DHA and EPA inhibited migration and invasion of MDA-MB-231 breast cancer cells *in vitro*. The aberrant increase in CD44 expression led to the generation of CSCs and contribute to metastasis. DHA and EPA were found to decrease mRNA as well as protein expression of CD44 in tumors of mice compared to control mice.[210] Song et al. have shown that DHA and EPA treatment to pancreatic cancer cells (SW1990 and Panc-1) resulted in cell growth inhibition and enhanced cell death along with reduction in the expression of β-catenin, decreased T-cell factor/lymphoid-enhancing factor reporter activity, and induced complex formation of β-catenin/Axin/GSK-3β. Immunohistochemical study showed upregulation and enhanced nuclear staining of β-catenin in tumors from pancreatic cancer patients. The implantation of mouse pancreatic cancer (PANC02) cells into fat-1 transgenic mice expressing ω3 desaturases and hence expressing elevated levels of endogenous ω3-PUFAs showed reduced tumor β-catenin levels and enhanced apoptosis compared to control mice.[211] Suzuki et al. have studied the inhibitory effects of unsaturated fatty acids including oleic acid (OA), DHA, and EPA alone or in combination on lung metastatic colonies in CDF1 mice bearing subcutaneous implants of the highly metastatic colon carcinoma. In the case of DHA, cotreatment with linoleic acid (LA) or EPA reduced its inhibitory activity, while DHA treatment showed very low potential to form colonies in lung after intravenous injection, which was partially reversed by cotreatment with EPA. OA treatment showed reduction in 57kDa isoform of type-IV collagenase (MMP-2), while DHA, but not DHA plus LA or EPA treatment, resulted in the reduction of 92kDa isoform (MMP-9), suggesting that the metastasis inhibitory potential of these compounds is mediated by the inhibition of type-IV collagenase activity.[212]

22.3.7 Miscellaneous

22.3.7.1 Bitter Melon

Bitter melon or bitter gourd (*Momordica charantia*) is commonly used in traditional Indian and Chinese system of medicine. The plant is cultivated worldwide such as in the tropical areas of Asia, Amazon, east Africa, and the Caribbean and utilized as vegetables as well as folk medicine for the treatment of diabetes and colic.[213,214] Almost all parts of the plant especially the fruit are usually consumed alone as well as cooked with other vegetables and stuffed or used in small quantities in soups to give little bitter taste and flavor. The plant is widely studied and reported to exhibit several biological activities including antioxidant,[215] anti-inflammatory,[216] anticancer,[217] and antidiabetic activities. In our group,[218,219] we have evaluated the potential of methanolic extracts of bitter melon against colon cancer stem as well as progenitor cells. We found that whole fruit (BMW) and skin (BMSk) extracts of bitter melon fruit exhibited a significant reduction of cell growth and colony formation, with BMW showing better activity. It caused cell cycle arrest in S-phase-induced LC3B cleavage and no caspase3/7, suggesting that cells were undergoing autophagy and not apoptosis. The treatment reduced Bcl-2 expression and enhanced Beclin-1, Atg-7, and Atg-12. BMW treatment also reduced the levels of cellular ATP and activated protein kinase, whereas exogenous addition of ATP resulted in cell proliferation. The treatment with BMW induced dose-dependent decrease in size and number of colonospheres and reduced the expression of stem cell markers such as DCLK1 and Lgr-5. Hsu et al. have shown that *M. charantia* leaf extract induces caspase- and mitochondria-dependent apoptosis in lung cancer cells by upregulating Wnt-2 expression, affecting migratory and invasive ability of CL1 cells via suppression of enzymatic activities of MMP-2 and MMP-9. The MCME treatment also inhibited migration by decreasing Src and focal adhesion kinase (FAK) activation as well as reducing the Akt, β-catenin, and MMPs' expression.[220]

22.3.7.2 Tomato

Lycopene is a bright-red carotene phytochemical present in tomatoes and other vegetables such as watermelon and red carrots. Tomatoes are used as a base ingredient in several vegetable recipes, pizza, ketchup, sauce, as well as soup/juice, which is very popular all over the world. Lycopene is one of the major ingredients in tomatoes and is reported for various biological activities including antioxidant,[221] anti-inflammatory,[222] and anticancer[223–225] activities. Lycopene was found to induce apoptosis and growth inhibition in prostate cancer cells.[226] Preet et al. have evaluated the effects of quinacrine (QC) and inhibitory potential on Wnt–TCF signaling in breast cancer cell lines. QC treatment affected Wnt–TCF signaling cascade components by increasing APC, DAB2, GSK-3β, and axin levels and reducing the β-catenin, p-GSK3β (Ser 9), and CK1 levels. It also decreased the activity of Wnt transcription factor TCF/LEF and expression of cyclin D1 and c-MYC. QC treatment also increased APC expression and induced apoptosis and DNA damage. Lycopene induced synergistic activity with QC and inhibited Wnt–TCF signaling pathway in breast cancer cells without altering the MCF-10A normal breast cell line.[227]

22.3.7.3 Blueberries

Blueberries are flowering plants with indigo-colored berries and occur perennially. Blueberry fruits are popular worldwide and are common ingredients of salads. It is consumed as such and in the form of juice. It occurs in different varieties like *Vaccinium caesariense* (known as New Jersey blueberry), *Vaccinium elliottii* (Elliott blueberry), and *Vaccinium formosum* (southern blueberry), in different areas. Blueberries due to its popularity are available commercially worldwide and hence studied for biological activities such as antioxidant,[228] anti-inflammatory,[229] and anticancer[230–233] activities. Blueberry extracts were found to inhibit colon cancer growth by inducing apoptosis and DNA fragmentation in HT-29 and Caco-2 cells.[234] Kanaya et al. have shown that whole blueberry powder reduced the metastasis of triple-negative breast cancer in female MDA-MB-231 tumor-bearing mice fed with a high-fat Western diet.[233] Jeyabalan et al. have shown that blueberry blend possessed inhibitory activity against 7β-estradiol-mediated mammary tumorigenesis as assessed by delayed tumor latency and reduced tumor multiplicity and volume. The results were found to be mediated via downregulation of CYP1A1, gene expression of ER-α, and modulation of miR-18a and miR-34c levels.[232] Pterostilbene, a natural stilbene compound found in blueberries, was studied for its potential CSCs/tumor-associated macrophage (TAM)-modulating activity in breast cancer. The treatment with pterostilbene resulted in M2 TAM-induced enhancement of CSCs and metastatic ability of breast cancer cells. Pterostilbene reduced the expression of NF-κB, Twist1, and vimentin and increased the expression of E-cadherin. The study also showed that pterostilbene-induced NF-κB downregulation was correlated with enhanced amount of microRNA 448. The tumor suppression and metastasis inhibitory activity of pterostilbene was also confirmed in mice bearing M2 TAM cocultured with MDA-MB-231 tumor. The study suggested the potential of pterostilbene as anti-CSC agent for the treatment of cancer.[235] Montales et al. have demonstrated that sera of adult mice on a diet containing soy isoflavone genistein and blueberry altered the stem/progenitor cells population as assessed by their self-renewal ability and spheroid-forming capacity *in vitro*. Genistein treatment alters the mammosphere formation and targets the basal stem CD44+/CD24-/ESA+ and the luminal progenitor CD24+ subpopulations in estrogen receptor–positive MCF-7 and estrogen receptor–negative MDA-MB-231 human breast cancer cell lines, while the blueberry phenolic acids were only active in MDA-MD-231 cells and its CD44+/CD24-/ESA+ subpopulation.[236]

22.3.7.4 Kokum

Garcinol is a polyisoprenylated benzophenone derivative isolated from the fruit rind of *Garcinia indica* that belongs to mangosteen family and is commonly known as kokum. The fruits are usually used for edible as well as for therapeutic purpose. It is usually included in various food recipes to give sour taste or used as a juice to drink. The plant is studied for antioxidant,[237] anti-inflammatory,[238] and anticancer[239,240] activities. Garcinol was found to be a major compound present in the extract, which is studied for anticancer activity. Tsai et al. have studied the protective effect of garcinol against DSS-induced colitis/inflammation and AOM/DSS-induced inflammation-related colon tumorigenesis in male Institute for Cancer Research (ICR) mice. Garcinol treatment prevented DSS-stimulated shortening of the colon length, aberrant crypt foci formation, and improved inflammation score in mice, while it also reduced DSS-induced nitric oxide synthase, cyclooxygenase-2, and expression of proliferating cell nuclear antigen protein. Garcinol treatment reduced the tumor size and incidence in the colon of mouse, while treatment downregulated the expression of cyclooxygenase-2, cyclin D1, and vascular endothelial growth factor via suppression of the extracellular signal–regulated protein kinase 1/2, phosphatidylinositol 3 kinase/Akt/p70 ribosomal S6 kinase, and Wnt/β-catenin signaling cascade.[238] Garcinol was found to efficiently reverse EMT in triple-negative MDA-MB-231 and BT-549 breast cancer cells, which was associated with the upregulation of E-cadherin, miR-200, and let-7 family microRNAs and downregulation of vimentin, ZEB-1, and ZEB-2. Garcinol treatment showed increased β-catenin phosphorylation, thereby reducing its nuclear translocation. Garcinol treatment in xenograft mouse model also inhibited the expression of NF-κB, miRNAs, vimentin, and nuclear β-catenin.[241]

22.3.7.5 Pepper

Piperine is an alkaloid isolated from black pepper and long pepper and is responsible for pungent taste. It has been used as a traditional medicine as well as an insecticide. Pepper is widely used in several food recipes to give pungent taste and popularly on pizza and other bakery products. Kakarala et al. have evaluated the effects of curcumin and piperine on mammosphere formation, expression of ALDH, and Wnt signaling pathway. Both curcumin and piperine reduced mammosphere formation, serial passaging, and percent of ALDH+ cells by half at 5 μM and completely at 10 μM concentration in normal as well as malignant breast cells. Both compounds inhibited Wnt signaling and inhibited the self-renewal of breast stem cell without causing toxicity to differentiated cells.[242]

22.4 CONCLUSION

Cancer remains a major problem due to treatment failure, metastasis, and recurrence after surgery. Recently, CSCs are reported as a small cell population exhibiting resistance to anticancer drugs and tumor recurrence. Studies identified several CSCs' signaling pathways that are responsible for acquiring and maintaining CSC in tumor mass. Current anticancer chemotherapy has failed to target CSCs. Hence, there is a need of novel compounds, which can target CSCs to achieve complete elimination of cancer. Functional foods and their active constituents, which are consumed regularly as dietary components, have been implicated in inhibiting these signaling pathways and target CSCs. Hence, the study of functional foods and their components may be helpful in the treatment of cancer in combination with current anticancer therapy to achieve complete resection of cancer. These active constituents will also serve as lead compounds for future anticancer drug discovery targeting CSC as well as other types of cancer cells.

Conflict of Interest

The authors declare that the research was conducted in the absence of any commercial or financial relationships that could be construed as a potential conflict of interest.

ACKNOWLEDGMENTS

We thank members of the Anant Laboratory for their discussion during the course of this study. This work was supported by National Institutes of Health Grants CA109269 and CA135559 and also Thomas O'Sullivan Foundation and The University of Kansas Cancer Center P30CA168524-01 grants. S. Anant is an Eminent Scientist of the Kansas Bioscience Authority.

REFERENCES

1. Siegel, R.L., Miller, K.D., and Jemal, A. Cancer statistics, 2015. *CA Cancer J Clin* 65, 5–29 (2015).
2. Vogelstein, B. and Kinzler, K.W. Cancer genes and the pathways they control. *Nat Med* 10, 789–799 (2004).
3. Gomez-Veiga, F. et al. Brachytherapy for the treatment of recurrent prostate cancer after radiotherapy or radical prostatectomy. *BJU Int* 109 Suppl 1, 17–21 (2012).
4. Meguid, R.A. et al. Recurrence after neoadjuvant chemoradiation and surgery for esophageal cancer: Does the pattern of recurrence differ for patients with complete response and those with partial or no response? *J Thorac Cardiovasc Surg* 138, 1309–1317 (2009).
5. Watt, F.M. and Driskell, R.R. The therapeutic potential of stem cells. *Philos Trans R Soc Lond B Biol Sci* 365, 155–163 (2010).
6. Carpentino, J.E. et al. Aldehyde dehydrogenase-expressing colon stem cells contribute to tumorigenesis in the transition from colitis to cancer. *Cancer Res* 69, 8208–8215 (2009).
7. Clarke, M.F. et al. CSCs—Perspectives on current status and future directions: AACR workshop on CSCs. *Cancer Res* 66, 9339–9344 (2006).
8. Dalerba, P. and Clarke, M.F. CSCs and tumor metastasis: First steps into uncharted territory. *Cell Stem Cell* 1, 241–242 (2007).

9. Ricci-Vitiani, L. et al. Identification and expansion of human colon-cancer-initiating cells. *Nature* 445, 111–115 (2007).
10. Huang, W.Y. et al. Microsomal epoxide hydrolase polymorphisms and risk for advanced colorectal adenoma. *Cancer Epidemiol Biomarkers Prev* 14, 152–157 (2005).
11. Dandawate, P., Padhye, S., Ahmad, A., and Sarkar, F.H. Novel strategies targeting CSCs through phytochemicals and their analogs. *Drug Deliv Transl Res* 3, 165–182 (2013).
12. Torre, L.A. et al. Global cancer statistics, 2012. *CA Cancer J Clin* 65, 87–108 (2015).
13. Tarraga Lopez, P.J., Albero, J.S., and Rodriguez-Montes, J.A. Primary and secondary prevention of colorectal cancer. *Clin Med Insights Gastroenterol* 7, 33–46 (2014).
14. Young, W.F., McGloin, J., Zittleman, L., West, D.R., and Westfall, J.M. Predictors of colorectal screening in rural Colorado: Testing to prevent colon cancer in the high plains research network. *J Rural Health* 23, 238–245 (2007).
15. Thenappan, A. et al. New therapeutics targeting colon CSCs. *Curr Colorectal Cancer Rep* 5, 209 (2009).
16. Subramaniam, D., Ramalingam, S., Houchen, C.W., and Anant, S. CSCs: A novel paradigm for cancer prevention and treatment. *Mini Rev Med Chem* 10, 359–371 (2010).
17. Sureban, S.M. et al. Selective blockade of DCAMKL-1 results in tumor growth arrest by a Let-7a MicroRNA-dependent mechanism. *Gastroenterology* 137, 649–659, 659 el–e2 (2009).
18. Nakanishi, Y. et al. Dclk1 distinguishes between tumor and normal stem cells in the intestine. *Nat Genet* 45, 98–103 (2013).
19. Metcalfe, C. and de Sauvage, F.J. A tumor-specific stem cell. *Nat Genet* 45, 7–9 (2013).
20. Westphalen, C.B. et al. Long-lived intestinal tuft cells serve as colon cancer-initiating cells. *J Clin Invest* 124, 1283–1295 (2014).
21. Rahib, L. et al. Projecting cancer incidence and deaths to 2030: The unexpected burden of thyroid, liver, and pancreas cancers in the United States. *Cancer Res* 74, 2913–2921 (2014).
22. Nieto, J., Grossbard, M.L., and Kozuch, P. Metastatic pancreatic cancer 2008: Is the glass less empty? *Oncologist* 13, 562–576 (2008).
23. Duffy, J.P. and Reber, H.A. Pancreatic neoplasms. *Curr Opin Gastroenterol* 19, 458–466 (2003).
24. Petrelli, F., Borgonovo, K., Ghilardi, M., Cabiddu, M., and Barni, S. What else in gemcitabine-pretreated advanced pancreatic cancer? An update of second line therapies. *Rev Recent Clin Trials* 5, 43–56 (2010).
25. Koutsounas, I., Giaginis, C., and Theocharis, S. Histone deacetylase inhibitors and pancreatic cancer: Are there any promising clinical trials? *World J Gastroenterol* 19, 1173–1181 (2013).
26. Kawasaki, B.T., Hurt, E.M., Mistree, T., and Farrar, W.L. Targeting CSCs with phytochemicals. *Mol Interv* 8, 174–184 (2008).
27. Rasheed, Z., Wang, Q., and Matsui, W. Isolation of stem cells from human pancreatic cancer xenografts. *J Vis Exp* 43, e2169 (2010).
28. Fitzgerald, T.L. and McCubrey, J.A. Pancreatic CSCs: Association with cell surface markers, prognosis, resistance, metastasis and treatment. *Adv Biol Regul* 56, 45–50 (2014).
29. Sureban, S.M. et al. DCLK1 regulates pluripotency and angiogenic factors via microRNA-dependent mechanisms in pancreatic cancer. *PLoS One* 8, e73940 (2013).
30. Bailey, J.M. et al. DCLK1 marks a morphologically distinct subpopulation of cells with stem cell properties in preinvasive pancreatic cancer. *Gastroenterology* 146, 245–256 (2014).
31. Jemal, A. et al. Global cancer statistics. *CA Cancer J Clin* 61, 69–90 (2011).
32. Sotiriou, C. et al. Breast cancer classification and prognosis based on gene expression profiles from a population-based study. *Proc Natl Acad Sci USA* 100, 10393–10398 (2003).
33. Sorlie, T. et al. Gene expression patterns of breast carcinomas distinguish tumor subclasses with clinical implications. *Proc Natl Acad Sci USA* 98, 10869–10874 (2001).
34. Isakoff, S.J. Triple-negative breast cancer: Role of specific chemotherapy agents. *Cancer J* 16, 53–61 (2010).
35. Chang, W.W. et al. Expression of Globo H and SSEA3 in breast CSCs and the involvement of fucosyl transferases 1 and 2 in Globo H synthesis. *Proc Natl Acad Sci USA* 105, 11667–11672 (2008).
36. Velasco-Velazquez, M.A., Homsi, N., De La Fuente, M., and Pestell, R.G. Breast CSCs. *Int J Biochem Cell Biol* 44, 573–577 (2012).
37. Velasco-Velazquez, M.A., Popov, V.M., Lisanti, M.P., and Pestell, R.G. The role of breast CSCs in metastasis and therapeutic implications. *Am J Pathol* 179, 2–11 (2011).
38. Al-Hajj, M., Wicha, M.S., Benito-Hernandez, A., Morrison, S.J., and Clarke, M.F. Prospective identification of tumorigenic breast cancer cells. *Proc Natl Acad Sci USA* 100, 3983–3988 (2003).
39. Lawson, J.C., Blatch, G.L., and Edkins, A.L. CSCs in breast cancer and metastasis. *Breast Cancer Res Treat* 118, 241–254 (2009).

40. Charafe-Jauffret, E. et al. Aldehyde dehydrogenase 1-positive CSCs mediate metastasis and poor clinical outcome in inflammatory breast cancer. *Clin Cancer Res* 16, 45–55 (2010).
41. Vaughan, S. et al. Rethinking ovarian cancer: Recommendations for improving outcomes. *Nat Rev Cancer* 11, 719–725 (2011).
42. Godwin, A.K., Testa, J.R., and Hamilton, T.C. The biology of ovarian cancer development. *Cancer* 71, 530–536 (1993).
43. McCluggage, W.G. Morphological subtypes of ovarian carcinoma: A review with emphasis on new developments and pathogenesis. *Pathology* 43, 420–432 (2011).
44. Desale, S.S., Soni, K.S., Romanova, S., Cohen, S.M., and Bronich, T.K. Targeted delivery of platinum-taxane combination therapy in ovarian cancer. *J Control Release* 220, 651–659 (2015).
45. Liu, S. et al. Prognostic value of cancer stem cell marker aldehyde dehydrogenase in ovarian cancer: A meta-analysis. *PLoS One* 8, e81050 (2013).
46. Zhou, Q. et al. Prognostic value of cancer stem cell marker CD133 in ovarian cancer: A meta-analysis. *Int J Clin Exp Med* 8, 3080–3088 (2015).
47. Tjhay, F. et al. CD44 variant 6 is correlated with peritoneal dissemination and poor prognosis in patients with advanced epithelial ovarian cancer. *Cancer Sci* 106, 1421–1428 (2015).
48. Oing, C. et al. Second-line chemotherapy for advanced and metastatic urothelial carcinoma—Vinflunine and beyond: A comprehensive review of the current literature. *J Urol* 195, 254–263 (2015).
49. Keymoosi, H., Gheytanchi, E., Asgari, M., Shariftabrizi, A., and Madjd, Z. ALDH1 in combination with CD44 as putative cancer stem cell markers are correlated with poor prognosis in urothelial carcinoma of the urinary bladder. *Asian Pac J Cancer Prev* 15, 2013–2020 (2014).
50. Radtke, F. and Clevers, H. Self-renewal and cancer of the gut: Two sides of a coin. *Science* 307, 1904–1909 (2005).
51. Ugolini, F. et al. Differential expression assay of chromosome arm 8p genes identifies Frizzled-related (FRP1/FRZB) and Fibroblast Growth Factor Receptor 1 (FGFR1) as candidate breast cancer genes. *Oncogene* 18, 1903–1910 (1999).
52. Curtin, J.C. and Lorenzi, M.V. Drug discovery approaches to target Wnt signaling in CSCs. *Oncotarget* 1, 552–566 (2010).
53. Brabletz, T. et al. Variable beta-catenin expression in colorectal cancers indicates tumor progression driven by the tumor environment. *Proc Natl Acad Sci USA* 98, 10356–10361 (2001).
54. Brabletz, S., Schmalhofer, O., and Brabletz, T. Gastrointestinal stem cells in development and cancer. *J Pathol* 217, 307–317 (2009).
55. Salama, P. and Platell, C. Colorectal CSCs. *ANZ J Surg* 79, 697–702 (2009).
56. Salama, P. et al. Tumor-infiltrating FOXP3+ T regulatory cells show strong prognostic significance in colorectal cancer. *J Clin Oncol* 27, 186–192 (2009).
57. Reguart, N. et al. The role of Wnt signaling in cancer and stem cells. *Future Oncol* 1, 787–797 (2005).
58. Taipale, J. and Beachy, P.A. The Hedgehog and Wnt signalling pathways in cancer. *Nature* 411, 349–354 (2001).
59. Oniscu, A. et al. Expression of Sonic hedgehog pathway genes is altered in colonic neoplasia. *J Pathol* 203, 909–917 (2004).
60. Yoshikawa, K. et al. Sonic hedgehog relates to colorectal carcinogenesis. *J Gastroenterol* 44, 1113–1117 (2009).
61. Bian, Y.H. et al. Sonic hedgehog-Gli1 pathway in colorectal adenocarcinomas. *World J Gastroenterol* 13, 1659–1665 (2007).
62. You, S. et al. PTCH1, a receptor of Hedgehog signaling pathway, is correlated with metastatic potential of colorectal cancer. *Ups J Med Sci* 115, 169–175 (2010).
63. Mazumdar, T. et al. Hedgehog signaling drives cellular survival in human colon carcinoma cells. *Cancer Res* 71, 1092–1102 (2011).
64. Varnat, F. et al. Human colon cancer epithelial cells harbour active HEDGEHOG-GLI signalling that is essential for tumour growth, recurrence, metastasis and stem cell survival and expansion. *EMBO Mol Med* 1, 338–351 (2009).
65. Ruiz i Altaba, A., Mas, C., and Stecca, B. The Gli code: An information nexus regulating cell fate, stemness and cancer. *Trends Cell Biol* 17, 438–447 (2007).
66. Ji, Z., Mei, F.C., Xie, J., and Cheng, X. Oncogenic KRAS activates hedgehog signaling pathway in pancreatic cancer cells. *J Biol Chem* 282, 14048–14055 (2007).
67. Bai, C.B., Auerbach, W., Lee, J.S., Stephen, D., and Joyner, A.L. Gli2, but not Gli1, is required for initial Shh signaling and ectopic activation of the Shh pathway. *Development* 129, 4753–4761 (2002).

68. Thiyagarajan, S. et al. Role of GLI2 transcription factor in growth and tumorigenicity of prostate cells. *Cancer Res* 67, 10642–10646 (2007).
69. Eichberger, T. et al. Overlapping and distinct transcriptional regulator properties of the GLI1 and GLI2 oncogenes. *Genomics* 87, 616–632 (2006).
70. Lauth, M., Bergstrom, A., Shimokawa, T., and Toftgard, R. Inhibition of GLI-mediated transcription and tumor cell growth by small-molecule antagonists. *Proc Natl Acad Sci USA* 104, 8455–8460 (2007).
71. Chen, J.K., Taipale, J., Cooper, M.K., and Beachy, P.A. Inhibition of Hedgehog signaling by direct binding of cyclopamine to Smoothened. *Genes Dev* 16, 2743–2748 (2002).
72. Berman, D.M. et al. Medulloblastoma growth inhibition by hedgehog pathway blockade. *Science* 297, 1559–1561 (2002).
73. Allenspach, E.J., Maillard, I., Aster, J.C., and Pear, W.S. Notch signaling in cancer. *Cancer Biol Ther* 1, 466–476 (2002).
74. Katoh, M. and Katoh, M. Integrative genomic analyses on HES/HEY family: Notch-independent HES1, HES3 transcription in undifferentiated ES cells, and Notch-dependent HES1, HES5, HEY1, HEY2, HEYL transcription in fetal tissues, adult tissues, or cancer. *Int J Oncol* 31, 461–466 (2007).
75. Ronchini, C. and Capobianco, A.J. Induction of cyclin D1 transcription and CDK2 activity by Notch(ic): Implication for cell cycle disruption in transformation by Notch(ic). *Mol Cell Biol* 21, 5925–5934 (2001).
76. Rangarajan, A. et al. Notch signaling is a direct determinant of keratinocyte growth arrest and entry into differentiation. *EMBO J* 20, 3427–3436 (2001).
77. Oswald, F., Liptay, S., Adler, G., and Schmid, R.M. NF-kappaB2 is a putative target gene of activated Notch-1 via RBP-Jkappa. *Mol Cell Biol* 18, 2077–2088 (1998).
78. Palomero, T. et al. NOTCH1 directly regulates c-MYC and activates a feed-forward-loop transcriptional network promoting leukemic cell growth. *Proc Natl Acad Sci USA* 103, 18261–18266 (2006).
79. Konishi, J. et al. Gamma-secretase inhibitor prevents Notch3 activation and reduces proliferation in human lung cancers. *Cancer Res* 67, 8051–8057 (2007).
80. Kukar, T.L. et al. Substrate-targeting gamma-secretase modulators. *Nature* 453, 925–929 (2008).
81. Wolfe, M.S. Intramembrane-cleaving proteases. *J Biol Chem* 284, 13969–13973 (2009).
82. Stanger, B.Z. Quit your YAPing: A new target for cancer therapy. *Genes Dev* 26, 1263–1267 (2012).
83. Tapon, N. et al. salvador Promotes both cell cycle exit and apoptosis in *Drosophila* and is mutated in human cancer cell lines. *Cell* 110, 467–478 (2002).
84. Avruch, J., Zhou, D., Fitamant, J., and Bardeesy, N. Mst1/2 signalling to Yap: Gatekeeper for liver size and tumour development. *Br J Cancer* 104, 24–32 (2011).
85. Steinhardt, A.A. et al. Expression of Yes-associated protein in common solid tumors. *Hum Pathol* 39, 1582–1589 (2008).
86. Avruch, J. et al. Protein kinases of the Hippo pathway: Regulation and substrates. *Semin Cell Dev Biol* 23, 770–784 (2012).
87. Overholtzer, M. et al. Transforming properties of YAP, a candidate oncogene on the chromosome 11q22 amplicon. *Proc Natl Acad Sci USA* 103, 12405–12410 (2006).
88. Cordenonsi, M. et al. The Hippo transducer TAZ confers cancer stem cell-related traits on breast cancer cells. *Cell* 147, 759–772 (2011).
89. Radisky, D.C. and LaBarge, M.A. Epithelial-mesenchymal transition and the stem cell phenotype. *Cell Stem Cell* 2, 511–512 (2008).
90. Hanahan, D. and Weinberg, R.A. Hallmarks of cancer: The next generation. *Cell* 144, 646–674 (2011).
91. Thomas, S.J., Snowden, J.A., Zeidler, M.P., and Danson, S.J. The role of JAK/STAT signalling in the pathogenesis, prognosis and treatment of solid tumours. *Br J Cancer* 113, 365–371 (2015).
92. Kiger, A.A., Jones, D.L., Schulz, C., Rogers, M.B., and Fuller, M.T. Stem cell self-renewal specified by JAK-STAT activation in response to a support cell cue. *Science* 294, 2542–2545 (2001).
93. Hernandez-Vargas, H. et al. Methylome analysis reveals Jak-STAT pathway deregulation in putative breast CSCs. *Epigenetics* 6, 428–439 (2011).
94. Neradugomma, N.K. et al. Prolactin signaling enhances colon cancer stemness by modulating Notch signaling in a Jak2-STAT3/ERK manner. *Carcinogenesis* 35, 795–806 (2014).
95. Porta, C., Paglino, C., and Mosca, A. Targeting PI3K/Akt/mTOR Signaling in Cancer. *Front Oncol* 4, 64 (2014).
96. Steelman, L.S. et al. Contributions of the Raf/MEK/ERK, PI3K/PTEN/Akt/mTOR and Jak/STAT pathways to leukemia. *Leukemia* 22, 686–707 (2008).
97. Xia, P. and Xu, X.Y. PI3K/Akt/mTOR signaling pathway in CSCs: From basic research to clinical application. *Am J Cancer Res* 5, 1602–1609 (2015).

98. Zhou, J. et al. Activation of the PTEN/mTOR/STAT3 pathway in breast cancer stem-like cells is required for viability and maintenance. *Proc Natl Acad Sci USA* 104, 16158–16163 (2007).
99. Huang, E.H. et al. Aldehyde dehydrogenase 1 is a marker for normal and malignant human colonic stem cells (SC) and tracks SC overpopulation during colon tumorigenesis. *Cancer Res* 69, 3382–3389 (2009).
100. Chai, X. et al. Metformin increases sensitivity of pancreatic cancer cells to Gemcitabine by reducing CD133(+) cell populations and suppressing ERK/P70S6K signaling. *Sci Rep* 5, 14404 (2015).
101. Vyas, A., Dandawate, P., Padhye, S., Ahmad, A., and Sarkar, F. Perspectives on new synthetic curcumin analogs and their potential anticancer properties. *Curr Pharm Des* 19, 2047–2069 (2013).
102. Dandawate, P.R. et al. Inclusion complex of novel curcumin analogue CDF and beta-cyclodextrin (1:2) and its enhanced in vivo anticancer activity against pancreatic cancer. *Pharm Res* 29, 1775–1786 (2012).
103. Padhye, S. et al. Perspectives on chemopreventive and therapeutic potential of curcumin analogs in medicinal chemistry. *Mini Rev Med Chem* 10, 372–387 (2010).
104. Abdel-Daim, M.M. and Abdou, R.H. Protective effects of diallyl sulfide and curcumin separately against thallium-induced toxicity in rats. *Cell J* 17, 379–388 (2015).
105. Subbaramaiah, K. et al. Dietary polyphenols suppress elevated levels of proinflammatory mediators and aromatase in the mammary gland of obese mice. *Cancer Prev Res (Phila)* 6, 886–897 (2013).
106. Subramaniam, D. et al. Curcumin induces cell death in esophageal cancer cells through modulating Notch signaling. *PLoS One* 7, e30590 (2012).
107. Abd Allah, E.S. and Gomaa, A.M. Effects of curcumin and captopril on the functions of kidney and nerve in streptozotocin-induced diabetic rats: Role of angiotensin converting enzyme 1. *Appl Physiol Nutr Metab* 40, 1061–1067 (2015).
108. Betts, J.W. and Wareham, D.W. In vitro activity of curcumin in combination with epigallocatechin gallate (EGCG) versus multidrug-resistant Acinetobacter baumannii. *BMC Microbiol* 14, 172 (2014).
109. Prasad, S. and Tyagi, A.K. Curcumin and its analogues: A potential natural compound against HIV infection and AIDS. *Food Funct* 6, 3412–3419 (2015).
110. Shakibaei, M. et al. Curcumin chemosensitizes 5-fluorouracil resistant MMR-deficient human colon cancer cells in high density cultures. *PLoS One* 9, e85397 (2014).
111. Lin, L. et al. STAT3 signaling pathway is necessary for cell survival and tumor sphere forming capacity in ALDH(+)/CD133(+) stem cell-like human colon cancer cells. *Biochem Biophys Res Commun* 416, 246–251 (2011).
112. Nautiyal, J., Kanwar, S.S., Yu, Y., and Majumdar, A.P. Combination of dasatinib and curcumin eliminates chemo-resistant colon cancer cells. *J Mol Signal* 6, 7 (2011).
113. Lin, L. et al. Targeting colon CSCs using a new curcumin analogue, GO-Y030. *Br J Cancer* 105, 212–220 (2011).
114. Yu, Y. et al. Elimination of colon cancer stem-like cells by the combination of curcumin and FOLFOX. *Transl Oncol* 2, 321–328 (2009).
115. Kanwar, S.S. et al. Difluorinated-curcumin (CDF): A novel curcumin analog is a potent inhibitor of colon cancer stem-like cells. *Pharm Res* 28, 827–838 (2011).
116. Sun, X.D., Liu, X.E., and Huang, D.S. Curcumin reverses the epithelial-mesenchymal transition of pancreatic cancer cells by inhibiting the Hedgehog signaling pathway. *Oncol Rep* 29, 2401–2407 (2013).
117. Wang, Z., Zhang, Y., Banerjee, S., Li, Y., and Sarkar, F.H. Notch-1 down-regulation by curcumin is associated with the inhibition of cell growth and the induction of apoptosis in pancreatic cancer cells. *Cancer* 106, 2503–2513 (2006).
118. Bao, B. et al. Anti-tumor activity of a novel compound-CDF is mediated by regulating miR-21, miR-200, and PTEN in pancreatic cancer. *PLoS One* 6, e17850 (2011).
119. Bao, B. et al. Curcumin analogue CDF inhibits pancreatic tumor growth by switching on suppressor microRNAs and attenuating EZH2 expression. *Cancer Res* 72, 335–345 (2012).
120. Li, Y. et al. Inactivation of AR/TMPRSS2-ERG/Wnt signaling networks attenuates the aggressive behavior of prostate cancer cells. *Cancer Prev Res (Phila)* 4, 1495–1506 (2011).
121. Mohapatra, P. et al. Resveratrol and curcumin synergistically induces apoptosis in cigarette smoke condensate transformed breast epithelial cells through a p21(Waf1/Cip1) mediated inhibition of Hh-Gli signaling. *Int J Biochem Cell Biol* 66, 75–84 (2015).
122. Chung, S.S. and Vadgama, J.V. Curcumin and epigallocatechin gallate inhibit the cancer stem cell phenotype via down-regulation of STAT3-NFkappaB signaling. *Anticancer Res* 35, 39–46 (2015).
123. Mukherjee, S. et al. Curcumin inhibits breast cancer stem cell migration by amplifying the E-cadherin/beta-catenin negative feedback loop. *Stem Cell Res Ther* 5, 116 (2014).
124. Mahmoud, A.M., Yang, W., and Bosland, M.C. Soy isoflavones and prostate cancer: A review of molecular mechanisms. *J Steroid Biochem Mol Biol* 140, 116–132 (2014).

125. Jacobsen, B.K., Knutsen, S.F., and Fraser, G.E. Does high soy milk intake reduce prostate cancer incidence? The Adventist Health Study (United States). *Cancer Causes Control* 9, 553–557 (1998).
126. Avci, C.B. et al. Genistein-induced mir-23b expression inhibits the growth of breast cancer cells. *Contemp Oncol (Pozn)* 19, 32–35 (2015).
127. Dastjerdi, M.N. et al. Inhibitory effect of genistein on PLC/PRF5 hepatocellular carcinoma cell line. *Int J Prev Med* 6, 54 (2015).
128. Papaj, K., Rusin, A., Szeja, W., and Grynkiewicz, G. Absorption and metabolism of biologically active genistein derivatives in colon carcinoma cell line (Caco-2). *Acta Pol Pharm* 71, 1037–1044 (2014).
129. Montales, M.T., Rahal, O.M., Nakatani, H., Matsuda, T., and Simmen, R.C. Repression of mammary adipogenesis by genistein limits mammosphere formation of human MCF-7 cells. *J Endocrinol* 218, 135–149 (2013).
130. Yu, D. et al. Genistein attenuates cancer stem cell characteristics in gastric cancer through the downregulation of Gli1. *Oncol Rep* 31, 673–678 (2014).
131. Iftikhar, H. and Rashid, S. Molecular docking studies of flavonoids for their inhibition pattern against beta-catenin and pharmacophore model generation from experimentally known flavonoids to fabricate more potent inhibitors for Wnt signaling pathway. *Pharmacogn Mag* 10, S264–S271 (2014).
132. Zhang, Y., Li, Q., and Chen, H. DNA methylation and histone modifications of Wnt genes by genistein during colon cancer development. *Carcinogenesis* 34, 1756–1763 (2013).
133. Hirata, H. et al. Genistein downregulates onco-miR-1260b and inhibits Wnt-signalling in renal cancer cells. *Br J Cancer* 108, 2070–2078 (2013).
134. Wang, H., Li, Q., and Chen, H. Genistein affects histone modifications on Dickkopf-related protein 1 (DKK1) gene in SW480 human colon cancer cell line. *PLoS One* 7, e40955 (2012).
135. Zhang, Y., Li, Q., Zhou, D., and Chen, H. Genistein, a soya isoflavone, prevents AOM-induced up-regulation of WNT/beta-catenin signalling and reduces colon pre-neoplasia in rats. *Br J Nutr* 109, 33–42 (2013).
136. Zhang, Y. and Chen, H. Genistein attenuates WNT signaling by up-regulating sFRP2 in a human colon cancer cell line. *Exp Biol Med (Maywood)* 236, 714–722 (2011).
137. Xia, J. et al. Genistein inhibits cell growth and induces apoptosis through up-regulation of miR-34a in pancreatic cancer cells. *Curr Drug Targets* 13, 1750–1756 (2012).
138. Bao, B. et al. Over-expression of FoxM1 leads to epithelial-mesenchymal transition and cancer stem cell phenotype in pancreatic cancer cells. *J Cell Biochem* 112, 2296–2306 (2011).
139. Pan, H. et al. Genistein inhibits MDA-MB-231 triple-negative breast cancer cell growth by inhibiting NF-kappaB activity via the Notch-1 pathway. *Int J Mol Med* 30, 337–343 (2012).
140. Fan, P. et al. Genistein decreases the breast cancer stem-like cell population through Hedgehog pathway. *Stem Cell Res Ther* 4, 146 (2013).
141. Ning, Y., Luo, C., Ren, K., Quan, M., and Cao, J. FOXO3a-mediated suppression of the self-renewal capacity of sphere-forming cells derived from the ovarian cancer SKOV3 cell line by 7-difluoromethoxyl-5,4′-di-n-octyl genistein. *Mol Med Rep* 9, 1982–1988 (2014).
142. Ning, Y.X., Li, Q.X., Ren, K.Q., Quan, M.F., and Cao, J.G. 7-difluoromethoxyl-5,4′-di-n-octyl genistein inhibits ovarian cancer stem cell characteristics through the downregulation of FOXM1. *Oncol Lett* 8, 295–300 (2014).
143. Soleas, G.J., Diamandis, E.P., and Goldberg, D.M. Resveratrol: A molecule whose time has come? And gone? *Clin Biochem* 30, 91–113 (1997).
144. Cheng, L. et al. Resveratrol attenuates inflammation and oxidative stress induced by myocardial ischemia-reperfusion injury: Role of Nrf2/ARE pathway. *Int J Clin Exp Med* 8, 10420–10428 (2015).
145. Joraholmen, M.W., Skalko-Basnet, N., Acharya, G., and Basnet, P. Resveratrol-loaded liposomes for topical treatment of the vaginal inflammation and infections. *Eur J Pharm Sci* 79, 112–121 (2015).
146. Yaylali, A., Ergin, K., and Cecen, S. Effect of resveratrol on leptin and Sirtuin 2 expression in the kidneys in streptozotocin-induced diabetic rats. *Anal Quant Cytopathol Histpathol* 37, 243–251 (2015).
147. Yu, X.D., Yang, J.L., Zhang, W.L., and Liu, D.X. Resveratrol inhibits oral squamous cell carcinoma through induction of apoptosis and G2/M phase cell cycle arrest. *Tumour Biol* 37, 2871–2877 (2015).
148. Kim, S.H. et al. Resveratrol attenuates norepinephrine-induced ovarian cancer invasiveness through downregulating hTERT expression. *Arch Pharm Res* 39, 240–248 (2015).
149. Trung, L.Q. et al. Resveratrol selectively induces apoptosis in malignant cells with the JAK2V617F mutation by inhibiting the JAK2 pathway. *Mol Nutr Food Res* 59, 2143–2154 (2015).
150. Siddiqui, I.A., Sanna, V., Ahmad, N., Sechi, M., and Mukhtar, H. Resveratrol nanoformulation for cancer prevention and therapy. *Ann N Y Acad Sci* 1348, 20–31 (2015).

151. Siddiqui, A. et al. Novel Aza-resveratrol analogs: Synthesis, characterization and anticancer activity against breast cancer cell lines. *Bioorg Med Chem Lett* 23, 635–640 (2013).
152. Ronghe, A. et al. Differential regulation of estrogen receptors alpha and beta by 4-(E)-{(4-hydroxyphenylimino)-methylbenzene,1,2-diol}, a novel resveratrol analog. *J Steroid Biochem Mol Biol* 144 Pt B, 500–512 (2014).
153. Liu, Y.Z. et al. The PTEN/PI3K/Akt and Wnt/beta-catenin signaling pathways are involved in the inhibitory effect of resveratrol on human colon cancer cell proliferation. *Int J Oncol* 45, 104–112 (2014).
154. Ji, Q. et al. Resveratrol inhibits invasion and metastasis of colorectal cancer cells via MALAT1 mediated Wnt/beta-catenin signal pathway. *PLoS One* 8, e78700 (2013).
155. Zhang, W. et al. Fluorinated N,N-dialkylaminostilbenes for Wnt pathway inhibition and colon cancer repression. *J Med Chem* 54, 1288–1297 (2011).
156. Gao, Q., Yuan, Y., Gan, H.Z., and Peng, Q. Resveratrol inhibits the hedgehog signaling pathway and epithelial-mesenchymal transition and suppresses gastric cancer invasion and metastasis. *Oncol Lett* 9, 2381–2387 (2015).
157. Qin, Y., Ma, Z., Dang, X., Li, W., and Ma, Q. Effect of resveratrol on proliferation and apoptosis of human pancreatic cancer MIA PaCa-2 cells may involve inhibition of the Hedgehog signaling pathway. *Mol Med Rep* 10, 2563–2567 (2014).
158. Mo, W. et al. Resveratrol inhibits proliferation and induces apoptosis through the hedgehog signaling pathway in pancreatic cancer cell. *Pancreatology* 11, 601–609 (2011).
159. Shankar, S. et al. Resveratrol inhibits pancreatic cancer stem cell characteristics in human and KrasG12D transgenic mice by inhibiting pluripotency maintaining factors and epithelial-mesenchymal transition. *PLoS One* 6, e16530 (2011).
160. Fu, Y. et al. Resveratrol inhibits breast cancer stem-like cells and induces autophagy via suppressing Wnt/beta-catenin signaling pathway. *PLoS One* 9, e102535 (2014).
161. Tsai, J.H. et al. 3,5,4′-Trimethoxystilbene, a natural methoxylated analog of resveratrol, inhibits breast cancer cell invasiveness by downregulation of PI3K/Akt and Wnt/beta-catenin signaling cascades and reversal of epithelial-mesenchymal transition. *Toxicol Appl Pharmacol* 272, 746–756 (2013).
162. Seino, M. et al. Differential contribution of ROS to resveratrol-induced cell death and loss of self-renewal capacity of ovarian CSCs. *Anticancer Res* 35, 85–96 (2015).
163. Zhong, L.X. et al. Inhibition of STAT3 signaling as critical molecular event in resveratrol-suppressed ovarian cancer cells. *J Ovarian Res* 8, 25 (2015).
164. Lubecka-Pietruszewska, K. et al. Sulforaphane alone and in combination with clofarabine epigenetically regulates the expression of DNA methylation-silenced tumour suppressor genes in human breast cancer cells. *J Nutrigenet Nutrigenomics* 8, 91–101 (2015).
165. Fofaria, N.M., Ranjan, A., Kim, S.H., and Srivastava, S.K. Mechanisms of the anticancer effects of isothiocyanates. *Enzymes* 37, 111–137 (2015).
166. Sestili, P. and Fimognari, C. Cytotoxic and antitumor activity of sulforaphane: The role of reactive oxygen species. *Biomed Res Int* 2015, 402386 (2015).
167. Ali Khan, M. et al. Sulforaphane reverses the expression of various tumor suppressor genes by targeting DNMT3B and HDAC1 in human cervical cancer cells. *Evid Based Complement Alternat Med* 2015, 412149 (2015).
168. Peng, X. et al. Sulforaphane inhibits invasion by phosphorylating ERK1/2 to regulate E-cadherin and CD44v6 in human prostate cancer DU145 cells. *Oncol Rep* 34, 1565–1572 (2015).
169. Khor, T.O. et al. Pharmacogenomics of cancer chemopreventive isothiocyanate compound sulforaphane in the intestinal polyps of ApcMin/+ mice. *Biopharm Drug Dispos* 27, 407–420 (2006).
170. Hu, R. et al. Cancer chemoprevention of intestinal polyposis in ApcMin/+ mice by sulforaphane, a natural product derived from cruciferous vegetable. *Carcinogenesis* 27, 2038–2046 (2006).
171. Kallifatidis, G. et al. Sulforaphane increases drug-mediated cytotoxicity toward cancer stem-like cells of pancreas and prostate. *Mol Ther* 19, 188–195 (2011).
172. Rodova, M., Fu, J., Watkins, D.N., Srivastava, R.K., and Shankar, S. Sonic hedgehog signaling inhibition provides opportunities for targeted therapy by sulforaphane in regulating pancreatic cancer stem cell self-renewal. *PLoS One* 7, e46083 (2012).
173. Li, S.H., Fu, J., Watkins, D.N., Srivastava, R.K., and Shankar, S. Sulforaphane regulates self-renewal of pancreatic CSCs through the modulation of Sonic hedgehog-GLI pathway. *Mol Cell Biochem* 373, 217–227 (2013).
174. Li, Q., Eades, G., Yao, Y., Zhang, Y., and Zhou, Q. Characterization of a stem-like subpopulation in basal-like ductal carcinoma in situ (DCIS) lesions. *J Biol Chem* 289, 1303–1312 (2014).
175. Li, Q. et al. Downregulation of miR-140 promotes cancer stem cell formation in basal-like early stage breast cancer. *Oncogene* 33, 2589–2600 (2014).

176. Li, Y. et al. Sulforaphane, a dietary component of broccoli/broccoli sprouts, inhibits breast CSCs. *Clin Cancer Res* 16, 2580–2590 (2010).
177. Li, Q. et al. Sulforaphane inhibits mammary adipogenesis by targeting adipose mesenchymal stem cells. *Breast Cancer Res Treat* 141, 317–324 (2013).
178. Forster, T. et al. Sulforaphane counteracts aggressiveness of pancreatic cancer driven by dysregulated Cx43-mediated gap junctional intercellular communication. *Oncotarget* 5, 1621–1634 (2014).
179. Weng, J.R., Tsai, C.H., Kulp, S.K., and Chen, C.S. Indole-3-carbinol as a chemopreventive and anti-cancer agent. *Cancer Lett* 262, 153–163 (2008).
180. De Kruif, C.A. et al. Structure elucidation of acid reaction products of indole-3-carbinol: Detection in vivo and enzyme induction in vitro. *Chem Biol Interact* 80, 303–315 (1991).
181. Leem, S.H., Li, X.J., Park, M.H., Park, B.H., and Kim, S.M. Genome-wide transcriptome analysis reveals inactivation of Wnt/beta-catenin by 3,3′-diindolylmethane inhibiting proliferation of colon cancer cells. *Int J Oncol* 47, 918–926 (2015).
182. Semov, A. et al. Diindolilmethane (DIM) selectively inhibits CSCs. *Biochem Biophys Res Commun* 424, 45–51 (2012).
183. Frei, B. and Higdon, J.V. Antioxidant activity of tea polyphenols in vivo: Evidence from animal studies. *J Nutr* 133, 3275S–3284S (2003).
184. Tipoe, G.L., Leung, T.M., Hung, M.W., and Fung, M.L. Green tea polyphenols as an anti-oxidant and anti-inflammatory agent for cardiovascular protection. *Cardiovasc Hematol Disord Drug Targets* 7, 135–144 (2007).
185. Sabu, M.C., Smitha, K., and Kuttan, R. Anti-diabetic activity of green tea polyphenols and their role in reducing oxidative stress in experimental diabetes. *J Ethnopharmacol* 83, 109–116 (2002).
186. Mukherjee, S. et al. Gold-conjugated green tea nanoparticles for enhanced anti-tumor activities and hepatoprotection—Synthesis, characterization and in vitro evaluation. *J Nutr Biochem* 26, 1283–1297 (2015).
187. Cerezo-Guisado, M.I. et al. Implication of Akt, ERK1/2 and alternative p38MAPK signalling pathways in human colon cancer cell apoptosis induced by green tea EGCG. *Food Chem Toxicol* 84, 125–132 (2015).
188. Afzal, M., Safer, A.M., and Menon, M. Green tea polyphenols and their potential role in health and disease. *Inflammopharmacology* 23, 151–161 (2015).
189. Wang, X. et al. EGCG enhances cisplatin sensitivity by regulating expression of the copper and cisplatin influx transporter CTR1 in ovary cancer. *PLoS One* 10, e0125402 (2015).
190. Tyagi, T., Treas, J.N., Mahalingaiah, P.K., and Singh, K.P. Potentiation of growth inhibition and epigenetic modulation by combination of green tea polyphenol and 5-aza-2′-deoxycytidine in human breast cancer cells. *Breast Cancer Res Treat* 149, 655–668 (2015).
191. Moses, M.A., Henry, E.C., Ricke, W.A., and Gasiewicz, T.A. The heat shock protein 90 inhibitor, (−)-epigallocatechin gallate, has anticancer activity in a novel human prostate cancer progression model. *Cancer Prev Res (Phila)* 8, 249–257 (2015).
192. Umeda, D., Yano, S., Yamada, K., and Tachibana, H. Involvement of 67-kDa laminin receptor-mediated myosin phosphatase activation in antiproliferative effect of epigallocatechin-3-O-gallate at a physiological concentration on Caco-2 colon cancer cells. *Biochem Biophys Res Commun* 371, 172–176 (2008).
193. Pahlke, G. et al. Impact of quercetin and EGCG on key elements of the Wnt pathway in human colon carcinoma cells. *J Agric Food Chem* 54, 7075–7082 (2006).
194. Bose, M. et al. Inhibition of tumorigenesis in ApcMin/+ mice by a combination of (−)-epigallocatechin-3-gallate and fish oil. *J Agric Food Chem* 55, 7695–7700 (2007).
195. Kim, J. et al. Suppression of Wnt signaling by the green tea compound (−)-epigallocatechin 3-gallate (EGCG) in invasive breast cancer cells. Requirement of the transcriptional repressor HBP1. *J Biol Chem* 281, 10865–10875 (2006).
196. Mineva, N.D., Paulson, K.E., Naber, S.P., Yee, A.S., and Sonenshein, G.E. Epigallocatechin-3-gallate inhibits stem-like inflammatory breast cancer cells. *PLoS One* 8, e73464 (2013).
197. Tang, S.N. et al. Inhibition of sonic hedgehog pathway and pluripotency maintaining factors regulate human pancreatic cancer stem cell characteristics. *Int J Cancer* 131, 30–40 (2012).
198. Appari, M., Babu, K.R., Kaczorowski, A., Gross, W., and Herr, I. Sulforaphane, quercetin and catechins complement each other in elimination of advanced pancreatic cancer by miR-let-7 induction and K-ras inhibition. *Int J Oncol* 45, 1391–1400 (2014).
199. Tabei, S.M. et al. Effect of vitamins A, E, C and omega-3 fatty acids supplementation on the level of catalase and superoxide dismutase activities in streptozotocin-induced diabetic rats. *Bratisl Lek Listy* 116, 115–118 (2015).

200. Zhang, W. et al. Omega-3 polyunsaturated fatty acids in the brain: Metabolism and neuroprotection. *Front Biosci (Landmark Ed)* 16, 2653–2670 (2011).
201. Lee, J.H., O'Keefe, J.H., Lavie, C.J., Marchioli, R., and Harris, W.S. Omega-3 fatty acids for cardioprotection. *Mayo Clin Proc* 83, 324–332 (2008).
202. Maroon, J.C. and Bost, J.W. Omega-3 fatty acids (fish oil) as an anti-inflammatory: An alternative to nonsteroidal anti-inflammatory drugs for discogenic pain. *Surg Neurol* 65, 326–331 (2006).
203. Wu, J.H. et al. Omega-3 fatty acids and incident type 2 diabetes: A systematic review and meta-analysis. *Br J Nutr* 107 Suppl 2, S214–S227 (2012).
204. Laviano, A., Rianda, S., Molfino, A., and Rossi Fanelli, F. Omega-3 fatty acids in cancer. *Curr Opin Clin Nutr Metab Care* 16, 156–161 (2013).
205. Piazzi, G. et al. Eicosapentaenoic acid free fatty acid prevents and suppresses colonic neoplasia in colitis-associated colorectal cancer acting on Notch signaling and gut microbiota. *Int J Cancer* 135, 2004–2013 (2014).
206. Vasudevan, A. et al. Omega-3 fatty acid is a potential preventive agent for recurrent colon cancer. *Cancer Prev Res (Phila)* 7, 1138–1148 (2014).
207. De Carlo, F., Witte, T.R., Hardman, W.E., and Claudio, P.P. Omega-3 eicosapentaenoic acid decreases CD133 colon cancer stem-like cell marker expression while increasing sensitivity to chemotherapy. *PLoS One* 8, e69760 (2013).
208. Algamas-Dimantov, A. et al. Amelioration of diabesity-induced colorectal ontogenesis by omega-3 fatty acids in mice. *J Lipid Res* 53, 1056–1070 (2012).
209. Xue, M. et al. Docosahexaenoic acid inhibited the Wnt/beta-catenin pathway and suppressed breast cancer cells in vitro and in vivo. *J Nutr Biochem* 25, 104–110 (2014).
210. Mandal, C.C., Ghosh-Choudhury, T., Yoneda, T., Choudhury, G.G., and Ghosh-Choudhury, N. Fish oil prevents breast cancer cell metastasis to bone. *Biochem Biophys Res Commun* 402, 602–607 (2010).
211. Song, K.S. et al. Omega-3-polyunsaturated fatty acids suppress pancreatic cancer cell growth in vitro and in vivo via downregulation of Wnt/Beta-catenin signaling. *Pancreatology* 11, 574–584 (2011).
212. Suzuki, I. et al. Inhibitory effects of oleic and docosahexaenoic acids on lung metastasis by colon-carcinoma-26 cells are associated with reduced matrix metalloproteinase-2 and -9 activities. *Int J Cancer* 73, 607–612 (1997).
213. Satyawati, G.V., Gupta, A.K., and Tandon, N. *Medicinal Plants of India.* (Indian Council of Medical Research, New Delhi, India, 2014).
214. Yesilada, E., Gurbuz, I., and Shibata, H. Screening of Turkish anti-ulcerogenic folk remedies for anti-Helicobacter pylori activity. *J Ethnopharmacol* 66, 289–293 (1999).
215. Lu, K.H. et al. Wild bitter gourd protects against alcoholic fatty liver in mice by attenuating oxidative stress and inflammatory responses. *Food Funct* 5, 1027–1037 (2014).
216. Xu, J. et al. Bitter gourd inhibits the development of obesity-associated fatty liver in C57BL/6 mice fed a high-fat diet. *J Nutr* 144, 475–483 (2014).
217. Kaur, M. et al. Bitter melon juice activates cellular energy sensor AMP-activated protein kinase causing apoptotic death of human pancreatic carcinoma cells. *Carcinogenesis* 34, 1585–1592 (2013).
218. Kwatra, D. et al. Methanolic extracts of bitter melon inhibit colon CSCs by affecting energy homeostasis and autophagy. *Evid Based Complement Alternat Med* 2013, 702869 (2013).
219. Kwatra, D. et al. Bitter melon extracts enhance the activity of chemotherapeutic agents through the modulation of multiple drug resistance. *J Pharm Sci* 102, 4444–4454 (2013).
220. Hsu, C. et al. Wild bitter melon (Momordica charantia Linn. var. abbreviata Ser.) extract and its bioactive components suppress Propionibacterium acnes-induced inflammation. *Food Chem* 135, 976–984 (2012).
221. Rao, A.V. and Agarwal, S. Role of antioxidant lycopene in cancer and heart disease. *J Am Coll Nutr* 19, 563–569 (2000).
222. Palozza, P., Parrone, N., Catalano, A., and Simone, R. Tomato lycopene and inflammatory cascade: Basic interactions and clinical implications. *Curr Med Chem* 17, 2547–2563 (2010).
223. Pisipati, S.V. et al. Lycopene: Redress for prostate cancer. *J Basic Clin Pharm* 3, 261–264 (2012).
224. Ono, M., Takeshima, M., and Nakano, S. Mechanism of the anticancer effect of lycopene (Tetraterpenoids). *Enzymes* 37, 139–166 (2015).
225. Kim, M.J. and Kim, H. Anticancer effect of lycopene in gastric carcinogenesis. *J Cancer Prev* 20, 92–96 (2015).
226. Renju, G.L., Muraleedhara Kurup, G., and Bandugula, V.R. Effect of lycopene isolated from *Chlorella marina* on proliferation and apoptosis in human prostate cancer cell line PC-3. *Tumour Biol* 35, 10747–10758 (2014).

227. Preet, R. et al. Lycopene synergistically enhances quinacrine action to inhibit Wnt-TCF signaling in breast cancer cells through APC. *Carcinogenesis* 34, 277–286 (2013).
228. Zhao, M. et al. The chemoprotection of a blueberry anthocyanin extract against the acrylamide-induced oxidative stress in mitochondria: Unequivocal evidence in mice liver. *Food Funct* 6, 3006–3012 (2015).
229. Huang, W.Y., Liu, Y.M., Wang, J., Wang, X.N., and Li, C.Y. Anti-inflammatory effect of the blueberry anthocyanins malvidin-3-glucoside and malvidin-3-galactoside in endothelial cells. *Molecules* 19, 12827–12841 (2014).
230. Bunea, A. et al. Anthocyanin determination in blueberry extracts from various cultivars and their antiproliferative and apoptotic properties in B16-F10 metastatic murine melanoma cells. *Phytochemistry* 95, 436–444 (2013).
231. Diaconeasa, Z., Leopold, L., Rugina, D., Ayvaz, H., and Socaciu, C. Antiproliferative and antioxidant properties of anthocyanin rich extracts from blueberry and blackcurrant juice. *Int J Mol Sci* 16, 2352–2365 (2015).
232. Jeyabalan, J. et al. Chemopreventive and therapeutic activity of dietary blueberry against estrogen-mediated breast cancer. *J Agric Food Chem* 62, 3963–3971 (2014).
233. Kanaya, N., Adams, L., Takasaki, A., and Chen, S. Whole blueberry powder inhibits metastasis of triple negative breast cancer in a xenograft mouse model through modulation of inflammatory cytokines. *Nutr Cancer* 66, 242–248 (2014).
234. Yi, W., Fischer, J., Krewer, G., and Akoh, C.C. Phenolic compounds from blueberries can inhibit colon cancer cell proliferation and induce apoptosis. *J Agric Food Chem* 53, 7320–7329 (2005).
235. Mak, K.K. et al. Pterostilbene, a bioactive component of blueberries, suppresses the generation of breast CSCs within tumor microenvironment and metastasis via modulating NF-kappaB/microRNA 448 circuit. *Mol Nutr Food Res* 57, 1123–1134 (2013).
236. Montales, M.T. et al. Repression of mammosphere formation of human breast cancer cells by soy isoflavone genistein and blueberry polyphenolic acids suggests diet-mediated targeting of cancer stem-like/progenitor cells. *Carcinogenesis* 33, 652–660 (2012).
237. Jing, Y. et al. Protective effects of garcinol in mice with lipopolysaccharide/D-galactosamine-induced apoptotic liver injury. *Int Immunopharmacol* 19, 373–380 (2014).
238. Tsai, M.L., Chiou, Y.S., Chiou, L.Y., Ho, C.T., and Pan, M.H. Garcinol suppresses inflammation-associated colon carcinogenesis in mice. *Mol Nutr Food Res* 58, 1820–1829 (2014).
239. Li, F. et al. Garcinol sensitizes human head and neck carcinoma to cisplatin in a xenograft mouse model despite downregulation of proliferative biomarkers. *Oncotarget* 6, 5147–5163 (2015).
240. Liu, C. et al. Garcinol: Current status of its anti-oxidative, anti-inflammatory and anti-cancer effects. *Cancer Lett* 362, 8–14 (2015).
241. Ahmad, A. et al. Garcinol regulates EMT and Wnt signaling pathways in vitro and in vivo, leading to anticancer activity against breast cancer cells. *Mol Cancer Ther* 11, 2193–2201 (2012).
242. Kakarala, M. et al. Targeting breast stem cells with the cancer preventive compounds curcumin and piperine. *Breast Cancer Res Treat* 122, 777–785 (2010).

23 Packaging
A Noteworthy Feature in Food Safety

Raquel Sendón and Ana Sanches-Silva

CONTENTS

23.1 INTRODUCTION

Consumers are conscious of the importance of healthy and safe food in their health status. Packaging plays an essential role in the protection of goods (Marsh and Bugusu, 2007). However, packaging is also one of the most noteworthy features that can affect food safety. Migration, the transfer of chemicals from the packaging to foods, is the food packaging interaction that affects food safety the most. Many factors can influence migration, such as the type of polymer, the properties of the migrants, the food itself (e.g., physical state, fat content), and the time and temperature conditions of the contact (Hotchkiss, 1997; Arvanitoyannis and Bosnea, 2004). Indeed, the high fat content of the food and the high temperature of contact favor the release of substances from the packaging (Cruz et al., 2008; Silva et al., 2008). Migration is not a health issue *per se*; however, there is a need for establishing a list of acceptable materials, authorized substances, and their limits (Arvanitoyannis and Bosnea, 2004). Several topics make the establishment of legislation a complex issue, namely, the number of compounds that can be used in the manufacture of packaging, the scarcity or lack of information on the potential migrants (for instance, toxicological data), and the reduction or not suitable methods of analysis (most of food matrices are complex and most of the potential migrants are present at low levels, requiring low detection limits and the need for confirmatory techniques to avoid false positives or false negatives) (Sanches-Silva et al., 2005; Sendón García et al., 2006; Gallart-Ayala et al., 2013; Grob, 2014). Moreover, in some cases due to the ubiquitous contamination, it is difficult to obtain blank samples (e.g., phthalates, bisphenol A [BPA], and related compounds). The knowledge on the food contamination levels with these compounds is important to understand the potential risk to human health. Table 23.1 resumes the compounds that are more prone to migration from different food contact materials (FCMs); these include additives that are used to achieve a certain technological function in the packaging and

TABLE 23.1
Most Used Packaging Materials, Their Major Components, and Compounds More Prone to Migration from Each Material

Packaging Material	Major Components	Compounds More Prone to Migration
Plastic	Polymers (5000–1 million Da)	Molecules with <1000 Da: monomers and other starting substances, additives, catalysts, and solvents.
Paper and board	Pulp from different vegetable sources	Fillers, starch and derivatives, wet strength sizing agents, retention aids, biocides, fluorescent whitening agents, and grease-proofing agents.
Metal cans	Tin plate (steel coated with tin, steel coated with chromium and chromium oxides) or aluminum	Most cans are coated with a polymeric layer; therefore, the layer in contact with food is a lacquer.
Glass	Silica, sodium, and calcium oxides	Major components are natural constituents of foods; therefore, there is no food safety concern.
Ceramics	Silica and clays (alumina, silica, and water)	Glazes, printing inks, lead, and cadmium.

they include plasticizers, antioxidants, light and thermal stabilizers, lubricants, antistatic agents, antiblocking agents, dyes and pigments, and slip additives (Poças and Hogg, 2007). Figure 23.1 compiles the most representatives European Union (EU) legislation regarding FCMs. Nowadays, in order to obtain the best mechanical and barrier properties, most of the FCMs are multilayer or multilayer and multimaterial.

In this chapter, the most important contaminants of food with origin in food packaging were reviewed, namely, ultraviolet (UV)-ink photoinitiators, BPA, primary aromatic amines (PAAs), phthalates, alkylphenols and phenols, UV filters, perfluorochemicals, antioxidants and NIAS. All these chemicals could be food packaging contaminants, but not all known contaminants related to FCMs are here included; for example, nanomaterials (polymer nanocomposites), mineral oil

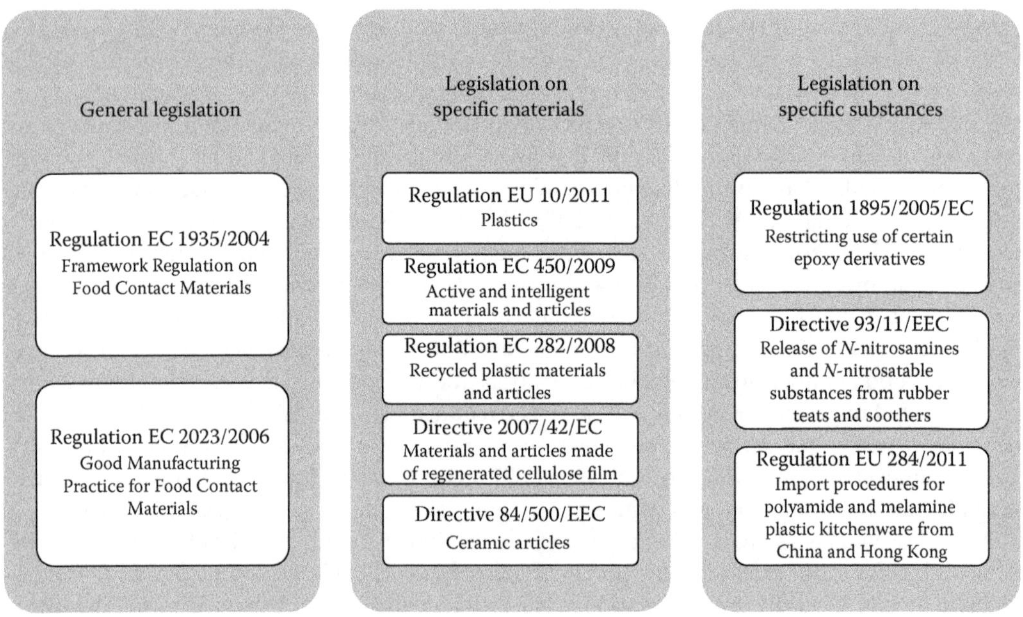

FIGURE 23.1 The most representative European Union legislation on food packaging materials.

hydrocarbons (mineral oil saturated hydrocarbons [MOSHs] and mineral oil aromatic hydrocarbons [MOAHs]) are topics of interest but are not addressed in this work.

Special attention was given to the analytical methods used to determine these substances in food and their regulatory control. Future trends in the area of the evaluation of food packaging safety were also addressed.

23.2 POSSIBLE FOOD PACKAGING CONTAMINANTS

23.2.1 Plasticizers

Plasticizers, namely, phthalate and nonphthalate (sebacate, citrate, adipate, and phosphate esters) plasticizers, are commonly used in plastic products due to enhanced material properties such as flexibility, elasticity, stretchability, softness, durability, and workability (Bošnir et al., 2007; Kueseng et al., 2007; Gallart-Ayala et al., 2013). Several studies have been made regarding plasticizers (Fankhauser-Noti et al., 2006; Bošnir et al., 2007; Kueseng et al., 2007; Lopez-Espinosa et al., 2007; Gärtner et al., 2009; Poças et al., 2010; Fasano et al., 2012; Fierens et al., 2012; Sendón et al., 2012; Demir and Ulutan, 2015). Although some plasticizers present low toxicity (e.g., sebacates, citrates, adipates), phthalates (diesters of ortho-phthalic acid) are considered to show a carcinogenic, mutagenic, and estrogenic effect and have also been associated with low fertility, asthma, and allergic symptoms in children (Lau and Wong, 2000; Bornehag et al., 2004; Gallart-Ayala et al., 2013). Due to their human health threats, several studies have concerned the toxicity of phthalate esters (Autian, 1973) and surveys have been carried out in the United Kingdom, Canada, and Hong Kong regarding the occurrence of plasticizers in food and food packaging due to their impact on health (Survey, 1987; Food Surveillance Sheet, 1995; Page and Lacroix, 1995).

Phthalates are considered as endocrine disruptors, that is, they are chemicals that alter the functioning of the endocrine system causing adverse health effects (Rudel et al., 2003; Mezcua et al., 2012). A plastic material may constitute up to 45% of its mass of phthalates, depending on the type and purpose of the product (Bošnir et al., 2007). Phthalates are mainly used as plasticizers of polymers such as polyvinyl chloride (PVC), but they can also be used in printing inks, lacquers, adhesives, and recycled pulp. Food ingestion is the main exposure pathway for these chemicals (Fierens et al., 2012). Therefore, it is important to monitor these chemicals in different foods to carry out up-to-date surveys, in order to assess human exposure.

Regarding the potential adverse effects of phthalates on human health, there are those who consider that the use of these compounds should be reduced due to their possible negative implications and those who consider that the risk associated with the exposure to phthalates is negligible because the studies on experimental animals that showed negative effects used levels of phthalates several times higher than the level of exposure of humans to these compounds, even in the worst scenario (Bošnir et al., 2007).

Di-2-ethylhexyl phthalate (DEHP), di-*n*-butyl phthalate (DBP), diisodecyl phthalate (DIDP), and di-isononyl phthalate (DINP) are the most commonly used phthalates with toxicological concerns (Gallart-Ayala et al., 2013). High-molecular-weight phthalates such as DEHP, DiNP, and DiDP are mostly used as plasticizers and low-molecular-weight phthalates (DEP, DBP, butyl-benzyl phthalate [BBzP]) as solvents to maintain color and scent of personal care products (Cao, 2010). Table 23.2 compiles the physicochemical information of some phthalates. The European Food Safety Authority (EFSA) has specified tolerable daily intakes (TDIs) for DnBP (0.01 mg/kg bw), DEHP (0.05 mg/kg bw), benzylbutyl phthalate (BBP) (0.50 mg/kg bw), and sum of DiDP and DiNP (0.15 mg/kg bw) (EFSA, 2005a,b,c,d,e,f). Plastic solid waste exposed to environmental conditions undergoes degradation. Therefore, phthalates may migrate to underground layers and water flows and contaminate waters that might be used by humans (Bošnir et al., 2007).

Several factors can affect the migration of phthalates to food. A study carried out on soft drinks and mineral water revealed that phthalates migrated several folds higher to soft drinks than to

TABLE 23.2
Chemical Properties and Other Information of Selected Phthalates

Phthalate	Abbreviation	CAS	Ref. No.	Formula	MW	Chemical Structure	Log K_{ow}[a]
Butyl-benzyl phthalate	BBzp	85-68-7	74560	$C_{19}H_{20}O_4$	312.36	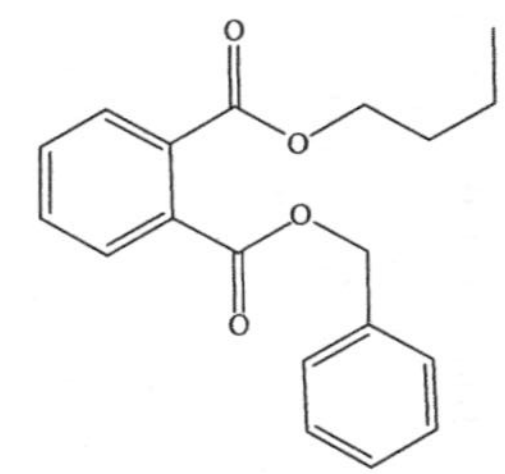	4.84
Dibutyl phthalate	DBP	84-74-2	74880	$C_{16}H_{22}O_4$	278.34	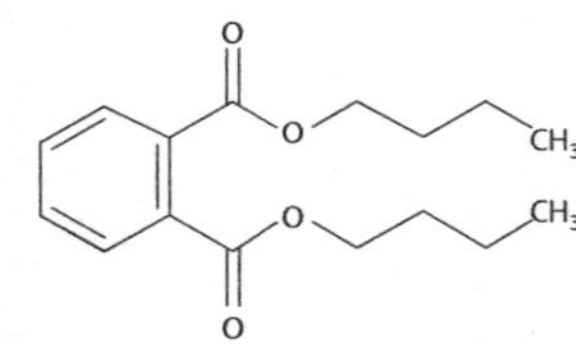	4.61
Bis(2-ethylhexyl) phthalate	DEHP	117-81-7	74640	$C_{24}H_{38}O_4$	390.56	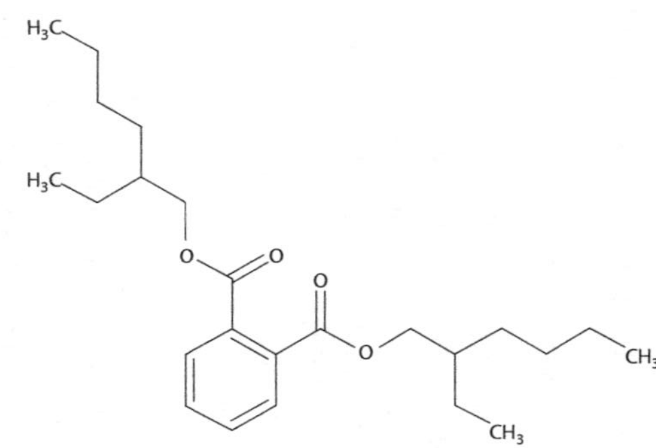	8.39

(Continued)

TABLE 23.2 (*Continued*)
Chemical Properties and Other Information of Selected Phthalates

Phthalate	Abbreviation	CAS	Ref. No.	Formula	MW	Chemical Structure	Log K_{ow}[a]
Diisodecyl phthalate	DiDP	68515-49-1	75105	$C_{28}H_{46}O_4$	446.66	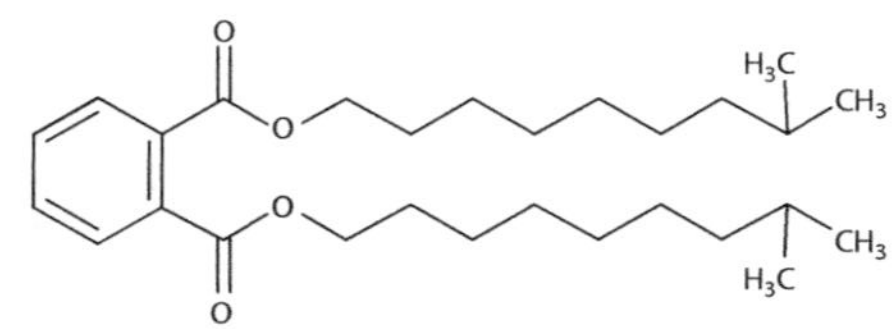	10.36
Di-isopropyl phthalate	DIP	605-45-8	—	$C_{14}H_{18}O_4$	250.29	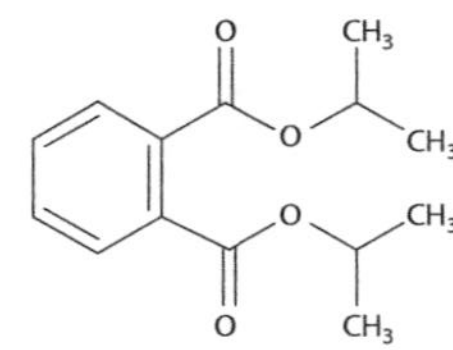	3.48
Di-isononyl phthalate	DiNP	28553-12-0	75100	$C_{26}H_{42}O_4$	418.61	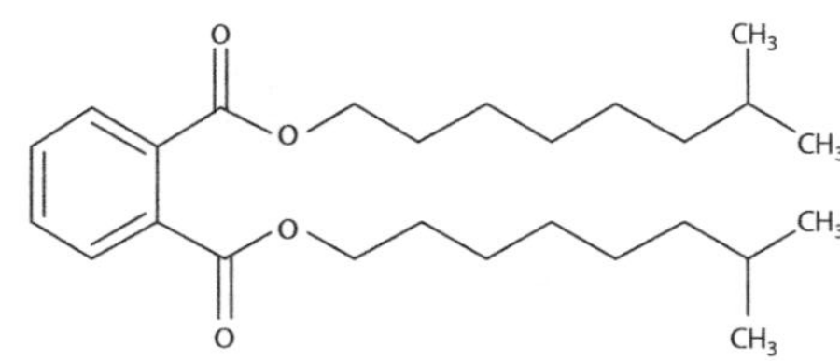	15.40

(*Continued*)

TABLE 23.2 (*Continued*)
Chemical Properties and Other Information of Selected Phthalates

Phthalate	Abbreviation	CAS	Ref. No.	Formula	MW	Chemical Structure	Log K_{ow}[a]
Dimethyl isophthalate	DMIP	1459-93-4	19210	$C_{10}H_{10}O_4$	194.18	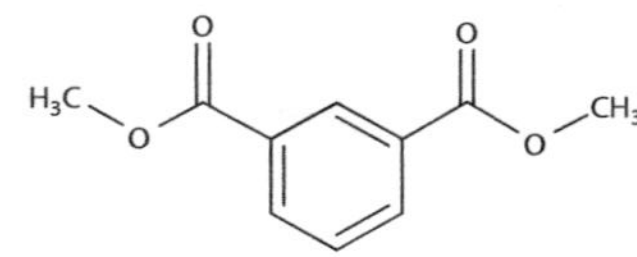	1.66
Diallyl phthalate	DAP	131-17-9	23230	$C_{14}H_{14}O_4$	246.26	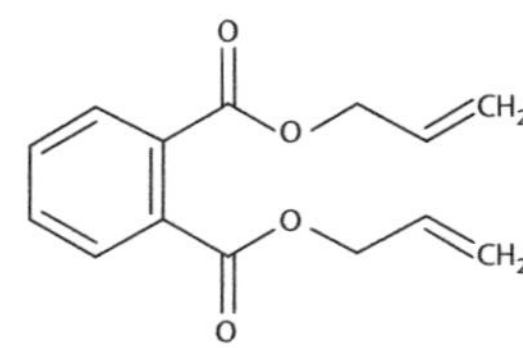	3.36
Diphenyl phthalate	DPP	84-62-8	—	$C_{20}H_{14}O_4$	318.32	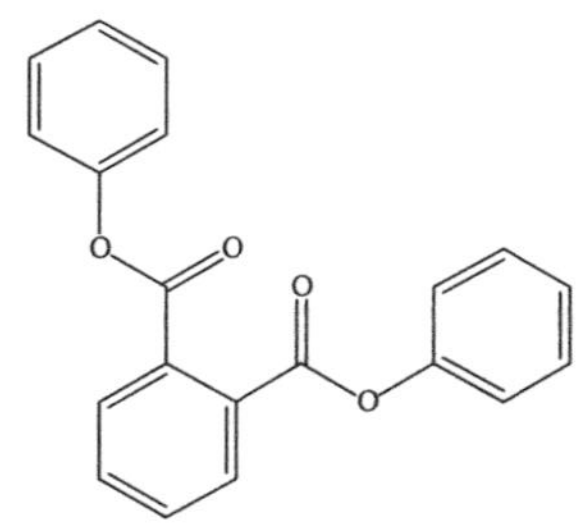	4.10

[a] Estimated with EPI Suite™ v 4.0.

mineral water, indicating that the acidic pH stimulated the migration of phthalates from plastic containers (Bošnir et al., 2007). Another effect that also influenced migration was the preservative used in soft drinks. It is important to know the potential factors that affect the migration of phthalates in order to be able to better design or select the packaging according to the food. Besides these factors, the composition of food is also important. Table 23.3 compiles the results of some studies that determined phthalates in FCMs or in foods after contact with these materials. A national survey (400 samples) analyzed phthalates in food products and packaging materials of the Belgian market (Fierens et al., 2012). This study has developed and validated extraction techniques for different matrices (high-fat foods, low-fat foods, aqueous-based beverages, and packaging materials). The lowest phthalate levels were found in baby foods and aqueous-based beverages, while DEHP was found in almost all the studied groups. Disobutyl phthalate (DiBP) was mainly responsible for cardboard phthalate contamination (Fierens et al., 2012). Sioen et al. (2012) reported after the dietary intake assessment of eight phthalates for the Belgian population that BBP, DnBP, and DEP were far below the TDIs. However, preschool children were in the worst case, reaching almost 80% of the TDI for DEHP.

Regarding the methods to determine phthalates, gas chromatography (GC)-MS is the most common method followed by LC-MS. However, in many cases, the detection limits of phthalates are high due to the background levels of contamination of the laboratory (plastic materials, glassware, reagents, and laboratory air) (Cao, 2010; Guo and Kannan, 2012).

23.2.2 UV Ink Photoinitiators

Photocured materials are obtained by photoinduced hardening or cross-linking of various monomers, oligomers, and polymer-based compositions. These materials could be cured, among others, by UV light and are commonly used as coatings, inks, and adhesives. Increasing demands for higher productivity and lower emission of volatile organic chemicals (VOCs) continue to support the expansion of photocured materials, especially important in case of printing inks and coatings in packaging applications and even more significant in food packaging where there is a continued need for safer, faster, and cheaper inks, functional coatings, and overprint varnishes (Polykarpov and Tiwari, 2015).

Thus, the main advantage of UV photocuring over other curing methods is the speed of the process mainly due to the high polymerization rates reached under UV irradiation. Therefore, the transformation of liquid resins into solid and insoluble polymers occurs in short times; moreover, it is a process carried out at room temperature, and unlike the classical solvent-based inks, the printing process occurs in the relative absence of residues (Lago et al., 2015a).

A UV curing ink basically contains a mixture of monomers, prepolymers (oligomers), pigments, and photoinitiators. The latter are highly photoactive compounds, usually of low molecular weight. There are two main types of photoinitiators used in the food contact industry: free-radical and cationic photoinitiators. Additionally, free-radical ones are divided into Type I and Type II photoinitiators. All these groups are very heterogeneous and with different reaction mechanisms. In case of free-radical photoinitiators, the UV energy produces free radicals that initiate a polymerization reaction: Type I photoinitiators involve one component that generates free radicals by fragmentation of cleavage (hydroxyacetophenones, benzylketals, benzoin derivatives, α-aminoalkylacetophenones, and phosphine oxides). Type II photoinitiators need an H-donor molecule, usually an amine synergist, to generate a ketyl radical and the donor radical (benzophenones and derivatives, and thioxanthones). Therefore, they are considered two-component photoinitiators. However, the cationic photoinitiators can contain sulfur, iodine, or phosphorous and an anion of type BF_4, PF_6, or AsF_6; they generate strong Brönsted acids that are very active initiators for the ring-opening polymerization. By far, most of the information found in the literature is related to the free-radical photoinitiators (Aparicio and Elizalde, 2015; Lago et al., 2015a).

TABLE 23.3
Compilation of Some Studies That Determined Phthalates and/or Adipates in Packaged Food

Phthalates/Adipates	Food/Food Simulant Tested	Technique Used	Results	Reference
DiDP, DINP, DEHP, DEHA	Food samples (158) packed in glass jars with metal closures	GC-MS	Twenty-two percent lids contained phthalates as the main additive and 6% DEHA. Concentrations in analyzed food are as follows: 270 mg/kg DINP, 740 mg/kg DIDP, 825 mg/kg DEHP, and 180 mg/kg DEHA.	Fankhauser-Noti et al. (2006)
DMP, DEP, DBP, BBP, DEHP, DOP	Soft drinks and mineral water packed in PET	GC-ECD	The pH level of the drink and the type of preservative influenced migration. DMP showed the highest migration in soft drinks. DBP and DEHP showed the highest migration into mineral water.	Bošnir et al. (2007)
DEHP and DEHA	Curry paste	GC-FID	DEHP found in the levels of 4 ng/g and DEHA in the range of 0.61 ng/g.	Kueseng et al. (2007)
DBP and DEHR	Takeaway food packed with paper and cardboard	HPLC-DAD and GC-MS	DBP ranged from 0.10 to 10,774 ng/g and DEHP from 0.52 to 61,013 ng/g.	Lopez-Espinosa et al. (2007)
DiBP, DEHP, DOP, BBP, DnBP, 2,6-DIPN	Infant food (milk powders, cereal flakes, semolina powders) packed with recycled paperboard containers	GC-MS	Only food packed in paper bags was highly contaminated with DiBP, DnBP, and DIPN. Paper does not provide an appropriate barrier against migration of semipolar compounds such as phthalates.	Gärtner et al. (2009)
DBP, DEHP, DiBP, DEP, DINP, DBP	Foods (cookies and biscuits, flour and sugar, cereals and dry pasta, butter, frozen ice cream, and chocolate) packed with paper and paperboard	GC-MS	DEHP was omnipresent in all packaging materials and in some cases migrated at levels close to safety limits. DiBP and DEHP varied between not detected and 360 and 2200 μg/kg fresh weight, respectively.	Poças et al. (2010)
DMP, DEP, DiBP, DnBP, BBP, DEHP, DCHP, and DnOP	Food products (fruits and vegetables, milk and dairy products, cereal and cereal products, meat and meat products, fish and fish products, fat and oils, snacks, condiments and sauces, baby food, beverages)	GC-EI-MS	DEHP was the phthalate with the highest concentration in almost all the studied groups (81% of samples), followed by DiBP (75%), DnBP (69%), and BBP (58%).	Fierens et al. (2012)

(*Continued*)

TABLE 23.3 (*Continued*)
Compilation of Some Studies That Determined Phthalates and/or Adipates in Packaged Food

Phthalates/Adipates	Food/Food Simulant Tested	Technique Used	Results	Reference
DMP, DBP, BBP, DEHP, and DEHA	Cans, marmalade cap, yogurt packaging, polystyrene dish, teat, bread bag, film, baby bottle, tetrapack and plastic wine top	GC-MS	Concentration of DEHP > DEHA > DBP > BBP > DMP. Migration was lower than SML and OML established.	Fasano et al. (2012)
BBP, DBP, DiNP, DiDP	Cork stoppers	LC-MS/MS	None of the cork stopper sampler analyzed for migration into ethanol 12% gave quantifiable levels of phthalates.	Sendón et al. (2012)

PET, polyethylene terephthalate.

Although these UV inks are applied in the outer surface of food packaging, they were responsible, in the last decade, of several notifications reported through Rapid Alert System for Food and Feed (RASFF) in the EU. The first one arose in 2005, and it was related to the presence of isopropylthioxanthone (ITX) in baby milk; benzophenone and 4-methyl-benzophenone were the focus of the EFSA in 2009 (Aparicio and Elizalde, 2015; Lago et al., 2015a). An overview of these notifications (around 145 during the last 10 years) shows different compounds found in a great variety of foods as a consequence of contact with different types of packaging (cartonboard, paperboard, different plastics), thus indicating that the problem of photoinitiators from FCMs is difficult to address (Lago et al., 2015b).

As it has been explained in the introduction, the term *migration* refers to the transfer of chemicals compounds from the package to the food. When photoinitiators are taken into account, how these compounds reach the food becomes a very important issue. There are two main routes: indirect and direct transfer. In the first case, the photoinitiators reach the food through the vapor, and semivolatile compounds are generally involved, as benzophenone and derivatives.

When direct transfer occurs, photoinitiators can follow several ways: diffusion and set-off. The low-molecular-weight molecules can diffuse through the different layer(s) of the package (from the outer side to the inner), but this phenomenon will be slower for components of high molecular weight. The set-off effect can be defined as the transfer of printing ink components from the external layer to the inner one, when they are in contact during storage of packages in reels or stacks until final assembling (Lago et al., 2015). Nevertheless, several mechanisms described earlier can be involved once these compounds are detected in foods.

Moreover, several other factors have been shown to affect the migration of photoinitiators such as its vapor pressure, molecular weight, polarity, and lipophilicity; time and temperature of storage; moisture content (when Tenax is employed as a food simulant); and fat content and porosity of the packaged food (Snedeker, 2014; Aparicio and Elizalde, 2015).

In order to ensure the safety of food, the presence of photoinitiators should be checked. But some difficulties could be found due to several reasons. As it was mentioned earlier, UV inks are complex mixtures and photoinitiators are a very heterogeneous group of molecules with very diverse physicochemical properties, so analytical approaches will modify as a function of the molecule or group investigated. Moreover, different by-products can be generated by photopolymerization and photocuring reactions. This leads to the probable presence of NIAS (see Section 23.2.8), making the analysis even more complex.

In the literature, liquid chromatography (LC) and GC are the most commonly used techniques to analyze food and/or FCM. Table 23.4 summarizes high-performance LC (HPLC)-based methods

TABLE 23.4
HPLC-Based Methods Found in the Literature

Technique	Photoinitiators	Sample	Extraction	References
LC-DAD/FLD	2-ITX	Food and FCM	LLE (QuEChERS)	Rothenbacher et al. (2007)
	MMMP, ITX, and EDB	Food	LLE (QuEChERS)	Jung et al. (2010)
HPLC-DAD	HCPK, DMPA, MMMP, BP, ITX, and EDB	Food and FCM	LLE	Sanches-Silva et al. (2008a)
	HCPK, DMPA, MMMP, BP, ITX, and EDB	Food simulants	Direct injection	Sanches-Silva et al. (2009)
	BP	Food and FCM	LLE	Pastorelli et al. (2008)
	HCPK, DMPA, MMMP, BP, ITX, EHA, BIS, THIO	Food simulants	SPE	Papilloud and Baudraz (2002a,b)
	HCPK, BP, ITX, EDB, and EHA	Food	LLE and SPE	Sagratini et al. (2008)
	HCPK, DMPA, MMMP, BP, ITX, and EHA	Food and FCM	LLE	Sanches-Silva et al. (2008b,c)
	2-HBP, 4-HBP, 4-MBP, MBB, DEAB, PBZ, BP	Food and FCM	LLE	Rodríguez-Bernaldo de Quirós et al. (2009)
	II: MK, DEAB, DMAB	FCM	LLE	Castle et al. (1997)
	2-HBP, 4-HBP, 4-MBP, MBB, DEAB, PBZ, BP, BPAcr, and EHA	FCM	LLE	Koivikko et al. (2010)
	BP	Food simulant	LLE	Li et al. (2012)
	HCPK, 4-MBP, PBZ, MBB, 4-HBP, DEAB, MK, DMAB, EDB, EHA	FCM	LLE	Jung et al. (2013)
LC-MS	HCPK, DMPA, MMMP, BP, ITX, and EHA	Food and FCM	LLE	Sanches-Silva et al. (2008a)
	ITX	Food	LLE	Benetti et al. (2008)
	HCPK, BP, ITX, EHA, EDB	Food	LLE and SPE	Sagratini et al. (2008)
	ITX	Food	PLE	Sagratini et al. (2006)
	4-MBP, PBZ, MBB, 4-HBP, DEAB, MK, DMAB, EDB, EHA	Food	LLE	Jung et al. (2013)
LC-MS/MS	HCPK, DMPA, HMPP, BP, PBZ, DEAB, 2-ITX, 4-ITX, DETX, EDB, EHA	Food and FCM	LLE and SPE (QuEChERS)	Gallart-Ayala et al. (2011)
	HCPK, BP, ITX, EHA, EDB	Food and FCM	LLE and SPE	Sagratini et al. (2008)
	2-ITX, EHA	Food	PLE and LLE	Gil-Vergara et al. (2007)
	ITX	Food	PLE	Sagratini et al. (2006)
	2-ITX	Food and FCM	LLE and SPE	Sun et al. (2007)
	2-ITX, 4-ITX	Food	LLE	Bagnati et al. (2007)
	2-ITX, 4-ITX	Food	LLE and SPE	Gallart-Ayala et al. (2008)
	HCPK, BDMB, MMMP, TPO, BP, ITX, EHA	Food and FCM	LLE and SPE	Shen et al. (2009)
	HCPK, 4-MBP, PBZ, MBB, 4-HBP, DEAB, MK, DMAB, EDB, EHA	Food	LLE	Jung et al. (2013)
	HCPK, DMPA, BP, 4-MBP, 2-ITX, MBB, PBZ, EDB, EHA	Food and FCM	LLE and SPE (QuEChERS)	Biedermann et al. (2013)

Source: Adapted from Lago, M.A et al., *Food Addit. Contam. A*, 32, 779–798, 2015a.

TABLE 23.5
GC-Based Methods Found in the Literature

Technique	Photoinitiators	Sample	Extraction	References
GC-MS	BP	Food and FCM	LLE	Pastorelli et al. (2008)
	BP, 4-MBP	Food	LLE and SPE	Van Hoeck et al. (2010)
	BP	Food and FCM	LLE	Anderson and Castle (2003)
	BP	Food and FCM	LLE	Papilloud and Baudraz (2002a,b)
	HCPK, BP, ITX, EHA, EDB	Food	LLE and SPE	Sagratini et al. (2008)
	2-ITX and EHA	Food and FCM	PLE and LLE	Gil-Vergara et al. (2007)
	HCPK, DMPA, MMMP, BP, ITX, EHA	Food and FCM	LLE	Sanches-Silva et al. (2008b)
	MK, DEAB, DMAB	FCM	LLE	Castle et al. (1997)
	HCPK, DMPA, MMMP, BP, ITX, EHA	Food simulants	SPE	Papilloud and Baudraz (2002a,b)
	2-HBP, 4-HBP, 4-MBP, MBB, DEAB, PBZ, BP, BPAcr; Amine syn: EHA	FCM	LLE	Koivikko et al. (2010)
	HCPK, BDMB, DMMB, MMMP, OMN 910, TPO, BP, PBZ, ITX, SP 7005, SP 7010, EHA	FCM	LLE	Bentayeb et al. (2013)
	HCPK, DMPA, BP, 4-MBP, ITX, EHA, EDB	Food and FCM	LLE	Negreira et al. (2010)
	HCPK, DMPA, MMMP, BP, 2-MBP, 3-MBP, 4-MBP, PBZ, 2-HBP, 4-HBP, MBB, 2-ITX, 4-ITX, DETX, EHA, EDB	Food and FCM	LLE	Bradley et al. (2013)
GC-MS/MS	ITX	Food	LLE and SPE	Allegrone et al. (2008)
GC-FID	HCPK, DMPA	Food simulants	LLE	Wang et al. (2009), Huang et al. (2012)

Source: Adapted from Lago, M.A et al., *Food Addit. Contam. A*, 32, 779–798, 2015a.

found in the literature (Lago et al., 2015a). LC is the elected technique most of the times due to the low volatility of photoinitiators; when coupled to a diode array detector (DAD) and/or fluorescence detection (FLD), it is the most extensively used for quantification purposes, and when coupled to mass spectrometry (MS or MS/MS), it can be used for confirmatory purposes or both identification and confirmation. However, LC-MS and LC-MS/MS are now commonly used to detect various substances. GC is very common for determining semivolatile photoinitiators such as benzophenone and derivatives. Moreover, different detectors can be used, among which the MS detector is the most common (Table 23.5; Lago et al., 2015b).

Existing information on risk assessment on photoinitiators is largely limited to studies on BP and 4-MBP. Even so limited evidence is available on the toxicology of BP and very little information other than EFSA-mandated genotoxicity studies is currently available on 4-MBP (Snedeker, 2014).

A recent study suggests that 2,2-dimethoxy-2-phenylacetophenone (2,2-DMPAP), 2-ethylhexyl-4-(dimethylamino) benzoate (2-EHDAB), 1-hydroxycyclohexyl-phenyl ketone (1-HCHPK), and methyl-2-benzoylbenzoate (MBB) induced cytotoxicity, whereas 2-ITX did not. Furthermore, 2-methyl-4′-(methylthio)-2-morpholinopropiophenone (MTMP) caused a slight increase in cytotoxicity. It is concluded that the combination of 1,2 dichloropropanone (the solvent used as a paint remover) and photoinitiators may increase the risk of respiratory diseases (Kawasaki et al., 2015).

Reitsma et al. (2013) evaluated the effects of 2-ITX, 4-ITX, 2,4-diethylthio-xanthone (2,4-diethyl-TX), 2-chlorothioxanthone (2-chloro-TX), and 1-chloro-4-propoxythioxanthone (1-chloro-4-propoxy-TX). None of the compounds showed androgenic or estrogenic activities, but clear antiandrogenic and antiestrogenic activities were observed for 2-ITX, 4-ITX, and 2,4-diethyl-TX, whereas 2-chloro-TX showed only antiandrogenic activity. They concluded that thioxanthone derivatives could act as potential endocrine disruptors at the level of both nuclear receptor signalling and steroid hormone production.

However, the photoinitiators mentioned earlier are just only a few of the many photoinitiators used in UV inks; however, except for benzophenones (which are subject to the restrictions set out in Regulation EU 10/2011 [European Commission, 2011]), most of them are not fully evaluated. EFSA, in the last meeting of the Food Ingredients and Packaging (FIP) Unit, expressed that printing inks (among others) have no specific EU regulation, whereas they are regulated at the national level, as is the case of Switzerland and Germany (not yet in force). The Swiss ordinance is based on the Council of Europe Resolution ResAP (2005) 2 on packaging inks and was a result of a collaboration between the Swiss Federal Office of Public Health (FOPH) and the European Printing Ink Association (EuPIA) (EFSA, 2014). The substances considered here are divided into two parts: evaluated and nonevaluated substances, which must not be detectable in a migration test at a detection limit not exceeding 0.01 mg/kg foodstuff or food simulant. Moreover, exclusion criteria are also applied (ILSI, 2011).

The EFSA Scientific Cooperation (ESCO) working group published an external scientific report in which one suggested strategies for the evaluation of substances when limited or that no information on toxicity is available, which could be based on a human exposure threshold approach. This approach uses structural information based on Cramer classification, which categorizes substances according to their potential toxicity into three types: class I, class II, and class III. The annex of the report shows the classification of some photoinitiators used in printing inks for FCMs (EFSA, 2012; Lago et al., 2015a). Available information on the risk assessment on photoinitiators is largely limited to studies on BP and 4-MBP (Snedeker, 2014).

23.2.3 Bisphenol A

Bisphenol A is one of the most common industrial chemicals produced in the world and is synthesized by the condensation of acetone with two equivalents of phenol (Schug and Birnbaum, 2014). Some important applications from the FCM point of view are as follows: It is the monomer used in the manufacture of polycarbonate, a thermoplastic polymer with high transparency, low weight, and high heat and impact resistance (Maia et al., 2009). Polycarbonate was used in returnable beverage containers and infant baby bottles among other applications (López-Cervantes and Paseiro-Losada, 2003). Moreover, BPA is one of the products used, jointly with epichlorohydrin, for the synthesis of the diglycidyl ether of bisphenol A (BADGE), which is the monomer employed in epoxy resins and widely used as coatings inside food cans or big storage vessels and/or as additive for vinyl and acrylic resins (Paseiro-Losada et al., 1993). BPA was also associated with cardboard take-out containers and other paper products, especially if they are made from recycled materials (Vandenberg, 2014).

BPA is considered an endocrine-disrupting chemical with estrogenic activity and nowadays is ubiquitous in our environment, and it is found in our bodies, and so epidemiological studies regarding BPA exposure and human health effects can be found elsewhere (Schug and Birnbaum, 2014).

But exposure to BPA is a very controversial issue and is even more disputable after the ban of its use in polycarbonate infant baby bottles in 2011 (European Commission, 2011). A recent report published by the EFSA stated that "the dietary exposure to BPA for the highest exposed groups, which include infants children and adolescents, is below the t-TDI of 4 μg/kg bw per day, indicating that there is no health concern for BPA at the estimated levels of exposure. These conclusions also apply to prenatally exposed children and to the elderly." Additionally, further research is recommended in order to minimize the uncertainties surrounding this risk assessment (EFSA, 2015).

Regarding the analysis of BPA, hundreds of works can be found in the literature with focus on different matrices, but some reviews include many of them. As an FCM migrant, BPA can be analyzed in a raw material, in a food simulant, and/or in a food; but generally, it could be extracted from food with polar solvents immiscible with fat. Moreover, due to its physicochemical characteristics, that is, its very good response in fluorescence, HPLC with FLD (HPLC-FLD) is the appropriate determination technique for BPA in food. Also, HPLC and GC with MS or MS/MS detectors are also widely used (Sendon García et al., 2006). More recently, the determination of BPA in food was addressed by Ballesteros-Gomez and coworkers paying particular attention to the available strategies for sample treatment (Ballesteros-Gómez et al., 2009).

23.2.4 Primary Aromatic Amines

PAAs are chemical compounds, well known as by-products of chemical manufacturing and contaminants of dyes, rubber, and textiles; they are also considered environmental pollutants as a consequence of gasoline and coal combustion and are present in tobacco smoke (Palmiotto et al., 2001).

Regarding FCMs, PAAs can occur in these articles as a consequence of the presence/use of diisocyanates (the polyurethane adhesives employed in laminates) or azo-dyes (employed, for instance, as colorants in nylon utensils).

Polyurethanes can be used as adhesives not only in laminated films but also in printing inks and lacquers; this type of adhesives is formed by the polymerization of polyols and diisocyanate monomers. If the ratio between the monomers is well balanced and the reaction is complete, all monomers disappear to form the polymeric network. But if free isocyanates are present (e.g., due to incomplete curing) and reach the food, the presence of water leads to the formation of PAAs (Aznar et al., 2009; Sendón et al., 2010).

Nylon kitchen utensils are another possible source of PAA contamination. Nylon is the generic name of polyamides that have characteristic amide groups in the backbone chain. Because these amide groups are very polar and interact with each other by hydrogen bonds and the polyamide backbone is regular and symmetrical, nylons are often crystalline and make very good fibers. There are two classes of synthetic nylons: one formed from cyclic monomers, for example, nylon 6 (manufactured from ε-caprolactam) and nylon 12 (from the lactam of 12-amino dodecanoic acid), and the second class involves various diamines and diacids, for example, nylon 6,6 (made from hexamethylene diamine and adipic acid) (Gupta, 1989). In the manufacture of these nylon kitchen utensils (usually black), azo dyes (synthetic organic colorants with characteristic chromophoric azo groups) could be incorporated where PAAs are the starting substances (Sendón et al., 2010).

Consequently, if the FCMs mentioned earlier are not well manufactured or made with inappropriate raw material, PAAs can contaminate the food.

PAAs have been implicated in several alerts, mainly related to PAA migration from kitchen utensils made of nylon imported mainly from China. The European legislation specifies that the levels of PAAs (total PAAs expressed as aniline) in food or in food simulants should not be detectable (i.e., 0.01 mg/kg), except for those that are listed in Regulation 10/2011 (European Commission, 2011). Though the first RASFF notification concerning PAA migration from cooking utensils was in 2004, the issue seems unresolved since during the last few years in the RASFF database, several notifications were found (Figure 23.2) (RASFF, 2015).

For screening purposes, a spectrophotometric method is available (based on diazotization of free PAA), but if PAA levels are above 2 μg aniline equivalent/kg, a confirmation method should be used (JRC, 2011); but this method is not selective, that is, no information about which PAA is obtained. So since then, numerous analytical methods that attain better selectivity (and also reach the required detection limit of 10 μg/kg) have been published. LC with UV detection has been used for the analysis of aniline in plastic materials subjected to repeat testing (Brede and Skjeurak, 2004) and for the determination of eight PAAs in the aqueous migration solution of laminates (Ellendt et al., 2003). Furthermore, GC coupled to MS was employed before (Brede and Skjeurak, 2003), but because a

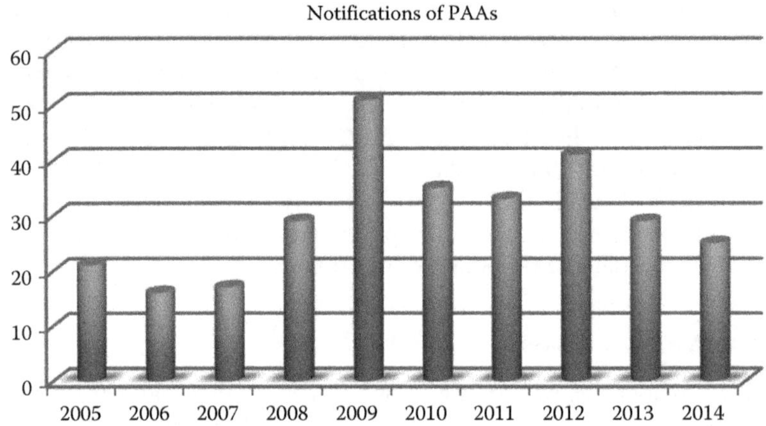

FIGURE 23.2 Notifications related to primary aromatic amines through the Rapid Alert System for Food and Feed system during the last few years.

prior step of derivatization is needed, this analytical tool has been replaced mainly by LC coupled to MS/MS techniques (Mortensen et al., 2005, Aznar et al., 2009, Sendón et al., 2010).

Recently, an LC-MS method was validated for 22 PAAs and its suitability was checked in food contact multilayer plastic laminates and paperboard materials (Lambertini et al., 2014); the same PAAs were also tentatively determined using an LC-Orbitrap full-scan HRMS by direct injection of the food simulant acetic acid 3% (w/v) (Mattarozzi et al., 2013). In line with this, Paseiro-Cerrato and coworkers have demonstrated the potential of using direct analysis in real-time mass spectrometry (DART-MS) as a tool for determining the presence of these compounds in food contact utensils (Paseiro-Cerrato et al., 2014).

Most of the methods published lately achieve the limit of detection (LOD) below the one recommended by Regulation (EU) 10/2011 (10 μg/kg food simulant); therefore, due to the toxicology concern related to PAAs, the suggested LOD for PAAs could be decreased to 2 ppb (EFSA, 2014).

Some of these PAAs are classified as "carcinogenic to humans" (Group 1), "probably carcinogenic to humans" (Group 2A), and "possibly carcinogenic to humans" (Group 2B) by the International Agency for Research on Cancer (IARC) (Group 3 corresponds to substances "not classifiable as to its carcinogenicity to humans," which often means that further research is needed [IARC]). Some examples are given in Table 23.6.

23.2.5 Alkylphenols: Nonylphenol and Octylphenol

Alkylphenols include *p*-nonylphenols, octylphenols, and their ethoxylated derivatives (Figure 23.3). The concern regarding these compounds is based on their estrogenic properties (Loyo-Rosales et al., 2004).

Alkylphenols are used for plastic production. In fact, around 10% of nonylphenol (NP) is used to produce tris(nonylphenyl)phosphate (TNPP), an antioxidant and heat stabilizer (which contributes for processing stability and performance integrity) used in plastics (Fernandes et al., 2008; Muncke, 2009). Several studies have reported the study of the migration of alkylphenols from packaging to food (Kawamura, 2000; Fernandes et al., 2003; Loyo-Rosales et al., 2004; Fernandes et al., 2008). Some studies revealed that several sources/pathways may contribute to the contamination of food with alkylphenols, once the amount of NP found did not correlate with packaging. Bottled water consumption can contribute to alkylphenol intake (Layo-Rosales et al., 2004). Moreover, different routes may contribute to human exposure such as inhalation, dermal absorption, or ingestion

TABLE 23.6
IARC Classification of Some PAAs Related to Food Contact Materials (IARC)

Name	Abbreviation	CAS No.	Structure	IARC Classification Group
m-Phenylenediamine	m-PDA	108-45-2	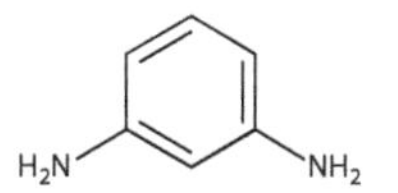	3
2,4-Toluenediamine	2,4-TDA	95-80-7	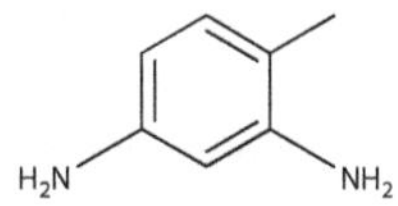	2B
1,5-Diaminonaphthalene	1,5-DAN	2243-62-1	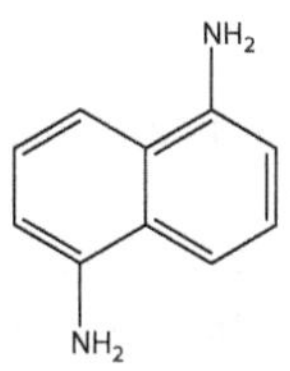	3
Aniline	ANL	62-53-3	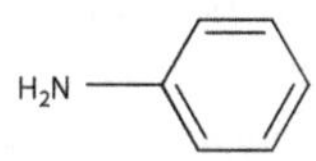	3
o-Toluidine	o-T	95-53-4	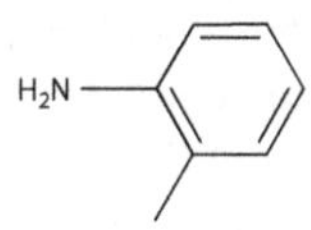	1
4,4′-Diaminodiphenylether	4,4′-DPE	101-80-4	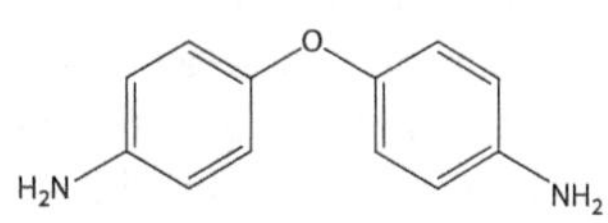	2B
Benzidine	BNZ	92-87-5	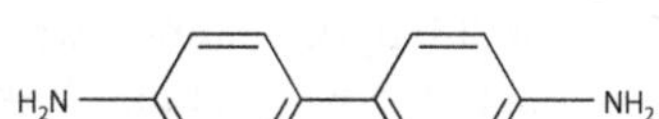	1
4,4′-Methylenedianiline	4,4′-MDA	101-77-9	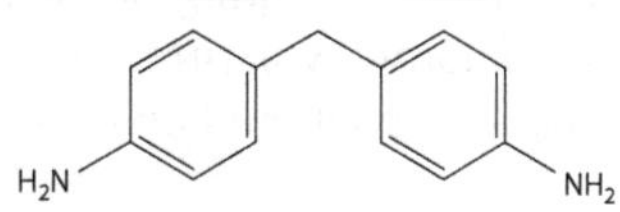	2B
3,3′-Dimethylbenzidine	3,3′-DMB	119-93-7	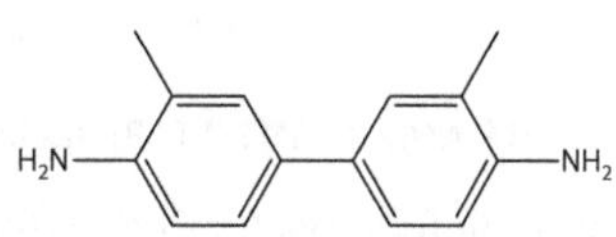	2B

of contaminated food (Mezcua et al., 2012). An interesting study revealed that 100% of indoor air samples detected the presence of 4-nonylphenol, although lower concentrations were found in outdoor air samples (Rudel et al., 2003).

23.2.6 Perfluorinated Compounds

Perfluorinated compounds have a wide range of applications in textile, household, automotive, and electrical wiring industries (Begley et al., 2005, 2008). Regarding FCMs, perfluorinated compounds are mainly used in nonstick coating (polytetrafluoroethylene [PTFE]) for cookware and in paper coating to improve resistance to oil and moisture (Trier et al., 2011). Fluorochemicals

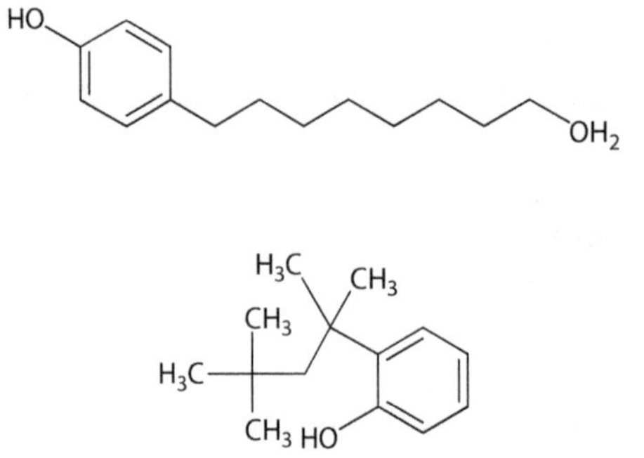

p-nonylphenol

CAS: 104-40-5
MW: 220.354

Octylphenol

CAS: 27193-28-8
MW: 206.327

FIGURE 23.3 Chemical structure of *p*-nonylphenol and octylphenol.

have been associated with adverse health side effects such as inhibition of male sex hormone synthesis (Rosenmai et al., 2013).

PTFE is a very resistance and stable material. In fact, it has a very high melting point (327°C).

Some papers (e.g., popcorn bag paper) used in contact with high-fat-content foods are treated with fluorotelomer or fluorotelomer-based paper coatings (mixtures of C6, C8, C10, and C12 perfluorochemicals or high-molecular-weight polymers with fluorometer-based side chains) to avoid oil stains or oil soak through paper (Begley et al., 2005).

Perfluorochemicals are nonvolatile and lack a sensitive chromophore; therefore, conventional techniques such as GC-MS or LC-UV are not suitable for most of these compounds. LC-MS/MS is required to allow the quantification of these compounds (Begley et al., 2005).

Two of the perfluorochemicals that have been subject of studies include perfluorooctane sulfonate (PFOS) and perfluorooctanoic acid (PFOA). PFOS is a residual impurity of paper coating, while PFOA is a processing aid in the manufacture of PTFE (Begley et al., 2005). The study carried out by Begley and coauthors reported residual amounts of PFOA (in low range of μg/kg) in PTFE cookware and PFOA of 300 μg/kg in microwave popcorn bag paper (Begley et al., 2005).

In a study carried out by Powley et al. (2005), salts of PFOA were used in the manufacture of fluoropolymers, applied to the coating of metal cookware. PFOA was analyzed by LC-MS/MS. None of the samples (fluoropolymer-treated cookware) showed detectable levels of PFOA after extraction under simulated cooking conditions (detection and quantification limits were as low as 100 pg cm^{-2}).

23.2.7 Antioxidants and Light Stabilizers

Antioxidants are common food packaging additives used to avoid degradation caused by oxidation. One of the main primary antioxidants is Irganox 1076 (which is used to intercept and stabilize free radicals), while Irgafos 168 is one of the main secondary antioxidants (which used to avoid further formation of free radicals) (Helmrothy et al., 2002; Reinas et al., 2012). Several papers have studied the migration of these compounds in food simulants or foods (Marcato et al., 2003; Lundbäck et al., 2006; Sanches-Silva et al., 2007). Regarding light stabilizers, these are used as radical scavengers having high molecular weight, which limits their mobility and consequently their loss, during processing and use. Tinuvin 622 and Chimassorb 944 are two of the most used hindered amine light stabilizers (HALSs) (Coulier et al., 2005; Su et al., 2015). Table 23.7 compiles some studies carried out with common antioxidants and/or light stabilizers.

Dopico-García et al. (2012) have carried out an extensive review regarding the analytical procedures to determine the additives of polyolefins, namely, antioxidants, UV light stabilizers, antistatic agents, and colorants.

TABLE 23.7
Compilation of Some Studies That Determined Antioxidants and/or Light Stabilizers Used in the Processing of Polymeric FCMs

Antioxidants/Light Stabilizers	FCM	Extraction Technique	Chromatographic Technique	Results	References
BHA, BHT, Irganox 1010, Ethanox 330, Irgafos 168, Irganox 1076[a]	LDPE	LLE (n-hexane)	RP-HPLC-DAD	Antioxidants were not found under the legislation limits, although Ethanox 330 and Irgafos 168 have been found in trace levels.	Dopico-García et al. (2003)
Tinuvin 770, Chimassorb 944, Tinuvin 622, Chimassorb 119, Chimassorb 2020	PP and HDPE	Dissolution (toluene)/ precipitation (methanol)	GC-MS and HPLC-UV/ELSD	Pyrolysis GC-MS developed for fast and straightforward identification. HPLC-UV/ELSD developed for quantification. Analytical performance was satisfactory.	Coulier et al. (2005)
Irganox 1076	LDPE	Dissolution (toluene)/ precipitation (methanol)	HPLC-DAD	The objective of the study was to obtain the concentration evolution of the antioxidant in the food simulants to predict its migration based on a mathematical model.	Galotto et al. (2011)
Irganox 1076	LDPE	LLE (*fatty solid foods,* hexane; nonfatty solid foods, acetonitrile and hexane; liquid foods, hexane)	GC-MS	Results showed that the migration of Irganox 1076 increased with fat content and storage temperature. Kinetics in simulants was higher than in selected foodstuffs.	Beldì et al. (2012)
Irgafos 168 and Irganox 1076	LDPE	LLE (isooctane)	GC-MS	Migration into Tenax was higher than into rice, showing that the food simulant tends to overestimate migration values.	Reinas et al. (2012)
Tinuvin 622, Cyasorb UV 9, Chimassorb 81, Isonox 129, Tinuvin 326, Irganox 3114, Tinuvin 327, Tinuvin 328, Vitamin E, Irganox 1010, Irganox 1330, Naugard DLTDP, Irganox 1076, Irgafos 168, among others	Polymers	SPE	UHPLC-MS/MS	A method was optimized and validated to monitor the extraction of additives and additive degradation products from polymer packaging.	Pouech et al. (2014)
Tinuvin P and Irganox 3114	HDPE	Food simulants, direct injection after dilution; milk, LLE (tetrahydrofuran)	LC-MS/MS	The food simulant gave a close estimation for Tinuvin P in the case of milk samples, while for Irganox 3114 it gave an overestimation with a factor of minimum 3.5.	Bodai et al. (2015)

[a] DBP, a degradation product of Irgafos 168, was also determined.

Su et al. (2015) reported a study in which the effect of antioxidants and light stabilizers on the migration of nanosilver–polyethylene composite packaging films into food simulants is evaluated. The results revealed that the addition of selected additives could decrease the migration of silver compared with the composite film without additives.

23.2.8 Non–Intentionally Added Substances

Non–intentionally added substances (NIAS) comprise reaction and degradation compounds, impurities, and substances from printing inks, adhesives, and solvents. The lack of information (e.g., identity, physicochemical properties, biological activity) is the greatest concern regarding NIAS, while the greatest difficulty is their identification due to the lack of information regarding the ingredients used to manufacture FCMs due to confidential agreements, which only declare major components (Poças and Hogg, 2007; ILSI, 2015). Moreover, NIAS can also negatively influence the organoleptic properties of packed foods and contribute to an unacceptable migration from the packaging to the food. However, due to the low levels of these substances, it is unlikely that a health concern can be generated. The complexity of FCMs tends to increase, with different layers, materials, and additives; therefore, the number of NIAS is increasing (Nerín et al., 2013). In fact, Regulation 10/2011 (European Commission, 2011) defines NIAS and recognizes their formation during manufacture and use of plastic materials and articles. Recent concerns regarding NIAS identification and safety assessment had led to the publication of different studies regarding this subject (Honkalampi-Hämäläinen et al., 2010; Félix et al., 2012; Bach et al., 2013, Isella et al., 2013; Nerín et al., 2013; Koster et al., 2014). Table 23.8 compiles some examples of NIAS according to their origin (degradation processes, impurities, neoformed compounds, and contaminants).

In order to assess food safety, it would be important to have a sensitive and selective tool that allowed detecting alert chemical structures in complex mixtures (Nerín et al., 2013). Overall, migration or specific migration tests cannot cover the evaluation of NIAS; therefore, improved screening methods such as techniques with MS detectors and chemical databases such as headspace GC-MS can be valuable tools for their control (Franz and Welle 2008). Greater attention is given to compounds with less than 1000 Da once it is generally accepted, namely, by the EFSA, that compounds with higher molecular weight are not absorbed by the body and they can be excluded from migration and exposure calculations (EFSA, 2008). Nerín et al. (2013) proposed valuable analytical protocols and decision tree for the identification of NIAS. In sum, for volatile substances, GC-MS is the most used technique due to the possibility of using mass spectra libraries. However, when substances are not volatile, LC-MS is recommended although libraries are not available. Recent LC-MS advances such as the possibility of alternating ionization modes in a single run and powerful softwares for data treatment allow identification of NIAS.

23.3 TRENDS IN THE USE OF ADDITIVES

Today, there is a major concern regarding the investigation for sustainable goods and sustainable food packaging is not an exception (Russell, 2004). In fact, there is a great attempt of replacing non-bio-based FCMs by biodegradable and cost-effective materials. Biopolymers are very promising materials, but in general, they present poor barrier and mechanical properties; therefore, there is the need of using additives. The use of natural additives, which are not associated with adverse health side effects, such as natural plasticizers (epoxidized triglyceride vegetable oils from soybean oil, linseed oil, castor oil, sunflower oil, and fatty acid esters [FAEs]) (Vieira et al., 2011) or natural antioxidants, is attracting great attention both from the industry and from the academia. However, more investigation and further developments are required in this area until it is possible to effectively replace the majority of non-bio-based materials by workable biopolymers.

TABLE 23.8
Examples of NIAS according to Their Origin (Degradation Processes, Impurities, Neoformed Compounds, and Contaminants)

Origin	Where	Cause	Example	References
Degradation processes	Polymer itself	High temperatures	Formaldehyde and acetaldehyde from PET.	Mutsuga et al. (2007)
		High irradiation energies		Deschênes et al. (1995), Jeon et al. (2007)
	Additives added to polymer	Antioxidants	Irganox 1010 and Irgafos 168 can originate different NIAS after microwave heating such as 2,4-ditertbutyl-phenol (2,4-DTB), 2,6-ditertbutyl-p-benzoquinone (2,6-DTBQ), and 3,5-ditertbutyl-4-hydroxyphenylpropionic acid.	Pouech et al. (2014)
			Tris(nonylphenyl)phosphite is used in PVC, polyolefins, and acrylics, and degradation originates NP or OP.	Howe et al. (2001)
		Surfactant	Polyethoxylated NP can originate NP or OP	Inoue et al. (2001)
		Stabilizer	Chimassorb 944 can originate (Z)-9-octadecenamide, 2,2,6,6-tetramethyl-4-aminopiperidine, 2,4-ditertbutyl-6-nitro-phenol and 2-cyclohexene-1-dione, 3,5-dimethyl, or o-methyloxime.	Haider and Karlsson (2001)
	Additives added to adhesives, coatings, or inks		Abietic acid, a tackifier in hot-melt adhesives for food packaging, can originate 4b-8-dimethyl-2-isopropylphenanthrene, dehydroabietin 1-methyl-10,18-bisnorabieta-8,11,13-triene, retene, dehydroabietal, and dehydroabietic acid methyl ester.	Castle et al. (1997), Ozaki et al. (2004, 2006)
Impurities			N2-Dodecanoyl-L-arginine (LAS), an impurity of ethyl lauroyl-arginate (LAE), which is an active agent of PET.	Aznar et al. (2013)
			3-Methyl-3-(4-methyl-3-pentenyl)-2-oxiranecarbaldehyde, an impurity of citral.	Aznar et al. (2012)
			1-Hexanol-2-ethyl and 2-ethylhexylacetate, impurities of 2-ethylhexylacrylate, which is a monomer of acrylic adhesives.	Canellas et al. (2010)
Neoformed compounds			PAAs formed from aromatic isocyanates when in contact with water.	Brede et al. (2003)
			Hydrolysis products (BAGE · H_2O and BADGE · $2H_2O$), chlorohydroxy derivatives (BADGE · HCl and BADGE · 2HCl), and oxidized form of BADGE (BAMGE · H_2O), used in epoxy based of food cans.	Lane et al. (2015)
			ESBO chlorohydrins from thermal degradation of PVC with ESBO.	Suman et al. (2010)
Contaminants			Found in recycled materials coming from previously packaged food or misuse by the consumer before discarding the packaging.	Coulier et al. (2007)

PET, polyethylene terephthalate.

23.4 CONCLUDING REMARKS

Due to the tendency to use biopolymers and natural additives, it is likely that the most common food packaging contaminants will change drastically in the new few years. Although newer and more sophisticated processing and analytical techniques are emerging, the task of identifying and quantifying contaminants of food packaging will continue to be a huge challenge due to a variety of compounds implicated, the low levels of migration, the reaction and degradation products (within the food packaging and resulting from the contact of FCMs with food), and the increasing complexity of materials, namely, multilayer food packaging, multilayer and multimaterial food packaging, nano-based food packaging, and recycled food packaging. Each of these materials adds new concerns regarding the evaluation of FCM safety. The European legislation on FCMs needs special attention; currently, only plastic materials, ceramics, regenerated cellulose films, and active and intelligent packaging are specifically regulated of the 17 specific materials referred in the framework regulation (European Commission, 2004). In response to the continuous changes in the FCMs' industry and in order to guarantee consumer safety, the implantation of newer legislation on other specific groups of FCMs is expected in the near future, besides the constantly update of the already existing.

ABBREVIATIONS

2,6-DIPN	2,6-Di-isopropyl naphthalene
2-DMEB	2-(Diethylamino)ethylbenzoate
2-HBP	2-Hydroxybenzophenone
2-MBP	2-Methyl-benzophenone
3-MBP	3-Methyl-benzophenone
4-HBP	4-Hydroxybenzophenone
4-MBP	4-Methyl-benzophenone
BADGE	Bisphenol A diglycidyl ether
BBP	Benzylbutyl phthalate
BBzP	Butyl-benzyl phthalate
BDMB	2-Benzyl-2-dimethylamino-4-morpholino butyrophenone
BHA	Tert-butyl-hydroxyanisole
BHT	2,6-Di-tert-butyl-*p*-cresol
BIS	Bis(4-diphenylsulfonium)phenylsulfide-bis(hexafluorophosphate)
BP	Benzophenone
BPA	Bisphenol A
BPAcr	Benzophenone acrylate: polymeric photoinitiator (59626-79-8)
Chimassorb 119	*N,N*-[1,2-ethanediylbis[[[4,6-bis-[butyl(1,2,2,6,6-pentamethyl-4-piperidinyl) amino]-1,3,5-triazine-2-yl]imino]-1,3-propanediyl]]bis[*N,N*-dibutyl-*N,N*-bis(1,2,2,6,6-pentamethyl-4-piperidinyl)-1,3,5-triazine-2,4,6-triamine
Chimassorb 2020	1,6-Hexanediamine, *N,N*-bis(2,2,6,6-tetramethyl-4-piperidinyl)-polymer, reaction products with *N*-butyl-1-butanamine and *N*-butyl-2,2,6,6-tetramethyl-4-piperidinamine
Chimassorb 81	2-Hydroxy-4-(octyloxy)benzophenone
Chimassorb 944	Poly-{6-[1,1,3,3-tetramethylbutyl]-imino]-1,3,5-triazine-2,4-diyl}{2-(2,2,6,6-tetramethylpiperinyl)-imino}
CPQ	D,L-Camphorquinone
Cyasorb UV 9	2-Hydroxy-4-methoxybenzophenone
DAD	Diode array detector
DART	Direct analysis in real time
DBA	9,10-Dibutoxyanthracene

DBP	Dibutyl phthalate
DCHP	Dicyclohexyl phthalate
DEAB	4,4′-Bis(diethylamino)-benzophenone
DEHA	Di-2-ethylhexyl adipate
DEHP	Diethylhexyl phthalate
DEHR	Di-2-ethylhexyl phthalate
DEP	Diethyl phthalate
DETX	2,4-Diethyl-thioxanthone
DiBP	Disobutyl phthalate
DiDP	Diisodecyl phthalate
DiNP	Di-isononyl phthalate
DMAB	4-(Dimethylamino)benzophene
DMMB	2-(4-Methylbenzyl)-2-dimethylamino-1-(4-morpholinophenyl)-1-butanone
DMP	Dimethyl phthalate
DMPA	2,2-Dimethoxy-2-phenyl acetophenone
DnBP	Di-n-butyl phthalate
DOP	Dioctyl phthalate or DnOP-di-n-octyl phthalate
EDB	Ethyl-4-(dimethylamino)benzoate
EFSA	European Food Safety Authority
EHA	2-Ethylhexyl-4-dimethylamino benzoate
ELSD	Evaporative light scattering detection
ESBO	Epoxidized soybean oil
ESCO	EFSA Scientific Cooperation
Ethanox 330	1,3,5-Trimethyl-2,4,6-tris(3,5-di-tert-butyl-4-hydroxybenzyl)benzene
EuPIA	European Printing Ink Association
FCM	Food contact materials
FLD	Fluorescence detector
FOPH	Swiss Federal Office of Public Health
GC	Gas chromatography
GC-ECD	Gas chromatography with an electron capture detector
GC-FID	Gas chromatography with a flame ionization detector
HALS	Hindered amine light stabilizers
HCPK	1-Hydroxycyclohexyl-phenyl-ketone
HMPP	2-Hydroxy-2-methyl propiophenone
HPLC	High-performance liquid chromatography
HRMS	High-resolution mass spectrometry
IARC	International Agency for Research on Cancer
Irgafos 168	Tris(2,4-di-tert-butylphenyl)phosphate
Irganox 1010	Pentaerythritol tetrakis [3-(3,5-di-tert-butyl-4-hydroxyphenyl]propionate
Irganox 1076	Octadecyl-3-(3,5-di-tert-butyl-4-hydroxyphenyl)-propionate
Irganox 1330	1,3,5-Trimethyl-2,4,6-tris(3,5-di-tert-butyl-4-hydroxybenzyl)benzene
Irganox 3114	1,3,5-Tris(3,5-di-tert-butyl-4-hydroxybenzyl)-1,3,5-triazine-2,4,6(1H,3H,5H)-trione
Isonox 129	2,2′-Ethylidene-bis(4,6-di-tert-butylphenol)
ITX	Mixture of two isomers: 2- and 4-isopropyl thioxanthone
LC	Liquid chromatography
LC-MS/MS	Liquid chromatography–tandem mass spectrometry
LLE	Liquid–liquid extraction
LOD	Limit of detection
MBB	Methyl-2-benzoylbenzoate
MBF	Methylbenzoyl-formate

MK	Michler's ketone: 4,4′-bis(dimethylamino) benzophenone
MMMP	Methyl-1-(4-methylthio)phenyl-2-morpholinopropan-1-one)
MPBP	4-(4-Methylphenylthio)benzophenone
MS	Mass spectrometry
Naugard DLTDP	Didodecyl3,3′-thiodipropionate
NIAS	Non–intentionally added substances
NP	Nonylphenol
o-cl-HABI	2,2-Bis-(2-chlorophenyl)-4,4′,5,5′-tetraphenyl-1,2-biimidazolyl
OML	Overall migration limit
OMN 910	Omnipol 910®: polymeric photoinitiator (886463-10-1)
OPAc	Oxyphenylacetic acid 2-[2-hydroxy-ethoxy]ethyl ester
PAAs	Primary aromatic amines
PBZ	4-Phenylbenzophenone
PET	Polyethylene terephthalate
PFOA	Perfluorooctanoic acid
PFOS	Perfluorooctane sulfonate
PTFE	Polytetrafluoroethylene
PVC	Poly(vinyl)chloride
RASFF	Rapid Alert System for Food and Feed
RHO 2074	Rhodorsil 2074®: [4-(1-methylethyl)phenyl](4-methylphenyl)-iodonium tetrakis(pentafluorophenyl) borate
RP	Reverse phase
SMLs	Specific migration limits
SP 7005	Speedcure 7005®: polymeric photoinitiator 1182753-56-5
SP 7010	Speedcure 7010®: polymeric photoinitiator 1182755-49-2
TDI	Tolerable daily intake
Tenax®	Modified polyphenylene oxide
THIO	Diphenyl[(phenylthio)phenyl]sulfonium hexafluorophosphate
Tinuvin 326	2-Tert-butyl-6-(5-chloro-2H-benzotriazol-2-yl)-4-methylphenol
Tinuvin 327	2,4-Di-tert-butyl-6-(5-chloro-2H-benzotrizol-2-yl)phenol
Tinuvin 328	2-(2H-benzotriazol-2-yl)-4,6-di-tert-pentylphenol
Tinuvin 622	Poly-(*N*-*b*-hydroxyethyl-2,2,6,6-teramethyl-4-hydroxy-piperdiyl succinate)
Tinuvin 770	Bis(2,2,6,6-tetramethyl-4-piperidinyl)sebacate
Tinuvin P	Phenol, 2-(2H-benzotriazol-2-yl)-4-methyl-
TNPP	Tris(nonylphenyl)phosphate
TPO	Diphenyl-(2,4,6-trimethylbenzoyl)phosphine oxide
UV	Ultraviolet
VOC	Volatile organic chemicals

REFERENCES

Allegrone, G., Tamaro, I., Spinardi, S., Grosa, G., 2008. Development and validation of a solid-phase extraction and gas chromatography-tandem mass spectrometry method for the determination of isopropyl-9H-thioxanthen-9-one in carton packaged milk. *J. Chromatogr. A*, 1214:128–133.

Anderson, W. C., Castle, L., 2003. Benzophenone in cartonboard packaging materials and the factors that influence its migration into food. *Food Addit. Contam.*, 20: 607–618.

Aparicio, J. L., Elizalde, M., 2015. Migration of photoinitiators in food packaging: A review. *Packag. Technol. Sci.*, 28: 181–203.

Arvanitoyannis, I. S., Bosnea, L., 2004. Migration of substances from food packaging materials to foods. *Crit. Rev. Food Sci. Nutr.*, 44: 63–76.

Autian, J., 1973. Toxicity and health threats of phthalate esters: Review of the literature. *Environ. Health Perspect.*, 4: 3–26.

Aznar, M., Canellas, E., Nerín, C., 2009. Quantitative determination of 22 primary aromatic amines by cation-exchange solid-phase extraction and liquid chromatography–mass spectrometry. *J. Chromatogr. A*, 1216: 5176–5181.

Aznar, M., Gómez-Estaca, J., Vélez, D., Devesa, V., Nerín, C., 2013. Migrants determination and bioaccessibility study of ethyl lauroyl arginate (LAE) from a LAE based antimicrobial food packaging material. *Food Chem. Toxicol.*, 56: 363–370.

Aznar, M., Rodriguez-Lafuente, A., Alfaro, P., Nerin, C., 2012. UPLC-Q-TOF-MS analysis of non-volatile migrants from new active packaging materials. *Anal. Bioanal. Chem.*, 404: 1945–1957.

Bach, C., Dauchy, X., Severin, I., Munoz, J.-F., Etienne, S., Chagnon, M.-C., 2013. Effect of temperature on the release of intentionally and non-intentionally added substances from polyethylene terephthalate (PET) bottles into water: Chemical analysis and potential toxicity. *Food Chem.*, 139: 672–680.

Bagnati, R., Bianchi, G., Marangon, E., Zuccato, E., Fanelli, R., Davoli, E., 2007. Direct analysis of isopropylthioxanthone (ITX) in milk by high-performance liquid chromatography/tandem mass spectrometry. *Rapid Commun. Mass Spectr.*, 21: 1998–2002.

Ballesteros-Gómez, A., Rubio, S., Pérez-Bendito, D., 2009. Analytical methods for the determination of bisphenol A in food. *J. Chromatogr. A*, 1216: 449–469.

Begley, T. H., Hsu, W., Noonan, G., Diachenko, G., 2008. Migration of fluorochemical paper additives from food-contact paper into foods and food stimulants. *Food Addit. Contam. A*, 25: 384–390.

Begley, T. H., White, K., Honigfort, P., Twaroski, M. L., Neches, R., Walker, R., 2005. Perfluorochemicals: Potential sources of and migration from food packaging. *Food Addit. Contam.*, 22: 1023–1031.

Beldì, G., Pastorelli, S., Franchini, F., Simoneau, C., 2012. Time- and temperature-dependent migration studies of Irganox 1076 from plastics into foods and food simulants. *Food Addit. Contam. A*: *Chem. Anal. Control Expo Risk Assess.*, 29: 836–845.

Benetti, C., Angeletti, R., Binato, G., Biancardi, A., Biancotto, G., 2008. A packaging contaminant: Isopropylthioxanthone (ITX) in dairy products. *Anal. Chim. Acta*, 617: 132–138.

Bentayeb, K., Ackerman, L. K., Lord, T., Begley, T. H., 2013. Non-visible print set-off of photoinitiators in food packaging: Detection by ambient ionization mass spectrometry. *Food Addit. Contam. A*, 30: 750–759.

Biedermann, M., Ingenhoff, J. E., Zurfluh, M., Richter, L., Simat, T., Harling, A., Altkofer, W., Helling, R., Grob, K., 2013. Migration of mineral oil, photoinitiators and plasticisers from recycled paperboard into dry foods: A study under controlled conditions. *Food Addit. Contam. A*, 30: 885–898.

Bodai, Z., Kirchkeszner, C., Novák, M., Nyiri, Z., Kovács, J., Magyar, N., Iván, B., Rikker, T., Eke, Z., 2015. Migration of Tinuvin P and Irganox 3114 into milk and the corresponding authorised food stimulant. *Food Addit. Contam. A*, 32: 1358–1366.

Bornehag, C.-G., Sundell, J., Weschler, C. J., Sigsgaard, T., Lundgren, B., Hasselgren, M., Hägerhed-Engman, L., 2004. The association between asthma and allergic symptoms in children and phthalates in house dust: A nested case-control study. *Environ. Health Perspect.*, 112: 1393–1397.

Bošnir, J., Puntarić, D., Galić, A., Škes, I., Dijanić, T., Klarić, M., Grgić, M., Čurković, M., Z. Šmit, 2007. Migration of phthalates from plastic containers into soft drinks and mineral water. *Food Technol. Biotechnol.*, 45: 91–95.

Bradley, E. L., Stratton, J. S., Leak, J., Lister, L., Castle, L., 2013. Printing ink compounds in foods: UK survey results. *Food Addit. Contam. B*, 6: 73–83.

Brede, C., Skjeurak, I., 2004. Migration of aniline from polyamide cooking utensils into food simulants. *Food Addit. Contam.*, 21: 1115–1124.

Brede, C., Skjevrak, I., Herikstad, H., 2003. Determination of primary aromatic amines in water food simulant using solid-phase analytical derivatization followed by gas chromatography coupled with mass spectrometry. *J. Chromatogr. A*, 983: 35–42.

Canellas, E., Nerín, C., Moore, R., Silcock, P., 2010. New UPLC coupled to mass spectrometry approaches for screening of non-volatile compounds as potential migrants from adhesives used in food packaging materials. *Anal. Chim. Acta*, 666: 62–69.

Cao, X. L., 2010. Phthalate esters in foods: Sources, occurrence, and analytical methods. *Compress. Rev. Food Sci. Food Saf.*, 9: 21–43.

Castle, L., Damant, A. P., Honeybone, C. A., Johns, S. M., Jickells, S. M., Sharman, M., Gilbert, J., 1997. Migration studies from paper and board food packaging materials. Part 2. Survey for residues of dialkylamino benzophenone UV-cure ink photoinitiators. *Food Addit. Contam.*, 14, 45–52.

Coulier, L., Kaal, E. R., Tienstra, M., Hankemeier, T., 2005. Identification and quantification of (polymeric) hindered-amine light stabilizers in polymers using pyrolysis-gas chromatography-mass spectrometry and liquid chromatography-ultraviolet absorbance detection-evaporative light scattering detection. *J. Chromatogr. A*, 1062: 227–238.

Coulier, L., Orbons, H. G. M., Rijk, R., 2007. Analytical protocol to study the food safety of (multiple-) recycled high-density polyethylene (HDPE) and polypropylene (PP) crates: Influence of recycling on the migration and formation of degradation products. *Polym. Degrad. Stab.*, 92: 2016–2025.

Cruz, J. M., Sanches-Silva, A., Sendón García, R., Franz, R., Paseiro Losada, P., 2008. Studies of mass transport of model chemicals from packaging into and within cheeses. *J. Food Eng.*, 87: 107–115.

Demir, A. P. T., Ulutan, S., 2015. Degradation kinetics of PVC plasticized with different plasticizers under isothermal conditions. *J. Appl. Polym. Sci.*, 132: 41579.

Deschênes, L., Arbour, A., Brunet, F., Court, M. A., Doyon, O. J., Fortin, J., Rodrigue, N., 1995. Irradiation of a barrier film: Analysis of some mass transfer aspects. *Radiat. Phys. Chem.*, 46: 805–808.

Dopico-García, M. S., López-Vilariño, J. M., González-Rodríguez, M. V., 2003. Determination of antioxidant migration levels from low-density polyethylene films into food simulants. *J. Chromatogr. A*, 1018: 53–62.

Dopico-García, M. S., Noguerol-Cal, R., Castro-López, M. M., Cela-Pérez, M. C., Piñón-Giz, E., López-Vilariño, J. M., González-Rodríguez, M. V., 2012. Determination of polyolefin additives by reversed-phase liquid chromatography. *Cent. Eur. J. Chem.*, 10: 585–610.

Ellendt, K., Gutsche, B., Steiner, G., 2003. Analysis of laminates—Determination of isocyanate residues and primary aromatic amine migration. *Dtsch. Lebensm.-Rundsch.*, 99: 131–136.

European Commission, 2004. Regulation EC No 1935/2004 of the European Parliament and of the Council of 27 October 2004 on materials and articles intended to come into contact with food. *Off. J. Eur. Union*, OJ L 338; and amendments.

European Commission, 2011. Commission Regulation 10/2011 on plastic materials and articles intended to come into contact with food. *Off. J. Eur. Comm.*, 12: 1–89; and amendments.

European Food Safety Authority (EFSA), 2005a. Opinion of the scientific panel on food additives, flavourings, processing aids and material in contact with food (AFC) on a request from the commission related to di-butylphthalate (DBP) for use in food contact materials. Question No EFSA-Q-2003-192. *EFSA J.*, 242: 1–17.

European Food Safety Authority (EFSA), 2005b. Opinion of the scientific panel on food additives, flavourings, processing aids and materials in contact with food (AFC) on a request from the commission related to butylbenzylphthalate (BBP) for use in food contact materials. Question No EFSA-Q-2003-190. *EFSA J.*, 241: 1–14.

European Food Safety Authority (EFSA), 2005c. Opinion of the scientific panel on food additives, flavourings, processing aids and materials in contact with food (AFC) on a request from the commission related to di-isononylphthalate (DINP) for use in food contact materials. Question No EFSA-Q-2003-194. *EFSA J.*, 244: 1–18.

European Food Safety Authority (EFSA), 2005d. Opinion of the scientific panel on food additives, flavourings, processing aids and materials in contact with food (AFC) on a request from the commission related to di-isodecylphthalate (DIDP) for use in food contact materials. *EFSA J.*, 245: 1–14.

European Food Safety Authority (EFSA), 2005e. Opinion of the scientific panel on food additives, flavourings, processing aids and materials in contact with food (AFC) on a request from the commission related to bis(2-ethylhexyl)phthalate (DEHP) for use in food contact materials. Question No EFSA-Q-2003-191. *EFSA J.*, 243: 1–20.

European Food Safety Authority (EFSA), 2005f. Opinion of the scientific panel on food additives, flavourings, processing aids and materials in contact with food (AFC) on a request from the commission on the application of a total reduction factor of 5 for di(2-ethylhexyl)adipate used as plasticizer in flexible PVC food packaging films. Question No EFSA-Q-2003-071. *EFSA J.*, 217: 1–5.

European Food Safety Authority (EFSA), 2008. Food Contact Materials: Note for guidance for petitioners presenting an application for the safety assessment of a substance to be used in food contact materials prior to its authorisation. *EFSA J.*, Issue 7: 1–125.

European Food Safety Authority, 2012. External Scientific Report. Report of ESCO WG on non-plastic Food Contact Materials. Available from: http://www.efsa.europa.eu/en/supporting/doc/139e.pdf. Accessed on November 13, 2015.

European Food Safety Authority (EFSA), 2014. FIP Scientific Network on Food Contact Materials. Minutes of the 1st meeting Held on 12–14 November 2014, Parma. Available from: http://www.efsa.europa.eu/sites/default/files/event/141112a-m.pdf. Accessed on November 13, 2015.

European Food Safety Authority, 2015. Opinion on BPA: Executive summary. *EFSA J.*, 13 (1): 3978.

Fankhauser-Noti, A., Biedermann-Brem, S., Grob, K., 2006. PVC plasticizers/additives migrating from the gaskets of metal closures into oily food: Swiss market survey June 2005. *Eur. Food Res. Technol.*, 223: 447–453.

Fasano, E., Bono-Blay, F., Cirillo, T., Montuori, P., Lacorte, S., 2012. Migration of phthalates, alkylphenols, bisphenol A and di(2-ethylhexyl)adipate from food packaging. *Food Control*, 27: 132–138.

Félix, J. S., Isella, F., Bosetti, O., Nerín, C., 2012. Analytical tools for identification of non-intentionally added substances (NIAS) coming from polyurethane adhesives in multilayer packaging materials and their migration into food stimulants. *Anal. Bioanal. Chem.*, 403: 2869–2882.

Fernandes, A., Costley, C., Rose, M., 2003. Determination of 4-octylphenol and 4-nonylphenol congeners in composite foods. *Food Addit. Contam.*, 20: 846–852.

Fernandes, A. R., Rose, M., Charlton, C., 2008. 4-Nonylphenol (NP) in food-contact materials: Analytical methodology and occurrence. *Food Addit. Contam. A: Chem. Anal. Control. Expo. Risk Assess.*, 25: 364–372.

Fierens, T., Servaes, K., Van Holderbeke, M., Geerts, L., Henauw, S. De, Sioen, I., Vanermen, G., 2012. Analysis of phthalates in food products and packaging materials sold on the Belgian market. *Food Chem. Toxicol.*, 50: 2575–2583.

Food Surveillance Sheet No. 6, Ministry of Agriculture Fisheries and Food, London, U.K., 1995.

Franz, R., Welle, F., 2008. Investigation of non-intentionally added substances (NIAS) in PET bottles and closures. Poster Presentation at the *Fourth International Symposium on Food Packaging*, No. 85354, November 2008.

Gallart-Ayala, H., Moyano, E., Galceran, M. T., 2008. Liquid chromatography/tandem mass spectrometry (highly selective selected reaction monitoring) for the analysis of isopropylthioxanthone in packaged food. *J. Chromatogr. A*, 1208: 182–188.

Gallart-Ayala, H., Núñez, O., Lucci, P., 2013. Recent advances in LC-MS analysis of food-packaging contaminants. *TrAC Trends Anal. Chem.*, 42: 99–124.

Gallart-Ayala, H., Núñez, O., Moyano, E., Galceran, M. T., 2011. Analysis of UV ink photoinitiators in packaged food by fast liquid chromatography at sub-ambient temperature coupled to tandem mass spectrometry. *J. Chromatogr. A*, 1218: 459–466.

Galotto, M. J., Torres, A., Guarda, A., Moraga, N., Romero, J., 2011. Experimental and theoretical study of LDPE: Evaluation of different food simulants and temperatures. *Food Res. Int.*, 44: 3072–3078.

Gärtner, S., Balski, M., Koch, M., Nehls, A., 2009. Analysis and migration of phthalates in infant food packed in recycled paperboard. *J. Agric. Food Chem.*, 57: 10675–10681.

Gil-Vergara, A., Blasco, C., Picó, Y., 2007. Determination of 2-isopropyl thioxanthone and 2-ethylhexyl-4-dimethylaminobenzoate in milk: Comparison of gas and liquid chromatography with mass spectrometry. *Anal. Bioanal. Chem.*, 389: 605–617.

Grob, K., 2014. Work plans to get out of the deadlock for the safety assurance of migration from food contact materials? A proposal. *Food Control*, 46: 312–318.

Guo, Y., Kannan, K., 2012. Challenges encountered in the analysis of phthalate esters in foodstuffs and other biological matrices. *Anal. Bioanal. Chem.*, 404: 2539–2554.

Gupta, S. K., 1989. Nylon polymerization. In: *Handbook of Polymer Science and Technology*, Vol. 1, ed. N. P. Cheremisinof, p. 211. Marcel Dekker, New York.

Haider, N., Karlsson, S., 2001. Loss of Chimassorb 944 from LDPE and identification of additive degradation products after exposure to water, air and compost. *Polym. Degrad. Stab.*, 74: 103–112.

Helmrothy, I. E., Dekker, M., Hankemeiery, T., 2002. Influence of solvent absorption on the migration of Irganox 1076 from LDPE. *Food Addit Contam.*, 19: 176–183.

Honkalampi-Hämäläinen, U., Bradley, E. L., Castle, L., Severin, I., Dahbi, L., Dahlman, O., Lhuguenot, J.-C. et al., 2010. Safety evaluation of food contact paper and board using chemical tests and in vitro bioassays: Role of known and unknown substances. *Food Addit. Contam. A: Chem. Anal. Control. Expo. Risk Assess.*, 27: 406–415.

Hotchkiss, J. H., 1997. Food-packaging interactions influencing quality and safety. *Food Addit. Contam.*, 14: 601–607.

Howe, S. R., Surana, P., Jakupca, M. R., Borodinsky, L., 2001. Potential dietary exposure to p-nonylphenol from food-contact use of tris(nonylphenyl)phosphite (TNPP). *Food Addit. Contam.*, 18: 1021–1039.

Huang, X. L., Wang, Z. W., Hu, C. Y., Zhu, Y., Wang, J., 2012. Factors affecting migration of contaminants from paper through polymer coating into food simulants. *Packag. Technol. Sci.*, 26: 26–31.

Inoue, K., Kondo, S., Yoshie, Y., Kato, K., Yoshimura, Y., Horie, M., Nakazawa, H., 2001. Migration of 4-nonylphenol from polyvinyl chloride food packaging films into food simulants and foods. *Food Addit. Contam.*, 18: 157–164.

International Agency for Research on Cancer (IARC). Monograph on the evaluation of carcinogenic risks to humans, available on line at: http://monographs.iarc.fr/ENG/Classification/index.php. Accessed on November 15, 2015.

International Life Sciences Institute (ILSI), 2011. Packaging Materials 8. Printing inks for food Packaging Composition and Properties of Printing inks.

International Life Sciences Institute (ILSI), 2015. Guidance on best practises on the risk assessment of non-internationally added substances (NIAS) in food contact materials and articles.

Isella, F., Canellas, E., Bosetti, O., Nerin, C., 2013. Migration of non-intentionally added substances from adhesives by UPLC-Q-TOF/MS and the role of EVOH to avoid migration in multilayer packaging materials. *J. Mass Spectrom.*, 48: 430–437.

Jeon, D. H., Park, G. Y., Kwak, I. S., Lee, K. H., Park, H. J., 2007. Antioxidants and their migration into food simulants on irradiated LLDPE film. *LWT - Food Sci. Technol.*, 40: 151–156.

JRC, 2011. Technical guidelines on testing the migration of primary aromatic amines from polyamide kitchenware and of formaldehyde from melamine kitchenware.

Jung, T., Simat, T. J., Altkofer, W., 2010. Mass transfer ways of ultraviolet printing ink ingredients into foodstuffs. *Food Addit. Contam. A*, 27: 1040–1049.

Jung, T., Simat, T. J., Altkofer, W., Fügel, D., 2013. Survey on the occurrence of photo- initiators and amine synergists in cartonboard packaging on the German market and their migration into the packaged foodstuffs. *Food Addit. Contam. A*, 30: 1993–2016.

Kawamura, Y., 2000. Nonylphenol in food contact plastics and toys. *J. Food Hyg. Soc. Jpn.*, 41: 212–218.

Kawasaki, Y., Tsuboi, C., Yagi, K., Morizane, M., Masaoka, Y., Esumi, S., Kitamura, Y., Sendo, T., 2015. Photoinitiators enhanced 1,2-dichloropropane-induced cytotoxicity in human normal embryonic lung fibroblasts cells in vitro. *Environ. Sci. Pollut. Res.*, 22: 4763–4770.

Koivikko, R., Pastorelli, S., Rodríguez-Bernaldo de Quirós, A., Paseiro-Cerrato, R., Paseiro- Losada, P., Simoneau, C., 2010. Rapid multi-analyte quantification of benzophenone, 4-methylbenzophenone and related derivatives from paperboard food packaging. *Food Addit. Contam. A*, 27: 1478–1486.

Koster, S., Rennen, M., Leeman, W., Houben, G., Muilwijk, B., Acker, F., Van, Krul, L., 2014. A novel safety assessment strategy for non-intentionally added substances (NIAS) in carton food contact materials. *Food Addit. Contam. A: Chem. Anal. Control. Expo. Risk Assess.*, 31: 422–443.

Kueseng, P., Thavarungkul, P., Kanatharana, P., 2007. Trace phthalate and adipate esters contaminated in packaged food. *J. Environ. Sci. Health. B*, 42: 569–576.

Lago, M. A., Rodríguez-Bernaldo de Quirós, A., Sendón, R., Bustos, J., Nieto, M. T., Paseiro, P., 2015a. Photoinitiators: A food safety review. *Food Addit. Contam. A*, 32: 779–798.

Lago, M.A, Sendón, R., Rodríguez-Bernaldo de Quirós, A., 2015b. Analytical methods for determining photoinitiators in food-contact materials. In: *Photocured Materials*, eds. A. Tiwari and A. Polykarpov, pp. 1–9. The Royal Society of Chemistry, Cambridge, U.K.

Lambertini, F., Di Lallo, V., Catellani, D., Mattarozzi, M., Careri, M., Sumana, M., 2014. Reliable liquid chromatography-mass spectrometry method for investigation of primary aromatic amines migration from food packaging and during industrial curing of multilayer plastic laminates. *J. Mass Spectrom.*, 49: 870–877.

Lane, R. F., Adams, C. D., Randtke, S. J., Carter, R. E., 2015. Chlorination and chloramination of bisphenol A, bisphenol F, and bisphenol A diglycidyl ether in drinking water. *Water Res.*, 79: 68–78.

Lau, O. W., Wong, S. K., 2000. Contamination in food from packaging material. *J. Chromatogr. A*, 882: 255–270.

Li, H., Guan, H., Dai, H., Tong, Y., Zhao, X., Qi, W., Majeed, S., Xu, G., 2012. An amperometric sensor for the determination of benzophenone in food packaging materials based on the electropolymerized molecularly imprinted poly-o-phenylenediamine film. *Talanta*, 99: 811–815.

López-Cervantes, J., Paseiro-Losada, P., 2003. Determination of bisphenol A in, and its migration from, PVC stretch film used for food packaging. *Food Addit. Contam.*, 20: 596–606.

Lopez-Espinosa, M.-J., Granada, A., Araque, P., Molina-Molina, J.-M., Puertollano, M.-C., Rivas, A., Fernández, M. et al., 2007. Oestrogenicity of paper and cardboard extracts used as food containers. *Food Addit. Contam.*, 24: 95–102.

Loyo-Rosales, J. E., Rosales-Rivera, G. C., Lynch, A. M., Rice, C. P., Torrents, A., 2004. Migration of nonylphenol from plastic containers to water and a milk surrogate. *J. Agric. Food Chem.*, 52: 2016–2020.

Lundbäck, M., Hedenqvist, M. S., Mattozzi, A., Gedde, U. W., 2006. Migration of phenolic antioxidants from linear and branched polyethylene. *Polym. Degrad. Stab.*, 91: 1571–1580.

Maia, J., Cruz, J. M., Sendón, R., Bustos, J., Sanchez, J. J., Paseiro, P., 2009. Effect of detergents in the release of bisphenol A from polycarbonate baby bottles. *Food Res. Int.*, 42: 1410–1414.

Marcato, B., Guerra, S., Vianello, M., Scalia, S., 2003. Migration of antioxidant additives from various polyolefinic plastics into oleaginous vehicles. *Int. J. Pharm.*, 257: 217–225.

Marsh, K., Bugusu, B., 2007. Food packaging roles, materials, and environmental issues. *J. Food Sci.*, 72: R39–R55.

Mattarozzi, M., Lambertini, F., Suman, M., Careri, M., 2013. Liquid chromatography–full scan-high resolution mass spectrometry-based method towards the comprehensive analysis of migration of primary aromatic amines from food packaging. *J. Chromatogr. A*, 1320: 96–102.

Mezcua, M., Martínez-Uroz, M. A., Gómez-Ramos, M. M., Gómez, M. J., Navas, J. M., Fernández-Alba, A. R., 2012. Analysis of synthetic endocrine-disrupting chemicals in food: A review. *Talanta*, 100: 90–106.

Muncke, J., 2009. Exposure to endocrine disrupting compounds via the food chain: Is packaging a relevant source? *Sci. Total Environ.*, 407: 4549–4559.

Mutsuga, M., Tojima, T., Kawamura, Y., Tanamoto, K., 2007. Survey of formaldehyde, acetaldehyde and oligomers in polyethylene terephthalate food-packaging materials. *Food Addit. Contam.*, 22: 783–789.

Negreira, N., Rodríguez, I., Rubí, E., Cela, R., 2010. Solid-phase microextraction followed by gas chromatography-mass spectrometry for the determination of ink photo-initiators in packed milk. *Talanta*, 82: 296–303.

Nerin, C., Alfaro, P., Aznar, M., Domeño, C., 2013. The challenge of identifying non-intentionally added substances from food packaging materials: A review. *Anal. Chim. Acta*, 775: 14–24.

Ozaki, A., Ooshima, T., Mori, Y., 2006. Migration of dehydroabietic and abietic acids from paper and paperboard food packaging into food-simulating solvents and Tenax TA. *Food Addit. Contam.*, 23: 854–860.

Ozaki, A., Yamaguchi, Y., Fujita, T., Kuroda, K., Endo, G., 2004. Chemical analysis and genotoxicological safety assessment of paper and paperboard used for food packaging. *Food Chem. Toxicol.*, 42: 1323–1337.

Page, B. D., Lacroix, G. M., 1995. The occurrence of phthalate ester and di-2-ethylhexyl adipate plasticizers in Canadian packaging and food sampled in 1985–1989: A survey. *Food Addit. Contam.*, 12: 129–151.

Palmiotto, G., Pieraccini, G., Moneti, G., Dolara, P., 2001. Determination of the levels of aromatic amines in indoor and outdoor air in Italy. *Chemosphere*, 43: 355–361.

Papilloud, S., Baudraz, D., 2002a. Analysis of food packaging UV inks for chemicals with potential to migrate into food stimulants. *Food Addit. Contam.*, 19: 168–175.

Papilloud, S., Baudraz, D., 2002b. Migration tests for substrates printed with UV ink systems in aqueous simulants. *Prog. Org. Coatings*, 45: 231–237.

Paseiro-Cerrato, R., Noonan, G. O., Begley, T. H., 2014. Development of a rapid screening method to determine primary aromatic amines in kitchen utensils using direct analysis in real time mass spectrometry (DART-MS). *Food Addit. Contam. A*, 31: 537–545.

Paseiro-Losada, P., Simal-Lozano, J., Paz Abuín, S., Lopez-Mahia, P., Simal Gándara, J., 1993. Kinetics of the hydrolysis of bisphenol A diglycidyl ether (BADGE) in water-based food simulants. *J. Anal. Chem.*, 345: 527–532.

Pastorelli, S., Sanches-Silva, A., Cruz, J. M., Simoneau, C., Paseiro, P., 2008. Study of the migration of benzophenone from printed paperboard packages to cakes through different plastic films. *Eur. Food Res. Technol.*, 227: 1585–1590.

Poças, M. F., Hogg, T., 2007. Exposure assessment of chemicals from packaging materials in foods: A review. *Trends Food Sci. Technol.*, 18: 219–230.

Poças, M. F., Oliveira, J. C., Pereira, J. R., Hogg, T., 2010. Consumer exposure to phthalates from paper packaging: An integrated approach. *Food Addit. Contam. A: Chem. Anal. Control. Expo. Risk Assess.*, 27: 1451–1459.

Polykarpov, A., Tiwari, A., 2015. Photocured materials: A general perspective. In: *Photocured Materials*, eds. A. Tiwari and A. Polykarpov, pp. 1–9. The Royal Society of Chemistry, Cambridge, U.K.

Pouech, C., Lafay, F., Wiest, L., Baudot, R., Léonard, D., Cren-Olivé, C., 2014. Monitoring the extraction of additives and additive degradation products from polymer packaging into solutions by multi-residue method including solid phase extraction and ultra-high performance liquid chromatography-tandem mass spectrometry analysis. *Anal. Bioanal. Chem.*, 406: 1493–1507.

Powley, C. R., Michalczyk, M. J., Kaiser, M. A., Buxton, L. W., 2005. Determination of perfluorooctanoic acid (PFOA) extractable from the surface of commercial cookware under simulated cooking conditions by LC/MS/MS. *Analyst*, 130: 1299–1302.

Rapid Alert System in Food and Feed [RASFF], 2015. European Commission. Brussels (BE). [Internet]. [Cited 10 November 2015]. Available from: https://webgate.ec.europa.eu/rasff-window/portal/.

Reinas, I., Oliveira, J., Pereira, J., Machado, F., Poças, M. F., 2012. Migration of two antioxidants from packaging into a solid food and into Tenax n®. *Food Control*, 28: 333–337.

Reitsma, M., Bovee, T. F. H., Peijnenburg, A. A. C. M., Hendriksen, P. J. M., Ron, L. A. P., Hoogenboom, R. L. A. P., Rijk, J. C. W., 2013. Endocrine-disrupting effects of thioxanthone photoinitiators. *Toxicol. Sci.*, 132: 64–74.

Rodríguez-Bernaldo de Quirós, A., Paseiro-Cerrato, R., Pastorelli, S., Koivikko, R., Simoneau, C., Paseiro-Losada, P., 2009. Migration of photoinitiators by gas phase into dry foods. *J. Agric. Food Chem.*, 57: 10211–10215.

Rosenmai, A. K., Nielsen, F. K., Pedersen, M., Hadrup, N., Trier, X., Christensen, J. H., Vinggaard, A. M., 2013. Fluorochemicals used in food packaging inhibit male sex hormone synthesis. *Toxicol. Appl. Pharmacol.*, 266: 132–142.

Rothenbacher, T., Baumann, M., Fügel, D., 2007. 2-Isopropylthioxanthone (2-ITX) in food and food packaging materials on the German market. *Food Addit. Contam.*, 24: 438–444.

Rudel, R. A., Camann, D. E., Spengler, J. D., Korn, L. R., Brody, J. G., 2003. Phthalates, alkylphenols, pesticides, polybrominated diphenyl ethers, and other endocrine-disrupting compounds in indoor air and dust. *Environ. Sci. Technol.*, 37: 4543–4553.

Russell, D. A. M., 2014. Sustainable (food) packaging—An overview. *Food Addit. Contam. A*, 31: 396–401.

Sagratini, G., Caprioli, G., Cristalli, G., Giardiná, D., Ricciutelli, M., Volpini, R., Zuo, Y., Vittori, S., 2008. Determination of ink photoinitiators in packaged beverages by gas chromatography-mass spectrometry and liquid chromatography-mass spectrometry. *J. Chromatogr. A*, 1194: 213–220.

Sagratini, G., Mañes, J., Giardina, D., Picó, Y., 2006. Determination of isopropyl thioxanthone (ITX) in fruit juices by pressurized liquid extraction and liquid chromatography-mass spectrometry. *J. Agric. Food Chem.*, 54: 7947–7952.

Sanches-Silva, A., André, C., Castanheira, I., Cruz, J. M., Pastorelli, S., Simoneau, C., Paseiro, P., 2009. Study of the migration of photoinitiators used in printed food-packaging materials into food simulants. *J. Agric. Food Chem.*, 57: 9516–9523.

Sanches-Silva, A., Cruz, J. M., Sendón García, R., Paseiro-Losada, P., 2007. Determination of butylated hydroxytoluene in food samples by high-performance liquid chromatography with ultraviolet detection and gas chromatography/mass spectrometry. *J. AOAC Int.*, 90: 277–283.

Sanches-Silva, A., Cruz Freire, J. M., Franz, R., Paseiro Losada, P., 2008. Time-temperature study of the kinetics of migration of diphenylbutadiene from polyethylene films into aqueous foodstuffs. *Food Res. Int.*, 41: 138–144.

Sanches-Silva, A., Pastorelli, S., Cruz, J. M., Simoneau, C., Castanheira, I., Paseiro-Losada, P., 2008a. Development of a method to study the migration of six photoinitiators into powdered milk. *J. Agric. Food Chem.*, 56: 2722–2726.

Sanches-Silva, A., Pastorelli, S., Cruz, J. M., Simoneau, C., Castanheira, I., Paseiro, P., 2008b. Development of an analytical method for the determination of photoinitiators used for food packaging materials with potential to migrate into milk. *J. Dairy Sci.*, 91: 900–909.

Sanches-Silva, A., Pastorelli, S., Cruz, J. M., Simoneau, C., Paseiro-Losada, P., 2008c. Development of a multimethod for the determination of photoinitiators in beverage packaging. *J. Food Sci.*, 73: C92–C99.

Sanches-Silva, A., Sendón-García, R., López-Hernández, J., Paseiro-Losada, P., 2005. Determination of triclosan in foodstuffs. *J. Sep. Sci.*, 28: 65–72.

Schug, T. T., Birnbaum, L. S., 2014. Human health effects of bisphenol A. In: *Toxicants in Food Packaging and Household Plastics*, ed. S. M. Snedeker, pp. 1–29. Springer-Verlag, London, U.K.

Sendón, R., Bustos, J., Sánchez, J. J., Paseiro, P., Cirugeda, M. E., 2010. Validation of a liquid chromatography–mass spectrometry method for determining the migration of primary aromatic amines from cooking utensils and its application to actual samples. *Food Addit. Contam.*, 27: 107–117.

Sendón, R., Sanches-Silva, A., Bustos, J., Martín, P., Martínez, N., Cirugeda, M. E., 2012. Detection of migration of phthalates from agglomerated cork stoppers using HPLC-MS/MS. *J. Sep. Sci.*, 35: 1319–1326.

Sendón García, R., Silva, A. S., Cooper, I., Franz, R., Losada, P. P., 2006. Revision of analytical strategies to evaluate different migrants from food packaging materials. *Trends Food Sci. Technol.*, 17: 354–366.

Shen, D., Lian, H., Ding, T., Xu, J., Shen, C., 2009. Determination of low-level ink photoinitiator residues in packaged milk by solid-phase extraction and LC-ESI/MS/MS using triple-quadrupole mass analyzer. *Anal. Bioanal. Chem.*, 395: 2359–2370.

Silva, A. S., Freire, J. M. C., Franz, R., Losada, P. P., 2008. Mass transport studies of model migrants within dry foodstuffs. *J. Cereal Sci.*, 48: 662–669.

Sioen, I., Fierens, T., Holderbeke, M. Van Geerts, L., Bellemans, M., Maeyer, M. De, Servaes, K., Vanermen, G., Boon, P. E., S. De Henauw, 2012. Phthalates dietary exposure and food sources for Belgian preschool children and adults. *Environ. Int.*, 48: 102–108.

Snedeker, S. M., 2014. Benzophenone UV-Photoinitiators used in food packaging: Potential for human exposure and health risk considerations. In: *Toxicants in Food Packaging and Household Plastics*, ed. S. M. Snedeker, pp. 151–176. Springer-Verlag, London, U.K.

Su, Q.-Z., Lin, Q.-B., Chen, C.-F., Wu, Y.-M., Wu, L.-B., Chen, X.-Q., Wang, Z.-W., 2015. Effect of antioxidants and light stabilizers on silver migration from nanosilver-polyethylene composite packaging films into food simulants. *Food Addit. Contam. A*, 32: 1561–1566.

Suman, M., Dominicis, E. De, Commissati, I., 2010. Trace detection of the chlorohydrins of epoxidized soybean oil in foodstuffs by UPLC–ESI–MS/MS. *J. Mass Spectrom.*, 35: 996–1002.

Sun, C., Chan, S. H., Lu, D., Lee, H. M. W., Bloodworth, B. C., 2007. Determination of isopropyl-9H-thioxanthen-9-one in packaged beverages by solid-phase extraction clean-up and liquid chromatography with tandem mass spectrometry detection. *J. Chromatogr. A*, 1143: 162–167.

Survey of Plasticiser Levels in Food Contact Materials and in Foods, 1987. The Report of the Steering Group on Food Surveillance: The Working Party on Chemical Contaminants from Food Contact Materials, Food Surveillance Paper No. 21, Ministry of Agriculture Fisheries and Food, London, U.K.

Trier, X., Granby, K., Christensen, J. H., 2011. Polyfluorinated surfactants (PFS) in paper and board coatings for food packaging. *Environ. Sci. Pollut. Res.*, 18: 1108–1120.

Van Hoeck, E., De Schaetzen, T., Pacquet, C., Bolle, F., Boxus, L., Van Loco, J., 2010. Analysis of benzophenone and 4-methylbenzophenone in breakfast cereals using ultrasonic extraction in combination with gas chromatography-tandem mass spectrometry (GC-MSn). *Anal. Chim. Acta*, 663: 55–59.

Vandenberg, L. N., 2014. Hazards of food contact material: Bisphenol A and endocrine disruption. In: *Encyclopedia of Food Safety*, Vol. 2, ed. Y. Motarjemi, pp. 424–429. Elsevier, Amsterdam, the Netherlands.

Vieira, M. G. A., Silva, M. A. Da, Santos, L. O. Dos, Beppu, M. M., 2011. Natural-based plasticizers and biopolymer films: A review. *Eur. Polym. J.*, 47: 254–263.

24 Role of Gut Microbiome in the Modulation of Environmental Toxicants and Therapeutic Agents

Ashfaque Hossain, Godfred Antony Menezes, Mohammad Al-Mogbel, and Ibraheem Ashankyty

CONTENTS

24.1 TERMINOLOGY

See Table 24.1.

24.2 INTRODUCTION

Our entire body is home to hundreds of trillions of diverse population of microorganisms, known as microbiome, mostly impacting us positively and contributing toward keeping us healthy. However, many recent scientific studies have demonstrated a link between dysbiosis (imbalance in microbiota) and a variety of metabolic diseases such as obesity, diabetes, cancer, and pulmonary diseases (Sekirov et al., 2010; The Human Microbiome Project Consortium, 2012; Bultman, 2014).

The role of microorganisms residing in the intestine (gut microbiota) in human health and disease has evolved into an area of active research due to its importance to the health and well-being of the human race. Humans can now be considered as "superorganisms" because in addition to our own genome, we are endowed with gene pool of resident microbiota that is many times larger than our own genome (Zhu et al., 2010; van Duynhovena et al., 2011; Kramer and Bressan, 2015). Gut microbiome is considered as an additional organ of our system with a metabolic capacity as robust as that of liver; which can be modulated or even replaced through a variety of means (O'Hara and Shanahan, 2006). It has been suggested that our microbiota develops with us and alters its own structure, function, and gene expression in response to changing environmental conditions (Ley et al., 2008; DiBaise et al., 2012). The gut microbiome exists in physiological homeostasis with the human host; the modulation of the gut microbiome by a variety of agents has the potential to improve health and prevent diseases (Sekirov et al., 2010). The colon (large intestine) is naturally the most complex ecosystem in the body. It contains an estimated 10^{14} microorganisms weighing about 1 kg, more than 99% of which are bacterial species (Lepage et al., 2012). More than 1300 species of bacteria so far have been reported to be present in the gut microbiota. However, according to culture-independent metagenomic studies, the human gut microbiome consists of at least 1,800 genera and approximately 15,000–36,000 species of bacteria (Frank et al., 2007). With an average length of 1.50 m and a diameter of 6.4 cm, the human large intestine is one of the largest interfaces where host–microbe interactions occur (Cummings and Macfarlane, 1991).

The Human Microbiome Project (HMP) of the National Institutes of Health (NIH) is an initiative of the NIH Roadmap for Biomedical Research (http://nihroadmap.nih.gov) launched to better understand the role of microbiome in human health with four objectives: to take advantage of the new, high-throughput technologies to characterize the human microbiome, (1) to determine whether individuals share a core microbiome by studying samples from multiple body sites using at least 250 "normal" volunteers, (2) to gain insight into the associations between changes in the microbiome and health/disease states by studying several medical conditions, (3) to develop genomic and bioinformatic tools and to provide standardized data resources to enable scientific communities to undertake such studies, and (4) to address the legal, social, and ethical implications of human microbiome research (Turnbaugh et al., 2007; Peterson et al., 2009; Betts, 2011). Important knowledge regarding the commonalities and differences among healthy individuals in microbial diversity in the microbiome and functional pathways is being generated, which is vast with wide-ranging implications (D'Argenio and Salvatore, 2015). More than 70% of all the microbes in the human body are found in the colon (Schippa and Conte, 2014), and microbiome significantly contributes to the metabolic processes of the human host to the extent that up to 36% of small molecules circulating in the human blood originate from gut microbiome (Hood, 2012).

TABLE 24.1
Terminology

Term	Definition
Microbiota	Microorganisms such as bacteria, fungi, archaea, protozoa, and viruses harbored by normal, healthy individuals.
Microbiome	Total microbial genomes associated with the human body. Currently, however, microbiome is being used to denote two different concepts: (1) the collective microbial genome and (2) the sum of all microscopic life forms, viz., microbes within an environment (microbiota), i.e., host-inhabiting microbes and their genes and genomes.
Metagenomics	Genomic analysis of DNA extracted directly from a clinical or environmental sample; it can assess community diversity (through 16 rRNA metagenomics) or fuller complexity (through whole genome shotgun metagenomics).
Pan-genome	The collection of genes found across all members of a species.
Metabolome/ Metabonome	Total metabolites found in a tissue or a sample. This profile represents a snapshot in time of what chemicals are present in the sample.
Metabonomics	The quantitative measurement of the dynamic, multiparametric metabolic response of living systems to pathophysiological stimuli or genetic modification, applying systems biology to the study of metabolic processes.
Dysbiosis	Imbalance in the intestinal microbiota.
Enterotype	Classification of human gut microbiome according to the dominant microorganism present. The dominant bacteria present in Type-1, Type-2 and Type-3 enterotypes are *Bacteroides*, *Prevotella*, and *Ruminococcus*, respectively, (Arumugam et al., 2011).
Extirpations	The loss of species in a locality (e.g., an individual host).
Allelopathy	A phenomenon in which a microbe uses chemical means to aid its competition within a group of microbes. Allelopathy may involve manipulation of third parties (e.g., the host) to favor competition.
Amphibiont	An organism (e.g., a microbe) that may have a pathogenic or symbiotic relationship with another organism (e.g., its host), depending on context. This is a more specific term than commensal.
Commensals	Organisms (e.g., microbes) that are involved in the form of symbiosis in which one organism derives a benefit while the other is unaffected.
Gnotobiotic	Describes an animal that is colonized solely by known strains of bacteria or other microorganisms. The term also describes germ-free animals, as the status of their microbial communities is known.
Virome	The genomes of all the viruses (bacteriophages, DNA and RNA viruses) that inhabit a particular environment, such as human intestine.
Gutome	Nutritional systems biology of gut microbiome and host–microbiome interactions.
Sterolbiome	The genetic potential of the gut microbiome to produce endocrine molecules from endogenous and exogenous steroids in the mammalian gut.
Exposome	It is the measure of all the exposures (both external such as environment, diet, lifestyle, and microbiome and internal such as physiology, genetics, and epigenetics) of an individual in a lifetime and how those exposures relate to health.
Probiotics	Living microorganism (bacteria and yeast) that can inhabit the gut and contribute to the health of the host.
Prebiotics	Food ingredient that is nondigestible by humans but promotes the growth of beneficial microorganisms (probiotics) in the intestines.
Immunobiotics	Bacteria that promote health through driving mucosal immune mechanisms, compared to those with strictly local effects.
Cobiotics	Unlike prebiotics that are used by probiotics, these are food ingredients that are utilized by probiotics and also by the host.
Synbiotics	Synergistic combinations of probiotics and prebiotics.
Postbiotic	A metabolic by-product generated by a probiotic microorganism that influences the host's biological functions.
Allochthonous	Describing organism that originates from a place other than that in which it is found. The organism is usually a transient member of a community.
Autochthonous	Describing an organism that is native to the place in which it is found.

24.3 HUMAN GUT MICROBIOME

24.3.1 Microbiome and Microbiota

Out of the 55 bacterial divisions, only two divisions of bacteria, Bacteroidetes and Firmicutes, are prominent in the mammalian gut microbiota, representing approximately 90% of the intestinal microorganisms present (Bäckhed et al., 2005). The less abundant phyla include *Actinobacteria*, *Verrucomicrobia*, and *Proteobacteria* (Ley et al., 2005, 2006a,b; Table 24.2). It is interesting to note that oral and gastrointestinal microbiota are distinctly different although the two organs are connected and oral bacteria pass into the intestine along with food (Gerritsen et al., 2011; Table 24.3). The human microbiota also contains other microorganism such as archaea, viruses, fungi, yeasts, and other Eukarya (such as Blastocystis and Amoebozoa) (Mai and Draganov, 2009; Lozupone et al., 2012; Table 24.2).

How a child is born (natural delivery or caesarean section) and how that child is fed (breastfeeding or bottle feeding) strongly influence the rate and trajectory of acquisition and development of gut microbiota. In addition, the duration of gestation and the living environment of the mother also influence the structure of gut microbiota in infants (Penders et al., 2006; Dogra et al., 2015). In a longitudinal study of 75 infants, Dogra et al. (2015) demonstrated high level of interindividual variability in the rate of microbiota progression. Normally delivered babies develop a gut microbiome high in *Prevotella*, *Lactobacillus*, and *Atopobium*, whereas babies delivered by cesarean section acquire gut microbiota whose composition is similar to that of the maternal skin community (Li and Jia, 2013).

Normally delivered babies acquired a microbiota high in *Bifidobacterium* and *Collinsella* within 6 months on an average; on the other hand, babies delivered by caesarean section and those born following a shorter duration of gestation took a longer time to acquire such microbiota (Dogra et al., 2015). It is a well-known fact that the intestine of a fetus is devoid of any microorganisms but is rapidly colonized by different microorganisms during and after the birth of the infant (Mackie et al., 1999). As babies are considered to be sterile at birth and the total microbiota outnumbers the total number of cells in an adult human in the ratio of 10:1, it is said that we are born 100% human, but at death we are 90% bacterial (Goodacre, 2007). However, recent evidences show that fetal intestinal colonization is initiated before birth due to ingestion of microbe-containing amniotic fluid by the fetus (Mshvildadze and Neu, 2010). Although considered sterile, the human placenta has recently been found to possess a unique microbiome. Culture-independent metagenomic analysis of population-based cohort of placental specimens collected under sterile conditions from 320 subjects revealed that a large variety of microorganisms are present in the placenta (Aagaard et al., 2014).

Important factors in determining the structure of gut microbiota of infants also include hygiene levels practiced and the medications used. The gut microbiota is usually considered fully developed by the age of 4 and bacteria initially colonizing new born babies are important in determining the

TABLE 24.2
Profile of Gut Microbiota

Total Number of Phylotypes	More than 1300
Four major phylotypes	*Firmicutes, Bacteroidetes, Proteobacteria, Acinetobacteria*
Two minor phylotypes	*Fusobacteria, Verrucomicrobia*
Major obligate anaerobes	*Bacteroides, Bifidobacterium, Clostridium, Eubacterium, Fusobacterium, Peptococcus, Peptostreptococcus, Ruminococcus*
Major facultative anaerobes	*Escherichia, Enterobacter, Enterococcus, Klebsiella, Lactobacillus, Proteus*

Sources: Adapted from Sekirov, I. et al., *Physiol. Rev.*, 90, 859, 2010; The Human Microbiome Project Consortium, *Nature*, 486, 207, 2012.

final composition of the microbiota in adults. An adult human carries approximately 2.5–3 pounds of bacteria (about the same weight as the brain) in the gut, and the bacterial content of our feces is approximately 60% (Guarner and Malagelada, 2003; Vanhoutte et al., 2004; Vrieze et al., 2010).

The spatial organization of the intestinal microbiota influences both the autochthonous (resident) and allochthonous (transient) microbes. Autochthonous microbes usually are closely associated with the intestinal mucosa that overlies the intestinal epithelium, while the allochthonous microbes are located usually in the central lumen as part of the fecal stream (Nava and Stappenbeck, 2011).

Gut microbiota is the most complex ecosystem and is composed of highly diverse microbes, which varies significantly between individuals. Even in any individual, the microbiota exhibits temporal variations, especially when there is a change in food habits and when one is affected by disease (Lozupone et al., 2012). The microbiota of a person is reported to be more similar at two different time points than to microbiota of another person (Huttenhower et al., 2012). Gut microbiota has been reported to be remarkably stable over time, with only little changes occurring in a 4-year study; more abundant strains remain more stable (Faith et al., 2013). An important property of gut microbiota is reduction in the overall diversity or microbiome gene count in several diseases and health conditions such as obesity and rheumatoid arthritis and old age (Turnbaugh et al., 2009; Lynch et al., 2015). Although individual host may experience extirpations of organisms such as *Helicobacter pylori* following the use of medications, human microbiome exhibits strong resilience and usually returns to pretreated composition after antibiotic-induced perturbation (Cho and Blaser, 2014).

Marked intraindividual variation in microbiome exists in different parts of the gastrointestinal tract (GIT), and also there is marked interindividual variation in the same anatomic site (Costello et al., 2009; Cho and Blaser, 2012). Revealed through high-throughput sequencing, such site-specific differences and individual-specific variations provide important basis for determining the beneficial and pathological significance of a particular microbiome composition (Table 24.3).

Metagenomic studies have brought to light an enormous richness and diversity of human gut microbiota compositions. Microorganisms colonize different parts of the GIT, and bacterial population density varies along the GIT (Guarner and Malagelada, 2003; Tappenden and Deutsch, 2007; Romano-Keeler et al., 2014). There is a qualitative and quantitative increase in complexity from the stomach to the colon (Table 24.3). In addition, there is variation in the composition of the flora along the GIT in terms of surface adherent and luminal bacteria. Although the ratio of anaerobes to aerobes is lower at the mucosal surface than in the lumen, the anaerobes outnumber aerobes and facultative anaerobes by two to three orders of magnitude in the overall count (Sekirov et al., 2010). A significant interplay exists between the metabolism of microbiota and the host as a comparison of germ-free and normal mice demonstrated that microbiota are responsible for a significant proportion of the metabolites that are detected in plasma (Wikoff et al., 2009).

TABLE 24.3
Compositional Differences in the Microbiome in the Different Parts of the Gastrointestinal Tract

Organ	Major Microbiome Members
Mouth	*Firmicutes, Proteobacteria, Bacteroides, Actinobacteria, Fusobacteria, Cyanobacteria*
Esophagus	*Firmicutes, Bacteroides, Actinobacteria, Proteobacteria, Fusobacteria*
Stomach (*H. pylori* negative)	*Actinobacteria, Firmicutes, Bacteroides, Proteobacteria, Fusobacteria*
Stomach (*H. pylori* positive)	*Helicobacter pylori, Firmicutes, Actinobacteria, Bacteroides*
Colon	*Firmicutes, Bacteroides, Actinobacteria, Proteobacteria*

Sources: Adapted from Costello, E.K. et al., *Science*, 326, 1694, 2009; Cho, I. and Blaser, M.J., *Nat. Rev. Genetics*, 13, 260, 2012.

24.3.2 Core Microbiome and Variable Microbiome

The core human microbiome is the microbiome present in a given habitat in all or the vast majority of humans. Habitat can refer to the entire body or a specific surface area, such as the gut or a region within the gut. The variable human microbiome is the microbiome present in a given habitat in a smaller subset of humans (Tremaroli and Bäckhed, 2012). The core microbiota comprises 50–100 bacterial species when the frequency of abundance at the phylotype level is not considered, and a core microbiome harboring more than 6000 functional gene group is present in the majority of the human guts surveyed (Zhu et al., 2010).

The various factors that influence the nature of variable microbiome include immunity, pathology (disease/health status), lifestyle, and environment of the host in addition to host's genetics (Turnbaugh et al., 2007; Franzosa et al., 2015). In order to determine whether such variation is sufficient to identify individuals within large populations or stable enough to identify them over time, stool samples were analyzed on day 1 and after 30–300 days. It was observed that gut microbiome was exceptionally stable and could be identified in >80% of the individuals, demonstrating the feasibility of microbiome-based identifiability, a result with important ethical implications for microbiome study (Franzosa et al., 2015).

24.3.3 Human Virome

Viruses affect the physiology of the host, that is, they cause acute diseases in the host, and hence, the virome is considered to be a component of the microbiome (Cadwell, 2015). Virome is reported to contain bacteriophages, pathogenic and nonpathogenic DNA and RNA viruses, prophages, newly discovered giant viruses, and interestingly several types of plant viruses (Scarpellini et al., 2015). Recent and continued advancements in sequencing and bioinformatic technologies have enabled researchers to reconstitute the entire viral genomes from the genetic material spread in the human gut, hence opening new perspectives on the understanding of the gut virome composition and possible future therapeutic use of gut virome modulation for the benefit of human health (Cadwell, 2015).

Humans are colonized by a large population of viruses and, which according to metagenomic studies, are mostly unique to each individual. The extreme interpersonal diversity of human gut viruses is predicted to derive from persistence of a small portion of the global virome within the gut of each individual and rapid evolution of some long-term virome members (Minota et al., 2013). Viruses infecting microbial and human cells can have profound effect on human health, directly causing acute or chronic infections or indirectly modifying the microbiome. Human virome-wide exploration revealed that viral infection can affect host immunity by infecting immune cells and interplay between virome and host immune system is implicated in several diseases such as type 1 diabetes, asthma, inflammatory bowel disease (IBD), and chronic gastritis (Foxman and Iwasaki, 2011; Chia et al., 2015). In addition to diet, prebiotics, probiotics, and other such recently developed supplements such as cobiotics, synbiotics, and immunobiotics (Gourbeyre et al., 2011; Hossain et al., 2015) can potentially modulate the gut virome. Exploration of the virome, its evolution and diversity, and its interactions with prokaryotes and with the host is emerging as an interesting frontier of research (Reyes et al., 2012; Blaser, 2014; Cadwell, 2015).

24.3.4 Human Gutome and Sterolbiome

The term *gutome* has been coined to describe the nutritional systems biology of gut microbiome and microbiome–host interactions (Dimitrov, 2011). The dynamic balance among gut microbiome, food nutrients, and host physiology and metabolism generates crucial model of systems biology, contributing to human health and disease (Brussow and Parkinson, 2014). Bile acids and their metabolites are now considered as hormones with both hepatic and extrahepatic effects on many host metabolic processes. The gut microbiota has the capacity to produce bile acid metabolites distinct from the liver, and the bile acid metabolites found in humans are produced by the combined action of enzymes encoded by the mammalian genome and also by the intestinal sterolbiome (Swann et al., 2011). The term *sterolbiome* describes the genetic potential of the gut microbiome to produce various bile acid metabolites from endogenous and exogenous steroids in the mammalian gut (Ridlon and Bajaj, 2015).

Sterolbiome enzyme bile salt hydrolases (BSHs) deconjugate the bile salts as they enter the terminal ileum and the proximal colon. Functional metagenomic screening demonstrated that BSH is present in 30% of Firmicutes, 14.4% of *Bacteroides*, and 8.9% of Actinobacteria. Members of the genus *Clostridium* are major producer of the secondary BAs, deoxycholic acid (DCA; 3α, 12α), and lithocholic acid (3α). These findings cumulatively indicate that the gut microbiome is capable of producing a cocktail of bile acid metabolites that can modulate host physiology through enterohepatic circulation. In addition, these bile acid metabolites can also modulate the structure and function of the gut microbiome (Swann et al., 2011; Ridlon and Bajaj, 2015). The BA composition observed in mammals is the result of equilibria between enzymes encoded in the mammalian genome and the gut sterolbiome. Bacterially produced secondary bile acids play an important role in the infection process of the enteric pathogen *Clostridium difficile.* While primary bile acids (produced by the host) induce germination of *C. difficile* spores, secondary bile acids (produced by gut microbiota), whose formation is often inhibited by antibiotics, suppress *C. difficile* germination (Ridlon and Bajaj, 2015).

24.4 TOOLS FOR STUDYING MICROBIOME

The vast number and variety of microorganisms present in the microbiota and the fact that more than 80% of the microbial species are not culturable in conventional growth media make the task of characterization of the microbiota a great challenge (Cho and Blaser, 2012). Recent advancement in culture-independent metagenomic and bioinformatic tools to analyze microbial population makes it possible to comprehensively explore the gut microbiome. Through characterization of the compositional range of the "normal" microbiome of healthy individuals leading to the analysis of the complex metabolism and microbe–host interactions that take place at the interface between microbes and host intestinal epithelia, the HMP made a great leap forward in the functional analysis of microbiome (The Human Microbiome Project Consortium, 2012). HMP focuses on the generation of a catalog of microbial genes present in the gut, which is being generated using several approaches including sequencing, assembling, and characterizing nonredundant microbial genes from fecal samples and whole genome sequencing of reference microbial species (Nelson et al., 2010; Qin et al., 2010; Wang et al., 2015).

Metagenomic approach involves assigning unassembled sequence generated by high-throughput shotgun sequencing to the NCBI nonredundant clusters of orthologous groups (Tringe et al., 2005; Andersson et al., 2008) and facilitating the assessment of interactions occurring with the microbiome and potentially the host–microbe interactions that occur at the interface between microbiome and the gut epithelial cells of the host. Another approach uses catalogs of known genes to identify functional clusters in a sample, which corresponds to the proposed taxonomic enterotypes (Arumugam, et al., 2011). High-throughput sequencing revealed that many microbial clades have large pan-genomes in contrast to the size of their core genomes. Any one *Escherichia coli* isolate typically contains ~4700 genes, but only some 2000 of these are found in all *E. coli*, and the pan-genome selected from all strains in the species now greatly exceeds 8000 genes (Huang et al., 2014).

Metatranscriptomics, metaproteomics, and metabolomics represent enormous complements to the understanding of the human gut microbiome using metagenomics, which can identify novel functional genes, microbial pathways, antibiotic resistance genes, and functional dysbiosis of the intestinal microbiome and determine interactions and coevolution between microbiota and host (Wang et al., 2015).

24.5 GUT MICROBIAL ACTIVITY IN HEALTH AND DISEASE

24.5.1 Gut Microbial Activity Beneficial to Humans

The enormous gene pool of the gut microbiome provides various bioactive molecules and enzymes, which communicate with the host and affect the host in numerous ways, both beneficial and harmful, acting locally and systemically (Tremaroli and Bäckhed, 2012; Hossain et al., 2015).

One of the major functions of gut microbiome that is beneficial to the host is fermentation of nondigestible carbohydrates such as resistant starches, cellulose, hemicellulose, gums, and pectins, resulting in the generation of short-chain fatty acids (SCFAs; acetic acid, propionic acid, butyric acid [Figure 24.1]), which serve as energy source to colonic epithelial cells. In addition, SCFAs stimulate differentiation and proliferation of intestinal epithelial cells (Cummings and Macfarlane, 1991; Sharma and Devi, 2014). Interaction between the gut bacteria and the host is important in the differentiation and proliferation of intestinal epithelial cells and the development of competent immune system and generation of appropriate immune and inflammatory responses (Yu et al., 2012). Studies with germ-free animals have revealed that gut microbiota plays an important role in the functional development of the mucosal immune system (Mazmanian et al., 2005; Tlaskalová-Hogenová et al., 2011). Also, gut bacteria stimulate the intestinal endocrine cells to produce hormones, which in turn enter circulation and modulate host function (Sekirov et al., 2010). In addition, gut microbiota communicates directly with immune cells of the gut-associated lymphoid tissue and terminal of visceral afferent nerves (Diaz-Heijtz et al., 2011). Gut microbiota also plays nonimmune, protective roles by directly blocking adhesion of intestinal pathogenic microbes to intestinal epithelial cells and enhancing mucosal integrity via epithelial cell stimulation (Sun et al., 2007; Hossain et al., 2015; Figure 24.2).

A state of homeostasis is maintained between human host and their gut microbiota as it dictates the health and disease states of the host, depending upon the relative proportion of beneficial bacteria and harmful bacteria (Kaminogawa, 2010). There are several excellent reviews on the beneficial role of microbiome (Guarner and Malagelada, 2003; Sekirov et al., 2010; Sommer and Backhed, 2013; Shreiner et al., 2015), so the discussion on this aspect is kept to a minimum as this is beyond the scope of this chapter.

24.5.2 Dysbiosis Leading to Disease States

Gut microbiota is a complex ecosystem with enormous diversity that exists in a state of homeostasis with the human host for the benefit of both the host and the microbes under normal circumstances. Any disruption in the state of homeostasis (dysbiosis) may enable the members of microbiota to overcome the protective host responses and exert pathological effects (Wallace et al., 2011; Brown et al., 2012; Carding et al., 2015).

Gut microbiome is being linked to a variety of diseases and health conditions with increasing frequency; more than 25 diseases and syndromes have been linked to dysbiosis (de Vos and de Vosm, 2012; Marchesi et al., 2015). Either excess or inadequate stimulation of the immune system can disrupt microbiome–host homeostasis and can lead to dysbiosis (Hardy et al., 2013). The intestinal mucosal immune system is equipped with mechanisms for eliminating the pathogenic microbes while showing tolerance for the gut microbiota (Schuijt et al., 2013). In addition, several indirect or associative support mechanisms of the mucosal immunity are involved in maintaining the state of

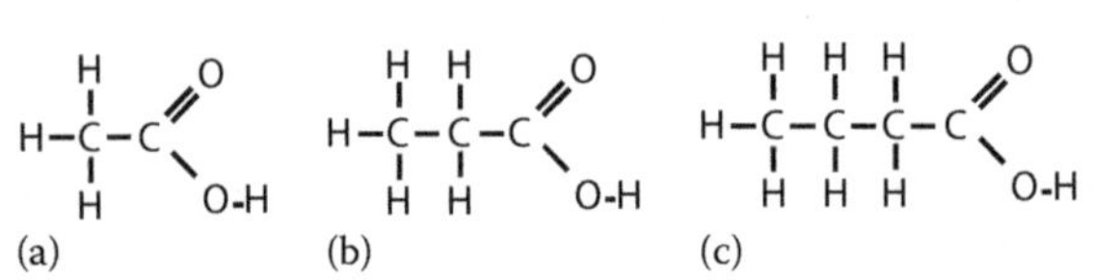

FIGURE 24.1 Short-chain fatty acids. (a) acetic acid, (b) propionic acid, and (c) butyric acid.

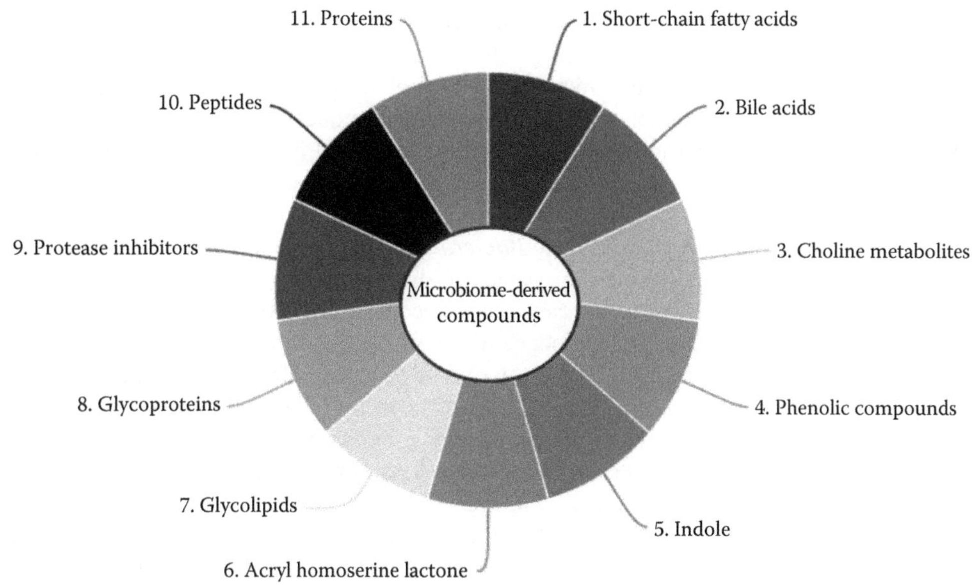

FIGURE 24.2 Microbiome-derived compounds that offer benefits to the host. (1) Energy supply to colonic epithelial cells. (2) Enteric absorption of nutrients; (3) Modulation lipid and glucose metabolism. (4) Aid in xenobiotic metabolism. (5) Protection against enteric stress. (6) Intra- and interspecies bacterial communication for host colonization. (7) Gut barrier function and integrity. (8) Immune tolerance/immune modulation. (9) Anti-inflammatory activity. (10) Transcription regulation/antimicrobial activity. (11) Inhibition of pathogen adhesion and colonization. Electrolyte absorption. (Adapted from Wikoff, W.R. et al., *Proc. Natl. Acad. Sci. USA*, 106, 3698, 2009; Tremaroli, V. and Bäckhed, F., *Nature*, 489, 242, 2012; Lukeš, J.et al., *PLoS Pathogens*, 11, e1005039, 2015; Clemente et al. 2012).

host–microbe homeostasis such as innate defense mechanisms, ensuring proper functioning of the mucosal barrier, secretory immunoglobulin A secretion, and existence of unique types of lymphocytes (Maranduba et al., 2015).

Human gut microbiota converts L-carnitine into trimethylamine (TMA) and trimethylamine-N-oxide, which can promote cardiovascular risk in humans (Zhu et al., 2014). Several cellular components of bacteria (e.g., LPS, flagella, peptidoglycans, teichoic acids, bacterial DNA, heat shock proteins, and superantigens) and the metabolic products of the gut microbiota are able to stimulate host immunity (both innate and adaptive) (Holmes et al., 2011, 2012). Low-grade inflammation may develop due to chronic host immune activation in response to signals generated by various gut bacterial components from gut microbiota. Studies have shown that low-grade inflammation can be a predisposing factor for a variety of multifactorial, multigenic complex diseases including diabetes, obesity, cardiovascular disease, allergy, asthma, IBDs, rheumatoid arthritis, autism, colon cancer, ageing, and several other infectious, inflammatory, autoimmune, and neoplastic diseases (Chang and Li, 2011; Hormannsperger et al., 2012; Guarner, and Rubio-Ruiz, 2014). Studies have shown that modulation of structure and function of gut microbiota through the use of probiotics, prebiotics, synbiotics, cobiotics, and immunobiotics can be considered as realistic therapeutic and preventive options for these diseases (Hossain et al., 2015).

24.5.3 Microbiome and Cancer

A study using a germ-free mouse model, which was selectively colonized by specific bacteria, revealed that microbiota can modulate susceptibility to cancer through a variety of mechanisms such as by modulating inflammation, and influencing the genomic stability of host cells, through epigenetic regulation of host gene expression (by producing metabolites that function as histone deacetylase inhibitors) (Bultman, 2014).

Zackular et al. (2014) analyzed gut microbiome in stool samples of patients from three clinical groups representing the stages of colorectal cancer development: healthy, adenoma, and carcinoma. The results showed both an enrichment and depletion of several bacterial populations that were associated with adenomas and carcinomas, hence demonstrating the possibility of using the gut microbiome composition to detect the presence of precancerous and cancerous lesions. Oncogenic potential of bacterial pathogens *H. pylori* and *Bacteroides fragilis* provides evidence regarding the carcinogenic potential of a single bacterium. Anaerobic bacterium *B. fragilis* is a classic example of an amphibiont organism, that is, it has the capacity to become either a symbiont or a cancer-causing pathogen, depending on whether it will produce an enterotoxin or not (Sears, 2009). The addition of enterotoxigenic *B. fragilis* (ETBF) to the microbiota induced colon tumor formation in mice, thus providing strong support for the disease-inducing capacity of single microbes. ETBF murine models confirmed the *in vitro* activities of *Bacteroides fragilis* toxin (BFT) to induce NF-κB and Wnt signaling, reactive oxygen species production, DNA damage, and carcinogenesis (Sears et al., 2014). Microbial dysbiosis has been linked with pancreatic cancer and breast cancer (Xuan et al., 2014; Zambirinis et al., 2014); on the other hand, the impact of gut microbiota in eliciting innate and adaptive immune responses beneficial to the host in the context of effective therapies against cancer has been shown recently; microbiome has been shown to play a role in generating anticancer immune responses in the host (Viaud et al., 2015).

24.5.4 Carcinogenic and Anticarcinogenic Activities of Gut Microbiota

Microbiome may promote carcinogenesis and can also mediate anticancer activity both directly (as an infectious agent, e.g., *H. pylori* causing gastric carcinoma) and indirectly through a variety of mechanisms such as modifying dietary components in ways that are either beneficial or detrimental to the human host (Hullar and Fu, 2014). Microbiome exists in a state of homeostasis with the host, and any shift in the balance can lead to a disease state: (1) microbial factors may induce changes in gut permeability, leading to bacterial translocation, (2) bacterial genotoxins such as cytolethal distending toxin and colibactin actively induce DNA damage in host cells in direct contact with the microbiome, and (3) various metabolites such as H_2S produced by the microbiota may be xenotoxic. Pathogen-associated molecular pattern interacts with Toll-like receptors of different cell types of immune system, leading to inflammation. Metabolic activities of the microbiome may result in the activation of various genotoxic substances such as dietary nitrosamine, acetaldehyde, and other carcinogens (Zackular et al., 2013).

Microbiome may also act to prevent carcinogenesis. Microbiome mediates anticarcinogenic activity directly by inactivating various carcinogens (Ou et al., 2013; Hagland and Søreide, 2015). The SCFAs produced by the microbiota (Figure 24.1) are versatile mediators of host metabolism. Microbiome metabolism also activates a variety of phytochemicals, which act as cancer-preventing agents. Microbial metabolites can also modulate gene expression in the host involving epigenetic mechanisms. Epigenetic mechanisms involve heritable change of gene expression without change in DNA via addition to or removal of methyl groups to nucleotides, acetylation, or deacetylation of histone, leading to either activation or deactivation of target genes (Huang et al., 2011; Hullar and Fu, 2014).

24.5.5 Dysbiosis and Restoration of Healthy Microbiota

Studies have demonstrated that various factors such as diet, disease, and antimicrobial therapy can alter gut microbiota, leading to dysbiosis; restoration of healthy microbiota can be achieved by the intake of probiotics and prebiotics and by fecal transplantation (Figure 24.3). Diet is a major factor influencing the composition and metabolism of the gut microbiota. The main dietary macronutrients

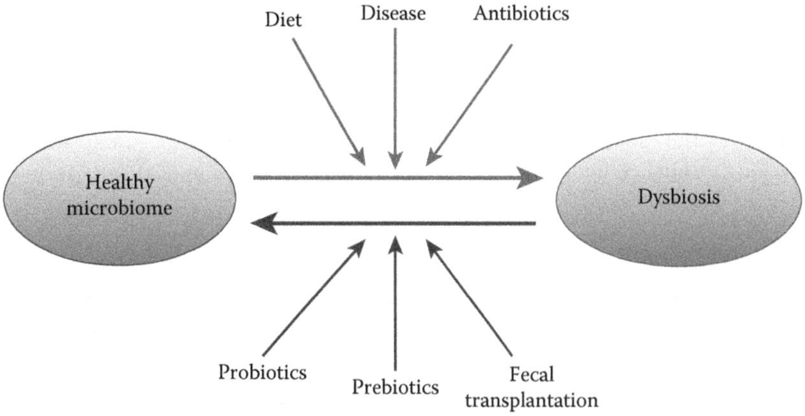

FIGURE 24.3 Factors influencing bidirectional transition between healthy microbiome and dysbiosis (altered microbiome).

(carbohydrates, proteins, and fats), their amount, and type have a great impact on the gut microbiota. A number of recent studies have shown that dietary factors alter the microbial community, which translates into biological changes in the host (summarized by Brown et al. (2012). High-fat diet leads to reduction in *Bifidobacteria* spp., which plays a vital role in protecting gut barrier. While changes in genetics resulted in 12% of the variation in the gut microbiota, the diet accounted for 57% of the total structural variation in a mouse model (Zhang et al., 2014). Diet rich in refined sugar leads to the overgrowth of opportunistic pathogens such as *C. difficile* and *C. perfringens*, while complex carbohydrates mediated an increase in the levels of beneficial *Bifidobacteria* spp. such as *B. longum* subspecies *longum*, *B. breve*, and *B. thetaiotaomicron* (Pokusaeva et al., 2011; Berg et al., 2012). The proportion of one macronutrient to the total energy intake affects the proportion of other macronutrients to the overall energy intake. Consequently, the changes induced in the gut microbiota by macronutrient intake usually result from the combinatory effect of all macronutrients (Graf et al., 2015). The connection between dysbiosis and disease and the effect of antibiotics on gut microbiota have been discussed in Sections 24.5.2 and 24.8.2, respectively.

Probiotics is defined by WHO as "a live organism which provides a benefit to the host when provided in adequate quantities," and a large number of bacteria meet this criterion. Various strains of lactic acid bacteria, *Bifidobacter* spp. and *E. coli* strain Nissle 1917, are commercially produced for human consumption (Soccol et al., 2010). The effects of the probiotics *Lactobacillus paracasei* or *Lactobacillus rhamnosus*, and two galactosyl-oligosaccharide prebiotics, were studied in germ-free mice inoculated with human baby flora. When the therapies were combined, populations of *Bifidobacterium longum* and *Bifidobacterium breve* increased, and *Clostridium perfringens* decreased. These microbial effects were associated with modulation of a range of host metabolic pathways indicated by changes in lipid profiles, gluconeogenesis, amino acid, and methylamine metabolism and were associated with fermentation of carbohydrates by different bacterial strains (Kinross et al., 2011). Hierarchical principal component analysis also permitted the visualization of multicompartmental transgenomic metabolic interactions that were also resolved at the compartment and pathway levels. More novel approaches that incorporate parasites that interact with the gut microbiome have also been suggested, such as the use of helminthes in IBD (Guarner, 2009). Probiotic therapies have now been proposed for a large variety of gut-related disorders such as IBD, inflammatory bowel syndrome, and pouchitis after surgery for IBD. However, there are more novel areas of potential application (Kinross et al., 2011).

Prebiotics are defined as "nondigestible food ingredients that beneficially affect the host by selectively stimulating the growth of one or a limited number of bacterial species in the colon, such as *Bifidobacteria* and *Lactobacilli*, which have the potential to improve host health" (Hardy et al., 2013). The enhancement of the number of beneficial organisms such as *Lactobacilli* and *Bifidobacteria* is considered as a marker of intestinal health and as a possible screening marker for prebiotics. Restoration of healthy microbiota can be achieved by the intake of probiotics and prebiotics and various combinations of these two ingredients (such as cobiotics, synbiotics, and immunobiotics) developed to enhance their spectrum of activity in host (Slavin, 2013; Hossain et al., 2015).

24.5.6 Fecal Microbiota Transplantation

Fecal microbiota transplantation (FMT, also known as fecal bacteriotherapy) is the process in which a fecal preparation from a carefully screened, healthy stool donor is transplanted via colonoscopy, nasoenteric tube, and rectal enema or in the form of capsules into the colon of a patient (Youngster et al., 2014). First used in 1958 in patients with pseudomembranous enterocolitis (Eiseman et al., 1958), FMT is being used with increasing frequency to treat recurrent *C. difficile* infection (CDI) (Cammarota et al., 2015).

The basic pathogenic mechanism of recurrent CDI is not completely understood. The salient features of the CDI are normal gut microbiota, which keeps the expansion of the *C. difficile* populations in check is disrupted due to antibiotic treatment. *C. difficile* spores (which are resistant to antibiotic treatment) germinate back into vegetative forms after antibiotic treatment has been discontinued. If residual normal intestinal microbiota cannot restrain the infection, *C. difficile* proliferate and produce toxins that cause destruction of colonic epithelial cells, thus leading to inflammation with resultant disease symptoms (Tang-Feldman et al., 2003). Recurrent CDI occurs in as many as 15%–30% of patients after an initial CDI, and up to 65% of patients who experience one recurrence will have subsequent recurrences after antibiotic therapy is stopped (McFraland et al., 2002; Bakken et al., 2011).

In a randomized clinical trial, FMT and vancomycin treatment were compared for the treatment of recurrent CDI. Ninety percent of the patients treated by FMT exhibited resolution of *C. difficile*–associated diarrhea. In FMT, five of the seven patients with pseudomembranous colitis also reported a resolution of diarrhea. On the other hand, resolution of CDI occurred in 26% patients in vancomycin-treated group ($P < 0.0001$), demonstrating that FMT using colonoscopy to infuse feces was significantly more effective than vancomycin regimen for the treatment of recurrent CDI (Cammarota et al., 2015). Youngster et al. (2014) used oral, capsulized, frozen FMT for the treatment of relapsing *C. difficile* infection. Resolution of diarrhea was achieved in 70% of the patients after a single capsule-based FMT; resolution level increased to 90% following a second capsule injection. Taken together, these studies have demonstrated that FMT is a useful procedure in treating relapsing or refractory CDI. However, practical barriers and safety concern issues need to be addressed before its widespread use. Larger studies are needed to confirm these results and to evaluate long-term safety and effectiveness (Cammarota et al., 2015).

24.6 HUMAN GUT MICROBIOME AND ENVIRONMENTAL TOXICANTS

Extensive studies on human microbiome have led to the birth of a novel concept: microbiome as the interphase between the external environment and our mucosal epithelial cells which serves as a mediator between the external agents to which we are exposed and our body systems. Chemicals that pass through the GIT and interact with microbiome may undergo major changes in bioavailability and in toxicity. The role played by microbiome as the interphase between the external environment and our body is so vast and varied that one of the objectives of HMP was to study how microbiome modulates the external agents to which we are exposed and how the microbiome is also influenced by the agents to which we are exposed. The study on such bidirectional interaction between the external agents and microbiome, transformative powers of the microbiome, and its implications for chemical toxicology and risk assessment has emerged as a major area in the field of toxicology.

24.6.1 Microbiome and Heavy Metals

The gut microbiota plays an essential role in maintaining intestinal homeostasis, and it also plays a critical role in the metabolism of xenobiotics. Exposure to environmental toxicant causes dysbiosis (alteration of gut microbiota), which may lead to disease states and affects the health of the host. High concentrations of heavy metal (HM) ions that are not absorbed have a detrimental impact on the gut microbiota, hence disrupting the homeostasis that exists between the microbiota and the host. Intestinal absorption of HMs accumulates in specific target organs and causes detrimental effects on human health (Upreti et al., 2004). The cellular mechanism involved in toxicity is not completely known, but excessive local and systemic oxidative stress or immune deregulation has been shown to be involved. Nonabsorbed HMs have a direct impact on the gut microbiota (Fazeli et al., 2011). Studies have shown that depletion of luminal iron alters the gut microbiota and chronic ingestion of lead and cadmium alters the bioavailability of essential metal ions and HMs (Breton et al., 2013a).

Germ-free mouse model provided vital experimental data highlighting the role of gut microbiota in limiting the burden of HMs. When compared with conventional-specific pathogen-free mice, significant increase in lead and cadmium absorption and dissemination was noted in germ-free mice. Metallothioneins are proteins that can bind various xenobiotic metal ions and play a role in the regulation of physiological metal ions and provide protection against xenobiotic metal ions (Coyle et al., 2002). The genes encoding metallothioneins and various enzymes and proteins involved in metal transport and oxidative pathways were differentially regulated in the germ-free mice in comparison to the control mice illustrating complex microbe–host interactions that take place when microbiome is challenged with HMs (Breton et al., 2013b,c). Investigations focusing on the complex interaction that involves environmental toxicants, such as HMs, the microbiome, and the host, may provide insights into "exposome," which takes into account both the direct effect of the toxicants on the microbiome and the host and also the resultant indirect effects such as oxidative stress, inflammation, modulation of gene expression, and metabolic processes that cumulatively lead to chronic diseases (Rapport and Smith, 2010; Brenton et al., 2013c).

24.6.2 Microbiome and Polycyclic Aromatic Hydrocarbons

Polycyclic aromatic hydrocarbons (PAHs) are a group of more than 100 chemicals containing multiple aromatic rings fused together lacking branching substitutions. It is one of the most widespread organic pollutants generated by incomplete burning of coal, oil, and gas, garbage, or other organic substances. People can be exposed to PAHs from many sources, including contaminated food and soil and charcoal-broiled meats (Baird et al., 2005; Haritash and Kaushik, 2009). PAHs are considered pollutants of great concern due to their mutagenic, carcinogenic, and also teratogenic nature with potential adverse health impacts and also due to their presence at significant levels over a period of time in human diets. PAHs are absorbed through intestinal walls of the human host and are metabolized into compounds with carcinogenic potential. Estrogens are a group of steroid hormones that regulate the expression of genes involved in the development and functioning of the female and male reproductive organs (Choi et al., 2006; Billiard et al., 2008). A study by Van de Wiele et al. (2005) showed that microbiome can convert PAHs into metabolites with estrogenic side effects. The authors established a model system that re-creates the digestive processes that occur along the length of the human GI tract and demonstrated that gut microbiome can convert PAHs into metabolites that were structurally and functionally similar to estrogen. The enzymes produced by gut epithelial cells were excluded, which could obscure the effects of the microbiome due to their own metabolic function. Interestingly, when PAH-contaminated soils were allowed to interact with gut microbiota, metabolites having estrogenic activity were detected at levels comparable to that seen with pure PAHs (Brinkmann et al., 2014). Microbiome modified PAH metabolites are usually hydroxylated derivatives. Investigation on monohydroxylated metabolites of PAHs revealed that the metabolites selectively induced estrogen receptor beta (ERβ) transcriptional activity at the concentrations tested but do not activate ERα estrogen receptors. In addition, monohydroxylated PAHs competitively bound ERβ, induced ERβ homodimers, and regulated ERβ target genes in human breast cancer cells and potentially interfere with ERβ signaling pathways (Sievers et al., 2009).

24.6.3 Microbiome and Deoxycholic Acid, a Microbial Cocarcinogen

Studies have shown that senescence-associated secretory phenotype (SASP) has a critical role in promoting obesity-associated hepatocellular carcinoma (HCC) (Kuilman and Peeper, 2009). DCA is a gut bacterial metabolite known to cause DNA damage. Obesity has been reported to alter gut microbiota so that increased level of DCA is produced. Enterohepatic circulation of DCA induces SASP phenotype in the hepatic stellate cells (HSCs), resulting in the secretion of a large amount of various inflammatory and tumor-promoting cytokines by these cells in the liver. This culminates in the development of HCC in mice after an exposure to chemical carcinogens. Reduction in gut bacteria that produce DCA prevented HCC development in murine model. Moreover, in patients with nonalcoholic steatohepatitis, SASP was observed in HSCs, indicating the involvement of a similar pathway in humans (Yoshimoto et al., 2013).

24.6.4 Microbiome and Arsenic

Cumulative evidences generated from recent studies suggest that microbiome plays a critical role in the transformation of various environmental toxicants, such as arsenic, which is one of the major toxicants. The contribution of gut microbiome in toxicant metabolism is so staggering that scientists compare it to genetic polymorphism regulating metal transformations in human body (Betts, 2011).

To a toxicologist, a toxicant enters the cell or tissue or circulates in the blood causing harmful effects in the host, but microbiome may carry out biologically relevant transformation before it crosses into the internal environment of the human body. For example, microbiome carries out methylation of arsenic (at least in part) prior to the entry of the toxicant to our body. However, it needs to be appreciated that toxicant–microbiome interaction is bidirectional (Mani et al., 2014). Although the metabolism of toxicants by gut microbiome is important, how the xenobiotics influence the structure and function of the microbiome is equally important. In addition to changes that a toxicant may induce into the microbiome, it may serve as a substrate for microbiome-encoded enzymes carrying out the biotransformation of a drug into its active or inactive metabolites (Ursell and Knight, 2013). Moreover, xenobiotics and their metabolites may modulate host drug metabolism spectrum by regulating the expression of genes in the host involved in drug metabolism. For example, cytochrome P450 (Cyp) enzymes such as Cyp2c29 and Cyp2c293a3a11 and host xenobiotics sensors pregnane X receptor and constitutive rostane receptor are significantly upregulated in germ-free mice in comparison to the conventional mice (Mani et al., 2014).

The gut microbiome metabolism of arsenic generates several intermediate compounds with different levels of toxicity, some with higher or lower toxicity than arsenic itself (Van de Wiele et al., 2010). In an interesting study, metagenomic profiling of gut microbiome of mice was compared before and after exposure to 10 ppm of arsenic for 4 weeks and with untreated mice. Metabolomic profiles of serum urine and feces of these mice were analyzed by comparing liquid chromatography/mass spectroscopy. Microbiome analysis revealed that Bacteroidetes and Firmicutes families were predominant in control mice, while in arsenic-treated mice, Firmicutes family was significantly reduced and Bacteroidetes population remained the same. Metabolite profile analysis showed that in arsenic-exposed mice, the concentration of 224 metabolites decreased and the concentration of 146 metabolites increased. Metabolites linked to insulin resistance, obesity, and cardiovascular disease were found to be altered (DiBaise et al., 2012; Lu et al., 2014). In an another similar study using mice model, Dheer et al. (2015) reported that chronic exposure to arsenic in drinking water resulted in time- and concentration-dependent alterations in Bacteroidetes and Firmicutes populations of the microbiome. Other changes included abundance of bacterial spores and alterations in microbiome/host nitrogen metabolism that support disease-enhancing pathogenic phenotypes. Taken together, these findings indicate that arsenic exposure leads to marked alterations in the composition of the microbiome; moreover, the profile of host metabolites also changed, leading to disease states. Incorporating the microbiome into the environmental health paradigm will increase our understanding of multiple-level interactions within the microbiome and the interaction between the microbiome and the different organ systems of the host, especially when exposed to exogenous toxicants (Dietert and Silbergeld, 2015).

24.6.5 Microbiome and Mercury

Mercury affects many metabolic processes and organ systems in humans and experimental animals. Reexamining environmental exposures to mercury factoring in microbiome is providing more insights into how toxicity of mercury can be modulated by the gut microbiome (Skurnik et al., 2010). Methyl-mercury can be demethylated by intestinal bacteria, which can lead to high-level exposure to inorganic mercury. Exposure of microbiome to mercury led to the development of mercury resistance in gut bacterial species, and the development of mercury resistance is often associated with the development of resistance to antimicrobial agents in bacterial population (Liebert et al., 1997; Skurnik et al., 2010). Mercury-induced changes in gut microbiota have been linked to the development of autism. Mercury exposure in a pregnant woman can lead to a negative ramification in her child's microbiome—loss of beneficial bacteria and abundance of pathogens. The ultimate effects are chronic infections, molecular mimicry, altered development of the nervous system, mitochondrial dysfunction, neurologic inflammation, neuropsychiatric symptoms, and autism (Conrick, 2014). Consumption of probiotics-supplemented yogurt was found to lower mercury levels in at-risk populations such as pregnant women in Tanzania (Bisanz et al., 2014).

24.7 MICROBIOME AND THERAPEUTIC AGENTS

Recent studies have demonstrated that the metabolic cross talk between the host and gut microbiota modulates the pharmacokinetics and pharmacodynamics of drug. Gut microbiota has the capacity to transform various organic chemicals such as nutrients, pollutants, toxins, and drugs using endogenous enzymes (Figure 24.4). Most drugs are biotransformed into active or inactive form by

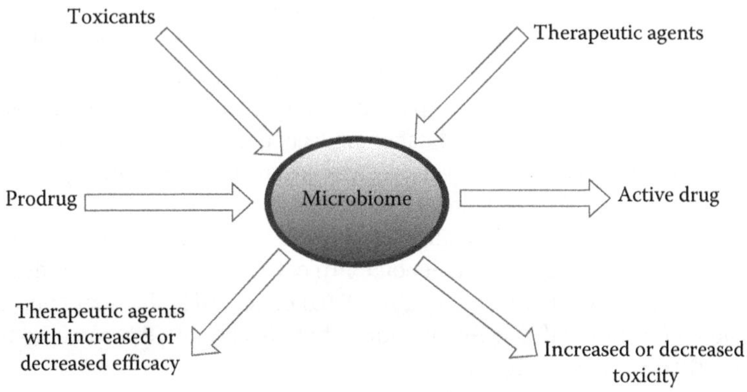

FIGURE 24.4 Microbiome metabolism of environmental toxicants and therapeutic agents. Various enzymatic reactions are involved: (1) reduction, (2) oxidation, (3) hydrolysis, (4) acetylation, (5) deacetylation, (6) denitration, (7) hydroxylation, (8) dehydroxylation, (9) conjugation, (10) deconjugation, and (11) proteolysis (Table 24.4).

endogenous enzymes and/or conjugation reactions, which facilitate their elimination via various routes such as through urine, feces, or bile (Li and Jia, 2013). The value of pharmacogenomics in predicting the response of the patients to the drugs is highlighted by the recommendation of the U.S. Food and Drug Administration to genotype patients prior to treatment with irinotecan, clopidogrel, trastuzumab, and carbamazepine, due to the differential efficacy and toxicity of these drugs in stratified patient cohorts (Yip and Chan, 2015). However, the important role played by the gut microbiome in drug metabolism and hence bioavailability and efficacy need to be factored in to fully recognize the interindividual differences in drug metabolism. The effect of drug treatments may vary in different individuals due to genetic factors; however, environmental factors such as age, disease, nutritional status, other drug use, and importantly individual microbiome can also influence individual metabolic phenotype, which modulates metabolism, efficacy, and toxicity of drugs (Clyton et al., 2009; Wilson, 2009).

Pharmacometabonomics is a new branch of science, defined as "the prediction of the effects of a drug on the basis of a mathematical model of pre-dose metabolite profiles" (Everett et al., 2013). Pharmacometabonomics promises to be an important new approach to the delivery of personalized medicine (PM) to improve both drug efficacy and safety for patients in the future. Pharmacometabonomics is particularly powerful as it is sensitive to both genetic and environmental factors such as diet, drug intake, and, most importantly, a person's microbiome.

There exists a strong connection between an individual's metabolic phenotype, in the form of a predose urinary metabolite profile, and the metabolic fate of a standard dose of the widely used drugs (Clayton et al., 2009). Certain human gut microbiome cometabolite can serve as predictor of drug efficacy. Microbiome-influenced drug explains for the variable responses induced by many different drugs and xenobiotics in different individuals. These findings indicate that assessing the effects of microbiome activity should be an integral part of drug development and of personalized health care and targeted manipulations of gut microbiome to reduce adverse drug reactions and to improve the efficacy of drug will appear as an option in the future (Clayton et al., 2009).

24.8 HUMAN GUT MICROBIOME AND THERAPEUTIC AGENTS

24.8.1 Microbiome and Therapeutics

All patients do not show the same response to the same treatment; there is considerable variation depending on their genotypes and/or phenotypes. This anticipates more tailored approach, which is already in progress with acceleration in PM, though its assurance has not always been translated into upgrading in patient care due to the complexities involved (Jeremy et al., 2012; Schork, 2015). Until recent times, PM was linked almost entirely to pharmacogenomics. In order to optimize the potential of PM (adjust the drug therapy to patient uniqueness), we have to seriously consider the contribution of pharmacometabonomics. Metabonomics concentrates on the study of metabonome, representing all communications between various characteristics of genome, microbiome, parasites, environment, etc. (Gutiu et al., 2010; Dhanasekaran et al., 2015).

There is an enormous rush in human microbiome research presently around the world. Succeeding the completion of the Human Genome Project, there has been significant development in recent years in the HMP (Rizkallah et al., 2010; Saad et al., 2012). Human microbiome not only does modulate human health and disease conditions but also affects the outcome of drug therapy (Rizkallah et al., 2010; Saad et al., 2012).

These advances open up new prospects for learning how compositional and functional variations in microbiome influence the efficiency of drugs and their toxicity (pharmamicrobiomics), especially in the gut. This takes account of research regarding how the microbiome works together with human metabolic enzymes in the intestine and liver. Eventually, we must identify better the future suggestions of HMP on PM and drug therapeutics (Saad et al., 2012).

During the last decade, novel molecular methods have been introduced to reveal the hidden truths about the role of microbiome in human life. We are the hosts to this population of microbes (including bacteria, viruses, and yeasts) living on us and in us, and the diversity of the human "microbiota" is enormous. As we begin to decipher the large number of functions that this microbiota performs, it is becoming clearer that our microbes play a far more significant and central role in the maintenance of our health than we ever thought before (Human Microbiome Project, 2012). As a result, it is clear that we are at a crucial point in our outlook toward microbes in a medical context.

The progress in HMP helps new drug therapies that improve health status by interacting with the human microbiome in specific ways. It can lead to the development of microbiome modulator products designed to alter bacterial populations and their environment in the GI tract to prevent or treat health conditions, such as prediabetes, diabetes, and obesity. It paves the way for unprecedented opportunity to develop novel, prescription products that influence this information to benefit human health (Lehouritis et al., 2015).

The composition of the intestinal microbiome is primarily determined by environmental and genetic factors although external factors as well as host immune reactions lead to the adjustment of the delicate equilibrium in both health and disease states. The metabolism of drugs by both intestinal bacteria and further by enterocytes leading to their systemic absorption deserves further consideration and may offer valuable insights into presystemic drug metabolism, delivery, and toxicity (Grundmann, 2010; Lehouritis et al., 2015).

More than 30 commercially available drugs are "biotransformed" by the gut microbiome (Sousa et al., 2008; Table 24.4). This processing can be implemented in various ways. For instance, microbes can secrete substances that serve as substrates for host enzymes processing a particular drug and thus reduce its efficiency. Further, this process may lead to drug toxicity due to extended circulation in the blood. On the contrary, some antibiotics can influence the shape and diversity of the microbiota (Freilich et al., 2009; Borenstein, 2012). It is of interest to note that transplanted microbiota by itself can be used as a therapy to treat various diseases, such as diabetes, diarrhea, obesity, acute gastroenteritis, and IBD (Virgin et al., 2011; Cammarota et al., 2015).

The gut microbiota is a chief contributor to individual differences in the bioavailability of orally administered drugs (Nicholson et al., 2005; Garber, 2015). As an essential and acquired organ of the body, the gut microbiota provides an extensive variety of valuable functions, such as secretion of antimicrobial products, which through the development of colonization resistance negatively select against pathogenic bacteria. These functions could be compromised due to the use of broad-spectrum antibiotics, leading to disorder of gut microbiota. The consequence of this may lead to conditions like improper development of the gut immune system and the development of *C. difficile* induced antibiotic-associated colitis, respectively (Sun and Chang, 2014). There is increasing awareness of the importance of the gut microbiome research in identifying novel microbiome modulator drugs (Holmes et al., 2011; Garber, 2015).

24.8.2 Microbiome and Antibiotics

The indiscriminate use of antibiotics has transformed a portion of their proposed targets into superbugs, selecting for those mutants and strains of bacteria that have the capacity to survive large doses of antibiotic drugs. Blaser and Falkow (2009) put forward that the recent increase in allergic and other diseases without any apparent explanation (e.g., asthma and metabolic diseases) is an outcome of the disappearing microbiota. The factors such as smaller families, cleaner water, delivery by caesarian section, and extensive antibiotic use (particularly in young children) have all resulted in the disappearance of our microbiota, which have evolved and coexisted with their hosts for millions of years. The genomic studies have been useful in demonstrating the effects of antibiotics on the human microbiota (Raymond et al., 2015). In a longitudinal study of the gut microbiota in three antibiotic-treated adults, ciprofloxacin treatment was associated with decreased bacterial diversity (Dethlefsen et al., 2008). Most bacterial taxa recovered after four weeks, but some taxa did not recover even after

TABLE 24.4
Microbiome-Induced Alteration of Potency and Toxicity of Drugs and Toxicants

Drug/Reaction	Action of Gut Microbiota	Mechanism/Animal Model
Digoxin/reduction	Drug converted to inactive metabolites, leading to reduced therapeutic efficacy	Gut bacterium *Eggerthella lenta* converts digoxin to dihydrodigoxin or dihydrodigoxigenin. Germ-free mice did not form these metabolites, but they did when they were colonized with *E. lenta.*
Prontosil/reduction	Conversion of prodrug into active drug	Azoreductase enzymes convert the azo bond containing prodrug into active drug, 5-aminosalicyclic acid.
Levodopa or L-Dopa (L-3,4-dihydroxyphenylananine)/ dehydroxylation	Alteration of pharmacokinetics of the drug	Drug converted into m-tyramine and m-hydroxyphenylacetic acid. In germ-free animals and antibiotic-treated animals (to eliminate gut flora), L-Dopa is not dehydroxylated.
Lactulose (stool softener)/ hydrolysis	Therapeutically active molecules produced	*Lactobacillus*, *Bacteroides*, and *Clostridium* spp. carry out hydrolysis; lactic acid and acetic acid are produced, which stimulate water secretion into the lumen by lowering the pH of the intestinal environment.
Sorivudine/hydrolysis	Increase in toxicity of another drug	*Bacteroides eggerthii* and *Bacteroides vulgatus* convert the drug bromovinyluracil, which suppresses the catabolism of 5-fluorourcil (5-FU), another antiviral drug, resulting in toxic levels of 5-FU.
Acetaminophen/competition	Enhancement of availability of drug	*C. difficile* produces *p*-cresol that competes with the drug for hepatic detoxification, thus reducing its elimination.
5-Aminosalicylic acid/ acetylation	Inactivation of the drug	Bacterial *N*-acetyltransferase enzyme acetylates the drug to acetylated 5-aminosalisylic.
Phenacetin/deacetylation	Production of toxic compounds	Bacterial enzymes cause deacetylation, producing toxic p-phenetidine, which causes methemoglobinemia and nephritis.
Ranitidine/cleavage of N-oxide bond	Poor bioavailability	Gut bacterial enzyme cleaves N-oxide bond.
Glyceryl trinitrate/denitration	Reduction in activity	Generation of dinitrated and mononitrated glycerol.
Insulin/proteolysis	Loss of activity	Gut bacterial protease digests insulin.
Levamisole/thiazole ring opening	Generation of metabolites levametabol-I, levametabol-II, and levametabol-III	*Bacteroides* spp. and *Clostridium* spp. carry out ring opening.
Chloramphenicol/amine formation and hydrolysis of amide linkage	Enhancement of toxicity	Metabolized to p-aminophenyl-2-amino-1,3-propanediol that causes aplasia in the bone marrow.

Sources: Adapted from Clayton, T.A. et al., *Proc. Natl. Acad. Sci. USA*, 106, 14728, 2009; Kinross, J.M. et al., *Genome Med.*, 3, 14, 2011; Haiser, H.J. et al., *Science*, 341, 295, 2013; Lehouritis, P. et al., *Sci. Rep.*, 5, 14554, 2015; Yip, L.Y. and Chan, E.C.Y., *Drug Metab. Disposit.*, 43, 1619, 2015.

six months, demonstrating the enormous effects of antibiotic treatment on the gut microbiota. As per a study in mice, the antibiotic treatment completely displaced the normal gut microbiota, allowing exogenously administrated vancomycin-resistant *Enterococcus* to completely invade the gut (Ubeda et al., 2010). The emergence of multidrug-resistant bacterial strains depends on reservoirs for antibiotic resistance genes that may be accessible to clinically relevant pathogens. The lateral gene transfers are principally responsible for the propagation of antibiotic resistance genes and virulence genes.

The lateral gene transfer can occur by direct uptake of environmental DNA (such as transformation of plasmids), bacteriophage-mediated transduction, conjugation, or transposition. Further, soil has also been identified as a potential source of antibiotic resistance genes (Dantas et al., 2008), which is not astonishing because several clinically relevant antibiotics are derived from soil actinomycetes. The functional characterization of the antibiotic resistance reservoir (culture independent) of human gut and saliva microbiota revealed that most resistance genes have not been previously identified and were evolutionarily distant from known resistance genes (Sommer et al., 2009). Moreover, most of the resistance genes identified in cultured isolates of the gut microbiota were identical to known resistance genes in clinically significant pathogens. These findings raise many questions about indiscriminate use of antibiotics leading to commensal human microbiota forming reservoirs for antibiotic resistance genes (Grice et al., 2012; Barriere, 2015).

24.8.3 Gut Microbiota: A Potential New Territory for Drug Targeting

The major involvement of the gut microbiota in human health and disease recommends that manipulation of microbiome through combinations of antibiotics, probiotics, and prebiotics could be a novel therapeutic approach. A systems viewpoint is required to understand the complex host–bacteria interactions and their association with pathophysiological phenotypes so that alterations in the composition of the gut microbiota in disease states can be reversed (Jia et al., 2008; Graf et al., 2015).

To date, engineering the ecologies (the living environments outside and inside of our bodies) has not been that successful because of an incomplete understanding of the systems involved. Currently, we still have limited mechanistic knowledge regarding how indigenous microorganisms have shaped our genome and biology in postnatal development. Emerging technologies, such as metabonomics and metagenomics, have provided information indicating that symbiotic microbes affect metabolic phenotypes and enzyme induction states, which ultimately affect the outcome of interventions (Dumas et al., 2006). With the progress of metagenomics in conjunction with other "omics" technologies, the composition and specific biological functions of the human gut microbiota will be gradually uncovered, opening up an entirely new approach to drug discovery and therapy (Jia et al., 2008).

24.8.4 Predicting Patient Response to Therapy (Drug Metabolism, Toxicity, and Efficacy) through Pharmacometabonomics

Pharmacometabonomics has particular value for studying the host–microbiome interactions and for evaluating new therapies that target the gut bacteria. Pharmacometabonomics has at present been used to know and predict the interventional outcome of drugs (Clayton et al., 2006; Everett et al., 2013). The majority of metabonomics studies have used nuclear magnetic resonance or mass spectrometry. To begin with, the influence of three structurally diverse hepatotoxins in rats (galactosamine, allyl alcohol, and paracetamol), which act via different mechanisms, was considered. It was demonstrated that predose urinary profiles carried information about the degree of postdosing toxicity. In case of paracetamol, information about the interanimal variation in drug metabolism was also demonstrated (Clayton et al., 2006). Consequently, the first demonstration of pharmacometabonomics in humans showed a clear association between an individual's predose urinary metabolic phenotype and the metabolic outcome of a standard dose of the extensively used analgesic acetaminophen (paracetamol) (Clayton et al., 2009). The metabolic outcome of the drug was obtained from postdose urine samples, determining the extent of the various drug metabolites excreted by each subject, this being known to show significant intersubject variations. It is important since information about how a particular drug is metabolized and excreted by each individual can have a major influence on its efficacy and safety. It is widely expected that metabonomics and analytical and statistical technologies will continue to develop and possibly find application in real clinical situations, even in monitoring patients in real time (Jeremy et al., 2012).

24.8.5 PharmacoMicrobiomics: Effects of Human Microbiome on Drugs

Pharmacogenomics inspects how variations within the human genome affect the action and disposition of drugs as well as drug tolerance (Nebert et al., 2008). However, variations within the human genome do not fully account for the tremendous phenotypic variations observed between individuals. Human-associated microbes significantly contribute to the effective human gene pool, and their combined genomes (known as the human microbiome) have gained attention recently. The HMP investigated the tremendous diversity of human-associated microbial communities residing in different human tissues and studied the effect of microbial genes and genomes on human health and disease (Turnbaugh et al., 2007; Peterson et al., 2009). However, the effect of these microbes on drugs remains largely unexplored. Since microbes have complex metabolism, including an extraordinary ability to metabolize xenobiotics (Clayton et al., 2009; Wilson, 2009; Wilson and Nicholson, 2009), they are expected to play a pivotal role in modulating the action, disposition, and toxicity of drugs with which they interact in different subecosystems within the human body (Rizkallah et al., 2010; Yip et al., 2015).

24.8.6 Therapeutic Implications of Gut Microbiome Modulation

The potential therapeutic role of the gut microbiota for human health has led to therapeutic approaches such as bacteriotherapy (Borody et al., 2004) and bioecological control (Bengmark, 2005). Broadly, these theories argue that modulation of intestinal floral populations either by the premorbid gut microbiota of the host or by prebiotics, probiotics, and synbiotics may be beneficial to the human health (Gibson and Roberfroid, 1995). Several studies now suggest that fecal transplantation may be beneficial in case of antibiotic-associated diarrhea or *C. difficile* infection (Khortus et al., 2010). However, the long-term functional or metabolic consequences for the host of microbial modulation are poorly understood. New insights are being provided through the analysis of probiotic and prebiotic studies using global metabolic profiling techniques, where evidence suggests that the effect may be cumulative (Turroni et al., 2009).

24.9 CONCLUSIONS AND FUTURE PERSPECTIVES

There has been a remarkable progress in the past few years on the understanding of gut microbiome and its immense impact on human biology. Although we have the necessary crucial tools, experimental proofs are required for an in-depth analysis of the role of microbiome in metabolism of xenobiotics and therapeutic agents. Gut microbiome possesses multiple pathways for xenobiotic metabolism, and the reason for the evolution of such diversity in key biotransformations remains one of the fascinating and thought-provoking questions. Possibly some of these pathways may have evolved in the microbiome to avoid the toxicity of environmental toxicants or host-generated inhibitory compounds. Alternatively, it is also possible that the enzymes involved in the metabolism of compounds of dietary, microbial, or human origin evolved to cometabolize xenobiotics.

Studies directed toward defining the metabolic pathways of utilization of various xenobiotics as growth substrate, with the aim of minimizing toxicity, will provide insights into the selection pressure inducing propagation of these genes. Obtaining biologically relevant samples of microbiome–toxicant interactions, careful annotation of metadata, and meticulous clinical phenotyping are essential in generating relevant microbiome data sets that would lead to meaningful insights into modulation of toxicants/therapeutic agents by the microbiome. An overarching challenge in the field of microbiome–toxicant interaction is to explore the complex, multitier analyses resulting from the eventual integration of metagenomic and human genetics data sets. The gut microbiota cumulatively possesses extremely complex and diverse metabolic power and hence enormous ability to metabolize xenobiotics. Hence, it is anticipated that eventual integration of metagenomic and human genetics data sets will provide valuable insights into the role of microbiome in modulating the action, toxicity, and disposition of therapeutic agents in the human body.

REFERENCES

Aagaard, K., J. Ma, K. M. Antony, R. Ganu, and J. Versalovic (2014). The placenta harbors a unique microbiome. *Science Translational Medicine* 6: 237ra65.

Andersson, A. F., M. Lindberg, H. Jakobsson, F. Bäckhed, P. Nyrén, and L. Engstrand (2008). Comparative analysis of human gut microbiota by barcoded pyrosequencing. *PLoS One* 3:e2836.

Arumugam, M., J. Raes, E. Pelletier, D. Le Paslier, T. Yamada, D. R. Mende et al. (2011). Enterotypes of the human gut microbiome. *Nature* 473: 174–180.

Bäckhed, F., R. E. Ley, J. L. Sonnenburg, D. A. Peterso, and J. I. Gordon (2005). Host-bacterial mutualism in the human intestine. *Science* 307: 1915–1920.

Baird, W. M., L. A. Hooven, and B. Mahadevan (2005). Carcinogenic polycyclic aromatic hydrocarbon-DNA adducts and mechanism of action. *Environmental and Molecular Mutagenesis* 45: 106–114.

Bakken, J. S., T. Borody, L. J. Brandt, J. V. Brill, D. C. Demarco, M. Franzos et al. (2011). Treating *Clostridium difficile* infection with fecal microbiota transplantation. *Clinical Gastroenterology and Hepatology* 9: 1044–1049.

Barriere, S. L. (2015). Clinical, economic and societal impact of antibiotic resistance. *Expert Opinion on Pharmacotherapy* 16: 151–153.

Bengmark, S. (2005). Bioecologic control of the gastrointestinal tract: The role of flora and supplemented probiotics and synbiotics. *Gastroenterology Clinics of North America* 34: 413–436.

Berg, A. M., C. P. Kelly, and F. A. Farraye (2013). *Clostridium difficile* infection in the inflammatory bowel disease patient. *Inflammatory Bowel Disease* 19: 194–204.

Betts, K. S. (2011). A study in balance: How microbiomes are changing the shape of environmental health. *Environmental Health Perspectives* 119: a340–a346.

Billiard, S. M., J. N. Meyer, D. M. Wassenberg, P. V. Hodson, and R. T. Di Giulio (2008). Nonadditive effects of PAHs on early vertebrate development: Mechanisms and implications for risk assessment. *Toxicological Sciences* 105: 5–23.

Bisanz, J. E, M. K. Enos, J. R. Mwanga, J. Changalucha, J. P. Burton, G. B. Gloor, and G. Reid (2014). Randomized open-label pilot study of the influence of probiotics and the gut microbiome on toxic metal levels in Tanzanian pregnant women and school children. *mBio* 5(5):e01580-14. doi:10.1128/mBio.01580-14.

Blaser, M. J. (2014). The microbiome revolution. *Journal of Clinical Investigation* 124: 4162–4165.

Blaser, M. J. and S. Falkow (2009). What are the consequences of the disappearing human microbiota? *Nature Reviews Microbiology* 7: 887–894.

Borenstein, E. (2012). Computational systems biology and *in silico* modeling of the human microbiome. *Briefings in Bioinformatics* 13: 769–780.

Borody, T. J., E. F. Warren, S. M. Leis, R. Surace, O. Ashman, and S. Siarakas (2004). Bacteriotherapy using fecal flora: Toying with human motions. *Journal of Clinical Gastroenterology* 38: 475–483.

Breton, J., C. Daniel, J. Dewulf, S. Pothion, N. Froux, M. Sauty et al. (2013a). Gut microbiota limits heavy metals burden caused by chronic oral exposure. *Toxicology Letters* 222: 132–138.

Breton, J., K. Le Clère, C. Daniel, C. Sauty, M. Nakab, L., Chassat, T. et al. (2013b). Chronic ingestion of cadmium and lead alters the bioavailability of essential and heavy metals, gene expression pathways and genotoxicity in mouse intestine. *Archives in Toxicology* 87: 1787–1795.

Breton, J., S. Massart, P. Vandamme, E. De Brandt, B. Pot, and B. Foligné (2013c). Ecotoxicology inside the gut: Impact of heavy metals on the mouse microbiome. *BMC Pharmacology and Toxicology* 14: 62. doi: 10.1186/2050-6511-14-62.

Brinkmann, M., S. Maletz, M. Krauss, K. Bluhm, S. Schiwy, J. Kuckelkorn et al. (2014). Heterocyclic aromatic hydrocarbons show estrogenic activity upon metabolization in a recombinant transactivation assay. *Environmental Science and Technology* 48: 5892–5901.

Brown, K., D. DeCoffe, E. Molcan, and D. L. Gibson (2012). Diet-induced dysbiosis of the intestinal microbiota and the effects on immunity and disease. *Nutrients* 4: 1095–1119.

Brüssow, H. and S. J. Parkinson (2014). You are what you eat. *Nature Biotechnology* 32: 243–245.

Bultman, S. J. (2014). Emerging roles of the microbiome in cancer. *Carcinogenesis* 35: 249–255.

Cadwell, K. (2015). Expanding the role of the virome: Commensalism in the gut. *Journal of Virology* 89: 1951–1953.

Cammarota, G., L. Masucci, G. Ianiro, S. Bibbò, G. Dinoi, G. Costamagna et al. (2015) Randomised clinical trial: Faecal microbiota transplantation by colonoscopy vs. vancomycin for the treatment of recurrent *Clostridium difficile* infection. *Alimentary Pharmacology and Therapeutics* 41: 835–843.

Carding, S., K. Verbeke, D. T. Vipond, B. M. Corfe, and L. J. Owen (2015). Dysbiosis of the gut microbiota in disease. *Microbial Ecology in Health and Disease* 26: 1–9.

Chang, S. and L. Li. 2011. Metabolic endotoxemia: A novel concept in chronic disease pathology. *Journal of Medical Sciences* 31: 191–209.

Chia, J. K., A. Y. Chia, D. Wang, and R. El-Habbal (2015) Functional dyspepsia and chronic gastritis associated with enteroviruses. *Open Journal of Gastroenterology* 5: 21–27.

Cho, I. and M. J. Blaser (2012). The human microbiome: At the interface of health and disease. *Nature Reviews Genetics* 13: 260–270.

Choi, H., W. Jedrychowski, J. Spengler, D. E. Camann, R. M. Whyatt, V. Rauh et al. (2006). International studies of prenatal exposure to polycyclic aromatic hydrocarbons and fetal growth. *Environmental Health Perspectives* 114: 1744–1750.

Clayton, T. A., D. Baker, J. C. Lindon, J. R. Everett, and J. K. Nicholson (2009). Pharmacometabonomic identification of a significant host-microbiome metabolic interaction affecting human drug metabolism. *Proceedings of the National Academy of Sciences USA* 106: 14728–14733.

Clayton, T. A., J. C. Lindon, O. Cloarec, H. Antii, C. Charuel, Hanton, G. et al. (2006). Pharmaco-metabonomic phenotyping and personalized drug treatment. *Nature* 440: 1073–1077.

Conrick, T. (2014). Age of autism. http://www.ageofautism.com/2014/04/autism-does-mercury-modulate-the-microbiome.html.

Costello, E. K., C. L. Lauber, M. Hamady, N. Fierer, J. I. Gordon, and R. Knight (2009). Bacterial community variation in human body habitats across space and time. *Science* 326: 1694–1697.

Coyle, P., J. C. Philcox, L. C. Carey, and A. M. Rofe (2002). Metallothionein: The multipurpose protein. *Cellular and Molecular Life Sciences* 59: 627–647.

Cummings, J. H. and G. T. Macfarlane. 1991. The control and consequences of bacterial fermentation in the human colon. *Journal of Applied Bacteriology* 70: 443–459.

D'Argenio, V. and F. Salvatore (2015). The role of the gut microbiome in the healthy adult status. *Clinical Chimica Acta* 451: 97–102.

Dantas, G., M. O. Sommer, R. D. Oluwasegun, and G. M. Church (2008). Bacteria subsisting on antibiotics. *Science* 320: 100–103.

de Vos, W. M. and E. A. de Vos. 2012. Role of the intestinal microbiome in health and disease: From correlation to causation. *Nutrition Reviews* 70: S45–S56.

Dethlefsen, L., S. Huse, M. L. Sogin, and D. A. Relman (2008). The pervasive effects of an antibiotic on the human gut microbiota, as revealed by deep 16S rRNA sequencing. *PLoS Biology* 6: e280.

Dhanasekaran, A. R., J. L. Pearson, B. Ganesan, and B. C. Weimer (2015). Metabolome searcher: A high throughput tool for metabolite identification and metabolic pathway mapping directly from mass spectrometry and using genome restriction. *BMC Bioinformatics* 16: 62. doi: 10.1186/s12859-015-0462-y.

Dheer, R., J. Patterson, M. Dudash, E. N. Stachler, E. N. Bibby, K. J. Stolz et al. (2015). Arsenic induces structural and compositional colonic microbiome change and promotes host nitrogen and amino acid metabolism. *Toxicology and Applied Pharmacology* 289: 397–408.

Diaz Heijtz, R., S. Wang, F. Anuar, Y. Qian, B. Björkholm, A. Samuelsson, et al. (2011). Normal gut microbiota modulates brain development and behavior. *Proceedings of the National Academy of Sciences of the United States of America* 108: 3047–3052.

DiBaise, J. K., D. N. Frank, and R. Mathur (2012). Impact of the gut microbiota on the development of obesity: Current concepts. *American Journal of Gastroenterology Supplements* 1: 22–27.

Dietert, R. R. and E. K. Silbergeld (2015). Biomarkers for the 21st century: Listening to the microbiome. *Toxicological Sciences* 144: 208–216. doi: 10.1093/toxsci/kfv013.

Dimitrov, D. V. (2011). The human gutome: Nutrigenomics of the host-microbiome interactions. *OMICS* 15: 419–430.

Dogra, S., O. Sakwinska, S. Soh, C. Ngom-Bru, W. M. Brück, W. M. Berger et al. (2015). Dynamics of infant gut microbiota are influenced by delivery mode and gestational duration and are associated with subsequent adiposity. *mBio* 6: e02419–e024114.

Dumas, M. E., R. H. Barton, A. Toye, O. Cloarec, C. Blancher, A. Rothwell et al. (2006). Metabolic profiling reveals a contribution of gut microbiota to fatty liver phenotype in insulin-resistant mice. *Proceedings of the National Academy of Sciences of the United States of America* 103: 12511–12516.

Eiseman, B., W. Silen, G. S. Bascom, and A. J. Kauvar (1958). Fecal enema as an adjunct in the treatment of pseudomembranous enterocolitis. *Surgery* 44: 854–859.

Everett, J. R., R. L. Loo, and F. S. Pullen (2013). Pharmacometabonomics and personalized medicine. *Annals of Clinical Biochemistry* 50: 523–545.

Faith, J. J., J. L. Guruge, M. Charbonneau, S. Subramanian, H. Seedorf, A. L. Goodman et al. (2013). The long-term stability of the human gut microbiota. *Science* 341: 1–12.

Fazeli, M., P. Hassanzadeh, and S. Alaei (2011). Cadmium chloride exhibits a profound toxic effect on bacterial microflora of the mice gastrointestinal tract. *Human and Experimental Toxicology* 30: 152–159.

Foxman, E. F. and A. Iwasaki (2011). Genome–virome interactions: Examining the role of common viral infections in complex disease. *Nature Reviews Microbiology* 9: 254–264.

Frank, D. N., A. L. Amand, R. A. Feldman, E. C. Boedeker, N. Harpaz, and N. R. Pace (2007). Molecular-phylogenetic characterization of microbial community imbalances in human inflammatory bowel diseases. *Proceedings of the National Academy of Sciences of the United States of America* 104: 13780–13785.

Franzosa, E. A., K. Huang, J. F. Meadow, D. Gevers, K. P. Lemon, B. J. Bohannan et al. (2015). Identifying personal microbiomes using metagenomic codes. *Proceedings of the National Academy of Sciences of the United States of America* 112: E2930–E2938.

Freilich, S., A. Kreimer, E. Borenstein, N. Yosef, R. Sharan et al. (2009). Metabolic-network-driven analysis of bacterial ecological strategies. *Genome Biology* 10: R61.

Garber, K. (2015). Drugging the gut microbiome. *Nature Biotechnology* 33: 228–231.

Gerritsen, J., H. Smidt, G. T. Rijkers, and W. M. de Vos (2011). Intestinal microbiota in human health and disease: The impact of probiotics. *Genes and Nutrition* 6: 209–240.

Gibson, G. R. and M. B. Roberfroid (1995). Dietary modulation of the human colonic microbiota: Introducing the concept of prebiotics. *Journal of Nutrition* 125: 1401–1412.

Goodacre, R. (2007). Metabolomics of a superorganism. *Journal of Nutrition* 137: 259S–266S.

Gourbeyre, P., S. Denery, and M. Bodinier (2011). Probiotics, prebiotics, and synbiotics: Impact on the gut immune system and allergic reactions. *Journal of Leukocyte Biology* 89: 685–695.

Graf, D, R. Di Cagno, F. Fa˚ k, H. J. Flint, M. Nyman, S. M. Saarela, and B. Watzl (2015). Contribution of diet to the composition of the human gut microbiota. *Microbial Ecology in Health and Disease* 26: 26164: 1–11.

Grice, E. A. and J. A. Segre (2012). The human microbiome: Our second genome. *Annual Reviews in Genomics and Human Genetics* 13: 151–170.

Grundmann, O. (2010). The gut microbiome and pre-systemic metabolism: Current state and evolving research. *Journal of Drug Metabolism and Toxicology* 1: 105.

Guarner, F. (2009). Prebiotics, probiotics and helminths: The 'natural' solution? *Digestive Diseases* 27: 412–417.

Guarner, F. and J. R. Malagelada (2003). Gut flora in health and disease. *The Lancet* 360: 512–519.

Guarner, V. and M. E. Rubio-Ruiz (2014). Low-grade systemic inflammation connects aging, metabolic syndrome and cardiovascular disease. In: A. I. Yashin and S. M. Jazwinski (eds.), *Aging and Health—A Systems Biology Perspective. Interdisciplinary Topics in Gerontology.* Karger, Basel, Switzerland. Vol. 40, pp. 99–106. doi: 10.1159/000364934.

Gutiu, I. A., A. Andries, C. Mircioiu, F. Radulescu, A. M. Georgescu, and D. Cioaca (2010). Pharmacometabonomics, pharmacogenomics and personalized medicine. *Romanian Journal of Internal Medicine* 48: 187–191.

Hagland, H. R. and K. Søreide (2015). Cellular metabolism in colorectal carcinogenesis: Influence of lifestyle, gut microbiome and metabolic pathways. *Cancer Letters* 356: 273–280.

Haiser, H. J., D. B. Gootenberg, K. Chatman, G. Sirasani, E. P. Balskus, and P. J. Turnbaugh (2013). Predicting and manipulating cardiac drug inactivation by the human gut bacterium *Eggerthella lenta. Science* 341: 295–298.

Hardy, H., J. Harris, E. Lyon, J. Beal, and A. D. Foey (2013). Probiotics, prebiotics and immunomodulation of gut mucosal defenses: Homeostasis and immunopathology. *Nutrients* 5: 1869–1912.

Haritash, A. K. and C. P. Kaushik (2009). Biodegradation aspects of polycyclic aromatic hydrocarbons (PAHs): A review. *Journal of Hazardous Materials* 169: 1–15.

Hattori, M. and T. D. Taylor (2009). The human intestinal microbiome: A new frontier of human biology. *DNA Research* 16: 1–12.

Holmes, E., J. Kinross, G. R. Gibson, R. Burcelin, W. Jia, S. Pettersson, and J. K. Nicholson (2012). Therapeutic modulation of microbiota-host metabolic interactions. *Science Translational Medicine* 4: 1–11. doi: 10.1126/scitranslmed.3004244.

Holmes, E., J. V. Li, T. Athanasiou, H. Ashrafian, and J. K. Nicholson (2011). Understanding the role of gut microbiome–host metabolic signal disruption in health and disease. *Trends in Microbiology* 19: 349–359.

Hood, L. (2012). Tackling the microbiome. *Science* 336: 1209.

Hormannsperger, G., T. Clavel, and D. Haller (2012). Gut matters: Microbe–host interactions in allergic diseases. *Journal of Allergy and Clinical Immunology* 129: 1452–1459.

Hossain, A., S. Akhtar, and Y. Kabir (2015). Diet, microbiome and human health. In: Tamang, J. (ed.), *Health Benefits of Fermented Foods*. Taylor & Francis Publications, Boca Raton, FL.

Huang, J., C. Plass, and C. Gerhauser (2011). Cancer chemoprevention by targeting the epigenome. *Current Drug Targets* 12: 1925–1956.

Huang, K., A. Brady, A. Mahurkar, O. White, D. Gevers, C. Huttenhower, and N. Segata, N. (2014). MetaRef: A pan-genomic database for comparative and community microbial genomics. *Nucleic Acids Research* 42: D617–D624.

Hullar, M. A. and B. C. Fu (2014). Diet, the gut microbiome, and epigenetics. *Cancer Journal* 20: 170–175.

Human Microbiome Project (2012). Structure, function and diversity of the healthy human microbiome. *Nature* 486: 207–214.

Huttenhower, C., D. Gevers, R. Knight, S. Abubucker, J. H. Badger, A. T. Chinwalla et al. (2012). Structure, function and diversity of the healthy human microbiome. *Nature* 486: 207–214.

Nicholson, J. K., J. R. Everett, and J. C. Lindon (2012). Longitudinal pharmacometabonomics for predicting patient responses to therapy: Drug metabolism, toxicity and efficacy. *Expert Opinion on Drug Metabolism and Toxicology* 8: 2012.

Jia, W., H. Li, L. Zhao, and J. K. Nicholson (2008). Gut microbiota: A potential new territory for drug targeting. *Nature Reviews Drug Discovery* 7: 123–129.

Kaminogawa, S. (2010). Effects of food components on intestinal flora, intestinal immune system and their mutualism. *Bioscience Microflora* 29: 69–82.

Khan, S. and G. Jena (2015). The role of butyrate, a histone deacetylase inhibitor in diabetes mellitus: Experimental evidence for therapeutic intervention. *Epigenomics* 7: 669–680.

Khoruts, A., J. Dicksved, J. K. Jansson, and M. J. Sadowsky (2010). Changes in the composition of the human fecal microbiome after bacteriotherapy for recurrent *Clostridium difficile*-associated diarrhea. *Journal of Clinical Gastroenterology* 44: 354–360.

Kinross, J. M., A. W. Darzi, and J. K. Nicholson (2011). Gut microbiome-host interactions in health and disease. *Genome Medicine* 3: 14.

Kramer, P. and P. Bressan (2015). Humans as superorganisms. *Perspectives on Psychological Science* 10: 464–481.

Kuilman, T. and D. S. Peeper (2009). Senescence-messaging secretome: SMS-ing cellular stress. *Nature Reviews* Cancer 9: 81–94.

Lehouritis, P., J. Cummins, M. Stanton, C. T. Murphy, F. O. McCarthy, G. Reid et al. (2015). Local bacteria affect the efficacy of chemotherapeutic drugs. *Scientific Reports* 5: 14554 doi: 10.1038/srep14554; www.nature.com/scientificreports.

Lepage, P., M. C. Leclerc, M. Joossens, S. Mondot, H. M. Blottière, J. Raes et al. (2012). A metagenomic insight into our gut's microbiome. *Gut* 62: 146–158.

Ley, R., F. Bäckhed, P. Turnbaugh, C. Lozupone, R. Knight, and J. Gordon (2005). Obesity alters gut microbial ecology. *Proceedings of the National Academy of Sciences USA* 102: 11070–11075.

Ley, R. E., M. Hamady, C. Lozupone, P. Turnbaugh, R. R. Ramey, J. S. Bircher et al. (2008). Evolution of mammals and their gut microbes. *Science* 320: 1647–1651.

Ley, R. E., D. A. Peterson, and J. I. Gordon (2006a). Ecological and evolutionary forces shaping microbial diversity in the human intestine. *Cell* 124: 837–848.

Ley, R. E., P. J. Turnbaugh, S. Klein, and J. I. Gordon (2006b). Microbial ecology: Human gut microbes associated with obesity. *Nature* 444: 1022–1023.

Li, H. and W. Jia (2013). Implications for drug metabolism and drug-induced toxicity. *Clinical Pharmacology and Therapeutics* 94: 574.

Liebert, C. A., J. Wireman, T. Smith, and A. O. Summers (1997). Phylogeny of mercury resistance (mer) operons of gram-negative bacteria isolated from the fecal flora of primates. *Applied and Environmental Microbiology* 63: 1066–1076.

Lozupone, C. A., J. I. Stombaugh, J. I. Gordon, J. K. Jansson, and R. Knight (2012). Diversity, stability and resilience of the human gut microbiota. *Nature* 489: 220–230.

Lu, K., R. P. Abo, K. A. Schlieper, M. E. Graffam, S. Levine, J. S. Wishnok et al. (2014). Arsenic exposure perturbs the gut microbiome and its metabolic profile in mice: An integrated metagenomics and metabolomics analysis. *Environmental Health Perspectives* 122: 284–291.

Lukeš, J., C. R. Stensvold, K. Jirků-Pomajbíková, and L. W. Parfrey (2015). Are human intestinal eukaryotes beneficial or commensals? *PLoS Pathogens* 11: e1005039. doi:10.1371/journal.ppat.100.

Lynch, D. B., I. B. Jeffery, and P. W. O'Toole (2015). The role of the microbiota in ageing: Current state and perspectives. *WIREs Systems Biology and Medicine* 7: 131–138.

Mackie, R. I., A. Sghir, and H. R. Gaskins (1999). Developmental microbial ecology of the neonatal gastrointestinal tract. *American Journal of Clinical Nutrition* 169: 1035S–1045S.

Mai, V. and P. V. Draganov (2009). Recent advances and remaining gaps in our knowledge of associations between gut microbiota and human health. *World Journal of Gastroenterology* 15: 81–85.

Mani, S., U. A. Boelsterli, and M. R. Redinbo (2014). Understanding and modulating mammalian-microbial communication for improved human health. *Annual Review of Pharmacology and Toxicology* 54: 559–580.

Maranduba, C. M., S. B. R. De Castro, G. T. de Souza, C. Rossato, F. C. da Guia, M. A. S. Valente et al. (2015). Intestinal microbiota as modulators of the immune system and neuroimmune system: Impact on the host health and homeostasis. *Journal of Immunology Research* 2015 Article ID 931574: 1–14.

Marchesi, J. R., D. H. Adams, F. Fava, G. D. A. Hermes, G. M. Hirschfield, G. Hold et al. (2015). The gut microbiota and host health: A new clinical frontier. *Gut* 65: 330–339. doi:10.1136/gutjnl-2015–309990.

Mazmanian, S. K., C. H. Liu, A. O. Tzianabos, and D. L. Kasper. (2005). An immunomodulatory molecule of symbiotic bacteria directs maturation of the host immune system. *Cell* 122: 107–118.

McFarland, L. V., G. W. Elmer, and C. M. Surawicz (2002). Breaking the cycle: Treatment strategies for 163 cases of recurrent *Clostridium difficile* disease. *American Journal of Gastroenterology* 97: 1769–1775.

McNally, L. and S. P. Brown (2015) Building the microbiome in health and disease: Niche construction and social conflict in bacteria. *Philosophical Transactions of the Royal Society B* 370: 20140298. http://dx.doi.org/10.1098/rstb.2014.0298.

Minota, S., A. Brysona, C. Chehouda, G. D. Wub, J. D. Lewis, and F. D. Bushmana (2013). Rapid evolution of the human gut virome. *Proceedings of the National Academy of Sciences of the United States of America* 110: 12450–12455.

Mshvildadze, M. and J. Neu (2010). The infant intestinal microbiome: Friend or foe? *Early Human Development* 86: 67–71.

Nava, G. M. and Stappenbeck, T. S. (2011). Diversity of the autochthonous colonic microbiota. *Gut Microbes* 2: 99–104.

Nebert, D. W., G. Zhang, and E. S. Vesell (2008). From human genetics and genomics to pharmacogenetics and pharmacogenomics: Past lessons, future directions. *Drug Metabolism Reviews* 40: 187–224.

Nelson, K. E., G. M. Weinstock, S. K. Highlander, K. C. Worley, H. H. Creasy, J. R. Wortman et al. (2010). A catalog of reference genomes from the human microbiome. *Science* 328: 994–999.

Nicholson, J. K., E. Holmes, and I. D. Wilson (2005). Gut microorganisms, mammalian metabolism and personalized health care. *Nature Reviews Microbiology* 3: 431–438.

O'Hara, A. M. and F. Shanahan (2006). The gut flora as a forgotten organ. *EMBO Reports* 7: 688–693.

Ou, J., F. Carbonero, E. G. Zoetendal, J. P. DeLany, M. Wang, and K. H. Newton (2013). Diet, microbiota, and microbial metabolites in colon cancer risk in rural Africans and African Americans. *American Journal of Clinical Nutrition* 98: 111–120.

Penders, J., C. Thijs, C. Vink, F. F. Stelma, B. Snijders, I. Kummeling et al. (2006). Factors influencing the composition of the intestinal microbiota in early infancy. *Pediatrics* 118: 511–521.

Peterson, J., S. Garges, M. Giovanni, P. McInnes, L. Wang, J. A. Schloss et al. (2009). The NIH human microbiome project. *Genome Research* 19: 2317–2323.

Pokusaeva, K., G. F. Fitzgerald, and D. van Sinderen (2011). Carbohydrate metabolism in bifidobacteria. *Genes and Nutrition* 6: 285–306.

Qin, J., R. Li, J. Raes, M. Arumugam, K. S. Burgdorf, C. Manichanh et al. (2010). A human gut microbial gene catalogue established by metagenomic sequencing. *Nature* 464: 59–65.

Rapaport, S. M. and M. T. Smith (2010). Epidemiology environment and disease risks. *Science* 330: 460–461.

Raymond, F., A. A. Ouameur, M. Déraspe, N. Iqbal, H. Gingras, and B. Dridi (2015). The initial state of the human gut microbiome determines its reshaping by antibiotics. *International Symposium on Microbial Ecology Journal* 10: 707–720. doi: 10.1038/ismej.2015.148.

Reyes, A., N. P. Semenkovich, K. Whiteson, F. Rohwer, and J. I. Gordon (2012). Going viral: Next-generation sequencing applied to phage populations in the human gut. *Nature Reviews Microbiology* 10: 607–617.

Ridlon, J. M. and J. S. Bajaj (2015). The human gut sterolbiome: Bile acid-microbiome endocrine aspects and therapeutics. *Acta Pharmaceutica Sinica B* 5: 99–105.

Rizkallah, M. R., R. Saad, and R. K. Aziz. (2010). The human microbiome project, personalized medicine and the birth of pharmacomicrobiomics. *Current Pharmacogenomics and Personalized Medicine* 8: 182–193.

Romano-Keeler, J., D. J. Moore, C. Wang, R. M. Brucker, C. Fonnesbeck, J. C. Slaughter et al (2014). Early life establishment of site-specific microbial communities in the gut. *Gut Microbes* 5: 192–201.

Saad, R., M. Rizkallah, and R. Aziz (2012). Gut pharmacomicrobiomics: The tip of an iceberg of complex interactions between drugs and gut-associated microbes. *Gut Pathogens* 4: 16.

Scarpellini, E., G. Laniro, F. Attili, C. Bassanelli, A. De Santis, and A. Gasbarrini (2015). The human gut microbiota and virome: Potential therapeutic implications. *Digestive and Liver Disease*. 12: 1007–1012. http://dx.doi.org/10.1016/j.dld.2015.07.008.

Schippa, S. and M. P. Conte (2014). Dysbiotic events in gut microbiota: Impact on human health. *Nutrients* 6: 5786–5805.

Schuijt, T. J., T. Poll, W. M. de Vos, and W. J. Wiersinga (2013). Human microbiome: The intestinal microbiota and host immune interactions in the critically ill. *Trends in Microbiology* 21: 221–229.

Sears, C. L. (2009). Enterotoxigenic *Bacteroides fragilis*: A rogue among symbiotes. *Clinical Microbiology Reviews* 22: 349–369.

Sears, C. L., A. L. Geis, and F. Housseau (2014). *Bacteroides fragilis* subverts mucosal biology: From symbiont to colon carcinogenesis. *Journal of Clinical Investigation* 124: 4166–4172.

Sekirov, I., S. L. Russell, C. M. Antunes, and B. B. Finlay (2010). Gut microbiota in health and disease. *Physiology Review* 90: 859–904.

Sharma, M. and M. Devi (2014). Probiotics: A comprehensive approach toward health foods. *Critical Reviews in Food Sciences and Nutrition* 54: 537–552.

Shreiner, A. B., J. Y. Kao, and V. B. Young (2015). The gut microbiome in health and in disease. *Current Opinion in Gastroenterology* 31: 69–75.

Sievers, C. K., E. K. Shanle, C. A. Bradfield, and W. Xu (2009). Differential action of monohydroxylated polycyclic aromatic hydrocarbons with estrogen receptors α and β. *Toxicology Science* 132: 359–367.

Skurnik, D., R. Ruimy, D. Ready, E. Ruppe, C. Bernède-Bauduin, F. Djossou et al. (2010). Is exposure to mercury a driving force for the carriage of antibiotic resistance genes? *Journal of Medical Microbiology* 2010: 804–807.

Slavin, J. (2013). Fiber and prebiotics: Mechanisms and health benefits. *Nutrients* 5: 1417–1435.

Soccol, C. R., L. P. D. S. Vandenberghe, M. R. Spier, A. B. P. Medeiros, C. T. Yamaguishi, and J. D. D. Lindner (2010). The potential of probiotics. *Food Technology and Biotechnology* 48: 413–434.

Sommer, F. and F. Bäckhed (2013). The gut microbiota—Masters of host development and physiology. *Nature Reviews Microbiology* 11: 227–238.

Sommer, M. O., G. Dantas, and G. M. Church (2009). Functional characterization of the antibiotic resistance reservoir in the human microflora. *Science* 325: 1128–1131.

Sousa, T., R. Paterson, V. Moore, A. Carlsson, B. Abrahamsson, and A. W. Basit (2008). The gastrointestinal microbiota as a site for the biotransformation of drugs. *International Journal of Pharmaceutics* 363: 1–25.

Sun, J. and E. B. Chang (2014). Exploring gut microbes in human health and disease: Pushing the envelope. *Genes Diseases* 1: 132–139.

Sun, J., G. W. Le, Y. H. Shi, and G. W. Su (2007). Factors involved in binding of *Lactobacillus plantarum* Lp6 to rat small intestinal mucus. *Letters in Applied Microbiology* 44: 79–85.

Swann, J. R., E. J. Want, F. M. Geier, K. Spagou, I. D. Wilson, J. E. Sidaway et al. (2011). Systemic gut microbial modulation of bile acid metabolism in host tissue compartments. *Proceedings of the National Academy of Sciences of the United States of America* 108: 4523–4530.

Tang-Feldman, Y., S. Mayo, J. Silva, and S. H. Cohen (2003). Molecular analysis of *Clostridium difficile* strains isolated from 18 cases of recurrent *Clostridium difficile*-associated diarrhea. *Journal of Clinical Microbiology* 41: 3413–3414.

Tappenden, K. A. and A. S. Deutsch (2007). The physiological relevance of the intestinal microbiota—Contributions to human health. *Journal of American College of Nutrition* 26: 679S–683S.

The Human Microbiome Project Consortium (2012). Structure, function and diversity of the healthy human microbiome. *Nature* 486: 207–214.

Tlaskalová-Hogenová, H., R. Štěpánková, H. Kozáková, T. Hudcovic, L. Vannucci, L. Tučková et al. (2011). The role of gut microbiota (commensal bacteria) and the mucosal barrier in the pathogenesis of inflammatory and autoimmune diseases and cancer: Contribution of germ-free and gnotobiotic animal models of human diseases. *Cellular and Molecular Immunology* 8: 110–120.

Tremaroli, V. and F. Bäckhed (2012). Functional interactions between the gut microbiota and the host metabolism. *Nature* 489: 242–249.

Tringe, S. G., C. von Mering, A. Kobayashi, A. A. Salamov, K. Chen, H. W. Chang, M. Podar et al. (2005). Comparative metagenomics of microbial communities. *Science* 308: 554–557.

Turnbaugh, P. J., M. Hamady, T. Yatsunenko, B. L. Cantarel, A. Duncan, R. E. Ley et al. (2009). A core gut microbiome in obese and lean twins. *Nature* 457: 480–484.

Turnbaugh, P. J., R. E. Ley, M. Hamady, C. M. Fraser-Liggett, R. Knight, and J. I. Gordon (2007). The human microbiome project. *Nature* 449: 804–810.

Turroni, F., J. R. Marchesi, E. Foroni, M. Gueimonde, F. Shanahan, A. Margolles et al. (2009). Microbiomic analysis of the bifidobacterial population in the human distal gut. *International Symposium on Microbial Ecology Journal* 3: 745–751.

Ubeda, C., Y. Taur, R. R. Jenq, M. J. Equinda, T. Son, M. Samstein et al. (2010). Vancomycin-resistant *Enterococcus* domination of intestinal microbiota is enabled by antibiotic treatment in mice and precedes bloodstream invasion in humans. *Journal of Clinical Investigation* 120: 4332–4341.

Upreti, R. K., R. Shrivastava, and U. C. Chaturvedi (2004). Gut microflora & toxic metals: Chromium as a model. *Indian Journal of Medical Research* 119: 49–59.

Ursell, L. K. and R. Knight (2013). Xenobiotics and the human gut microbiome: Metatranscriptomics reveal the active players. *Cell Metabolism* 17: 317–318.

Van de Wiele, T., C. M. Gallawa, K. M. Kubachka, J. T. Creed, N. Basta, E. A. Dayton et al. (2010). Arsenic metabolism by human gut microbiota upon *in vitro* digestion of contaminated soils. *Environmental Health Perspective* 118: 1004–1009.

Van de Wiele, T., L. Vanhaecke, C. Boeckaert, K. Peru, J. Headley, W. Verstraete, and S. Siciliano (2005). Human colon microbiota transform polycyclic aromatic hydrocarbons to estrogenic metabolites. *Environmental Health Perspective* 113: 6–10.

van Duynhovena, J., E. E. Vaughana, D. M. Jacobsa, R. A. Kempermana, E. J. J. van Velzena, G. Grossa et al. (2011). Metabolic fate of polyphenols in the human superorganism. *Proceedings of the National Academy of Sciences of the United States of America* 108: 4531–4538.

Vanhoutte, T., G. Huys, E. Brandt, and J. Swings (2004). Temporal stability analysis of the microbiota in human feces by denaturing gradient gel electrophoresis using universal and group-specific 16S rRNA gene primers. *FEMS Microbiology Ecology* 248: 437–446.

Viaud, S., R. Daillère, I. G. Boneca, P. Lepage, P. Langella, M. Chamaillard et al. (2015). Gut microbiome and anticancer immune response. *Cell Death and Differentiation* 22: 199–214.

Virgin, H. W. and J. A. Todd (2011). Metagenomics and personalized medicine. *Cell* 147: 44–56.

Vrieze, A., F. Holleman, E. G. Zoetendal, W. M. de Vos, J. B. Hoekstra, and M. Nieuwdorp (2010). The environment within: How gut microbiota may influence metabolism and body composition. *Diabetologia* 53: 606–613.

Wallace, T. C., F. Guarner, K. Madsen, M. D. Cabana, G. Gibson, E. Hentges, and M. I. Sanders (2011). Human gut microbiota and its relationship to health and disease. *Nutrition Reviews* 69: 392–403.

Wang, W., S. Xu, Z. Ren, L. Tao, J. Jiang, and S. Zheng (2015). Application of metagenomics in the human gut microbiome. *World Journal of Gastroenterology* 21: 803–814.

Wikoff, W. R., A. T. Anfora, J. Liu, P. G. Schultz, S. A. Lesley, E. C. Peters et al. (2009). Metabolomics analysis reveals large effects of gut microflora on mammalian blood metabolites. *Proceedings of the National Academy of Sciences of the United States of America* 106: 3698–3703.

Wilson, I. D. (2009). Drugs, bugs, and personalized medicine: Pharmacometabonomics enters the ring. *Proceedings of the National Academy of Sciences of the United States of America* 106: 14187–14188.

Wilson, I. D. and J. K. Nicholson (2009). The role of gut microbiota in drug response. *Current Pharmaceutical Design* 15: 1519–1523.

Xuan, C., J. M. Shamonki, A. Chung, M. L. Dinome, M. Chung, P. A. Sieling et al. (2014). Microbial dysbiosis is associated with human breast cancer. *PLoS One* 9:e83744.

Yip, L. Y. and E. C. Y. Chan (2015). Investigation of host–gut microbiota modulation of therapeutic outcome. *Drug Metabolism and Disposition* 43: 1619–1631.

Yoshimoto, S., T. M. Loo, K. Atarashi, H. Kanda, S. Sato, S. Oyadomari et al. (2013). Obesity-induced gut microbial metabolite promotes liver cancer through senescence secretome. *Nature* 499: 97–101.

Youngster, I., G. H. Russell, C. Pindar, T. Ziv-Baran, J. Sauk, and E. L. Hohmann (2014). Oral, capsulized, frozen fecal microbiota transplantation for relapsing *Clostridium difficile* infection. *Journal of American Medical Association* 312: 1772–1778.

Yu, L. C., J. Wang, S. Wei, and Y. Ni (2012). Host-microbial interactions and regulation of intestinal epithelial barrier function: From physiology to pathology. *World Journal of Gastrointestinal Pathophysiology* 3: 27–43.

Zackular, J. P., N. T. Baxter, K. D. Iverson, W. D. Sadler, J. F. Petrosino, G. Y. Chen, and P. D. Schloss (2013). The gut microbiome modulates colon tumorigenesis. *mBio* 5;4(6):e00692–e006913. doi: 10.1128/mBio.00692-13.

Zackular, J. P., M. A. Rogers, M. T. Ruffin, and P. D. Schloss (2014). The human gut microbiome as a screening tool for colorectal cancer. *Cancer Prevention Research* 7: 1112–1121.

Zambirinis, C. P., S. Pushalkar, D. Saxena, D., and G. Miller (2014). Pancreatic cancer, inflammation, and microbiome. *Cancer Journal* 20: 195–202.

Zhang, C., M. Zhang, S. Wang, R. Han, Y. Cao, W. Hua et al. (2010). Interactions between gut microbiota, host genetics and diet relevant to development of metabolic syndromes in mice. *International Symposium on Microbial Ecology Journal* 4: 232–241.

Zhu, B., X. Wang, and L. Li (2010). Human gut microbiome: The second genome of human body. *Protein Cell* 1: 718–725.

Zhu, Y., E. Jameson, M. Crosatt, H. Schäfer, K, Rajakumar, T. D. Bugg, and Y. Chen (2014). Carnitine metabolism to trimethylamine by an unusual Rieske-type oxygenase from human microbiota. *Proceedings of the National Academy of Sciences of the United States of America* 111: 4268–4273.

25 The Future of GM Foods or GM Foods of the Future

Where Is the Biotech Revolution Heading?

Stuart J. Lucas, Aydın Türkeç, and Elif Karlık

CONTENTS

25.1 INTRODUCTION

Ensuring food security for the world is one of the pressing scientific and political challenges of the twenty-first century. Currently, approximately 800 million people worldwide are living in chronic hunger (FAO, IFAD, and WFP 2014). Furthermore, the Food and Agriculture Organization of the United Nations (FAO) has stated that 70% more food needs to be produced by 2050 to feed a population projected to grow to over 9 billion (Godfray et al. 2010). It is also now clear that climate change will have an increasingly negative effect on crop yields as this century progresses (IPCC 2015), recognized by the recent COP21 international agreement to limit the rise in global temperature to less than 2°C by 2050.

Over the past 30 years, genetically modified (GM) varieties of a number of major crop plants have been developed and commercialized. In classical plant breeding, crop varieties are crossbred and selected for desired characteristics such as a higher yield or disease resistance. This is a lengthy process (8–10 years) requiring several generations of progeny to be grown and extensive phenotyping to test whether they perform better than the existing varieties. GM greatly accelerates this process, allowing a gene conferring a desired trait to be introduced and selected for in a single generation.

However, increasing public awareness and opposition to the use of genetically modified organisms (GMOs), particularly in foods, have made GM crops a highly contentious issue. Proponents of GM crops argue that, by allowing increased yields with fewer agricultural inputs, they offer the best prospect for meeting the challenge of food security outlined earlier (James and ISAAA 2014). However, opponents raise concerns about potential health, environmental, and socioeconomic hazards of widespread GM crop use. With some countries (such as the United States, China, and Brazil) investing heavily in GM crops while others (notably in the European Union) have chosen to ban their cultivation, the debate has become politicized and increasingly complex; a recent review found >32,000 papers in the scientific literature relating to GM crops, food, and feed (Nicolia et al. 2014).

Against this backdrop, in the last 5 years, a new set of techniques collectively referred to as "genome editing" technologies have been attracting attention as a method that is more precise and poses fewer safety risks than those used in existing GMOs. In this chapter, we focus on the food safety aspects of GM crops, by exploring the methods by which GM plants are generated, the risks that these modifications may present, and methods for detection and monitoring of GM ingredients in foodstuffs. In each section, we review existing GM foods before discussing the contrasting opportunities and challenges represented by genome editing for the next generation of biotech crops.

25.2 CURRENT AND EMERGING TECHNOLOGIES FOR GENETIC ENGINEERING OF CROP PLANTS

In GMO development, efficacy and performance are the most significant measures of success: does the genetic modification impart the desired phenotype? Assuming it does, the next most important considerations for food safety are related to the molecular characteristics of the GMO. Ideally, engineered DNA should be present at a single locus and as a single copy, with no sequence from the transformation vector backbone. Moreover, endogenous genes should not have been disrupted or turned into chimeric fusion proteins, and there should be no rearrangements of the engineered locus, as these increase the possibility of unexpected off-target effects. It is not always possible to obtain these ideal conditions (Ladics et al. 2015), which largely depend on the transformation techniques used. Most current GM crops were generated using either *Agrobacterium*-mediated transformation or biolistics to transfer DNA into plant cells, each of which has its own advantages and disadvantages (Table 25.1).

25.2.1 *Agrobacterium*-Mediated Transformation

Agrobacterium tumefaciens has served as an essential tool for the production of GM crops, owing to its ability to inject its Ti plasmid into plant cells, from which the "T-DNA" (transferred DNA) is inserted into the plant genome (Newell 2000). *Agrobacterium* can be used to transform cultured plant cells or tissues, or reproductive organs in plants, to generate transgenic plants. Three genetic elements of *A. tumefaciens* are required for this activity: virulence genes (*chv*) located in the chromosomal DNA, further virulence genes (*vir*) located on the Ti plasmid, and border sequences that delimit the T-DNA. *Agrobacterium*-based plant transformation is carried out by replacing most of the bacterial T-DNA with the transgenes to be inserted, leaving just the border regions intact. The large size of the Ti plasmid initially made it difficult to manipulate; however, efficient and convenient *Agrobacterium*-mediated transformation was achieved using "binary vectors" such as pBin19 (Bevan 1984), the main features of which are typical of most subsequent *Agrobacterium* transformation constructs. The key development made by pBin19 was realizing that T-DNA located on a different plasmid from the *vir* genes could still be integrated into the plant genome. Therefore, the plasmid containing the T-DNA can be small and easy to manipulate in *Escherichia coli*; it is then transformed into a "disarmed" *A. tumefaciens* strain containing a Ti plasmid from which the bacterial T-DNA has been deleted. The T-DNA from the vector inserts at a seemingly random location and may cause small deletions or rearrangements at the insertion site (Mayerhofer et al. 1991).

TABLE 25.1
Comparison of Conventional Genetically Modified and Genome Editing Technologies

Technology	*Agrobacterium* Vectors	Biolistics	ZFNs/TALENs	CRISPR/Cas9 System
DNA insert size	3–50 kb	Not limited[a]	No insert *or* not limited[a]	No insert *or* not limited[a]
Target range	*Agrobacterium* host range	All plants	All plants	All plants
Efficiency	High	Low	Low (ZFNs) Moderate to high (TALENs)	Moderate to high
Insertion site	Random	Random; can also be in organellar genomes	Specific nucleotide sequence of 12 bp (ZFNs) or 19 bp (TALENs)	Specific nucleotide sequence of 20 bp
Copy number	Single or few copies	Single or few copies	Single	Single
Optimized for plant codons	Yes	Yes	Yes	Yes
Cost	Low	Moderate	Low to moderate	Low to moderate
Vector design	Easy	Easy	Moderate to difficult	Easy
Applications	Transgene insertion	Transgene insertion	Gene disruption Gene modification Transgene insertion	Gene disruption Gene modification Transgene insertion

[a] The size of DNA fragment that can be inserted is not limited by the method itself. However, very large constructs are unlikely to be incorporated into the genome intact. The maximum practical insert size has not been determined.

Advantages of this method include (1) its simplicity; (2) the ends of the transgenic DNA are clearly defined; (3) the ability to transfer multiple expression cassettes (trait genes, marker genes, etc.); (4) efficient, usually single-copy insertion; (5) infrequent transgene silencing; (6) long DNA sequences (>150 kb) can be transferred; and (7) lack of requirements for specialized equipment. The host range of *A. tumefaciens* is relatively wide, in that most dicotyledonous and gymnosperm families and a few monocotyledonous families are susceptible to *Agrobacterium* infection (Cleene and Ley 1976). The major limitation of *Agrobacterium*-mediated transformation was initially considered to be transferring DNA into cereals. However, in 1993, a pioneering study succeeded in efficiently transforming rice using *Agrobacterium* (Chan et al. 1993); subsequently, transformation protocols for other major cereal crops including wheat, maize, sorghum, and barley were also developed, for example by manipulating plant tissue culture conditions or artificially activating *Agrobacterium* virulence (Newell 2000).

25.2.2 Direct Transformation Methods

Besides *Agrobacterium*-mediated transformation, DNA can be directly inserted into plant cells by electroporation, microinjection, and microprojectile bombardment. The last of these is the most extensively used in plants and is also referred to as biolistics or the gene gun. Biolistics was developed in the late 1980s for the transformation of species recalcitrant to *Agrobacterium* infection (Klein et al. 1992). DNA-coated gold particles are uniformly bombarded over reproductive or regenerative plant cells or tissues, where a minority of particles will become lodged in the nucleus and some of their DNA will be incorporated into the genome by the cell's own DNA repair mechanisms. The advantages of biolistics are as follows: (1) plants resistant to *A. tumefaciens* can be transformed; (2) no vector DNA sequences are required; (3) multiple constructs can be inserted simultaneously by co-bombardment; and (4) false-positive reporter gene expression is avoided.

The major disadvantage of biolistics is that its shotgun nature can cause cell damage, and may lead to the integration of multiple copies of transgenes into the plant genome. These can cause altered expression levels, the silencing/disruption of endogenous genes, or genetic instability (Travella et al. 2005). Nevertheless, biolistics has been used successfully to produce a number of GM crops, notably pest-resistant "Bt" maize.

Direct transformation methods developed for other organisms have also been reported to be applicable to plants, including microinjection (de la Peña et al. 1987), Polyethylene glycol (PEG)-mediated chemical transformation, and electroporation (Fromm et al. 1986). However, the transformation efficiencies obtained using these methods are typically much lower than that can be achieved using *Agrobacterium* or biolistics. Also, chemical transformation and electroporation can only be carried out on plant protoplasts (isolated cells from which the cell wall has been removed). This requires extended plant tissue culture, and for many crop species, it is not currently possible to regenerate a fertile plant from protoplasts. As a result, these techniques are not generally used in GM crop development.

25.2.3 Transformation Constructs and Their Safety Aspects

From a food safety perspective, the specific source of potential health risks in GM crops is the non-endogenous DNA construct or vector introduced during transformation, along with novel proteins expressed from the transgene(s).

The pBin19 backbone (Bevan 1984) contains origins of replication for both *E. coli* and *A. tumefaciens* along with a kanamycin resistance gene, enabling it to be maintained and selected in both species. Generally, these sequences are not integrated into the plant genome as they lie outside the T-DNA borders; in any case, it is straightforward to screen new GMOs to confirm that these sequences are absent.

Between the T-DNA border sequences are found a multiple cloning site within a *lacZ* gene (so that successful transgene insertion can be confirmed in *E. coli* by blue-white selection) and a chimeric nopaline synthase–neomycin phosphotransferase (*npt*) gene at the right T-DNA border, enabling successfully transformed plants to be selected by their resistance to the antibiotics kanamycin and neomycin. Later, binary vectors made it easier to recover transformants with full-length inserts by adding a reporter gene adjacent to the left T-DNA border or replaced *npt* with other selectable markers conferring resistance to antibiotics (*bla*, *aadA*) and/or herbicides (*bar*, *pat*, *gox*) (Jiang et al. 2014).

However, the inclusion of these bacterial antibiotic resistance genes in GM crop genomes has raised biosafety fears; it is argued that they could become incorporated into existing soil or gut bacteria by horizontal gene transfer, potentially causing a spread of antibiotic-resistant pathogens (Kleter et al. 2005). While this would be a very unlikely event, efforts have been made to generate vectors that do not include antibiotic resistance genes. Instead, "environmentally safer" selection is based on positive selection of transformants, either by enhancing growth or by expressing a visual marker. In the first category, bacterial genes that enhance plant glycometabolism have been employed, such as *pmi* (Wright et al. 2014), *xylA* (Haldrup et al. 1998), *manA*, and *tps1* (Jiang et al. 2014). Glycometabolism selectable marker genes are environmentally safe and cost-effective, but their efficacy varies depending on the carbohydrate ratios of each plant species, limiting their utilization. Other approaches to enhance growth include adding the phytohormone-related *ipt* gene (Endo et al. 2001) or drought/salt stress tolerance genes such as *SOS1* and *DREB* (Shi et al. 2003), which have been utilized to transform tomato (Jiang et al. 2014). In the second category, markers confer color (*gus*, anthocyanin) or fluorescence (*gfp*, *rfp*) to transformed cells, permitting simple visual selection of transformants. As an alternative to using safer selectable markers, several approaches have been tested to try to delete the selectable marker genes from the GMO following initial transformation. For example, the pCLEAN vector has been used to produce marker-free transformants in tobacco (Thole et al. 2007).

A secondary biosafety concern is the possibility of expressing unexpected partial or fused proteins from vector sequences or disruption of plant genes at or near the insertion site. Over the last 30 years, the following transformation constructs have been refined as follows: (1) minimizing the presence of undesired open reading frames by inserting stop codons or altering a translational start codon, (2) including the optimal promoter, Kozak sequence, and optimized codons for transgene expression (Joshi et al. 1997), (3) eliminating sequence identity or high homology to known toxins/allergens (Ladics et al. 2011), (4) removing direct or inverted repeat sequences from the construct to avoid the generation of unpredicted small RNA transcripts that could affect gene expression (Goodarzi et al. 2012), and (5) adding of enhancer-blocking sequences to prevent interference between regulatory sequences of adjacent endogenous genes with those found in the transgene cassette (Singer et al. 2011).

GM crops constructed using these safer vector constructs are now beginning to enter the market. However, in the meantime, a new methodology has emerged that promises to remove the need for a transformation vector altogether.

25.2.4 Genome Editing Methodology

The field of targeted genome engineering, also known as genome editing, has recently emerged as a novel transgenic technique to improve crop plants. These new approaches have significant advantages over conventional GM methods (Table 25.1), the most important being that specific plant genes can be modified *in situ*. Genome editing uses "programmable" nucleases such as zinc finger nucleases (ZFNs), TAL effector nucleases (TALENs), or clustered regularly interspaced short palindromic repeat (CRISPR)-associated endonucleases to modify existing plant genes or introduce new trait genes at a determined location (Sprink et al. 2015). All of these proteins have a recognition domain that can be engineered to target specific DNA sequences, and a nuclease activity, which is provided by the FokI domain in both ZFNs and TALENs, and Cas9 in CRISPR systems. The sequence-specific nuclease induces a double-stranded break (DSB) in the target DNA sequence (Bogdanove and Voytas 2011), which is then repaired by the cell's own nonhomologous end-joining (NHEJ) or homologous recombination (HR) mechanisms (Sprink et al. 2015). The amino acid sequence of the ZFN/TALEN directly determines the DNA recognition site, meaning the engineered nucleases can catalyze DSBs at the desired location with high specificity (Bogdanove and Voytas 2011). The simplest application of genome editing is to induce two DSBs a short distance apart, resulting in precise deletion of the intervening nucleotides, but new DNA can also be inserted by providing a donor DNA fragment homologous to the sequences flanking the DSBs (Voytas and Gao 2014). For example, one commercial research group used conventional *Agrobacterium*-mediated transformation of maize to insert "trait landing pads"—target sites for specific ZFNs that could then be used to stack additional traits by genome editing (Ainley et al. 2013).

The precise design of DNA-binding sites for ZFNs is restricted due to protein structural limitations; however, these limitations do not apply to TALENs, which can theoretically target any DNA sequence. Diverse plant genomes including *Arabidopsis*, *Brachypodium*, tobacco, tomato, rice, and barley have been successfully edited using TALENs (Sprink et al. 2015).

Although ZFNs/TALENs have been successful in editing plant genomes, the nucleases themselves are relatively complex to engineer. Recently, the CRISPR/Cas9 system has been proposed as a simpler technique for crop transformation (Shan et al. 2013). This system is based on a bacterial ribonucleoprotein complex consisting of two RNA species, CRISPR RNA (crRNA) and trans-activating CRISPR RNA (tracrRNA), and the Cas9 endonuclease. When the crRNA hybridizes with the target DNA sequence, cleavage occurs between locations determined by complementarity between the target and a ~20 bp guide sequence in the crRNA. Significantly, cleavage can also be directed by a synthetic single guide RNA (sgRNA) formed by fusing of the crRNA and tracRNA sequences (Jinek et al. 2012). Engineering the sgRNA to target a desired sequence is relatively simple, meaning that the Cas9 nuclease can easily be programmed to cleave specific genes. The CRISPR/Cas9

TABLE 25.2
Examples of Successful Genome Editing in Crop Plants

Crop	Genes Disrupted	SSN Type	Phenotypic Effect	Reference
Tobacco	*SuRA and SuRB*	ZFN	Herbicide resistance	Townsend et al. (2009)
Maize	*IPK1*	ZFN	Reduction of antinutrients	Shukla et al. (2009)
Rice	*OsSWEET14*	TALEN	Bacterial blight resistance	Li et al. (2012)
Soybean	*FAD2–1A and 1B*	TALEN	Improved oil quality and shelf-life	Haun et al. (2014)
Wheat	*TaMLO* genes	TALEN/CRISPR	Powdery mildew resistance	Wang et al. (2014)
Rice	*ALS*	CRISPR	Herbicide resistance	Sun et al. (2016)

system also has the ability to carry out multiple cleavages simultaneously by including multiple guide RNAs. Using this strategy, genomic fragment deletions of 140 bp and 1.8 kb have been made in tomato (Brooks et al. 2014) and tobacco (Gao et al. 2014), respectively.

Genome editing by deleting a gene fragment almost always results in loss of function of the targeted gene, in contrast to conventional GM in which the inserted transgene(s) confer a new function. However, it is still possible to introduce agronomically important traits through such loss-of-function mutations (Table 25.2). ZFNs were the first genome editing technique applied to crop plants, being used to confer herbicide tolerance in tobacco by disrupting the acetolactate synthase gene that the herbicide acts on (Townsend et al. 2009), and they improved the nutritional value in maize by disrupting one of the enzymes involved in the synthesis of phytic acid (Shukla et al. 2009).

Similarly, using TALENs, pathogen resistance was improved in rice by modifying the promoter of a bacterial blight susceptibility gene (Li et al. 2012). In soybean, TALEN-mediated disruption of fatty acid desaturase genes increased the content of monosaturated oils, resulting in improved oil quality and shelf-life (Haun et al. 2014). There are also reports of CRISPR/Cas9 being used successfully to edit genes of many crop plants including tobacco, rice, sorghum, wheat (Belhaj et al. 2013), and tomato (Brooks et al. 2014). As this is a relatively newer technology, most of these initial studies were carried out in cultured cells or tissues to optimize the method and establish its applications. However, reports of varieties of engineered traits such as herbicide tolerance using CRISPR/Cas9 are beginning to emerge (Sun et al. 2016).

Perhaps the most dramatic demonstration to date of the potential provided by genome editing technologies for crop improvement was the simultaneous disruption (using a combination of TALENs and CRISPR/Cas9) of all three homologs of the disease susceptibility locus *TaMLO* in bread wheat, conferring resistance to powdery mildew (Wang et al. 2014). The authors further suggested that, as no foreign DNA or proteins are present in the engineered plants, and the mutations introduced could in theory have arisen naturally, such crops should not be considered as GMOs. In this case, they would be exempt from the regulations and risk assessments applied to conventional GM crops (Kanchiswamy et al. 2015).

25.3 ASSESSING THE RISKS OF GENETICALLY ENGINEERED FOOD PRODUCTS

The adoption of GM crops by 18 million farmers in 28 countries, with a global planting area of 181.5 million ha. by 2013, has made GM arguably the fastest adopted novel crop technology (James and ISAAA 2014). However, this same rapid adoption combined with public suspicion of GMOs has made GM food safety a pressing issue. The speed of development of new techniques such as genome editing represents a significant challenge for authorities responsible for risk assessment and approval of newly engineered crops. Since the publication of the first guidelines for GM food safety

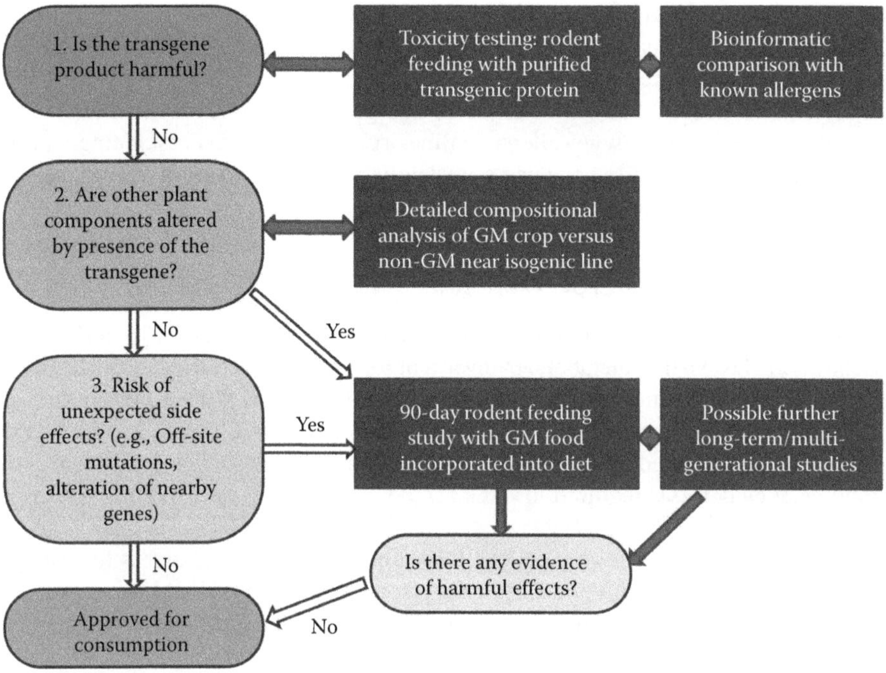

FIGURE 25.1 Flowchart showing the stages of current food safety risk assessments for novel genetically modified crops. White arrows indicate decisions and shaded arrows indicate information from experiments.

(WHO and FAO 1991), a risk assessment strategy has been developed that is applied in some form by most regulatory bodies (e.g., EFSA 2011), the major components of which are summarized in Figure 25.1.

25.3.1 Possible Causes of Toxicity in Conventional GMOs

The only direct health risk of GM foods is that the transgenic components, introduced to confer, for example, herbicide or insect resistance, may be somehow harmful when ingested by humans. As there is no known mechanism by which nucleic acids could cause toxicity on ingestion, this concern is focused on the expression of novel proteins, particularly of nonplant origin, which could be toxic and/or allergenic. To test for toxicity, high doses of transgenic protein (at least 1000 times the concentration that would be ingested in GM food) can be administered to rodents in a 28-day acute toxicology study; the European Food Safety Authority (EFSA) requires such a study for novel GM proteins following OECD guidelines (OECD 2008). To date, numerous acute toxicology studies using the proteins expressed in current GM crops have reported no adverse effects (Delaney 2015).

Allergenicity is much harder to determine than toxicity in food, as it is an abnormal immune response to a food ingredient that is harmless in the majority of individuals. Food allergy is reported to be found in 3%–8% of the population in Western countries (Bøgh and Madsen 2015). All known food allergens are proteins, but only a small proportion of food proteins are associated with allergy (Boyce et al. 2010). It is not yet known why some proteins are more allergenic than others, so any proteins newly introduced into GM crops could present an allergy risk. The potential for allergenicity in GM foods may be through coincidentally increased expression of endogenous allergens or unexpected expression of a *de novo* allergen from the novel protein(s) (Ladics et al. 2011). For example, transgenic expression of a soybean protein in pea was reported to produce a structurally altered protein, which had a proinflammatory effect when administered intragastrically to mice

(Prescott et al. 2005). It was argued that these effects could potentially lead to an allergic reaction, although the probability of such a reaction occurring in a normal feeding regime was not evaluated.

Bioinformatics provides the primary tool for evaluating the risk of allergenicity by comparing the novel GM protein with all known or suspected allergens, the sequences of which are available in a searchable online database (www.allergenonline.org). Any GM food containing a protein that already causes allergy in specific consumers is clearly a significant health hazard, although the degree of similarity to known allergens that presents a genuine risk is not clear (Ladics et al. 2011).

25.3.2 Effects on the Host Crop: Fault Lines in the Ongoing Debate

Assuming the transgenic proteins are not directly harmful, an indirect risk is that the presence of the transgene alters the existing metabolic pathways of the crop or that unexpected additional mutations were induced during transformation. While it is relatively straightforward to establish that a specific GMO contains no known toxins or allergens (the first step in Figure 25.1), it is extremely difficult to rule out unexpected effects altogether. Therefore, the most appropriate risk assessment strategy, and what evidence is required for a GMO to be considered safe, remains a hotly debated topic.

Numerous analytical studies have produced laboratory data aiming to assess the safety of GM crops and foods for human health and the environment. Proponents of GM argue that the great majority of studies have found no harmful effect from foods derived from GM crops (Delaney 2015), while sceptics contest that this is an oversimplistic view and that all novel GM crops should be rigorously assessed on a case-by-case basis (Hilbeck et al. 2015).

International guidelines for the risk assessment of GM crops have adopted the principle of "substantial equivalence" (WHO and FAO 1991). This recognizes that it is impossible to guarantee that any food is completely risk-free but that many foods have been consumed safely for many years. Therefore, if a GM crop is found to have no significant changes in composition compared with a non-GM equivalent, and any proteins expressed by the GM cassette itself are not harmful, it should be considered safe. However, as with other aspects of the GM debate, what constitutes "substantial equivalence" has become a contentious issue. Current safety assessment regulations (e.g., EFSA 2011) require compositional analysis of nutrients, antinutrients, secondary metabolites, and toxins in a GM crop alongside a non-GM near-isogenic variety. Accordingly, industry sources cite over 80 published compositional studies that, following these regulations, found GM crop varieties to be substantially equivalent to their non-GM comparators; it is also noted that compositional differences were often less than the natural variation produced by classical breeding methods (Herman and Price 2013). However, critics argue that many of these studies did not measure important secondary factors. For example, unlike non-GM varieties, glyphosate-tolerant GM crops are frequently sprayed with this herbicide by farmers, resulting in the accumulation of glyphosate and its metabolites in the crop (Bøhn et al. 2014).

Animal feeding trials are a second approach for assessing the safety of GM foods. However, it should be noted that most animal feeding study protocols were developed for toxicology testing of individual chemical compounds rather than whole foods containing a GM component. Therefore, in cases where any novel GM protein has been shown to be nontoxic (see Section 25.3.) and compositional equivalence has been established, feeding trials are not required (WHO and FAO 1991; EFSA 2011). Feeding trials are indicated if the composition of the GM crop has changed significantly (e.g., due to insertion of a gene intended to increase nutritional value) or if there is a high probability of unexpected effects. Furthermore, the design of the trial (animal species, diet, dosage, group size, controls, etc.) must be appropriate for a test of whole foods rather than chemical compounds; the recommended design is a 90-day feeding study in laboratory rats, with the GM food administered at several levels up to 33% of the diet, in accordance with OECD Guideline 408 (OECD 1998).

A large number of rodent feeding studies that were carried out according to these guidelines have indicated no evidence for adverse effects of current GM crops (EFSA GMO Panel Working

Group 2008). Initial results of the European GRACE (GMO Risk Assessment and Communication of Evidence) project, of which animal feeding trials are a major component, also found that two different GM maize varieties were not harmful to rats (Zeljenková et al. 2014). It should be noted that 90-day trials are not appropriate for detecting allergenicity or possible effects on reproductive health (EFSA GMO Panel Working Group 2008). For these purposes, longer term or multigenerational feeding studies may be carried out. Once again, over 20 such studies have reported no evidence of adverse effects of extended dietary exposure to foods derived from GM crops (Snell et al. 2012). Apart from rodents, feeding studies carried out with poultry have also detected no evidence of nutritional inferiority in GM feeds (McNaughton et al. 2014).

Although a large number of feeding trials carried out as part of risk assessments have found no effect of GM foods on animal health, a handful of studies have reported fine molecular and histological changes in the gut, pancreas, and liver of mice (Malatesta et al. 2008, and references therein) and detected DNA fragments from GM plants in some tissues of rabbits (Tudisco et al. 2006) after feeding with GM soybean. Additionally, a small number of animal feeding studies have reported adverse effects such as signs of toxicity in the kidney and liver of rats (de Vendômois et al. 2009) or increased the incidence of stomach inflammation (Carman et al. 2013) in GM-fed pigs compared with controls. One controversial long-term animal feeding study concluded that glyphosate-treated GM maize induced cancer in rats with increased incidence of tumors, higher death rates, and associated liver and kidney pathologies (Séralini et al. 2012). This paper has been widely criticized for significant scientific weaknesses, and a retraction of the findings was published by the journal a year after the original paper (Elsevier 2013).

In summary, while a great majority of the literature supports the assertion that current GM foods are as safe as their non-GM equivalents, a small number of negative results continue to raise safety concerns that critics believe the GM industry has not addressed (Carman et al. 2013). Here, again the polarized nature of the discussion is evident—for example, GM proponents point to a review published by the Royal Society of Medicine, which concluded that GM foods have been eaten for more than 15 years in some parts of the world with no reported negative effects on human health (Key et al. 2008). However, sceptics argue that no statistically rigorous epidemiological studies have assessed the effects of GM food on human health, so it is impossible to know whether they have had an impact (Hilbeck et al. 2015). While further research may resolve some of these issues, it should be remembered that proving any food (GM or otherwise) to be risk-free is an unattainable goal. A more useful way to frame the ongoing debate might be to ask whether the potential benefits outweigh the risks.

25.3.3 Are New Technologies Less Risky Than Conventional GM?

As noted earlier, foods and crops produced by conventional GM methods are far from globally accepted and must go through lengthy and expensive risk assessments. The first generated crops produced using genome editing (cf Section 25.2.4) are likely to enter the market in the near future. This technology allows modification of plant genes *in situ* without inserting a foreign transgene; the ZFN, TALEN, or CRISPR/Cas9 nucleases are transiently introduced into the plant cell and eliminated in subsequent generations. Therefore, it is suggested that the resulting crops should be considered as non-GM (Belhaj et al. 2013). Another emerging technology that may reduce risk is RNA interference (RNAi), in which conventional GM methods are used to insert a construct that expresses a regulatory RNA sequence (Ramon et al. 2014). In this case, although a nonendogenous construct is inserted into the crop genome, no transgenic proteins are produced.

Considering the risk assessment strategy summarized in Figure 25.1, both genome editing and RNAi do not introduce transgenic proteins, eliminating the risk of direct toxicity (step 1). Furthermore, the mutations created by genome editing have no qualitative differences from natural mutations that are exploited in conventional breeding. Similarly, RNAi uses the endogenous host-induced gene silencing mechanism that is part of the plant cell's defense against pests and pathogens.

Therefore, it seems highly unlikely that the nutrient composition of crops would be accidentally altered by either technique. However, the possibility of unexpected off-target mutations remains a risk. CRISPR/Cas9-mediated genome editing has been observed to cause off-target DSBs and chromosomal translocations (Cho et al. 2014), although these were reduced to undetectable levels by careful choice of a unique target sequence and optimization of the guide RNA and Cas9 sequences. RNAi activity is based on sequence homology, so other genes with sufficiently similar sequences to the target gene can also be silenced. Off-target gene silencing could occur in the GM plant itself, or conceivably in organisms that ingest GM material, although in all probability the RNA molecules would be digested before they have any effect (Lundgren and Duan 2013).

It is to be hoped that genome editing and RNAi live up to their potential to improve crop yields while avoiding the safety controversies that restrict conventional GMOs. However, thorough and well-communicated research is required to assess the potential risks associated with off-target effects (Ramon et al. 2014) and ideally to eliminate them. Meanwhile, risk assessors will need to establish new criteria for fairly evaluating these new technologies.

25.3.4 Regulation of Risk Assessments for GM Crops

GM crops have induced widespread debate over the health, environmental, and economic risks and the potential benefits of GMOs. Based on recommendations and guidelines from scientific and regulatory authorities (e.g., WHO and FAO 1991), several international policies and standards have been developed (WTO, Codex Alimentarius Commission, and the Cartagena Protocol on Biosafety) to deal with trade and regulation of GM foods and crops. Furthermore, contrasting public or political responses to GM crops have resulted in the adoption of quite different risk assessment and regulatory strategies and GMO trade conflicts between different countries.

In particular, a sharp contrast has arisen between the United States and the EU, likened by some to a "GM Cold War" (Meijer and Stewart 2004). For complex reasons, the EU has decided not to invest in GM crops and adopted a restrictive approach to the use of GMOs in foods, whereas the United States has more liberal rules that facilitate the development of GM crop technologies. The US–EU conflict has powerfully shaped the global debate over GMO trade regulatory policies. For instance, in the toxicological risk assessment of GM crops, according to the US regulatory agencies including the FDA (Food and Drug Administration) and EPA (Environmental Protection Agency), the novel proteins and GM product should each be evaluated as described in Section 25.3.2. If the developer followed the standard risk assessment process, the product can be released to the market. In contrast, regulatory bodies in Europe (e.g., EFSA and the European Commission) assess each event on a case-by-case basis and sometimes require additional studies addressing potential adjuvanticity, alteration of fertility, or the potential to induce cancers. Nevertheless, many GM products are permitted for use in animal feed in EU countries, even when they are not yet approved for human consumption.

With regard to the new crop genome engineering technologies, according to the definition of a "living modified organism" in the Cartagena Biosafety Protocol, genome-edited crops, but not those transformed with RNAi constructs, may fall outside the scope of current GMO regulations if lines are produced that lack any transgenic insert (Kanchiswamy et al. 2015). No genome-edited crops have yet come into the market, but the future regulatory status of such crops is under consideration in several countries (Araki and Ishii 2015). Some of the preliminary decisions and opinions are summarized in Table 25.3. Here, there is a contrast between countries that have product-based GMO regulations (where a product is regulated only if it contains a GM component) with those that have process-based regulations (where any substance produced using a GM process is regulated). For example, according to the U.S. Department of Agriculture, plants containing targeted deletions generated by the cell's self-repair mechanisms contain no GM elements; hence, genome-edited crops containing only deletions would be excluded from GMO regulations.

TABLE 25.3
Preliminary Regulatory Positions on Genome-Edited Crops

	EU	United States	Argentina	Australia	New Zealand
GMO regulation type	Process based	Product based	Product based	Process based	Process based
NHEJ pathway, one or a few base pairs altered	No opinion expressed	Not regulated	Not regulated	Not regulated	Regulated[a]
HR pathway with short repair template, resulting in a small insertion but no transgene	No opinion expressed	Not regulated	Regulated if a new coding sequence is present	Not regulated	No opinion expressed
HR pathway used to insert a transgene	No decision	Regulated	Regulated	Regulated	No opinion expressed
Relevant legislation for regulated GMOs	European Commission (2003a)	7 CFR Part 340	National Biosafety Framework	Gene Technology Act 2000	Hazardous Substances and New Organisms Act 1998

[a] Initially, the EPA of New Zealand ruled that crops edited by the NHEJ pathway would not be regulated as GMOs, but this ruling was reversed by a High Court decision (Araki and Ishii 2015).

However, the EU has yet to reach a decision on whether mutations made by genome editing fall within their process-based regulations (Hartung and Schiemann 2014). Interestingly, Australia, which also has process-based GM regulations, has adopted the position that crops edited by the NHEJ mechanism are not GMOs; however, a similar ruling in New Zealand was overturned in the High Court (Araki and Ishii 2015). Plants containing transgenes introduced by genome editing generally are regulated as GMOs, although the EU review expressed the opinion that the site-specific transgene insertion provided by this approach minimizes the risks associated with conventional GM. International discussions as to the status of genome-edited crops are ongoing, the outcome of which will have a major effect on their future development.

25.4 CHALLENGES FOR MONITORING OF GENETICALLY ENGINEERED MATERIAL IN FOODS

25.4.1 Consumer/Regulatory Requirement for Monitoring and Labeling

In the last 20 years, the number and variety of GM crops have grown enormously, and hundreds of millions of people worldwide have consumed GM foods without reported ill effects (Key et al. 2008). Nevertheless, fierce public debate continues as to whether GM foods present unusual risks for health (Romeis et al. 2013). Consequently, with the aim of providing consumer choice, the European Union introduced labeling laws for foods containing GM material in 1997, and over 40 other countries have followed suit (Gruère and Rao 2007). These policies vary widely from country to country; the EU policy is one of the most stringent, requiring mandatory labeling of food and feed produced using any GM ingredients above a certain threshold (>0.9% total GMO content), even if the end product contains no detectable DNA or protein (e.g., refined oil from GM canola) (European Commission 2003a,b). Furthermore, the EU has recently adopted a zero tolerance policy toward the low-level presence of unauthorized GMOs in foodstuffs (European Commission 2011). At the other end of the spectrum, countries such as the United States and Canada have introduced a standard for voluntary labeling based on the detection of >5% GM material in the final product. The U.S. Congress recently passed a bill blocking any mandatory labeling of foodstuffs containing

GM crops by individual states (Pompeo 2015). The labeling issue remains contentious, with proponents of GM objecting that, owing to implementation costs and impossibility of preventing cross-contamination on production lines, food producers simply stop using GM ingredients in markets with strict labeling requirements. Nevertheless, wherever labeling regulations exist, there is a need for reliable GMO detection methods to apply them.

25.4.2 Methods for GMO Detection in Foods

Over the last two decades, hundreds of procedures have been developed for detecting GM elements in food, the large majority of which are summarized in the GMO detection methods' database (GMDD; Dong et al. 2008). These methods rely on the identification of proteins or DNA sequences that are only found in specific GMOs.

1. *Protein-based methods* are designed to detect foreign proteins that have been introduced into GM crops, such as the CP4-EPSPS enzyme from *A. tumefaciens* (found in glyphosate-tolerant GMOs) and the crystal proteins from *Bacillus thuringiensis* (conferring insect resistance in "Bt" GMOs). The proteins are detected using specific monoclonal antibodies (mAb), in the form of either a chromatographic protein strip test (PST; reviewed in Van Den Bulcke et al. 2007) or an enzyme-linked immunosorbent assay (ELISA, described in detail by Ermolli et al. 2006). The PST is a simple dipstick test that, after suitable extraction of the sample, gives a result in 2–5 minutes with no additional equipment. The ELISA format is more complex but offers higher sensitivity than the PST and the potential for quantification of the protein detected. However, protein-based methods have major limitations; first, a different mAb must be developed and a separate test is carried out for every new transgenic protein, which becomes increasingly costly as the number and variety of GMOs increase. Second, they cannot discriminate between different GM events that express the same protein. Finally, many of the steps involved in food production break down or denature proteins, making them unsuitable for detection in cooked or processed foods. Therefore, although a number of kits for the detection of widely used transgenic proteins are commercially available—some of which are approved for use by regulatory bodies (ISO 2013)—in most cases, detection of GM-specific DNA sequences, which are much more stable than proteins, is used instead.
2. *Polymerase chain reaction (PCR)* is by far the most widely used method for the detection of GM elements in food, with PCR assays comprising more than 90% of the methods in the GMDD (Dong et al. 2008). PCR is a routine method used in laboratories around the world, adaptable to any DNA sequence of interest, relatively rapid and inexpensive, and highly sensitive. For the detection of DNA from GM crops, element- or construct-specific PCRs amplify sequences found in transgenes and GM vectors, respectively, and so can be used to screen samples for multiple GMOs that contain common elements. On the other hand, event-specific PCRs target the junction where the GM construct integrated into the plant genome. Current GM constructs integrate at random locations, so junction sequences are unique to each GM event and can be used to discriminate between them. As a result, a large number of PCR protocols optimized for different target sequences have been validated by the European Union for use in GM monitoring (Bonfini et al. 2012) and are made publically available at http://gmo-crl.jrc.ec.europa.eu/gmomethods/. Standard endpoint PCRs are highly sensitive and can detect GM DNA at concentrations well below those that trigger labeling requirements; such low concentrations could easily arise from cross-contamination in food production lines (Turkec et al. 2016). Therefore, quantitative real-time PCR (qPCR) is also essential to establish whether labeling is necessary. Typically, a reference gene from the plant genome is amplified in parallel with the GM sequence, and the ratio of the amplification signal from each is used to calculate the

relative GMO concentration (Turkec et al. 2016). However, the specialist equipment and fluorescent probes used in real-time PCR also make it more expensive than qualitative PCR tests; furthermore, each individual qPCR assay must be carefully optimized (Holst-Jensen et al. 2003).

PCR, qPCR in particular, is a currently used method of choice for GMO detection in foods, but it also has limitations. Although DNA is much more stable than protein, it can be degraded during food processing, making detection more difficult in highly processed foods. A more serious issue is that PCRs can be inhibited by other common food ingredients such as salts, sugars, and oils. Therefore, isolation of DNA from the food sample is a critical step, and different foodstuffs require different isolation methods to obtain DNA of optimal purity (Turkec et al. 2015a).

3. *High-throughput DNA-based methods.* The other major limitation of PCR is the need for a separate test for each target sequence. Multiplex PCR strategies have been developed to monitor multiple GMOs simultaneously, but interference and competition between PCR primers restrict them to amplify a handful of targets in a single reaction. For this reason, systems designed to screen for most known GMOs by PCR typically require a large number of single reactions (Querci et al. 2009) or a moderate number of multiplex PCRs to be carried out in parallel (Cottenet et al. 2013), leading to high cost and labor per sample. Other researchers have addressed the multiplex problem by physically dividing PCRs into microdroplets (Guo et al. 2011) or across a microwell chip (Shao et al. 2014). In these approaches, a high-throughput analysis method is then used to characterize the pooled PCR products.

The DNA microarray is a well-established high-throughput analysis technology, allowing sequence-specific, parallel detection of a large number of genetic elements from complex DNA samples. Accordingly, microarray analysis has also been applied to GMO detection (Zhang and Guo 2011). To date, several methods based on multiplex PCR followed by hybridization of the PCR products to a low-density DNA microarray have been reported for the detection of GM events, including one that has been developed into a commercial product (Leimanis et al. 2008). However, the limited multiplexing capacity of PCR still restricts the number of different GM events that can be detected in parallel. PCR amplification prior to microarray detection also makes reliable quantification problematic, as different samples may amplify with different efficiencies. Therefore, alternative methods have been developed that modify or substitute the PCR amplification step with ligation-based probes that amplify more consistently and can then be detected on a microarray (Ujhelyi et al. 2012). Microarray analysis has also been used to characterize PCR products amplified in parallel on a microwell chip (Shao et al. 2014).

In a few cases, a microarray-based method for GMO detection has been described in which target DNA is hybridized to the array without an initial amplification step. In principle, this has potential for GMO quantification, as the amount of DNA bound to the microarray should be proportional to the copy number of the element under scrutiny. We explored this possibility by designing a microarray for direct detection of 16 GMOs approved for feed use in Turkey using a modified comparative genome hybridization microarray platform (Baumbusch et al. 2008) and found that the GMOs could be detected without amplification and quantified (Turkec et al. 2015b). Importantly, the array system has the capacity to detect hundreds more GMOs simultaneously, once optimal hybridization probes are developed.

All of the PCR and microarray hybridization methods mentioned earlier require prior knowledge of the DNA sequence of the GMO to be detected. As the insertion site of the transgenic construct is random in conventional GM, this means that every single transformation event creates a new GMO, which requires unique probes to provide unequivocal identification. Addressing this issue, a microarray probe set representing all possible sequences has been proposed as a means of detecting unknown GM events (Nesvold et al. 2005). This concept was further refined by Tengs et al. using overlapping probe sequences covering all the vectors and the majority of trait genes used in conventional GM plants. It was demonstrated that this approach could be used accurately to determine the inserted GMO cassette, even for an event where no prior sequence data were available (Tengs et al. 2009); however, this required undiluted GMO DNA, so it may not be applicable to food samples containing an unknown GM contaminant.

Next-generation DNA sequencing (NGS) perhaps presents the best opportunity for the characterization of unknown GMOs, with its extremely high throughput and steadily decreasing cost. Accordingly, NGS has been applied to GMO characterization. In some cases, GM regions are amplified by PCR and then sequenced—this is similar to the multiplex PCR screens mentioned earlier, with the added benefit that the complete sequence of the amplified region is verified (Song et al. 2014). Alternatively, it is possible to resequence the entire genome of a suspected GMO. In a proof-of-principle study, the GM rice variety LLRICE62 was resequenced using the published sequence of *Oryza sativa* spp. *japonica* as a reference and found to contain a single intact GM cassette in the location reported by the developer (Wahler et al. 2013). Conversely, re-sequencing of the GM cassettes two other GM rice events, TT51-1 and T1c-19, revealed additional unintended insertions not detected by other techniques (Yang et al. 2013). While NGS clearly has an important role to play particularly in the detection of unknown GMOs, its high cost and degree of expertise needed to analyze the data mean that it is still currently impractical for routine GM monitoring (Fraiture et al. 2015).

25.4.3 Challenges for GM Monitoring Presented by Genome Editing

While it is not yet clear whether genome-edited crops will be considered as GMOs by regulatory authorities, it is clear that they present a major problem for current detection strategies. A GMO that contains no nonendogenous sequence is likely to be very difficult to discriminate from wild type by endpoint PCR, especially if mixed with unmodified DNA from the same species. Where a known deletion has taken place, it may be feasible to detect the deletion using a hybridization probe (for qPCR or on a microarray) that bridges the deletion site; however, even this would be unlikely to discriminate 1–2 base changes sufficient to knockout a protein by deleting its start codon or causing a frameshift. Technologies currently used for genotyping single-nucleotide polymorphisms in crops, such as TaqMan and competitive allele-specific PCR (KASP), could potentially be adapted for this purpose, although it is unknown whether they are sensitive enough to analyze mixed samples, and unique probes would have to be developed for each GM event. Next-generation sequencing has the power to distinguish even small modifications and rule out the presence of off-target mutations in species where the complete genome sequence is already known (Cho et al. 2014); however, many important crop species such as bread wheat have very large and complex genomes for which only draft sequences are available (Lucas et al. 2014). One possible solution to this problem is requiring genome-edited crops to add a DNA tag to the site of modification specifically to facilitate detection (Araki and Ishii 2015). This would mean that the plant was no longer completely free of nonendogenous DNA, but a carefully chosen detection tag would present far fewer risks than a complete transgenic construct.

25.5 CONCLUSION

For the last 20 years, GM crops have been seen both as the best opportunity to meet the agricultural challenges of the twenty-first century and as a technology fraught with unknown risks and dangers. The controversy between these two viewpoints does not seem likely to abate in the foreseeable future. In the authors' experience, the great majority of the scientific community believe that the benefits outweigh the potential risks. This is particularly true for genome-edited crops, which can be engineered much more subtly and with greater precision than conventional GMOs; with NGS analysis, it may also be possible to eliminate off-target mutations in many cases. However, the past GMO controversies show that patience and clear communication are required to achieve public and political acceptance of these technologies before they are likely to be accepted in food products.

REFERENCES

Ainley, W.M., L. Sastry-Dent, M.E. Welter, M.G. Murray, B. Zeitler, R. Amora, D.R. Corbin et al. 2013. Trait stacking via targeted genome editing. *Plant Biotechnology Journal* 11 (9): 1126–1134. doi:10.1111/pbi.12107.

Araki, M. and T. Ishii. 2015. Towards social acceptance of plant breeding by genome editing. *Trends in Plant Science* 20 (3): 145–149. doi:10.1016/j.tplants.2015.01.010.

Baumbusch, L.O., J. Aarøe, F.E. Johansen, J. Hicks, H. Sun, L. Bruhn, K. Gunderson et al. 2008. Comparison of the agilent, ROMA/NimbleGen and illumina platforms for classification of copy number alterations in human breast tumors. *BMC Genomics* 9 (1). BioMed Central: 379. doi:10.1186/1471-2164-9-379.

Belhaj, K., A. Chaparro-Garcia, S. Kamoun, and V. Nekrasov. 2013. Plant genome editing made easy: Targeted mutagenesis in model and crop plants using the CRISPR/Cas system. *Plant Methods* 9 (1): 39. doi:10.1186/1746-4811-9-39.

Bevan, M. 1984. Binary agrobacterium vectors for plant transformation. *Nucleic Acids Research* 12 (22): 8711–8721. http://www.pubmedcentral.nih.gov/articlerender.fcgi?artid=320409&tool=pmcentrez&rendertype=abstract.

Bogdanove, A.J. and D.F. Voytas. 2011. TAL Effectors: Customizable proteins for DNA targeting. *Science* 333 (6051): 1843–1846. doi:10.1126/science.1204094.

Bøgh, K.L. and C.B. Madsen. 2015. Food allergens: Is there a correlation between stability to digestion and allergenicity? *Critical Reviews in Food Science and Nutrition* 56 (9): 1545–1567. doi:10.1080/10408398.2013.779569.

Bøhn, T., M. Cuhra, T. Traavik, M. Sanden, J. Fagan, and R. Primicerio. 2014. Compositional differences in soybeans on the market: Glyphosate accumulates in roundup ready GM soybeans. *Food Chemistry* 153 (June): 207–215. doi:10.1016/j.foodchem.2013.12.054.

Bonfini, L., M.H. van den Bulcke, M. Mazzara, E. Ben, and A. Patak. 2012. GMOMETHODS: The European union database of reference methods for GMO analysis. *Journal of AOAC International* 95 (6): 1713–1719. doi:10.5740/jaoacint.12-050.

Boyce, J.A., S.M. Jones, L. Rock, H.A. Sampson, S.F. Cooper, S. Boyce, and E.T. Al. 2010. Guidelines for the diagnosis and management of food allergy in the United States: Report of the NIAID-sponsored expert panel. *Journal of Allergy and Clinical Immunology* 126. doi:10.1016/j.jaci.2010.10.007.

Brooks, C., V. Nekrasov, Z. Lippman, and J. Van Eck. 2014. Efficient gene editing in tomato in the first generation using the CRISPR/Cas9 system. *Plant Physiology* September, 114.247577. doi:10.1104/pp.114.247577.

Carman, J.A., H.R. Vlieger, L.J. Ver Steeg, E. Verlyn, G.W. Robinson, C.A. Clinch-Jones, I. Julie, and J.W. Edwards. 2013. A long-term toxicology study on pigs fed a combined genetically modified (GM) soy and GM maize diet. *Journal of Organic Systems* 8 (1): 38–54.

Chan, M.T., H.H. Chang, S.L. Ho, W.F. Tong, and S.M. Yu. 1993. Agrobacterium-mediated production of transgenic rice plants expressing a chimeric alpha-amylase promoter/beta-glucuronidase gene. *Plant Molecular Biology* 22 (3): 491–506. http://www.ncbi.nlm.nih.gov/pubmed/8392395.

Cho, S.W., S. Kim, Y. Kim, J. Kweon, H.S. Kim, S. Bae, and J.S. Kim. 2014. Analysis of off-target effects of CRISPR/Cas-derived RNA-guided endonucleases and nickases. *Genome Research* 24 (1): 132–141. doi:10.1101/gr.162339.113.

Cleene, M. and J. Ley. 1976. The host range of crown gall. *The Botanical Review* 42 (4): 389–466. doi:10.1007/BF02860827.

Cottenet, G., C. Blancpain, V. Sonnard, and P.F. Chuah. 2013. Development and validation of a multiplex real-time PCR method to simultaneously detect 47 targets for the identification of genetically modified organisms. *Analytical and Bioanalytical Chemistry* 405 (2013): 6831–6844. doi:10.1007/s00216-013-7125-5.

de la Peña, A., H. Lörz, and J. Schell. 1987. Transgenic rye plants obtained by injecting DNA into young floral tillers. *Nature* 325 (6101): 274–76. doi:10.1038/325274a0.

de Vendômois, J.S., F. Roullier, D. Cellier, and G.E. Séralini. 2009. A comparison of the effects of three GM corn varieties on mammalian health. *International Journal of Biological Sciences* 5 (7): 706–726. doi:10.7150/ijbs.5.706.

Delaney, B. 2015. Safety assessment of foods from genetically modified crops in countries with developing economies. *Food and Chemical Toxicology: An International Journal Published for the British Industrial Biological Research Association* 86 (October): 132–143. doi:10.1016/j.fct.2015.10.001.

Dong, W., L. Yang, K. Shen, B. Kim, G.A. Kleter, H.J.P. Marvin, R. Guo, W. Liang, and D. Zhang. 2008. GMDD: A database of GMO detection methods. *BMC Bioinformatics* 9 (1): 260. doi:10.1186/1471-2105-9-260.

EFSA. 2011. Guidance for risk assessment of food and feed from genetically modified plants. *EFSA Journal* 9 (5): 2150. doi:10.2903/j.efsa.2011.2150.

EFSA GMO Panel Working Group. 2008. Safety and nutritional assessment of GM plants and derived food and feed: The role of animal feeding trials. *Food and Chemical Toxicology: An International Journal Published for the British Industrial Biological Research Association* 46 Suppl 1 (March): S2–S70. doi:10.1016/j.fct.2008.02.008.

Elsevier. 2013. Elsevier Announces Article Retraction from Journal Food and Chemical Toxicology. https://www.elsevier.com/about/press-releases/research-and-journals/elsevier-announces-article-retraction-from-journal-food-and-chemical-toxicology. Accessed on July 4, 2016.

Endo, S., T. Kasahara, K. Sugita, E. Matsunaga, and H. Ebinuma. 2001. The isopentenyl transferase gene is effective as a selectable marker gene for plant transformation in tobacco (*Nicotiana tabacum* Cv. Petite Havana SRI). *Plant Cell Reports* 20 (1): 60–66. doi:10.1007/s002990000279.

Ermolli, M., A. Fantozzi, M. Marini, D. Scotti, B. Balla, S. Hoffmann, M. Querci, C. Paoletti, and G. Van Den Eede. 2006. Food safety: Screening tests used to detect and quantify GMO proteins. *Accreditation and Quality Assurance* 11: 55–57. doi:10.1007/s00769-005-0027-2.

European Commission. 2003a. Regulation (EC) No 1829/2003 of the European Parliament. *Official Journal of the European Union* L268: 1–23. http://eur-lex.europa.eu/legal-content/EN/TXT/?uri=uriserv:OJ.L_.2003.268.01.0001.01.ENG.

European Commission. 2003b. Regulation (EC) No 1830/2003 of the European Parliament. *Official Journal of the European Union* L268: 24–28. http://eur-lex.europa.eu/legal-content/EN/TXT/?uri=uriserv:OJ.L_.2003.268.01.0024.01.ENG.

European Commission. 2011. Commission Regulation (EU) No 619/2011. *Official Journal of the European Union* L166: 9–15. http://eur-lex.europa.eu/legal-content/EN/TXT/?uri=uriserv:OJ.L_.2011.166.01.0009.01.ENG.

FAO, IFAD, and WFP. 2014. *The State of Food Insecurity in the World 2014. Strengthening the Enabling Environment for Food Security and Nutrition.* Rome, Italy. http://www.fao.org/3/a-i4037e.pdf.

Fraiture, M.-A., P. Herman, I. Taverniers, M. De Loose, D. Deforce, and N.H. Roosens. 2015. Current and new approaches in GMO detection: Challenges and solutions. *BioMed Research International* 2015 (Table 1): 1–22. doi:10.1155/2015/392872.

Fromm, M.E., L.P. Taylor, and V. Walbot. 1986. Stable transformation of maize after gene transfer by electroporation. *Nature* 319 (6056): 791–793. doi:10.1038/319791a0.

Gao, J., G. Wang, S. Ma, X. Xie, X. Wu, X. Zhang, Y. Wu, P. Zhao, and Q. Xia. 2014. CRISPR/Cas9-mediated targeted mutagenesis in *Nicotiana tabacum. Plant Molecular Biology* 99–110. doi:10.1007/s11103-014-0263-0.

Godfray, H.C.J., J.R. Beddington, I.R. Crute, L. Haddad, D. Lawrence, J.F. Muir, J. Pretty, S. Robinson, S.M. Thomas, and C. Toulmin. 2010. Food security: The challenge of feeding 9 billion people. *Science (New York, N.Y.)* 327 (5967): 812–818. doi:10.1126/science.1185383.

Goodarzi, H., H.S. Najafabadi, P. Oikonomou, T.M. Greco, L. Fish, R. Salavati, I.M. Cristea, and S. Tavazoie. 2012. Systematic discovery of structural elements governing stability of mammalian messenger RNAs. *Nature* 485 (7397): 264–268. doi:10.1038/nature11013.

Gruère, G.P. and S.R. Rao. 2007. A review of international labeling policies of genetically modified food to evaluate India's proposed rule. *AgBioForum*. 10 (1): 51–64.

Guo, J., L. Yang, L. Chen, D. Morisset, X. Li, L. Pan, and D. Zhang. 2011. MPIC: A high-throughput analytical method for multiple DNA targets. *Analytical Chemistry* 83 (5): 1579–1586. doi:10.1021/ac103266w.

Haldrup, A., S.G. Petersen, and F.T. Okkels. 1998. The xylose isomerase gene from thermoanaerobacterium thermosulfurogenes allows effective selection of transgenic plant cells using D-xylose as the selection agent. *Plant Molecular Biology* 37: 287–296. doi:10.1023/A:1005910417789.

Hartung, F. and J. Schiemann. 2014. Precise plant breeding using new genome editing techniques: Opportunities, safety and regulation in the EU. *The Plant Journal: For Cell and Molecular Biology* 78 (5): 742–752. doi:10.1111/tpj.12413.

Haun, W., A. Coffman, B.M. Clasen, Z.L. Demorest, A. Lowy, E. Ray, A. Retterath et al. 2014. Improved soybean oil quality by targeted mutagenesis of the fatty acid desaturase 2 gene family. *Plant Biotechnology Journal* 12 (7): 934–940. doi:10.1111/pbi.12201.

Herman, R.A. and William, D.P. 2013. Unintended compositional changes in genetically modified (GM) crops: 20 years of research. *Journal of Agricultural and Food Chemistry* 61 (48). American Chemical Society: 11695–11701. doi:10.1021/jf400135r.

Hilbeck, A., R. Binimelis, N. Defarge, R. Steinbrecher, A. Székács, F. Wickson, M. Antoniou et al. 2015. No scientific consensus on GMO safety. *Environmental Sciences Europe* 27 (1): 4. doi:10.1186/s12302-014-0034-1.

Holst-Jensen, A., S.B. Rønning, A. Løvseth, and K.G. Berdal. 2003. PCR technology for screening and quantification of genetically modified organisms (GMOs). *Analytical and Bioanalytical Chemistry* 375 (8): 985–993. doi:10.1007/s00216-003-1767-7.

IPCC. 2015. *Climate Change 2014: Synthesis Report. Contribution of Working Groups I, II and III to the Fifth Assessment Report of the Intergovernmental Panel on Climate Change.* Geneva, Switzerland. http://www.ipcc.ch/report/ar5/syr/. Accessed on July 4, 2016.

ISO. 2013. *ISO 21572:2013 Foodstuffs—Molecular Biomarker Analysis—Protein-Based Methods.* Geneva, Switzerland. http://www.iso.org/iso/home/store/catalogue_ics/catalogue_detail_ics.htm?csnumber=51005.

James, C. and ISAAA. 2014. *Global Status of Commercialized Biotech/GM Crops: 2014—ISAAA Brief 49.* The International Service for the Acquisition of Agri-biotech Applications (ISAAA). http://www.isaaa.org/resources/publications/briefs/49/default.asp. Accessed on July 4, 2016.

Jiang, Y., X. Hu, and H. Huang. 2014. Recent patents on biosafety strategies of selectable marker genes in genetically modified crops. *Recent Patents on Food, Nutrition & Agriculture* 6 (1): 3–15. doi:10.2174/2212798406666141024104637.

Jinek, M., K. Chylinski, I. Fonfara, M. Hauer, J.A. Doudna, and E. Charpentier. 2012. A programmable dual-RNA—Guided DNA endonuclease in adaptive bacterial immunity. *Science* 337 (August): 816–822. doi:10.1126/science.1225829.

Joshi, C.P., H. Zhou, X. Huang, and V.L. Chiang. 1997. Context sequences of translation initiation codon in plants. *Plant Molecular Biology* 35 (6): 993–1001. doi:10.1023/A:1005816823636.

Kanchiswamy, C.N., M. Malnoy, R. Velasco, J.-S. Kim, and R. Viola. 2015. Non-GMO genetically edited crop plants. *Trends in Biotechnology* 1–3. doi:10.1016/j.tibtech.2015.04.002.

Key, S., J.K.-C. Ma, and P.M.W. Drake. 2008. Genetically modified plants and human health. *Journal of the Royal Society of Medicine* 101 (6): 290–298. doi:10.1258/jrsm.2008.070372.

Klein, R.M., E.D. Wolf, R. Wu, and J.C. Sanford. 1992. High-velocity microprojectiles for delivering nucleic acids into living cells. 1987. *Biotechnology (Reading, Mass.)* 24: 384–386. doi:10.1038/327070a0.

Kleter, G.A., Ad.A.C.M. Peijnenburg, and H.J.M. Aarts. 2005. Health considerations regarding horizontal transfer of microbial transgenes present in genetically modified crops. *Journal of Biomedicine & Biotechnology* 2005 (4): 326–352. doi:10.1155/JBB.2005.326.

Ladics, G.S., A. Bartholomaeus, P. Bregitzer, N.G. Doerrer, A. Gray, T. Holzhauser, M. Jordan et al. 2015. Genetic basis and detection of unintended effects in genetically modified crop plants. *Transgenic Research* 587–603. doi:10.1007/s11248-015-9867-7.

Ladics, G.S, R.F. Cressman, C. Herouet-Guicheney, R.A. Herman, L. Privalle, P. Song, J.M. Ward, and S. McClain. 2011. Bioinformatics and the allergy assessment of agricultural biotechnology products: Industry practices and recommendations. *Regulatory Toxicology and Pharmacology* 60 (1): 46–53. doi:10.1016/j.yrtph.2011.02.004.

Leimanis, S., S. Hamels, F. Nazé, G. Mbongolo Mbella, M. Sneyers, R. Hochegger, H. Broll et al. 2008. Validation of the performance of a GMO multiplex screening assay based on microarray detection. *European Food Research and Technology* 227 (6): 1621–1632. doi:10.1007/s00217-008-0886-y.

Li, T., B. Liu, M.H. Spalding, D.P. Weeks, and B. Yang. 2012. High-efficiency TALEN-based gene editing produces disease-resistant rice. *Nature Biotechnology* 30 (5): 390–392. doi:10.1038/nbt.2199.

Lucas, S.J., B. Akpınar, H. Šimková, M. Kubaláková, J. Doležel, and H. Budak. 2014. Next-generation sequencing of flow-sorted wheat chromosome 5D reveals lineage-specific translocations and widespread gene duplications. *BMC Genomics* 15 (1): 1080. doi:10.1186/1471-2164-15-1080.

Lundgren, J.G. and J.J. Duan. 2013. RNAi-based insecticidal crops. *BioScience* 63 (8): 657–665. doi:10.1525/bio.2013.63.8.8.

Malatesta, M., F. Boraldi, G. Annovi, B. Baldelli, S. Battistelli, M. Biggiogera, and D. Quaglino. 2008. A long-term study on female mice fed on a genetically modified soybean: Effects on liver ageing. *Histochemistry and Cell Biology* 130 (5): 967–977. doi:10.1007/s00418-008-0476-x.

Mayerhofer, R., Z. Koncz-Kalman, C. Nawrath, G. Bakkeren, A. Crameri, K. Angelis, G.P. Redei, J. Schell, B. Hohn, and C. Koncz. 1991. T-DNA integration: A mode of illegitimate recombination in plants. *The EMBO Journal* 10 (3): 697–704. http://www.pubmedcentral.nih.gov/articlerender.fcgi?artid=452704&tool=pmcentrez&rendertype=abstract.

McNaughton, J., M. Roberts, D. Rice, B. Smith, B. Hong, B. Delaney, and C. Iiams. 2014. Comparison of broiler performance and carcass yields when fed diets containing genetically modified canola meal from event DP-Ø73496-4, near-isogenic canola meal, or commercial canola meals. *Poultry Science* 93 (7): 1713–1723. doi:10.3382/ps.2013-03645.

Meijer, E. and R. Stewart. 2004. The GM cold war: How developing countries can go from being dominos to being players. *Review of European Community and International Environmental Law* 13 (3): 247–262. doi:10.1111/j.1467-9388.2004.00406.x.

Nesvold, H., A.B. Kristoffersen, A. Holst-Jensen, and K.G. Berdal. 2005. Design of a DNA chip for detection of unknown genetically modified organisms (GMOs). *Bioinformatics (Oxford, England)* 21 (9): 1917–1926. doi:10.1093/bioinformatics/bti248.

Newell, C.A. 2000. Plant transformation technology developments and applications. *Molecular Biotechnology* 16 (1): 53–65. doi:10.1385/MB:16:1:53.

Nicolia, A., A. Manzo, F. Veronesi, and D. Rosellini. 2014. An overview of the last 10 years of genetically engineered crop safety research. *Critical Reviews in Biotechnology*, February. http://www.tandfonline.com/doi/abs/10.3109/07388551.2013.823595.

OECD. 1998. Test No. 408: Repeated dose 90-day oral toxicity study in rodents. In *Guidelines for the Testing of Chemicals, Section 4*. Paris, France: OECD Publishing. doi: http://dx.doi.org/10.1787/9789264070707-en.

OECD. 2008. Test No. 407: Repeated dose 28-day oral toxicity study in rodents. In *Guidelines for the Testing of Chemicals, Section 4*. Paris, France: OECD Publishing. doi:http://dx.doi.org/10.1787/9789264070684-en.

Pompeo, M. 2015. H.R.1599—114th Congress (2015–2016): Safe and Accurate Food Labeling Act of 2015. https://www.congress.gov/bill/114th-congress/house-bill/1599. Accessed on July 4, 2016.

Prescott, V.E., P.M. Campbell, A. Moore, J. Mattes, M.E. Rothenberg, P.S. Foster, T.J.V. Higgins, and S.P. Hogan. 2005. Transgenic expression of bean??-Amylase inhibitor in peas results in altered structure and immunogenicity. *Journal of Agricultural and Food Chemistry* 53: 9023–9030. doi:10.1021/jf050594v.

Querci, M., N. Foti, A. Bogni, L. Kluga, H. Broll, and G. van den Eede. 2009. Real-time PCR-based ready-to-use multi-target analytical system for GMO detection. *Food Analytical Methods* 2: 325–236. doi:10.1007/s12161-009-9093-0.

Ramon, M., Y. Devos, A. Lanzoni, Y. Liu, A. Gomes, A. Gennaro, and E. Waigmann. 2014. RNAi-based GM plants: Food for thought for risk assessors. *Plant Biotechnology Journal* 12 (9): 1271–1273. doi:10.1111/pbi.12305.

Romeis, J., M. Meissle, S. Brunner, D. Tschamper, and M. Winzeler. 2013. Plant biotechnology: Research behind fences. *Trends in Biotechnology* 31 (4): 222–224. doi:10.1016/j.tibtech.2013.01.020.

Séralini, G.E., E. Clair, R. Mesnage, S. Gress, N. Defarge, M. Malatesta, D. Hennequin, and J.S. de Vendômois. 2012. Long term toxicity of a roundup herbicide and a roundup-tolerant genetically modified maize. *Food and Chemical Toxicology* 50 (11): 4221–4231. doi:10.1016/j.fct.2012.08.005.

Shan, Q., Y. Wang, J. Li, Y. Zhang, K. Chen, Z. Liang, K. Zhang et al. 2013. Targeted genome modification of crop plants using a CRISPR-Cas system. *Nature Biotechnology* 31 (8): 686–688. doi:10.1038/nbt.2650.

Shao, N., S.M. Jiang, M. Zhang, J. Wang, S.J. Guo, Y. Li, H.W. Jiang et al. 2014. MACRO: A combined microchip-PCR and microarray system for high-throughput monitoring of genetically modified organisms. *Analytical Chemistry* 86 (2): 1269–1276. doi:10.1021/ac403630a.

Shi, H., B.-H. Lee, S.-J. Wu, and J.-K. Zhu. 2003. Overexpression of a plasma membrane Na+/H+ antiporter gene improves salt tolerance in *Arabidopsis thaliana*. *Nature Biotechnology* 21 (1): 81–85. doi:10.1038/nbt766.

Shukla, V.K., Y. Doyon, J.C. Miller, R.C. DeKelver, E.A. Moehle, S.E. Worden, J.C. Mitchell et al. 2009. Precise genome modification in the crop species *Zea mays* using zinc-finger nucleases. *Nature* 459 (7245): 437–441. doi:10.1038/nature07992.

Singer, S.D., K.D. Cox, and Z. Liu. 2011. Enhancer-promoter interference and its prevention in transgenic plants. *Plant Cell Reports* 30 (5): 723–731. doi:10.1007/s00299-010-0977-7.

Snell, C., A. Bernheim, J.B. Bergé, M. Kuntz, G. Pascal, A. Paris, and A.E. Ricroch. 2012. Assessment of the health impact of GM plant diets in long-term and multigenerational animal feeding trials: A literature review. *Food and Chemical Toxicology* 50 (3–4): 1134–1148. doi:10.1016/j.fct.2011.11.048.

Song, Q., G. Wei, and G. Zhou. 2014. Analysis of genetically modified organisms by pyrosequencing on a portable photodiode-based bioluminescence sequencer. *Food Chemistry* 154: 78–83. doi:10.1016/j.foodchem.2014.01.001.

Sprink, T., J. Metje, and F. Hartung. 2015. Plant genome editing by novel tools: TALEN and other sequence specific nucleases. *Current Opinion in Biotechnology* 32: 47–53. doi:10.1016/j.copbio.2014.11.010.

Sun, Y., X. Zhang, C. Wu, Y. He, Y. Ma, H. Hou, X. Guo, W. Du, Y. Zhao, and L. Xia. 2016. Engineering herbicide resistant rice plants through CRISPR/Cas9-mediated homologous recombination of the acetolactate synthase. *Molecular Plant*, January. doi:10.1016/j.molp.2016.01.001.

Tengs, T., A.B. Kristoffersen, H. Zhang, K.G. Berdal, M. Løvoll, and A. Holst-Jensen. 2009. Non-prejudiced detection and characterization of genetic modifications. *Food Analytical Methods* 3 (2): 120–128. doi:10.1007/s12161-009-9111-2.

Thole, V., B. Worland, J.W. Snape, and P. Vain. 2007. The pCLEAN dual binary vector system for agrobacterium-mediated plant transformation. *Plant Physiology* 145 (4): 1211–1219. doi:10.1104/pp.107.108563.

Townsend, J.A., D.A. Wright, R.J. Winfrey, F. Fu, M.L. Maeder, J.K. Joung, and D.F. Voytas. 2009. High-frequency modification of plant genes using engineered zinc-finger nucleases. *Nature* 459 (7245): 442–445. doi:10.1038/nature07845.

Travella, S., S.M. Ross, J. Harden, C. Everett, J.W. Snape, and W.A. Harwood. 2005. A comparison of transgenic barley lines produced by particle bombardment and agrobacterium-mediated techniques. *Plant Cell Reports* 23 (12): 780–789. doi:10.1007/s00299-004-0892-x.

Tudisco, R., P. Lombardi, F. Bovera, D.d'Angelo, M.I. Cutrignelli, V. Mastellone, V. Terzi, L. Avallone, and F. Infascelli. 2006. Genetically modified soya bean in rabbit feeding: Detection of DNA fragments and evaluation of metabolic effects by enzymatic analysis. *Animal Science* 82 (02): 193–199. doi:10.1079/ASC200530.

Turkec, A., H. Kazan, A. Baykut, and S.J. Lucas. 2015a. Evolution of DNA extraction methods in order to monitor genetically modified materials in soy foodstuffs and feeds commercialised in Turkey by multiplex real-time PCR. *Journal of the Science of Food and Agriculture* 95 (2): 386–392. doi:10.1002/jsfa.6731.

Turkec, A., S.J. Lucas, B. Karacanli, A. Baykut, and H. Yuksel. 2015b. Assessment of a direct hybridization microarray strategy for comprehensive monitoring of genetically modified organisms (GMOs). *Food Chemistry*, 194: 399–409. doi:10.1016/j.foodchem.2015.08.030.

Turkec, A., S.J. Lucas, and E. Karlık. 2016. Monitoring the prevalence of genetically modified (GM) maize in commercial animal feeds and food products in Turkey. *Journal of the Science of Food and Agriculture*, 96: 3173–3179. doi:10.1002/jsfa.7496.

Turkec, A., S.J. Lucas, and E. Karlik. 2016. Monitoring the prevalence of genetically modified (GM) soybean in Turkish food and feed products. *Food Control* 59 (January): 766–772. doi:10.1016/j.foodcont.2015.06.052.

Ujhelyi, G., J.P. van Dijk, T.W. Prins, M.M. Voorhuijzen, A.M. Angeline Van Hoef, H.G. Beenen, D. Morisset, K. Gruden, and E.J. Kok. 2012. Comparison and transfer testing of multiplex ligation detection methods for GM plants. *BMC Biotechnology* 12 (1): 4. doi:10.1186/1472-6750-12-4.

Van Den Bulcke, M., A. De Schrijver, D. De Bernardi, Y. Devos, G. Mbongombella, A.L. Casi, W. Moens, and M. Sneyers. 2007. Detection of genetically modified plant products by protein strip testing: An evaluation of real-life samples. *European Food Research and Technology* 225 (1): 49–57. doi:10.1007/s00217-006-0381-2.

Voytas, D.F. and C. Gao. 2014. Precision genome engineering and agriculture: Opportunities and regulatory challenges. *PLoS Biology* 12 (6): 1–6. doi:10.1371/journal.pbio.1001877.

Wahler, D., L. Schauser, J. Bendiek, and L. Grohmann. 2013. Next-generation sequencing as a tool for detailed molecular characterisation of genomic insertions and flanking regions in genetically modified plants: A pilot study using a rice event unauthorised in the EU. *Food Analytical Methods* 6 (6): 1718–1727. doi:10.1007/s12161-013-9673-x.

Wang, Y., X. Cheng, Q. Shan, Y. Zhang, J. Liu, C. Gao, and J.-L. Qiu. 2014. Simultaneous editing of three homoeoalleles in hexaploid bread wheat confers heritable resistance to powdery mildew. *Nature Biotechnology* 32 (9): 947–951. doi:10.1038/nbt.2969.

WHO and FAO. 1991. *Strategies for Assessing the Safety of Foods Produced by Biotechnology: Report of a Joint FAO/WHO Consultation [held in Geneva from 5 to 10 November 1990]*. Geneva, the Switzerland: World Health Organization. http://www.who.int/iris/handle/10665/41465.

Wright, M., J. Dawson, E. Dunder, J. Suttie, J. Reed, C. Kramer, Y. Chang, R. Novitzky, H. Wang, and L. Artim-Moore. 2014. Efficient biolistic transformation of maize (*Zea mays* L.) and wheat (*Triticum aestivum* L.) using the phosphomannose isomerase gene, Pmi, as the selectable marker. *Plant Cell Reports* 20 (5): 429–436. doi:10.1007/s002990100318.

Yang, L., C. Wang, A. Holst-Jensen, D. Morisset, Y. Lin, and D. Zhang. 2013. Characterization of GM events by insert knowledge adapted re-sequencing approaches. *Scientific Reports* 3: 2839. doi:10.1038/srep02839.

Zeljenková, D., K. Ambrušová, M. Bartušová, A. Kebis, J. Kovrižnych, Z. Krivošíková, M. Kuricová et al. 2014. Ninety-day oral toxicity studies on two genetically modified maize MON810 varieties in Wistar Han RCC rats (EU 7th Framework Programme Project GRACE). *Archives of Toxicology* 88 (12): 2289–2314. doi:10.1007/s00204-014-1374-8.

Zhang, D. and J. Guo. 2011. The development and standardization of testing methods for genetically modified organisms and their derived products. *Journal of Integrative Plant Biology* 53 (7): 539–551. doi:10.1111/j.1744-7909.2011.01060.x.

Index

D

E

F

G

H

I

J

K

L

M

N

O

Q

R

S

T

U

V

W

X

Y

Z